Engineering your future: an Australasian guide

FIFTH EDITION

David Dowling

Roger Hadgraft

Anna Carew

Tim McCarthy

Doug Hargreaves

Caroline Baillie

Sally Male

WILEY

Fifth edition published 2025 by
John Wiley & Sons Australia, Ltd
310 Edward Street, Brisbane Qld 4000

Typeset in 10/12pt Times LT Std

ISBN 9781394267217 (print); 9781394267224 (e-text)

A catalogue record for this book is available from the National Library of Australia

Wiley acknowledges the Traditional Custodians of the land on which we operate, live and gather as employees, and recognise their continuing connection to land, water and community. We pay respect to Elders past, present and emerging.

Creators
David Dowling (Author), Roger Hadgraft (Author), Anna Carew (Author), Tim McCarthy (Author), Doug Hargreaves (Author), Caroline Baillie (Author), Sally Male (Author)

Wiley
Mark Levings (Publishing Manager, Tertiary and Professional), Jess Carr (Manager, Higher Education Content Management), Rebecca Campbell (Senior Production Editor), Tara Seeto (Senior Publishing Coordinator), Renee Bryon (Copyright & Image Research), Delia Sala (Cover Design), Laura Davies (Copyeditor)

Cover images: © SergeyBitos / Adobe Stock Photo; © newb1 / Adobe Stock Photo; © LOVE A Stock / Adobe Stock Photo; © TSANI / Adobe Stock Photo

Typeset in India by diacriTech

Printed in Singapore
M130003_190824

BRIEF CONTENTS

CONTENTS

ABOUT THE AUTHORS

David Dowling

DipLSurv, ARMIT, BAppSci, MSurvMap
Honorary Professor of Engineering Education
University of Southern Queensland

David Dowling is passionate about facilitating student learning and helping engineering students to achieve their career goals. Consequently, much of his work and research has been concentrated on working with industry representatives to develop practice-based curricula and to enhance teaching and learning environments. In particular, his focus has been on: facilitating student transition to university; identifying and addressing factors that influence success at university; assessing workplace learning; working with practitioners to define graduate attributes; embedding graduate attributes into program curricula; and engineering technician education.

David worked as a surveyor for 12 years prior to accepting a lecturing position at the University of Southern Queensland (USQ) in 1978. David was appointed Head of Surveying in 1989 and worked intensively with industry organisations to design, develop and gain accreditation for three new distance education programs in surveying and GIS — the first in Australia.

In 1995, he accepted the role of Associate Dean (Academic) in the Faculty of Engineering and Surveying; a position he held until 2009. His major achievement in this role was the successful 2001 accreditation of the first Australian Bachelor of Engineering to be offered by distance education. In 2005, David developed the content, structure and study materials for the innovative Master of Engineering Practice program. This distance education program is accredited by Engineers Australia and enables experienced Engineering Technologists to become Professional Engineers by using their workplace learning to demonstrate their competence.

From 2009 to 2015, David led two major projects funded by the federal Office of Learning and Teaching and was a member of the project team on three other projects. He has been a member of Engineers Australia's National Articulation Committee since 2006 and a member of the Australasian Association for Engineering Education (AAEE) since 1998, serving as President during 2005 and 2006.

In 2006, David received the AAEE Excellence in Engineering Education Award for Inclusive Teaching, and in 2008 he received an Australian Learning and Teaching Council (ALTC) Citation which read: *For sustained leadership in using graduate attributes to design and deliver programs, courses and resources that enhance students' learning and their achievement of career goals.*

When David retired in 2015, he accepted an appointment as an Honorary Professor at the University of Southern Queensland. This allows him to continue his research in engineering education topics that enable students to achieve their career goals. He has also consulted with engineering and other organisations to help them define and develop the personal and professional capabilities their employees need to have a successful career.

He is currently completing a 20-year longitudinal study of the educational pathways 300 engineering students have followed to achieve their career goals, after commencing an Associate Degree in Engineering in the early 2000s. Preliminary results are very encouraging as they indicate that overall, 62% of the students graduated from at least one engineering program, with a further 7% graduating from a program offered by another faculty.

Roger Hadgraft

BE(Hons), MEngSc, DipCompSc, PhD
Visiting Professor, University of Technology Sydney

Roger Hadgraft is a civil engineer with more than 30 years of involvement in improving engineering education. He has published many papers on problem- and project-based learning (PBL), and the use of student-centred learning to meet the needs of engineering employers. He was instrumental in introducing a project-based curriculum in civil engineering at Monash University and in civil, chemical and environmental engineering at RMIT. He has led change at University of Technology Sydney (UTS) since 2016, transforming the engineering curricula using studios to deliver innovative, design-able graduates. He has consulted on PBL to universities both nationally and internationally.

Anna Carew

BSc(Hons), PhD
Agricultural and Environmental Consultant, RM Consulting Group

Dr Anna Carew trained as a water microbiologist and fell sideways into engineering as an early career researcher investigating novel microbial water quality indicators. Working alongside engineers, she came to appreciate engineering as a powerful and fascinating field that impacted the way people lived every day. While Anna was working in sustainable water and waste management with the Institute for Sustainable Futures, UTS, she observed some poor outcomes from engineering decisions that failed to take account of the people and ecosystems they impacted. A passion to address this problem led to a PhD at the University of Sydney, investigating the teaching and learning of sustainability in engineering (2005) and subsequent work with engineering academics to infuse the curriculum with sustainability tools and ideas. In 2011, Anna was awarded an Australian Learning and Teaching Council Citation for her outstanding contribution to the teaching of graduate attributes in engineering.

Seeking a new challenge, Anna joined the Tasmanian Institute for Agriculture at the University of Tasmania (UTAS) to refocus on technical research and completed a second PhD, this time on the marvellous fusion of microbiology, chemistry and bioprocess engineering that is pinot noir winemaking (2015). For a time, Anna worked in wine science and managed a teaching team delivering short courses, Diplomas and Associate Degrees in applied technologies at UTAS. She then stepped out into 'the real world' with roles in the Tasmanian public service, with a construction sector training board, and thence into the high-energy, rapid-response world of private consulting.

Anna is now buried neck-deep in a range of projects that draw on her diverse experiences. On this particular work day, Anna will finalise a slidedeck to teach agricultural producers better business skills, she will review some GIS maps for an emergency plan to protect endangered species and ecosystems in Victoria, and then she'll reshape her CV for a bid to deliver a zero municipal waste strategy needed by a council in New South Wales.

Over the 15 years Anna has contributed to this textbook, she has lived a life of learning. Life-long learning is a privilege. It keeps the grey matter in top condition, greases the transition between jobs, and allows people with curious minds to contribute to their communities in a myriad of ways aligned with the values they hold dear.

Tim McCarthy

BE, MSc, PhD, MIEI
Professor of Sustainable Buildings, University of Wollongong

Professor Tim McCarthy has been the Director of the Sustainable Buildings Research Centre at the University of Wollongong (UOW) since 2020. Prior to that he was Professor of Structural Engineering at the School of Civil Mining and Environmental Engineering, also at UOW. Tim's specialisations include sustainable buildings, engineering education research, integrated design systems and steel structure design, and he has supervised and co-supervised over 80 PhDs, MPhil and MSc theses. In 2010, he received an Australian Learning and Teaching Citation for Outstanding Contribution to Student Learning for leadership in curriculum and space design that fosters collaborative learning. Tim is also the author of the best-selling textbook *AutoCAD Express*. He has taken UOW students to two Solar Decathlon world finals. In 2013, Team UOW won the Solar Decathlon China while in 2018 they achieved the overall silver medal in the Middle East contest in Dubai.

Doug Hargreaves

AM, PhD, MSc, BEng, EngExec, Hon (IEAust)

Professor Doug Hargreaves is an Emeritus Professor in mechanical engineering at Queensland University of Technology (QUT). He has spent his professional life in a mixture of academic and industry practice. He was National President of the peak professional body for engineers, Engineers Australia, which had more than 100 000 members in 2010. He was Head of School of Engineering Systems at QUT for seven years, leading about 145 staff members. He returned to teaching over 1000 first-year engineering students a unit called 'Engineering and Sustainability', which effectively taught what engineering graduates do in the real world. He has published over 50 papers on the topic of engineering education and over 100 on his discipline of tribology. He is co-author of a leadership book called *Values-Driven Leadership*. He was awarded a member of the Order of Australia in the 2014 Queen's Birthday Honours list for his significant contribution to engineering education and to the community. He is the Executive Officer for the Australian Council of Engineering Deans (ACED), represents Engineers Australia (EA) on the Federation of Engineering Institutions of Asia and the Pacific (FEIAP), chairs EA's International Advisory Panel and was also Chair of the technical committee for the World Engineering Convention held in Melbourne in November 2019.

Caroline Baillie

BSc(Hons), MHEduc, PhD
Professor Integrated Engineering and Director of MESH (MS Engineering, Sustainability and Health), University of San Diego

Caroline Baillie is a Professor at the University of San Diego and Founding Director of a transdisciplinary masters program on engineering, sustainability and health. She brings over 30 years of experience in teaching engineering across multiple engineering disciplines and countries, research and development in engineering and education as well as community development and social justice work. Previously she was Chair of Engineering Education for the Faculty of Engineering, Computing and Mathematics at University of Western Australia, Chair of Engineering Education at Queens University, Canada, and educational developer and materials lecturer at Imperial College, UK and the University of Sydney, as well as holding a three-year position founding and running the Materials Engineering Subject Centre in the UK. Baillie's research considers socio-technical processes and systems, which enhance social justice, and educational systems that promote these. Baillie brings lessons learnt from these studies and practices into the curriculum and the classroom to facilitate the transformation of future generations of engineers. In 2006, Professor Baillie founded Waste for Life (WFL) (https://www.wasteforlife.org) to 'practise what she preached' as a socially just engineer. WFL 'socialises knowledge' about materials engineering with communities wishing to transform waste into composite material products for income generation. Her most recent program, 'Standing People Together' (https://www.wasteforlife.org/spt) adapts forest school pedagogy in support of grassroots community development. Baillie also co-founded the Engineering, Social Justice and Peace network in 2004 (https://esjp.org). Professor Baillie has published 28 scholarly books, an edited series of books on 'Engineers, Technology and Society' and over 200 book chapters and peer-reviewed journal and conference papers.

Sally A Male

Professor, BE(Hons), PhD
Director, Teaching and Learning Laboratory, Faculty of Engineering and IT, The University of Melbourne

Sally A Male is a Fellow of Engineers Australia. She holds qualifications in electrical engineering and a PhD on employability in engineering from The University of Western Australia. Sally is Professor of Engineering and Technology Education, and Director of the Teaching and Learning Laboratory in the Faculty of Engineering and IT at The University of Melbourne. Sally has served on the Governance Board of the Engineering Institute of Technology and held the Chair in Engineering Education at The University of Western Australia, where she is now an Adjunct Professor. Sally has led competitively funded research projects in engineering education and higher education, focusing on curriculum development, inclusion and work integrated learning. She is Editor-in-Chief of the *Australasian Journal of Engineering Education* and a former Associate Editor of *Journal of Engineering Education*. The World Federation of Engineering Organizations selected Sally as the Laureate to receive the 2023 Medal for Excellence in Engineering Education. In 2017 and 2021, Sally received Engineers Australia Medals for contributions to the engineering profession.

PREFACE

The 1996 Review of Engineering Education in Australia found that '*engineering education must become more outward looking, more attuned to the real concerns of the communities. Courses should promote environmental, economic and global awareness, problem-solving ability, engagement with information technology, self-directed learning and lifelong learning, communication, management and teamwork skills, but on a sound base of mathematics and engineering technology*.'[1] The report contained a series of recommendations that changed the way engineering was taught and learned over the following decade.

One of the key changes was the adoption by Engineers Australia of an outcomes-focused accreditation system for undergraduate degrees, based on a set of graduate attributes that Engineers Australia defined in consultation with industry at that time. During the early 2000s the graduate attributes were further developed to form a detailed set of competencies. Since then engineering schools have adapted their curriculum to ensure that engineering students have opportunities to acquire these competencies, in addition to the competencies defined by their own university. Many of these competencies are introduced in first-year subjects and students then practise and enhance those skills in subjects and projects in the later years of their programs.

The consultations undertaken for the Engineers for the Future project, commissioned by the Australian Council of Engineering Deans (ACED) in 2008, found that industry supports this explicit focus on graduate attributes. It also reported on engineering-specific graduate outcomes and attributes. They formed the view from their consultations that '*engineers do their work by having knowledge and skills in varying combinations of the following thematic areas: the engineering life-cycle of concept, design, implementation, operation, maintenance and retirement (with increasing emphasis on dealing with uncertainty and risk assessment; systems thinking, and integrating ideas and technologies); managing complex engineering projects …; mathematical modelling …; and scientific knowledge of established and emerging areas …*'[2]

More recently, ACED commissioned the Engineering Futures 2035 project to review how engineering education could address the anticipated changes to the profession by the year 2035. The project was conducted between 2018 and 2021 and two major findings were as follows.

- Industry wants to see a re-balancing of the theory-practice components of professional engineering education, with a greater emphasis on practice, including the human dimensions of engineering.
- Potential students are motivated to solve 'real-world' problems and want to see that engineering practices addresses societal needs.[3]

ACED established the Engineering Futures initiative in 2022 to '*lead, drive, and host the activities necessary for progressive implementation of the Engineering Futures 2035 recommendations at a national level*'.[4] Phase 1 of the implementation process commenced in March 2023.

Since the first edition, this text has been designed to provide first-year engineering students with a solid and practical grounding in many of the engineering and generic graduate attributes and skills needed to succeed, as well as many of the tools and techniques that facilitate the application of those skills in real engineering work and study. Subsequent editions continued this alignment with industry requirements for engineering education, including this fifth edition, which is aligned with the Engineering Futures 2035 initiative.

Numerous historic and contemporary Australian, New Zealand and international examples are used to illustrate the principles that are discussed in the text, and to highlight many of the important innovations that have built the reputation of Australian and New Zealand engineers. The examples are drawn from a range of current engineering disciplines, from emerging disciplines, and from a range of organisations and projects, large and small. These examples will enable students to explore engineering and how it is practised in Australasia, as well as the approaches used by Australasian engineers, who have a reputation

1 Institution of Engineers Australia 1996, *Changing the culture: Engineering education into the future*, IEAust, Canberra, p. 4.

2 King, R 2008, *Engineers for the future: Addressing the supply and quality of Australian engineering graduates for the 21st century*, Australian Council of Engineering Deans, Sydney, p. 61, https://www.engineersaustralia.org.au/sites/default/files/content-files/ACED/engineers_for_the_future.pdf

3 Australian Council of Engineering Deans 2021, *Engineering change: The future of Engineering Education in Australia*, Engineering 2035 Project, ACED, p. 5.

4 Australian Council of Engineering Deans 2023, *2035 Implementation Project — Phase 1*, March, ACED, p. 1.

for being flexible and adaptive.[5] The Australasian focus and context of the text will also assist students to formulate their future career preferences.

Although this text is designed to be used as the text for one or more first-year courses, the authors see it as a resource that students can use throughout their program.

The chapters are arranged in six sections to facilitate student learning. The first section provides an introduction to engineering and the engineering method. This is followed by a section on engineering in society, which includes sustainable engineering, professional responsibility and ethics. This is followed by two sections that provide students with the opportunity to acquire some of the key skills they will need to be successful in their first year at university, such as self-management, teamwork and communication. The fifth section provides an overview of each of the steps engineers use when they apply the engineering method: information and research skills, design, evaluating solutions, making decisions, managing engineering projects and communicating outcomes. The final chapter provides information about the engineering profession, as well as existing and emerging specialisations — information that will help students to refine their career choices.

The authors wish to thank the engineering academics who provided feedback on the previous four editions of this text. Their comments helped to shape the content and the structure of the fourth and fifth editions, as well as the focus and content of individual chapters.

The support for the first four editions of this text, as well as what it is achieving for undergraduate engineering education in Australia and New Zealand, means a lot to us. In many ways, this text and its accompanying extensive resource package should be seen as a resource generated for all engineering schools in Australia and New Zealand. The authors would therefore welcome constructive feedback from academic staff and students so that future editions continue to meet the needs of first-year engineering students. This includes information about innovative engineering projects that may be suitable for inclusion in future editions of the text.

Many people have contributed information that was incorporated directly into the body of the text, or in one of the many practical engineering 'spotlight' features. The authors acknowledge the important contribution of the following people to the development of this fifth edition: Armando Apan, Madhu Bhaskaran, Gunilla Burrowes, Caroline But, Ian Cameron, Miles Cattach, Susan Conrad, Cheryl Desha, Jason Eshraghian, Peter Fagan, Tim Gale, Peter Gibbings, Sue Murphy, Andrew Guzzomi, Chris McAlister, Nicole Hahn, Charlie Hargroves, Prue Howard, Ali Kharrazi, Peter Knights, Ilsa Kuiper, Nelson Lam, Julia Lamborn, Traci Nathans-Kelly, Christine Grohowski Nicometo, Katherine Nguyen, Sharon Nightingale, Timothy Pfeiffer, Matthew Preston, Carl Reidsema, Philip Rubie, John Russell, David Sampson, Warren Sharpe, Lori Sowa, Geoff Spinks, Clive Stack, Peter Stasinopoulos, Andrew Schroder and Tim Wark.

We would like to thank the team at Wiley for their assistance in the development of this text and its associated resources.

We would also like to acknowledge the members of our families who have lived the highs and lows of this project with us, some for the fifth time. We know the many sacrifices you made to help us meet the tight deadlines that accompany a project of this nature.

David Dowling
Roger Hadgraft
Anna Carew
Tim McCarthy
Doug Hargreaves
Caroline Baillie
Sally Male
May 2024

5 Institution of Engineers Australia 1996.

ENGINEERING SPOTLIGHTS: AT A GLANCE

Chapter	Spotlight	Civil	Mechanical/ manufacturing	Electrical/ electronics
1	Engineering vision through artificial intelligence			X
	Aerospace engineering and surf safety		X	X
	Engineering better flood modelling	X		
	Software engineer wins top student award	X		
	3D-printed heart valves		X	X
	Learning journeys	X	X	X
	Early Aboriginal engineering	X		
	Engineering breakthroughs in early New Zealand history		X	
	Consumerism and the growth of e-waste		X	X
2	Wind energy	X	X	X
	Water recycling	X		
	Smartwatches			X
	Design for earthquake-resistant buildings	X	X	
	Engineers Australia and Engineering New Zealand — continuing professional development	X	X	X
	Driverless vehicles will change our lives	X	X	X
	Finding the unasked question	X	X	X
3	Lake Pedder hydro-electric scheme	X		X
	NZ ETS prompts metal manufacturer to upgrade and save money		X	X
	Australasian students shine on the world stage	X	X	X
	Energy tips: landfill gas		X	X
	Wine carbon footprinting		X	X
	Unintended consequences of engineering breakthroughs		X	
	Multi-criteria evaluation: recreation and tourism in Victoria	X		
	Least cost planning in the 'sunburnt country'	X		

Environmental	Chemical (includes petroleum)	Mining	Other
			Software; hardware; mechatronics
			Aeronautical
X			
			Software
			Biomedical; materials
X	X	X	All other disciplines
X			Agricultural; aeronautical; maritime
			Maritime
X			Computer
X			Aerospace/aviation
X	X		
			Software; hardware
X			Materials
X	X	X	All other disciplines
X	X	X	All other disciplines
X			
X			Materials
X	X	X	All other disciplines
X			
X	X		
X	X		
X			
X			

Chapter	Spotlight	Civil	Mechanical/ manufacturing	Electrical/ electronics
4	Shared paths and the role of engineers	X		
	Creating recycling facilities in a post-conflict country	X	X	
	Design for the dump		X	X
	Learning from nature's design principles	X	X	X
	Reporting a leaky pipe			
	James Hardie and asbestos-related disease	X	X	
	Admitting failure	X	X	X
	Whistleblower slams Japan nuclear regulation	X	X	X
	Free prior and informed consent	X		
	Monsanto penalised for bribery	X	X	X
5	The Competencies of Engineering Graduates Project	X	X	X
	A civil engineering graduate's perspective	X		X
	Putting passion into practice 1			X
	Putting passion into practice 2	X		
	Hydrographic surveys	X		
	Life-long learning as a career develops			X
	A 3D printed car		X	X
	The law of the pendulum	X	X	X
	Promoting a culture of life-long learning among engineering staff	X	X	X
	A reflection: working on large projects in isolated areas	X		
6	The collaboration capabilities for two engineering disciplines	X	X	X
	Team innovation and success: why we should fight at work	X	X	X
	Collaborating and colliding: when alliances go wrong	X	X	X
	Etiquette tips for virtual meetings	X	X	X
	Negotiating tight spaces	X	X	X

Environmental	Chemical (includes petroleum)	Mining	Other
X			Geomatic
X			
X			
X			
X	X		
X		X	
X	X	X	All other disciplines
X	X	X	Chemical; nuclear
X			
X	X	X	Agricultural
X	X	X	All other disciplines
X			
X			
X			Geomatic
X	X	X	All other disciplines
X			
X		X	Geomatic
X	X	X	All other disciplines
X	X	X	All other disciplines
X	X	X	All other disciplines
X	X	X	All other disciplines
X			

Chapter	Spotlight	Civil	Mechanical/ manufacturing	Electrical/ electronics
7	Catastrophic communication breakdown	X	X	X
	Releasing Fletcher Aluminium's invisible handbrake		X	
	Developing, managing and communicating our brand	X	X	X
	Environmental engineering in Alaska			
	Communicating from space			X
	Swarm communication		X	X
	The plant visit		X	
	Who is the gatekeeper?	X	X	
8	Reporting on data breaches	X	X	X
	Communicating good and bad news	X	X	X
	Working with rubbery figures	X	X	X
9	Smart technologies: how accurate are their measurements?	X	X	X
	Exploring the accuracy of spatial location devices	X	X	X
	Flood protection for a mine tailings slurry system	X	X	
	WestConnex stakeholders seek compensation	X	X	
	Surf, dive or scan?	X	X	X
	The ethics of data management	X	X	X
	3D buildings in a 3D world	X		
	Do you need better cybersecurity?	X	X	X
10	The good news about plastic waste			
	Living Building Challenge	X		
	Formula E — high-performance electric cars		X	X
	Rethinking timber: a story of a long life	X		
	Qantas Q Bag Tag		X	X
	The Gateshead Millennium Bridge	X	X	X
11	Upgrade of concentrate thickeners		X	
	Water jet propulsion — HamiltonJet	X	X	
	Software checking	X	X	X

Environmental	Chemical (includes petroleum)	Mining	Other
X	X	X	All other disciplines
X	X	X	All other disciplines
X			
			Geomatic; telecommunications
			Geomatic; telecommunications
X			Biomedical; materials
X	X	X	All other disciplines
X	X	X	All other disciplines
X	X	X	All other disciplines
X	X	X	All other disciplines
X		X	Geomatic
		X	Minerals/metallurgical
X	X	X	Geomatic
X	X	X	
X	X	X	All other disciplines
X	X	X	All other disciplines
X			Materials
X			
X	X		Materials
X			
			Telecommunications; software
X			Structural
	X		
X			
X	X	X	Software

Chapter	Spotlight	Civil	Mechanical/ manufacturing	Electrical/ electronics
12	Understanding complex systems	X	X	X
	Dubai's Burj Khalifa	X		
	Designing a natural air-conditioning system		X	
	SMART decisions for bridge maintenance	X	X	
	Flood emergency DSS for the Gold Coast	X		
13	Winning project management in engineering	X	X	
	Lean management and Fast Jet		X	
	Project management oversights and the Australian eCensus fail			
14	When public information becomes unintelligible to the public	X	X	X
	Developing a communication plan for a student project	X	X	X
	What does ChatGPT say?	X	X	X
	The differences between academic and practitioner writing	X	X	X
	Planning a proposal	X	X	X
	When things go wrong: a legal perspective	X	X	X
	Legal games at Wembley Stadium	X	X	X
	Death by lack of communication and lack of design	X	X	X
	Slide rules	X	X	X
	Using $3D^{+}$ models to communicate design and operational information	X	X	
	Using virtual mines to communicate the complexity of mining operations to students			
15	Digital diagnosis: how your smartphone or wearable device could forecast illness			X
	Climate change adaptation and maritime engineering	X	X	X
	On mining, poverty and development			
	Microscope in a needle		X	X
	Mimicking the retina			X
	Weta Workshop and Weta Digital		X	X

Environmental	Chemical (includes petroleum)	Mining	Other
X	X	X	
X			
			Structural; computer systems
X			Geomatic; software
	X		Aeronautical
			Aerospace
			Software
X	X	X	All other disciplines
X	X	X	All other disciplines
X	X	X	All other disciplines
X	X	X	All other disciplines
X	X	X	All other disciplines
X	X	X	All other disciplines
			Structural
X	X	X	All other disciplines
X	X	X	All other disciplines
X			
		X	
X			Coastal/ocean/marine
		X	
			Mechatronic; software

PART 1

INTRODUCTION TO ENGINEERING

CHAPTER 1

What is engineering?

'Engineering is intellectual and practical — understanding social needs, conceiving solutions and predicting how they'll work . . .'

James Trevelyan (mechatronic engineer)

LEARNING OBJECTIVES

After studying this chapter, you should be able to:

1.1 describe the roles of an engineer

1.2 identify the major engineering disciplines

1.3 list the core skills and attributes of an engineer

1.4 identify some of the fundamentals of engineering science

1.5 explain the impact engineering has had on society over time

1.6 explain the need for professionalism and ethics in engineering.

Introduction

In this chapter you will explore what engineers *know, think* and *do* in practising engineering. Perhaps you have chosen to study engineering with clear ideas about engineers being people who design technological solutions to problems. You may be motivated by job security and a good salary. You might want to make a difference in the world.

You will learn about the many dimensions of engineering work. The emergence of the main engineering disciplines will be explained. You will learn how engineering innovation has coloured people's lives throughout history, and how the fundamental concepts of engineering science underpin engineering. You will also discover some of the core skills of an engineer (many of which are explored in depth in the coming chapters). But first, what is engineering?

In his article 'Engineering a future', published in the *New Zealand Education Review* (2007), journalist John Gerritsen asked leading professionals and academics for their take on 'what engineering is'. The article won Gerritsen a New Zealand Engineering Excellence Award in Engineering Journalism. Consider the following extract from this article (p. 8):

> 'The difference between engineering and science', says the chair of the Council of Engineering Deans, Bob Hodgson, 'is that engineering has to work'. After all, there's not a lot of use for bridges that test interesting principles but can't be driven on, or electricity grids that black out.
>
> The Institution of Professional Engineers New Zealand [now Engineering New Zealand] executive director Andrew Cleland agrees. Overseas experience indicates businesses are stronger if engineers are involved in their management and governance. Andrew suggested engineering is divided into two broad types — infrastructural or 'civil' engineering, and wealth creation. The former is concerned with the systems on which modern society depends, such as roads, and the latter with new products or systems, such as those developed in the field of information technology.

Natasha McCarthy, policy advisor to the Royal Academy of Engineering in the United Kingdom, states in her book (2012, p. 15):

> Engineering is inherently creative — it takes what nature provides in order to make something new, or to make things behave in novel ways. Engineers should never be satisfied with the way things are; the successful engineer will always want to intervene, to change things so that they work better.

In Australia and New Zealand, professional bodies accredit engineering courses and register engineers to practice. These bodies are Engineers Australia and Engineering New Zealand. There are nine general areas of practice recognised by Engineers Australia: aerospace, biomedical, chemical, civil, electrical, environmental, ITEE (information, telecommunications and electronics engineering), mechanical and structural (Engineers Australia 2018a). Many fields of engineering sit within the nine general areas of practice. For example, mining engineering, mechatronic engineering and medical engineering are named engineering courses at Australian and New Zealand universities. Graduates from those courses would register under the most closely related general area of practice.

In addition to general areas of practice, it is possible for practising engineers to register their expertise and competence to undertake specialist engineering work under a 'special area of practice'. In Australia, there are 13 recognised special areas of practice including amusement ride inspection, fire safety, heritage and conservation management, leadership and management, petroleum engineering and subdivisional geotechnics.

These areas of practice and specialisation change over the years in response to industry and societal demand. Engineering is about innovation and there are a range of views on what the future holds for engineers and engineering specialisations. We will examine these in more detail as this chapter progresses and in subsequent chapters of this text.

Engineers Australia offers a fairly clear description of what engineering is from an Australian perspective. For the sake of brevity, the following summarises the key aspects of Engineers Australia's description of engineering from their Stage 1 Competency Standards (Engineers Australia 2019).

Engineers take responsibility for engineering projects and programs in the most far-reaching sense. This includes:

- reliable functioning of all materials and technologies used and the systems created
- interactions between the technical system and the social and physical environment where it functions
- understanding the requirements of clients and of society as a whole

- working to optimise social, environmental and economic outcomes over the lifetime of the product or program
- interacting effectively with the other disciplines, professions and people involved
- ensuring that the engineering contribution is properly integrated into the totality of the project or program.

Engineers Australia emphasises the engineer's role in explaining technological possibilities to society, business and government; bringing knowledge to bear from multiple sources to develop solutions to complex problems and issues; and managing risk. These statements from Engineers Australia emphasise the importance of technical excellence and a capacity to consider and take into account the full social and environmental context in which engineering work takes place. This is the career you have chosen and the coming chapters in this text will start you on your journey into the rich, complex and important world and work of the professional engineer.

1.1 What is the role of an engineer?

LEARNING OBJECTIVE 1.1 Describe the roles of an engineer.

The role of an engineer is constantly evolving. Market demands, community expectations and advances in technology are constantly reshaping the practice of engineering. Engineering is pervasive in that it influences and is evident in a wide array of our day-to-day activities. This means that engineers are everywhere and their roles are many and diverse. Today's engineers work in a variety of roles:

- choosing to be *specialists* who design or develop technological solutions to problems
- thriving in *management* or *project management* roles where they oversee budgets and organise the work of others
- devoting their working lives to *researching* radical new applications of the fundamental concepts of engineering and physics
- getting '*hands on*', working with operators and tradespeople to maintain equipment in heavy industry
- moving into *policy* and specialising in creating the regulatory or legislative frameworks to ensure engineering solutions take account of the broader economic, social and political factors that engineering work sits within
- opting to work *alone* and in the background, or being '*out there*' meeting clients and working with people.

Some students may even graduate and elect not to work as an engineer. Engineers Australia reported that approximately 44 per cent of people with an engineering qualification work outside the field of engineering (Engineers Australia 2023). This means a substantial proportion of engineering graduates move out of engineering and apply their problem-solving skills in non-traditional fields such as medical logistics, political activism or merchant banking. Ultimately, all of these different roles are important to the transformation of natural capital into built and human capital, and to the prosperity and functioning of our society. They may all lead towards satisfying, productive careers.

We will now look at how the role of the engineer has evolved over time. Different skills and attributes have become increasingly important in engineering — a reflection of the diverse job responsibilities of today's graduates and professionals.

KEY POINT

All engineers should be technically competent; beyond that there are vast arrays of ways to be, and to practise as, an engineer.

A historical perspective

The engineer, as we understand the role, began to emerge during the Industrial Revolution of the eighteenth and nineteenth centuries in the United Kingdom and Europe. These engineers relied heavily on precedent, rules of thumb and experimentation to execute technical projects. The engineer was considered as much an artisan or master craftsperson as they were a scientist. During the 1900s, significant discoveries in mathematics and physics and the advent of computing paved the way for engineering approaches based on mathematical modelling and the application of fundamental principles known as engineering science. (An overview of engineering science is given later in the chapter.) This science-based approach to engineering enabled many of the technological breakthroughs of the twentieth century. Two examples of technological innovation that have affected the work and role of engineers are early advances in polymer chemistry and advances in glass purity and lasers, which allowed development of optical fibre technology. We will now briefly explain how these two examples came about and how they influence the work and lives of contemporary engineers.

During the 1920s and 1930s, a brilliant chemist called Wallace Carothers worked at the newly formed DuPont chemical company in the United States in the frontier field of polymers. Carothers and his team are credited with kick-starting the commercial development and manufacture of synthetic rubber (neoprene) and synthetic silk (nylon). Contemporary engineers rely on these polymers for a range of uses. Neoprene is commonly used for safety gloves and face masks, corrosion-resistant gaskets and hoses, water-resistant housings for consumer electronics (e.g. tablet devices, smartphones) and for personal buoyancy and insulation gear used by engineers working offshore and in the maritime industry. Nylon has evolved from its earliest, most recognisable application as a replacement for the silk that was commonly used to manufacture ladies' hosiery. Now engineers use nylon to achieve engineering outcomes in a range of applications such as in the form of injection moulded rocker covers and gaskets in car manufacturing; as strong, porous, chemically stable matting for soil erosion control; and for the biomedical engineering of prosthetic limbs and organs (e.g. bladders, heart valves).

Biomedical engineers use nylon in the design of prosthetic limbs.

As a result of the advances in glass purity and lasers, engineers (and most of the rest of society) came to rely heavily on dependable, high-speed, high-fidelity telecommunications based on fibre optics. Fibre optics allowed engineers to, for example, track and digitise human movement to develop new kinds of movie graphics, and develop *distributed control systems* by combining fibre optics, infra-red detectors and signal-processing electronics to safely and remotely monitor and control systems in chemical engineering and manufacturing. Optical fibres came about as a result of two breakthroughs. The first was a breakthrough in physics by Charles Townes and his colleagues at Columbia University based on Einstein's discoveries on *stimulated emission* (the capacity of electrons to absorb and emit photons). The Columbia team developed microwave amplification by stimulated emission of radiation or 'maser'. The maser was a precursor to light amplification by stimulated emission of radiation, which is now known as 'laser'.

The second breakthrough paving the way for society's current ubiquitous reliance on optic fibre was the development in the 1970s of high-purity glass fibres. To make optical fibre communications viable, these fibres needed to be of sufficient purity for light travelling through the glass to suffer *attenuation* of less than 20 dB/km. Attenuation is the loss of signal strength over distance and is expressed in units of decibels (dB) per kilometre (km) travelled. Much modern-day fibre optic technology utilises lasers as the most effective and efficient light source. Australia's largest ever infrastructure project, the National Broadband Network (NBN), which commenced roll-out in 2010, relies on fibre optic technology. The NBN was built to support growth of the digital economy through high-definition, high-speed communications. Much of our life, work and study now relies on this fibre optic supported internet access. Low Earth orbit (LEO) satellites (e.g. Starlink) offer a new approach to connectivity, particularly for regional and remote users, but most of us will continue to rely on fibre optics for the foreseeable future.

As you will discover in the coming chapters, engineering science, engineering innovation and engineering work have evolved over time. With this evolution has come the need for engineering practitioners to adapt their knowledge, skill sets, perspectives and attitudes.

A contemporary perspective

The skills, knowledge and attitudes employers and society expect of engineers have changed over time. Literacy in maths, physics and computer modelling have long been considered fundamental to the work of professional engineers (Stephan 2001), but engineers are expected to know and do more than that. Mitchell et al. (2004) suggested the main role of engineers is to apply technical competence as a means of transforming 'natural capital', such as water, solar energy or mineral deposits, into 'human and built capital', such as energy or metal goods. Strong technical competence is the requisite basis for designing this transformation; however, as many engineering practitioners are now discovering, technical competence is not the only requisite skill for professional practice.

According to prominent UK engineer Roland Clift, an engineer must be not only technically competent, but must understand that technology often has diverse consequences. The job of engineers is to seek an understanding of, and take responsibility for, the broader social, environmental and economic impacts of the work they do. They need to consider the impact of their work through the whole life cycle of an engineering product or process — from cradle (extraction or acquisition of raw materials) through to production and use and then to grave (decommissioning and disposal or recycling) — in terms of possible impacts on future generations (Clift 1998).

In a survey involving 300 Australian engineers, participants were asked to identify which skill, attribute or area of knowledge was most lacking in recent engineering graduates (Male et al. 2010). The survey participants rated the following competencies as the most lacking: communication skills, self-management, attitude, problem solving and teamwork. In line with these findings, the UK Henley Report (Henley Management College 2006) and the Australian King Report (2008) suggested that engineers needed the skills to fit into one of five broad categories. These are outlined in figure 1.1.

FIGURE 1.1 Engineer roles in the future

- *Technical specialists.* Engineers at the leading edge of their discipline, who work with those in other disciplines to conceive, create and implement highly innovative engineered systems.
- *Integrators.* Engineers skilled at working across technical, disciplinary and organisational boundaries in complex business, social and political environments.
- *Change agents.* Engineers who apply creativity, innovation and leadership developed through engineering studies and practice to solve high-profile issues, such as ecological sustainability and climate change.
- *Project managers.* Engineers who deploy highly developed interpersonal, organisational and financial management skills to execute engineering projects.
- *Asset maintainers.* Engineers who are responsible for understanding, maintaining, adapting and upgrading existing major engineering infrastructure, equipment and systems.

Source: Adapted from the UK Henley Report (2006) and the Australian King Report (2008).

Augustine (1994, p. 3) observed that, 'Engineers must become as adept in dealing with societal and political forces as they are with gravitational and electromagnetic forces'. A contemporary perspective on the role of an engineer is that today's engineers must possess more than technical competence; they must have the skills and inclination to understand and take account of the people and contexts that surround and give purpose to all engineering endeavours.

SPOTLIGHT

Engineering vision through artificial intelligence

Engineering innovation can change the way people live and create opportunities for many who face challenges. Ms Marita Cheng, who has been recognised as one of Australia's most innovative engineers, is using sensing and analytics to help people stuck at home with disability or illness to participate in school and work. She was listed in 2018 as one of *Forbes* magazine's World's Top 50 Women in Tech. Ms Cheng is a roboticist and co-founder of robotics start-up Aubot. Aubot manufactures a telepresence robot — basically a tablet on wheels — controlled by a remote user. The tablet displays the user's face and allows people to participate more fully in school or work, when they are unable to be there in real life. The platform is height adjustable, has mechanisms to avoid getting stuck or running into people, and can drive itself back to a charging dock.

Ms Cheng also co-founded Aipoly, a company that developed a smartphone app allowing people who are blind, vision impaired or colourblind to identify objects and colours. Aipoly Vision can recognise up to three objects per second in real time using a smartphone. The app recognises over 1000 different food items and dishes, and can tell 1400 colours apart and name them. The app offers simple colour naming if the user is not familiar with unusual colour names like 'celadon', 'malachite' or 'seafoam blue' (Engineers Australia 2016).

According to Ms Cheng:

> All you have to do is hold your phone, pass it over the various objects, and in real time it recognises chairs, the floor, tables, different colours . . . A blind person [can] have a much richer experience of the world through this kind of technology (Engineers Australia 2016).

The pathway to working as an AI, robotics and technology entrepreneur has been interesting and satisfying for Ms Cheng. She grew up in a Queensland housing commission suburb in a single-parent family. After graduating from high school in 2006, Ms Cheng studied for a Bachelor of Engineering (Mechatronics) and Bachelor of Computer Science at the University of Melbourne. Marita became aware that there were very few female students in her engineering classes so she and her fellow engineering students went to schools to teach girls robotics. While on academic exchange at Imperial College London, Ms Cheng expanded her program Robogals to London and then throughout Australia, the UK, the US and Japan. Today Robogals continues, running robotics workshops, career talks and community activities to introduce young women to engineering (Cheng n.d.).

CRITICAL THINKING

It is clear that Ms Cheng is strongly motivated to develop innovations and take actions that enable people who are challenged by vision impairment, illness or gender stereotypes. In your future work as an engineer, what might motivate you and shape your direction as a professional? It could be that you want status and financial security, or you may be driven to protect the environment, support social justice or develop radical new ways to do everyday things. Take the time to draft a clear statement of the personal mission guiding your career direction.

1.2 Engineering disciplines

LEARNING OBJECTIVE 1.2 Identify the major engineering disciplines.

The importance of engineering in the day-to-day lives of Australians and New Zealanders should not be underestimated. The work of all engineering disciplines is important. Engineers and the engineering work they do influence how people live, work, play, recuperate, eat, breathe and communicate. Beyond influencing lives, it is no exaggeration to say that engineering also saves lives, and mistakes and oversights in engineering can cost lives.

Here are a few examples of engineering in our day-to-day lives.

- We eat and drink fermented, cultured, cooked, canned and cured foodstuffs (e.g. yoghurt, beer, cheese, salami, canned tomatoes, smoked salmon), the safe mass-production of which is often designed and overseen by chemical, manufacturing, agricultural, process and biochemical engineers.
- We drive cars or ride in trains designed by mechanical, electrical and rolling stock engineers, which conform to stringent safety standards ensuring that only the unlucky few fail to make it to their destination. The work of traffic engineers and software engineers ensures that the speed and routing of cars and trains are controlled to reduce commuter frustration, delays and road and rail accidents.
- We consume therapeutic medicines and vitamin supplements that have been mass-produced with the aid of biochemical, chemical and manufacturing engineers, and distributed by road, air and rail by engineers who manage pharmaceutical supply chains. Some of these medications are administered to us via technology designed and maintained by biomedical and maintenance engineers who work in the health and hospital sector.
- We stay warm in winter and cool in summer increasingly through electrical power from renewable energy systems developed by mechanical and electrical engineers who work in photovoltaic, hydro and wind energy production. We also rely on clever, passive solar design by civil and mechanical engineers.
- We breathe air that is maintained within safe limits through the work of environmental engineers, who monitor and mitigate industrial air pollution. Environmental, chemical and civil engineers also ensure that wastes discharged to water (e.g. sewage, stormwater, industrial liquid wastes) are monitored to ensure waters remain safe for recreational use (e.g. swimming, boating) and as healthy ecosystems (e.g. maintenance of biodiversity).
- Access to reliable engineered telecommunications devices and geospatial services allows us to conduct business internationally, maintain relationships with loved ones interstate and overseas, and to call for emergency services when things go wrong. We can also find out where we are in any city and where the nearest restaurant is.

This list of how engineering touches our everyday lives could go on and on, and in the coming chapters the numerous spotlight sections will provide many further insights into the pervasive nature and importance of engineering work. You may like to think through the wide range of engineering technologies and engineered services that lead you safely and comfortably through each day. Figure 1.2 shows some of the engineering disciplines that may contribute to the lives and experiences of two typical undergraduate students, in terms of the design, manufacture and delivery or distribution of each of the products or services they are using.

About three years after graduation it becomes important to understand the wide range of engineering disciplines and specialisations. At that point, an engineer can seek to register as a Professional Engineer on the National Engineering Register (NER). Administered by Engineers Australia, it includes the nine general areas of practice under which a new engineer can nominate to practise; these are listed in the introduction to this chapter.

In New Zealand, the registration authority for engineers is called Engineering New Zealand and graduates can apply to become Chartered Professional Engineers (CPEng). Nominees to CPEng in New Zealand undertake an assessment task to demonstrate their capacity to apply specialist knowledge to complex engineering problems. The assessment is evaluated by a panel of experts and those nominees who pass attain CPEng status. To retain registration within New Zealand, CPEngs are required to demonstrate New Zealand-specific technical experience and be reassessed every six years (Engineering New Zealand 2024a).

In the following sections, we will consider in a bit more detail some of the recognised and emerging fields of engineering including electrical, electronics and telecommunications engineering; mechanical engineering; aerospace and aviation engineering; chemical engineering; civil engineering; environmental engineering; materials engineering; and mining engineering. The discussion in the following discipline-specific paragraphs draws on information from Engineers Australia (2018b) and Engineering New Zealand (2024b).

KEY POINT

Having a clear idea about what each of the engineering disciplines does will help you settle on the right specialisation for you.

FIGURE 1.2 The engineering behind everyday life

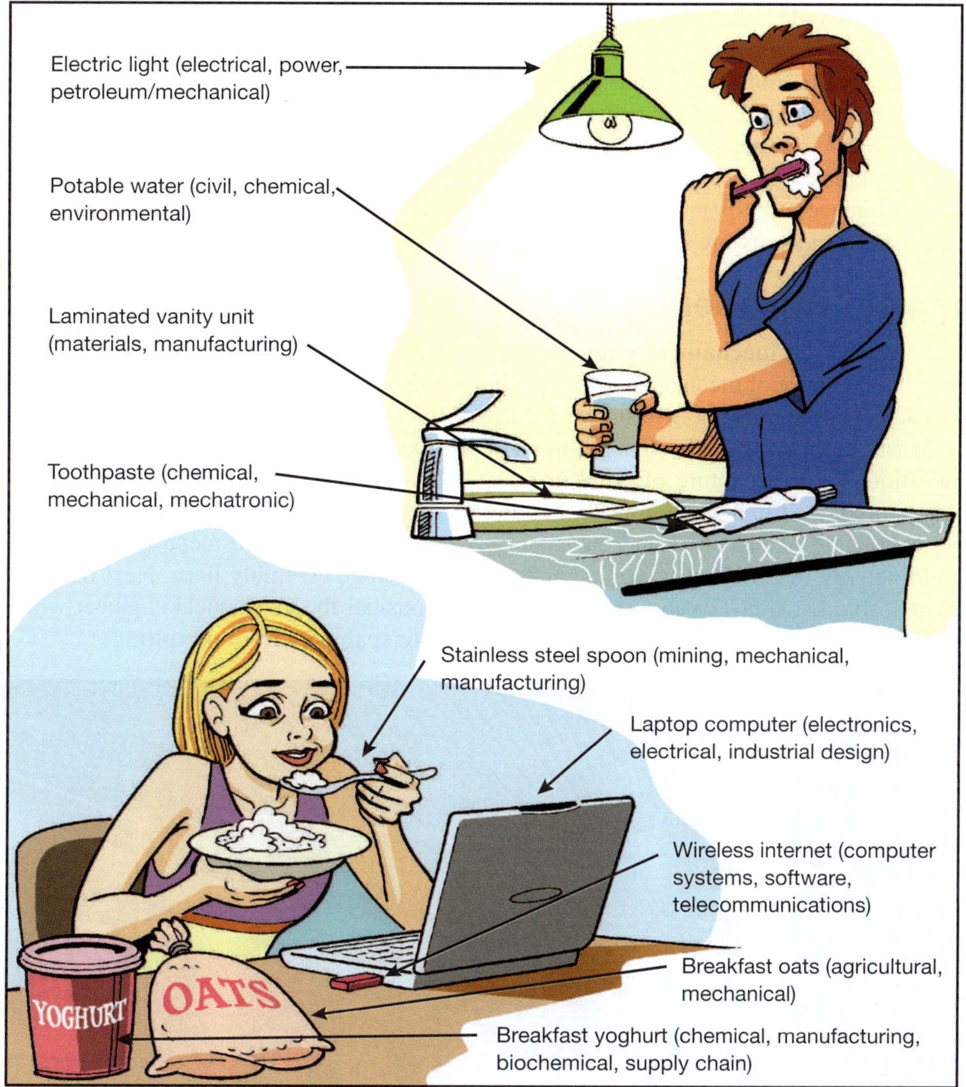

Electrical, electronics and telecommunications engineering

These three related fields of engineering account for those engineers responsible for the creation, transmission and use of electrical energy. **Electrical engineering** is concerned with the way electrical energy is produced and used in homes, the community and industry. Electrical engineers design and build the systems and machines that generate, transmit, measure, control and use the electrical energy essential to modern life. One specialisation within electrical engineering is power generation and distribution engineering, which is concerned with planning, developing, testing, installing, using and maintaining power plants or stations, and dealing with the transmission of that power to where it is needed — cities, towns, railway lines, large businesses and industry. Some power engineers also conduct research on developing alternative power sources that utilise solar and wind energy.

Electronics and telecommunications engineering are viewed as clearly defined fields that sit alongside electrical engineering. Electronics and telecommunications engineers design devices and systems that use small amounts of electrical energy to analyse, transmit and store information for communications, broadcasting, aviation, defence, robotics, in computers, for medical applications or in meteorology. Transmission of electronic signals forms the basis of communications in the information technology industry, and includes the field of microelectronics and the use of silicon chip technology. These engineers are responsible for technologies like mobile and satellite communications systems, optical fibre technologies and computer systems that allow for real-time communication between people who may be located on opposite sides of the globe. They also contribute to the development of electronics, sensors and telemetry (data collection and transmission) systems to support health applications such as the bionic ear, heart pacemakers and life support systems.

Electrical engineering has spawned an array of engineering specialisations and fields, such as software engineering (designing, modifying and supporting the use of computer programs and packages) and mechatronics (combining aspects of mechanical and electrical engineering to design processes and systems that, for example, control machines or simulate human thought and responses).

Computer systems engineering is another field related to electronics engineering that, in combination with computer science, is responsible for the analysis, design, development and operation of interlinked computer hardware and software systems (Engineers Australia 2014).

Mechanical engineering

Engineers who work in **mechanical engineering** mostly specialise in the design, optimisation, trial, creation and maintenance of mechanical systems and processes, and systems and processes that deal with heat. A mechanical system is one with moving parts or 'mechanisms' and much of the work mechanical engineers do with heat involves the generation and use of various forms of energy to power mechanisms (e.g. combustion of fuels, capture of wind or wave energy). These mechanical, energy-using systems pervade our everyday existence, from aircraft, car and ship engines to the mechanical devices used for the mass manufacture of industrial and consumer goods, such as extruded metal pipes, toothpaste (via automated toothpaste tube filling machines) and bottled beverage assembly lines. Next time you enter a cool room on a summer's day, or choose to take the lift, consider the mechanical engineer who designed the air-conditioning system or the elevator that has helped to spare you from perspiring.

Mechanical engineering assists the mass manufacture of consumer goods, such as the assembly line production of bottled beverages.

A lot of the work of mechanical engineers relies on the concepts that fall under the headings of 'dynamics', 'thermodynamics' and 'fluid dynamics'. All these concepts are about how force, energy and motion apply in different settings. Dynamics is broadly the study of the force and energy required to set solid members in motion, to stop them or to change their trajectory. Take a look at the pedal and braking system of a child's bike for a simple version of the application of dynamics in mechanical engineering. Thermodynamics allows mechanical engineers to understand and control how heat moves within materials and the conversion of heat from one energy form to another (e.g. from potential energy to momentum). Mechanical engineers can, for example, use the principles of thermodynamics to design and size car engine cooling systems to optimise engine performance and minimise fuel consumption. Fluid dynamics concerns the way that liquids, gases and mass particulates move under different conditions (e.g. under pressure, on a surface or in conduits). Mechanical engineers use their knowledge of fluid dynamics to understand how, for example, changes in hydraulic pressure can affect the performance of shock absorbers.

A specialisation closely aligned with the work of civil engineers is building services engineering. Specialists in this field often have an undergraduate degree in mechanical or electrical engineering. They design heating, air-conditioning, electric lighting and power, and communications systems for commercial and industrial premises. They also design water and gas supply, plumbing and drainage systems, and fire safety and security systems to service the needs of businesses and individuals who live and work in large commercial buildings.

Some mechanical engineers are leading the way in the capture of renewable sources of energy, and hence in fighting global climate change. For example, mechanical engineers are involved in the design of wave power generators, optimising the transfer of energy from source (wave) to the paddles, which in turn power energy-generating turbines. The effectiveness of wave power generators depends upon mechanical engineers calculating the fluid dynamics (such as drag and flow) of the sea water, the interaction of these dynamics with the paddle's shape and its surface qualities (aerodynamics, material properties). By optimising these (and other) variables, mechanical engineers contribute to the harnessing of a lucrative and environmentally sound source of energy.

Aerospace and aviation engineering

Engineers who work in **aerospace and aviation engineering** design, construct and monitor the operation of aircraft, aerospace vehicles and propulsion systems for planes, jets, helicopters, gliders, missiles and spacecraft. They are also involved in researching, developing and testing new materials for the aerospace and aviation industry as well as researching engines, body shapes and structures in the quest to increase the efficiency, speed and strength of air and spacecraft. They are also responsible for planning aircraft safety and maintenance programs and might develop space and aircraft automatic control and communication systems.

The Australian Civil Aviation Safety Authority employs aeronautical engineers whose main task is to ensure that aircraft are airworthy. This covers the certification of aircraft and involves the assessment of manufacturers' data from within Australia and overseas. In this role, they may have to assess mechanical systems, flight characteristics and aircraft performance. This is usually done through test flights, measuring take-off distances, rate of climb, stall speeds, manoeuvrability and landing capacities, and then comparing results with safety standards.

Another specialist role that some aerospace engineers carry out is investigating irregularities in the performance of air system components, faulty engines or failures in safety or communications procedures that affect, or have the potential to affect, the safety of air travel, air travellers or aircraft operators (Engineers Australia 2014).

SPOTLIGHT

Aerospace engineering and surf safety

Australia's beach culture is part and parcel of summer for locals and tourists alike. Over a million tourists visit our beaches annually and countless more Australians flock to the seaside each year for fun in the sun (Mackay 2020). This heavy visitation and over 20 000 kilometres of coastline, however, pose a substantial safety challenge (Mackay 2020). The range of dangers to manage along the coastline include rip currents, sharks and hazardous objects, swimmers in distress and dangerous surf conditions that may be less visible to the human eye.

Keeping Australia's coastline safe is a potential resourcing and logistical nightmare but aerospace engineers have put forward a partial solution (Mackay 2020).

Unmanned Aerial Vehicles (UAVs), or 'drones', have begun to be used by lifesavers in Australia. Surf Lifesaving Australia (2023, p. 2) has said that a drone:

> allows lifesavers to monitor large areas of the ocean quickly and efficiently. With the help of high-definition cameras and thermal imaging technology, UAVs can . . . provide a bird's-eye view of the ocean for better search and rescue operations.

Drones not only reduce response time and reduce risk to lifesavers, but they have also already saved lives.

An Australian company founded in 2015 called The Ripper Group has developed drones that can monitor beaches and drop emergency care packages to people in need (Mackay 2020). The Little Ripper drone uses onboard artificial intelligence (AI) to identify hazards (e.g. rips, crocodiles) to an accuracy of 90 per cent (University of Technology Sydney 2020). Since September 2020, UAV operators have been flying shark-spotting drones at beaches in South-East Queensland (Ripper Corp. 2021). Dr Nabin Sharma from the University of Technology Sydney's Centre for Artificial Intelligence has said:

> [This] is a great example of how an AI application can help humans, as it has significantly higher rates of visual accuracy in shark detection than people (University of Technology Sydney 2020).

This clever application of aerospace engineering has saved lives. In 2018, the lifesaving drone was key to the rescue of two teenagers off the coast of Lennox Head in New South Wales, when it was used to send an inflatable rescue pod to keep them afloat until surf lifesavers retrieved them (Mackay 2020). This use of drone technology was a world first, and resulted in the Little Ripper's inclusion in the National Maritime Museum as a 'significant piece of Australia's maritime history' (Ripper Corp. 2021).

This lifesaving technology combines AI with aerospace engineering, and demonstrates how the creative engineer can unlock opportunities in supposedly unrelated fields, enhancing human capacity, saving lives and generating new business opportunities (Mackay 2020).

CRITICAL THINKING

Having a passion for surf lifesaving or shark spotting aren't likely to be listed as 'essential selection criteria' in many of the engineering jobs you'll see advertised when you graduate. These elements of an engineer's background, however, might allow them to provide evidence of job-relevant personal qualities, experience and professional characteristics beyond engineering technical skill. What sporting, work, volunteering or other activities do you undertake that might show a future employer that you stand out from others in the employment game?

Chemical engineering

Engineers who work in **chemical engineering** design and manage technology and processes for the large-scale conversion of raw materials into useful and commercial products. This is chemistry on a grand scale. These engineers might choose to spend their professional lives researching raw materials and their properties. They might work in design and develop equipment and operating processes to extract and refine raw materials, or in manufacturing to produce food, petrol, plastics, paints, cosmetics, pharmaceuticals, paper, ceramics, minerals and more. Extracting and processing raw materials safely and without harming the environment is a major specialisation of chemical engineers. This means that clean production, hazard analysis, work health and safety, industrial ecology, environmental protection and the reclamation or clean-up of contaminated sites account for a substantial part of the employment market for chemical engineers.

Engineering fields that are closely related to chemical engineering include: combustion and petroleum engineering (design and operation of large-scale combustion chemistry for fuel refining and energy generation), smelting engineering (heat extraction of metals from bulk metal-rich ores), and water and wastewater treatment engineering (physical, chemical and biological treatment of water, wastewater and sewage sludge) (Engineers Australia 2014).

Civil engineering

Engineers working in the field of **civil engineering** undertake civil works that contribute to how people live their everyday lives. Their work focuses on the physical infrastructure of both urban and rural environments. In other words, when you walk around outside you will see civil engineering everywhere: buildings, roads and traffic control systems, water supply, stormwater and wastewater treatment, bridges, ports, dams and so on.

Much of the work of civil engineers relies on the concepts that fall under the headings of 'structural mechanics', 'fluid mechanics' and 'geomechanics'. Structural mechanics is the study of the behaviour and distortion of solid elements placed under various forms of load; for example, a fully laden Mack truck parked on a wooden bridge. Many civil engineering graduates work as structural engineers, designing the frameworks to support buildings and bridges. Fluid mechanics allows civil engineers to understand the forces that operate within, and are caused by, the movement of water and wind; for example, in the design of a wastewater treatment plant or in a desalination plant or a bridge pier in a fast-flowing river. Geomechanics is a specialised application of structural and fluid mechanics. It explains how soil and rock behave under pressure and in relation to water; for example, in the design of a retaining wall and drainage system to prevent land slip under conditions of extreme rainfall or saturation.

While some civil engineers specialise in designing big structures, there are also numerous roles in this field devoted to organising construction works, making policy and planning decisions related to transport and water, and consulting with the community and other users of urban and rural environments on their needs and aspirations. Many civil engineering graduates actually work as environmental engineers.

Civil engineers have played a lead role in the design, construction and maintenance of some of Australia's and New Zealand's most beautiful and iconic bridges, shown in figure 1.3.

FIGURE 1.3 Bridging form and function

Te Rewa Rewa Bridge, Taranaki, New Zealand
Modelled on the ribcage of a whale, this carefully planned and constructed bridge acts as the perfect frame for Mount Taranaki, seen here in the distance. The steel arch bridge has a 70-metre-long pedestrian and cycle path.

Gladesville Bridge, NSW
This bridge, with a 300-metre arch span, spans the Parramatta River. Construction took five years to complete and at the time of its opening, in 1964, it was the largest, longest single-span concrete arch ever constructed.

Webb Bridge, Melbourne Victoria
A collaboration between the artist Robert Owen and architects Denton Corker Marshall, this cycle and pedestrian link sits in the heart of the city. It crosses the Yarra River in a 200-metre tunnel of metal webbing designed to mimic an Aboriginal fish trap.

Story Bridge, Brisbane, Queensland
Connecting the two halves of the city, the landmark bridge opened in 1940. It is the longest cantilever bridge in Australia, constructed of 12 000 tonnes of steelwork and 22 storeys high. It was designed by John Bradfield, who was also responsible for the Sydney Harbour Bridge.

Hokitika Gorge Swing Bridge, Hokitika, New Zealand
This single-span swing footbridge crosses the Hokitika Gorge. It provides breathtaking views across the azure waters of the gorge. The main supporting cables are steel wire and it is a popular tourist destination.

Environmental engineering

Engineers who work in **environmental engineering** are responsible for protecting the environment by assessing the impact a technology, project or process will have on ecosystems, the air, water, soil and noise levels in its vicinity. This is done by investigating and analysing engineering works and designing operating procedures and processes to minimise adverse effects on the environment.

Environmental engineers predict what problems may be caused by accidents, such as oil or chemical spills, and work to assess likely long-term environmental effects. Environmental engineers are also involved in removing or remediating problems caused by past human and industrial activity. This might include deciding how to clean up contaminated industrial land so it can be used for housing, deciding on a plan to manage heavy metal contamination of sediments in harbours and fishing ports, or working out whether an old landfill site is safe to be used as a sporting field. They also plan and design equipment and processes for the treatment and safe disposal of waste material from municipalities (e.g. domestic solid waste and sewage sludge) and industry (e.g. biological hazards, construction and demolition waste), and they direct the conservation and wise use of natural resources. Some environmental engineers are involved in research and development of alternative energy sources, processes and procedures for water reclamation, and development of innovative waste treatment and recycling technologies and procedures.

Environmental engineers need to maintain their knowledge of the legislative and regulatory frameworks that govern the environmental performance of engineering work. They also keep up-to-date with environmental standards such as those governing emissions of gaseous and liquid-borne wastes to the atmosphere or waterways (Engineers Australia 2014).

SPOTLIGHT

Engineering better flood modelling

Constantly evolving technology can help local authorities track and plan for the impact of natural disasters such as flooding. According to Ms Monika Balicki, Director at Queensland consultancy WMS, 'Australia is one of the best places to see the advantages of graphics processing unit (GPU) modelling' for effective river and catchment management (Engineers Australia 2016).

Balicki's work included GPU modelling in Toowoomba as part of the region's 2015 planning scheme, a massive undertaking that covered nearly 13 000 square kilometres, including the Condamine River floodplain in Queensland's south-east. This work was highly significant as 33 people died across Toowoomba, Lockyer Valley and Ipswich due to catastrophic flooding over the summer of 2010–11 (Haffenden 2020).

The Condamine stretches across four local government jurisdictions and the councils there saw a need to update modelling of how water moved across this vast area. They were keen to use new technologies to improve flood-mapping accuracy.

Light detection and ranging (LiDAR) survey data and GPU technology made it possible for Balicki to develop a comprehensive map of flood elevation surfaces, velocities and depths, as well as flood hazard and hydraulic categories for a full set of modelled events. Using GPU 1D and 2D flexible mesh, Balicki was able to adjust the resolution for greater detail in areas of interest, such as towns (Engineers Australia 2016).

Balicki relied heavily on her engineering knowledge acquired through a Master of Science (Environmental Engineering) to improve councils' understanding of how development and management of the Condamine River floodplain would enable greater environmental sustainability and improve safety during flooding events.

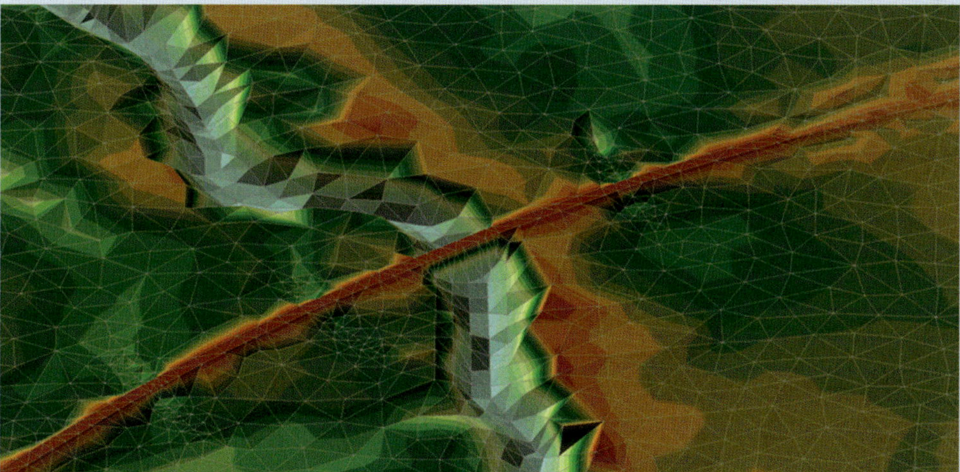

'Approach and outcomes: The hydraulic modelling of the Upper Condamine River catchment was undertaken using a coupled 1D/2D MIKE FLOOD Flexible Mesh model. The model covered an area of approximately 2230 km^2 with mesh elements ranging from 3 m^2 to 3600 m^2 in area. Over 30 bridges and culverts were represented in a 1D MIKE 11 model and coupled to the mesh topography. Comprehensive mapping showing modelled surface elevations, velocities, depths, flood hazard and hydraulic categories for the full set of modelled events was also produced to inform future Planning Scheme amendments' (Water Modelling Solutions 2017).

CRITICAL THINKING

Consider how the model and the associated information systems may be used by state government agencies and local government bodies to prepare for future flood events.

Materials engineering

Engineers who work in the field of **materials engineering** are employed in industries that turn raw materials into finished products. Materials engineering can be thought of as fitting in between mining or chemical engineering (where raw materials are produced) and manufacturing (where final products are made). Materials engineers design and select materials and materials-processing methods to fit a particular product or application. Put simply, they define the best material for the job. Some engineers in this field research and develop new, improved materials. If a suitable material does not yet exist, materials engineers will imagine, design and make a brand new one. Their work impacts every field of engineering, including construction (e.g. the development of lightweight, insulating concrete panels for energy efficient buildings), transport (e.g. the production of light, high-strength alloys for vehicles to reduce fuel consumption and increase safety), minerals processing (e.g. using plasma to produce hard surfaces on mining equipment so it lasts longer), electronics (e.g. the substitution of rare earth metals in mobile technology) and biomedical devices (e.g. the fabrication of materials for implants that promote regrowth of natural bone or tissue). Because they design and specify materials for such a range of different uses, materials engineers usually work in teams that involve other engineering disciplines, as well as many other specialists and stakeholders.

Materials engineers understand how the structure and properties of a material can be controlled by its composition and processing. Their work relies on knowledge of chemistry, physics and mathematics as well as statics, dynamics, mechanics of materials, fluids and heat flow, and mass transfer. As you'll find later in the chapter, this knowledge foundation will become very familiar to you over the next few years — regardless of your discipline of engineering — because these foundational engineering science ideas underpin the practice of all kinds of engineers.

Materials engineering is increasingly concerned with minimising the environmental impact of our modern society. Current developments include increasing the use of recycled materials, redesigning processes to minimise greenhouse gas emissions, designing nanotechnology for environmental remediation, and developing 'energy materials' for improved solar cells and hydrogen-based systems.

Mining engineering

Engineers who work in the field of **mining engineering** work with geologists to investigate and carry out the extraction of ore bodies and mineral deposits, as well as the extraction of non-metallic ores and fuels such as coal and uranium. They are responsible for planning the safest and most cost-effective way of removing minerals from the ground, rivers or the sea bed. They may be involved with designing, installing and supervising the use of explosives, ventilation equipment, and mining machinery and equipment, and for inspecting the progress of mining operations.

Computerised techniques are often used in the development and operation of mines. Mining engineers are responsible for protecting conditions for both people and the environment in the vicinity of mines. They work both on mining sites and in the head and regional offices of mining companies. Experienced mining engineers have a wide range of career options including mine planning and design, operations management, technical specialism, contracting, consulting, the mines inspectorate, investment analysis and advice, research, tertiary education and general management. Mining engineers use knowledge from other disciplines in their work (e.g. civil, electrical, mechanical engineering) (Engineers Australia 2014).

Other engineering disciplines

As will be apparent from the coming chapters and from Engineers Australia's long list of specialisations, there are many, many different branches or fields of engineering beyond those profiled here. Some are subsets of more established disciplines that have grown in size or complexity. For example, some suggest the field of environmental engineering grew out of civil engineering's focus on water provision, and water and wastewater treatment. Others see environmental engineering as having emerged from the manufacturing and chemical industry's need to meet the emissions standards that were introduced under various Clean Air Acts legislated in the United States, UK and elsewhere from the early 1900s.

Some fields of engineering have resulted from an intersection between established engineering fields (e.g. mechanical + electrical = mechatronic), or between established engineering fields and other disciplines (e.g. mechanical + electrical + human movement studies = biomechanical prosthetics engineering). Other engineering disciplines have emerged as a result of technological innovation (power engineering, software engineering) or shifts in social order. For example, agricultural engineering boomed as a result of the so called 'Green Revolution' of the 1960s that shifted production of food staples like rice and wheat to an industrial scale. Plant breeding, intensive use of synthetic fertilisers and wide-scale irrigation underpinned the need for industrial-scale engineering of irrigation networks, fertiliser distribution mechanisms and mass automated harvesting equipment (International Food Policy Research Institute 2002); hence, a heyday for the field of agricultural engineering.

SPOTLIGHT

Software engineer wins top student award

Victoria University of Wellington graduand Jack Robinson won the [Engineering New Zealand (ENZ)] Ray Meyer Medal for Excellence in Student Design.

The [now 29-year-old] was awarded the prize for his final year project. His project simplifies the creation of traffic management plans for roadwork sites.

'Whenever a contractor or event organiser would like to conduct work or hold an event on, or near a road they have to complete a Temporary Traffic Management Plan (TTMP). These are ten-page documents often with hand drawn diagrams of the site,' says Jack.

'My project moves the whole process online and makes it much easier. Users can specify work sites on a Google Maps-like page and generate a fully contextual work site.'

The Ray Meyer Medal is [ENZ's] top award for students and aims to encourage a new generation of innovative engineering designers. Jack is the first software engineering student to win the award.

The judges said Jack stood out as a clear winner of this award, with a well-presented project that had excellent commercial potential. They also commented on the good level of user-testing that Jack had engaged in.

Jack's project was supervised by Professor Dale Carnegie and Dr David Pearce.

'Jack was very motivated from the beginning and worked hard throughout the project,' says Professor Carnegie. 'The award is great recognition of the work he has done — one that solves a real-world problem. Jack displays all of the qualities we hope to instil in our Engineering graduates.'

Jack [graduated] with his Bachelor of Engineering with Honours in Software Engineering in [2017].

Source: Victoria University of Wellington (2017).

CRITICAL THINKING

All engineering students complete a final year research project as part of their degree program. Consider five engineering innovations you would like to research.

While students have many different motivations for commencing an undergraduate degree in engineering, the prospect of a good salary is a strong motivator. On average, Australian engineering graduates can initially expect to earn between \$70 000 and \$84 000 per annum (Glassdoor 2024a) and in the most senior roles of Managing Director or CEO of an engineering firm, salaries can range from \$220 000 to \$290 000 (Glassdoor 2024b).

As the next section demonstrates, engineers can work in many ways and can benefit from a wide range of skills.

1.3 The core skills and attributes of an engineer

LEARNING OBJECTIVE 1.3 List the core skills and attributes of an engineer.

Earlier in this chapter, it was highlighted that approximately 44 per cent of engineering graduates work in non-engineering occupations (Engineers Australia 2023). Engineering graduates are highly prized as employees in many fields of work outside engineering. Why is this? Engineering know-how and technical competence are important, but the skills and attributes of engineers have developed and changed in response to the changing role of engineers. Some consider that today's engineers are required to be more versatile than the engineers of yesteryear. Engineers need to look at their careers as an opportunity for life-long learning through the ongoing pursuit of professional learning (*continuing professional development* [CPD]). Engineers are expected to have, maintain and develop a range of what are sometimes termed *non-technical skills*, including (Clift 1998; Henley 2006; King 2008; Mitchell et al. 2004):

- being able to work with others
- being able to communicate technical ideas in non-technical language
- possessing business acumen
- having common sense
- desiring to contribute to sustainable development
- being able to think creatively and having intellectual independence
- possessing the ability to organise and manage budgets and logistics.

Demand for these skills has raised the benchmark for engineering students, who are expected to develop proficiency in a range of areas during their course. Undergraduate students should apply themselves to mastering technical knowledge and skills, and also to developing their generic or non-technical skills.

- *Basic skills.* These include written and oral communication skills, numeracy, computer and information literacy (ability to find, evaluate, use and create information) skills. Communication and information literacy skills are discussed in more detail in the chapters on understanding communication, communication skills and communicating information.
- *Practical skills.* These include interpersonal skills, basic budgeting skills, project management skills, and having an awareness of ethical frameworks and workplace health and safety laws (covered in the chapters on professional responsibility and ethics, understanding the problem and managing engineering projects).
- *Higher order thinking skills.* These include having a capacity for critical analysis, creativity and systems thinking, and the ability to reflect on individual work (known as reflective practice). These concepts are discussed in the chapters on self-management, communication skills, engineering design, evaluating options and engineering decision making.

In this section we will explore the need for, and nature of, some of these skills. Engineers Australia and Engineering New Zealand have both outlined key characteristics and attributes for professional engineers. These requirements form the basis for accreditation of engineering courses and membership of the professional body.

KEY POINT

Engineering graduates with a capacity for technical analysis and design, combined with well-developed non-technical skills, will be valuable to a wide range of employers.

The Engineers Australia competency framework

In the mid-1990s, Engineers Australia commissioned a review of engineering education in Australia. The report from the review was called *Changing the culture: Engineering education into the future* (Engineers Australia 1996). This report called for a bold shift in the direction of Australian undergraduate engineering education. The authors recommended that — as well as being competent in their discipline — undergraduate students needed to be introduced to innovation and creativity, and to have an 'awareness of when analysis should be supplanted by synthesis', at the 'earliest possible stages' of engineering courses (p. 7). Engineers Australia (1996) also stated this new breed of engineer needed to be a sophisticated ethical agent — someone able to 'inform themselves, their clients and employers, of the social, environmental, economic and other possible consequences which may arise from their actions' (p. 2).

The report made it clear that future engineers would operate in highly complex, dynamic and uncertain settings and that this would require engineers to be educated in key skills, known as graduate attributes. These skills were listed in the 1996 report and have become the foundation of the accreditation system for engineering programs in Australia. Engineers Australia requires accredited undergraduate engineering programs to build students' graduate capabilities and professional attributes, as laid out in the Stage 1 Competency Standard (Engineers Australia 2019). Stage 1 Competency corresponds to the completion of a four- or five-year Engineers Australia-accredited Bachelor of Engineering. The competencies represent the level of preparation necessary for a person to begin practising as a graduate engineer. A person with a non-standard educational qualification and appropriate engineering experience is also eligible to apply directly to Engineers Australia for assessment at the level of Stage 1 Competency.

Engineers Australia's Stage 1 Competency specifies a range of competencies or attributes that are listed in one of three categories: knowledge and skill base; engineering application ability; and professional and personal attributes.

Each category has several elements of competency.

1. KNOWLEDGE AND SKILL BASE

1.1 Comprehensive, theory based understanding of the underpinning natural and physical sciences and the engineering fundamentals applicable to the engineering discipline.

1.2 Conceptual understanding of the mathematics, numerical analysis, statistics, and computer and information sciences which underpin the engineering discipline.

1.3 In-depth understanding of specialist bodies of knowledge within the engineering discipline.

1.4 Discernment of knowledge development and research directions within the engineering discipline.

1.5 Knowledge of engineering design practice and contextual factors impacting the engineering discipline.

1.6 Understanding of the scope, principles, norms, accountabilities and bounds of contemporary engineering practice in the specific discipline.

2. ENGINEERING APPLICATION ABILITY

2.1 Application of established engineering methods to complex engineering problem solving.

2.2 Fluent application of engineering techniques, tools and resources.

2.3 Application of systematic engineering synthesis and design processes.

2.4 Application of systematic approaches to the conduct and management of engineering projects.

3. PROFESSIONAL AND PERSONAL ATTRIBUTES

3.1 Ethical conduct and professional accountability.

3.2 Effective oral and written communication in professional and lay domains.

3.3 Creative, innovative and pro-active demeanour.

3.4 Professional use and management of information.

3.5 Orderly management of self, and professional conduct.

3.6 Effective team membership and team leadership (Engineers Australia 2019).

In order for undergraduate students and their teachers to understand what learning needs to take place at Stage 1 Competency, Engineers Australia lists 'indicators of attainment' under each element. The indicators are not mandated competencies, but provide examples of the quality of thinking that would prove a graduate was competent in the specified element. The indicators of attainment for the communications competency (element 3.2) are listed as 3.2(a) and (b).

3.2 Effective oral and written communication in professional and lay domains.

(a) Is proficient in listening, speaking, reading and writing English, including:
- comprehending critically and fairly the viewpoints of others
- expressing information effectively and succinctly, issuing instruction, engaging in discussion, presenting arguments and justification, debating and negotiating — to technical and non-technical audiences and using textual, diagrammatic, pictorial and graphical media best suited to the context
- representing an engineering position, or the engineering profession at large to the broader community
- appreciating the impact of body language, personal behaviour and other non-verbal communication processes, as well as the fundamentals of human social behaviour and their cross-cultural differences.

(b) Prepares high quality engineering documents such as progress and project reports, reports of investigations and feasibility studies, proposals, specifications, design records, drawings, technical descriptions, and presentations pertinent to the engineering discipline (Engineers Australia 2019).

These communication indicators have been outlined here because communication skills are a key attribute according to engineering employers and professionals (Male et al. 2010). The need for engineers to develop good communication skills will be looked at in more depth in the chapters on understanding communication, communication skills and communicating information.

The assessment of graduating engineers' mastery of the Stage 1 Competencies is embedded in the routine assessment regime of the undergraduate degrees accredited by Engineers Australia. If you are studying an

Engineers Australia-accredited program, an important part of your faculty's role is reporting to Engineers Australia on how undergraduate students are taught and assessed to assure attainment of graduate attributes. This means that you will be unlikely to have to sit an exam or test at the end of your degree to prove your knowledge of the competencies for graduation. What is likely, however, is that some part of your formal assessment in many of your engineering subjects will be devoted to assessing your attainment of the Engineers Australia competencies. Engineers Australia does not expect that graduates will demonstrate a high level of attainment in every single detail of the knowledge, and competency described in the Standard, but they must demonstrate 'at least the substance of each element' (Engineers Australia 2013).

One of the reasons accreditation is so important is that holding a degree from an accredited course strengthens a graduate engineer's ability to work internationally. Engineers trained in Australia and New Zealand have their qualifications recognised internationally through multilateral agreements like the **Washington Accord**.

The Washington Accord is an agreement between professional engineering accreditation bodies (such as Engineers Australia and Engineering New Zealand) that they will offer mutual recognition of graduates' qualifications. This means that graduates of courses accredited by Engineers Australia or Engineering New Zealand may apply for graduate-level membership of other signatory bodies without having to undertake further study. Thus, graduates can attain work internationally and are very likely work with engineers who have trained in the countries listed in the accord (see figure 1.4).

FIGURE 1.4 Signatory and provisional signatory countries to the Washington Accord, which allows mutual recognition of engineering qualifications

Washington Accord

Signatories

- Australia — Represented by Engineers Australia (EA) (1989)
- Canada — Represented by Engineers Canada (EC) (1989)
- Ireland — Represented by Engineers Ireland (EI) (1989)
- New Zealand — Represented by Engineering New Zealand (EngNZ) (1989)
- United Kingdom — Represented by Engineering Council United Kingdom (ECUK) (1989)
- United States — Represented by Accreditation Board for Engineering and Technology (ABET) (1989)
- Hong Kong China — Represented by Hong Kong Institution of Engineers (HKIE) (1995)
- South Africa — Represented by Engineering Council South Africa (ECSA) (1999)
- Japan — Represented by Japan Accreditation Board for Engineering Education (JABEE) (2005)
- Singapore — Represented by Institution of Engineers Singapore (IES) (2006)
- Korea — Represented by Accreditation Board for Engineering Education of Korea (ABEEK) (2007)
- Chinese Taipei — Represented by Institute of Engineering Education Taiwan (IEET) (2007)
- Malaysia — Represented by Board of Engineers Malaysia (BEM) (2009)
- Turkey — Represented by Association for Evaluation and Accreditation of Engineering Programs (MÜDEK) (2011)
- Russia — Represented by Association for Engineering Education of Russia (AEER) (2012)
- Sri Lanka — Represented by Institution of Engineers Sri Lanka (IESL) (2014)
- India — Represented by National Board of Accreditation (NBA) (2014)
- China — Represented by China Association for Science and Technology (CAST) (2016)
- Pakistan — Represented by Pakistan Engineering Council (PEC) (2017)
- Peru — Represented by Instituto de Calidad y Acreditacion de Programas de Computacion, Ingeneria y Technologia (ICACIT) (2018)
- Costa Rica — Represented by Colegio Federado de Ingenieros y de Arquitectos de Costa Rica (CFIA) (2020)
- Mexico — Represented by Consejo de Acreditación de la Enseñanza de la Ingeniería (CACEI) (2022)
- Indonesia — Represented by Indonesian Accreditation Board for Engineering Education (IABEE) (2022)

Provisional signatories

- Bangladesh — Represented by The Institution of Engineers Bangladesh (IEB)
- Philippines — Represented by Philippine Technological Council (PTC)
- Chile — Represented by Agencia Acreditadora Colegio De Ingenieros De Chile S A (ACREDITA CI)
- Thailand — Represented by Council of Engineers Thailand (COET)
- Myanmar — Represented by Myanmar Engineering Council (MEngC)
- Saudi Arabia — Represented by Education and Training Evaluation Commission (ETEC)
- Nigeria — Represented by Council for the Regulation of Engineering in Nigeria (COREN)

Source: International Engineering Alliance (2024).

The Engineering New Zealand framework

Engineering New Zealand is a signatory to the Washington Accord and mandates that engineering programs develop undergraduate engineers' skills in ten areas that align with the Washington Accord Graduate Competency Profiles. The ten Engineering New Zealand areas are summarised as follows.

- 1.3.1 Apply knowledge of mathematics, natural science, engineering fundamentals and an engineering specialization . . . to the solution of complex engineering problems.
- 1.3.2 Identify, formulate, research literature and analyse complex engineering problems reaching substantiated conclusions using first principles of mathematics, natural sciences and engineering sciences.
- 1.3.3 Design solutions for complex engineering problems and design systems, components or processes that meet specified needs with appropriate consideration for public health and safety, cultural, societal, and environmental considerations.
- 1.3.4 Conduct investigations of complex problems using research-based knowledge and research methods including design of experiments, analysis and interpretation of data, and synthesis of information to provide valid conclusions.
- 1.3.5 Create, select and apply appropriate techniques, resources, and modern engineering and IT tools, including prediction and modelling, to complex engineering problems, with an understanding of the limitations.
- 1.3.6 Apply reasoning informed by contextual knowledge to assess societal, health, safety, legal and cultural issues and the consequent responsibilities relevant to professional engineering practice and solutions to complex engineering problems.
- 1.3.7 Understand and evaluate the sustainability and impact of professional engineering work in the solution of complex engineering problems in societal and environmental contexts.
- 1.3.8 Apply ethical principles and commit to professional ethics and responsibilities and norms of engineering practice.
- 1.3.9 Function effectively as an individual, and as a member or leader in diverse teams and in multi-disciplinary settings.
- 1.3.10 Communicate effectively on complex engineering activities with the engineering community and with society at large, such as being able to comprehend and write effective reports and design documentation, make effective presentations, and give and receive clear instructions (Engineering New Zealand 2020).

The Engineering New Zealand and Engineers Australia graduate competency statements look very similar, but each country has highlighted certain skills that feature more or less strongly in these statements. For example, the Engineering New Zealand statement makes it clear that engineering graduates from Engineering New Zealand programs will be skilled in 'carrying out experiments', whereas the Engineers Australia standard does not mention experimentation. The two frameworks also describe the 'communication' competency in different ways, with the Engineers Australia description being longer, and more detailed.

In addition to generic attributes, it is important engineering students build a strong knowledge of the basics of engineering science. These are discussed in the next section, and provide the foundation for the role engineers play in cooperating with other professionals to create technologies and processes intended to improve society. Many of these are aligned with the Engineers Australia and Engineering New Zealand competencies.

The following spotlight shows how interesting and varied engineering work can be, and the multicultural and multidisciplinary teams involved.

SPOTLIGHT

3D-printed heart valves

Australia's population is ageing and, as at 2021, 65 per cent of the adult population was overweight or obese (Australian Institute of Health and Welfare 2023). Ageing and obesity have both been associated with higher incidence of aortic stenosis. Aortic stenosis is a narrowing of the aortic valve that restricts blood flow from the left ventricle to the aorta, leading to increased angina and breathlessness (Perkins 2023). If left untreated, it can lead to heart failure, with up to 50 per cent of patients not surviving beyond 2–3 years (Perkins 2023). Around 1 in 8 elderly Australians has aortic stenosis (Perkins 2023).

Engineers, however, have contributed to a significant and dramatic breakthrough in cardiology. A Western Australian biomedical engineering and medical research team are developing next-generation artificial heart valves. The team's 3D-printed heart valve scaffolds have the potential to tackle this pervasive heart condition and save lives around the world (Becerra Mellet 2023).

Senior lecturer in biomedical engineering Dr Elena Juan Pardo, University of Western Australia (UWA) and the Harry Perkins Institute of Medical Research are working with cardiologist and expert in heart valve intervention Dr Abdul Ihdayhid to take this groundbreaking innovation to its next stage of development (Perkins 2023). According to Dr Juan Pardo:

> Heart valves are very complex organs — they are strong and flexible and thin. For a replacement heart valve, it's a harsh environment. Our 3D-printed valves are flexible yet strong [and] they have the same moving flaps, or leaflets (Becerra Mellet 2023).

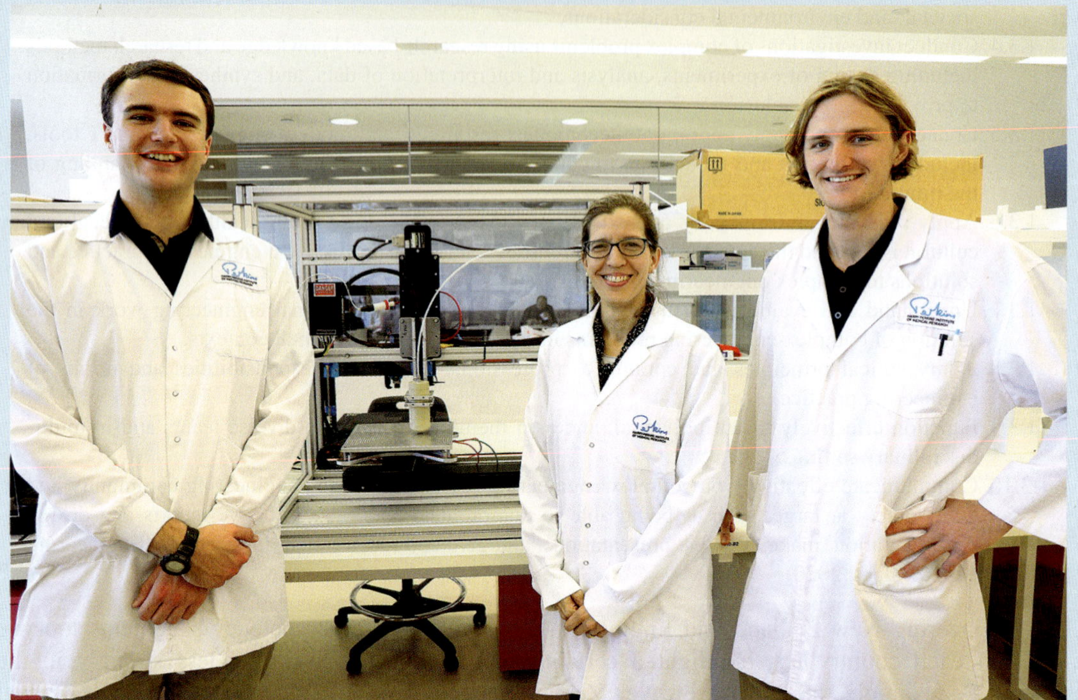

Dr Juan Pardo (centre) with PhD Students Christopher Lamb (L) and Michael Vernon (R) with their 3D Printer in the T3mPLATE lab.

The team is evaluating biopolymer-based valves as an alternative to current animal-derived or carbon-fibre mechanical valves. Current biological heart valves developed from animals can fail within 5 years; Dr Juan Pardo explains:

> Using a 3D-printing technique called Melt Electrowriting (MEW), we have designed and tested a patented biopolymer 3D-printed heart valve scaffold which exquisitely and uniquely mimics the mechanical properties of native human valves (Perkins 2023).

The research team is working to develop a 3D-printed heart valve scaffold that accurately mimics human valves and will endure for the rest of a patient's life (Becerra Mellet 2023). The importance of a valve's effective life cannot be understated; every time an implanted heart valve needs to be replaced, the patient undergoes significant surgery with significant concomitant risk and cost (Perkins 2023).

The global market for aortic valve replacement is estimated to grow from US$4.4 billion in 2021 to around US$13.4 billion by 2030 (Perkins 2023). The inspiring work of Dr Juan Pardo and her team is likely to help save lives, and will also reduce the financial cost to patients and the health system due to our currently increasing national incidence of heart failure from aortic stenosis.

CRITICAL THINKING

Excellent design and research skills are only one part of devising new medical technologies. Brainstorm some of the other skills and attributes, beyond technical skills in biomedical engineering, that the engineering team in the spotlight might need to develop a 3D-printed heart valve.

1.4 Engineering science

LEARNING OBJECTIVE 1.4 Identify some of the fundamentals of engineering science.

How does an engineering masterpiece come about? How does an established engineering discipline devise a new way to solve a problem? Thomas Edison is credited with the view that 'genius is 1 per cent inspiration and 99 per cent perspiration'. In engineering, *ingenious* innovation results in part from understanding the context of the problem, in part from the application of creativity, imagination and perseverance, and in part from good 'engineering know-how'. Engineering know-how involves a deep understanding of the interactions between the mass of facts, laws, ideas and traditions you will learn as an undergraduate student. Some of your early studies in maths, physics and chemistry might seem unrelated to solving practical problems. The basic concepts and laws of engineering science may seem abstract or overly theoretical. However, as you will discover, all of this fundamental knowledge provides useful building blocks for creating your own innovations as an engineer. Increasingly, it is as important to have engineering *know-why* as it is to *know-what* and *know-how*.

Some of the facts, laws and ideas that are fundamental to engineering have been around for hundreds of years. These facts serve as the fundamental pieces of knowledge for the profession, and are used by engineers of every discipline. They include:

- *modelling* — how complex systems can be represented and mathematically optimised
- *statics and structural mechanics* — how force affects solid objects
- *dynamics* — what happens when objects move
- *fluid dynamics* — the behaviour of liquids and gases
- *thermodynamics* — how heat moves
- *electromagnetics* — the interaction and utilisation of electrical and magnetic fields
- *quantum theory* — the operating principle of electronic devices
- *control* — ensuring engineered systems have predictable, controllable behaviour.

You will learn these facts, laws and ideas, and how to apply them, during your first few years of studying engineering science in subjects such as physics, statics or mechanics, dynamics, chemistry, circuit theory and thermodynamics. The following sections explain how some of these fundamental ideas in engineering science from your first and second year of study will form the basis for later learning and problem solving that is much more discipline-specific, concrete and applied.

KEY POINT

Basic concepts and laws of maths, physics and chemistry provide the building blocks for engineering work.

SPOTLIGHT

Learning journeys

Fluid mechanics learning journey

Fluid mechanics is a fundamental set of ideas in engineering. This area of study allows aerospace engineers to understand how solid objects, such as aircraft wings, move through the fluid medium of air. It underpins the work of chemical engineers in optimising heat and mixing conditions in processing systems full of reacting liquid — from electrolysis tanks in zinc metal processing to fermentation tanks full of beer or yoghurt. Understanding fluid mechanics allows civil engineers to plan for flood mitigation, and even to understand how wet concrete behaves during the process of pumping, distribution and settlement.

Basic concepts in physics and fluid properties, such as compressibility and shear stress, are pre-requisite to understanding the forces that fluids exert when they are at rest (known as *fluid statics*). These ideas, in turn, underpin concepts related to how fluids behave in terms of their energy and mass when they are in motion (*fluid dynamics*). Basic concepts such as compressibility and shear stress can be viewed as the first steps in a fluid mechanics learning path, which leads students to an understanding of fluid dynamics. The fluid mechanics learning path diverges after this point, into many sub-fields and applications (e.g. free surface flow, compressible/incompressible flow, laminar and turbulent flow, flow within porous media).

Structural mechanics learning journey

You may study structural mechanics early in your program. This set of topics is sometimes known as 'mechanics' or 'statics'. Structural mechanics is fundamental for many engineering disciplines in Australia and New Zealand, and usually comprises around 10 per cent of the first year of engineering studies, and up to 25 per cent of second-year engineering studies. This suite of ideas, laws and equations is based on requisite skills and knowledge from physics, and allows engineers to estimate the effects of different types and magnitudes of load on static elements and structures.

One concept in physics that underpins subsequent learning and the application of structural mechanics knowledge in civil and mechanical engineering is *equilibrium*. A bridge stays where we put it because the structure obeys Newton's laws of motion and all the forces and torsions balance out. Concepts from physics such as force, pressure and energy form the basis for understanding how steel I-beams react to loads placed upon them. Foundation concepts in physics, and the fundamental concepts from engineering science, underpin many applications of statics in the disciplines of civil and mechanical engineering.

Understanding how static or rigid objects behave under load is crucial for engineers who wish to accurately and economically size objects in a design. Once they understand stress and strain, and related concepts like shear force and bending moment, a civil engineering student is equipped to begin learning how to design and analyse real physical structures like bridges, roads and buildings.

CRITICAL THINKING

At this stage, the next few years of study might look like a series of unconnected (and sometimes incomprehensible) subject names. Get together with some classmates and use the handbook for your degree or discipline to make a map of which subjects link together into a path (you can start by looking at pre-requisites). Make your map more sophisticated by reading subject outlines or learning objectives and document where concepts build from year to year.

1.5 The impact of engineering on society and national identity

LEARNING OBJECTIVE 1.5 Explain the impact engineering has had on society over time.

Engineering is an age-old discipline. Many books refer to the substantial and sophisticated civil and mechanical engineering work that promoted the Egyptian, Roman, Persian and Chinese Empires through periods of domination during the pre-Christian age. Ancient engineering feats and modern engineering feats have both had a lasting and substantive impact on the world. Figure 1.5 lists some examples of significant advances in engineering over time from various countries. It includes some Australian and New Zealand engineering feats that have contributed to life, prosperity and national identity in this part of the world as well as many others that have contributed to improving quality of life on the international stage.

FIGURE 1.5 Selected engineering advances over time

- Invention of the printing press (China 1040, Germany 1440)
- Alexander Graham Bell's invention of the telephone (UK/USA 1876)
- Construction of the North Island Main Trunk Railway (New Zealand 1908)
- Sydney Harbour Bridge completion (Australia 1932)
- First mass-production of antibiotic Penicillin (UK 1945)
- Discovery of the double helix shape of DNA (USA/UK 1953)
- Vostok 1 — first manned spaceflight (Russia 1961)
- Kevlar — lightweight super strong fibre (USA 1965)
- Apollo 11 Lunar Module — first manned lunar landing (USA 1969)
- Invention of the portable telephone handset (USA 1973)
- Construction of Sydney Opera House completed (Australia 1973)
- Birth of the world wide web (Switzerland 1989)
- Wireless LAN (wifi) (Australia 1996)
- World land-speed record set in the car Thrust SSC — 1227.985 kilometres per hour (UK 1997)

- Oresund Bridge (Sweden/Denmark 2000)
- Maiden flight of the double-decker Airbus 380 (France 2005)
- First Total Artificial Heart transplanted to patient in Southern Hemisphere (Australia 2010)
- Implantation of the world's first 3D printed tibia (Brisbane, Australia 2017)
- Opening of the 55-kilometre Hong Kong–Zhuhai–Macao Bridge (a bridge–tunnel system) (Hong Kong 2018)

In the next section, we look at the impact of engineering from a historical perspective. We will examine early Australian and New Zealand engineering breakthroughs and consider the impact of engineering in modern times. Limitations to engineering progress and some of the unintended negative consequences of engineering will also be discussed.

KEY POINT

Over time, engineering innovations have triggered turning points in nations' histories, fostering economic prosperity and a sense of national identity.

A historical perspective

Engineering innovation has supported human development since the first cave-dweller attached a sharp chip of stone to a stick to create a hunting spear. Imagine the improvement in the efficiency and effectiveness of hunting food that a spear would create for people who had previously needed to chase, catch and club their evening meal. These cave-dwellers were not 'engineers' by modern standards, but the capacity to create technical solutions to resolve social problems has been at the heart of human development from the dawn of civilisation.

Engineering has its origins in the dual influences of scientific advance and social pressure. The widespread modern use of the term 'engineer' is based on the Latin word *ingeniator*, and appears to have started in the eleventh century. *Ingeniator* was applied to those who built 'ingenious devices and fortifications' such as the Tower of London (Auyang 2004, p. 14). The term achieved widespread use during the Renaissance (1300–1600), which was a period in European history of prolific change and advancement in many fields, including music, science, philosophy and technology. Auyang (2004) describes several phases in the emergence of modern engineering. The emergence of each of these overlapping phases came about in response to the social and political environments of different times, as well as because of advances in science, materials and technical ability. The phases overlap and are not clearly defined in time because change in engineering (and society generally) emerges as an often amorphous culmination of cultural, social, scientific and political pressures. In science, these revolutions are termed 'paradigm shifts' and are defined as radical change within a discipline, from one set of assumptions, concepts, values or practices to another (Kuhn 1970). Paradigm shifts, or in Auyang's terms 'revolutions', result in dramatic changes in the thinking and practice within a field or discipline. Looking at Auyang's work allows us to see when and why the various disciplines of engineering came into effect, and how they have impacted society. Auyang's treatise has been condensed into three distinct phases in the development of engineering.

Phase 1. The Scientific and Industrial Revolutions (1540–1890)

The **Scientific Revolution**, which began in the mid-1500s, heralded an era in which advances in science caused people to question some of the most fundamental beliefs and assumptions of their time. Advances in scientific thinking included the idea that matter was made of *atoms* rather than the *elements* (earth, air, water, fire and ether); that the sun was the centre of the universe rather than the earth; and that motion was governed by the concept of *inertia* rather than by the need for continued compulsive action. The Scientific Revolution laid the philosophical and scientific foundation for the first **Industrial Revolution**, which began in the 1700s. The first Industrial Revolution was a phase in (largely UK) history, when a substantial portion of the economy and labour force moved from small-scale, decentralised agricultural production, to large-scale, specialised industrial processing and manufacturing.

Pivotal to the first Industrial Revolution was work by the British mechanical engineer James Watt, who had improved on the design and fuel efficiency of early steam engines to such an extent that their widespread use prompted a period of massive economic growth in the UK. During this period, industrial processes — which had originally relied on the physical labour of individual workers — were increasingly driven by coal-fired steam engines. Reliance on steam engines meant a parallel reliance on the engineers who designed and maintained this crucial technology. Along with economic growth, the first Industrial Revolution caused a period of massive social unrest as droves of factory workers lost their jobs and incomes because of the widespread use of engineering technologies that substituted for human labour and skill. This was a turbulent time for underprivileged labourers in the UK, many of whom moved for work into cities and production centres that had neither the civic infrastructure (e.g. sewers, clean water supply) nor the social services (e.g. schools, hospitals) to meet their needs. Major manufacturing centres rapidly became overcrowded and public health suffered. The bacterial water-borne diseases typhoid fever and cholera killed thousands of people in the UK during this period and the environmental impact on air and water quality in some regions was severe. The Great Stink of London and its associated cholera epidemic convinced parliament in 1858 to fund what was then the largest municipal engineering project of all time: the London sewer system. A civil engineer, Joseph Bazalgette, was entrusted with the £3 million project that began in 1859 and was opened in 1865. This major civil engineering work not only rid London of its septic open sewers and polluted river, but it also eliminated the cholera and laid the foundation for London to become one of the greatest cities of its time (Cadbury 2004).

The steam engine was a major catalyst for the late 18th–early 19th century Industrial Revolution.

While the social and political impact of industrialisation in the UK was problematic, engineering played a profoundly significant role in creating the first Industrial Revolution. Engineers were the inventors, designers and operators of the ingenious and powerful technologies that transformed the UK and other parts of the world from decentralised, human labour-driven economies to economically powerful, centralised, manufacturing-focused economies. As a result, engineering became firmly established as a legitimate professional field, with the disciplines of mining, metallurgical, mechanical and civil engineering gaining traction in society by the turn of the nineteenth century.

Phase 2. The dissemination and diversification of engineering (1900–1945)

The first half of the twentieth century was a phase of significant scientific endeavour, which provided a platform for engineering to grow and diversify to serve needs and fill niches in all conceivable facets of early twentieth century life. During this time, advances in the fields of chemistry, physics and maths provided the platform for new products and services that were supported and disseminated by emerging engineering disciplines. The work of the Wright brothers introduced a new mode of transport for the very wealthy, spawning the study and growing practice of aeronautical engineering. The mass-production and distribution of chemicals, pharmaceuticals and food goods was supported by the emergence of chemical engineering and engineers who planned and oversaw growing national and international transport networks. Mechanical engineers, such as American Henry Ford, mass-produced the optimised 'internal-combustion engine', packaging it as a convenient mode of individual transport for the masses. The *Titanic* was a (short-lived) mascot for the considerable interest in — and the sophisticated technology that evolved in response to — mass transportation by sea. The public wanted electricity, and the disciplines of electrical and electronic engineering developed to assist in the generation, distribution and consumption of this new commodity.

In the early 20th century, the Model T Ford became the first mass-produced automobile. The founder of the Ford Motor Company, Henry Ford, was an engineer.

Phase 3. The Information Revolution (1945–2000)

The period after World War II has been described as the Information Revolution. This was a time when advances in micro-electronics and information systems brought computers into everyday usage. Consider the evolving size and memory capacity of computers, from the massive first computer — the Electronic Numerical Integrator and Computer (ENIAC) developed in 1946 — to the vast memory storage capacity and compact size of today's laptop and tablet devices and the evolving micro and nano-scale computing technology.

This period has also seen spectacular growth in space exploration and research. Aerospace engineering enabled the first moonwalk in 1969 and has laid the groundwork for the telecommunications satellites that underpin the proliferation of information and communication technologies we enjoy today. During this time, materials scientists developed, refined and mass-produced radical new polymers and materials such as plastics, nylon, Kevlar and lightweight metal alloys. Materials engineering supported the mass-production, downstream processing and production of consumer goods made from these new materials. These new materials revolutionised the work and capacity for innovation in many engineering disciplines. Consider the contrast in maritime engineering and naval architecture between the first half of the twentieth century, when much of the boat-building industry still relied on timber and steel construction, and today, where fibreglass, carbon fibre and modern metal alloys have changed the properties of hulls (e.g. drag and mass) and led to complementary changes in the business and efficiency of seafaring industries.

The Electronic Numerical Integrator and Calculator (ENIAC) was a forerunner to modern-day computers. It filled a 6-by-12 metre room, weighed 30 tonnes and used more than 18 000 vacuum tubes.

It was during the time of the Information Revolution that developments in the atomic sphere provided a knowledge basis for growth in the specialised field of nuclear engineering. This field continues to attract attention because the refinement and use of uranium for energy production and nuclear medicine creates waste products with highly toxic qualities that degrade extremely slowly. In the wake of the 2011 Fukushima earthquake and tsunami, which caused the meltdown of the Fukushima Daichi nuclear power plant, many countries are reviewing their dependence on such power. The German government announced in 2011 that it would phase out nuclear power. Initially, the Japanese government shut down all of their nuclear power stations but has gradually brought most of them back on stream. The continuing debate demonstrates the interconnectivity between engineering, social and environmental decision making.

Castells (2010) highlights how the information revolution has continued to impact during the first decade of the new millennium. For example, widespread, secure, fast internet has affected society both positively and negatively, including in:

- global trade and financial transactions (from eBay, internet shopping and cryptocurrencies to money laundering)
- improved communication between international police forces, which has led to intra and international coordinated action against insidious crime networks involved in such activities as drug trafficking, paedophilia and the illegal arms trade
- mobile communications and live streaming, which have helped shape political revolution and transition towards democracy — such as in the Egyptian people's overthrow of president Hosni Mubarak during 2011
- radical change in what we understand by 'socialising' and 'community', as citizens of the internet meet and interact in ways never before seen or possible but for the connective power of the world wide web and high-speed information transmission
- the rapid rise (and fall) of internet media sensations such as Myspace, Facebook, Instagram, TikTok and X (formerly known as Twitter)

- new categories of crime, defence and evidence such as cyberbullying and cyberstalking, ransomware, phishing, penetration testing and digital forensics.

Early Australasian engineering

Engineering innovation has a proud history in this part of the world, with early engineering changing the lives, wellbeing and destiny of our earliest peoples. Early engineering innovations by Aboriginal Australians and the Māori and Moriori peoples in New Zealand are discussed in the following spotlights. These achievements took place before white settlement in both countries and illustrate how thinking like an engineer has underpinned significant advances and development in our cultures and countries.

The nature of the engineering carried out by these ancient peoples relates more to discovery and craft than detailed calculations. Nevertheless, the achievements of the original inhabitants and the refinements they made from one generation to the next mirror the development of engineering in other parts of the world.

SPOTLIGHT

Early Aboriginal engineering

The earliest known engineering in Australia was undertaken by Aboriginal Australians. The Gunditjmara people of Victoria's Lake Condah (Tae Rak) area built and operated a complex aquaculture system comprising weirs, channels and holding ponds to regulate and aid the cultivation and trapping of eels and other freshwater fish (Gunditjmara People & Wettenhall 2022). This innovation involved the design, operation and maintenance of constructed watercourses that stretched over kilometres. It is estimated that up to 3000 cubic metres of soil and stone were excavated in the construction of these weirs (Johnston et al. 1995). This was a monumental task, as the people undertaking the excavation relied on hand tools fashioned from wood and stone. Today, the design and construction of watercourses is the responsibility of civil engineers, who have the benefits of sophisticated surveying equipment to set out earthworks and heavy machinery that can be used in excavations.

Excavations at Lake Condah reveal the construction history of fishing traps used in the nineteenth century, such as Muldoon's Trap Complex, seen here in 2008. Subsequent radiocarbon dating has determined that construction of the pictured fish trap/aquaculture system began 6600 years ago (McNiven et al. 2021).

A second example of early engineering innovation in Australia is from the field of aeronautical engineering (which, as outlined earlier in this chapter, is a branch of aerospace engineering). The boomerang is a deceptively simple-looking piece of aeronautical engineering. A boomerang is a curved wooden stick that is designed to fly through the air from the hand of a thrower, eventually returning to the thrower via an elliptical path. Traditionally, hunting boomerangs were used to strike and kill animals such as kangaroos and wallabies. Consider the exquisite craft involved in selecting and honing this lethal and efficient hunting tool. Modern aeronautical engineers work at length — often for years — perfecting aerodynamics for air transport. Aboriginal Australians spent a considerable amount of time and used great ingenuity in designing, testing and refining the boomerang.

Possibly the best example of an engineered Aboriginal artefact involving buoyancy is the bark canoe traditionally used along the Murray River valley. Red river gumtrees were tested for their suitability using yam sticks. A section of the bark of a suitable tree was then carved and peeled off using smaller sticks. The canoes varied in size with some scarred trees displaying sizes up to 4.5 metres long and 0.9 metres wide. The bark was fashioned into the canoe shape while still supple and tied at both ends. These boats did not have a long service life and were used for fishing in calm rivers and estuaries and for river crossings (Edwards 1972).

CRITICAL THINKING

Some consider that Western engineering only began with the advent of the Industrial Revolution and the move from a craft-based trial and error engineering tradition to the scientific engineering method based on calculated results. Do an internet search for the Brewarrina Fish Traps and one for the Roman aqueduct at Nîmes. Compare these projects to modern engineering methods for accurate surveying, levelling and construction. How do you describe the engineering demonstrated in these projects?

SPOTLIGHT

Engineering breakthroughs in early New Zealand history

Early maritime engineering allowed Polynesian settlers to discover and colonise mainland New Zealand and the Chatham Islands (about 850 kilometres due east of Christchurch). These early travellers, who became known as the Māori and Moriori peoples, designed, constructed and navigated canoes. These vessels transported them across the vast stretches of water separating Polynesia from New Zealand and the Chatham Islands.

An early account of the ocean canoes indicates how remarkable the engineering behind them was. Skinner (1919, p. 67) callously said the craft was 'designated by courtesy a canoe', but that it was 'in reality a . . . raft, able to be propelled or steered by oar or paddles. It was not in the slightest way watertight, and when fully laden must have been waterlogged to the seats'. Although his comments appear disparaging, Skinner was clearly impressed by the skill and bravery of those who built and sailed the canoes, and went on to describe the construction of these early engineering marvels in detail.

The Moriori had no access to metal-fastening technology such as nails, screws and rivets. Instead, all of the component parts of the ocean canoes were lashed together with string made from flax. Each canoe relied on two wooden *keels* (longitudinal beams that extended vertically into the water to provide lateral stability), short sections of the Ake-ake and Supplejack shrubs, and dual wooden poles for structural strengthening. Relatively soft, lightweight materials, such as the flower stems of the flax plant, fern stalks and small lengths of Matipo — a tough shrub — were also used in the construction process. Larger models of these vessels accommodated between 60 and 70 travellers (Skinner 1919). Considering the materials used, the distance covered and the probable seafaring conditions, these canoes were a substantial achievement in early maritime engineering.

CRITICAL THINKING

Use the internet to search your favourite period in history; you might search the 'Roman Empire', 'Pompeii', the 'first Emperor of China, Qin Shihuang' or 'Gorm, the first Viking king'. How did engineering skills and thinking influence people's lives during the period you selected?

After European colonisation, settlers in Australia and New Zealand relied on maritime engineering to transport supplies and people between the United Kingdom and colonial outposts (Australian Academy of Technological Sciences and Engineering [AATSE] 2000). Settlers spread out across the landscape from the ports where they arrived. These ports became capital cities in some states of Australia and regions of New Zealand. Transport networks still spread out from them into the hinterlands. Early settlement in Australia was beset with problems to do with fresh water supply (Johnston et al. 1995). Building aqueducts and sewer systems (examples of civil engineering) was an urgent priority for the fledgling colonies in both countries. As the settlers learnt how to farm in unfamiliar climates, demand for a means to transport crops and livestock to trading posts and ports increased. Advances in civil engineering — including the development of extensive road (and, later, rail) networks — allowed for the overland transportation of agricultural products and opened the way for timber and minerals exploitation (AATSE 2000). During this time, the growing wealth of the colonies paid for civil structures of increasing size, complexity and grandeur. Some of these, including Australia's oldest bridge (built by convicts in Richmond, Tasmania between 1823 and 1825) and Old Saint Paul's Cathedral in Wellington, New Zealand (a wooden structure built in the 1860s), remain a testament to Australia and New Zealand's early civil engineering heritage.

These early engineering endeavours helped establish viable settlements in Australia and New Zealand. In doing so, they set each country on a course towards independent nationhood. Over time, engineering innovations have triggered turning points in each nation's history, fostering economic prosperity and the development of a sense of national identity. Some examples of iconic engineering include New Zealand's Roxburgh Dam and North Island Main Trunk Railway, and Australia's Sydney Harbour Bridge and Snowy Mountains Hydro-electric Scheme.

Construction of the Snowy Mountains Hydro-electric Scheme saw over 100 000 migrants from more than 30 countries descend upon the Snowy Mountains to be greeted by the harshest of conditions. The majority of the project was carried out underground, with work in wet and often dangerous surroundings resulting in the deaths of more than 120 workers during the project's 25-year lifespan.

The Snowy Mountains Hydro-electric Scheme, completed in 1974, was a major engineering feat in Australia's history.

At a cost of more than $820 million, the workers managed to complete the project, which commenced in 1949 and ceased in 1974, in the time frame given. To this day it remains the largest engineering project carried out in Australia.

The entire scheme spans approximately 5124 square kilometres of the mountain range in southern New South Wales, yet only 2 per cent of the total construction can be seen above the ground. With more than 200 kilometres of tunnels, pipelines and aqueducts, sixteen major dams, seven power stations and a pumping station, it is acknowledged as one of the most complex hydro-electric schemes anywhere in the world (Australian Government 2008a).

The completion of the 680-kilometre North Island Main Trunk Railway (NIMT) between Wellington and Auckland in 1908 marked an era of enhanced trade and ease of travel between the major North Island regions. However, political and geographic challenges delayed the development, which took more than 36 years to complete (Pierre 1981).

The majestic scenery of New Zealand is well known. Wild mountain ranges are a feature of the terrain, which is also scored with chasms, rivers and glacial valleys. Although beautiful, the dramatic landscape tested the abilities of the engineers and tradespeople who were given the task of building the railway. The surveyor behind the NIMT, John Rochfort, also faced difficulties during the project. He entered the *Rohe Pōtae* (a region in New Zealand's North Island) during the 1880s, at a time when relations between the Māori and the *Pākehā* (New Zealand's white inhabitants) were tenuous. Rochfort's communication, negotiation and diplomacy skills were credited with paving the way for the construction of large sections of the railway (McLintock 1966).

The completed line, which still operates today, passes through 26 tunnels and along eight major bridges or viaducts. One section of the railway climbs 635 metres in 51 kilometres, negotiating gradients of up to 1:50 in a spectacular display of engineering brilliance. The NIMT also has another celebrated engineering feat, the Raurimu Spiral. At this stage, the railroad overcomes difficult terrain with a complete circle, two tunnels and three horseshoe bends (Pierre 1981).

The NIMT made a tangible difference to the way business was conducted between the cities of Wellington and Auckland. It was also symbolic for New Zealanders, particularly for those people who felt a profound need to subdue the strident landscape in which they were living.

In Australia, the much-photographed Sydney Harbour Bridge is an iconic engineering project. It opened in 1932 and remains the world's widest long-span bridge, with the capacity for eight lanes of car traffic and two railway lines (Guinness World Records 2009). The completion of the Harbour Bridge project was credited in most part to an engineer named Dr J.C.C. Bradfield. After much reworking of the original design, Bradfield settled on the two-hinged steel arch primarily because of its durability. The construction took eight years and involved 53 000 tonnes of steel, and 6 000 000 hand-driven rivets. Sixteen workers lost their lives in the process and numerous injuries ensued as a result of the perilous nature of the work. Skilled labourers including stonemasons and ironworkers were brought in from other countries to assist with the project. The Sydney Harbour Bridge gained international recognition as a unique and visionary structure. This recognition was vital to Australia, arriving at a time when the nation was attempting to overcome negative perceptions about its convict past (Australian Government 2008b).

The iconic Sydney Harbour Bridge is the world's widest long-span bridge.

A contemporary perspective

Engineering continues to change the way people across the globe are born, live their lives, and eventually die. Our lives are so interwoven with technology it can be hard to imagine how people survived without many of the innovations we take for granted. Consider the engineering technology that it takes to mass-produce, package and distribute antibiotics. Prior to the widespread availability of antibiotics, people in Australia and New Zealand commonly died from some injuries or diseases that we now think of as mere inconveniences. Biomedical, biochemical and chemical engineers design and maintain the massive *fermentation tanks* (large vessels, sometimes called 'bioreactors', which contain liquid media for the mass growth of micro-organisms like bacteria, yeasts and moulds) and associated equipment in which antibiotic-producing moulds are cultured under optimal conditions of temperature and nutrient availability.

Growing the antibiotic-producing mould is only one of the many engineered steps in antibiotic production. To fulfil their life-saving function, antibiotics rely on additional engineered technologies and processes overseen by engineers who specialise in separation science, mass product drying, pill-pressing, bottle capping, labelling, packaging, and transport and distribution logistics. Throughout this text you will find numerous examples of the ways in which engineering work innovates and evolves, with profound effects on human health and wellbeing.

The challenges of modern life mean that new fields of engineering continually emerge. They emerge as a result of both societal pressures and advances in technical aptitude. For example, the implications of over a century of industrial production are now weighing heavily on industrialised nations.

There has been a rapid growth in the discipline of environmental engineering involving the study of remediating, alleviating, managing or preventing pollution of the air, land and waters. Mitigating the detrimental environmental effects of polluting industries and activities like mine sites, manufacturing plants, municipal waste dumps, sewage treatment works and chemical processing plants is now big business.

One big business for environmental engineers intersects with the work of petroleum and maritime engineers — environmental engineers are sometimes called on to remediate damage to the environment caused by crude oil spills. In 2010, 11 workers were killed and an estimated 4.9 million barrels (670 000 tonnes) of crude oil was released into the Gulf of Mexico, United States, following an explosion on BP's Deepwater Horizon oil drilling platform in the Northern Gulf (Hu et al. 2011). The Deepwater Horizon spill took a tense three months to cap and called on the skills of structural, drilling and petroleum engineers as part of the multidiscipline and multi-institutional team tasked to stem the spill. The spill also mobilised a huge and rapid containment and remediation response that drew heavily on the knowledge and skills of environmental engineers. The company held responsible, BP, paid out approximately US$65 billion in clean-up and compensation costs (Vaughan 2018).

A bioreactor designed by biochemical engineers and used in commercial production of monoclonal antibodies. Photo courtesy of Professor Yusuf Chisti, Massey University, New Zealand.

Although on a smaller scale than these international spills, Australia and New Zealand have not been spared from such events. The 2009 blowout of the Montara Wellhead Platform resulted in oil spill and gas leakage in the Timor Sea north of Western Australia. The leak continued for 74 days, and a commission of inquiry highlighted numerous failures that led to the spill, including irregularities in the methods used to cement wells, and in the use of cement in stabilising and sealing off wells (Hunter 2010). A bad year for Australian marine environments, 2009 also saw up to 250 tonnes of oil spilt off Moreton Island in Queensland by the Hong Kong ship the *Pacific Adventurer*. During this incident, approximately 620 tonnes of ammonium nitrate also fell into the sea. Subsequent contamination to beaches on Moreton Island, Bribie Island and the Sunshine Coast required a massive clean-up operation, and the environmental damage to ecosystems in the region was predicted to be felt for years to come (Elks 2009). In October 2011 in New Zealand, an oil spill from the *MV Rena* container ship off the North Island was declared the country's worst maritime pollution disaster. It was reported that up to 350 tonnes of heavy fuel oil leaked into the Bay of Plenty (Australian Broadcasting Corporation [ABC] 2011). The clean-up operation cost over NZ$500 million (Davison 2016). Nowadays, more than a decade after the disaster, the wreck of the *MV Rena* has broken up and been colonised by fish and algae; it is now a popular-but-challenging scuba diving attraction.

These spill events strongly illustrate how the work of many engineering disciplines intersect in such complex situations: naval architects and off-shore engineers design boats and oil rigs; civil and mechanical engineers cooperate with petroleum engineers to fix drilling and piping to undersea oil and gas resources and effect extraction; petroleum and chemical engineers develop processes and plants to refine oil; and chemical and environmental engineers design clean up and remediation for when things go horribly wrong.

Engineering innovations

In the coming chapters, some of the most dynamic feats of engineering in modern times will be described and consideration given to how they were conceived, designed, implemented and operated. Consideration will also be given to the impact of engineering innovation on society. When you think of memorable engineering feats, many grand ventures are likely to spring to mind. The Eiffel Tower, constructed in 1889 in celebration of the French Revolution, was designed as a temporary structure; however, it has become one of the most iconic structures ever built.

There has been a lot of commentary about the innovative mechanical, electrical and computer engineering that allowed NASA's Mars Rovers to traverse the surface of Mars to collect samples to help us understand how the Earth was formed. As we discussed earlier, hundreds of thousands of lives have been saved through the mass-production of antibiotics. The internet, broadband, mobile telecommunications and associated hardware and software are also marvels of electrical and software engineering, which have substantially changed the way we conduct business and our relationships.

Limitations of engineering

Most engineers have the best of intentions when they design and implement new processes and technologies to solve problems or improve standards of living; however, even actions based on the best intentions can have unexpected or unpredicted negative outcomes. Examples of these include built-in obsolescence, which means a product or technology is designed for disposal rather than reuse or repurposing; large financial commitment to infrastructure development that encourages users into avoidable, undesirable, expensive or polluting behaviours (e.g. high-cost, fossil fuel powered seawater desalination to maintain a generous supply of potable, or drinking, water); or failure to provide adequate supporting services to ensure safe and sustainable use of the engineered solution (e.g. a new housing subdivision with inadequate public transport, employment opportunities or schools, causing residents to use excessive amounts of private car transportation to commute to work).

In the next spotlight, the drawbacks of consumer electronics — a rapidly growing electrical engineering speciality — are examined in more detail.

SPOTLIGHT

Consumerism and the growth of e-waste

The design and manufacture of cheap consumer electronic goods is an example of an engineering field that has both positive and negative effects. Cheap, lightweight consumer electronics allow even those on relatively low incomes to enjoy the benefits of home computing and internet access, quality sound systems, home theatres and mobile communications technologies.

The downside of this is the vast quantity of waste created. These cheap electronic goods are generally not designed to be repaired if they break down and, because they contain mixed components, they are difficult to recycle. They are mostly thrown away by users who then purchase a replacement model. The resulting refuse is called e-waste. As well as valuable substances, e-waste contains hazardous materials. These include lead, cadmium, mercury and persistent organic pollutants (POPs). This is one of the fastest growing forms of compound and potentially toxic waste in the developed world.

In 2019, Australia generated 511 000 tonnes of e-waste. By 2030 the national total is projected to rise by nearly 30 per cent, to 657 000 tonnes (Department of Climate Change, Energy, the Environment and Water 2023). New Zealanders are estimated to create an average of 19 kilograms of e-waste per year, which equates to 89 000 tonnes nationally per annum (Dreaver 2018).

Australia and New Zealand are parties to the Basal Convention, which concerns the international transport of hazardous waste. In 2025, signatories to the Basal Convention will adopt tighter controls on export of e-waste. In Australia, this offers challenges and opportunities that engineers will be central to realising. A discussion paper from consulting firm Clayton Utz (Smith et al. 2022) offered two e-waste scenarios, outlined as follows.

1. Increase recycling of e-waste to 80 per cent, and increase high-efficiency recycling systems
   This scenario would result in an e-waste recycling rate increase from 54 per cent to 80 per cent from 2019 to 2030, based on the use of high-efficiency recycling processes (where e-waste is dismantled to recover all materials). Approximately 380 000 tonnes of waste would go to high-efficiency recycling, leaving 160 000 tonnes (or 30 per cent of total recycling) to be processed by low-efficiency recycling

(where items are shredded). Landfill would account for 128 000 tonnes. This would mean an additional 140 000 tonnes of material being recycled, and more than 340 000 additional tonnes of e-waste being dismantled for high-value recycling.

The value of materials recovered would increase from $160 million to $600 million, avoiding approximately 2.7 million tonnes of CO_2 by 2030.

2. Extend e-product life spans by 10 per cent
Increasing the lifespan of all products by 10 per cent, starting in 2021, would directly reduce e-waste by 9 per cent, with fewer products introduced to the market. This would avoid 60 000 tonnes of waste, particularly in the category of solar photovoltaic (PV) and battery storage products. Annual consumption would be reduced by 19 per cent.

This strategy is projected to save 1.2 million tonnes of CO_2.

CRITICAL THINKING

What would it take to make 'repairability' an important design criterion for consumer electronics?

It can be easy or tempting to get so wrapped up in the pleasure of engineering design work or the implementation of technology that the usefulness, practicality or wider impacts of an engineered solution are easy to miss. It can also be tempting to dismiss the responsibility to think beyond the technical as 'someone else's problem' or 'outside my field of expertise'. The fact is that, as professionals, it is imperative engineers consider the broader impacts and possible negative ramifications of developing new processes and technologies. Where there are uncertainties about the likely impacts of an engineered solution, an engineer needs the confidence and vision to call a halt to their work and to seek input from others with appropriate expertise.

As well as considering the impact of the work they do, engineers need to take into account the importance of professionalism in the workplace. Ethical practice in engineering is promoted by Engineers Australia and Engineering New Zealand.

1.6 Professionalism, certification and ethical practice in engineering

LEARNING OBJECTIVE 1.6 Explain the need for professionalism and ethics in engineering.

Professions such as medicine, law and engineering are afforded a high degree of trust by the societies in which they operate. These societies allow professions like engineering to *self-accredit* (establish and administer their own review processes for the purpose of checking and maintaining quality). This means professional associations such as Engineers Australia or Engineering New Zealand hold the responsibility for checking and ensuring the quality of training for the next generation of engineers. If an individual engineer is judged by professional organisations such as Engineers Australia or Engineering New Zealand to be fully qualified, they have the legal right to practise independently as a Professional Engineer.

If you are studying for a bachelor degree in engineering, you have embarked on the first step that could lead to you becoming a Chartered Professional Engineer (CPEng). There are two stages to achieve certification to practise in Australia (see figure 1.6). First, as outlined earlier in this chapter, the candidate must successfully complete an undergraduate degree that is accredited by Engineers Australia as meeting the Stage 1 Competencies (Engineers Australia 2013). Graduates are eligible to become members of Engineers Australia at the level of Graduate Engineer (Engineers Australia 2024a).

After completing three or four years of relevant experience a Graduate Engineer may apply to become a full member of Engineers Australia, that is, a Professional Engineer (Engineers Australia 2024a). While gaining this experience the Graduate Engineer collects evidence and builds a portfolio to demonstrate competency in the criteria specified under Engineers Australia's Stage 2 Competency Standards — the level required for Chartered Engineer status. Currently there are 16 Elements of Competency covering four areas: personal commitment, obligation to the community, value in the workplace and technical proficiency (Engineers Australia 2024b).

The process of becoming a CPEng begins with the candidate conducting a self-assessment by rating themselves as either developing, functional, proficient or advanced in each of the 16 elements and in one (or more) of the listed Chartered Areas of Practice, such as Electrical Engineering (Engineers Australia 2024b). The self-assessment is undertaken online and may be completed several times until the candidate is ready to complete the next stage of the process — the Industry Review — where a short, written statement is provided to demonstrate competency in each of the 16 elements at the functional or higher level. The submission is assessed by Assessors from Engineers Australia, who may seek clarification or additional information.

FIGURE 1.6 Engineers Australia's pathways to membership categories and registration

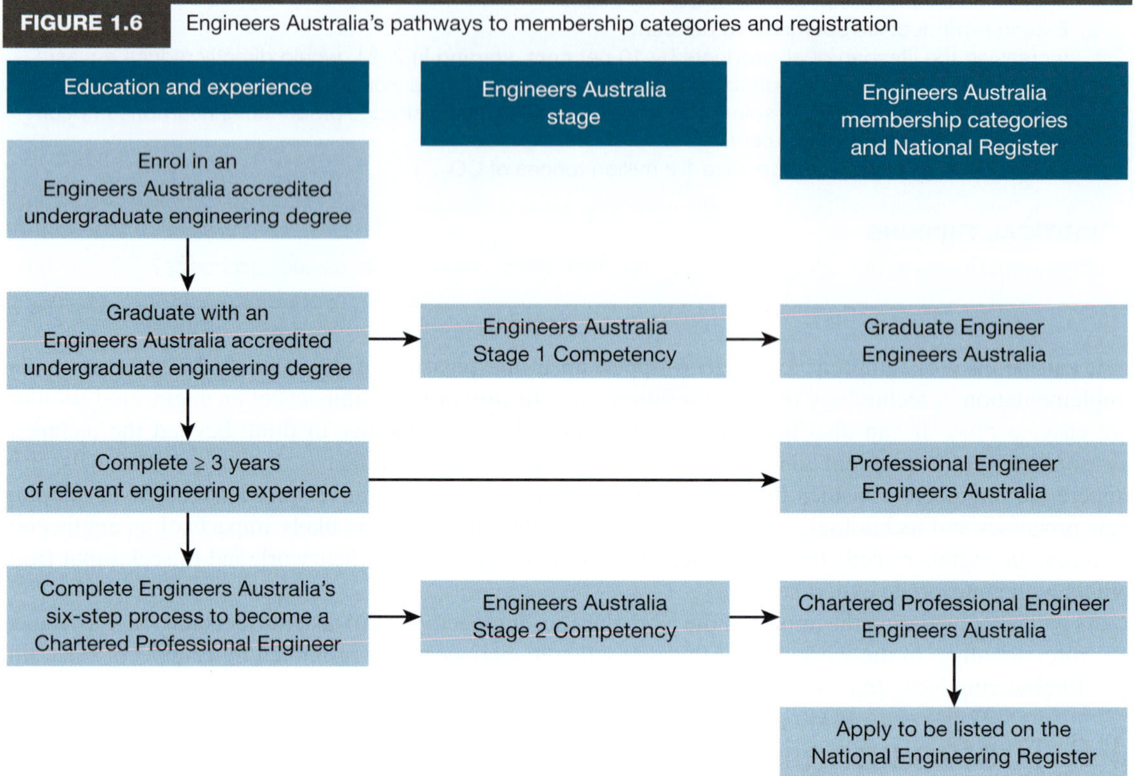

Source: Adapted from Engineers Australia (2024a, 2024b, 2024c).

Once this process has been satisfactorily completed a professional interview will be arranged and, if successful, the candidate will be awarded Chartered status in one or more areas of practice. They may then undertake independent practice under the general codes and frameworks set out by Engineers Australia, and other organisations and government agencies.

Chartered Professional Engineers are also eligible to be listed on the National Engineering Register and, if practising in Queensland, to apply for registration as a Registered Professional Engineer Queensland (RPEQ) (Engineers Australia 2024c).

In addition to the need for professional certification in some fields of engineering, there are legislative requirements that govern engineering work. Engineering ethics underpin both certification and compliance with legislative requirements. Breaches of legislative requirements can result in the threat of personal liability or imprisonment. Professional engineers know that in their position of trust as a professional, the first point of responsibility in engineering — to do the right thing technically and ethically — rests with the individual engineer. This places a high standard of behaviour on practising engineers. They need to uphold society's trust in the profession, being practitioners who are careful and responsible enough to operate independently without undue restriction and oversight.

Ethics in engineering and the codes of ethics that guide engineering work will be discussed in detail in the chapter on professional responsibility and ethics.

KEY POINT

A clear sense of personal and professional ethics is essential for independent practice as a registered engineer.

SUMMARY

In this chapter, we have looked at the basics of engineering. We have examined the role of an engineer, looked at different engineering disciplines and considered the core skills and attributes of an engineer. The impact of engineering on society over time has been discussed, while reviewing major phases of engineering growth and focusing on early Australasian engineering advances. We have considered the role engineering has to play in improving society and looked at some of the greatest engineering innovations. The importance of ethical practice in engineering has also been highlighted. We will now briefly revisit the learning objectives from this chapter.

1.1 Describe the roles of an engineer.

The role of an engineer is constantly evolving. Market demands, community expectations and advances in technology are constantly reshaping the practice of engineering. There are a range of different roles for an engineer and these roles require skills beyond just basic technical competence.

1.2 Identify the major engineering disciplines.

While there are more than 100 fields of engineering recognised by Engineers Australia, professional registration comes under the nine general areas: aerospace, biomedical, chemical, civil, electrical, environmental, ITEE (information, telecommunications and electronics engineering), mechanical and structural.

1.3 List the core skills and attributes of an engineer.

Engineers need both technical and non-technical skills to deploy their engineering know-how effectively and appropriately. The range of non-technical skills identified by Engineers Australia and Engineering New Zealand include being able to work in teams; being able to communicate technical ideas in non-technical language; possessing business acumen; designing experiments; a desire to contribute to sustainable development; being able to think creatively and having intellectual independence; and the ability to organise budgets and logistics.

1.4 Identify some of the fundamentals of engineering science.

Fundamental concepts of engineering science are applied to develop new processes and technologies. Engineering students can think about learning in different key areas of engineering as a learning path: from maths and physics foundations, through engineering science fundamentals, to application in design and problem solving. Engineers deploy the fundamental concepts of engineering in a range of different ways to understand and resolve problems in the various fields of engineering.

1.5 Explain the impact engineering has had on society over time.

Engineers have a profound impact on the lives, livelihoods and wellbeing of people — historically, now and in the future. Early engineering in Australia and New Zealand and iconic developments in both countries have had an impact on the national psyche of each. Engineering has helped shape the economic prosperity, quality of life and national identity of Australians and New Zealanders. Engineering innovation can occur on any scale and in any field of endeavour.

1.6 Explain the need for professionalism and ethics in engineering.

Engineers have professional responsibilities. Mechanisms to support professionalism include accreditation of undergraduate programs, registration to practice and professional codes of ethics. A clear sense of personal and professional ethics is essential for dealing with ethical dilemmas that may arise in the practice of engineering.

As you work your way through subsequent chapters, you will continue to develop your sense of what it means to work as an engineer and your understanding of the technical and non-technical skills that are essential to be an engineer. This will assist you in remaining motivated and focused on your goal of future practice as an engineer.

KEY TERMS

aerospace and aviation engineering The field of engineering concerned with the design, construction and operation of aircraft, aerospace vehicles and the propulsion systems for planes, jets, helicopters, gliders, missiles and spacecraft.

chemical engineering The field of engineering concerned with the design and management of technology and processes for large-scale conversion of raw materials into useful and commercial products.

civil engineering The field of engineering concerned with the undertaking of civil works that contribute to how people live their everyday lives, mainly focusing on the physical infrastructure of the urban and rural environments.

electrical engineering The field of engineering concerned with the way electrical energy is produced and used in homes, the community and industry.

environmental engineering The field of engineering concerned with the protection of the environment by assessment of the impact of technology, projects or processes on ecosystems, the air, water, soil and noise levels.

Industrial Revolution A phase in (largely UK) history, when a substantial portion of the economy and labour force moved from small-scale, specialised industrial processing and manufacturing to large-scale, specialised industrial processing and manufacturing.

materials engineering The field of engineering concerned with designing, producing and selecting the best material for a particular application. If a suitable material doesn't exist, materials engineers research, develop and produce new, improved materials.

mechanical engineering The field of engineering concerned with design, optimisation, trial, creation and maintenance of mechanical systems and processes, and systems and processes that deal with heat.

mining engineering The field of engineering concerned with the investigation and extraction of ore bodies and mineral deposits, as well as the extraction of non-metallic ores and fuels such as coal and uranium.

Scientific Revolution An era where advances in science caused people to question some of the most fundamental beliefs and assumptions of their time.

Washington Accord An agreement between professional engineering accreditation bodies that offers mutual recognition of engineering graduates' qualifications.

EXERCISES

1 Explain the difference between a craft or trade, such as carpentry, and engineering.

2 Explain how the role of an engineer has changed over time.

3 Name five engineering disciplines and explain clearly what they are. How are they different? Give at least two examples for each engineering discipline.

4 Identify the skills you will need to develop to win the engineering job you really want, considering the Engineers Australia and Engineering New Zealand competencies and graduate attributes outlined in this chapter. Think about the changing role of engineers as you formulate your response. Outline the skills you think will be more important for engineers in your field in the future.

5 Obtain a list of the graduate attributes defined by your university. Compare them to the Engineers Australia/Engineering New Zealand competencies. Are all the competencies included in the university list?

6 What do you think can be learnt about engineering from studying history? Split into groups in class or online and discuss the statement, 'History cannot teach the engineering student anything new'. Then either support or oppose this statement in a debate. Consider the impact of different engineering disciplines on society over time in your response. What are the reasons for and against this claim?

7 Discuss what you know about the fundamentals of engineering science with others. Why is engineering science important?

8 Consider an innovative engineering feat in recent history. What was special about this feat? How did it benefit society? Did it use any new technology or require creative thinking? Were there any setbacks to the project? What were the limitations of the project? Did it have any critics and, if so, why? Do you think this project will be considered innovative in 100 years' time?

9 Consider the following list of engineering technologies.
   - Optical fibres
   - The Cochlear implant
   - Pencils
   - Space shuttles
   - The internet
   - Composting toilets
   - Teabags
   - Solar water desalination

- Laptop computers
- Lawnmowers
- Cars
- Nuclear power
- Refrigeration
- The OPAL Reactor
- Antibiotics
- Robots
- Telecommunications
- Satellites
- Hydro-electricity
- Greenhouse gas reduction targets
- The Hill's Hoist
- Automatic teller machines
- Macadamia nut processing plant
- Electricity
- The Three Gorges Dam (China)

Which of these do you think are innovative? Why or why not? Do you think some of them have made more of a contribution to society than others? The list here is not comprehensive and some of the items may seem odd — the point of the list is to provoke you to discuss, debate, think through and justify your response.

PROJECT ACTIVITY

Your client is Engineers Australia. They are keen to promote *the real work of engineers* to high school students. The Director of Engineers Australia's Outreach program has asked you to prepare an *engaging* and *accurate* essay, explaining what engineers do in one field of engineering (e.g. civil, mechanical or environmental engineering). Choose the field of engineering you are most interested in.

- What do you know about its origins and history?
- Is it an ancient form of engineering like civil engineering or mechanical engineering, or has it emerged recently (for example, aerospace engineering or biomechanical engineering)?
- What do engineers in this field of study do?
- What different roles do they have and what are some of the obstacles they are likely to encounter 'on the job'?
- How does this type of engineering affect society?

Your essay should be between 2000 and 3000 words, word processed, with correct spelling, grammar and referencing employed (see the chapter on communication skills for ideas on collecting and presenting information).

Mention modern examples of this type of engineering and consider any trends in the field. Make sure you justify your response, explaining why you believe this is an exciting area of engineering. The information in this chapter should provide a good starting point for your research.

REFERENCES

Augustine, NR 1994, 'Socioengineering', *The Bridge*, vol. 24, no. 3, pp. 3–14.

Australian Academy of Technological Sciences and Engineering 2000, 'Technology in Australia 1788–1988: A condensed history of Australian technological innovation and adaptation during the first two hundred years', https://www.atse.org.au

Australian Broadcasting Corporation 2011, 'NZ says oil spill is country's worst', *ABC News*, 12 October, https://www.abc.net.au/news/2011-10-11/nz-says-oil-spill-is-countrys-worst/3497542

Australian Government 2008a, 'The snowy mountains scheme', https://www.pm.gov.au

—— 2008b, 'Sydney harbour bridge', https://www.pm.gov.au

Australian Institute of Health and Welfare 2023, 'Overweight and obesity', https://www.aihw.gov.au/reports/overweight-obesity/overweight-and-obesity/contents/overweight-and-obesity

Auyang, AY 2004, *Engineering — an endless frontier*, Harvard University Press, Massachusetts.

Becerra Mellet, G 2023, 'Harry perkins institute on the cusp of medical breakthrough with 3D printed heart valve', *PerthNow*, 21 September, https://www.perthnow.com.au/local-news/perthnow-western-suburbs/harry-perkins-institute-on-the-cusp-of-medical-breakthrough-with-3d-printed-heart-valve-c-11959594

Cadbury, D 2004, *Seven wonders of the industrial world*, Harper Perennial, London.

Castells, M 2010, *End of millennium: The information age: Economy, society and culture*, Wiley-Blackwell, Chichester.
Cheng, M n.d., 'Bio', https://www.maritacheng.com/bio.html
Clift, R 1998, 'Engineering for the environment: The new model engineer and her role', *Transactions of the Institution for Chemical Engineering*, vol. 76, no. 2, pp. 151–160.
Davison, I 2016, 'MV Rena owners will cover ongoing costs', *The New Zealand Herald*, 27 February, https://www.nzherald.co.nz/nz/mv-rena-owners-will-cover-ongoing-costs/IWICYV227EBS2OZGKLX6ZA4TPY
Department of Climate Change, Energy, the Environment and Water 2023, 'E-stewardship in Australia', https://www.dcceew.gov.au/environment/protection/waste/e-waste
Dreaver, C 2018, 'The fight against e-waste continues', *Radio New Zealand*, 6 February, https://www.rnz.co.nz/news/national/349753/the-fight-against-e-waste-continues
Edwards, R 1972, *Aboriginal bark canoes of the murray valley*, Rigby for the South Australian Museum, Adelaide.
Elks, S 2009, 'Oil spill ship owners deny "lying"', *The Australian*, 15 March, https://www.theaustralian.com.au
Engineering New Zealand 2020, *Requirements for accreditation of engineering education programmes (ACC 02)*, version 3.1, https://www.engineeringnz.org/documents/123/ACC_02_Accreditation_Criteria_V3.1_20201030.pdf
—— 2024a, 'Chartered professional engineer', https://www.engineeringnz.org/join-us/cpeng
—— 2024b, 'Working in engineering', https://www.engineeringnz.org/our-work/engineeringyourfuture/working-engineering
Engineers Australia 1996, *Changing the culture: Engineering education into the future*, Engineers Australia, Canberra.
—— 2013, *Guide to assessment of eligibility for membership (stage one competency)*, Engineers Australia, Canberra.
—— 2014, *Engineering disciplines*, Engineers Australia, Canberra.
—— 2016, 'Australia's most innovative engineers', *Create Digital*, 21 July.
—— 2019, *Stage 1 Competency Standard for Professional Engineers*, https://www.engineersaustralia.org.au/publications/stage-1-competency-standard-professional-engineers
—— 2018a, 'Engineering areas of practice', https://www.engineersaustralia.org.au
—— 2018b, 'What is engineering?', https://www.engineersaustralia.org.au
—— 2023, *The engineering profession: A statistical overview*, 15th edn, https://www.engineersaustralia.org.au/publications/engineering-profession-statistical-overview-15th-edition
—— 2024a, 'Membership', https://www.engineersaustralia.org.au/membership
—— 2024b, 'Chartered', https://www.engineersaustralia.org.au/credentials/chartered
—— 2024c, 'Registration', https://www.engineersaustralia.org.au/credentials/registration
Gerritsen, J 2007, 'Engineering a future', *New Zealand Education Review*, 15 June, p. 8.
Glassdoor 2024a, 'Graduate engineer salaries', https://www.glassdoor.com.au/Salaries/graduate-engineer-salary-SRCH_KO0,17.htm
Glassdoor 2024b, 'Director of engineering salaries', https://www.glassdoor.com.au/Salaries/director-of-engineering-salary-SRCH_KO0,23.htm
Guinness World Records 2009, *Guinness: World records 2009*, Guinness World Records Limited, UK.
Gunditjmara People & Wettenhall, G 2022, *The people of Budj Bim: Engineers of aquaculture, builders of stone house settlements and warriors defending Country*, PRESS Publishing, Ballarat.
Haffenden, D 2020, 'Queensland floods of 2011: The crisis that killed 33 people and devastated communities', *7 News*, 11 January, https://7news.com.au/news/qld/queensland-floods-of-2011-the-crisis-that-killed-33-people-and-devastated-communities-c-634419
Henley Management College 2006, 'Report for the royal academy of engineering', in N Spinks, N Silburn & D Birchall, *Educating engineers for the 21st century: The industry view*, 8 March, The Royal Academy of Engineering, UK.
Hu, C, Weisberg, RH, Liu, Y, Zheng, L, Daly, KL, English, DC, Zhao, J & Vargo, GA 2011, 'Did the northeastern Gulf of Mexico become greener after the deepwater Horizon oil spill?', *Geophysical Research Letters*, vol. 38, no. 9, pp. 1–5, https://doi.org/10.1029/2011GL047184
Hunter, T 2010, 'The Montara oil spill and the marine oil spill contingency plan: Disaster response or just a disaster?', *Australian and New Zealand Maritime Law Journal*, vol. 24, no. 2, pp. 46–58.
International Engineering Alliance 2024, 'Washington Accord', https://www.ieagreements.org/accords/washington/signatories
International Food Policy Research Institute 2002, *Green revolution: Curse or blessing?*, IFPRI, Washington, DC.
Johnston, S, Gostelow, P, Jones, E & Fourikis, R 1995, *Engineering and society: An Australian perspective*, Longman, Melbourne.
King, R 2008, *Addressing the supply and quality of engineering graduates for the new century*, report on engineering education in Australia for the Carrick Institute for Teaching and Learning in Higher Education, Sydney.
Kuhn, T 1970, *The structure of scientific revolutions*, University of Chicago Press, Chicago.
Mackay, S 2020, 'How drones are saving lives in Australian oceans', *Engineering Institute of Technology*, 30 April, https://www.eit.edu.au/how-drones-are-saving-lives-in-australian-oceans-2
Male, SA, Bush, MB & Chapman, ES 2010, 'Perceptions of competency deficiencies of engineering graduates', *Australasian Journal of Engineering Education*, vol. 16, no. 1, pp. 55–68, https://doi.org/10.1080/22054952.2010.11464039
McCarthy, N 2012, *Engineering: A beginner's guide*, Oneworld Publications, London.
McLintock, AH (ed.) 1966, *An encyclopaedia of New Zealand*, New Zealand Ministry for Culture and Heritage, Wellington.
McNiven, IJ, Crouch, J, Richards, T, Gunditj Mirring Traditional Owners Aboriginal Corporation, Dolby, N & Jacobsen, G 2012, 'Dating Aboriginal stone-walled fishtraps at Lake Condah, southeast Australia', *Journal of Archaeological Science*, vol. 39, no. 2, pp. 268–286.
Mitchell, CA, Carew, AL & Clift, R 2004, 'The role of the professional engineer and scientist', in A Azapagic, S Perdan & R Clift, *Case studies for engineers and scientists*, John Wiley & Sons, Chichester.
Perkins (Harry Perkins Institute of Medical Research) 2023, '3D printed heat valves to be commercialised', *News & Events*, 25 May, https://perkins.org.au/3d-printed-heart-valves-to-be-commercialised
Pierre, WA 1981, *North Island main trunk: An illustrated history*, AH & AW Reed, Wellington.

Ripper Corp 2021, 'SLSQ invests in world-first-hi-tech drone rescue enterprise', 7 May, https://rippercorp.com/2021/05/slsq-invests-in-world-first-hi-tech-drone-rescue-enterprise

Skinner, HD 1919, 'Moriori sea-going craft', *Man*, 19 May, pp. 65–68.

Smith, C, Brennan, A & McFadyen, M 2022, 'Draft national e-waste stewardship strategy: Too good to waste', *Clayton Utz*, 3 February, https://www.claytonutz.com/insights/2022/february/draft-national-e-waste-stewardship-strategy-too-good-to-waste

Stephan, KD 2001, 'Is engineering ethics optional?', *IEEE Technology and Society Magazine*, Winter 2001–02, pp. 7–12.

Surf Lifesaving Australia 2023, 'Australian UAV service guide to surf life saving club use of UAVs', https://www.surflifesaving.com.au/wp-content/uploads/sites/2/2023/09/AUAVS-Guide-to-SLSC-Use-of-UAVs-Aug-2023.pdf

University of Technology Sydney 2020, 'Life-saving technology for Australian beaches', https://www.uts.edu.au/research-and-teaching/research/explore/impact/life-saving-technology-australian-beaches

Vaughan, A 2018, 'BP's Deepwater Horizon bill tops $65bn', *The Guardian*, 16 January, https://www.theguardian.com/business/2018/jan/16/bps-deepwater-horizon-bill-tops-65bn

Victoria University of Wellington 2017, 'Software engineer wins top student award', 10 April, https://www.wgtn.ac.nz/news/2017/04/software-engineer-wins-top-student-award

Water Modelling Solutions 2017, 'Condamine River flood study', https://wmseng.com.au/portfolio/condamine-river-flood-study

ACKNOWLEDGEMENTS

Photo: © AnnaStills / Adobe Stock Photo
Photo: © Roman Zaiets / Shutterstock
Photo: © Kadmy / Adobe Stock Photo
Photo: © Water Modelling Solutions
Photo: © Harry Perkins Institute of Medical Research
Photo: © razerzone23 / Adobe Stock Photo
Photo: © ClassicStock / Alamy Stock Photo
Photo: © Everett Collection Historical / Alamy Stock Photo
Photo: © Professor Ian McNiven
Photo: © PearlBucknall / Alamy Stock Photo
Photo: © SARAYUT RADOMKIT / Shutterstock
Photo: © pisaphotography / Shutterstock
Photo: © Yusuf Chisti
Photo: © phoenix021 / Adobe Stock Photo
Figure 1.3: © S Watson / Shutterstock; Taras Vyshnya / Shutterstock; Robyn Mackenzie / Shutterstock; Thomas Hansson / Shutterstock; Klanarong Chitmung / Shutterstock
Figure 1.4: © International Engineering Alliance 2024, 'Washington Accord', https://www.ieagreements.org/accords/washington/signatories
Figure 1.6: Adapted from Engineers Australia 2024.
Text: © Gerritsen, J 2007, 'Engineering a future', *New Zealand Education Review*, 15 June, p. 8.
Text: © Water Modelling Solutions 2017, 'Condamine River flood study', https://wmseng.com.au/portfolio/condamine-river-flood-study
Text: © Victoria University of Wellington 2017, 'Software engineer wins top student award', 10 April, https://www.wgtn.ac.nz/news/2017/04/software-engineer-wins-top-student-award
Text: © Engineers Australia 2019, *Stage 1 Competency Standard for Professional Engineers*, https://www.engineersaustralia.org.au/publications/stage-1-competency-standard-professional-engineers
Text: © Engineering New Zealand 2020, *Requirements for accreditation of engineering education programmes (ACC 02)*, version 3.1, https://www.engineeringnz.org/documents/123/ACC_02_Accreditation_Criteria_V3.1_20201030.pdf

CHAPTER 2

The engineering method

'The engineering method is the use of heuristics, to cause the best change, in a poorly understood situation, within the available resources.'

Koen (2003)

LEARNING OBJECTIVES

After studying this chapter, you should be able to:

2.1 describe the activities that constitute the engineering method and apply the method to an identified problem or challenge

2.2 identify a range of system definitions for a problem and use these definitions to present different solutions to a problem

2.3 apply basic project management principles to plan a project and maintain organised project documentation

2.4 describe the role of an engineer throughout the life cycle of an engineering asset, including the differences between conceptual design and detailed design

2.5 use critical thinking skills to scope and solve engineering problems.

Introduction

This chapter will consider how engineers approach problems or projects using the *engineering method* — a problem-solving and design process that can be applied to any problem. The context of engineering problem solving, including project management processes and procedures that surround the engineering process, will also be considered. These processes include the nature of the engineering project life cycle, using systems thinking, scheduling and resources, maintaining project documentation and, finally, reflective practice as a means for continuous improvement.

This is a first introduction to the engineering method, so the intention is to present the whole picture, rather than getting into too much detail, which is covered in subsequent chapters. *Systems engineering* is the formalised view of what we call here the engineering method (International Council on Systems Engineering n.d.).

Engineers work on problems at various stages of an engineering project. Some work on strategic planning — what energy supply will be required in 20 years' time? Some work in research and development — how can a bionic eye be made to work? Some work in conceptual or detailed design — determining the nature of the best solution and how it will be implemented. Some engineers work on the implementation itself — manufacture or construction. Others work on the commissioning of a new project and in operations — getting an oil refinery operating the first time and keeping it working throughout its lifetime, including regular maintenance processes. Finally, some engineers work in decommissioning — think how complex it would be to dismantle a nuclear power plant.

Now, consider a real engineering problem. Many cities in Australia are faced with declining water supplies, likely caused by climate change. Periods of drought sometimes reduce the city reservoirs below half their capacity. What can be done? Many capital cities have built a desalination plant (Perth, Sydney, Melbourne and Adelaide). Why? Who made these decisions and how did they do it? Why did they recommend such an expensive piece of infrastructure, which might only be used occasionally? What are the greenhouse gas implications?

A wind farm provides the energy for Perth's desalination plant and some other states have committed to buying renewable energy but have not indicated whether this energy will come from new or existing facilities. Are these good engineering choices? Brisbane has taken a different route, recycling wastewater as potable water should dams fall to critical levels; however, this is not the preferred option for many people. For example, the city of Toowoomba in Queensland rejected such an option in a referendum in 2006 (*The Sydney Morning Herald* 2006).

How can calm choices be made between these alternatives? What other alternatives are there? Melbourne has built a north–south pipeline, taking water formerly used for irrigation across the Great Dividing Range into city reservoirs. There are also plans to pipe recycled wastewater from the Eastern Treatment Plant at Carrum to Gippsland for use as cooling water for power stations, and to pipe the fresh water currently used in the power stations back to Melbourne. These projects are currently on hold.

Which of these projects should go ahead? How can decisions be made, and who will be affected by them? If you were the engineer in charge of water supply for your city and the reservoir is sitting at 20 per cent capacity with 8 months' supply, which options would you recommend? Engineers require a systematic process that will guide them through such difficult decisions. The engineering method is such a process.

The process starts with the need to be well informed. The problem needs to be well researched and carefully defined. This research needs to identify what *criteria* will be used for choosing the preferred option, including considerations of sustainability: social and environmental as well as economic measures of performance. This stage usually requires engaging with the stakeholders for the product or project to understand the social and environmental dimensions of the project.

Second, it is important to consider all the available options as well as their advantages and disadvantages. Third, there needs to be a way of evaluating the options against the criteria so that it can be determined which options best fit the criteria. Checks need to be performed ensuring all required criteria have been satisfied, and then a recommendation can be communicated to decision makers.

This problem-solving process is one part of a more complex process. Project management skills will be required to manage the process; systems thinking skills will be required to ensure the right problem is being solved; documentation management will be necessary to track the thousands of pages and files created during the process; and a continuous improvement strategy will be required. This involves reflective practice — constantly questioning whether this problem-solving process can be done better.

Throughout this chapter, we will discuss the engineering method in relation to a common scenario — buying a car. This scenario has been deliberately chosen because it is a scenario with which many university students are familiar. By applying the engineering method to a familiar problem, you should be able to grasp how you could apply this method to more challenging engineering situations.

Victorian desalination plant at Wonthaggi

2.1 The engineering method

LEARNING OBJECTIVE 2.1 Describe the activities that constitute the engineering method and apply the method to an identified problem or challenge.

Many engineering jobs start with a request of some kind from a client. The client could be an external body or from another part of your organisation. The client may have drafted a formal document and attempted to define the problem and the scope of work, perhaps including measures of performance (criteria) that will be important. Alternatively, the request could be a simple verbal request.

For example, a company that designs electronic control devices is approached by an automotive company to design an active cruise control module that includes emergency brake assist. This new module will be required to control the speed (accelerator) and brake (to control excess speed going downhill). It will also interface with a forward radar system to prevent nose to tail collisions by slowing the vehicle to maintain a safe separation distance. Another company is designing the radar sensor and the interface specification has been provided. The interface to the accelerator is well understood (such devices have been in use for at least 40 years). The interface to the braking system will need to be developed. This is a new feature that will require some new mechanical components as well as matching electronics. The device will likely be manufactured in China and there are three months in which to develop the prototype. Five thousand units per year will be manufactured.

In this case, the problem has been quite well defined, even though there are still some detailed design issues to be resolved, such as how the device will interface to other systems in each car — the device should operate across many different vehicles.

In many other jobs, the engineer may have to spend some time with the client helping them to define the problem and desired performance criteria, thinking of options the client has not mentioned. After all, this is why clients hire consultants — because they want expert advice. The process of clarifying the problem for a client will be considered later under systems thinking.

Regardless of how the request is made and how detailed it is, it initiates an **engineering project**, which is a task that is identified in response to an identified engineering problem. In response to a perceived problem, an engineer can apply the **engineering method**.

The engineering method has five basic steps.

1. *Explore* the problem. Gather data and knowledge (this is the research phase) and, in the process, define the problem, the performance criteria and any constraints that must be considered (collectively, the **requirements**).
2. *Explore* the available solutions to the problem.
3. *Evaluate* the alternative solutions against the requirements using various models (e.g. economic and technical models).
4. *Choose* the preferred solution, including monitoring, reviewing, and checking the outcomes (against the original specification and constraints).
5. *Communicate* the recommended solutions to the client.

Now we will consider these five steps of the engineering method in more detail.

KEY POINT

The engineering method is a systematic problem-solving process.

Step 1. Exploring the problem

The process starts with an expressed need, which is usually documented in some form of **client brief**. A client brief is a document that describes the *need* of the client as well as the **performance criteria** and **constraints** — the requirements — for an engineering project. If these are not included, the engineer needs to help the client articulate them.

Consider the following possible problems clients may want resolved: longer battery life in a mobile phone with minimal increase in weight, a new aircraft with sustainable fuel options, or a more efficient wind turbine with lower noise emissions. Each of these problems suggests requirements, for example, minimum weight, minimum fuel, renewable fuel, minimum noise or maximum efficiency. At this point sustainability goals and constraints can be clearly articulated and agreed — lower material usage, lower energy consumption, recyclability, and so on — and an experienced engineer begins to *plan* how to satisfy these needs. The following spotlight discusses the need for ongoing research into the production of wind energy.

SPOTLIGHT

Wind energy

Denmark leads the world in terms of the percentage of its electricity that is generated from wind (2.3 TWh, representing 68 per cent of national demand for the month of January 2022 with more than 100 per cent on many days) (State of Green 2022). The government's aim is to phase out fossil fuels by 2050 (Danish Energy Agency 2014).

Denmark is in the fortunate position of being able to export excess energy to Norway, Sweden, Germany and the United Kingdom, and to import hydropower from Norway and Sweden when the wind drops.

Denmark is also a leader in the development of the technology that is now ubiquitous, such as the three-bladed wind turbine (Shahan 2014). The history of electrical wind turbines covers more than a century, from James Blyth in Scotland, and Charles Brush in the US, both in the late nineteenth century, who developed turbines to charge the batteries in their homes, to the Danish school teacher and aerodynamicist Paul la Cour, who developed the three-bladed turbine that is so common today. He built a wind tunnel for his tests and used his turbines to produce hydrogen, which was stored and used to light the lamps in his school, sometimes with explosive consequences!

Modern research undertaken by electrical and mechanical engineers, in companies such as Vestas, Siemens and General Electric, is constantly improving both the electrical and mechanical efficiency of turbines, while reducing undesirable side effects such as noise to adjacent homes. Noise reduction has been achieved by paying more attention to insulating the mechanical components of the generating equipment (State of Green 2021). Quieter blades have also been developed.

An issue that has had special attention in Australia has been the effect of wind turbines on birds, including the orange-bellied parrot and the wedge-tailed eagle (Smales 2006). Noise impact on adjacent homes has also been and continues to be a concern in Australia, though recent research may have put this issue to rest (Flinders University 2023). Further research and community engagement will be necessary in the resolution of these issues and in the wider adoption of wind energy in Australia and New Zealand.

CRITICAL THINKING

If you were involved in the planning for a potential wind farm site, what data or other evidence would you seek to ensure that noise levels and other adverse effects are contained? Where might you find such evidence?

The engineering method is the basis of engineering problem solving and is wrapped within a project management process. This is illustrated in figure 2.1, where the engineering method is represented by the steps in the inner loop. The focus will be on explaining these steps, with a discussion on how the method is related to project management later in the chapter.

FIGURE 2.1 The engineering method and its relationship to the project management process

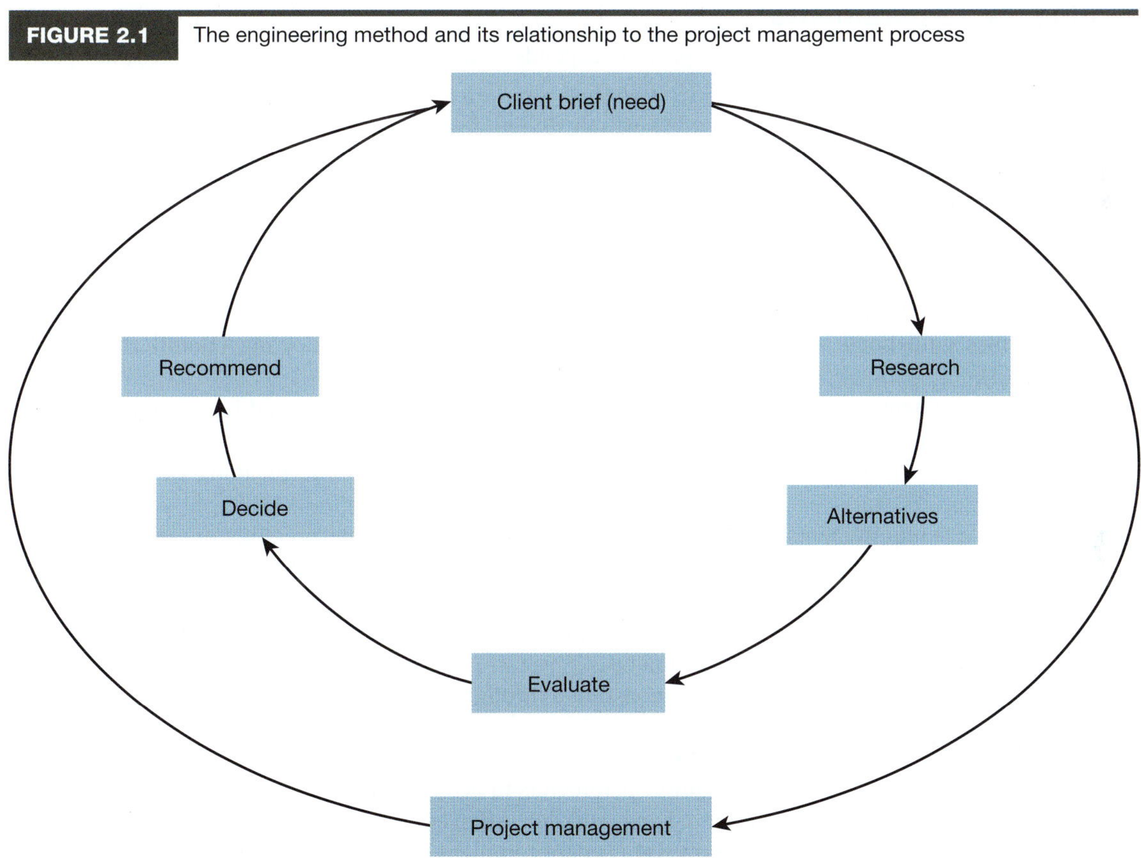

The engineering method will be explained by discussing how it is applicable to the modern scenario of buying a new (or used) car. Buying a car is a purchase decision many university students are faced with and the problem-solving process that is integral to the engineering method can be applied to this scenario. A similar situation is choosing a mobile phone. If you are choosing a phone for yourself, that might be easy enough. But what if you have to advise one of your parents which phone to choose? How would you go about that?

By looking at practical issues such as this, the process should become clearer. Consider the case study ‘Buying a car’. It should give you an idea of the roles of both the consultant and the client.

CASE STUDY

Buying a car

Imagine you have been asked to act as an adviser for a friend or a younger family member (in an engineering project, they would be known as your client). Your role is to help them to buy a car. The budget, the type of car and the requirements must be defined by them. They have told you that they want a small to medium-sized car, with good fuel economy, low kilometres, automatic transmission and a reasonably large luggage compartment (big enough for two suitcases). It must not be red and it has to cost less than $10 000. A towbar would be useful and leather seats would be nice as well. A good sound system would be another bonus.

Although this is not an engineering case, it is a great opportunity to apply the engineering method to a practical problem. Use the engineering method to choose the car. First, you should clarify the *requirements*. Second, consider several alternatives. Third, rank these against the criteria. And last, check and document your recommendation. This is the specification for the car. This specification would be a type of car, rather than a particular car.

If your client accepted your recommendation and took the project to implementation, you would look for the type of car matching the specification. During this process, you might modify your choice. In effect, the implementation phase of a proposed solution might change the design recommendation. This is quite common during the implementation of engineering projects, with unforeseen issues often influencing the construction process (e.g. the site conditions at a bridge may require a redesign of the foundations).

Writing the requirements

The successful completion of engineering projects depends on having good requirements, ones that specify what is to be delivered. Requirements are so important that they have their own syntax. Note the use of the word 'shall' in the following examples — not will or should, but *shall*.

- The car shall be no larger than medium size.
- The car shall be fuel efficient.

Do you think that these are clear and unambiguous? Could they be defended in a court of law? It is much better to have clear, measurable criteria; take, for instance, the following.

- The car shall have a mass of no more than 1200 kilograms.
- The car shall use less than 8 litres/100 kilometres in city driving.

Some further guidelines for writing good requirements include the following (Bahill et al. 1996; Lennard 2013).

- *Focus.* A requirement is expressed in terms of 'what' and 'why' or form, fit and function, not in terms of how to develop the products or the materials to be used.
- *Clarity.* A requirement is readily understandable without analysis of the meaning of the words or terms used.
- *Measurability.* A requirement is expressed so as to allow for only one interpretation of meaning (i.e. not defined by words or terms, such as 'excessive', 'sufficient' or 'resistant', that cannot be measured, as in the previous examples).
- *Singularity.* A requirement statement cannot be sensibly expressed as two or more requirements having different agents, actions, objects or instruments.
- *Testability.* A requirement can be only demonstrated through a finite and objective process.

Data, information and reliability

An engineering project almost always requires parameters to be quantified. *Data* needs to be collected for this purpose. **Primary data** is data observed and recorded, often collected in the field or from respondents (e.g. in a questionnaire survey). **Secondary data** is data compiled both inside and outside an organisation, to be used for the project at hand. Examples include national census data and meteorological data.

Depending on how they are collected, these primary or secondary data may include information such as the details of electricity consumption, battery life, temperatures, rainfall, solar energy input or community needs. Increasingly, many of these data sets are available online.

These types of generic data are very useful for feasibility studies. The data can often be accessed quickly, for free or at low cost, and it can be used in preliminary engineering models. For example, *Perry's chemical engineers' handbook* contains a vast collection of materials and performance data needed by chemical and process engineers (Green & Southard 2018). Almost every chemical engineer has one of these on their bookshelf for instant reference.

For more specific data, other professionals are paid to collect primary data for you, such as for a water quality assessment, wind tunnel testing of a prototype, land surveying or market research of desired new mobile phone features.

However you collect the data, you need to assess its *reliability*. Has the data come from a reliable source? Can it be trusted? Perry's handbook, for instance, is a highly reliable source. Now in its ninth edition, it was originally published in 1934 so has had many revisions to improve its reliability and is constantly peer reviewed.

Likewise, formal publications such as textbooks and journal articles are reliable sources. They have usually been peer reviewed for their accuracy. One of the problems with many articles on the internet is that they have not been reviewed by anyone other than their author. Consequently, you need to be very careful when using such unreviewed sources. Make sure that you can corroborate any data by finding other sources that support the original article.

Engineering is a conservative profession, and it is important to have good data as input to its processes. People's lives often depend on the design decisions engineers make. Think about this the next time you fly. Even the design of consumer electronics can be fatal if not performed adequately, with several examples of people killed by faulty phone chargers due to inadequate insulation (Howden 2014).

It is important not to underestimate the value of peers as an information source. In an engineering firm there are more-experienced employees who can help you with gaps in your knowledge. They have probably 'done it all before'. They may know where data sets are, and be aware of useful industry catalogues, old plans and specifications from other jobs. Industry peers may also be able to help by suggesting alternative solutions or ways of thinking about a problem.

Being a member and attending local meetings of an engineering institution — for example, Engineers Australia, Engineering New Zealand, Institution of Electrical and Electronics Engineers (IEEE), Institution of Chemical Engineers, a UK-based, international organisation (IChemE) or Institute of Engineering and Technology (IET) — helps build a professional network of contacts that can also be very useful.

Learning

An engineering project may require learning a new procedure or using a new technology. This might require researching information using various means, such as talking to someone else, doing some reading, mastering a new computer program, searching the internet or doing a short course.

Research normally reveals ways a problem can be solved. For example, if sufficient research is done, a chemical engineer will identify several ways of producing a particular product and a structural engineer will find various structural forms and materials that they can use to complete an engineering project. An electrical engineer would develop alternative solutions using alternative electronic components for circuits.

Typically engineering companies use a **collaboration environment** that helps engineers to share documents and expertise. It stores past documents and plans, specifications, tenders, contracts and reports. These systems resemble the learning management system or course management system you use at university (such as Teams, Canvas or Moodle).

A collaboration platform is especially important in team projects. Electronic forums allow for regular discussion about a project and are very useful when team members are geographically dispersed. Even if team members are not based in different locations, an online system is useful for tracking the evolution of ideas and for informal chats between formal meetings. A Facebook group, for instance, can serve this purpose.

All of these types of online tools are likely to be available via your university's learning management systems, which provide the capability for groups of students to use a shared learning space for storing files, having discussions and keeping in touch online. Ask your lecturer for assistance in making these facilities available to you or your class. Alternatively, you might want to use Cloud storage, Dropbox, Google Docs, and so on.

In a large company, a collaboration environment may also give you access to special interest groups (e.g. providing the contact details for environmental engineers or chemical engineers who specialise in fractional distillation). Multinational engineering firms may have access to these specialists from company offices across the globe. Such specialists may be able to provide project advice; however, it is important to do your research first before you ask others for assistance.

Taking the time to research and analyse the client's needs is an important part of the first step in the engineering method — exploring the problem (*research*). Careful needs analysis and a straightforward specification help limit misunderstandings between the consultant and the client on any project.

Suitable organisational tools for paper items are folders or ring binders, which allow you to group related material together, for example, using tabbed dividers. Electronic information can easily be dropped into separate folders on your computer, or you might use a tool such as OneNote or Evernote. It is much simpler to search electronic information. By doing a keyword search using a desktop search tool, you will be able to find information quickly on your computer. All the results of your research are collectively called a **design file** — discussed in more detail later in the chapter.

CASE STUDY

Discussing the client's needs

Returning to the car-buying scenario, some of the requirements that would need to be determined with the car-buying client include:

- a realistic *budget* for the car
- the *purpose* or purposes for the car
- *preferences* (such as size, colour, manufacturer, country of origin, age, engine, entertainment system, luggage space, etc.).

It is important to have defined clear requirements with your client. Have new questions arisen in your investigation that the client may not have thought about? For our car-buying scenario, for example, you could ask the client, 'How do you feel about a hybrid, a plug-in hybrid or an electric vehicle (EV)?'

After clarifying the client's requirements, as a consultant you could begin by making a preliminary list of all the cars you think will meet them; this could be a short or long list, depending on how the requirements are specified. If the client was prescriptive — for example, saying 'The car shall be a Ford' or 'The car shall be a convertible' — you may only be able to choose a particular brand or style of car. If the client gave a functional specification such as, 'The car shall be less than 1200 kilograms', 'The car shall use less than 8 litres/100 kilometres around town' or 'The car shall be less than ten years old', then you may have a wider set of choices.

After considering the client's requirements, you might make a list of places to find useful data, such as advertisements online from car dealers. When determining how you will find information, it is best to be specific. Consider which websites are likely to yield the most useful buying information for you.

Your list of suitable cars will likely get longer as you collect information. You may find types of cars you had not thought of and sources of information you were not aware of or had forgotten about. You may end up with a collection of newspaper advertisements, printouts from websites and notes from conversations with car dealers. By now, you are beginning to realise that you need to keep your data and information well organised, whether it is paper or electronic.

It is also useful to *summarise* information. This summary may be a single written page that shows related facts grouped together. A **mind map** — a hierarchical, two-dimensional graphical representation of a complex topic — is also a useful way of making sense of information. These short summaries are very useful as briefing documents for the rest of your team. They would normally be uploaded to your collaboration environment to keep everyone informed. For an elaborated view of the scope of mind maps, see resources from the inventors of mind mapping Buzan and Buzan (1993) or Buzan (2013), or the quick online guide at https://www.mindmappro.com/insights/how-to-make-a-mind-map. There are many good apps for making mind maps, for example, FreeMind.

For the purposes of the example, figure 2.2 shows some likely requirements for the car represented as a mind map. Note this is a *hierarchical* organisation of several details. The idea is not to have too many branches at one level — between three and seven is a good number. Once the number of branches exceeds this guideline, the map becomes confusing, due to our limited short-term memory — a human can remember 7 ± 2 items at once, as famously proposed by Miller (1956). This has important implications when people are under stress or are trying to learn new things. Miller's research focused on showing groups of objects to a series of volunteers. When the objects were removed from sight, the volunteers were asked to list how many they remembered. Miller found that most people could remember between five and nine items.

The handy thing about using software for making mind maps, as opposed to drawing them by hand, is that it lets you drag and drop ideas, shifting them from one part of the diagram to another. This helps get a large number of ideas organised quickly and easily, and allows you to see patterns in information that do not emerge easily from long lists.

This first step of the engineering method, research, gets you started on any project, no matter how daunting it may seem initially. The research phase will also reveal potential solutions to the problem, which will help you with step 2, generating alternative solutions. In fact, it is hard not to revisit step 1 as you embark on step 2.

FIGURE 2.2 Car specification mind map

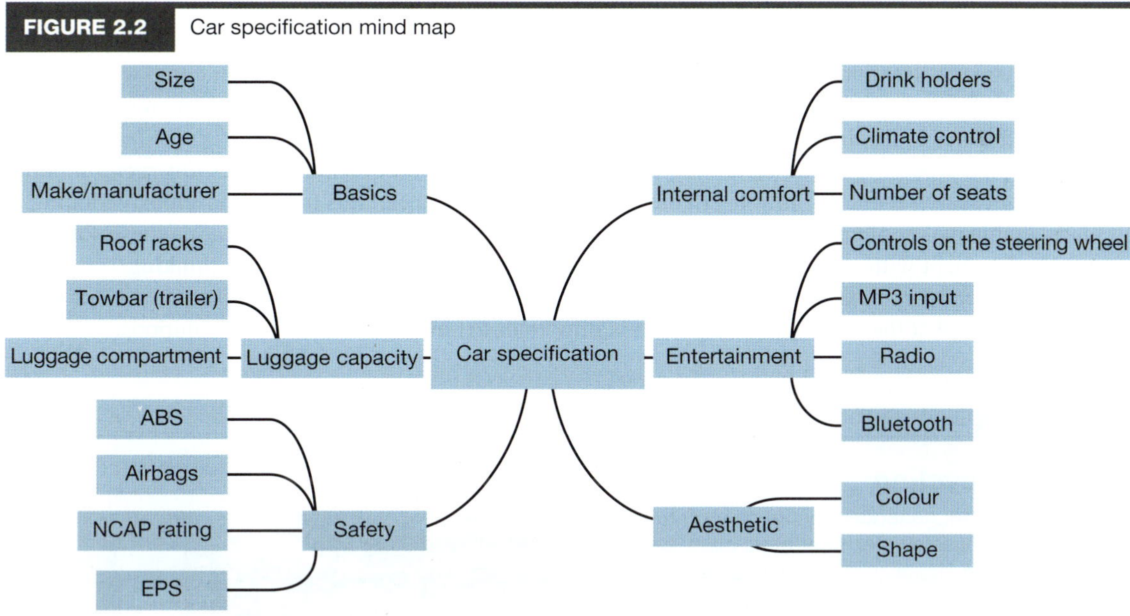

Step 2. Exploring alternative solutions

Once the information has been organised and the problem is well specified, it is time to think clearly about the next step in the engineering method, *exploring alternative solutions*. In the context of the example about selecting a car, every possible car could be listed as a potential solution. In the research phase, however, it would be normal to reject some possible solutions and to collect data about likely solutions. A comprehensive list of suitable vehicles could now be made. For each alternative, it will be necessary to collect further information. Table 2.1 shows a simple way of organising the collection of this information.

TABLE 2.1 **Alternatives: seeking further information**

Alternatives	What do I/we need to know?	Who will find out?	From where?	By when?
Holden Commodore SV6				
Ford Falcon XR6				
Subaru WRX				
Toyota Yaris				

Note that the vehicles in column one are generic solutions. The field of options has not yet been narrowed to specific vehicles; rather, classes of appropriate solutions (cars) for the client's needs are being identified.

Questions for the second column of table 2.1 could include fuel consumption, the price of specific vehicles and preferences such as the colour and age of the vehicle. The third column indicates who will find this information. If you were doing a project as part of a team, you could divide the work so that each person is allocated specific tasks. The likely source of this information is outlined in the fourth column. This might be a particular website, an industry catalogue, or an alternative source you would need to specify. A deadline (e.g. by the next team meeting) should also be listed. This is shown in the fifth column. A deadline is helpful because it helps keep everyone organised.

Because of the amount of content that is included, an alternative solutions document such as table 2.1 may grow very large. If you were working on the car scenario or on a real engineering project, it would be useful to keep your alternative solutions document as a spreadsheet.

After you have finished documenting all the solutions that match the client's needs, a summary that lists each viable alternative and its key characteristics can be produced.

Based on the information in your design file, how many possible solutions have you identified for the scenario? Are there any other possible solutions you have not gathered information about? Now it is time to identify cars that match your client's original specification. As you identify these alternatives, you may realise that you require more information about each option. Does a certain type of car come with an anti-lock braking system or electronic stability control? How many airbags does it have? How much luggage space does it have? Is it suitable for tall people?

Note that this process is about *expanding* the range of possible solutions; it requires *divergent* thinking and a constant search for new solutions. In the context of choosing a car, this means consciously looking at all the available makes of cars, which is a very long list. The next phase is about evaluating these solutions to provide the client with a short list of likely acceptable vehicles. This is *convergent* thinking.

In an engineering context, the following spotlight outlines a range of solutions to the problem of water recycling, outlined at the start of the chapter. It is important to consider all the viable solutions.

SPOTLIGHT

Water recycling

The wastewater generated by our cities is normally treated in a wastewater treatment plant before it is discharged, usually into the sea, as irrigation or into inland streams. Traditionally this treatment was implemented in three stages: primary treatment removes grit and other solids; secondary treatment removes organic material through biological digestion in ponds, trickle filters or through activated sludge methods; and tertiary treatment removes various forms of nutrients such as nitrogen and phosphorus, since these nutrients can have harmful effects on waterways, encouraging algal blooms.

Interior of a reverse osmosis water purification plant

In the last decade, as water shortages have become more acute, serious attention has shifted to further treatment processes that would render the resulting water suitable for human consumption or for specialist industry usage. These processes include micro-filtration, nano-filtration, reverse-osmosis, oxidation and ultraviolet disinfection. These processes are designed to remove any harmful organisms such as bacteria and viruses, which are not able to pass through the very fine membranes used to filter the water. These membranes also intercept a range of complex molecules such as hormones, which might also be harmful if allowed to return to the drinking water system.

Naturally there is considerable community unease about such systems. This may be an instinctive understanding that such systems would need to be foolproof — operated at very high levels of reliability. For example, how would a membrane filtration system cope with the sudden rupturing of one of the membranes, allowing the polluted wastewater to mix with the pure water on the downstream side? To manage this eventuality the engineers would need to create a control system that could immediately isolate the ruptured filtration module from the rest of the plant so that the cleaned water would not be contaminated by such a failure.

Such systems are already operating in the United States, particularly on the west coast, where population growth is quickly exceeding the available water supply from natural sources. In Singapore, the government funded the development of NEWater, where used water is treated by micro-filtration and reverse-osmosis before being transferred back to mix into existing water storages (PUB Singapore's National Water Agency 2018). This water is also supplied to some industry users, such as the semiconductor industry, who benefit from its very low level of dissolved solids.

City West Water (now Greater Western Water) in Melbourne opened a water recycling plant in Altona in 2011 that produces two grades of water for irrigation and industrial purposes (Greater Western Water 2023). The plant uses reverse osmosis to produce up to 9 ML/day or more than 3 GL/year. This reduces demand on Melbourne's drinking water supply by an equivalent amount. The plant cost $46 million and uses technology from Spain.

As the reliability of these systems becomes better understood, it is likely that such systems will operate routinely in Australian cities. These plants will be designed by chemical engineers who have the process engineering skills for the task, together with control engineers, electrical engineers, mechanical engineers, and civil engineers who will design the various subsystems.

CRITICAL THINKING

How would you convince attendees at a community meeting that water from such a treatment facility is safe to drink, and how would you present it?

Step 3. Evaluating alternative solutions

Having identified a range of solutions, either through research or through creative thinking, the next step is to evaluate each solution against the requirements.

As you gather information about each alternative solution on a project, you will be able to start identifying the attributes of each alternative. This will assist you in completing the third step of the engineering method for problem solving, which is *evaluating alternative solutions*.

For the car scenario, the attributes of each car may be different, including its fuel efficiency, noisiness, power rating and embodied energy. Some of these measures may require detailed calculations or careful investigation.

The attributes of each alternative can be documented in a table. An example of how this information might be shown is given in table 2.2. Many car magazines use similar tables to compare vehicles. You might like to analyse suitable alternatives for the engineering project now, using this table or your own comparative document. You can include this table or document in your project file.

TABLE 2.2 **Comparing the feasibility of alternatives**

Alternative	Pros (advantages)	Cons (disadvantages)	Feasibility (y/n)
Holden Commodore SV6			
Ford Falcon XR6			
Subaru WRX			
Toyota Yaris			

You will need to consider the *feasibility* (the need, value and practicality) of each alternative. Some alternatives may be viable, while you might reject others quickly for different reasons (such as cost or environmental impact). An option becomes *infeasible* when it fails to satisfy an essential design requirement. In the car scenario, if the client has specified they require a four-wheel-drive car, then a two-wheel-drive car will be an infeasible option. There is no point providing a collection of alternatives that do not match this essential requirement.

A feasibility check allows engineers to 'prune the solution tree', or to quickly eliminate several options by declaring them infeasible. This is a natural process that is very helpful. It allows engineers to deal with complex problems despite the limitations of short-term memory mentioned earlier in the chapter. If there are a small number of available solutions (say between five and nine) then these can be kept in an 'engineer's head' while they work. If there are many more, it will be likely that writing them all down and keeping careful records will be vital.

Grouping solutions is another way of dealing with short-term memory problems. The range of solutions could perhaps be grouped into small cars, medium cars and large cars. In each category, there might then be five to nine different car types. This reduced number makes it easier to think about or compare them and be less overwhelmed by the range of solutions. A mind map may be useful.

In modelling the alternative solutions to an engineering problem, it is important to rate feasible alternatives against decision criteria. This can be done using a spreadsheet that uses simple mathematical formulas to work out the best alternative. Table 2.3 demonstrates how this could be done for the car scenario.

Based on this analysis, the Subaru WRX seems to be better suited to the client's needs. However, the scores are all quite close together, suggesting that any of the choices might be suitable with the right options.

TABLE 2.3 **Rating feasible solutions: a hypothetical comparison**

Alternative	Size	Fuel	Audio	Safety	Doors	Value
Holden Commodore SV6	3	3	5	5	4	20
Ford Falcon XR6	3	3	5	5	4	20
Subaru WRX	5	4	4	5	5	23
Toyota Yaris	5	5	3	4	5	22

Key
Size: prefer small-medium, so small-medium = 5, large = 3
Fuel: consumption — less consumption means a higher rating
Audio: availability of sophisticated equipment as standard (higher rating indicates more likely)
Safety: Australasian New Car Assessment Program (ANCAP) rating
Doors: availability of hatchback option = 5; else = 4

The approach documented is simplistic. Most of us weight some factors more than others when we make decisions, even if we are not aware of doing so. (See the later chapter on decision making for more about how we make decisions.) For example, a client might be very concerned about fuel consumption for environmental reasons and about personal safety. Someone else might be more concerned about the make or colour of the car. An easy way of making decisions for a problem is to add a weighting factor to each criterion, in consultation with the client. The weights can be any magnitude. In our car-buying scenario, a weight factor of 0 to 5 has been used, with fuel efficiency having the greatest weight.

The final score, shown in table 2.4, is calculated by multiplying the appropriate weight by the rating and adding them together. In mathematics, this is called the *dot product*.

TABLE 2.4 **Rating alternatives with weights**

Alternatives	Size	Fuel	Audio	Safety	Doors	Score
Weights	4	4	3	5	3	—
Holden Commodore SV6	3	3	5	5	4	76
Ford Falcon XR6	3	3	5	5	4	76
Subaru WRX	5	4	4	5	5	88
Toyota Yaris	5	5	3	4	5	84

$score = \sum w_i r_i$
Where w_i = weight i, and
r_i = rating i

These calculations are easily handled in a spreadsheet program such as Microsoft Excel, particularly if you know how to use an 'array formula'.

Note that as a result of this process, the scores in our car-buying example are now more variable, and the WRX is a clear choice. However, let's check what happens if we change the weights. Let's make size fairly unimportant ($w_1 = 2$) if the fuel option is met ($w_2 = 5$) and if safety is met ($w_4 = 5$). Let's also make audio and doors fairly unimportant ($w_3 = w_5 = 2$). This changes the overall scores to 64, 64, 73 and 71 respectively, so the WRX is still the preferred option, with the Yaris a close second.

Having evaluated the range of solutions, it is now time to check that the client's needs have been met.

Step 4. Engineering decision making

At all stages of the engineering method, and particularly before the final report is released, it is important to monitor, check and review the recommendations. *Engineering decision making* ensures that obvious errors have not been made. For example, are the calculations correct? Is the formula in the spreadsheet correct? Is the answer realistic?

A quick calculation on one of the rows in table 2.4 reveals it is correct (for example, $4 * 5 + 4 * 5 + 3 * 3 + 4 * 5 + 3 * 5 = 84$). A check of the second row confirms the formula is also followed in this column (for example, $4 * 3 + 4 * 3 + 3 * 5 + 5 * 5 + 3 * 4 = 76$ is correct). The last row of the column is correct, which implies the first row should also be correct. This is an additional check on the outcomes.

In large projects, independent consultants check the calculations. This vital check helps protect the safety of both the public and the workforce; as such, it is part of the quality assurance process.

Ideally a co-worker should check your work on a project when you are employed as an engineer. Your manager might want a report on progress once or twice a week. You will want to check progress and direction with the client at strategic stages as well. Some strategic checkpoints include when the project is 15 per cent complete, 50 per cent complete, 85 per cent complete and finished. Hopefully the client will be 100 per cent satisfied with your efforts.

If you are working on a large project, you will need to report your progress to other team members, so their efforts complement your work. Engineering failures have occurred from team members not keeping each other informed. Once you have met the client's needs, then you can communicate your recommendations more formally.

Step 5. Communicating your recommendation

The fifth step in the engineering method involves *communicating recommended solutions*, for which you need access to the collected documentation for the engineering project, including the research, the alternative solutions proposed, evaluation of the alternatives and the recommended course of action.

Communication of the recommendations is often conveyed in multiple ways. First, a report could be drafted outlining the recommendations. This report would be followed by a meeting where the recommendations are presented and discussed with the client. There may then be negotiation around a final set of preferred options. This is a complex process requiring a range of communication skills — writing, speaking, listening and negotiating. These skills are discussed further in the chapters on understanding communication, communication skills and communicating information. Reaching agreement on a set of preferred solutions is the final step in the engineering method as it has been described here.

Surrounding this core problem-solving process is a range of project management processes, which will be discussed later in this chapter. Before considering project management, further consideration will be given to properly understanding the problem. Before we do that, let's compare the engineering method with the scientific method.

The scientific method

Science has transformed our world in the last 300 years, and it has certainly transformed engineering practice, particularly in the last 100 years. But what *is* science? Is it what we learn from textbooks, or is it a process of creating new knowledge in a systematic way? Since the seventeenth century, science has placed an emphasis on the empirical collection of *evidence* as its fundamental principle. This evidence must be observable, measurable and reproducible if we are to have any confidence in it.

How does the scientific method work? How is it similar to, and different from, the engineering method? Both methods start with a problem or a need. For example, the problem could be to explain our observations of rising global temperatures and sea levels. Scientists begin by forming *hypotheses* about the cause (of climate change in this case). Competing hypotheses could be (i) climate change is induced by human activity, such as burning fossil fuels and (ii) climate change is a naturally occurring process that has been observed throughout the life of the planet. This is similar to the 'generate alternative solutions' phase of the engineering method.

Scientists address this kind of problem by carefully collecting data such as temperature, tree growth rings, atmospheric CO_2 levels and so on. These data are correlated in an attempt to confirm one hypothesis and reject the other. This is the *evaluation* part of the process, using empirical data to test the alternative hypotheses. Although it is impossible to *prove* a theory, it is sometimes easier to disprove or reject a competing theory or hypothesis. Despite the work of thousands of climate scientists, there is still considerable doubt expressed, particularly by politicians, that the case for climate change has been proven.

Finally, conclusions must be drawn and communicated to the intended audience through reports, papers and presentations. So, the scientific method is quite similar to the engineering method, as shown in table 2.5.

TABLE 2.5 **Comparison of the engineering method and the scientific method**

Engineering method	Scientific method
Begins with a problem or a need	The problem or need is often a *question* to be answered
Requires *research* to better understand the problem and to define suitable criteria and constraints (the boundary) for a problem solution	Likewise requires a research phase to better understand the question and to define its boundaries
What are some suitable *solutions* to this problem? This can require creative and divergent thinking	What are some suitable *hypotheses* to answer the question? This can also require creative and divergent thinking
Evaluate the solutions against the criteria and constraints using available data and modelling tools	Evaluate the hypotheses using data and modelling tools, with data either collected first-hand or derived from other reliable sources
Choose one or more suitable solutions	Choose the hypothesis that is best supported by the evidence, preferably rejecting all but one of the hypotheses
Make a recommendation for further action, such as to implement one of the solutions	Make a recommendation for further action, which might also include the need for further data collection and analysis

2.2 Systems thinking

LEARNING OBJECTIVE 2.2 Identify a range of system definitions for a problem and use these definitions to present different solutions to a problem.

Throughout the discussion in this chapter, we have assumed the client understands the problem and needs an engineer to find the best solution; this is often not the case. It is often the engineer's role to help the client understand the problem and to identify appropriate requirements. This is the realm of systems thinking. **Systems thinking** is the process of identifying all the elements of a problem. It requires explicit attention to problem boundaries and the components of a problem; it also requires an understanding of the interactions across the boundary and interactions between the components of the system.

Systems thinking is an essential framework for handling complex problems. Some basic ideas in systems thinking include the following.

- *The boundary.* What is inside and what is outside the system of interest? A problem usually needs to be simplified in order to make progress in solving or managing it.
- *Components (subsystems).* What are the major elements of the system? These are also often systems, so they can be called subsystems. For example, the electrical system is a subsystem of a car (the total vehicle), which is in turn part of a transport system.
- *Interactions (interfaces).* What are the ways in which the components interact with each other? For example, a population will have a birth rate and a death rate. The rate of growth of the population is dependent on both these factors. It may also have immigration and emigration rates (across the system boundary), which may affect population size. Figure 2.3 shows the relationship between a system and its environment.

When the components and interactions of a system are understood, it may enable a mathematical model to be built and eventually a simulation of it on a computer. For example, the weather forecasts on the nightly news rely on complex climate models, which are run every day on supercomputers. As the capacity of these computers has increased, the models have become more sophisticated and seven- or even ten-day forecasts are now common.

KEY POINT

Systems thinking is essential in analysing complex problems.

FIGURE 2.3 A system in its environment with components, interactions and boundary

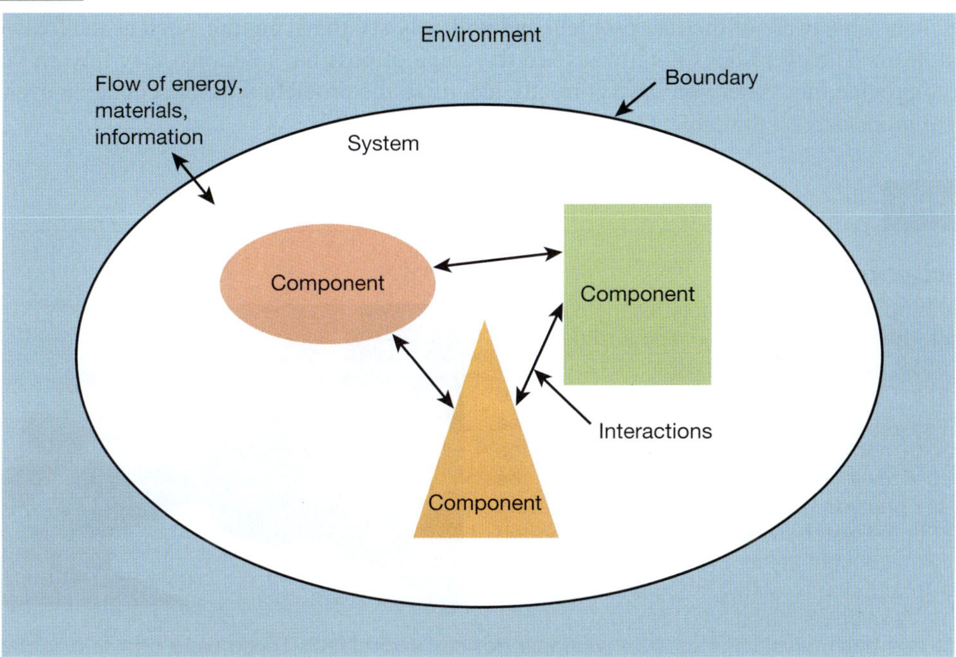

The system boundary

The system boundary is the key to solving the right problem. If it is too small, there is a risk of solving only a symptom of the real problem; if it is too large, the problem becomes so large it is intractable (that is, it is computationally intensive to the point of being unsolvable).

For example, looking at a freeway at eight o'clock in the morning and seeing the traffic is barely moving on a particular route into the city may lead to an assumption that more lanes are needed as the solution. This defines the problem as a road issue. Another view of the problem could suggest that the situation be improved by providing commuters with a more effective public transport system that takes people off the freeway, reducing its congestion. Now the problem is seen as a transport issue.

One might also wonder whether it is necessary for so many people to have to commute into the city for work; perhaps people could work from home (as has become commonplace since the beginning of the COVID-19 pandemic) to reduce the freeway congestion problem. As these perspectives show, by redefining the boundary of the problem, new solutions are possible. Figure 2.4 is a conceptual map that outlines the crossover of different systems boundaries for this problem.

FIGURE 2.4 Alternative system definitions

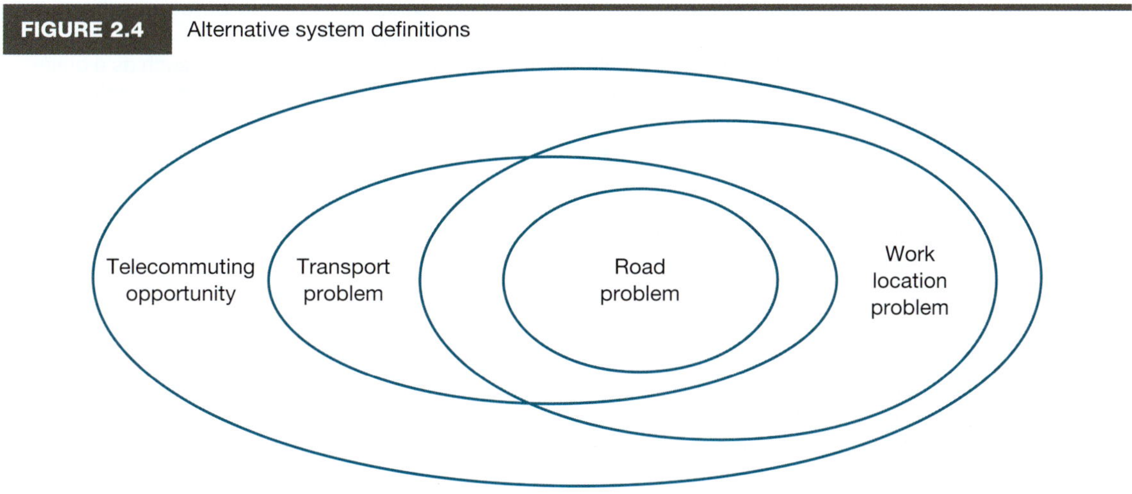

It is important to solve the right problem when you work as an engineer. Do not assume the client knows what the problem is and that they require you only to design and implement the most efficient solution; often the client is unaware of the real problem and will only see the symptom, such as the freeway being blocked at 8 am. The client does not always see the range of possible solutions; they rely on you as the engineer to provide this. After you have correctly identified the problem with the client, an even broader array of solutions may be available. Consider the following spotlight.

SPOTLIGHT

Smartwatches

The development of the wearable technology market has been increasing for some time now. The Pebble watch — developed in 2012, funded through a Kickstarter campaign — was one of the first smartwatches offering integration with a phone to provide notifications, calendar alerts, a range of apps and, of course, the time. However, the company shut down in December 2016, leaving all the Pebble watch owners with an unsupported watch. Fitbit bought the remnants of the company, mostly to acquire its software expertise and key software developers (Kovach 2016).

Fitbit had been developing its own wearable devices since 2007. These were slim and lightweight, intended for tracking steps, heart rate, stairs climbed and other fitness activities. The accompanying app allowed food intake to be entered, so that energy-in versus energy-out calculations could be performed. More recent Fitbit models include Bluetooth connection to the phone at all times, allowing notifications, alerts and the time to be displayed on the fitness device. Fitbit had taken a very different approach to the watchmakers, focusing instead on fitness as the 'killer app' that would drive sales. From 60 000 sales in 2010, Fitbit sales grew to US$2.2 billion in 2016 then plunged to US$1.2 billion in 2022 (Statista 2023). The fitness tracker seemed no longer to be the device that people wanted.

In the interim, Apple had released its first watch in 2015 to ambiguous reviews: it looked a bit chunky and it offered only 18 hours of battery life, meaning it needed to be charged every day. Nevertheless, it offered close integration with the iPhone, allowing the user to answer their phone through the watch, call Siri, read notifications and alerts, and run a range of apps. It had the advantage of Apple's market presence and a willing team of developers.

Reviewers of the watch wondered about Apple's strategy. The watch was comparatively expensive and seemed a gimmick for the well-to-do, particularly the gold edition, priced at US$15 000 — a peculiar product given that the internal technology would date very quickly. Did people really want a $400 watch that, yes, told the time and provided alerts, but had modest fitness capability? By comparison, Fitbit's focus on providing a fitness wearable that was fit-for-purpose, at half the price of the Apple Watch, might explain its popularity.

The Apple Watch, however, did have a heart rate sensor, which, apart from its obvious use in fitness circumstances, began to make a name for itself in detecting heart problems such as irregular heartbeats that were life threatening (Smith 2018). Apple Watch sales continued to grow, outselling all other smartwatches in 2017 (as well as the entire Swiss watch industry) (Statt 2020).

The latest Apple Watch, series 9, takes the health capability to a new level with features such as a built-in electrocardiogram, sleep tracking, temperature tracking, and blood oxygen and heart rate monitoring. It allows the watch to record the electrical activity for an episode of irregular heartbeats. This means it can collect initial data for conditions such as supraventricular tachycardia (SVT), a condition that causes the heart to race to 200 beats per minute or more. Apple has FDA (Food and Drug Administration) approval in the United States (Gurman 2018), which suggests a thorough regime of testing to ensure accurate performance. Of course, this is not a replacement for professional equipment at a doctor's surgery but could provide an initial data set to aid diagnosis before more thorough testing.

A personal view

As a smartwatch wearer (two Pebble watches) from 2012 to 2017, the initial value for this author was in reading notifications and alerts and taking calls (and being able to detect who was calling). Having owned an Apple Watch (series 3, then series 7) since 2017, I found its ability to act as an independent device particularly appealing. It has its own connection to the phone network, so it is not necessary to have my phone in my pocket. Calls can be both accepted and initiated from the watch. Text messages are received and can be sent by Siri — no typing necessary. Fitness capabilities are also much improved, and the three fitness circles are my challenge every day — active energy burned (in kJ), 30 exercise minutes to complete and 12 hours of some activity (to encourage not sitting for long periods). All these features encourage me to leave my phone on my desk rather than carry it all the time.

The 'killer app', however, for me, is touch-and-go payment with Apple Pay. Connect a credit or debit card to your iPhone and it is also connected to the watch. A double-click on the button brings the card onto the screen and the watch can be scanned by an EFTPOS terminal as if it was the actual card. Now I can leave my phone in my office, since I can pay for coffee or lunch without either my phone or wallet. Store cards are also moving into the Apple Wallet, further emptying one's own wallet of pieces of plastic. Smartwatches can also be used for public transport and will soon be available for security access for buildings.

It may not have been obvious in 2015 what the future of wearables was but it is becoming increasingly clear that with an Apple Watch, or similar, on your wrist, most of the routine functions of your phone and wallet can be carried out much more easily. (It's probably harder to lose too!) The next step could be the inclusion of a camera.

CRITICAL THINKING

What features of a smartwatch are most compelling for you? What apps would you find most useful? Can you suggest a feature or application that hasn't been thought of yet? Is there a chance for a start-up using your idea?

In this chapter, we have seen a systematic method for working through complex problems — the engineering method. We have looked at systems thinking as a way of ensuring we are solving the right problem. We will now explore some basic project management techniques to manage the work involved in the engineering method.

2.3 Project management

LEARNING OBJECTIVE 2.3 Apply basic project management principles to plan a project and maintain organised project documentation.

The engineering method provides a systematic problem-solving process, which is aided by **project management**. This process of managing projects — temporary endeavours designed to produce a specific product or service — is itself a small project, intended to meet the client's needs. Any project also needs adequate *resources* (time and money) and an assessment of the *risks* involved. In the example discussed earlier in the chapter, the outcome is to provide a recommended vehicle type. As a project, it is made up of several distinct activities, which must be *scheduled* adequately if the project is to be completed on time.

Now we will consider project management in a simplified way, with a more thorough treatment provided in the chapter on managing engineering projects. This discussion is deliberately focused on how to run student team projects.

Engineers Australia (2019) lists the required competencies for project management (the application of systematic approaches to the conduct and management of engineering projects) as follows.

- Contributes to and/or manages complex engineering project activity, as a member and/or as leader of an engineering team.
- Seeks out the requirements and associated resources and realistically assesses the scope, dimensions, scale of effort and indicative costs of a complex engineering project.
- Accommodates relevant contextual issues into all phases of engineering project work, including the fundamentals of business planning and financial management.
- Proficiently applies basic systems engineering and/or project management tools and processes to the planning and execution of project work, targeting the delivery of a significant outcome to a professional standard.
- Is aware of the need to plan and quantify performance over the full life-cycle of a project, managing engineering performance within the overall implementation context.
- Demonstrates commitment to sustainable engineering practices and the achievement of sustainable outcomes in all facets of engineering project work.

KEY POINT

Scheduling is a key component of project management.

Defining the scope

Possibly the most important step in project management is to define the scope of the project. The scope will identify the steps or tasks required to be completed, which will be defined formally in the contract of work to be conducted. Determine what needs to be delivered. Then break this down into system components, using your systems thinking skills. What are the individual parts of the product? Are these components readily available (e.g. a pump or a software library) or do they need to be developed from scratch? How much time is involved in developing or sourcing these components? How much time will be required for systems integration (bringing all the parts together to make a working system? Now think about whether these components can be developed in series or parallel. Do some parts or tasks need to be done before others? That will affect the scheduling of the tasks.

Scheduling

One of the most important aspects of any project is the **scheduling** of the tasks.

- What tasks are required?
- How long will each task take?
- How much time is available?

We will look at the **timeline** for a project first. A deadline — the final due date for a completed project — needs to be negotiated with the client and may contain **milestones** (intermediate deadlines) to be met within the project. This requires some overall estimate of total effort and the resources available. During undergraduate studies, the lecturer will usually set project deadlines, with milestones at certain weeks of the semester. The assessment may be in a traditional form, such as a report or presentation, and/or it may involve participation in an online forum, such as a blog. You may also be required to document your learning in a journal. This helps you become conscious of the skills you need to develop to become an engineer.

Once an overall project deadline is established, it is important to look at the plan as a whole to establish probable due dates for each activity in the project. This will require making estimates of the amount of time in which each activity can be completed and making whatever adjustments are necessary to fit all the activities into the time available before the deadline.

After this, a **Gantt chart** — a bar chart that illustrates a project schedule — or a similar chart, can be drafted to inform team members of due dates. Table 2.6 is an illustrative Gantt chart for a semester project. The first column lists all the tasks. The top row shows the overall timeline (12 weeks in this case) and the coloured rectangles show the start, finish and duration of each of the activities.

TABLE 2.6 Gantt chart for a semester project

	Wk 1	Wk 2	Wk 3	Wk 4	Wk 5	Wk 6	Wk 7	Wk 8	Wk 9	Wk 10	Wk 11	Wk 12
Client need	■											
Planning	■	■	■	■	■	■	■	■	■	■		
Problem definition		■	■									
Research		■	■	■	■							
Selection criteria					■	■						
Alternative solutions					■	■	■					
Analysis						■	■	■	■			
Decision (Choosing)									■	■	■	
Reporting											■	■

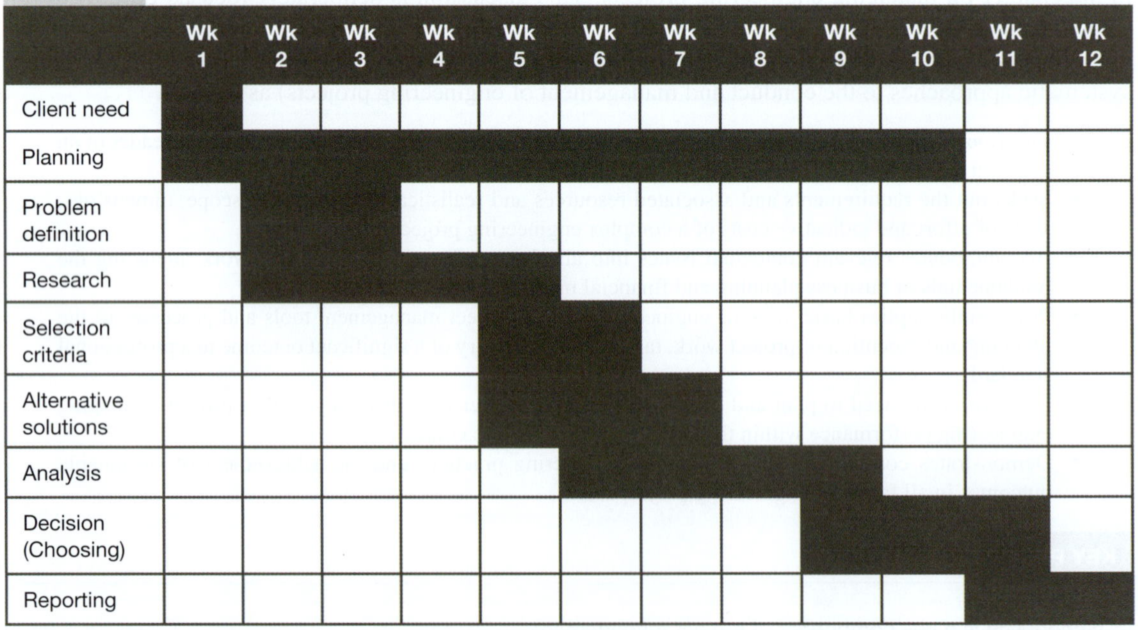

It is important that the plan is revisited regularly to make sure the project is on track. This means the plan can be adjusted, even if some tasks take longer than anticipated. For example, in the car-buying scenario, the requirements might have been identified during the research and problem definition phase, allowing for more effort to be allocated to other tasks earlier.

There are many project management packages available to help with this process. These packages will allow you to build Gantt charts and track existing projects. In the chapter on managing engineering projects, the One-Page Project Manager is described as a simplified approach that is suitable for managing larger projects.

Resources

Project management is a constant tug-of-war between the time available, the physical resources available, and the scope and quality of the work. When working with an external client, it is also necessary to consider the customer relationship. In this section, commonly available resources — time, money (budget), equipment, and people — will be considered.

Time, as with other resources, is always in limited supply. Frequently the time available depends on the client's needs to get a project completed by a particular date. At other times, the client is willing to negotiate because they may have no idea how long it will take to complete the project. There is usually a cost–time trade-off, with shorter completion times being possible by spending more money to accelerate tasks.

In all engineering projects, the budget available will determine what is delivered. Are there sufficient funds to deliver what has been committed? How much is reasonable to spend on this project? From whom will the money be borrowed?

In student team projects, the individuals who make up your team are the most valuable resource, together with the time that they have to devote to the task. Because of this, the needs and talents of team members must be taken into account as part of effective project management; take, for example, the following.

- What are the talents of individual team members? Are they creative thinkers, managers, implementers?
- How do you make the most of these talents?
- What aspects of the project do team members want to work on?
- What do team members want to learn in the process?
- Will any team members be away during any stage of the project?
- Risk management — what is the contingency plan if a team member becomes sick? What if a team member drops out or fails to contribute to the project? How will this affect the schedule and the team's performance?

Team skills are discussed further in the chapter on communication skills.

Access to resource equipment, including computers and the internet, is important for student projects. For engineering projects, specialised equipment may be required. For example, equipment for collecting data such as traffic counts, land surveys or soil samples might be necessary. Crash testing of vehicles to test passenger safety could be required. The availability of such equipment for the desired time frame needs to be ensured. Project management software allows you to schedule these resources. This will be discussed further in the chapter on managing engineering projects.

Documentation — the design file

Working on projects generates a considerable amount of documentation. This documentation includes meeting agendas and action plans as well as handwritten notes, sketches, calculations, photocopies of published material, and printouts from websites and other computer programs. Keeping track of all this information and being able to make it available to others, such as your manager and possibly the client, is important. This information needs to be collected in a design file, a standard practice for work as an engineer.

Keeping a design file may involve a significant shift from how you have worked in the past. For example, polished work (in the form of typed assignments and essays) may have been important in your earlier studies. You might have accumulated a lot of scrappy notes, which were discarded after you finished the assessment items.

In engineering, neat handwritten calculations, drawings, sketches, and notes are important project documents. These should be preserved for access by others. It is often easier to handwrite this material, than to use a computer or drawing package to improve the presentation of the work (that may come later for selected figures and tables for reports). Reports are word processed to make them easily read and understood by others.

Design files are standard practice in engineering organisations for many reasons. They allow you to hand a job to someone else to complete (you may have been assigned to other work or be moving to a new job). It allows a new team member to see what has already been done. It allows your manager to see what you have done. It shows the detailed assumptions and calculations behind a particular design decision. It might also be required in a court of law if something goes wrong. In this case you will need to show that you have followed 'best practice' in your work.

In the Stage 1 Competency Standard, Engineers Australia (2019) describes this competency as follows.

> Professional use and management of information.
>
> - Is proficient in locating and utilising information — including accessing, systematically searching, analysing, evaluating and referencing relevant published works and data; is proficient in the use of indexes, bibliographic databases and other search facilities.
> - Critically assesses the accuracy, reliability and authenticity of information.
> - Is aware of common document identification, tracking and control procedures.

A design file should include the following.

- *Records of your own work.* These include notes, sketches, library searches, meeting notes, agendas, and action plans.
- *Collection of all relevant documentation.* These include handouts, printouts from websites, photocopies, and plans.
- *Personal reflections on the process.* This is the stage at which you think out loud about the project or discuss what you have learned so far.

A ring binder with section separators is a useful design file for all your paper documents. A design file should always have a *table of contents* at the start describing its major sections. This will usually be a work in progress, as the information in the sections will change and be updated throughout the process. A design file needs to be logically organised. It might be arranged into major sections based on the five steps in the engineering method, or it might be organised by a major theme or by major components in the design. A design file needs to be able to be audited by project stakeholders at any time during the process. It may be audited according to an agreed project schedule. For example, a project manager may want to review the progress of a young engineer once a week or once a fortnight. Your lecturer may want to review your work during your classes.

As work is now largely computer-based, most information is kept electronically. The electronic information you have stored on your computer as documentation for an engineering project is the e-version of your design file. It is important that this electronic information is as organised as the hard copy of the design file. Project folders and sub-folders for the stages of a project can be created on your computer. This process is like the act of using dividers in a ring binder. If necessary, printed documents, such as sketches, can be scanned and stored as computer files. It may be useful to have a hard copy of the key documents of your design file for reference and as a backup of your electronic documents. Cloud storage services are a convenient means to keep your documents available anywhere and anytime.

Tablets such as the Apple iPad are viable alternatives to laptops for many purposes. As a result, spontaneous work such as sketches can now be captured easily in an e-format. Whiteboards support group discussions, allowing notes to be captured both on paper and electronically with the aid of a phone camera.

Mobile phone cameras are a valuable, readily available resource for collecting field data and information. Regardless of the electronic device being used, digital images can capture an overall arrangement (e.g. a pump installation) and specific details (such as its electrical supply or the electronic control module) for transmission to other group members. Apps such as Evernote and OneNote are excellent for storing notes and images as part of your design file.

Time–accuracy trade-off

It is important that engineers can produce answers to problems with an accuracy that fits the time available. A five-minute answer requires some clever simplifications and a 'back of the envelope' approach, sometimes without any calculator assistance. An appropriate level of accuracy may be to get within 25 per cent of a good or optimal answer. With five hours, you could use a more sophisticated method, a spreadsheet or other program and perhaps get within 10 per cent of a good answer. With five days, you could use better methods again and you might get within 5 per cent of a good answer. With five weeks or five months, even better methods are possible, and you might improve your answers by a few percentage points.

A useful way of expressing this accuracy is in terms of the uncertainty of your answer. If you have only a few minutes to estimate the likely cost of a project for instance, you might say 'half a million dollars, plus or minus two hundred thousand'. This would be an indicative figure. If you had time to do some conceptual design (say a day or week of work), then your new estimate might be '\$600 000 ± \$50 000'. If you are quoting on the job, then you might quote \$650 000. (Of course, there is a risk in any quotation, and this will be discussed later in the chapter.)

Is all this extra time spent on analysis worth it? This depends on the project. If you are working on a project that will cost \$1 million, an expenditure saving of 5 per cent is \$50 000. If your time developing the new plan costs the client an extra \$5000, the saving to the client is \$45 000. In this case, they are likely to be suitably grateful. Of course, if the project is worth \$1 billion, much more sophisticated methods can be used to create greater savings. It may be in the client's best interests to pay engineers to do calculations for several months to achieve a 1 per cent improvement in the cost of the \$1 billion job.

This ability to consider time versus accuracy is a crucial engineering skill. As you embark on each project or calculation, you should start with a quick estimate of what you think the answer will be. After all, engineers have been working for at least 5000 years and have had computers for only the last 60 years. A lot of the engineering works in the world today were designed with clever assumptions and very simple calculations.

Agile project management

Agile project management has become very popular in software development (software engineering). It assumes that many projects are too complex to use the standard scoping and scheduling approach. The client simply does not understand their problem well enough to undertake a proper breakdown of the problem into its components at the start.

Instead, the method relies on incremental 'scrums and sprints' to make progress. At the first scrum meeting, the challenge is to define the work for the first sprint — say to the end of the week. This might be to understand the problem as well as possible and to reflect that back to the client to develop a refined problem statement and a list of requirements. This might take more than one sprint before the problem is sufficiently defined for the next step.

A subsequent sprint might then be to consider solutions and technologies. Again, working with the client, the problem becomes better defined and suitable solutions identified. Each sprint is used to ensure quick and focused action, with scrum meetings once a week or even once a day to ensure a rapid pace is maintained.

As the solution emerges, this can be further refined into modules or components, with individuals and small teams assigned to work on each. Again, short sprints define clearly what is to be delivered and scrum meetings check these deliveries and set up the next stage of work for the following sprint.

As you can see, it is possible (and probably advisable) to adapt the traditional project methodology to an agile approach. Every project needs to proceed from a problem statement through exploration, solutions, evaluation, recommendation and then detailed design.

The agile version of this is well suited to student projects, given students are often time poor and need careful organisation to ensure work gets done. If you think of a sprint as a week (or even shorter) then it might be possible to keep all team members aligned and working enthusiastically with a sense of achievement. This might reduce the occurrence of social loafers not contributing to the project.

There are several software tools to support agile project management, including Trello, MS Planner, Podio and Basecamp.

Risk management

Most business decisions, including engineering ones, are driven by risk management. In some ways, engineers have always been conscious of risk. They have used probabilities of failure for structures or dam spillways and other large components since the beginning of the twentieth century. This became possible when scientific understanding of engineering systems allowed them to quantify system performance and to estimate the probability of the loads on these systems.

Engineering involves the conscious trade-off of risk versus consequence. The expected cost is the probability multiplied by the consequence. In 2010 and 2011, Christchurch in New Zealand was devastated by two serious earthquakes. How do engineers go about designing for this kind of risk? The following spotlight provides some insight into the development of design codes for earthquake loads on buildings.

SPOTLIGHT

Design for earthquake-resistant buildings

PROFESSOR NELSON LAM, UNIVERSITY OF MELBOURNE

A sudden change in the earth's crust thousands of metres deep generates waves in rock, which then propagate to the surface as an earthquake. As the ground shakes, buildings are brought into motion. By Newton's laws of motion, forces are generated from within the building by its own weight.

The destruction in Christchurch

There are well-established calculation procedures for determining the magnitude of forces that can be generated for a given ground motion intensity. Similarly, there are methods for determining if the soil would experience *liquefaction* or *spreading* when subject to a certain degree of shaking. Ground shaking intensity is mainly a function of the magnitude (size) of the earthquake, and distance measured from the building to the earthquake source.

Had it not been for major events, few experts would have predicted that the epicentre of a magnitude 6.3 earthquake would have originated so close (10 kilometres) to Christchurch, New Zealand. Likewise, experts from Japan itself would have considered the chance of having a magnitude 9 earthquake in northern Japan to be very remote. Clearly, the world is still on a steep learning curve when it comes to predicting future earthquake events. When a natural phenomenon is not well understood, experts resort to the use of statistics to make predictions in probabilistic terms.

In regions of high seismic activity, it is common practice to design a building for ground shaking corresponding to a *return period* of 500 years. This means that, *on average*, there is one such event every 500 years. So, there is a 1/500 chance of this level of ground shaking being exceeded in any given year, which means a roughly 10 per cent chance of it being exceeded at some time during a design lifespan of 50 years. While 500 years appears to be a very long period, ground motions recorded in Christchurch during the February 2011 earthquake were found to be more severe than that predicted for a 2500-year return period event, as per the design standard currently used in New Zealand. Thus, predictions based on current statistics were shown to be inadequate.

The probability of these occurrence values is based on a building in isolation and should not be confused with the probability of a country experiencing a disaster, which is characterised by much higher values. The level of seismic loading stipulated for the design of buildings in Christchurch will have to be increased in the future (as for many cities that have experienced a disaster). The philosophical question is do we simply increase design requirements when prompted by a disaster? It is preferable to prevent damage resulting from disasters before they occur, but what level of loading should we anticipate when data is lacking? Does Australia too have destructive earthquakes?

The most serious in Australia's history was in Newcastle in 1989 with a magnitude of 5.7, which resulted in loss of life. A 5.0 earthquake was recorded in Kalgoorlie, Western Australia, on 20 April 2010. A slightly higher magnitude earthquake occurred roughly a year later — 5.3 on 16 April 2011 in a remote part of northern Queensland — and a 5.4 earthquake in Gippsland, when there was extensive damage felt across Gippsland and in Melbourne, with an aftershock of 4.4 felt in Melbourne (Geoscience Australia 2021). Could these earthquakes be used as 'statistics' to predict when and where the next earthquake strikes? Common sense tells us this would be impossible.

Even more research is warranted in developing earthquake-resistant design of buildings that are also economical and practical for their environment. Improving design methods alone would not mitigate a future disaster; retrofitting existing infrastructure is just as important. A good knowledge of the relative risks of existing buildings is required to assist with prioritising, given that retrofitting every building would be very costly.

Since the Christchurch earthquake in February 2011, the New Zealand government has outlined the goal of developing an earthquake-resilient community. Accomplishing this goal requires substantial input from the disciplines of engineering, social science, financial planning and public health, as well as individual citizens. Engineers will play a pivotal role in making this major interdisciplinary undertaking a success. In the future, Christchurch can be used as a case study to guide other countries in formulating their own disaster mitigation plan.

CRITICAL THINKING

1. Can you think of other natural phenomena that are rare enough to require a national statistical approach (that is, it is not enough to rely on data from just a single site)?
2. Find some data on:
   (i) the value of current building stock in Christchurch's CBD
   (ii) the cost of repairing some comparable damaged buildings following the earthquake.
3. If you were a Christchurch CBD building owner, how could you decide whether retrofitting greater structural strength is worthwhile, in terms of the cost versus the risk?

Buildings are often designed for 50- or 100-year wind events or to survive a 500-year earthquake, as described above. If a building lasts for 200 years (this is not uncommon), there is a chance its design conditions will be exceeded. Fortunately, other excess capacity is built into most structures, so — even under these extreme conditions — the building is likely to survive.

If a building is designed to a 100-year wind standard, the probability of failure will be much less than 1 per cent, perhaps 0.1 per cent, given the excess capacity in most structures. This means the structure has a survival probability of 0.999 in any one year. Over a 200-year life, this equates to a survival probability of $0.999^{200} = 0.812$; or a 19 per cent chance of failure. Is this reasonable?

You may not be estimating probabilities for structural behaviour or other engineered systems just yet. However, completing a **risk assessment** of your own work is an important process, using these four steps.

1. Identify potential *impacts* (e.g. unavailability of key people, information, equipment or skills).
2. Assess the *likelihood* of each impact (e.g. are they almost certain, likely, possible, unlikely or rare?).
3. Compare the potential *consequences* of each impact on a scale (e.g. grouping them as insignificant, minor, moderate, major or catastrophic).
4. Calculate *risk* as the product of likelihood and consequence of each impact.

Reporting and documentation

Formal documentation is necessary as it marks key stages in the development of a project. The project starts with a document, usually a client brief. The research stage produces many documents, such as data sheets, spreadsheets, catalogue data and discussion summaries of various kinds.

The alternative solutions are documented through sketches, data sheets and summaries of advantages and disadvantages. The analysis and evaluation of each alternative solution stage will involve calculations, modelling and other assessments that are summarised in a design report. All the important processes and conclusions in the design report are revised to make sure the preferred alternatives meet the original client specification. These checks are carefully documented for future reference. This guarantees appropriate quality assurance processes are enacted.

The *final report* describes recommendations of what is required. This is usually a summary of everything achieved to date, which is written in sufficient detail for the client. This is normally presented to the client in a meeting; some follow-up work or revisions may be required.

Writing and speaking skills

In addition to needing access to documentation, engineers need to have good writing skills so they can communicate recommendations to the client and to senior management. Regular written reporting is required. As a professional, you will need to be able to write simply and concisely; simple sentences are usually more effective than longer sentences. Write in the active voice (e.g. 'I submitted my report to the client') rather than in the passive voice (e.g. 'My report was submitted to the client'). Define terms, abbreviations and acronyms that may not be familiar to your readers. Provide supporting evidence for your assertions. Writing skills are discussed in more detail in the chapters on communication.

As an engineer, you will be required to verbally present your work to clients and to discuss it. This will require personal confidence as well as presentation skills. The project work you do throughout your degree will provide you with opportunities to improve and develop your communication skills.

In the car-buying example, you would sit down with your client and outline the range of solutions you have identified, and the criteria you have used to assess them, and finally describe how you have arrived at your recommendations. It is quite possible that your client will then have new insights, which might lead to additional work for you. There may be new criteria to consider or different weights of the existing factors that could change the recommendations. When all such changes have been finalised, the final report, in written form, can be transmitted to the client.

Improving practice

On any project, most professionals review what they are doing and think about how they might improve their results. Sports coaches are experts at this. These coaches review how the team is playing and make recommendations, such as changing the players on the field and the plays made in games, as well as helping individuals improve their kicking, catching and throwing skills by encouraging them to train regularly.

Likewise, you need to get into the habit of being your own coach, both as a student and in your future career, and engage in thinking about your own practice as a means of personal and professional growth. A mentor can also be useful to help you improve.

A systematic way to engage in this kind of reflective practice is to keep a diary or a journal for jotting down ideas, so when you are employed as an engineer, you will be able to look up ideas you had last week or last month.

A formalised process of doing this is called **action learning**, which is the process of continual improvement or learning through action. There are four basic steps: plan, act, observe and reflect (plan to improve). For example, you may draft plans to improve the efficiency of a process in a chemical processing plant. Your plans are enacted through the installation of some new equipment. You then observe the performance of the process by collecting data over the following weeks and reflect on whether the changes have been successful. Perhaps the process needs further adjustment, such as an increase in temperature. This leads to a new set of plans, which might include adjustments to the key process settings for the equipment (temperature, flow rate, etc.).

Dick (2001) shows continuing cycles of action and reflection — action, reflection, action, reflection, and so on. As tasks are completed, reflection needs to take place. Can a project be completed more efficiently, more quickly, more cheaply or with less environmental impact? As better solutions are developed, they should be implemented so that even better solutions can be created, and the cycle continues. This process has allowed humans to move from caves to satellites in about 7000 years. The transition from the first powered flight to a moon landing took just 65 years — a truly remarkable achievement. Such is the power of continual reflection and improvement. You will learn more about reflective practice in the chapter on self-management.

Team management and improvement

When you work in teams, it is easy for things to go wrong — work is not done, members do not show up for meetings and you may be unsure of what needs to be done. You may begin to panic, with too much work to do and not enough time to do the work well to meet the deadline. Rather than procrastinate and proceed to panic, the purpose of project management is to 'plan, proceed and perform'. The project management processes discussed earlier in the chapter provide the tools for team organisation. Once a project is underway, regular progress checks are needed to keep the plan on track.

At each meeting, team members should check off what has been done and decide what will be done next. This process is documented in an **action plan**. These meetings are sometimes uncomfortable discussions. You may need to tell a colleague that they need to put in more effort or improve their attendance at meetings. A team member may be holding others up. Questions may need to be asked, such as 'Does everyone understand the task required? Is the project too large? Can someone else help? Can a group brainstorm be conducted to scope out the task (work out the real engineering problem) so it can be completed more quickly?'

In these discussions it is advisable to *focus on the task* rather than on the person. How can the task be moved forward? It may be necessary to re-allocate certain tasks, or for the whole team to provide input to the task. It is easy to get caught up in blaming others for not getting work done; however, there may be a range of reasons why tasks have not been completed. People respond to positive feedback much more than they do to negative feedback. If you can provide team members with a helping hand, it is much more likely that they will respond positively to you. Negative criticism tends to lead to team members putting in less work rather than more work. Team dynamics will be discussed in greater detail in the chapter on communication skills.

There are now several apps for organising team project activities, including Trello, Podio, Basecamp and MS Planner. Each allows tasks to be defined, assigned to individuals and completion recorded. These are very helpful tools to ensure that work gets done. They also provide evidence, if required, when team members fail to do their share.

In your engineering classes, you will likely have to review your team's performance, either informally each week or formally to score each other's performance. In your learning journal you are also likely to be asked to reflect on your performance at regular intervals to gain insight into your strengths and weaknesses and how you can continue to develop your engineering capabilities. Keep thinking of that first job interview: what skills will you demonstrate to make sure that the job is offered to you?

In addition to being aware of the importance of participating in reflective practice, you will need to engage in life-long learning.

Life-long learning

As a reflective engineering practitioner, you will always be learning and improving. This will be essential to the development of your career. In the Engineers Australia Stage 1 Competency Standard (2019), the need for professional development and life-long learning is expressed as follows.

> Orderly management of self, and professional conduct.
>
> - Demonstrates commitment to critical self-review and performance evaluation against appropriate criteria as a primary means of tracking personal development needs and achievements.
> - Understands the importance of being a member of a professional and intellectual community, learning from its knowledge and standards, and contributing to their maintenance and advancement.
> - Demonstrates commitment to life-long learning and professional development.
> - Manages time and processes effectively, prioritises competing demands to achieve personal, career and organisational goals and objectives.
> - Thinks critically and applies an appropriate balance of logic and intellectual criteria to analysis, judgement and decision making.
> - Presents a professional image in all circumstances, including relations with clients, stakeholders, as well as professional and technical colleagues across wide ranging disciplines.

The importance for engineers to undertake on-going professional development is discussed in the following spotlight.

SPOTLIGHT

Engineers Australia and Engineering New Zealand — continuing professional development

Whether you join Engineers Australia or Engineering New Zealand, or a similar professional body, such as IEEE or IChemE, continuing professional development (CPD) will be an essential component of your professional life. Engineers graduate from accredited programs at universities. These programs are designed to deliver a range of graduate capabilities that will prepare graduates to begin practice as a professional engineer.

The next step in professional development is to become a Chartered Professional Engineer (CPEng). In Australia, this requires documentation of achievement of a range of Stage 2 competencies through *career episode reports* — documentation linking a competency (such as 'manage a small technical team') and a project in which an individual has been involved. Many companies now provide explicit guidance and mentoring in this process. Individuals accumulate the required number of competencies usually over a period of two to three years, drawing on the range of projects in which they have been involved. Supportive employers make sure that they have the range of project opportunities to cover all the required competencies.

Once CPEng has been achieved, engineers are required to keep their professional skills up to date to enable them to:

- maintain technical competence
- retain and enhance their effectiveness in the workplace

- be able to help, influence and lead others by example
- successfully deal with changes in their career
- better serve the community (Engineers Australia 2014).

Professional bodies such as Engineering New Zealand and Engineers Australia provide a range of regular meetings, seminars and conferences of specialised interest groups and technical societies, which cater for this need to stay up to date. There is also a range of other international professional bodies such as the Australian Computer Society (ACS), IEEE, IET, ASME (American Society of Mechanical Engineers) and IChemE. Many of these have a broad collection of professional meetings and publications designed to keep members up to date with best practice, which is changing constantly. It is advisable to become a student member of one of these bodies and to attend their meetings, many of which are now conveniently online. Face-to-face meetings have the advantage of also being good networking opportunities, which might lead to an internship or your first job.

The engineering method and project management

Earlier in the chapter, the engineering method was shown to be the basis of engineering problem solving, wrapped within a project management process. This was shown in figure 2.1, where the engineering method was represented by the steps in the inner loop and the overall concept of project management was represented by the outer loop. The major principles of project management are shown in greater detail in figure 2.5, which includes scoping and scheduling tasks and resources for the project; arranging meetings and organising the work, personal and group documentation; accuracy and risk assessment (less accuracy leads to more risk, and vice versa); reflection and reporting. Project management will be discussed in greater detail in the chapter on managing engineering projects.

FIGURE 2.5 The engineering method and project management process

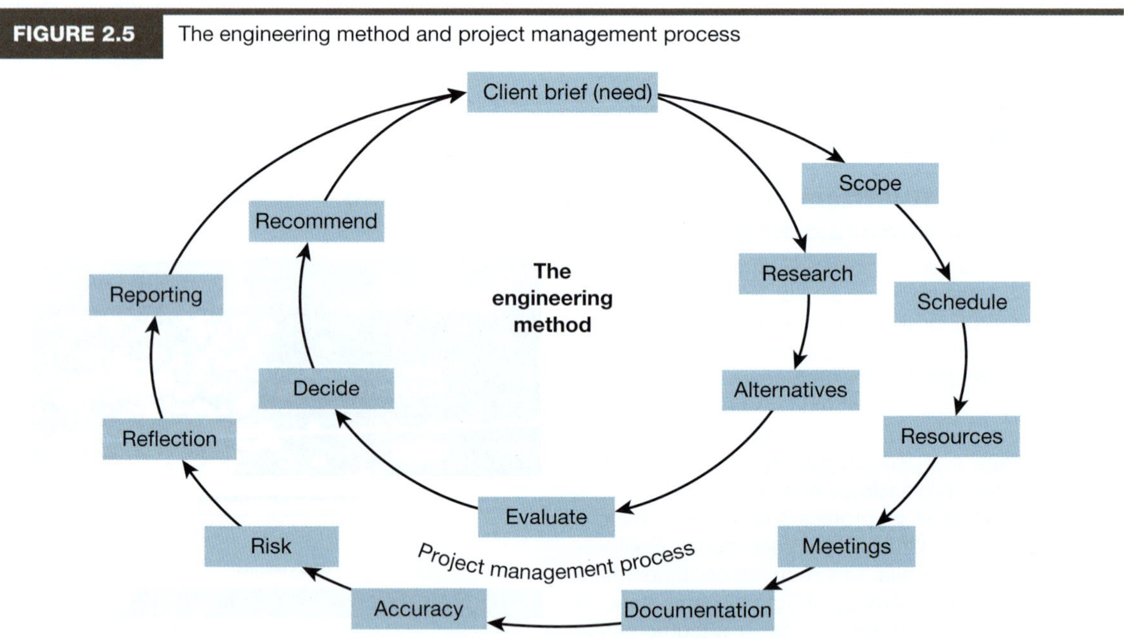

2.4 The life cycle of an engineering asset

LEARNING OBJECTIVE 2.4 Describe the role of an engineer throughout the life cycle of an engineering asset, including the differences between conceptual design and detailed design.

The preceding sections of this chapter have provided an overview of the engineering method (the basic problem-solving process that engineers use), and the need for a systems approach. Wrapped around the engineering method is a set of project management processes that ensure the project is delivered efficiently and effectively. This section of the chapter provides an overview of the life cycle of an engineering project from strategic planning through to decommissioning. It provides the context for engineering problem solving.

We have considered how problems can be defined in many different ways, resulting in quite different projects, even for the same client brief. Does all engineering work follow this pattern? Engineers are involved in all stages of the life of an engineering asset, whether it is a mobile phone, oil refinery, artificial heart or a power station. The seven common stages in the life cycle of an engineering asset are:

1. strategic planning — planning for anticipated problems at medium to long time frames, for example, 5–20 years
2. research and development — investigating novel or unusual problems that do not have standard solutions
3. conceptual design — identifying a preferred solution to a problem
4. detailed design — detailing the preferred solution so that it is ready to be implemented
5. implementation — construction or manufacturing of the preferred solution
6. commissioning, testing, and operation — getting the implemented solution started the first time (commissioning); making sure it is performing according to specification (testing); keeping the system operating, perhaps over an extended period (operations); and maintenance
7. decommissioning — closing a facility and possibly dismantling and recycling it.

Table 2.7 provides an example of these seven stages in the life cycle of an engineering asset.

TABLE 2.7 Stages in the life cycle of a hypothetical new mobile phone

Stage	Engineering questions, considerations and tasks
Strategic planning	What will mobile phones look like in five or ten years?
Research and development	What gaps in our understanding do we need to plug? What new services will be required? Where is battery technology going? What about screen resolution and networking standards? Do we need to collect new kinds of data to support our decisions?
Conceptual design	What range of solutions is available? What new technologies are emerging? How will we choose between them? What features do we need? What support will be required for them?
Detailed design	How should a particular solution be specified? How will it be integrated with other components? What are the power requirements? What manufacturing processes will we use? Where should it be built? Who are the stakeholders that need to be consulted?
Implementation	How will we build the preferred design? What project management and industrial engineering is required? What assembly equipment will be needed? What training is required for the manufacturing workforce? What government approvals are required?
Commissioning, testing and operation	How will the new model be tested (likely in secret)? How will users learn to operate the new device? What regular maintenance or support will be required (e.g. screen or battery replacement)?
Decommissioning	At some point, the phone will be inefficient or of limited capacity. It may need to be recycled or refurbished. What will be required? How will the materials be efficiently recovered for reuse?

The processes of strategic planning, research and development are quite specialised and often require either several years of practical experience or several years of specialised training after graduation. Likewise, construction, manufacturing, commissioning, testing, operation and decommissioning are complex activities that require a considerable amount of field knowledge and experience.

Design is more easily codified and is more easily taught at university. An understanding of how engineering systems are designed, no matter which branch of engineering is of interest, underlies the other processes. In this sense, design is at the heart of engineering and engineering is quite properly considered as a design profession. Design will be looked at in more detail next.

KEY POINT

Engineering projects arise across the life of an engineering asset, from the initial strategic planning of the asset right through to its eventual decommissioning and recycling.

Design

> Engineering design is a systematic, intelligent process in which designers generate, evaluate and specify designs for devices, systems or processes whose form(s) and function(s) achieve clients' objectives and users' needs while satisfying a specified set of constraints (Dym & Little 2009).

There are many definitions of how design takes place. For example, Dym and Little describe a three-phase process that includes **conceptual design**, **preliminary design** and **detailed design**. Conceptual design is the process of identifying a suitable solution to a problem by considering many alternatives. It involves making realistic estimates of costs with somewhat limited information. Preliminary design fills in some of the detail by sizing and specifying key subsystems in the conceptual design. Detailed design is the process of specifying all components so that they are ready for implementation. A detailed design gets down to the component level, while conceptual design works at the overall system level. In detailed design, how the engineering asset will be built needs to be considered. What size will all the components be? How will they be made?

Engineering deals with complex problems. It proceeds from top to bottom, from outside to inside, from the statement of a need to the statement of the problem to the statement of a likely solution or management strategy. It is unlikely a client will ask you for the final detailed design of an engineering project at your first meeting. They may not even be sure you can do the work. Initially, they may want assistance in understanding the problem and may look to you as someone who can provide them with answers through the conceptual design process.

Interestingly, the conceptual design stage has the greatest impact on the final cost of the project. This is because the most important decisions in a project are locked in at this stage. The basic technology is decided, usually based on imperfect estimates of final cost and environmental impact. These decisions are often difficult to alter at the detailed design stage. In detailed design, the emphasis is on establishing how to implement the technology chosen at the conceptual design stage and further refined through preliminary design.

Different engineers prefer different aspects of this work. Individuals with a preference for uncertainty and risk are attracted to conceptual design. They are often 'big picture' people. They can discuss complex ideas and revel in being bold and creative. Other engineers, with an attention to detail, prefer detailed design. At this stage, creativity is more limited and components are usually carefully optimised. Irrespective of personal preferences, design is an essential part of engineering.

Engineers Australia (2019) stresses the importance of design in its Stage 1 Competency Standard.

> Application of systematic engineering synthesis and design processes.
>
> - Proficiently applies technical knowledge, open ended problem-solving skills, appropriate tools and resources to design components, elements, systems, plant, facilities and/or processes to satisfy user requirements.
> - Addresses broad contextual constraints such as social, cultural, environmental, commercial, legal, political and human factors, as well as health, safety and sustainability imperatives as an integral part of the design process.
> - Understands and applies the whole systems design cycle approach. Is aware of the accountabilities of the professional engineer in relation to the 'design authority' role.

Why are these design skills important to an engineer? Engineers develop solutions to societal problems, as outlined in the 'What is engineering?' chapter. It is important that engineers make good use of the resources available to the society they serve and that they are mindful of the effect of their work on the environment and the society they serve. Engineering work is constrained in various ways and there is an expectation that engineers will produce the 'best' work they can, within these limits, or as the quote at the start of the chapter put it '*cause the best change, in a poorly understood situation, within the available resources*'. Consider now this issue of design constraints.

All design is constrained, be it by cost, by environmental impact, by weight, by power consumption, by size, and so on. The challenge of engineering is to produce excellence in design within the identified constraints, as this next spotlight shows.

SPOTLIGHT

Driverless vehicles will change our lives

There continues to be plenty of hype about driverless vehicles (DVs) (Walker 2023; Winton 2018) as well as the occasional, sobering story of fatalities associated with testing such vehicles (Greenemeier 2018). Maybe it will be some time before we can jump in our car and simply tell it to 'take me to the shops'. Maybe that will never happen. Perhaps the future might be quite different from that. An interesting article in *Medium* highlighted the potential disruptive changes that could occur with DVs. The author described more than 70 aspects of our lives that could change (Nesnow 2018).

Google's self-driving car

The first change, perhaps most obviously, is that more people will stop owning cars. This is already a trend in the inner parts of large cities where people use public transport, taxis and car share schemes (Clay 2014). Driverless vehicles will become a new form of taxi. It is no surprise that Uber is actively involved in the development of driverless vehicles (*Sky News* 2018).

The implication of this is that large tech companies, such as Google, Apple, Uber, and Amazon, will expand into the transport sector and, possibly, the energy sector, which could make use of the large battery capacity in thousands or millions of electric vehicles plugged in at any time (Mwasilu et al. 2014). The distributed use of energy in electric vehicles also supports the expanding use of renewable sources of energy (solar and wind primarily) with batteries providing load-levelling across the grid from daytime to night time. Oil companies will dwindle in importance.

The design of cars will also change in this process. If a car is no longer a status symbol, it can be optimised for performance rather than kerbside appeal. Cars may become much more uniform in appearance, perhaps painted in corporate colours (Facebook blue, Google's primary colours).

Driver licences will no longer be the default form of identification and other forms will be necessary in places like Australia that do not have a national identification card. Many young people are already delaying getting licences or rarely using them if they live in inner cities where public transport is more convenient.

Parking lots and service stations will become redundant and will be repurposed. Traffic police and parking inspectors, likewise, will be able to be used for other purposes. Traffic signals and signage will eventually disappear. Car chases will be a thing of the past. Drink-driving offences will disappear, so bars and restaurants might sell more alcohol and it will be easier to go out for a meal and a drink (or two).

Industries that support the car industry will also decline — car financing and insurance, car repairs, refuelling stations, advertising, and those involved with legal and medical issues associated with car accidents. There will be fewer ambulances required, fewer rehabilitation hospitals, fewer medical practitioners.

Many people will have much more money to spend on other purposes, likely driving shifts of employment from car-related industries into others, for example, entertainment. This may be a difficult shift for many, given that driving a vehicle is a relatively low-skilled occupation that is readily available to many with low levels of formal education.

Roads will be emptier because DVs can travel much closer together and they will likely carry multiple people. We will think of them like elevators. We will order a vehicle, which might already have someone riding in it along our route and it might pick up another one or two people along the way, for example, to the train station. There will be more use of public transport as it transports people further, faster and more cheaply.

Cities will become denser, with more people living close to transport hubs. This trend is already occurring (Infrastructure Australia 2013). Roads will take up less space as they become more efficient and on-street parking disappears or is substantially reduced. The walkable, cyclable, green city would become more common (former road space turned into green space and bike lanes).

There will be new marketplaces for services sold in the DVs. This will include entertainment and educational services. There will be less privacy, of course, with cameras and vehicle logs automatically recording our movements.

Mobility for seniors and people with disability will be greatly improved. Secure transport of children to schools might also be possible and encouraged. Transport of urgent goods could follow similarly — possibly as cheap as a courier over short distances.

Driveways and garages of detached homes will be repurposed. (Although many garages are already turned into family rooms as families get older.)

Cars can be full of sensors, collecting all kinds of climate and other data along their route, leading to new data services, like the way that Google can now advise of traffic congestion based on its many users' simultaneous data.

Cars will be full of advertising, and we might spend our time booking our next holiday as we commute.

Of course, these changes will likely not be uniform and will be seen first in the inner cities where car ownership is declining (or discouraged) (Clay 2014). Before we see fully autonomous vehicles, we will likely see growing autonomy — smart extensions to cruise control — frontal collision detection, already available in many vehicles, automatic parking and lane following (great for the highway), and capabilities such as Tesla's Autopilot and Enhanced Autopilot (Turner 2024). All these features are already available in many new vehicles. Other applications will include inner city shuttles (Sisson 2018) and street sweepers (Bellamy 2018).

CRITICAL THINKING

What do you see as the pros and cons of driverless vehicles? Would you use one? How? What features would make them more appealing to you? What do you see as the principal hazards for the user, the design engineer and the company operating them?

Engineers must be creative within a constrained environment of resource availability, data, information, knowledge and regulation. How do engineers deal with problems under these conditions? As well as the engineering method (the design process), which structures their problem-solving process, they must be able to make judgements about the alternative solutions. The rules applied in this process are a collection of **heuristics**. The opening quote for this chapter related to heuristics — a problem-solving technique in which the most appropriate solution is selected using rules.

Heuristics are rules of thumb. Before the twentieth century, engineering relied heavily on rules of thumb based on past engineering practices. For example, a rule of thumb when choosing a beam for a particular span is that the depth of the beam should equal the span divided by 20. So, a span of 20 metres requires a beam 1-metre deep.

We now have more complex models and heuristics based on science. The twentieth century saw a revolution in engineering through the application of the scientific principles of mathematics, physics, chemistry and, more recently, biology, as well as management tools such as economics and psychology. Engineering is sometimes described as 'applied science', though this text shows that engineering is much more complex than that and relies on an interdisciplinary combination of science, economics, management and social science.

Even science is a set of sophisticated heuristics (Koen 2003). Think about how science has changed over the last century, particularly since Albert Einstein's first paper on special relativity in 1905. Every new scientific advance means discarding or modifying old ideas that were considered 'true'. Before Einstein's work, Newtonian mechanics was considered an exact description of the laws of motion. We now know that it is a good approximation at small velocities — certainly adequate for day to day living or even getting a spacecraft to rendezvous with Pluto — but not very useful when designing particle accelerators that operate at 99.9999 per cent of the speed of light. Every scientific theory is a new heuristic that is trying to explain reality better than the last theory. Note also that science can't 'prove' theories. It can only stack up supporting evidence. It can, however, disprove theories, by finding contradicting evidence (Popper 1983).

It is important that engineers understand modelling with heuristics because their work is governed by codes of practice that specify minimum or maximum performance standards, such as temperature, pressure, stress, deflection, current, voltage, water quality and so on. These standards may be Australia–New Zealand standards or international standards and they will be discussed in more detail in subsequent chapters.

Engineering is also about trading off cost and risk. Engineers work within available resources. They are responsible for making the most of resources that might be better spent on another product or service, and constantly work with limited information. Engineers cannot predict future weather events or human events; therefore, engineers are often left to make predictions of future behaviour based on historical trends.

Engineering involves the choice of a best solution or a best change. What is best? Sometimes it is what is cheapest to build, or it might be what is cheapest over an engineering asset's total lifespan. It could be what produces the greatest economic return, or it may be the option with the least environmental impact as assessed by a life cycle assessment. Ultimately, engineers must work with communities to help define what is best for them.

It is important for engineers to be conscious of the models and the assumptions made (heuristics), what the client wants (what is best), what resources are available (usually money) and what is and is not understood about the problem of interest (limited information).

2.5 Critical thinking

LEARNING OBJECTIVE 2.5 Use critical thinking skills to scope and solve engineering problems.

One of the key skills that you will learn at university is *critical thinking*. Most universities include critical thinking in their list of attributes or qualities that all their graduates will acquire during their studies. This is because critical thinking can be applied to all facets of our lives and all our human endeavours. You will learn about, acquire, and apply critical thinking skills at university, and you will continue to develop these skills throughout your career.

Critical thinking is an essential component of good engineering thinking, as will be illustrated in this section. There are many definitions of critical thinking, from the simple to the complex (see, for example, Scriven & Paul 1987). A good starting point is a statement by Glaser (1941) that 'Critical thinking calls for a persistent effort to examine any belief or supposed form of knowledge in the light of the evidence'. This requires knowledge of the methods of logical inquiry and reasoning, and a predisposition and ability to apply them (Glaser 1941).

Critical thinking normally involves more than just the application of cognitive thinking skills to an issue or problem. It also involves the evaluation of the thinking processes being used. Paul and Elder (2006, p. xvii) include this dimension in their working definition of critical thinking.

> Critical thinking is the art of thinking about thinking while thinking in order to make thinking better. It involves three interwoven phases: it analyses thinking, it evaluates thinking, it improves thinking.

The term critical thinking is also used by social theorists to 'critique', question or comment on aspects of society from specific critical theoretical perspectives. The most important element of this genre of thinking is to *question assumptions* and to move from subjective reactive *opinions* to *well-informed evidence-based positions*.

In an engineering context, this means that while solving a problem, an engineer should also be thinking about the validity of the approaches, processes, principles, and engineering theories and practices in the specific social and environmental contexts that are being used. Thus, critical thinking involves not only the application of cognitive thinking skills, but also the evaluation of the thinking processes being used.

Paul and Elder (2006) neatly unpack the complexity of critical thinking by describing it as a process that involves three components:

1. *intellectual standards*, such as clarity, precision, accuracy, significance, relevance, completeness, logic, fairness, breadth and depth
2. *elements of reasoning*, such as purposes, inferences, questions, concepts, points of view, implications, information and assumptions
3. *intellectual traits*, such as intellectual humility, intellectual perseverance, intellectual autonomy, confidence in reason, intellectual integrity, intellectual empathy, intellectual courage and fair-mindedness.

They connect these three components by describing the process as follows: 'Critical thinkers routinely apply the intellectual standards to the elements of reasoning in order to develop intellectual traits' (Paul & Elder 2006, p. 54). Thus, our critical thinking skills are enhanced over time through the systematic application of intellectual standards and reasoning to create designs, investigate issues and solve problems.

Finally, it is important to note that critical thinking is complementary to creative thinking. Creative thinking is used to expand the set of options available to us, while critical thinking is used to systematically eliminate those options that do not match our requirements.

KEY POINT

Critical thinking skills are used by engineers to solve problems and develop design solutions.

Critical thinking in an engineering context

In an engineering context, critical thinking involves:

- clarifying the purpose of the task
- identifying the fundamental questions to be answered
- accessing and evaluating the information required to answer those questions
- noting any assumptions made, including the cultural and knowledge systems employed
- applying relevant concepts
- developing and evaluating alternative solutions
- making informed and wise decisions or recommendations
- accurately reporting findings or outcomes
- noting our own personal bias and those of other contributors, and questioning the impact of these on the appropriateness of the process and solution
- recognising the influence of personal preferences while being open to new ideas, concepts and approaches.

This is not always easy to do, particularly when faced with high workloads, tight budgets or rapidly approaching deadlines. In these situations, it is easy to fall into the trap of applying a familiar or routine solution, even when there are indications that it doesn't quite fit. While the resulting outcome may be acceptable, it may not be the best outcome or most efficient outcome in the long term.

Critical thinking skills are used throughout a project, not just in the design or problem-solving stage of the process. To emphasise this point, examples of the critical thinking questions that may be asked are provided for each stage of the engineering method. At the beginning of each task, particularly new tasks, the engineer should reflectively select the relevant standards and elements of reasoning they believe are required to complete the task. They would then apply those standards and elements of reasoning to the problem, adjusting their approach when new or unforeseen problems arise, all the while applying their intellectual traits to the task.

A real, and incredibly difficult, design challenge will be used to illustrate how the standards and elements of reasoning are used in each stage of the engineering method. In the Calscan challenge, a leading health and lifestyle company asked a medical electronics company to develop a concept and conduct a feasibility study for the development of a calorie scanner app for a mobile phone. The proposed app would be used to scan a food item, or a plate of food, estimate the food energy, and then use that information to enable the user to monitor calorie intake.

The current app requires the user to enter the type and amount of each food item consumed. It then uses that data and the information in a database of more than 400 000 food items to calculate the calorie intake. The app also allows the user to log personal characteristics such as height and weight, exercise, water intake, and other information, and then uses the information to calculate daily and weekly balances and healthy lifestyle plans. The proposed Calscan app would give the company a market edge, as the user would not have to estimate the content or amount of the food items they consume.

Stage 1. Client brief

The client brief is a document or verbal instruction describing the needs of the client, as well as the performance criteria or requirements for an engineering project. The brief may be developed by the client or by the client and the engineer during a briefing meeting. Either way, the engineer should always work through the brief with the client and help them to articulate their requirements. The following list provides examples of questions that the engineer might ask to understand the brief for the Calscan.

Intellectual standards

Clarity. Is the client's brief clear, and does it contain all the relevant details so that a detailed specification and/or contract can be drawn up? Is the client fully aware of the approach that will be used and the tasks that will be undertaken?

Breadth. Does the client's brief cover all aspects of the project?

Accuracy and precision. What accuracy is required from the calorie scanner? What precision is required?

Context. How and where does the client expect the Calscan to be used? Does the client expect to market the scanning technology in computer applications in addition to the mobile phone app? Should the client be the only one to decide how this product is made and what it will do?

Elements of reasoning

Purpose. Is the purpose of the Calscan app clearly spelt out in the brief?
Implication. What impact might Calscan have on eating habits? How and when might people typically use the app?
Feasibility. How technically feasible is the proposal, using available technology?
Risk. What are the health risks if the Calscan is misused or is inaccurate in its measurements?
Social impacts. What social impacts might there be?
Ethics. What are the ethical implications of Calscan?
Assumptions. What assumptions is the proposal based on?
Sustainability. What are the sustainability goals, such as energy and waste?

In answering these questions, it is important to apply the *intellectual traits*, such as intellectual humility, intellectual perseverance, intellectual autonomy, confidence in reason, intellectual integrity, intellectual empathy, intellectual courage and fair-mindedness.

Once the details are finalised and agreed upon, the detailed client brief can be drawn up and signed by all parties.

Stage 2. Research

During this phase of the project, the engineer analyses the client brief and undertakes a needs analysis to understand and clearly define the problem, often resulting in a detailed engineering specification. This understanding of the problem is used to determine what the data and information needs are for the project. This may involve:

- using existing data, designs or information
- gathering new data
- reviewing codes and manufacturers' handbooks
- researching new information
- seeking knowledge from colleagues.

The data and information are then evaluated and synthesised into a report that may be shared in a knowledge management system.

For the Calscan project, four initial problems were identified.

1. What food components will need to be scanned to assess calories in contemporary food, for example, carbohydrates, fats, fibres, proteins and nutrients?
2. What technologies could be used to scan and assess the calories in food items?
3. Can the camera systems commonly used in mobile phones be used as part of the scanning system?
4. What impact would the proposed technologies have on food items or live tissue during the scanning process?

Other problems may arise during the research process, and these would then become part of the project. The following critical thinking questions may be asked to assess the data and information gathered to answer the research questions.

Intellectual standards

Context. Are the findings reported in this article transferable to the Calscan project?
Assumptions. What assumptions are being made about ways of living and how the product might affect these?
Knowledge. What is the knowledge system being employed?
Relevance. How relevant is this data or information to the project?
Evidence. Are the findings supported by the evidence provided in the article?
Significance. What is the significance of the outcomes reported in the article?
Accuracy. What is the accuracy of the data? Is it sufficient for the project?
Precision. Is the precision reported for this machine suitable for the project?
Logic. Is the information reported in a logical manner? Is there any ambiguity?

Elements of reasoning

Scepticism. Is information from this source trustworthy?
Purpose. Was the purpose of the design reported in the article similar to the Calscan project?

Assumptions. What assumptions were made to undertake the reported work?
Concepts. What scientific and/or engineering principles was the design based on?
Inferences. What other inferences can be drawn from the evidence provided in the article?

Stage 3. Alternatives

Once the data and information has been gathered, analysed and synthesised, it is time to think clearly about the next step in the engineering method, *generating alternative solutions*. Careful needs analysis and a straightforward specification help limit misunderstandings between the consultant and the client on any project. The following questions would be considered during this stage of the Calscan project.

Intellectual standards

Concepts. What engineering methods should be used for this project?
Codes. What codes of practice and government regulations apply, for example, food hygiene, electromagnetic radiation, health and safety?
Accuracy and *precision*. How will the required accuracy and precision be achieved?
Completeness. How will we know when the design is complete?
Validity. Is this a valid approach to solve this problem?

Elements of reasoning

Systems. How will the developed technology, hardware and software be integrated into the next generation mobile phone?
Contexts. What user and locational characteristics need to be considered in the design process?
Sustainability. What approaches can be used to achieve the sustainability goals?
Open-mindedness. What other approaches could be used to achieve a better outcome?
Information. Has all the required information been gathered so that the conceptual design of the Calscan can be undertaken?
Assumptions. What assumptions have we made in developing this design?
Rigour. Is the solution based on rigorous scientific and engineering principles?

Stage 4. Evaluate

Once one or more alternative solutions have been developed, they are evaluated against the client brief, performance criteria and requirements, and the preferred alternative is then selected. The following questions should be considered.

Intellectual standards

Integrity. Does the design satisfy all the client's performance criteria?
Completeness. Is the design complete?
Codes. Does the design satisfy all relevant codes and government regulations?
Accuracy and precision. Will the design meet, or better, the client's accuracy and precision requirement for the Calscan? Are the projected figures realistic?

Elements of reasoning

Purpose. Does the design achieve the purpose specified in the client's brief?
Feasibility. Is the concept financially feasible?
Sustainability. Are the sustainability goals achieved?
Risk. What are the safety, financial and environmental risks associated with implementing the design?
Social impact. What impact might this design have on lifestyles?

Stage 5. Check

It is common practice for a project team to monitor, check and review all aspects of the project on a regular basis, normally at weekly meetings. This ensures that the team members have a shared understanding of the progress being made, current design issues, and key dates. These meetings also provide an opportunity for the team to review outcomes and ensure that errors, including obvious errors, have not been made.

Once the preferred alternative has been selected, it is important that all aspects of the design or solution are carefully checked before the outcomes are communicated to the client. Once this checking has been undertaken, some companies engage third parties, such as other engineering companies, to undertake final checks on their designs, calculations, plans and reports before they are delivered to the client. The questions used in this stage would be like those in the previous stage, but they would be asked by different people.

Stage 6. Recommend

Once the preferred alternative has been checked, and it is clear that it meets the client's brief, the outcomes are communicated to the client. This could be a two-stage process, in which case the client is given an opportunity to review and comment on the outcomes prior to them being formally presented at the end of the project. Normally, the outcomes of a project include a report, plans and other relevant documents.

The following questions may be considered while the report and recommendations are being prepared. These would be asked as part of the ten-step PCR planning process, discussed in the chapter on understanding communication.

Intellectual standards

Clarity. Will the client be able to understand the report and accompanying plans?
Relevance. Is all the information included in the report relevant?
Significance. How should the most significant outcomes of the project be highlighted?
Logic. Is the document written in a logical manner from the perspective of the client?
Ambiguity. Does the document contain any ambiguities?
Completeness. Does the report contain all of the required information from both the client's and company's perspectives?

Elements of reasoning

Purpose. Does the report simply and effectively communicate that the purpose of the project has been achieved?
Points of view. Does the report demonstrate that all of the relevant stakeholders have been consulted and their interests considered?
Assumptions. Does the report describe the key assumptions the conceptual design is based on?
Implications. Does the report clearly communicate the impacts and risks of implementing the proposed design?
Information. What media (i.e. tables, charts, plans, photos etc.) should be used to effectively communicate the recommended solution, and the data, information and evidence required to justify the design?

What a lot of questions! You can probably think of a lot more questions that would need to be answered if the Calscan app is to become a reality.

It is important to remember that the engineer and other team members would be applying their *intellectual traits* to the critical thinking tasks in the project and, because they are using their critical thinking skills in a systematic manner, they would be enhancing their intellectual traits.

The Calscan example shows the range of critical thinking skills that engineers use, and it highlights the importance of systematically thinking about the approaches, knowledge, skills and tools that will be applied to a problem before commencing work on the problem itself. This is difficult to do, as the temptation is always to 'jump in' and begin to solve the problem, particularly in fields we like working in, or have experience in.

It is just as important to stop at each stage of the design or problem-solving process and critically reflect on the outcomes achieved and the success of the approaches used, and to consider other approaches that could be used to produce a better solution or to validate the current solution. At these waypoints, it is also important to consider if the approaches that were previously selected for the next stage are still valid, and whether better approaches can be used. By spending a little more time to explore other options and approaches, a far better outcome may be achieved at minimal additional cost when considered in the context of the overall project budget.

Knowledge systems refer to the ways in which we think and know in our communities and cultures. Western ways are often considered to be very individualistic, and many other communities work in a more communal way. Knowledge, to these groups, is not something found in one person's head. This profoundly affects the way we view education, employment, and social norms such as eating and sharing and could affect the considerations when undertaking a project such as the Calscan one.

The key message from this critical thinking section is that you should strive to:

- continually improve the way you think and reason
- question your biases and assumptions
- think with greater rigour
- think with deeper empathy
- think your way through to more logical, balanced and valuable conclusions.

The following chapters include descriptions of some of the approaches, tools, techniques, and ways of thinking that engineers use to develop solutions for complex operational and design problems, and to manage projects.

The following spotlight highlights an important critical thinking skill: using thinking skills to find unasked questions. Russell Mineral Equipment (https://www.rmeglobal.com) is based in Toowoomba, Queensland, and employs more than 300 people worldwide. It is the technological leader and dominant supplier of specialist relining equipment for hard rock grinding mills. The business has been built from a zero-capital base to a debt free business, enjoying around $75 million in annual sales. John Russell and Julie McKerrow describe the journey in the following spotlight.

SPOTLIGHT

Finding the unasked question: a sign of professionalism

JOHN RUSSELL, MANAGING DIRECTOR, AND JULIE MCKERROW, MARKETING MANAGER, RME

Engineers are good, perhaps even gifted, at solving problems. However, big opportunities lie not in solving problems identified by other people, but in finding your own problems or opportunities. At Russell Mineral Equipment (RME), we call that 'finding the unasked question'.

Metso mill lining using an RME machine

RME's existence, and its success, is built on the identification of unasked questions. The unasked question I found during 1983 was 'Can relining mechanisation reduce concentrator plant shutdown durations?' The implication was that such a result would leverage the productivity of the entire site. As you will see, pursuing this simple question changed my life forever.

After graduating in 1979 with a mechanical engineering degree, I started work with Mt Isa Mines in north-west Queensland, and worked there until 1985. Mt Isa Mines is a significant producer of copper and silver, lead and zinc metals, and most of my time was spent in the 'concentrator' area — the processing plants that extract the mineralisation from the run of mine ore to produce a 'concentrate' suitable for transportation and subsequent smelting.

To extract the mineralisation from the host rock, all the run-of-mine ore must be blasted, crushed and finely ground to average particle size of around 75 microns. The grinding of ore occurs inside large rotating cylinders, known as grinding mills. These mills can be very large, up to 12.8 metres in diameter and 7.6 metres long, requiring up to 28 megawatts of power to rotate them. The structural shell of a mill would wear away in hours if it was not protected, so they are protected with very heavy steel alloy liner segments, referred to as mill liners. In the largest mills, there may be up to 2000 tonnes of liners, with each liner weighing up to five tonnes. Despite their massive size, these liners are sacrificial — designed to wear away and be replaced every four to six months.

This liner exchange task, which requires the concentrator to be shut down, used to be an intensely manual task, often accomplished without the benefits of mechanisation, particularly in smaller mills. It was dangerous, and it took a very long time. The combination of the number of mill liners, their life expectancy and their exchange rate therefore dictates concentrator plant availability, and concentrator plant availability dictates the throughput of the entire mine.

The answer to my first unasked question was 'yes', and today RME's Mill Relining System, a suite of individual technologies, has reduced typical large grinding mill reline from 160 hours to 60 hours when using RME Single Mill Relining Machine Systems, and to 40 hours with RME Twin Mill Relining Machine Systems.

This performance has liberated between 2 and 4 per cent of extra plant utilisation at our customer's sites, leveraging the entire mine-site investment (billions of dollars of investment for new projects these days).

So, after finding the first unasked question, I set about solving the problem, beginning with the next question: 'What does grinding mill relining mechanisation look like?' This second question has been asked again and again by RME engineers since the production of our first Mill Relining Machine in 1990. This thinking process has yielded a range of innovative technologies, such as:

- Russell 8 and Russell Twin 8 mill relining machines
- Thunderbolt Recoilless Hammers, which deliver high-energy, high-momentum recoilless blows to remove worn liners and liner bolts quickly and safely

- T-MAG, which is a guide for the Thunderbolt Hammer that eliminates the need for personnel to be in the hammer strike zone
- O-ZONE Liner Lifting Tools, which save one minute per liner while also improving safety
- Feed chute transportation technologies, which allow rapid extraction, transportation and replacement of mill feed chutes.

RME illustrates the importance of finding the unasked question. This thinking process ensures engineers not only build things right, but build the right things.

CRITICAL THINKING

Consider the world around you, and over the coming days and weeks think about what you hear, read, and see and try to find an unasked question that needs an engineering solution. This may be a business opportunity for your future!

Moving from having an opinion to taking a position

One of the most important skills for critical thinking is the ability to develop substantial evidence for your position. There is rarely a single right answer to a real-world challenge faced by engineers. There will always be multiple viewpoints and agendas influencing what gets designed or produced. Thinking critically will enable you to evaluate the problem and potential designs and work out and tell others *why they are better for this context.*

Everyone has their opinion, but they will not be able to appropriately defend their ideas unless they know what they are talking about and have researched the issue. It is not good enough to retell a story from the latest newspaper article and claim that you are the expert. It is not enough to have ten articles that support your view, if they are all reporting on projects funded by a biased source. You need to locate multiple independent sources that enrich your argument, helping you to move from a biased subjective opinion to a well-informed evidence-based position. You will still be biased — your ideas will never be 'objective' — but if you declare your bias and back up your arguments with sound evidence, you will be able to convince your audience much more effectively.

SUMMARY

This chapter has provided an overview of what it means to work as an engineer, using the engineering method (the design process), which begins by defining a problem and the requirements for its solution, generates alternatives, evaluates those alternatives against the requirements and recommends a solution. This basic problem-solving process is wrapped in project management processes to ensure that time, resources and other constraints are properly acknowledged. Within the engineering method, systems thinking is an essential skill to make sure that the correct problem is being solved. Any engineering project will progress through a life cycle from strategic planning to decommissioning. Each engineer needs to develop reflective practice skills to improve their performance as they move from one project to the next. We will now briefly revisit each of the chapter learning objectives.

2.1 Describe the activities that constitute the engineering method and apply the method to an identified problem or challenge.

The engineering method starts with the client's need, which is identified in a client brief. The engineering method requires the collection of *data* and other *research*, which yields *alternative solutions* to the problem. These solutions are evaluated against the *requirements* defined by the client to choose a preferred solution. After this, *checks* are made of all the important processes and conclusions and a *recommendation* is communicated to the client about the most appropriate solution to the problem. A car-buying scenario was used in this chapter to illustrate these problem-solving steps.

2.2 Identify a range of system definitions for a problem and use these definitions to present different solutions to a problem.

Systems thinking is required to make sure that the right problem is being solved. It defines a problem by seeking to identify its boundary, its components and the interactions between the components, as well as those between the system and its environment (across the boundary). Key questions to ask are as follows.

- Are you setting out to solve the right problem?
- Are there other ways of looking at the problem?
- Do you have all the facts?

A congested freeway example was used to illustrate how, by redefining the problem, a range of other solutions become possible.

2.3 Apply basic project management principles to plan a project and maintain organised project documentation.

Project management is the process of managing projects. A project is a temporary endeavour designed to produce a specific product or service. A key skill required in project management is scheduling. A Gantt chart is a simple way of mapping the tasks between the commencement of the project and its required completion date. Resources need to be mapped against the schedule. When and where are they required? Also, consider accuracy. How accurate an answer is required? How much effort is required to get that level of accuracy? Is this justified within the time and resources available? Risk is also an important consideration. What risks may impact your project? What is the likelihood of these risks occurring, and what are the possible consequences? What can you do about them?

2.4 Describe the role of an engineer throughout the life cycle of an engineering asset, including the differences between conceptual design and detailed design.

All engineering projects pass through distinct stages of development. These include big picture thinking such as strategic planning, and research and development. From here, design proceeds in two stages: (1) conceptual design — considering what solutions are available — and (2) detailed design — considering how the preferred solution will be implemented. Implementation in the field then requires some form of manufacturing or construction, followed by commissioning, testing and operation. At some point, the project will have outlived its usefulness and will be decommissioned. Conceptual design is the process of identifying a suitable solution to a problem by considering many alternatives. Detailed design is the process of taking a conceptual design and detailing its components, so they are ready for implementation.

2.5 Use critical thinking skills to scope and solve engineering problems.

While you are at university, you will have many opportunities to enhance and apply your critical thinking skills, both in engineering and other contexts. Critical thinking is a complex process that involves the application of intellectual skills, attitudes, dispositions and values to a task. It normally involves two parallel activities: first, applying thinking skills to the problem at hand and, second, thinking about the relevance and effectiveness of the thinking skills being used. Paul and Elder suggest that we apply intellectual standards and intellectual elements to think our way through engineering and other issues as we gradually enhance our intellectual traits. Critical thinking is the cornerstone of engineering; it is required for the successful completion of engineering projects.

KEY TERMS

action learning The process of continual improvement or learning through action. The four basic steps are: plan, act, observe and reflect (ready to make a new plan).
action plan A documented process of tasks that need to be completed, when and by whom.
client brief A document or verbal instruction describing the needs of the client: the basis of an engineering project.
collaboration environment An online environment for sharing documents and expertise.
conceptual design The process of identifying a suitable solution to a problem.
constraints A defined limit on the design, such as noise, cost, weight, environmental impact.
design file The personal documentation for each project, including notes and sketches, reflections, printed materials and electronic materials. It may be stored online and may be shared with other group members.
detailed design The process of taking a preliminary design and detailing all its components so they are ready for implementation.
engineering method A process that identifies the problem and the required performance criteria and constraints, considers a range of solutions, evaluates the solutions against the criteria and constraints, and recommends one or more 'best' solutions.
engineering project A task identified in response to a recognised engineering problem.
Gantt chart A simple bar chart showing the duration of project activities against a timeline.
heuristics A problem-solving technique in which the most appropriate solution is selected using rules.
milestones An intermediate checkpoint, usually indicating a major task is complete.
mind map A hierarchical, two-dimensional graphical representation of a complex topic.
performance criteria The values and goals of an engineering project.
preliminary design The process of specifying major subsystems in the conceptual design.
primary data Data directly observed and recorded, collected from the field or from respondents (e.g. a questionnaire survey).
project management A suite of processes used to manage a project to completion.
requirements The criteria and constraints that must be satisfied to deliver a satisfactory engineering product or project. These are usually written in functional terms.
risk assessment A process of identifying impacts, their likelihood and consequences. The risk is the product of likelihood and consequences.
scheduling The process of assessing what tasks are required, how long they will take and what their relationship is to other tasks.
secondary data Data compiled from both inside and outside an organisation for a particular project.
systems thinking The process of identifying all the elements of a problem. It requires explicit attention to the problem boundary and the components of the problem and an understanding of the interactions across the boundary and between the components within the system.
timeline The overall duration of the project, which is usually broken down into the major milestones required for project completion.

EXERCISES

1 In the car-buying scenario in this chapter, an attempt was made to identify suitable cars for a client to purchase. Outline other solutions available if you expand the system boundary for this problem (e.g. other transport solutions). Analyse what options are more cost effective and document the trade-offs to be made with personal convenience and time taken for each trip with the alternative solutions.
2 Repeat the car-buying activity using a mobile phone as an example. Ask a parent, sibling, friend or relative to act as the intended user. Start by helping them to specify their requirements and then proceed to investigate possible solutions. Did you find your preferred solution being at odds with your client's?

Could this happen in real engineering projects, where stakeholders have different views of acceptable solutions from you as engineer?

3 How much time do you have to devote to your studies each week? You are likely to have a timetable that shows all your scheduled classes. Work out how many hours you spend attending classes each week and the amount of study you need to do for all your subjects. You may be able to schedule some of these hours between your contact hours, using your library or other study spaces available to you. If you are a distance education student, you may find you spend hours online, keeping up to date with study requirements and communicating with course coordinators. With careful organisation, you can avoid having to spend every evening and much of the weekend studying. University study is a full-time job, which, with careful organisation, you should be able to do between 9 am and 5 pm, five days per week.
4 Use a reflective journal to think about all your studies this semester. Which subjects are going well? Which ones are not going so well? Which ones do you enjoy more or find easy? Why might this be? Identify some ways in which you could improve your performance. Are there skills that you need to learn (e.g. computing skills, oral presentation skills or library skills)? How will you learn these skills? Would a study group help, where you work with other students to develop these skills?
5 Choose an engineering project in your favourite discipline. Write a brief description of the type of work that would be done at each of the seven common stages of the engineering life cycle — strategic planning; research and development; conceptual design; detailed design; implementation; commissioning, testing and operations; and decommissioning. Which aspects of the work appeal to you most? What companies or organisations operate in these areas? How might you approach them for an internship?
6 Use the standards and elements of the reasoning approach, explained prior, to critically review the thinking processes you used to solve a difficult problem or to develop a design.

PROJECT ACTIVITY

Consider a problem of need or opportunity in your discipline. It could be an energy problem to reduce greenhouse gases, or a water, traffic, communication or biomedical challenge. Imagine that you work for an organisation that wishes to reduce this problem or develop an innovation that can be marketed to address the issue. Write a short hypothetical client brief based on this challenge to recruit talented engineers to work on the problem.

Now shift your focus to the engineer's perspective. Using the engineering method, identify the stakeholder views, sharpen your definition of the challenge to be addressed and find potential solutions. Write a list of functional requirements that must be satisfied. Prepare a conceptual design for the problem, generating a short list of solutions that are superior to other alternatives. Now conduct a detailed design analysis of a component of one of your preferred solutions to specify it ready for implementation. Prepare a report that is based on your recommendation. Your design file and communication skills should help with this.

REFERENCES

Bahill, AT, Bentz, B & Dean, FF 1996, 'Discovering system requirements', https://doi.org/10.2172/263004

Bellamy, D 2018, 'China gets its first driverless street sweeper', *Euronews*, 27 May, https://www.euronews.com/2018/05/27/china-gets-its-first-driverless-street-sweeper

Buzan, T 2013, *Mind map handbook: The ultimate thinking tool*, HarperCollins Publishers, Sydney.

Buzan, T & Buzan, B 1993, *The mind map book*, Penguin, New York.

Clay, M 2014, 'How Millennials are driving the shift away from cars', *ABC News*, 20 November, https://www.abc.net.au/news/2014-11-20/clay-millennials-are-driving-the-shift-away-from-cars/5906406

Danish Energy Agency 2014, *Danish climate policies*, https://ens.dk/en/our-responsibilities/energy-climate-politics/danish-climate-policies

Dick, B 2001, 'Action research: Action and research', in S Sankaran, B Dick, R Passfield & P Swepson (eds), *Effective change management using action learning and action research: Concepts, frameworks, processes, applications*, Southern Cross University, Lismore, NSW.

Dym, CL & Little, P 2009, *Engineering design: A project-based introduction*, John Wiley & Sons, Hoboken.

Engineers Australia 2014, 'Key CPD requirements', https://www.engineersaustralia.org.au/sites/default/files/content-files/2016-12/key_cpd_requirements_aug_2014.pdf

—— 2019, *Stage 1 Competency Standard for Professional Engineers*, https://www.engineersaustralia.org.au/publications/stage-1-competency-standard-professional-engineers

Flinders University 2023, 'Wind farm noise exposure doesn't wake people up more than road traffic', https://news.flinders.edu.au/blog/2023/06/29/wind farm noise-exposure-doesnt-wake-people-up-more-than-road-traffic
Geoscience Australia 2021, *Earthquakes@GA*, Commonwealth of Australia, Canberra, https://earthquakes.ga.gov.au
Glaser, E 1941, *An experiment in the development of critical thinking*, Teacher's College, Columbia University.
Greater Western Water 2023, 'Your water news (autumn/winter 2023)', 1 April, https://www.gww.com.au/about/news/your-water-news-autumn-winter-2023
Green, DW & Southard, MZ 2018, *Perry's chemical engineers' handbook*, 9th edn, McGraw-Hill, New York.
Greenemeier, L 2018, 'Uber self-driving car fatality reveals the technology's blind spots', *Scientific American*, 21 March, https://www.scientificamerican.com/article/uber-self-driving-car-fatality-reveals-the-technologys-blind-spots1
Gurman, M 2018, 'Apple gets FDA approval for new watch, touts health gains', *Bloomberg*, 12 September, https://www.bloomberg.com/news/articles/2018-09-12/apple-says-it-got-fda-approval-for-new-watch-touts-health-gains
Howden, S 2014, 'Warning over USB chargers after woman dies from apparent electrocution', *The Sydney Morning Herald*, 26 June, https://www.smh.com.au/national/nsw/warning-over-usb-chargers-after-woman-dies-from-apparent-electrocution-20140626-zsngd.html
Infrastructure Australia 2013, *Urban transport strategy*, Commonwealth of Australia, Canberra, https://www.infrastructureaustralia.gov.au/sites/default/files/2019-06/infrastructureaus_rep_urbanstrategy.pdf
International Council on Systems Engineering n.d., 'What is systems engineering?', https://www.incose.org/archived-pages/systems-engineering_old
Koen, B 2003, *Discussion of the method*, Oxford University Press, New York.
Kovach, S 2016, 'How the hot startup that stole Apple's thunder wound up in Silicon Valley's graveyard', *Business Insider*, 17 December, https://finance.yahoo.com/news/hot-startup-stole-apples-thunder-130000269.html
Lennard, M 2013, *MIET2380 Lecture notes on requirements*, RMIT University, Melbourne.
Miller, GA 1956, 'The magical number seven, plus or minus two: Some limits on our capacity for processing information', *The Psychological Review*, vol. 63, no. 2, pp. 81–97, https://psycnet.apa.org/doi/10.1037/h0043158
Mwasilu, F, Justo, JJ, Kim, E-K, Do, TD & Jung, J-W 2014, 'Electric vehicles and smart grid interaction: A review on vehicle to grid and renewable energy sources integration', *Renewable and Sustainable Energy Reviews*, vol. 34, pp. 501–516, https://doi.org/10.1016/j.rser.2014.03.031
Nesnow, G 2018, '73 mind-blowing implications of self-driving cars and trucks', *Medium*, 10 February, https://geoffnesnow.medium.com/73-mind-blowing-implications-of-a-driverless-future-58d23d1f338d
Paul, R & Elder, L 2006, *Critical thinking: Tools for taking charge of your learning and your life*, 2nd edn, Pearson Prentice Hall, Columbus, Ohio.
Popper, K 1983, 'On the non-existence of scientific method', in K Popper (ed.), *Realism and the aim of science*, Hutchinson, London.
PUB Singapore's National Water Agency 2018, 'NEWater', https://www.pub.gov.sg/Public/WaterLoop/OurWaterStory/NEWater
Scriven, M & Paul, R 1987, *Defining critical thinking*, Foundation of Critical Thinking, https://www.criticalthinking.org/pages/defining-critical-thinking/766
Shahan, Z 2014, 'History of wind turbines', *Renewable Energy World*, 21 November, https://www.renewableenergyworld.com/storage/grid-scale/history-of-wind-turbines
Sisson, P 2018, 'Small, autonomous shuttles seek to disrupt downtown transit', *Curbed*, 12 July, https://archive.curbed.com/2018/7/12/17566500/shuttle-autonomous-driverless-bus-transit
Sky News 2018 August 28, 'Uber and Toyota team up in race for driverless cars', https://news.sky.com/story/uber-nets-500m-from-toyota-ahead-of-driverless-car-venture-11484038
Smales, I 2006, 'Impacts of avian collisions with wind power turbines: An overview of the modelling of cumulative risks posed by multiple wind farms', in Biosis Research, *Wind farm collision risk for birds: Cumulative risks for threatened and migratory species*, Department of the Environment and Heritage, https://www.dcceew.gov.au/sites/default/files/documents/wind-farm-bird-risk.pdf
Smith, C 2018, 'Apple Watch saves the life of Florida teen with a life-threatening disease', *BGR*, 1 May, https://bgr.com/tech/apple-watch-saves-life-chronic-kidney-disease
State of Green 2021, 'The effects of wind turbine noise', 17 November, https://stateofgreen.com/en/news/the-effects-of-wind-turbine-noise
——2022, 'Danish wind energy breaks record in January', 4 February, https://stateofgreen.com/en/news/danish-wind-energy-breaks-record-in-january
Statista 2023, 'Fitbit Inc. revenue from 2010 to 2022', https://www.statista.com/statistics/472518/fitbit-revenue
Statt, N 2020, 'Apple now sells more watches than the entire Swiss watch industry', *The Verge*, 6 February, https://www.theverge.com/2020/2/5/21125565/apple-watch-sales-2019-swiss-watch-market-estimates-outsold
The Sydney Morning Herald 2006, 'Toowoomba says no to recycled water', 31 July, https://www.smh.com.au/national/toowoomba-says-no-to-recycled-water-20060731-gdo2hm.html
Turner, A 2024, 'Is it legal to use Tesla Autopilot in Australia?', *Drive*, 27 January, https://www.drive.com.au/caradvice/is-it-legal-to-use-tesla-autopilot-in-australia
Walker, J 2023, 'Much-hyped reality of the driverless car finally edges into view', *The Australian*, 23 December, https://www.theaustralian.com.au/inquirer/muchhyped-reality-of-the-driverless-car-finally-edges-into-view/news-story/c1c3e0dbabb2d18e5e39dea0a6033728
Winton, N 2018, 'Autonomous car hype is way ahead of reality', *Forbes*, 2 January, https://www.forbes.com/sites/neilwinton/2018/01/02/autonomous-car-hype-is-way-ahead-of-reality

ACKNOWLEDGEMENTS

Photo: © Nils Versemann / Shutterstock
Photo: © engel.ac / Adobe Stock Photo
Photo: © DenGuy / Shutterstock
Photo: © sitthiphong / Adobe Stock Photo
Photo: © Adwo / Shutterstock
Photo: © Engineering New Zealand
Photo: © Caryn Becker / Alamy Stock Photo
Photo: © Metso Corporation
Text: © Engineers Australia 2019, *Stage 1 Competency Standard for Professional Engineers*, https://www.engineersaustralia.org.au/publications/stage-1-competency-standard-professional-engineers
Text: © Engineers Australia 2014, 'Key CPD requirements', https://www.engineersaustralia.org.au/sites/default/files/content-files/2016-12/key_cpd_requirements_aug_2014.pdf

PART 2

ENGINEERING IN SOCIETY

CHAPTER 3

Sustainable engineering

'Sustainability is the art of living well, within the ecological limits of a finite planet.'

Professor Tim Jackson (2010)

LEARNING OBJECTIVES

After studying this chapter, you should be able to:

3.1 discuss the origins of sustainable engineering, what it is and why it is important

3.2 detail strategies for practising sustainable engineering and how to evaluate a solution using a triple bottom line analysis

3.3 discuss and critique various means for assessing environmental sustainability of engineered solutions

3.4 explain how to estimate the social impacts of engineered options using community/stakeholder communication and consultation

3.5 describe general approaches for estimating the economic sustainability of an engineered option.

Introduction

Engineering ingenuity has changed the way that many people live. In countries such as Australia and New Zealand, many people enjoy a life expectancy of 80 or more years, and largely take for granted cheap fresh water, access to good quality fruit and vegetables, and modern health-care technologies. Our societies regularly partake of consumer goods (e.g. luxury food and drink, electronic devices, entertainment options) that previous generations would have considered rare and extravagant, even as recently as 50 years ago.

However, the generous lifestyles that most of us enjoy do come at a cost. Human consumption results in the depletion of the planet's natural resources, and current rates of consumption are still depleting our natural resources at a rate that the planet cannot sustain. Engineering has helped to provide the technological means to consume at an extravagant rate; it also has a responsibility to help bring human consumption into line with a level the planet can support in the long term. This is the realm of sustainable engineering. Engineers in all fields need to develop the skills and knowledge to implement sustainability principles. To do this, engineers must understand the main principles of ecologically sustainable development (ESD), develop a commitment to practise in a sustainable way, and master the tools and analytical techniques to evaluate whether proposed engineering solutions will result in sustainable outcomes.

On 1 January 2016, the UN Sustainable Development Goals (SDGs) came into force. The 17 goals, shown in figure 3.1, include equity, health, responsible consumption, climate and sustainable communities. The SDGs have been adopted by many large corporations and governments as the chief framework for corporate and governmental social responsibility. Engineering sustainability and ecological sustainable development intersect across many of the SDGs.

FIGURE 3.1 UN Sustainable Development Goals

Source: United Nations (2019).

Australians generated approximately 76 million tonnes of waste in the 2018–19 financial year, which was a 10 per cent increase on 2016–17. Approximately 16 per cent of this total was municipal solid waste (MSW) (Australian Bureau of Statistics [ABS] 2020). In New Zealand, 3.2 tonnes of MSW per capita goes to landfill annually (NZ Ministry for the Environment 2021, as cited in New Zealand Infrastructure Commission 2021). MSW is made up of all the rubbish generated by you and your community just getting on with living. It comprises things such as plastic and food scraps that you throw in your bin at home, and gets combined with rubbish that your local pizza shop throws into its skip, demolition waste from the refurbishment of your university's lecture theatres, and furniture, clothes and shoes that are too far gone for a second-hand shop to on-sell.

Estimates of MSW to landfill do not account for wastes from industry, such as manufacturing, agriculture and mining; waste that is incinerated, such as some sewage sludge and hazardous medical wastes; liquid waste; waste discharged to waterways; and wastes that are recycled, such as newspapers, beverage containers and green waste. Based on the Organisation for Economic Co-operation and Development (OECD) data, the website Statista ranked Australia 14th in the world for per capita waste generation in 2022 (Statista 2024), and the ABS estimated the cost of waste services in Australia for the year 2018–19 to be A$17 billion (ABS 2020). On a positive note, Australia's resource recovery rate has reached 60 per cent (ABS 2020), which is below the world leader in MSW recycling, Germany, at approximately 68 per cent (OECD 2024). There is scope for greater MSW recycling in New Zealand with the recovery rate estimated at 35 per cent (Wilson et al. 2017).

The landfills in which we dump municipal solid waste emit greenhouse gases like carbon dioxide and methane from microbial digestion of organic waste. These greenhouse gases contribute to global climate change. In Victoria, the Environmental Protection Agency (EPA) requires that landfill gas be captured and, as you will learn later in this chapter, the methane fraction of this gas can be burned for energy generation. However, a challenge has faced some of the mechanical engineers tasked with designing Australia's first gas reciprocating generation engines to use landfill gas: the models available to estimate the volume of landfill gas were wildly inaccurate, and ranged from 90 per cent to 4000 per cent off the mark. An Australian environmental engineer, Professor Julia Lamborn, solved this problem by developing complex mathematical models that more accurately predicted gases typically emitted by landfill sites, and how the volume and gas ratios change over the life of a landfill (see figure 3.2).

FIGURE 3.2 Relative ratios of hydrogen, oxygen, nitrogen, methane and carbon dioxide emitted from a typical landfill over its 40+ year lifespan

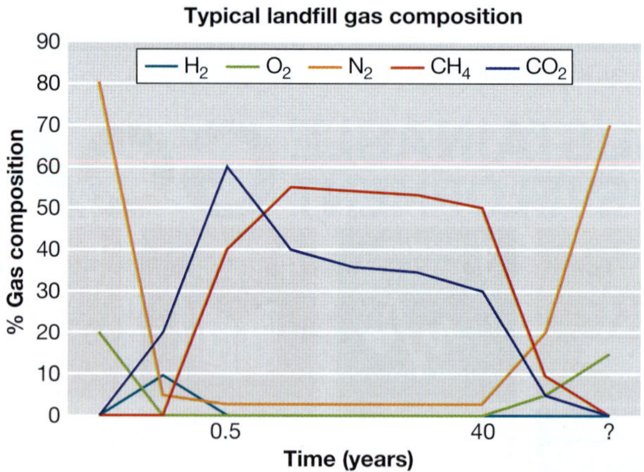

Source: Professor Julia Lamborn, Monash University.

Based on Professor Lamborn's (2012) work, we now know that an average urban landfill site, like Narre Warren in Victoria (3.5 million m^3), should be able to supply commercial volumes of gas for 40 to 50 years powering around 10 000 homes. Professor Lamborn's modelling means that energy utilities and the mechanical engineers who design, implement, and operate generation engines can tailor the engines to optimal generation from a given gas profile, and can adjust generation capacity over the active life of a landfill. This example of cooperation between engineering disciplines has resulted in an environmentally and economically sound outcome — less wasted generation capacity, less wasted methane and less emission of greenhouse gas — from something as environmentally unpromising as a big pile of rubbish (Derkley 2009).

We all have a responsibility to limit or reduce the amount of waste we generate. Professional engineers, however, have a far greater opportunity to make a big difference to the problem of waste generation and management. Environmental and civil engineers have a direct role in the siting, design, construction and management of landfills. Mechanical and electrical engineers are involved in designing waste handling and sorting, and waste-to-energy generation. Beyond these roles, however, engineers in all the professional fields are expected to be highly conscious of, and clever about, waste. Clever engineering design minimises the use and wastage of resources during construction, production or manufacture. Clever engineering considers waste generation and emissions during construction, manufacture and use. Clever engineering design reduces waste at the time of decommissioning, recycling or disposal.

In this chapter, we will look at a wide range of measures, particularly measures of social and environmental impact assessment, which allow engineers to evaluate the likely environmental, social and economic impact of their technical decisions.

Professor Julia Lamborn (Monash University) and the Narre Warren landfill gas extraction manifold

You will gain tools and insights to help you work as a sustainable engineer. These tools will also help you in situations in which you need to evaluate the work of other engineers, or perhaps remedy decisions and solutions made in the past that have caused unintended negative outcomes.

3.1 What is ecologically sustainable development (ESD)?

LEARNING OBJECTIVE 3.1 Discuss the origins of sustainable engineering, what it is and why it is important.

Economic, social and environmental analysis of engineering options or outcomes form what is called the **triple bottom line (TBL)**. For engineers, the TBL can be viewed as a set of design criteria that can be folded into the engineering method, sitting alongside technical measures, to inform sustainable design choices. In the chapter on the engineering method, you developed criteria to judge between the different makes of car a client could afford to buy, and explained the pros and cons of the different car options in terms of these criteria. The process of negotiating criteria with a client and explaining options in terms of these criteria is vital for producing outcomes that satisfy the client's needs. In straightforward problems such as choosing a car, optimising batch chemical conversions, or improving safety conditions on a building site, the criteria for a 'good' solution are relatively easy to agree upon.

But what happens when a client is not the only one in need of a result? What happens when engineering decisions must be made and judged using criteria imposed by society? What happens when a pro or con is uncertain, difficult or even impossible to judge? What happens when a problem you are engineering a solution for is complex and urgently needs to be resolved?

In these situations, *sustainable engineering* comes into play. Sustainable engineering has emerged from the general concept of ecologically sustainable development (ESD), and it allows engineers to tackle 'wicked problems' (Rittel & Webber 1973). Wicked problems are typically complicated and entangled in contested social and political situations (Funtowicz & Ravetz 1993). Often, there is confusion about what might be causing a wicked problem. There can be difficulty formulating a clear problem statement and even less agreement around how these sorts of problems might be resolved. Some current wicked problems in our society include the need to realise sustainable urban transport or development, the equitable management of inland waters and catchments, the maintenance of biodiversity, the positioning of municipal waste facilities, reducing the incidence of childhood obesity, and comprehending and responding to global climate change.

Before delving into what sustainable engineering entails, it is important to understand what ESD is and where the idea came from. Today, the term 'sustainable' is widely applied. A football coach might describe the team's losing streak as 'just not sustainable', or a politician might pledge 'a strong, sustainable health care system for all New Zealanders'. In these cases, the term is being used as an extension of one of the range of meanings attached to its base word 'sustain'. Sustain can mean 'to hold or bear up from below; bear the weight of; be the support of, as in a structure; to bear (a burden, charge, etc.); to undergo, experience, or suffer (injury, loss, etc.); endure without giving way or yielding; to keep (a person, the mind, the spirits, etc.) from giving way, as under trial or affliction; to uphold as valid, just, or correct, as a claim or the person making it' (Macquarie Dictionary 2024).

In the phrase 'ecologically sustainable development', the term 'sustainable' has evolved to have a more specific meaning than the range of meanings that are implied by the base word 'sustain'. It is important to grasp the meaning of ESD because there is now a strong expectation that engineers will implement ESD through their professional work.

KEY POINT

The concept of ESD requires an engineer to consider and take account of the broader environmental, social and economic consequences of their work.

Definitions of ESD

The Australian National Strategy for Sustainable Development, which was developed in 1992, acknowledges that ESD can have a range of meanings (Department of the Environment 1992). Here we present a few definitions of ESD and invite you to consider which of these definitions makes sense to you.

The quote opening this chapter — 'Sustainability is the art of living well, within the ecological limits of a finite planet' — comes from Tim Jackson, who is Professor of Sustainable Development at the University of Surrey, UK. It is a simple definition of ESD but requires some thought on what it means to 'live well', and whether all members of the world population might aspire to similar living conditions. Being able to enact this definition of ESD would depend on robust measures of the planet's 'ecological limits' and the capacity to model what impact various engineering options might have on those limits. Similar critical thinking about the following quotes and definitions will help you to make a critically informed choice about how you define (and eventually enact) ESD.

An early report by the World Commission on Environment and Development (1987), *Our Common Future* (also known as the *Brundtland Report*), defined ESD as 'development that meets the needs of the present without compromising the ability of future generations to meet their own needs'.

The New Zealand Business Council for Sustainable Development (2008) explained sustainable development as a holistic concept requiring the integration of economic growth, social equity and environmental management. Along similar lines, the Australian *National Strategy for Ecologically Sustainable Development* (Department of the Environment 1992) gave the following definition for ESD in Australia:

> using, conserving and enhancing the community's resources so that ecological processes, on which life depends, are maintained, and the total quality of life, now and in the future, can be increased.

Based on all these definitions, **ecologically sustainable development**, or **ESD**, could be described as development that integrates economic growth, social equity and environmental management, so that ecological processes (on which life depends) are maintained in a state sufficient to meet the quality of life needs of present and future generations.

The concept of ESD has its origins in increasing community discontent about the environmental impact of industrialisation following the second Industrial Revolution (1870–1914).

Rachel Carson's book *Silent spring* (1962) is recognised as an early and important catalyst in the modern environmental movement. In this book, Carson uncovered a trail of chemical poisoning that was traced back to use of the agricultural chemical DDT in the United States. The US author had noticed the spectacularly noisy 'dawn chorus' — in which thousands of birds sang together at the start of spring days — appeared to be diminishing. Her book was a wake-up call about the flow-on effects of broadscale use of bio-accumulating herbicides and pesticides.

In the 1970s, advances in mass media and telecommunications allowed environmental activists to bring their concerns about environmental damage to a national and international audience. The next spotlight discusses a dispute over Lake Pedder, Tasmania. Although the picturesque lake was dammed (and damned), the dispute increased public awareness about the importance of environmental issues.

SPOTLIGHT

Lake Pedder hydro-electric scheme

In 1967, the National Park status of the iconic Lake Pedder, a glacial lake with unique pink silicon sand 'beaches', was revoked by the Tasmanian government. Subsequently, the construction of dams on the Huon and Serpentine Rivers led to the flooding of the lake in 1972. During the 1960s, the Tasmanian government-owned Hydro-Electric Commission (HEC) (the electricity-generating division of which now operates as Hydro Tasmania) put forward a proposal to create large impoundments on the Huon and Serpentine Rivers. The hydro dams would increase the surface area of Lake Pedder from only 9 square kilometres to 242 square kilometres. The resulting impoundment provides 40 per cent of water used by the Gordon power station, Tasmania's largest hydropower producer. The Pedder–Gordon system produces a total of 432 megawatts (MW) of electricity (Hydro Tasmania 2014).

The Gordon Dam

However, many viewed the scheme as unnecessary and considered the environmental cost of submerging the existing lake and valleys as too great for the gain. There was significant opposition to the proposed decision to flood Lake Pedder, both within the Tasmanian community and elsewhere, leading to the formation of the world's first Green Party, the United Tasmania Group. The situation galvanised environmentalists in Tasmania and brought the issue of environmental protection into the living rooms of everyday Australians. However, the protests against the damming of Lake Pedder failed to stop the dam from being built (Simons 1992).

In 1978, the Tasmanian government and the HEC proposed building another hydro-electric dam, this time on the Franklin River. However, by then, protecting the environment had become a prominent social issue. The proposal was countered by a massive environmental campaign that was, for Tasmania, costly and divisive, and for its politicians, politically damaging. The then-Premier, Doug Lowe, was ousted from power by his own party during a referendum on the Franklin Dam. Lowe and another parliamentarian resigned from their party leading to the 1982 state election, which resulted in a change of government. On the federal political scene, the Prime Minister, Malcolm Fraser, failed to intervene in favour of the Tasmanian river and this was a contributing factor to his defeat in the 1983 election. These dramatic political events demonstrated that environmental concerns had the power to cause enormous consequences for those who chose to ignore them (Chen & Hay 2006; Peter Fagan 2009, pers. comm.).

CRITICAL THINKING

As an engineer designing a large dam like the one which flooded Lake Pedder, what information might you need, to check the project aligned with ESD (defined by Brundtland as 'development that meets the needs of the present without compromising the ability of future generations to meet their own needs')? On a broader note, just because the regulations allow something, does that make it the best thing to do?

Such early concerns and battles over specific environmental impacts and engineering projects, as outlined in the spotlight, paved the way for the emergence of the modern environmental movement during the 1980s. By the late 1980s, environmentalism was well established in some parts of the world. Independent state-based environmental protection authorities (EPAs) began to be established during the 1970s and 1980s — Victoria's EPA was established in 1970 — and successive countries toughened up on air and water pollution via changes to legislation (e.g. Clean Air and Clean Water Acts).

The emerging environmental movement inspired a pivotal meeting of world leaders in Rio De Janeiro, Brazil. The Rio Earth Summit was sponsored by the United Nations and took place in 1992. The Rio Declaration was drafted at this meeting and was signed by all attending nations, formally recognising the need for sustainable development, and agreed joint action on environmental imperatives.

During the 1990s, Australian and New Zealand governments began to enact legislation that required private enterprise to manage and evaluate the environmental impacts of their activities. These activities required action — for example, through the application of ISO 14000 environmental management standards and commissioning of environmental impact assessments (EIAs). An important piece of legislation along these lines was the NZ Resource Management Act (RMA) of 1991. It was amended through the Resource Legislation Amendment Act in 2017 and the Act remains the most significant legislation that details how New Zealand manages its environment.

In the twenty-first century, engineers in industry worked to interpret and implement sustainability in real industrial, community and consultancy settings. Government schemes and legislation supported incremental action towards more sustainable management of, for example, inland water resources, energy efficiency for industry and residential homes, and fisheries. The New Zealand government banned the construction of new fossil fuel power generation plants in 2007, and engineers responded by building capacity for renewable energy generation.

Australia's Biodiversity Conservation Strategy 2010–2030 (Department of the Environment 2010) set ten targets to measure environment protection, and software and environmental engineers collaborated to improve geographical information systems (GIS) tools that measure biodiversity. In 2015, the review of the first five years of the strategy concluded that it was not really working; it 'did not engage, guide, or communicate its objectives to all audiences in a useful way' (Department of the Environment and Energy 2016). The review also concluded that the strategy was 'generally concerned with the restoration and protection of natural environments and does not provide a framework for biodiversity conservation in built or production landscapes' (Department of the Environment and Energy 2016). More recently, Australia legislated a world-first framework for a national Nature Repair Market to reward biodiversity protection (Department of Climate Change, Energy, the Environment and Water [DCCEEW] 2024a). This, in combination with emerging carbon markets, will likely impact many engineering activities.

In 2011, the Australian Government legislated to put a tax on carbon dioxide emissions, the main human emission linked to climate change (Min et al. 2011). This proved to be a short-lived piece of legislation that was repealed by the next government in 2014. As at 2024, the establishment of targets and a Safeguard Mechanism, and the introduction of Australian Carbon Credit Units, are the main mechanisms for emissions abatement (DCCEEW 2024b). The New Zealand Emissions Trading Scheme, called the NZ ETS, was first legislated in 2008. It allows manufacturers to trade their emissions in NZU (New Zealand Units) (Leining & Kerr 2018) and may have contributed to stabilising New Zealand's emissions, as is apparent from the following spotlight.

SPOTLIGHT

NZ ETS prompts metal manufacturer to upgrade and save money

The introduction of the New Zealand Emissions Trading Scheme (NZ ETS) has sharpened the focus on technological innovation to reduce energy consumption and associated carbon emissions. One New Zealand company that has responded to calls for more energy-efficient operations is McKechnie (formerly MCK Metals), New Plymouth. McKechnie extrudes its own billet (metal bars) from scrap metal to produce finished components for a wide range of industry applications (Energy, Efficiency and Conservation Authority 2011). With the introduction of the NZ ETS, McKechnie saw an opportunity to improve the energy efficiency of the main furnace in their re-melt facility.

This large, gas-fired reverberatory furnace previously had roof-mounted burners that melted scrap metal but created a temperature gradient, leading to inefficient heating of metal located at the bottom of the furnace. The molten metal would be stirred using a rake attached to a forklift, but this allowed heat loss from the furnace and it took several hours for alloying additions, such as copper, magnesium and silicon, to mix and stabilise. The decision was made to upgrade the gas-fired furnace by the addition of a magnetic stirrer — the Z-Mag (Energy, Efficiency and Conservation Authority 2011).

The Z-Mag uses low-energy permanent magnet technology, and effectively performs the same function as a spoon in a cup of coffee, distributing heat more evenly throughout the vessel. One of the benefits of the Z-Mag stirrer for McKechnie has been that the furnace doors can be kept closed during mixing, thus conserving heat. Z-Mag offers other benefits. 'The stirrer not only reduces the melting time required, by improving the energy transfer — it also increases the amount of aluminium we can melt per hour', explained a McKechnie project engineer (Energy, Efficiency and Conservation Authority 2011). McKechnie evaluated the operation of the Z-Mag stirrer after its installation and, at monthly melt rates of 880 tonnes, found savings of 561 MJ per tonne. Table 3.1 shows the reduction in McKechnie's energy consumption with installation of the Z-Mag stirrer.

TABLE 3.1 Project figures	
Energy consumption (without stirrer)	8.26 GJ/tonne
Energy consumption (with stirrer)	7.70 GJ/tonne
Estimated reduction in annual energy usage	5049 GJ (i.e. 6.8% saving)

Source: Energy, Efficiency and Conservation Authority (2011).

These energy savings mean the Z-Mag reduces McKechnie's CO_2 emissions by 266 tonnes per year. The stirrer has delivered a host of other benefits: it is low maintenance, easy to operate has a large circulation capacity (> 50 tonnes per minute). Importantly, the stirrer is reducing waste and resulting in improved product quality. According to Terry Fitzgerald, works engineer at McKechnie, the stirrer has improved metal quality because the efficient mixing process means the molten metal doesn't get as hot, thus reducing the amount of ash and dross (floating impurities) produced (Energy, Efficiency and Conservation Authority 2011).

McKechnie's installation of the Z-Mag demonstrates how engineers and the companies they work for are acting to improve the sustainability of industry operations, and how the effectiveness of that action can be quantified by simple energy auditing (Energy, Efficiency and Conservation Authority 2011).

CRITICAL THINKING

What do you consider more important — the improvements in product quality or the reduction in CO_2 emissions? What things lead you to that consideration? How might you decide between a stirrer that *only* improved product quality and a stirrer that *only* reduced CO_2 emissions?

Figure 3.3 is a timeline showing how environmentalism has led to widespread political, social and industry acceptance of the need for ESD.

FIGURE 3.3 The origins of ecologically sustainable development (ESD). Widespread acceptance by government, industry and the community of the need for ecologically sustainable development started in the social margins with the early green movement in the 1970s.

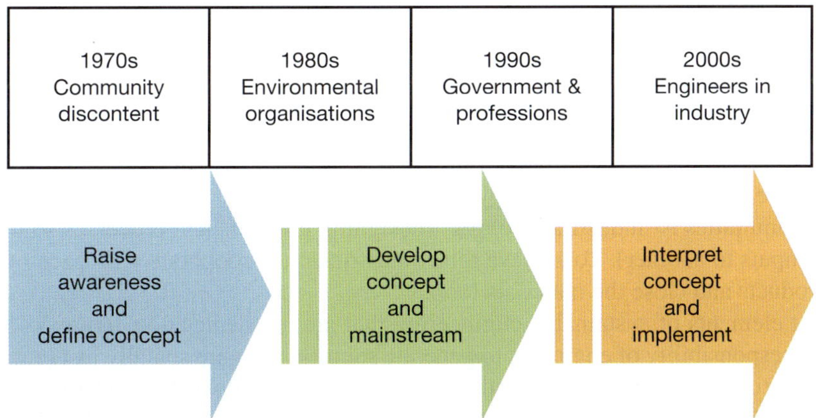

Now that ESD is a mainstream objective, engineers in industry are expected to implement it by interpreting and responding to legislation, and by anticipating the future impact of current threats. This might include considering the impact of predicted future extreme weather events on the performance of your technology, considering how a proposed design or solution will reduce or offset carbon emissions, or designing new infrastructure to accommodate projected changes in sea level and coastal storm surge events.

What is sustainable engineering?

Sustainable engineering is a desirable objective. There are different dimensions to enacting sustainable engineering, and this means that various engineers and engineering organisations define the term differently or add an emphasis to particular dimensions. Here are a few definitions, but perhaps you will come up with your own definition once you have spent more time on this chapter learning about how sustainable engineering is done. **Sustainable engineering** has been defined as:

- doing more with less — less resource input, less waste generated
- using energy and resources at a rate that will not compromise the environment or the health and welfare of current and future generations.

The first definition has its focus entirely on the use of materials and the generation of waste. In contrast, the second definition takes a broader view of engineers' responsibilities and includes reference to the environment and human health and welfare.

Engineers Australia considers sustainable engineering a fundamental competency for graduating engineers. Engineers Australia lists the capacity to consider sustainability throughout the Stage One Competency Standard: the mandatory competencies for every engineer graduating from an Engineers Australia-accredited undergraduate degree. For example, under the Element of Competency '*Knowledge* of engineering design practice and contextual factors impacting the engineering discipline', an engineering graduate would be expected to show they can identify and understand:

> the interactions between engineering systems and people in the social, cultural, environmental, commercial, legal and political contexts in which they operate, including both the positive role of engineering in sustainable development and the potentially adverse impacts of engineering activity in the engineering discipline (Engineers Australia 2019).

According to Engineers Australia's Element of Competency '*Understanding* of the scope, principles, norms, accountabilities and bounds of sustainable engineering practice in the specific discipline', a graduate engineer would be expected to demonstrate he or she 'Appreciates the social, environmental and economic principles of sustainable engineering practice' (Engineers Australia 2019).

The Element of Competency '*Application* of systematic approaches to the conduct and management of engineering projects' mandates that the graduate engineer 'Demonstrates commitment to sustainable engineering practices and the achievement of sustainable outcomes in all facets of engineering project work' (Engineers Australia 2019).

The Engineers Australia Stage One Competency elements and descriptors emphasise the possible breadth and complexity of sustainable engineering practice by linking the concept of sustainable development with other concepts, such as social, cultural, environmental, commercial, legal and political contexts; social, environmental and economic principles; and the need to enact sustainability in all facets of engineering project work.

Figure 3.4 shows the hierarchy of engineering activities to achieve a sustainable future as a progression from reducing harm caused by engineering processes through to improved processes, planning and developing a holistic, community or regional approach. The initial action is about establishing industrial pollution control to limit emissions to air and water (using methods that minimise environmental damage). The following two stages are 'process integration' and 'whole facility planning', which allow individual industries to operate at lower levels of energy and raw materials use. The final part of the path leading to sustainable communities is 'industrial ecology' (designing products and processes so that wastes from one are used as inputs to another). To achieve this, industries must cooperate with each other to use each other's waste products and close the materials loop.

Two important elements of sustainable engineering that are not depicted in figure 3.4 but are equally important are the responsibility of engineers to generate solutions that are socially just (e.g. using methods that provide sufficient food, water, shelter and mobility for a growing world population), and the need to think beyond the technical and financial (e.g. to incorporate broader environmental and social constraints into engineering decisions).

So, we could say sustainable engineering is engineering that minimises energy and resource use, minimises pollution and waste, and considers broader environmental, social and economic impacts on people, both now and in the future.

Why sustainable engineering?

The association between environmentalism and ESD is now well established. A societal shift towards exercising stewardship for the planet has led to the public exerting pressure (both directly and indirectly) on engineering professionals and the companies they work for. Growing acceptance of the phenomenon of human-induced global climate change (Min et al. 2011) has meant the public expects a higher level of accountability and stricter adherence to environmental standards from the engineering profession.

FIGURE 3.4 Engineering a bridge to a sustainable future: actions engineers could take to promote sustainable communities

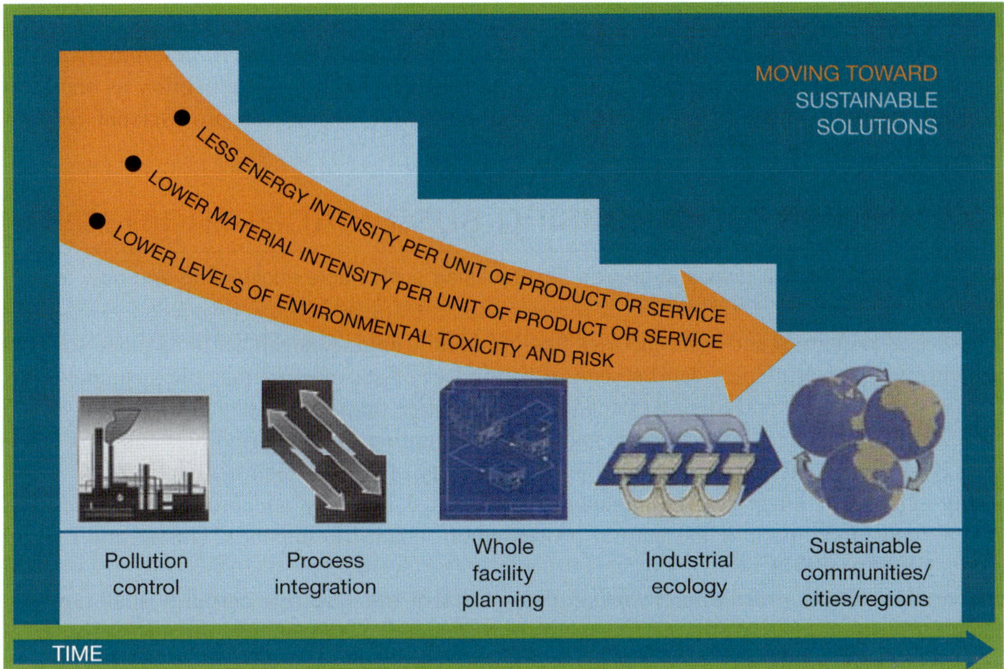

Source: National Science and Technology Council (1995, p. 20).

Legislation makes more sustainable engineering practice a necessity in some fields. Mechanical, civil, environmental and chemical engineers need to design processes and products that meet various national and state-based limits for releases of pollutants to air and water. For example, engineers working in the automotive industry work within the Australian Design Rules (ADRs), which set the fuel consumption labelling, emissions and noise standards that each vehicle model is required to meet, prior to its first supply to the market. The ADR was introduced by the Australian Department of Infrastructure and Transport in the early 1970s and has been progressively tightened since then (Department of Infrastructure and Regional Development 2018). Australian civil and environmental engineers working on water treatment need to meet the chemical, physical and microbiological limits set out in the Australian Drinking Water Guidelines (National Water Quality Management Strategy 2013).

Civil, construction and building services engineers in Australia and New Zealand work within building codes. New Zealand's building code is very stringent due to the threat of earthquake, and the Building Code of Australia has been substantially amended to include increased energy efficiency for residential and commercial/industrial buildings (Australian Building Codes Board 2016). In New Zealand, the introduction in 1991 of the Resource Management Act (RMA) (NZ Ministry for the Environment 2013) had a big impact on the practice of engineers, as proposed decisions and changes in projects (e.g. road development that will result in changed traffic volumes, the establishment of an industrial site with potential noise or odour impacts) now need to be reviewed by an appointed board of independent commissioners. In New Zealand, civil engineers and those involved in natural resource management spend a lot of time preparing and delivering submissions to comply with the requirements of the RMA.

Potential for financial reward is another reason there has been a shift towards sustainable engineering practice. During the late 1990s, there were increases in NSW in the cost per tonne to dump construction and demolition (C&D) waste to landfill. The cost increase was A$10 per tonne per annum plus CPI adjustments, reaching A$135 in NSW metropolitan areas by 2015–16 (Picken & Randell 2017). This levy improved the rate of recycling for C&D waste (Australian Council of Recyclers Inc 2012) and caused construction firms to begin adopting alternatives to landfill (e.g. on-site waste segregation/sorting and recycling of sorted streams — glazing; timber beams and boards; brick, tile and concrete rubble; steel members and roofing). The more financially favourable options tended to be those that involved minimal reprocessing of waste, along with reuse on-site (e.g. rubble for road base or backfilling and machining or chipping wood for reuse as structural members or landscaping mulch). In terms of future technologies, carbon, and water trading promise to deliver large financial dividends to those who can engineer more carbon- and water-efficient processes. We will look at some of these future technologies in more detail in the chapter on your engineering future.

If all those reasons are not compelling enough, all engineers registered to practise under Engineers Australia have an obligation to uphold the Engineers Australia *Code of Ethics and Guidelines on Professional Conduct* (Engineers Australia 2022), which lists promoting sustainability as one of the four ethical tenets. We will further explore the ethical tenets in the chapter on professional responsibility and ethics.

3.2 Strategies for practising sustainable engineering

LEARNING OBJECTIVE 3.2 Detail strategies for practising sustainable engineering and how to evaluate a solution using a triple bottom line analysis.

There are several principles and main theories that guide how engineers can act to advance sustainability. The following general principles can be seen as a mindset or set of aims that the sustainable engineer holds in the front of their mind to guide decisions and practice on a day-to-day basis. In the sections that follow, we will explain some of the theories and values that underpin these principles and look at tools and techniques engineers use to test or evaluate the sustainability of various technical options.

1. *Maintain ecological integrity.* Maintenance of ecological integrity and biodiversity is essential for the 'continued functioning of the natural processes on which life depends' (Institution of Engineers Australia [IEAust] 1997).
2. *Apply precaution ('the precautionary principle').* 'Lack of full scientific certainty shall not be used as a reason for postponing cost-effective measures to prevent environmental degradation' (United Nations General Assembly 1992).
3. *Consult stakeholders.* Those with a stake in the problem have a right to be consulted and have knowledge that will likely result in better technical solutions.
4. *Promote equity.* 'Ensure that all people have the opportunity to achieve economic, environmental and social well-being' (IEAust 1997).
5. *Conserve resources.* This includes reducing the amount of raw materials that a product or process requires (e.g. weight of steel in a beam, energy required to produce a tonne of concrete), disturbance to the natural environment that is required to produce the resources (e.g. logging old growth forests, open-cut mining in environmentally sensitive areas), and the durability or recyclability of the product.
6. *Limit emissions.* This includes all types of pollutants that will reduce the planet's overall capacity to handle, receive and store waste products. The main categories are liquid, such as polluted water and spent chemicals in solution; gaseous, such as methane, carbon dioxide and volatile organic carbons; and solid, such as waste packaging, slag and sludge, construction and demolition waste, biohazardous materials, and non-recyclable e-waste.
7. *Practise industrial ecology.* 'Reduce the industrial system's impact on the environment; in particular creating a closed industrial system, analogous to a natural ecosystem, where waste from one industry can be used as input for another' (Newton 2001). An extension of this thinking is termed '**circular economy**' (Fet & Deshpande 2023), which is reuse and regeneration of materials or products as part of sustainable, continuing production.

KEY POINT

Sustainable engineering invites engineers to design processes and products that create a good outcome and fit within environmental, social, economic and technical constraints.

SPOTLIGHT

Australasian students shine on the world stage

The Solar Decathlon is the world's largest competition for solar-powered homes in the world. The first Solar Decathlon was held in Washington DC in 2002 and has since been held roughly every two years. It is coordinated by the US Department of Energy (US Department of Energy 2019). University teams from around the world compete to get into the competition. The successful finalists are set the challenge to design, build and operate a fully solar-powered house. An added complication to the challenge is that the teams must transport their house across the world and assemble and fit it out within just a few days.

The Solar Decathlon consists of ten individual contests, each adding up to 1000 points. There are five measured contests where teams are benchmarked against relevant standards, including producing solar hot water and energy, and maintaining strict thermal and humidity comfort levels within the home. The other five are judged contests, each marked by a panel of experts in the relevant field during the two weeks of the on-site competition (US Department of Energy 2019).

Additional deliverables include project manuals, design portfolios, costings, simulations, safety and risk assessments — in fact, everything needed to fully document a complex industrial building (it is much more than a house) must be provided well in advance to the judges (US Department of Energy 2019).

In 2010, students from Victoria University of Wellington became the first Australasian team to gain a place in the finals of the competition. Led by architecture students and academic Guy Marriage, the New Zealand students began a two-year epic learning journey, designing a house that would be energy efficient, beautiful, and capable of being built and dismantled in very short lead times. In addition, the team wanted to produce a quintessential 'kiwi' house that would showcase the New Zealand lifestyle on the world stage (First Light Studio 2019).

First Light at Washington DC

The Victoria University students developed the concept of their First Light house as a bach (a typical New Zealand beach house), a place of refuge and entertainment where the occupants can experience the first light as a new day dawns. The long sea and land journey across the Pacific and continent of North America meant that the New Zealand team had to complete their work well before most of the other entrants. The re-build on the West Potomac Park on Washington DC's Mall went smoothly and the team weighed up the competition (First Light Studio 2019).

The 19 entrants came from all over the United States, Belgium, Canada, China and, of course, New Zealand (US Department of Energy 2019). After just seven days of building and commissioning and ten days of competition, First Light took gold in engineering, silver in architecture and bronze in market appeal (US Department of Energy 2019; First Light Studio 2019). They also had perfect scores in energy balance and hot water (US Department of Energy 2019). In overall standings, Team NZ won the bronze medal behind the University of Maryland and Purdue University — a podium finish for the first Australasian team in the so-called Energy Olympics (US Department of Energy 2019; First Light Studio 2019). A fantastic result — and all done by students, just like you.

The First Light house returned to New Zealand and now overlooks the sea at Waimarama where it is being run as a short-stay holiday accommodation. The core group of students have taken control of their futures and have set up their own architectural design studio (First Light Studio 2019). Visit them at https://firstlightstudio.co.nz.

Inspired by their NZ colleagues, Team UOW was established by a core group of Australian engineering students and two academics, Paul Cooper and Tim McCarthy. They worked on their application for the first Solar Decathlon competition to be held in China in 2013 (US Department of Energy 2019). The Wollongong students became the first Australian team to gain a final spot in a Solar Decathlon.

They began work on their house design, a sustainable retrofit of a 1960s Aussie fibro shack. Their aim was to demonstrate that even the humblest abode could be upgraded to the highest energy efficiency standards. The house, named the Illawarra Flame, took shape on the page and computer screen during 2012 and construction began in January 2013. It took the students of UOW and TAFE NSW 12 weeks to complete the first assembly. Four days later it had been completely disassembled and shipped across town for a practice re-build. The re-build was completed in seven days, giving the team confidence for China, where they were to be allowed two weeks for construction. Finishing touches were made and the house was once again disassembled and packed in seven shipping containers and put on the sea to Tianjin Port. After 45 days at sea the convoy began its 500-kilometre trek to Datong in Shanxi, Northern China, not far from the border with Mongolia. In mid-July 2013, the team of 50 students and teachers flew to Datong for their five-week stay in the Energy Olympic Village.

The teams came from all over the world: Australia, Belgium, China, Iran, Israel, Malaysia, Singapore, Sweden, the United Kingdom and the United States (US Department of Energy 2019). Team UOW were the first to complete construction and commissioning, well in advance of the deadline. Their practice build in Wollongong had paid off. The construction was followed by two weeks of intensive competition, monitoring, presentations and exhibition to the public.

Over 35 000 visitors came through the Australian house. Over 300 000 locals visited the competition site. The Australians had never seen such crowds. It was neck and neck with Team Sweden and South China University of Technology in the competition proper. On the last day, Team UOW was narrowly behind, in third place. The final announcement of the judged competitions gave UOW three gold medals in architecture, engineering and solar application and silvers in market appeal and communications. Team UOW's final score was 957.6/1000, beating South China who came second, with Chalmers University Sweden coming third (US Department of Energy 2019).

Illawarra Flame solar-powered house, Wollongong

The Illawarra Flame house returned to Australia and has been re-built on the University of Wollongong's Innovation Campus, where it is available for short-term rental and special events. It is also being operated by the Sustainable Buildings Research Centre as a laboratory of sustainable living. Many of Team UOW have embarked on successful careers and two of the team have set up their own home-building and retrofit businesses (Team UOW Australia 2013).

Team UOW re-formed in 2016 to design and build a dementia-friendly net-zero energy home for the Solar Decathlon Middle East 2018 contest. Their new house, named Desert Rose, was selected along with entries by 21 other universities from around the world to be showcased in Dubai in late 2018. While the process was the same, the cast was completely new — over 200 UOW and TAFE NSW students became involved. After two years of design, drawing and report writing, Desert Rose was a reality. With advanced predictive control of the HVAC, light gauge steel framing and building integrated photovoltaic (BIPV) roof tiles, this was a very complex construction project. In addition, Desert Rose had to perform flawlessly in the heat of the Dubai desert (Team UOW Australia 2019).

Shipped in eight containers, the modular house arrived in Dubai in October 2018. The team of 50 students and staff met up with these containers at the MBR Solar Park, the world's largest single site solar array and the site for the Solar Decathlon Middle East 2018. To compound the challenge, this site is 50 kilometres into the desert inland from Dubai. The nearest civilisation and accommodation is an hour away (Team UOW Australia 2019).

Desert Rose and Team UOW in Dubai, November 2018

The team had planned for building in the desert heat but it still took the allotted 15 days to complete the construction and commissioning of this state-of-the-art house. As happened in China, UOW was the first team to complete their build and get full sign-off. The others, including Futurehaus from Virginia Tech, were not far behind. When the whistle blew for the contest, all but two houses had been completed. There were ten scoring contests and three extra prizes on offer. In the non-scoring contests, Desert Rose won gold for interior design, silver for BIPV and silver (plus AE$50 000 — about A$20 000) for the best creative solution. Just as in China, Team UOW was in third place overall going into the final day and managed to jump into second place overall ahead of Team Baitykool from the University of Bordeaux with the final contest. Futurehaus was some 33 points out in front. Desert Rose achieved gold medals in innovation and comfort conditions, a silver in energy efficiency and bronzes in sustainability and house functioning (Team UOW Australia 2019).

The performance of the students from Victoria University of Wellington and the University of Wollongong have shown that Australian and New Zealand architecture and engineering students can match and lead the world.

CRITICAL THINKING

Student competitions create enormous learning opportunities beyond the normal scope of an undergraduate degree program. What inter-university competitions has your university been involved with? How can you get involved? The extracurricular work is not easy but is rewarding. Can you identify how being involved in such competitions can improve your employability and general experience in engineering?

Constraints of sustainable engineering practice

To recognise how engineers engage in sustainable engineering and to appreciate the tools and techniques of sustainable engineering, it is necessary to understand the constraints imposed on engineering work by ESD.

ESD is often explained as development that will fit within constraints imposed by three important subsystems (see figure 3.5):

1. the environment
2. society
3. the economy.

FIGURE 3.5 Dimensions of sustainability with technical options as the foundational constraint

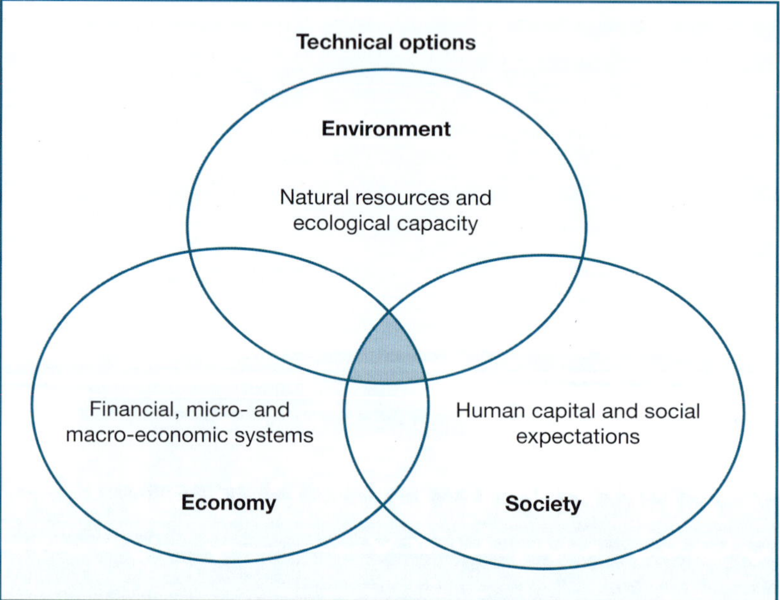

Source: Clift (1995), reproduced by permission of The Royal Academy of Engineering.

These subsystem constraints must be met within the possibilities created by technical constraints in sustainable engineering practice. This is not to say that the constraints of sustainable engineering are an afterthought. Rather, consideration of environmental, social and broader economic objectives need to run alongside and accompany the full design process right from the initial stage of the engineering method as described in the chapter on the engineering method.

Environmental subsystem

The term 'environmental subsystem' is applied broadly. It includes the natural environment (e.g. trees, the air, and oceans) and the built environment (roads, factories, cars, and so on). It is important to remember sustainability calls on engineers to consider the environment as it is today, as well as its potential future condition.

Societal subsystem

The societal subsystem includes individual people and groups of people who might affect or be affected by engineering. Socially sustainable engineering practice recognises groups who might suffer detrimental effects from engineering decisions. It also considers how to avoid or remedy such effects. In some cases, the greater societal benefit of new infrastructure needs to be weighed against the negative impact of the development. For example, a new freeway provides transport benefits and connects communities along its length, but it can cut off communities on either side. This calls for consideration of the cultural needs of groups in the community and an understanding of community group values, beliefs, priorities and perceptions of engineering work. Politics and legislation are part of the social sphere, as legislation is enacted by the public representatives to promote social order and meet community expectations.

Economic subsystem

Economic subsystem constraints include the financial costs that engineering options might impose on the person or organisation paying for the work (i.e. the proponent or client), and the subsequent return derived from this work. In sustainability thinking, it is important to also consider and quantify the financial costs and benefits for those who are affected by the work (i.e. stakeholders). The economic subsystem also draws attention to costs beyond the dollar. Costs and benefits can be financial, but they may also take other forms; including social good (e.g. the pleasure children derive from playing on a reserve; lost opportunity for recreation when public land is used for private ventures) or environmental impacts (e.g. reduced pollutant release from engineered improvements to operation of a smelting smokestack; loss of habitat for wildlife).

The interaction between the subsystems of sustainability can be depicted using an intersecting Venn diagram, with the central zone of overlap representing concurrent environmental, social, and economic sustainability. In engineering, an additional filter of technical constraints can be imposed. This is the range of options that are feasible, effective and safe, given the available engineering science, knowledge and materials. The interactivity of these subsystems is shown in figure 3.5.

Triple bottom line analysis (TBLA)

Triple bottom line analysis (TBLA) is an approach to cost–benefit analysis that may be used for evaluating the overall or general sustainability of corporate or industry operations, or for evaluating a range of options (e.g. impact of a new motor vehicle registration process, changes to a waste collection service). The TBLA attempts to evaluate the environmental, social, and economic costs and benefits of a process, product or activity. The concept was defined by John Elkington (1980) to focus companies:

> not just on the economic value they add, but also on the environmental and social value they add — and destroy. At its narrowest, the term 'triple bottom line' is used as a framework for measuring and reporting corporate performance against economic, social and environmental parameters.

The corporate sector tends to use TBLA as an approach to reporting on the sustainability performance of a company or department; however, the approach is equally useful as a means of evaluating and commenting on different proposals for solving engineering problems. The argument supporting this theory is that stating costs and benefits makes it easier for engineers to make decisions that fit reliably into *the zone of sustainability*, which is shown in the centre of figure 3.5.

The Group of 100 (G100) are an association of accounting and finance executives representing an array of government-owned enterprises and major companies in Australia. The G100 published a guide to TBL Reporting (G100 2003) that explained the qualities of good TBL indicators. As the G100 pointed out, indicators of environmental, social and economic sustainability can be either qualitative or quantitative; however, they should aim to be reliable, useful, consistently presented, reproducible and auditable, and demonstrate full disclosure.

The indicators selected and data sets gathered for each of the three bottom lines (environmental, social and economic) need to be well researched, honest and interpretable. It can be difficult to select indicators; however, while reviewing the TBL practices of Australian and international businesses, Suggett and Goodsir (2002) observed some common environmental indicators in use, including:

- the amount of energy consumed and its origin
- volume or mass of material resource and water use
- solid waste management
- emissions to air
- quantity and quality of effluents released.

Typical social indicators were the health and safety of workers or community members, and the extent of community involvement. Common economic indicators included the amount of taxation paid, and estimates of wealth created by the company. As is apparent from Suggett and Goodsir's research, calculating the TBL requires detailed information about a range of environmental, social and economic indicators.

3.3 Environmentally sustainable engineering

LEARNING OBJECTIVE 3.3 Discuss and critique various means for assessing environmental sustainability of engineered solutions.

There are several concepts and techniques that allow engineers to evaluate the environmental impact of their decisions, and a range of more complex and detailed strategies are available for assessing the sustainability of engineering. In the chapter on evaluating options, we will look at the technical and economic feasibility of an engineered solution for a wind farm. Technical and economic feasibility are important aspects of determining the merit of an engineered alternative. In this chapter, we will now look at methods for evaluating the environmental, social and broader economic impact of engineered solutions. The following tools and techniques will help you as you strive to understand and engage in sustainable practice, both during your studies and when you enter the workforce.

KEY POINT

There are a variety of auditing, analytical and reporting tools and processes that support an engineer in understanding the likely environmental impact of engineering decisions.

A global perspective

Through our engineering prowess, humans have gained the unprecedented capacity to modify the environment on a global scale. A current example of this is global climate change. As we will discover in this section, having this capacity for global environmental modification requires thought and consideration of impacts on the environment on a global scale. One aspect of this global perspective is to consider the global systems and timescales associated with the creation of resources (e.g. timber and coal), and how long it takes for the planet to absorb breakdown products from the use of resources, and to 'remanufacture' the resources.

The timescale associated with global remanufacturing of resources provides insights into whether resources are renewable or non-renewable. **Renewable resources** are those that are either continuously available in almost limitless quantity (e.g. sunshine, wind and wave action) or flow resources, which are finite but are able to be regenerated within a human timescale of days or years (e.g. timber and drinking water). **Non-renewable resources** are resources that rely on geological activity (i.e. take hundreds of thousands of years) to regenerate. Figure 3.6 provides examples of both renewable and non-renewable natural resources.

FIGURE 3.6 Natural resources can be classified as either renewable or non-renewable.

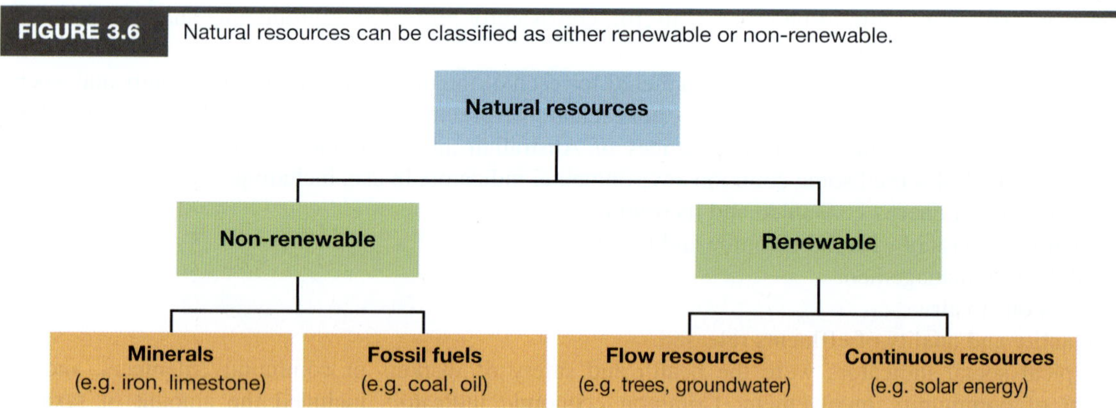

It can sometimes be difficult to distinguish between renewable and non-renewable resources. If a mineral resource can be recycled through human intervention, is it reasonable to describe it as renewable? How would the recycling (collection and reprocessing) need to be done to justify calling a mineral resource renewable? The question is somewhat philosophical, but an answer might require examination of the energy required for collection and reprocessing, and what the source of energy was (i.e. renewable or non-renewable).

Consider how the Earth functions as a system. We might think of the Earth operating as a series of resource, energy and waste flows between global systems (Mitchell et al. 2004). Global systems include the global water cycle, the global atmospheric system, energy inputs from the sun, and oceans and currents. Energy is used in the extraction and transformation of non-renewable resources such as mined ores, phosphate, and limestone from the Earth. Such non-renewable resources are transformed into consumer goods and services via industrial production and agriculture.

As figure 3.7 shows, there are energy inputs and emissions to land, air, or water at three transformation points: (1) extraction, (2) industrial and agricultural production and (3) consumption. Some of these emissions are cycled back to the Earth. Where does the energy come from to power these transformations? Before industrialisation, this energy came directly or indirectly from the sun. Since industrialisation, people have come to rely — almost completely in developed countries — on non-renewable fossil fuels, both for direct energy supply for industrial and domestic use and for producing the fertilisers used in intensive agriculture. Supplies of fossil fuels are finite, and when they are burned the carbon dioxide 'sequestered' when the coal or hydrocarbon was formed is put back into the atmosphere (Mitchell et al. 2004).

FIGURE 3.7 Resource flows in the human economy. The human economy depends on the Earth operating as a series of resource, energy, and waste flows between global systems such as the global water cycle, the global atmospheric system, energy inputs from the sun, and the oceans and currents.

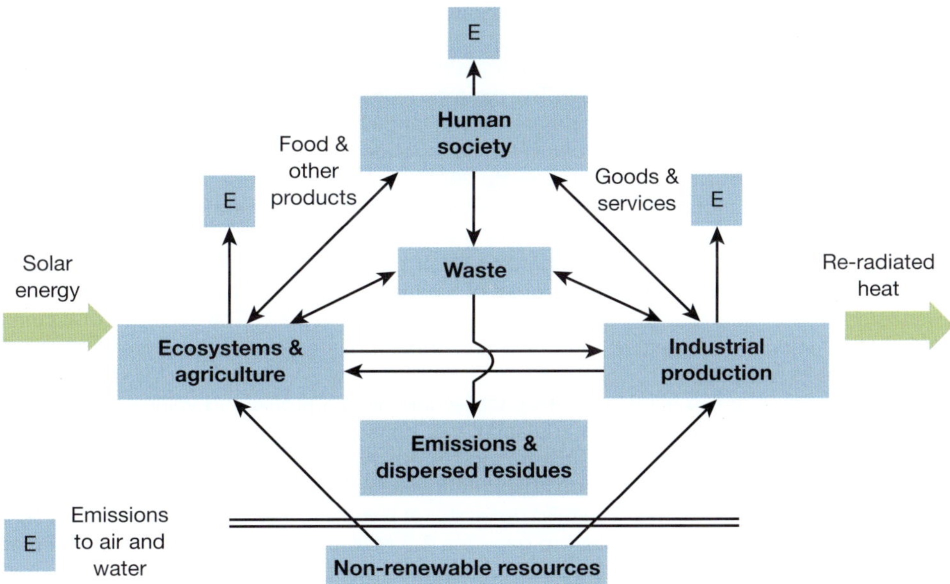

Source: Clift (1998).

The use of fossil fuels to power the transformations shown in figure 3.7 is a problem. The rate of carbon dioxide production from use of fossil fuels is now outstripping the capacity of our global waste sinks. The ocean, air and plants, which have always absorbed carbon dioxide production, are no longer able to cope with the volume of carbon dioxide in the atmosphere. The result is global climate change.

Gore (2006) has highlighted the problem of the increase in concentrations of atmospheric carbon dioxide over the last one hundred years. In the film and book, *An inconvenient truth*, Gore presents a graph of the concentration of atmospheric carbon dioxide over the last thousand years. This graph has come to be known as 'the hockey stick', because it shows concentration of atmospheric carbon dioxide climbing from a relatively flat base up to the 1800s, and then exponentially during the 1900s. Research published in the prestigious scientific journal *Nature* has demonstrated a link between human-induced climate change and extreme precipitation (rainfall, snowfall) events (Min et al. 2011). Imagine the engineering that will be needed to adapt to this problem.

Figure 3.7 should help you begin to conceptualise what might be done to protect the environment. Engineers can reduce the use of non-renewable resources including fossil fuels via the processes, systems and technologies they design.

SPOTLIGHT

Energy tips: landfill gas

The 3.2 MW landfill gas generator at Henderson, south of Perth, Western Australia

While landfill gas can seriously contribute to air pollution, it can generate a reliable and renewable source of energy. This gas is emitted from decomposing rubbish in a landfill site. As discussed at the outset of this chapter, landfill is an area designated to receive solid wastes, such as MSW, construction debris and sludges from sewage treatment and other processes. These landfill sites contain anaerobic bacteria that thrive in the oxygen-free environment under the surface of the landfill. The anaerobic bacteria decompose the landfill waste, which contains a significant number of organic compounds. During this anaerobic decomposition process, landfill gases such as methane and various dioxides are produced. Landfill gas consists of about 50 per cent methane (CH_4), a very potent greenhouse gas that is a key contributor to global climate change. However, methane is also the primary component of natural gas, a valuable fossil fuel (between 80 and 99 per cent of natural gas is made up of methane). When the methane is burned to produce energy, it gives off CO_2, another but significantly less potent greenhouse gas than CH_4.

Several Australian and New Zealand energy companies (large and small) have gone into partnership with organisations operating landfill services. The big growth in this sector occurred in the first decade of this century. Australian gas supplier AGL operates seven landfill electricity generation installations with a total output of 19.3 MW (AGL 2019). Envirowaste in NZ operates several landfill gas to electricity (LGTE) facilities generating 7 MW of electricity (Envirowaste 2017).

Western Australia has a successful history of developing landfill gas generators that convert methane from landfill into electricity. Eight LGTE and one sewage gas to electricity installations were operating in Perth in June 2010 and they were contributing 22 MW of power to the main grid — enough to supply 31 000 households with electricity (WA Department of Finance 2014).

The Henderson Renewable Energy Facility was opened on 7 June 2006. The initial 2.1 MW landfill gas generator at Henderson was projected to reduce greenhouse gas emissions by more than one million tonnes between 2006 and 2022.

As we learned in the opening section of this chapter, landfill gas profile and volume evolves over the active life of a landfill site. In response to additional gas generation at the Henderson site, Waste Gas Resources installed a third generator at Henderson (pictured). As is apparent from the picture, the landfill gas generation plant is designed to be modular so that these facilities can be up- and down-sized as the landfill's generation capacity changes. The third generator at Henderson has increased capacity by 33 per cent, allowing the methane extraction plant to generate 3.2 MW, powering more than 3300 homes. The new unit prevents an additional 35 000 tonnes of carbon dioxide-equivalent emissions each year (City of Cockburn 2011; Western Power 2008).

Western Australia is also set to be the home of the nation's first waste-to-energy incineration plant. The A$450 million Kwinana waste-to-energy plant is under construction and will produce up to 36 MW when it opens in 2021 (Phoenix Energy 2019).

CRITICAL THINKING

Biofuels, like methane, come from microbiological digestion of plant material and are considered 'renewable' energy sources. Is landfill gas a 'renewable' energy source? If not, why not? If so, should we create as much municipal solid waste as possible to ensure a generous ongoing supply? Incineration of waste for energy has caused much debate around the world. How do you think it rates in balancing the benefits against the damage in TBLA?

Measuring environmental impacts

There are various ways engineers can quantify the environmental impact of technical decisions. These fall under the two broad categories of environmental impact assessment (which directly assesses the impact of a development on the local environment), and measures of resource intensity (that quantify resources and energy required to create, use and dispose of engineered goods). Different countries and states have their own planning regulations and assessment requirements. In Australia, matters of national importance come under the *Environment Protection and Biodiversity Conservation Act 1999* (EPBC Act).

Generally, for medium impact developments, state and local government environmental reporting policies need to be complied with. The general method of discovering the environmental implications of a proposed development is to commission an impact study.

Environmental impact assessment (EIA)

An **environmental impact assessment (EIA)** is a tool for assessing or monitoring ecosystem health and is often used in engineering and development projects to determine the likely impact of a new engineering endeavour. An EIA is usually commissioned by the proponent of a development and is delivered to the government or regulatory body that will approve or monitor operations of the development. For example, the NSW planning approval for the NorthConnex nine-kilometre road tunnel in Sydney covers a huge range of environmental and Aboriginal heritage effects (NorthConnex 2015).

An EIA needs to define and report on indicators of impact. These indicators can be:

- *biological* — for example, estimated abundance of a range of animal and plant species, or species diversity
- *physical* — for example, water quality parameters such as turbidity, or soil quality parameters such as arability and aridity
- *chemical* — for example, nutrient load, concentration of heavy metals in soils, or dissolved oxygen in waters.

Good indicators are typically easy to measure, can be quantified using measuring techniques that provide reasonably reliable and accurate results, and should have some predictive properties. If an indicator is well chosen, an increase or decrease in the incidence or quantity of the indicator should tell you something about what is happening in the broader environment or to a broader range of species. Frogs, for example, are often described as a **bellwether species** because their decline can signal imminent ecosystem-wide problems that are likely to affect more robust species. Environmental impact assessments can be described as one-off or ongoing. We will look at the difference now.

One-off EIA

A *one-off EIA* is mostly undertaken when it is necessary to determine the value of an environment that might be subject to perturbation (e.g. being disturbed or damaged because of logging or mining) or a candidate for preservation (e.g. a marine system with rich species biodiversity that is used for fishing or maritime operations). A one-off EIA tells the proponent and those regulating the development what the environmental cost is for disturbing or destroying the proposed site. If the environmental cost is judged to be too high, the development is not approved, and if the environmental cost is low or can be mitigated through an environmental management plan, the development is likely to be approved, or approved subject to conditions.

Although the proponent for a project pays for an EIA, it is important EIAs are undertaken with integrity. Otherwise, the perception of a compromised process can lead to substantial community disquiet over the environmental impacts of a proposal.

Ongoing EIA

An *ongoing EIA* is designed to establish and support monitoring when there is a desire (by a company) or requirement (by a regulator) to maintain environmental health at a defined state over a period of time. For example, if there were community concerns that leachate from a local landfill could pollute local waterways, or that particulates and ultra-fine particulate emissions from a local waste incineration smokestack could affect the respiratory health of residents.

After indicators are defined and measured, an ongoing EIA often describes the regularity of measurement (a 'monitoring regime') and specifies a level for the indicators deemed 'acceptable'. Acceptable levels may be self-specified (e.g. by the company/proponent/organisation involved or by that party in conjunction with the local community) or outlined in local or national environmental protection legislation or guidelines. The defined levels outline what an acceptable emission is and the concentration, volume, flow rate and/or mass at which this emission becomes unacceptable (known as a *reference point*). An ongoing EIA should describe the actions that will be taken if reference points are breached.

A company may develop an environmental management plan (EMP) to manage the implementation of actions to address any conditions such as ongoing monitoring and reporting.

Resource intensity

Considering the non-renewable to renewable ratio of goods and services is one way to evaluate their environmental sustainability. Another way is to consider all the resources and energy required to deliver a service, or to create, use and dispose of the goods. This is described as **resource intensity**. Several analytical frameworks can be used to calculate resource intensity. We will investigate some of these in the following section.

Ecological footprint

Wackernagel and Rees (1996) created the idea of ecological footprinting. An **ecological footprint** is an estimate of the total surface area of arable land that would be required to provide renewable resources and energy for each person on Earth to maintain their current standard of living (Mitchell et al. 2004). This measure gives us an idea of the share of resources used by individuals in different societies and cultures. National Footprint Accounts are published each year by the Global Footprint Network (Lin et al. 2018).

Perdan (2004), Lin and colleagues (2018) and others have pointed out the current global level of resource consumption is far higher than the biocapacity the planet can sustain, and providing the level of resource consumption enjoyed by the developed nations would require several 'planet Earths' (Mitchell et al. 2004). The Global Footprint Network established the concept of Overshoot Day — the day during the year when global consumption exceeds the annual biocapacity of the planet, that is, the ability of the planet to replenish itself in one year (Global Footprint Network 2019). Overshoot Day occurred on 2 August in 2023 (Lin et al. 2023), meaning that by the end of the year we were using the equivalent of 1.7 Earths.

There are several footprinting accounting approaches that are based on similar theoretical assumptions. The National Footprint Account method developed by Wackernagel and Rees (1996) measures the land required to regenerate the resources needed for human activity. Another common measure used for ecological accounting is carbon footprinting. Carbon footprinting measures the mass of CO_2 emissions that a particular activity produces. Much human activity requires energy from fossil fuels that create direct CO_2 emissions. According to Lin et al. (2018) our current global carbon emissions make up 60 per cent of the total ecological footprint. Carbon dioxide is being pumped into the atmosphere at a greater rate than the planet can cope, leading to increases in the greenhouse gas effect. It has become an important skill for engineers to calculate carbon footprints with the introduction of emissions trading schemes and general awareness of climate change.

SPOTLIGHT

Wine carbon footprinting

The Australian wine industry produces around 1.4 billion litres of wine annually, about 60 per cent of which is exported, contributing about $2.6 billion dollars annually to the Australian balance of payments (Wine Australia 2019). The wine industry employs approximately 28 000 people and, as a large industry, has an important role to play in meeting Australia's greenhouse gas reduction targets.

Winemaking in Australia relies on an array of engineers; for example, chemical and bioprocess engineers who design the fermentation tanks that coax yeast to transform grape juice into wine; software and mechatronic engineers who design and maintain the distributed control systems that monitor ferments and flows in the winery; agricultural and mechanical engineers who design and maintain irrigation, trellising and pruning machinery for the vineyard; chemical and civil engineers who audit and improve water, wastewater, sanitation and energy use; and logistics engineers who specialise in supply-chain management to arrange the careful transport of grapes and finished wine, both of which must be treated with great care to ensure optimal quality in the glass.

The Australian wine industry has developed a comprehensive tool that allows wineries to measure their carbon footprint. The Australian Wine Carbon Calculator (AWCC) (Australian Wine Research Institute 2024) is an Excel worksheet with embedded formulas. The AWCC builds on an international version that has been in use since early 2008, and includes components specific to Australian needs, like government-endorsed emission factors. As an example, the calculator assists wineries in measuring their packaging footprint using a range of estimated emissions for all types of packaging material currently used in the Australian wine industry (i.e. glass bottles, aluminium cans, corks, synthetic stoppers, cardboard, plastic pallets). The calculator requires wine producers to enter the quantity of units (i.e. number of glass bottles) and weight per unit for each packaging item used, and then calculates total packaging emissions in tonnes of CO_2.

It is worth downloading the calculator (https://www.awri.com.au/industry_support/sustainable-wine growing-australia/carbon-calculator) to get a feel for the wide array of factors that need to be quantified to calculate a large industrial operation's carbon footprint.

CRITICAL THINKING

In your view, who is morally responsible for carbon emissions from the production of value-added products like wine: the suppliers of raw materials (e.g. farmers who grow the grapes), the company that converts the raw material (e.g. wine-making companies), the engineers who design the process where emissions occur (e.g. bioprocess engineers designing fermentation tanks), the consumer (e.g. you and I) or someone else?

Ecological rucksack

A product's **ecological rucksack (ER)** is the weight of natural material that is disturbed to generate a product, minus the weight of the product itself. The general term for this type of environmental impact evaluation is 'material intensity'. It represents the quantity of resources required for:

- extracting raw materials
- manufacturing
- transporting, using, and disposing.

In 2005, researchers associated with the Wuppertal Institute in Germany investigated the ER of 36-inch high-definition televisions (Aoe & Michiyasu 2005). The researchers calculated the ER for televisions produced in 1993 and in 2003. The ER of the 1993 product was found to be 19 tonnes, compared with 7.7 tonnes for the 2003 product. Perhaps you will look differently at your home entertainment set-up, imagining the tonnes of materials that have gone into its production and use.

The decline in ER between 1993 and 2003 showed that during a decade of technological refinement, the manufacturers had achieved a 60 per cent reduction in the material intensity of their televisions. Further investigation by Aoe and Michiyasu (2005) revealed that the decline in ER was due to improvements in the design of printed circuit boards and associated electronic components, the replacement of cathode ray tubes with liquid crystal and to plasma screens giving greater energy efficiency during use.

One thing that is noticeable in ER and related calculations is that high use of non-renewable resources such as driving a car, air travel, drinking bottled water and eating large amounts of grain-fed beef tends to come with a higher ER than activities that rely more on renewables (e.g. riding a bike, drinking tap water and eating vegetables). This suggests non-renewable resources should be avoided as much as possible. This applies if you want to reduce your personal ER, and if you want to create environmentally sustainable engineered solutions in your future professional life.

There are several ways in which engineers can reduce environmental impact when non-renewable ('big rucksack') resources are being used.

- *Substitute.* Use alternative, renewable or less scarce materials (e.g. laminated timber beams as opposed to rolled steel joists).
- *Conserve.* Increase 'material intensity', by using less to achieve the same service (e.g. using water efficient fittings and appliances).
- *Reuse/recycle.* Use again with no or minimal additional processing (e.g. utilise crushed, used bricks as road base).
- *Reprocess.* Salvage, transport and reprocess non-renewable resources (e.g. aluminium cans).
- *Redress impact.* Take action to minimise or make up for the impact of use (e.g. purchase carbon credits).

The ecological rucksack, which attempts to calculate the quantity of resources required to produce, use, and dispose of a particular product, is a measure of material intensity. Life cycle assessment (LCA) is another tool for quantifying the material intensity of a process or product.

Life cycle assessment (LCA)

Life cycle assessment (LCA) is a tool for quantifying the 'environmental impacts associated with a product, process or service throughout its life cycle, from the extraction of the raw materials through to processing, transport, use, re-use, recycling or disposal' (Department of the Environment 1992). It is used to identify and quantify the environmental impacts of an economic system from cradle (beginning) to grave (end). In this description, the term *economic system* means more than just a particular product (e.g. 1 kg of steel or a mobile phone), or a process (e.g. a lifestyle or a kilometre of private vehicle travel). In LCA the term economic system includes all the environmental inputs and outputs associated with a product or process over its entire life (Mitchell et al. 2004).

What is meant by the life of a product? Consider the life cycle of a bottle of water, illustrated in figure 3.8. At each stage of the transformation — from mineral resource extraction to landfill — there are inputs of energy (for transport and processing) and likely emissions of waste to land, air and/or water.

FIGURE 3.8 Life cycle of a bottle of water. As this diagram shows, even the life cycle of a bottle of water is complex and has important environmental implications.

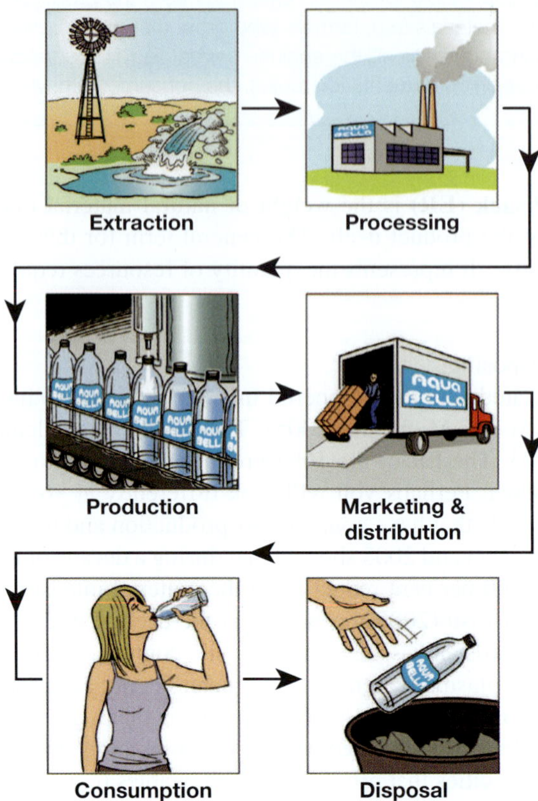

The life cycle of the economic system of a product or process can be placed inside a system boundary, shown in figure 3.9. This allows us to evaluate the total energy and material inputs over the whole life cycle of the product or process, and the total wastes and the functional (useful) outputs from the product or process during its life. Examples of functional outputs might include edible apples (from an LCA of the environmental effects of apple farming), traffic control at a particular intersection (from an LCA of a set of traffic lights) or square metres (m^2) of lawn cut (from an LCA of a residential lawn mower). At this stage of your course, you may be learning about concepts such as conservation of mass, conservation of energy, or mass balance. These are terms and concepts that are fundamental to the practice of engineering. They contribute some of the basic theoretical framework for LCA.

FIGURE 3.9 Life cycle conceptualisation of an economic system showing material and energy inputs, and functional and waste outputs

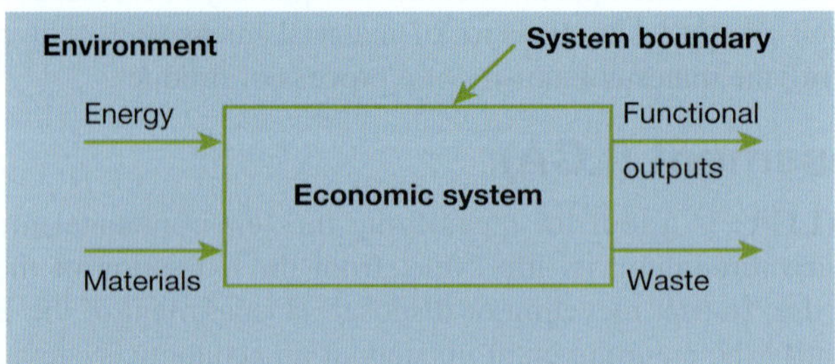

Because it is complicated, expensive and time-consuming to gather data on the wide array of inputs and outputs associated with even simple processes or products, important impact categories are negotiated

within sectors or industries and tend to be used across the sector so that LCA results can be compared. There are currently ten standards, published by the International Organization for Standardization (ISO) in the series ISO 14040–14049 that detail principles, frameworks, guidelines, data and documentation requirements and examples of LCA. Two of these have been adopted as Australian and New Zealand standards (ASNZ/ISO 14040 and AS/ISO 14048).

Engineers who use LCA to quantify the impact of engineered products or processes refer to these international standards to ensure the analysis they undertake is methodologically robust. LCA is often undertaken using proprietary software like SimaPro™ and impact factors are adapted to suit the sector. Students can download a demonstration of LCA from the SimaPro™ website. Another useful web-based LCA tool for the construction industry is the Australian eTool software. This open source tool is available from their website (https://cerclos.com/welcome). Should you require LCA certification, there is a subscription model.

LCA is a specialised skill in engineering that requires engineers to interpret and work across standards and databases. For example, materials, environmental or construction management engineers might use the ISO documents in conjunction with SimaPro™ and the AusLCI's *Best Practice Guide for Mid-Point Life Cycle Impact Assessment in Australia* (Renouf et al. 2018) to generate life cycle information on the impact factors specified by AUSLCI:

1. climate change
2. ozone depletion
3. ionising radiation
4. photochemical ozone formation
5. particulate matter formation
6. acidification
7. eutrophication (adding substances to and aquatic system, e.g. fertilisers ending up in streams)
8. toxicity
9. land stress (land use)
10. water stress (consumptive water use)
11. resource depletion — fossil fuels
12. resource depletion — minerals.

These factors are often grouped into three main areas of protection: human health, ecosystem quality, and natural resources.

LCA is becoming a more recognised and routine approach. However, there are limitations to the use and usefulness of LCA (Elghali et al. 2008; Greene 1995), such as:

- it can be time consuming and expensive to conduct a full LCA
- it can be difficult to compare and trade-off between various environmental impact categories (e.g. is moderate methane emission more or less detrimental than large carbon dioxide emissions?)
- system unknowns need to be covered using assumptions, and the accuracy of these can substantially influence the findings
- local factors are not always taken into account (e.g. high water consumption may be a non-issue in Wellington, New Zealand, but disastrous in Gympie, Australia)
- the usefulness of LCA may depend on the type of engineered system being evaluated (e.g. LCA is arguably more useful as a stand-alone approach for choice of site for an existing technology, rather than for choice of technology for a given site).

Despite these limitations, it is remarkable how shocking the results of a LCA can be for a person who has not thought about the difference between how much a product costs in dollars and how much it may cost in terms of environmental impact.

3.4 Socially sustainable engineering

LEARNING OBJECTIVE 3.4 Explain how to estimate the social impacts of engineered options using community/stakeholder communication and consultation.

Many companies now report annually on their community activity or corporate social responsibility. Society and shareholders are demanding higher standards of practice from companies, as corporate behaviours have the capacity to adversely affect the health and wellbeing of workers, consumers and the general community. One of the main principles of sustainability listed earlier in this chapter is that of promoting equity.

KEY POINT

Socially sustainable engineering invites engineers to move beyond 'do no harm' and become fully engaged with stakeholders from today's and tomorrow's societies.

Promoting intergenerational and intragenerational equity

While the environmental imperative of sustainable engineering is quite broadly understood and accepted, this social aspect of sustainable engineering is sometimes misunderstood or overlooked. Hence, we will now discuss **intergenerational equity** and **intragenerational equity** in more detail. The meaning of these two terms was foreshadowed in the *Brundtland Statement* (World Commission on Environment and Development 1987), which referred to our responsibility to ensure the resources and quality of life we enjoy do not adversely affect future generations' ability to access these resources and attain a similar quality of life.

The term 'intergenerational' invites us to think of the future generations of people who will inhabit the Earth — including our children, their children and the generations who will flow ahead through time using the Earth's resources to sustain their existence. Consider the exploitation of marine stocks that led to the collapse of cod fishing in the north-west Atlantic in the 1990s (Clark 2006). Cod was a prolific fish and provided food and income for generation upon generation of fishing families in that part of the world. The collapse of this fishery industry was an example of a failure to ensure intergenerational equity, because future generations will not have access to this food supply and the income that has traditionally been earned through its exploitation.

Intergenerational equity can also apply to the overuse of **waste sinks**. For example, the atmosphere and the oceans are waste sinks for excess carbon dioxide, providing storage for the compound before it is sequestered in the shells of marine crustaceans as calcium carbonate, or processed by plants through photosynthesis. Therefore, to achieve intergenerational equity, it would be necessary to leave waste sinks with enough capacity to process the carbon dioxide that is likely to be generated by future generations. Evidence suggests that this goal is not currently being met.

Can you think of other situations in which past and current activities have compromised intergenerational equity? Some examples include ozone depletion, increasing the risk of skin cancer for current and future generations; the depletion of fossil fuel reserves, affecting petrol and diesel prices; and salinity from poor irrigation and farming practices, reducing arable land area for current and future generations.

Intragenerational equity involves considering the current distribution and consumption of resources on a global scale. It is about ensuring the consumption of resources and distribution of wastes does not disproportionately favour one country, region or social group, causing disadvantage to another. For instance, the UN states that in 2022 '2.2 billion people still lacked safely managed drinking water, … 3.5 billion people lacked safely managed sanitation, … and 2 billion lacked a basic handwashing facility, including 653 million with no handwashing facility at all' (United Nations 2023). Goal 6 of the SDGs has a target that by 2030 there will be 'universal and equitable access to safe and affordable drinking water for all' (United Nations 2023). The current disproportionate availability of clean drinking water demonstrates intragenerational *inequity*.

Environmental justice is a field of study that is focused on intragenerational equity. It attempts to document, prove or refute allegations that a community or population is suffering disproportionately because of the actions of another community (e.g. a community may have loss of health or amenity because of the pollution caused by a factory or a waste treatment facility).

Pearce et al. (2006) have studied human exposure to air pollution from domestic heating in Christchurch, New Zealand. The researchers used an atmospheric dispersion model, concluding exposure to ambient air pollution and extreme pollution events was significantly higher in Christchurch's disadvantaged communities. These researchers concluded the communities experiencing the majority of the pollution were not responsible for producing most of it.

In this instance, a group benefited from a polluting activity — for example, having access to domestic heating in winter — while another group (the affected communities) was disproportionately disadvantaged, suffering from higher pollution levels during winter. This was a situation where environmental justice was lacking and intragenerational *inequity* was shown to exist between communities in Christchurch.

Native title and mining

Engineers who work in the mining sector are familiar with one of Australia and New Zealand's most tangible examples of intergenerational inequity: the dispossession of the First Nations peoples of Australia and Māori peoples of New Zealand when white colonisers claimed each country more than 200 years ago. Currently, some mining companies are at the forefront of efforts to redress the resultant cultural, social and economic damage.

The very act of dispossession limited economic opportunity for First Nations and Māori peoples who had lost their lands. This was perhaps more significant given each group has important cultural connections with the land. In both countries, First Nations and Māori peoples continue to suffer the economical and emotional impact of dispossession. They experience shorter average life expectancy than the population averages and substantial portions of their communities are trapped in a chronic cycle of unemployment, poor education outcomes and social disadvantage (Australian Human Rights Commission 2018).

In New Zealand, the signing of the *Treaty of Waitangi* in 1840 signalled some form of recognition of Māori peoples' original ownership of the land and right to citizenship in New Zealand. The *Treaty of Waitangi* also provided a precedent for legal recognition of some Māori legal systems and traditional practices. Nowadays this recognition is enshrined in the RMA, which requires engineering proponents of major developments to consult with, and account for, cultural impacts on local Māori communities.

In Australia, white settlers declared the land to be *terra nullius* (literally translated this means 'land belonging to no-one') in 1788 and the land claims of Aboriginal and Torres Strait Islander peoples were not legally recognised until the Mabo decision in 1992. The Mabo decision debunked the idea of Australia as *terra nullius* and ushered in legal recognition of First Nations land ownership — 'native title'.

One of the outcomes of native title has been that mining companies in Australia are now required to negotiate with traditional owners on land use, and to offer compensation for the loss of First Nations rights and interests in the land. These negotiations can lead to Indigenous Land Use Agreements (ILUA) that can include monetary payments to First Nations trust companies, community infrastructure and employment quotas for local First Nations peoples to be employed by the mining companies. However, ILUAs are still negotiated on an ad hoc basis (Australian Human Rights Commission 2016).

One way to make sense of the engineer's role in enacting social sustainability is to consider what aspects of human need are the responsibility of the engineer. Is an engineer merely responsible for ensuring their engineering activities do not cause anyone to die? Should an engineer feel responsible for extending human life, and for building human or community health? Is it the engineer's role to grow happiness, wealth or social cohesion? To become clear about the engineer's role in social sustainability, it is worth considering the array of human needs we might seek to fulfil.

SPOTLIGHT

Unintended consequences of engineering breakthroughs

Thomas Midgley Jr (1889–1944) was a mechanical engineer and inventor who worked for General Motors in their research division. His research into problems with high compression petrol engines found that 'knocking' was caused by the fuel mixture not burning evenly. He experimented with fuel additives, exploring different elements until in 1921 he discovered that tiny amounts of lead completely eliminated the engine knock. The use of lead introduced a new problem: namely, the build up of lead in the combustion chamber. In order to expel the lead, Midgley found that ethylene dibromide was a suitable extra additive to clean the engine. This resulted in the lead being ejected with the exhaust gases into the atmosphere (Giunta 2006).

This discovery created an enormous demand for bromine. Indeed, Midgley estimated that the demand would use up more than the complete world supply of bromine. Undeterred by the scale of this task, he is credited, along with Willard H Dow, with the development of new techniques to extract bromine from seawater. In 1933 he became vice-president of the Ethyl-Dow Chemical Company, producing bromine from seawater. With his inventions, Midgley was responsible for the worldwide use of lead in petrol from the 1930s until its eventual phase out in the mid-1990s. Midgley was awarded the Nichols Medal by the American Chemical Society (ACS) in 1923 for the 'use of anti-knock compounds in motor fuels' (ACS 2014). The toxic problems associated with lead were well known at the time of Midgley's inventions. Indeed, the engineer himself suffered from lead poisoning in 1923 and spent an extended period convalescing in Florida.

Although Midgley was responsible for introducing highly toxic lead into the atmosphere, he also worked on the elimination of poisons from industrial products. This was not necessarily due to altruism but was to improve the products and make them more marketable. In particular, he worked to find new compounds to use as refrigerants. Much work was done in the early decades of the twentieth century to develop refrigeration techniques and to move away from the need for ice boxes.

Early refrigerants included poisonous compounds such as methyl chloride, ammonia and sulphur dioxide. As well as being dangerous, these gases are malodorous and unsuitable for use in domestic environments. General Motors' Frigidaire division was keen to replace their sulphur dioxide–based technology with a more palatable gas — one that was safe and did not smell bad. Midgley was tasked with leading the team in this research.

In 1931 the team gave the first public description of a gas they named Freon 12, chemical name dichlorodifluoromethane, the first CFC. This was a popular invention and soon Freon 12 and other CFCs replaced the noxious and explosive refrigerants. The 'benign' gas made it possible to build domestic refrigerators and air conditioners, and for General Motors to develop air conditioners for automobiles. Midgley was awarded the American Chemical Society's top honour, the Priestley Medal, in 1941 (ACS 2014; Giunta 2006).

Midgley contracted polio in 1940, which left him disabled. He invented a system of pulleys and a harness so that he could hoist himself from bed into his wheelchair. Midgley was the President of the ACS in 1944 when he died suddenly. He became entangled in the ropes of his harness and was strangled. At the time, this was reported as an accident, though historians have speculated that the device he invented was intended to allow him to end his own life. Perhaps this was his third invention with unintended consequences (Encyclopaedia Britannica 2014).

While Thomas Midgley Jr could certainly have extrapolated the damage lead in petrol might do to air quality and human health, it is probably unfair of McNeill (2000) to dub him the 'Fritz Haber [inventor of chemical warfare] of the atmosphere'. CFCs were thought to be very stable and not reactive. It was only 30 years after Midgley's death that CFCs were linked to UV and damage to the ozone layer in the upper atmosphere.

CRITICAL THINKING

Can you think of other 'wonder compounds' that have subsequently been proven to create more damage than good? Are there any current inventions that might benefit from a precautionary approach so that we don't overuse them and discover unintended consequences?

Maslow's hierarchy of needs

Psychologist Abraham Maslow differentiated levels of human needs (Maslow 1987). Maslow described a hierarchy of human needs, arguing some human needs are more important than others. The hierarchy begins with basic physiological needs for food, drink and thermal stasis (e.g. staying warm enough to avoid death by hypothermia, and cool enough to avoid death by heat stroke). The next step involves the human need for safety from physical harm, fear or threat. The need for love, enfranchisement and belonging to a group (e.g. family and community) is identified as the next need stage. The fourth and fifth levels of the hierarchy are the need for self-esteem and the esteem of others, and the need for self-actualisation in the form of fulfilment of individual personal potential.

Maslow suggested humans work to satisfy the needs at the base of the hierarchy (e.g. food and safety), and after realising these needs, turn their efforts towards higher-order needs (e.g. belonging, self-esteem and fulfilment of potential). Maslow's hierarchy of needs is shown in figure 3.10, with basic physiological needs forming the base of the pyramid.

FIGURE 3.10 Maslow's hierarchy of needs

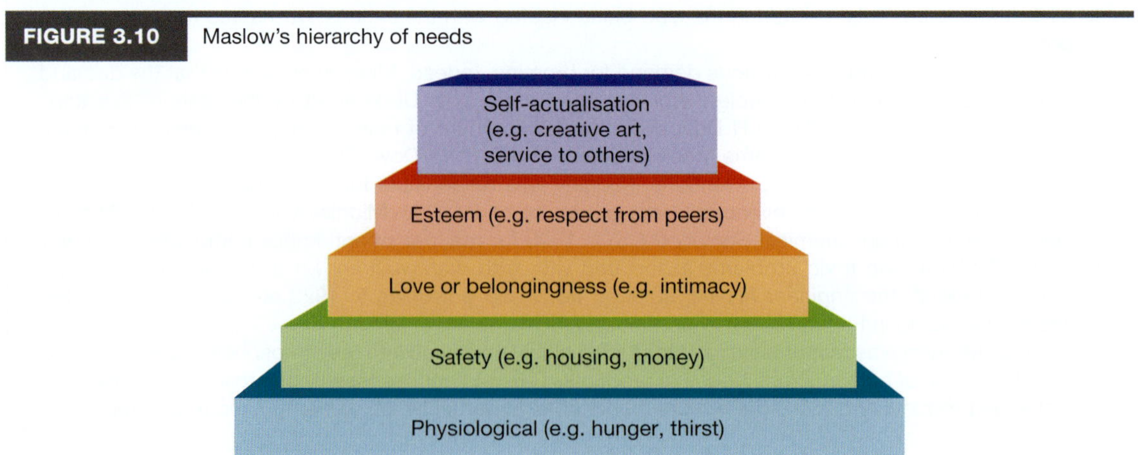

Maslow's work provides us with ideas about what it might mean for an engineer to move beyond the social responsibility of 'doing no harm' and into the realm of socially sustainable engineering. The idea of 'do no harm' holds engineers to the first two steps of Maslow's hierarchy. This leaves three steps in the hierarchy of needs. Are engineers responsible for enacting or promoting these human needs through their work? Interestingly, some engineering professionals and engineering companies are answering this question with a tentative 'yes' — and beginning to move their thinking beyond a 'do no harm' ethic. These companies are interested in ensuring the engineered solutions they produce consider and even contribute to achieving some of the higher-level needs described by Maslow.

Community communication and consultation

In a substantial or controversial engineering project, it is not sufficient, or good practice, for an engineer to think up and rate the importance of various costs and benefits to stakeholders. Engineers should gather real data on social costs and benefits by actively consulting members of the community about the projects that might have an impact on their lives, communities, cultures, and livelihoods. This promotes individuals' sense of belonging, cohesion and esteem within their community (levels 3 and 4 of Maslow's hierarchy).

This consultation is usually undertaken before engineering decisions are made and can be done as a way of clarifying the problem (in problem analysis). This approach is known as **community consultation**. It moves engineering practice beyond selling technical solutions to the public (sometimes called a 'public relations' approach) and even beyond informing or educating the community about technologies ('science communication' approach).

So how can you set up a community consultation process for an engineering project? Community consultation involves gathering stakeholders and providing them with a range of ideas, information and expertise, before facilitating discussion and debate among the group in order to generate comment on an engineering problem or proposed set of solutions. The facilitator in community consultation has a complex role. A **facilitator** organises and runs community consultation processes, paying attention to:

- who is invited to attend
- breaking the ice and establishing the process for debate and discussion
- what information is made available to participants
- making sure each of the participants gets a say
- dealing with conflict between participants in a constructive way
- guiding the participants towards a decision (if this is the desired outcome of consultation).

Community consultation can be a challenging and expensive process and it invites the community to take an active role in the engineering process. It should only be undertaken if the scale or importance of the problem warrants it. Dick (1997) has suggested some key questions and pointers for designing a community consultation process. These principles can be applied in engineering practice, as follows.

- Are you aiming to inform the community (e.g. about a new development such as the construction of a new off-ramp, or change in truck movements from your company's site)? If so, the use of mass media including newspaper or television advertising may be more appropriate than community consultation.
- Do you want to gather or exchange information on something (e.g. community acceptance of a new nanotechnology medical therapy, or explaining the workings and special features of a new weir that has been constructed on a local river)? In this case, a low-key managed approach, such as running information sessions or focus groups with a diverse cross-section of community members, might be a more appropriate approach.
- Is your intention to reach agreement or consensus on a design or solution for a particular engineered project? If so, full-scale community consultation is likely to be an effective approach. Agreement is likely to require that at least some stakeholders change their views and it is most likely to occur in face-to-face meetings in a consensual and supportive climate.

Scope

The scope of a community consultation refers to the range of issues stakeholders are being consulted on. Are they being consulted on a single issue, or on general community aims? For example, a local council may consult the community if they want to assess public sentiment about a commercial or retail development. This is single issue consultation. Dick (1997) suggests that if it is not handled carefully, this form of consultation can bring firmly held and differing views to the surface and may result in adversarial or divisive outcomes. Consultation on general community aims (e.g. planning for a more sustainable community or revising an urban transport system) is broader in scope, tending to raise less ire from individual participants. Therefore, single issue consultation may warrant more focused or controlled approaches than general or ongoing community consultation.

Time span

The time span for a consultation process is also important in informing how it should be conducted and the likelihood of a successful outcome. Dick (1997) explains, 'It is hard to maintain wide involvement over lengthy time spans, but brief consultation may raise more issues than it can address. Most planning authorities allow far too little time for effective consultation'.

Community consultation needs to be carefully planned and facilitated to make the most of stakeholders' time, to gain their enthusiasm for participating in good faith, and to allow for adequate coverage of the ideas, issues and concerns stakeholders hold or may develop about the issues or situations under review.

Multi-criteria decision analysis (MCDA)

Multi-criteria decision analysis (MCDA) is an approach that allows us to analyse the viewpoints that community consultation delivers. Researchers in Australia, South Africa, Europe and the United Kingdom have pioneered and are refining the approach (Azapagic et al. 2004). There are different 'schools of thought' in MCDA, including different ideas about how to conduct and interpret processes and outcomes. Essentially, MCDA attempts to use stakeholders' opinions or values to rank perceived environmental, social and economic impacts of engineered solutions.

The social impact of a solution is established by (1) asking stakeholders what they believe the benefits and setbacks of an option are and (2) getting stakeholders to rate these various benefits and setbacks. The relative importance stakeholders give to social costs and benefits is 'traded off' against other stakeholders' rankings to identify a preferred option (Elghali et al. 2008; Petrie et al. 2004). MCDA is often used in conjunction with LCA. For example, data on environmental impacts may be generated using LCA and stakeholders might be asked to rank these impacts using MCDA.

SPOTLIGHT

Multi-criteria evaluation: recreation and tourism in Victoria

The Goulburn Broken Catchment of Victoria, Australia, covers an area of 2.4 million hectares that stretches from just north of Melbourne in the south to the Murray River in the north. The catchment is characterised by myriad environmental problems, including soil salinity, rising water tables and poor water quality. About 200 000 people live in the catchment area.

The upper catchment is renowned as a tourist destination for residents of Melbourne (5 million people); however, the influx of tourists each year has caused serious environmental problems for the area. Many of these problems are related to water issues in the catchment, which have flow-on effects for users further downstream (Proctor & Drechsler 2003). The problems of how to address and solve the complex issues of tourism management and environmental impact in the upper catchment were investigated by Proctor and Drechsler (2003) using a deliberative process aided by multi-criteria evaluation (MCE), which is another name for multi-criteria decision analysis (MCDA).

With many public policy decisions, such as those concerning the environment, the objectives of the decision may conflict, and the criteria used to assess the effectiveness of different policy options may vary widely in importance. MCE is a technique to identify trade-offs in the decision-making process, with the goal of achieving compromise. It is also an important means by which structure and transparency can be imposed upon the decision-making process. The origins of MCE lie in the fields of mathematics and operations research, and it has been used by environmental and civil engineers, and public planners in such applications as the siting of health facilities, highways and energy-generation facilities (Proctor & Drechsler 2003).

In the case of the Goulburn Broken Catchment, a jury was chosen that comprised a group of natural resource managers (stakeholders), rather than randomly chosen members of the public (citizens). This group was termed a 'stakeholder jury' to distinguish it from the citizens' juries that are more commonly used in MCE/MCDA. The stakeholder jury was convened for a one-day deliberative workshop. At the outset of the workshop, members were asked to individually rank a list of 13 different assessment criteria for choosing between management options to reduce tourism impacts in the Goulburn Broken Catchment. The rankings were fed into ProDecX, a software package used to support MCE processes (Proctor & Drechsler 2003). Figure 3.11, generated by ProDecX, shows the lack of consensus between jury members at the outset of the day.

FIGURE 3.11 Ranking of criteria prior to participation in stakeholder jury; where a value of 1 represents the highest rank and a value of 13 represents the lowest

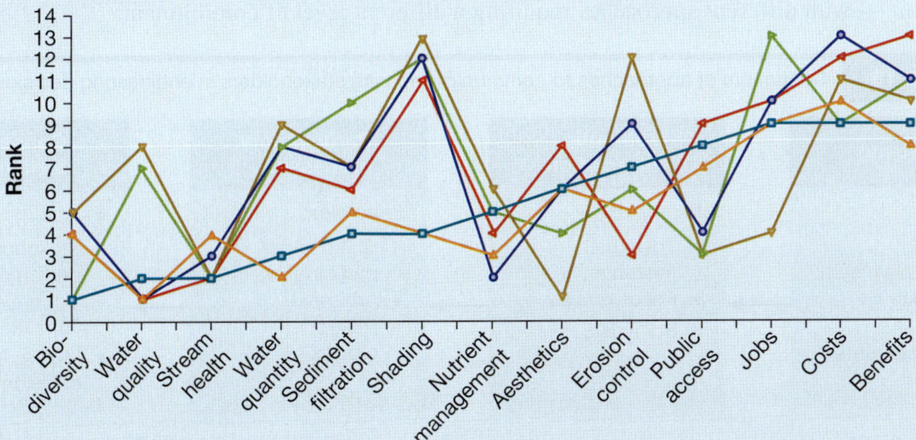

Source: Proctor and Drechsler (2003).

The jury members were then asked to consider information presented to them by expert witnesses and in the form of a pre-prepared Impact Matrix, and then deliberate on that information in terms of how their own professional perspective, needs and values matched the criteria. The jury were told that by the end of the day, they were expected to reach a unanimous decision on how useful the various assessment criteria were in deciding between management options to reduce tourism impacts. At the close of the workshop, jury members again individually ranked the assessment criteria, and a graph was generated.

Figure 3.11 shows how the facilitated process of MCE moved the stakeholder jury towards greater consensus. It should be noted that figure 3.12 has fewer categories on the x-axis than figure 3.11. This is because members of the stakeholder jury came to agree that only four criteria were needed to represent ecosystem services (ES), rather than the original proposal of nine ES criteria (Proctor & Drechsler, 2003).

FIGURE 3.12 Weighting of criteria after participation in stakeholder jury

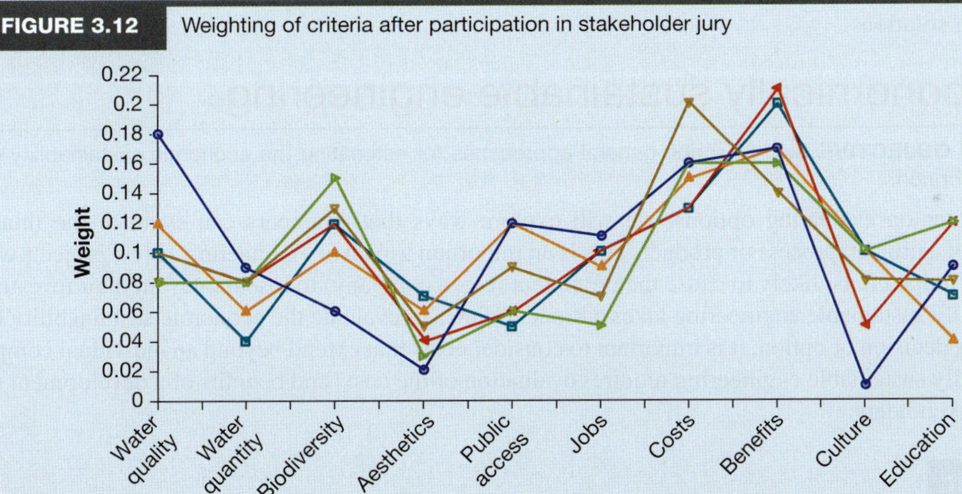

Source: Proctor and Drechsler (2003).

Engineers tasked with achieving social sustainability can use MCE or MCDA to generate consensus amongst stakeholders in an engineering decision and can also have some trust that metrics derived from MCE/MCDA accurately represent the views and aspirations of the community that will be impacted by a given engineering decision. The skills needed to run an effective MCE or MCDA often require the use of professional facilitators to assist the engineers and the stakeholders (Proctor & Drechsler 2003).

CRITICAL THINKING

As a professional engineer facilitating MCE or MCDA, how could you remain impartial in the decision-making process? When would you consider it necessary to employ a professional facilitator?

Engagement techniques

Mitchell et al. (2004) recognised engineering professionals can engage with stakeholders in several ways. These engagement techniques are shown in figure 3.13, which demonstrates how engagement is a continuum — with different approaches requiring a different level of commitment.

FIGURE 3.13 Continuum of approaches to communicating with stakeholders in engineering decisions

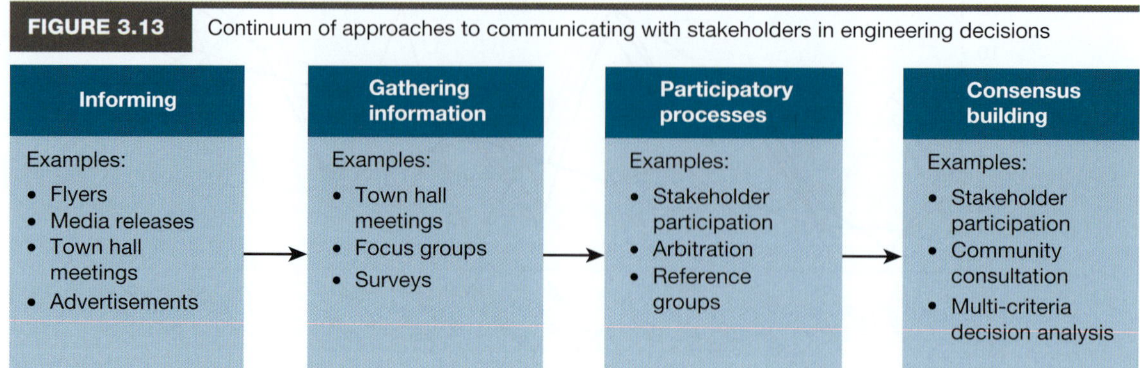

Informing involves *telling* the community what you are planning or have done. This requires little time and, more significantly, allows an engineer to maintain control and responsibility for decision-making processes and outcomes. In this approach, the engineer's main duty is to ensure stakeholders are kept up to date with project developments. At the other end of the consultation spectrum, approaches such as deliberation and consensus-making represent a challenge to engineers who aspire to retain control over the direction and outcome of a project (Mitchell et al. 2004).

The *consensus* approaches are more likely to recognise stakeholders as experts in their own needs and wants, and to give substantial power to stakeholders in guiding the decision-making process and determining preferred options. MCE/MCDA is a tool that has been used to build consensus. It has been suggested that the loss in control over decision making in consensus approaches is compensated by gains in public acceptance of outcomes and the production of more contextually appropriate solutions (Fawcett & Roberts 2001). We will now look at tools for understanding the economic sustainability of engineering solutions.

3.5 Economically sustainable engineering

LEARNING OBJECTIVE 3.5 Describe general approaches for estimating the economic sustainability of an engineered option.

In the chapter on evaluating options, we will explore ways that engineers can evaluate the financial feasibility of a project, process or product. This is an important aspect of engineering as few projects would go ahead if judged to be likely to cause the proponent (e.g. a company) to lose money on their investment. Economically sustainable engineering takes a broader view. To evaluate the economic sustainability of an engineering decision or option, it is important to consider costs that extend beyond an individual company. Economically sustainable engineering requires evaluation of the costs and benefits of a development to all affected stakeholders.

KEY POINT

Engineers may be asked to evaluate projects on costs and benefits that go beyond the direct financial returns to a proponent or shareholder group.

Costing

Cost–benefit analysis (CBA) is a common approach used to appraise projects in engineering. It is the assignment of a value to all the direct and indirect negative and positive outcomes that will likely result from a project, product or course of action. CBA provides a single positive or negative value and therefore makes decisions relatively simple. A positive value suggests a project's benefits will outweigh its costs and therefore should proceed.

In this framework, the costs and benefits of a project are expressed in terms of monetary value. Examples of costs and benefits considered in a CBA include the cost of raw materials and labour, and an estimate of income from the sale or rental of a product. A CBA considers all of the dollar costs and benefits of an engineering proposal over a set period of time (*planning horizon*). It uses them to calculate the net present value (NPV) (see the chapter on evaluating options).

A planning horizon can be any amount of time; however, it is usually determined by the amount of time an engineering project, or the production of a product, is expected to endure or take to payback the initial investment and be financially predictable (e.g. a typical planning horizon for a major civil engineering construction might be 25 years).

As illustrated in the wind turbine example in the chapter on evaluating options, NPV gives an estimate of the present value of all the invested money in a particular scheme or project over several years. This takes account of inflation and other factors that erode the value of money. The NPV equation considers the time value of money, recognising that an amount of money received in the future will be worth less than the same amount of money received now. It subtracts the initial investment from discounted future earnings and discounted future operating costs. If the total discounted earnings exceed the discounted operating costs and initial investments, the project or purchase will likely proceed. The NPV analysis compares future costs and benefits in terms of current dollar value.

This form of economic project appraisal is widely used but relies on all stakeholders agreeing on what costs and benefits sit within the boundary of the economic appraisal, and on all parties accepting the dollar values that are placed on the costs, benefits and likely environmental impact.

For example, it is relatively easy to apply a straight dollar value to the transport and landfill disposal of construction wastes (cost per kilometre of diesel for truck transport from construction site to landfill, plus cost per tonne in landfill charges); however, consider the difficulty of calculating costs in direct dollar terms of transporting construction waste to landfill, if the impact categories used in LCA are included (e.g. depletion of world resource reserves, ozone depletion, increase in eutrophication of waters and contribution to photochemical smog).

CBA and NPV are useful and widely used tools, but their use in evaluating the economic costs and benefits of engineering work rests with the boundary assumptions made by those who use CBA and NPV, such as the estimated future inflation and interest rates used in the analysis. Like the social and environmental parameters of sustainable engineering, the economics of sustainable engineering can be conceptualised using different frameworks. There are different approaches to thinking about the economics of engineering, including considering what things are really worth, what should be paid for, who pays for what, and the buyer's return on purchase.

Economic theories

The economic frameworks and tools (such as CBA) that are routinely used to judge the merit of engineered solutions tend to be based on a school of economic thought called neoclassical economics. In **neoclassical economics** it is assumed people have rational preferences for a particular outcome, people are fully informed and act independently of each other, and people seek to maximise their satisfaction in the consumption of various goods and services. These three assumptions underpin the general theoretical framework of neoclassical economics.

In this framework, some goods or services within an economic system are recognised as having a dollar value (e.g. timber), while it is assumed other goods and services (described as externalities) have no direct dollar value, or a dollar value that is impossible to accurately measure or estimate. Examples of externalities include ecosystem services such as the static physical and process physical services performed by trees, 'public good' such as the community cohesion and health contribution made by public sporting facilities, and social and environmental amenity, such as the visual pleasure and biodiversity value provided by a green, open public space.

Ecological economics is an alternative framework proposed in response to some of the shortcomings of neoclassical economics. It attempts to address three values that are supposedly in conflict under neoclassical economics: (1) allocation (efficiency), (2) distribution (justice), and (3) scale (sustainability) (Daly 1991). This framework was developed from the thinking of economist Georgescu-Roegen (Mayumi 2001).

According to ecological economics theory, goods and services (defined as 'externalities' in neoclassical economic theory) should be valued to address the values in conflict. Ecological economics proponents believe the value of externalities is intrinsically linked to an acknowledgement of the interconnections and interdependence between humans and nature. The value of externalities, such as environmental goods and services, affects engineering decision processes and outcomes as the costs and benefits of different technological options and products differ with an ecological economics approach (as opposed to a neoclassical approach) (Carew & Mitchell 2008). Emotions and values can influence engineering decisions when more than simple cost is considered.

There are several economic analysis approaches that attempt to enact an ecological economics mindset. Two approaches you may come across are *least cost planning* and *full cost accounting*. Consider the cost of a bottle of water. Do you agree the cost of a bottle of water might rise if the full cost of the environmental impact is factored into the price?

Least cost planning (LCP)

Least cost planning (LCP) is a costing tool used by engineers and others to measure the economic sustainability of a range of options. It allows, for example, a water service provider to identify which of a range of options provides their customers with the lowest cost water services. The term 'water services' highlights that customers don't want water for its own sake; rather, they want the services that water provides, such as personal hygiene, quenching of thirst, a green lawn and clean clothes (Turner et al. 2005).

LCP focuses on the services provided by resources — such as water, energy, carpet, or motor vehicles — rather than the resource itself. The tool is used for evaluating economic sustainability because it investigates the whole-of-society costs and benefits, rather than just the costs and benefits to the service provider (e.g. water utility, energy supplier, carpet factory or vehicle manufacturer). LCP allows us to quantify economic sustainability in dollar terms through the unit 'levelised cost'. Levelised cost is a measure of the capital and operating cost of supply over a set period, divided by present value unit cost (e.g. $0.30/kL of water).

SPOTLIGHT

Least cost planning in the 'sunburnt country'

Australia is a land of 'drought and flooding rains', which means managing its water supply requires clever engineering of physical systems, and clever analysis of how water is used and can be saved. The federal government and many state governments are reluctant to build new water supply dams, and so non-traditional water sources and water conservation have become a focus for ensuring security of water supply in times of drought.

The ACT government committed to a 12 per cent and 25 per cent reduction in per capita water demand by 2013 and 2023, respectively (based on 2003 levels of per-capita water consumption). As new dams cannot be built, two options are available: 'source substitution' and 'demand-side management'. Source substitution is using water other than traditional potable supply, such as recycled water, ocean water or stormwater.

Part of source substitution is working out how to match the lower quality of these waters with lower-quality applications (i.e. these would be inappropriate for drinking, but may be suitable for watering parks and gardens, or use in water features). Demand-side management involves reducing the amount of water people use to meet their water service needs; for example, installing water-efficient showerheads in residential buildings reduces the amount of water required to sanitise one human. Demand-side management usually involves a combination of technology upgrade and education for changed water-use behaviour (e.g. educating the public to take shorter showers).

To achieve the ACT's 2013 and 2023 demand management targets, Turner et al. (2005) outlined a range of options grouped under 'demand management' and 'source substitution'. The water savings of each option are presented as figure 3.14, which shows projected total water demand for Canberra under a range of scenarios. Most relevant in figure 3.14 are the:

- 'reference case demand', which shows how many megalitres of water Canberrans would consume up to 2053 under a business-as-usual scenario (i.e. with no intervention to modify current water use practices)
- black dots, which show the ACT government's per-capita water demand reduction targets for 2013 and 2023
- pink line, which shows water consumption projected if a range of demand-side measures are instituted.

In addition to modelling water savings from each option, Turner et al. (2005) used LCP to calculate the levelised cost for various demand-side and source-substitution options, and table 3.2 shows some of these. It is apparent that a program where householders are provided AAA-rated showerheads at substantial discount would cost 22c/kL saved, whereas a program to retrofit old homes with dual flush toilets would cost 59c/kL saved. The source-substitution options are generally more expensive, but LCP shows us that providing a rebate for new homes to install rainwater tanks would be more economically efficient (at $4.45/kL saved) than retrofitting existing homes with rainwater tanks (at $10.62/kL saved).

FIGURE 3.14 Suite of options developed to achieve water consumption targets for Canberra, Australia

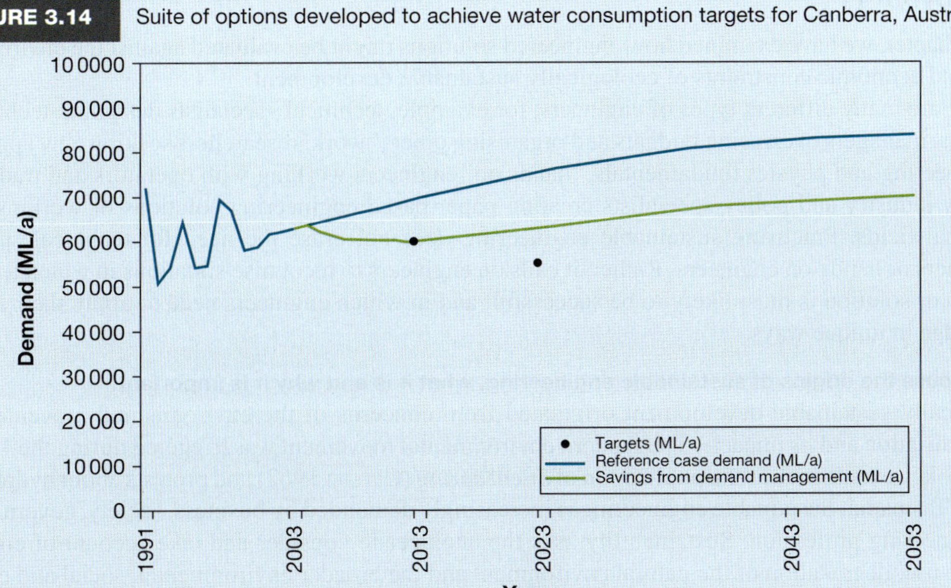

Source: Turner et al. (2005).

TABLE 3.2 **Option summary table**

Option	Levelised cost (A$/kL)	Savings assumptions
	Demand management	
AAA-rated showerhead rebate	0.22	16.5 kL/household/a
Dual-flush toilet program	0.59	37 and 23 kL/household/a for single and multi-residential households respectively
AAAA washing machine rebate	1.02	50 per cent reduction compared to top-loading machines
Active unaccounted for water control program	0.99	Conservative estimation of 750 ML/a
	Source substitution	
Rainwater tank rebates (existing houses)	10.62	35 kL/household/a for a 5 kL tank in a single residential household
Rainwater tank rebates (new houses)	4.45	55 kL/household/a for a 10 kL tank in a single residential household
Grey-water system rebates (new houses)	4.87	50 per cent outdoor use in a single residential household

Source: Turner et al. (2005).

Based on LCP analysis, engineers and managers at ACT Water were better placed to decide which options would best reduce ACT's demand for water at the lowest cost to the wider community, as noted in table 3.2. According to the ABS, the ACT was close to its target in 2015–16, with total water demand of 51 gigalitres (ABS 2017).

CRITICAL THINKING

Is an engineer responsible for costing projects in a way that shows costs for the whole community (e.g. all stakeholders) or just the costs to the proponent (e.g. the company commissioning the engineered works)?

SUMMARY

In this chapter, we have examined how engineered solutions might be evaluated against the environmental, social and economic constraints of ecologically sustainable development.

There are many different types of engineers, for example, technical specialists designing technological solutions, managers overseeing budgets and organising others' work, researchers seeking new applications of engineering and physics fundamentals, 'hands on' engineers working with operators and tradespeople in heavy industry and policy specialists creating paper-based engineering solutions or working in non-traditional fields. Practising sustainable engineering does not erase the need for technical specialists, researchers or hands-on engineers. Rather, it calls on engineers to recognise situations in which a technical decision or solution is most likely to be successful, and in which engineers need to apply their skills and knowledge in unique ways.

3.1 Discuss the origins of sustainable engineering, what it is and why it is important.

Ecologically sustainable development originated from concerns of the environmental movement about industrialisation and its impacts. The modern environmental movement was triggered during the 1960s and 1970s by key events such as the publication of *Silent spring* (Carson 1962) and protests about hydro-electric dams in Tasmania. Sustainable engineering is increasingly demanded by business, society, government and the engineering profession. Sustainability asks the engineer to consider and take account of energy and resource use, degradation of the natural environment and the broader environmental, social and economic consequences of their work.

3.2 Detail strategies for practising sustainable engineering and how to evaluate a solution using a triple bottom line analysis.

Strategies for practising sustainable engineering include maintaining ecological integrity, applying precaution, consulting stakeholders, promoting equity, conserving resources, limiting emissions and practising industrial ecology. The triple bottom line is an approach to integrating findings of environmental, social and economic evaluation of products or processes that supports sustainable decision making and corporate responsibility.

3.3 Discuss and critique various means for assessing the environmental sustainability of engineered solutions.

There are a variety of auditing, analytical and reporting tools and processes that support an engineer in understanding the likely environmental impact of engineering decisions. These include EIAs, which allow an engineer to assess or monitor ecosystem health. EIAs are often used in engineering to determine the likely impact of a new engineering endeavour; and in LCA (which quantifies all energy and resource inputs required to produce, use and dispose of an engineered economic system, and all functional outputs from the system and wastes generated). These approaches allow rational quantification of the likely impact of engineering work on the natural environment.

3.4 Explain how to estimate the social impacts of engineered options using community/stakeholder communication and consultation.

Engineers have a responsibility to promote intergenerational and intragenerational equity. These are measures that ensure the actions or activities of one set of people in society do not disadvantage or compromise the quality of life for another group in society (even if these people are not yet born). A continuum of communication and consultation approaches stretches from engineers informing the public of engineering work or solutions (e.g. advertisements, flyers, website information), through to engineers participating in consensus or consultation with potential stakeholders.

3.5 Describe general approaches for estimating the economic sustainability of an engineered option.

Engineers are often asked to evaluate projects on their financial feasibility (e.g. direct financial returns to a proponent or shareholder group). Economic sustainability asks engineers to move beyond simply calculating the costs and benefits of a project or proposal, or NPV (analysis of financial feasibility with discounting of future costs and benefits) for a client or company. Sustainable engineering invites the engineer to consider and account for externalities and alternate economic assumptions and theories. This takes engineering into the realm of least cost planning, where costs and benefits to all stakeholders are considered and count towards the evaluation of project viability.

KEY TERMS

bellwether species Species of plant or animal whose decline can signal imminent ecosystem failure. The word *bellwether* comes from shepherding. A wether (castrated ram) has a bell around his neck and guides the other sheep to the shearing shed or the sheep dip. When you hear the bellwether, you know the rest of the sheep are on their way.

circular economy An economic system that involves the reuse and regeneration of existing materials or products in order to extend their life cycle in a sustainable or environmentally friendly way (European Parliament 2023).

community consultation Actively consulting members of the community about the projects that might impact on their lives, communities, cultures, and livelihoods.

cost–benefit analysis (CBA) Assigns values to all costs and benefits to generate a single positive or negative value, and thereby show whether project costs will outweigh benefits.

ecological economics An alternative economic framework that attempts to address three values supposedly in conflict under neoclassical economics: (1) allocation (efficiency), (2) distribution (justice) and (3) scale (sustainability).

ecological footprint An estimate of the total surface area of arable land that would be required to provide renewable resources and energy for each person on Earth to maintain their current standard of living.

ecological rucksack (ER) The weight of natural material that is disturbed to generate a product, minus the weight of the product itself.

ecologically sustainable development (ESD) Development that integrates economic growth, social equity, and environmental management so that ecological processes (on which life depends) are maintained in a state sufficient to meet the quality of life needs of present and future generations.

environmental impact assessment (EIA) A tool for assessing or monitoring ecosystem health.

environmental justice Study that attempts to document, prove, or refute allegations a community or population is suffering disproportionately as a result of the actions of another community.

facilitator Organises and runs community consultation processes.

intergenerational equity Ensuring future generations have access to resources that would provide opportunity for a quality of life equal to or better than that enjoyed by current generations.

intragenerational equity Ensuring the consumption of resources and distributions of wastes does not disproportionately favour one country, region, or social group, causing disadvantage to another.

least cost planning (LCP) A costing approach for evaluating economic sustainability that investigates whole-of-society costs and benefits.

life cycle assessment (LCA) Environmental impacts associated with a product, process, or service throughout its life cycle, from the extraction of the raw materials through to processing, transport, use, re-use, recycling or disposal.

multi-criteria decision analysis (MCDA) Uses stakeholders' opinions or values to rank perceived environmental, social, and economic impacts of engineered solutions.

neoclassical economics An assumption that people have rational preferences for a particular outcome, people are fully informed and act independently of each other, and people seek to maximise their satisfaction in the consumption of various goods and services.

non-renewable resources Resources that rely on geological activity (i.e. take hundreds of thousands of years) to regenerate.

renewable resources Resources that are either continuously available in almost limitless quantity (e.g. sunshine, wind, and wave action) or flow resources, which are finite but are able to be regenerated within a human timescale of days or years (e.g. timber and drinking water).

resource intensity All of the resources and energy required to deliver a service, or to create, use and dispose of a product.

sustainable engineering Practices that promote environmental, social, and economic sustainability through greater resource efficiency, reduced pollution, and consideration of the wider social impacts of new technologies, processes and practices.

triple bottom line (TBL) A set of design criteria that can be folded into the engineering method, sitting alongside technical measures, to inform sustainable design choices.

triple bottom line analysis (TBLA) An approach to cost–benefit analysis commonly used for evaluating the sustainability of corporate or industry operations, or for evaluating a range of options (e.g. impact of new motor vehicle registration process; changes to a waste collection service).

waste sink A part of the environment that receives and accommodates waste products. The air and the ocean have traditionally acted as waste sinks for gaseous pollutants and gaseous and liquid wastes respectively.

EXERCISES

1 Consider this scenario: Janet and Jiang are civil engineers who have been asked to design and site a new sustainable wastewater treatment plant.* Janet interprets 'sustainable' to mean reducing the use of non-renewable resources, doing no harm to the community and ensuring the plant can operate within the council's 'budget allocation'. Jiang interprets 'sustainable' to mean not reducing biodiversity, improving cultural and equity outcomes for the community and maximising opportunities to return goods and services to the local community.

(a) Describe what environmental, social, and economic indicators Janet and Jiang might come up with, given their very different interpretations of 'sustainable'. Record your ideas in a table where the left-hand column lists the triple bottom line (environment, society, economy), the middle column lists Janet's selected indicators, and the right-hand column lists Jiang's selected indicators.

(b) Which engineer will have the most useful set of indicators for judging the sustainability of the engineering outcome? Why?

(c) Which engineer has a more difficult design task to fulfil and evaluate? Why?

*Different engineering scenarios that you may like to consider as alternatives to the water treatment example in this exercise include the:

- design of a sustainable voice-activated lift for the local library
- siting and operations of a sustainable tailings dam for a uranium mine
- sustainable herbicide selection and application regime for broad-acre broccoli farming
- sustainable side airbag safety standards for motor vehicles.

2 Access a corporate responsibility report, annual report or sustainability report from the website of a large mining firm or civil engineering construction company and find evidence of commitment to UN SDGs, triple bottom line analysis, and reporting. Provide a summary of the economic, environmental and social values that are reported by the organisation.

3 Look around you at a nearby building, either near where you live or at university. List the materials that the building is made from and identify which of these are renewable as opposed to those that would be classified as non-renewable. If there are composite materials such as concrete, split them up into their component parts if you know what they are made from. By running your eye down this list, you can make a rough estimate of what proportion of the materials used in the building are renewable. Outline how the building could have been constructed differently to reduce reliance on non-renewables.

4 Use the WWF Australia's footprint calculator (https://wwf.org.au/get-involved/ecological-footprint-calculator) or do an internet search for a 'footprint calculator'. Alternative search terms include 'ecological rucksack' or 'materials intensity calculator'. You will find quite a few websites, so choose one that seems reasonably easy to use. Calculate your own ecological footprint. Next, find a friend or someone in your class and encourage them to use a different internet footprint calculator. Compare what factors were considered in both calculators, and what factors were unique to one or other of the calculators. How much of a difference is there between the two footprints? Is the difference due to different factors and weightings in the footprint models, or due to differences in lifestyles or lifestyle choices?

5 Visit the website for your local municipal or city council. How have they been consulting with the local community about engineering solutions to local planning, development, social and environmental issues? Consider the following aims described by Dick (1997):

- informing
- gathering or exchanging information
- reaching agreement or consensus.

Can you find examples of each of these forms of communication between your council and the community? You might need to look in an Annual Report or State of the Environment Report (if your council produces one) or use alternative search terms such as 'stakeholder consultation'.

6 Use Google Earth, a street directory, or a local real estate guide to print out or draw up a map of 10 or 20 suburbs around your home on an A3 sheet of paper. Perform the following analyses.
   - Check your real estate guide to see how much an average three-bedroom home costs in each suburb. Another way to determine this is by going to one of the online services that provide prices and other relevant data for each suburb.
   - Use a street directory to identify where various municipal waste facilities are located within these suburbs (e.g. sewage treatment plants, waste recycling depots and landfill sites). Also, use Google Earth to scan the area and mark in any other industry or activities that may make a suburb more unpleasant, noisy, or potentially dangerous. Mark these on your map. You may also like to mark in sites with high volumes of traffic, such as major highways and shopping centres.
   - What does your map show about the relationship between the number of polluting, noisy or unpleasant industries or activities in a suburb, and the cost of houses? Are there any clear patterns?

7 Do some internet research on an engineering practice, technology or situation that has allegedly caused social health effects or discontent about potential health effects. Some examples might be:
   - copper mining at Ok Tedi
   - the smokestack at Port Kembla (find out about the environmental concerns relating to its demolition)
   - Union Carbide's operations in Bhopal, India
   - Nestlé's provision of powdered milk in Africa
   - the tailings dam operation at Jabiluka Uranium Mine
   - mobile phone use and cancer fears
   - volatile organic carbons and indoor office air quality
   - noise from wind farms
   - elevated blood lead levels in children living near Xstrata's Mt Isa mine.

   Imagine you live or work next to one of these operations or use one of the products and are concerned about health effects. After your internet research, write three or four questions you would want answered about the operations or product so that you could make a judgement about whether your health was at risk. Then, imagine you are an engineer whose job it is to answer these questions. How would you gather the information you need to provide honest, accurate answers to these questions? What if the answers went against the position put forward by your employer?

8 Your cousin who is studying economics at university reads the term 'economically sustainable engineering' on a website and says to you 'What does that mean and how do you calculate it?' Prepare a 250-word explanation and include examples to keep your cousin interested.

9 Where is the nearest landfill gas to electricity facility to you? Contact the operators and, if possible, arrange a visit for you and your classmates.

10 Compare Maslow's hierarchy of needs to the 17 UN SDGs. Where do Maslow's needs fit within the SDGs? Which of the SDGs do you think should be added as high priorities to Maslow's hierarchy?

PROJECT ACTIVITY

Engineering work takes place in complex social and political environments. Spend some time reviewing your local newspaper, online news source or a relevant trade magazine to find a proposed development that is both controversial and relevant to your discipline. It might be the Southern Hemisphere Synchrotron, a new hybrid fuel cell, desalination technology, genetic modification of food crops or farm animals, or software that will potentially aid those practising identity fraud. The proponent wants to commission a triple bottom line study and has asked you for an expression of interest to prepare the TBL.

Review the tools, ideas and techniques in this chapter and develop a proposal for how you could assess the triple bottom line for the controversial development. You will need to prepare a short verbal report with supporting written expression of interest to present during your upcoming meeting with the proponent.

REFERENCES

AGL 2019, 'Landfill gas and biogas', https://aglpt.digital.agl.com.au/about-agl

American Chemical Society 2014, 'Priestly medal', https://www.acs.org

Aoe, T & Michiyasu, T 2005, '"Ecological rucksack" of high definition TVs', *Material Transactions*, vol. 46, no. 12, pp. 2561–2566, https://doi.org/10.2320/matertrans.46.2561

Australian Building Codes Board 2016, 'National construction code', https://ncc.abcb.gov.au
Australian Bureau of Statistics 2017, *Water Account, Australia, 2016–17*, cat. no. 4610.0, ABS, Canberra.
——— 2020, *Waste Account, Australia, experimental estimates*, https://www.abs.gov.au/statistics/environment/environmental-management/waste-account-australia-experimental-estimates/latest-release
Australian Council of Recyclers Inc. 2012, *Australian Council of Recyclers submission on NSW Waste Levy Review*, https://acor.org.au/
Australian Human Rights Commission 2016, *Social justice and native title report 2016*, AHRC, Sydney.
——— 2018, *Close the gap — 10 year review (2018)*, AHRC, Sydney.
Australian Wine Research Institute (AWRI) 2024, 'Australian wine carbon calculator', https://www.awri.com.au/industry_support/sustainable-winegrowing-australia/carbon-calculator
Azapagic, A, Perdan, S & Clift, R 2004, *Sustainable development in practice: Case studies for engineers and scientists*, John Wiley & Sons, West Sussex, England.
Carew, AL & Mitchell, CA 2008, 'Teaching sustainability as a contested concept: Capitalizing on variation in engineering educators' conceptions of environmental, social, and economic sustainability', *Journal of Cleaner Production*, vol. 16, no. 1, pp. 105–115, https://doi.org/10.1016/j.jclepro.2006.11.004
Carson, R 1962, *Silent spring*, Houghton Mifflin, Boston.
Chen, HCL & Hay, P 2006, 'Defending island ecologies: Environmental campaigns in Tasmania and Taiwan', *Journal of Developing Societies*, vol. 22, no. 3, pp. 303–326, https://doi.org/10.1177/0169796X06068033
City of Cockburn 2011, 'Henderson Renewable Energy Facility', https://www.cockburn.wa.gov.au
Clark, CW 2006, *The worldwide crisis in fisheries economic models and human behavior*, Cambridge University Press, Cambridge.
Clift, R 1995, 'The challenge for manufacturing', in J McQuaid (ed.), *Engineering for sustainable development*, The Royal Academy of Engineering, London.
——— 1998, 'Engineering for the environment: The new model engineer and her role', *Process Safety and Environmental Protection*, vol. 76, no. 2, pp. 151–160, https://doi.org/10.1205/095758298529443
Daly, HE 1991, 'Elements of environmental macroeconomics', in R Costanza (ed.), *Ecological economics*, Columbia University Press, New York.
Department of Infrastructure and Regional Development 2018, *Australian design rules*, Australian Government, Canberra.
Department of the Environment 1992, *National strategy for ecologically sustainable development*, Australian Government, Canberra.
——— 2010, *Australia's Biodiversity Conservation Strategy 2010–2030*, Australian Government, Canberra.
Department of Climate Change, Energy, the Environment and Water 2024a, 'Nature repair market', https://www.dcceew.gov.au/environment/environmental-markets/nature-repair-market
——— 2024b, 'Safeguard mechanism', https://www.dcceew.gov.au/climate-change/emissions-reporting/national-greenhouse-energy-reporting-scheme/safeguard-mechanism
Department of the Environment and Energy 2016, *Report on the review of the first five years of Australia's Biodiversity Conservation Strategy 2010–2030*, Commonwealth of Australia, Canberra.
Derkley, K 2009, 'Profit forecast to fire up landfill gas', *Swinburne Magazine*, December.
Dick, B 1997, *Community consultation checklist*, Southern Cross University, Lismore NSW.
Elghali, L, Clift, R, Begg, KG & McLaren, S 2008, 'Decision support methodology for complex contexts', *Engineering Sustainability*, vol. 161, no. 1, pp. 7–22, https://doi.org/10.1680/ensu.2008.161.1.7
Elkington, J 1980, *The ecology of tomorrow's world*, Associated Business Press, London.
Encyclopaedia Britannica 2014, 'Thomas Midgley, Jr', https://www.britannica.com/biography/Thomas-Midgley-Jr
Energy, Efficiency and Conservation Authority 2011, 'Stirring up energy savings at MCK Metals Pacific Ltd', https://zmag.net/documents/mckmetals.pdf
Engineers Australia 2019, *Stage 1 Competency Standard for Professional Engineers*, https://www.engineersaustralia.org.au/publications/stage-1-competency-standard-professional-engineers
——— 2022, *Code of Ethics and Guidelines on Professional Conduct*, https://www.engineersaustralia.org.au/publications/code-ethics
Envirowaste 2017, 'Doing good with waste', 1 March, https://environz.co.nz
European Parliament 2023, 'Circular economy: Definition, importance and benefits', https://www.europarl.europa.eu/topics/en/article/20151201STO05603/circular-economy-definition-importance-and-benefits
Fawcett, A & Roberts, P 2001, 'The institution of engineers, Australia — at the turn, remaking for relevance', *Transactions of Multidisciplinary Engineering*, vol. GE25, pp. 51–63.
Fet, AM & Deshpande, PC 2023, 'Closing the loop: Industrial ecology, circular economy and material flow analysis', in AM Fet (ed.), *Business transitions: A path to sustainability*, Springer, Berlin, https://doi.org/10.1007/978-3-031-22245-0_11
First Light Studio 2019, 'The Meridian First Light House', https://firstlightstudio.co.nz
Funtowicz, S & Ravetz, J 1993, 'Science for the post-normal age', *Futures*, vol. 25, no. 7, pp. 739–755, https://doi.org/10.1016/0016-3287(93)90022-L
G100 2003, 'Available sustainability: A guide to triple bottom line reporting', https://group100.com.au
Giunta, CJ 2006, 'Thomas Midgley Jr and the invention of chlorofluorocarbon refrigerants: It ain't necessarily so', *Bulletin for the History of Chemistry*, vol. 31, no. 2, pp. 66–74.
Global Footprint Network 2019, 'Earth Overshoot Day 2018', https://www.overshootday.org/movethedate-live-stream
Gore, A 2006, *An inconvenient truth*, Guggenheim, D (dir.), film, Paramount Pictures.
Greene, D 1995, 'Towards sustainable engineering practice: Engineering frameworks for sustainability', resource folder prepared for the IEAust Task Force on Sustainable Development, Library & Learning Service.
Hydro Tasmania 2014, 'Gordon-Pedder', https://www.hydro.com.au/clean-energy/our-power-stations/gordon-pedder

Institution of Engineers Australia (IEAust) 1997, *Towards sustainable engineering practice: Engineering frameworks for sustainability*, report prepared by D Greene for the Task Force on Sustainable Development, Institution of Engineers, Australia, Canberra.

Jackson, T 2010, 'Keeping out the giraffes', in A Tickell (ed.), *Long horizons*, British Council, London, p. 20.

Lamborn, J 2012, 'Observations from using models to fit the gas production of varying volume test cells and landfills', *Waste Management*, vol. 32, no. 12, pp. 2353–2363, https://doi.org/10.1016/j.wasman.2012.06.013

Leining, C & Kerr, S 2018, *A guide to the New Zealand Emissions Trading Scheme*, Motu Economic and Public Policy Research, Ministry for the Environment, Wellington.

Lin, D, Hanscom, L, Murthy, A, Galli, A, Evans, M, Neill, E, Mancini, MS, Martindill, J, Medouar, F-Z, Huang, S & Wackernagel, M 2018, 'Ecological footprint accounting for countries: Updates and results of the national footprint accounts, 2012–2018', *Resources*, vol. 7, no. 3, p. 58, https://doi.org/10.3390/resources7030058

Lin, D, Wambersie, L & Wackernagel, M 2023, 'Estimating the date of Earth Overshoot Day 2023', https://www.overshootday.org/content/uploads/2023/06/Earth-Overshoot-Day-2023-Nowcast-Report.pdf

Macquarie Dictionary 2024, *The Macquarie Dictionary online*, Pan Macmillan Australia, https://www.macquariedictionary.com.au

Maslow, AH 1987, *Motivation and personality*, Longman, New York.

Mayumi, K 2001, *The origins of ecological economics*, Routledge, London.

McNeill, JR 2000, *Something new under the sun: An environmental history of the 20th century world*, WW Norton & Company, New York.

Min, SK, Zhang, X, Zwiers, FW & Hegerl, GC 2011, 'Human contribution to more-intense precipitation', *Nature*, vol. 470, no. 7334, pp. 378–381, https://doi.org/10.1038/nature09763

Mitchell, CA, Carew, AL & Clift, R 2004, 'The role of the professional engineer and scientist', in A Azapagic, S Perdan & R Clift (ed.), *Sustainable development in practice: Case studies for engineers and scientists*, John Wiley & Sons, London.

National Science and Technology Council 1995, *A bridge to a sustainable future: National environmental technology strategy*, US Government Printing Office, Washington DC.

National Water Quality Management Strategy 2013, *Australian drinking water guidelines 6, 2011 Version 2.0 (updated Dec 2013)*, National Health and Medical Research Council and Natural Resource Management Ministerial Council, Canberra.

Newton, PW 2001, *Australia state of the environment report 2001*, Commonwealth DEHWA, Canberra, Australia.

New Zealand Business Council for Sustainable Development 2008, 'Definition of sustainable development', https://nzbcsd.org.nz

New Zealand Infrastructure Commission 2021, *Draft New Zealand Infrastructure Strategy*, https://tewaihanga.govt.nz/media/454ps2vc/draft-new-zealand-infrastructure-strategy.pdf

NorthConnex 2015, 'Infrastructure approval SSI-6136', https://www.northconnex.com.au

NZ Ministry for the Environment 2013, *Resource Management Act 1991*, New Zealand Government, Wellington.

Organisation for Economic Co-operation and Development (OECD) 2024, *Environmental performance reviews: Germany 2023*, OECD Publishing, Paris, https://doi.org/10.1787/19900090

Pearce, J, Kingham, S & Zawar-Reza, P 2006, 'Every breath you take? Environmental justice and air pollution in Christchurch, New Zealand', *Environment and Planning*, vol. 38, no. 5, pp. 919–938, https://doi.org/10.1068/a37446

Perdan, S 2004, 'Introduction to sustainable development', in A Azapagic, S Perdan & R Clift (ed.), *Sustainable development in practice: Case studies for engineers and scientists*, John Wiley & Sons, London.

Petrie, J, Basson, L, Notten, P & Stewart, M 2004, 'Multi-criteria decision analysis: The case of power generation in South Africa', in A Azapagic, S Perdan & R Clift (ed.), *Sustainable development in practice: Case studies for engineers and scientists*, John Wiley & Sons, London.

Phoenix Energy 2019, https://www.phoenixenergy.com.au

Proctor, W & Drechsler, M 2003, *Deliberative multi-criteria evaluation: A case study of recreation and tourism options in Victoria Australia*, European Society for Ecological Economics, Frontiers 2 Conference, 11–15 February, Tenerife.

Renouf, MA, Grant, T, Sevenster, M, Logie, J, Ridoutt, B, Ximenes, F, Bengtsson, J, Cowie, A & Lane, J 2018, *Best practice guide for mid-point life cycle impact assessment in Australia, version 2*, Australian Life Cycle, Assessment Society.

Rittel, H & Webber, M 1973, 'Dilemmas in a general theory of planning', *Policy Sciences*, vol. 4, pp. 155–169, https://doi.org/10.1007/BF01405730

Simons, N 1992, 'Lake Pedder: The beginning of a movement', *Green Left Weekly*, 30 September, iss. 73, https://www.greenleft.org.au/content/lake-pedder-beginning-movement

Statista 2024, 'Average annual per capita municipal waste generated by OECD countries as of 2022', https://www.statista.com/statistics/478928/leading-countries-by-per-capita-generated-municipal-waste

Suggett, D & Goodsir, B 2002, 'Triple bottom line measurement and reporting in Australia, making it tangible', The Allen Consulting Group, Melbourne.

Team UOW Australia 2013, 'Illawarra Flame — house', https://www.illawarraflame.com.au/house.php

——— 2019, 'Desert rose: Team UOW Australia — Dubai', https://desertrosehouse.com.au

Turner, A, White, S & Bickford, G 2005, 'The Canberra least cost planning case study', International Conference on the Efficient Use and Management of Urban Water, Santiago, Chile, pp. 15–17.

United Nations 2019, 'Sustainable development goals', https://www.un.org/sustainabledevelopment

——— 2023, 'Sustainable development goals: Goal 6 — ensure access to water and sanitation for all', https://www.un.org/sustainabledevelopment/water-and-sanitation

United Nations General Assembly 1992, 'Rio declaration on environment and development', *Report of the United Nations Conference on Environment and Development*, https://www.un.org/en/conferences/environment/rio1992

US Department of Energy 2019, 'Solar Decathlon', https://www.solardecathlon.gov

Wackernagel, M & Rees, W 1996, 'Our ecological footprint: Reducing human impact on the earth', *Population and Environment*, vol. 17, pp. 195–215.

WA Department of Finance 2014, '2010 electricity generation tables', Government of Western Australian, Perth.

Western Power 2008, 'The facts about landfill gas', https://www.westernpower.com.au

Wilson, D, Chowdhury, T, Elliott, L, Elliott, T & Hogg, D 2017, *The New Zealand Waste disposal levy: Potential impacts of adjustments to the current levy rate and structure*, Eunomia Research & Consulting, Bristol.

Wine Australia 2019, 'Market insights — Australian wine exports', https://www.wineaustralia.com/market-insights/australian-wine-exports

World Commission on Environment and Development 1987, *Our common future*, Oxford University Press, Oxford.

ACKNOWLEDGEMENTS

Photo: © Professor Julia Lamborn

Photo: © chris / Adobe Stock Photo

Photo: © Andriano / Shutterstock

Photo: © First Light Associated Photographers

Photo: © SBRC UOW

Photo: © Alison Gillespie

Photo: © City of Cockburn

Photo: © Kwest / Adobe Stock Photo

Figure 3.1: © United Nations 2019, 'Sustainable development goals', https://www.un.org/sustainable development

Figure 3.2: © Professor Julia Lamborn, Monash University

Figure 3.4: © National Science and Technology Council 1995, *A bridge to a sustainable future: National environmental technology strategy*, US Government Printing Office, Washington DC.

Figure 3.5: © Clift, R 1995, 'The challenge for manufacturing', in J McQuaid (ed.), *Engineering for sustainable development*, The Royal Academy of Engineering, London.

Figure 3.7: © Clift, R 1998, 'Engineering for the environment: The new model engineer and her role', *Process Safety and Environmental Protection*, vol. 76, no. 2, pp. 151–160, https://doi.org/10.1205/095758298529443

Figures 3.11 and 3.12: © Proctor, W & Drechsler, M 2003, *Deliberative multi-criteria evaluation: A case study of recreation and tourism options in Victoria Australia*, European Society for Ecological Economics, Frontiers 2 Conference, 11–15 February, Tenerife.

Figure 3.14 and table 3.2: © Turner, A, White, S & Bickford, G 2005, 'The Canberra least cost planning case study', International Conference on the Efficient Use and Management of Urban Water, Santiago, Chile, pp. 15–17.

Text: © Engineers Australia 2019, *Stage 1 Competency Standard for Professional Engineers*, https://www.engineersaustralia.org.au/publications/stage-1-competency-standard-professional-engineers

CHAPTER 4

Professional responsibility and ethics

'Few are those who see with their own eyes and feel with their own hearts.'

Albert Einstein (1954)

LEARNING OBJECTIVES

After studying this chapter, you should be able to:

4.1 explain the difference between negligence, malpractice and commercial liability

4.2 detail the ways engineers can avoid, manage and remediate WHS risks, discuss how engineers can be held personally liable for their decisions and discuss the advantages and drawbacks of risk assessment frameworks when working globally

4.3 explain the influence of values on ethical decisions

4.4 give examples of how moral theories can be applied in ethics

4.5 identify the critical importance and the potential limitations of codes of ethics

4.6 describe macro ethics and how it is different from micro ethics

4.7 explain the role culture and corruption play in ethical decisions and perspectives across the globe.

Introduction

Engineers generally excel at using the principles of physics and the tools of mathematics to apply engineering science in crafting technical processes and designing technical products. They enjoy the comfort of finding the 'correct' answer and believe that maths and physics give them certainty about the problems they solve. What often comes as a bit of a shock to young graduates is that the world doesn't know disciplinary boundaries; engineering does not operate in a vacuum but is influenced by, and itself influences, the broader social and environmental world in which it is framed. In reality, the engineering problems that graduates work on in their jobs are messy ill-defined problems. They soon realise that even their technically 'correct' answers are at best calculated on the basis of an approximation of the context, that they will have to include margins of error, but worse they start to see that *they might actually not be working on the right problem to begin with*! To produce a good solution they will need to consider the social and environmental implications, as well as any legal ramifications, and consider the possibility that they have to start again with a new problem definition.

Here is an example that may encourage you to ponder the distinction between 'correct' and 'good'. In 1942, during World War II, the Manhattan Project was established in the United States. This project brought together some of the greatest physicists and engineers of the day to produce an atom bomb for the United States, including Director of the Manhattan Project, J Robert Oppenheimer. In August 1945, two of these bombs were dropped on the Japanese cities of Hiroshima and Nagasaki, resulting in the death of an estimated 200 000 people, the majority of whom were civilians. Japan surrendered under threat of further devastating atomic attacks, thus concluding World War II. There are public documents available online that show the choices made about which city should receive the bomb, one where it would result in sufficient death and destruction to cause attention. Arguments are often made in support of this decision related to the need to stop the war before more lives were lost; but what was the cost of ending the war? Can this technically correct engineering be considered 'good' engineering? If there is a potential for a technology does it always have to be developed? There is much debate about this case, including the consideration of why the German researcher Werner Heisenberg did not develop the bomb, whereas Oppenheimer led its development in Los Alamos. The 2023 film *Oppenheimer*, written and directed by Christopher Nolan, explores the motivations and ethical dilemma faced by Oppenheimer during his work on the Manhattan Project. The film shows Oppenheimer's eventual horror at the role he played in developing the weapons; he is said to have told President Truman 'I have blood on my hands' (Bella 2023). The Manhattan Project and films like *Oppenheimer* invite us to consider a profound question: does a scientist or engineer have the right, or do they have the professional responsibility, to decide if and how their knowledge is used?

An appreciation of your professional responsibilities and engineering ethics allows you to judge whether the engineered project or solution you are proposing goes beyond being technically correct. In other chapters, the environmental and financial aspects of engineering decisions were considered alongside technical aspects. In this chapter, we will consider how the solutions an engineer proposes might fit within the social, ethical and legal frameworks that exist around the engineering profession.

First, consider the following scenario. Mehreen is an environmental engineer working on a mine site in Peru for a large Australian company. She has been asked to visit the community to find out about recent claims that the mine site has been poisoning local fish and to prepare a plan of action for her manager. She is sent to speak to a representative of the community who explains that in the past weeks, the fish in the lake, where fishing usually occurs, have been dying. They believe it is due to the mine tailings pond leaking into the river. She takes photos of the dead fish and takes notes from the conversation. She also prepares herself by considering the following issues.

- The mining company has a good environmental record, but profits are down and the shareholders are applying pressure to cut costs. There are rumours the company is planning to minimise the budget allocated to reduce environmental impact.
- There have been violent clashes with communities in this area due to previous environmental concerns.
- One of the known impacts of the toxins created from mine tailings (mine waste) is fish dying in local lakes and rivers.
- There are rumours that the mine tailings are leaking from the tailings pond into the local river.

Consider what additional information Mehreen would need before going to her manager. She could explore the potential impact of tailings on fish. She could explore the issues that communities have complained about in the past. She could explore whether the tailings pond is leaking. These all take a lot of time and she wonders whether her boss will be sympathetic to her spending time on this and not other areas of her work. She is worried that if she takes this seriously and finds out the potential harm or social and health impacts that someone else will be asked to do the job instead of her, someone who will not ask questions. She could check with her boss to find out if her consultancy's reputation, legal liability and ethical culture would support her need to spend time assessing the situation before deciding on her action. She might even ask if there could be a team available to help her with this important assessment. If she gets the support and finds out that the fish are dying because of the tailings, she could then make her decision to report and walk away or stay and fight against this proposed plan. If she doesn't get the support, again she has the right to walk away or decide to stay and act against her own values, which might eventually cause her stress and health issues. When personal and professional values collide, it is very difficult to remain untarnished by the experience — we feel we are doing the wrong thing. Instinctively, you may believe it is the mining company's responsibility — and not Mehreen's — to respond to these issues. As professionals, however, engineers need to evaluate the situations in which they work from both an *ethical* and a *legal* perspective. Engineers are expected to apply greater insight and judgement about the likely impacts of their actions on the environment and on others. Engineers need to apply their skills and expertise to understanding and taking responsibility for the ethical ramifications of what they do, and to abiding by legal, regulatory and policy standards.

An engineer can face multiple ethical dilemmas when exercising their responsibilities as a professional. An understanding of professional responsibility can provide a structured way of recognising, understanding, analysing and deciding on a legal and ethical course of action. Throughout this chapter, different real-life dilemmas faced by engineers will be analysed. The legal frameworks and codes of ethics that govern the work of professional engineers will be highlighted, and some of the personal, professional and societal consequences of the ethics of engineers' decisions will be explored.

4.1 Professional responsibility: standards and professional liability

LEARNING OBJECTIVE 4.1 Explain the difference between negligence, malpractice and commercial liability.

In earlier chapters, some of the standards, codes, guidelines and legislation that form the regulatory framework within which engineers practise were touched on. Some of these regulations — for example, anti-discrimination legislation and duty of care for co-workers — apply to anyone who has a job in Australia and New Zealand. Some regulations apply directly to those with a position of power or trust in the workplace, such as ensuring privacy of personnel records and details, legislation precluding anti-competitive behaviour, legislation precluding insider-trading and mandatory reporting of health and safety breaches. Some regulations apply specifically to the form of work engineers are engaged in; for example, standards for emissions to air and water, structural strength requirements for cars, and building codes.

In the chapter on sustainable engineering, you read about the International Organization for Standardization (ISO) and the documents produced by this group. Standards are agreed specifications that are used to guide the design and assessment of products, processes, services or performance. Australia's national equivalent to the ISO is Standards Australia, which is the 'leading independent, non-governmental, not-for-profit standards organisation . . . [and] Australia's representatives of the International Organization for Standardization (ISO) and International Electrotechnical Commission (IEC), [who] are specialists in the development and adoption of internationally-aligned standards in Australia' (Standards Australia 2018).

In New Zealand, the ISO equivalent is called Standards New Zealand and they develop new standards using expert committees that actively seek public input. The majority of New Zealand standards are developed in partnership with Standards Australia, and take account of existing international standards. While standards and codes are only mandatory if specified by law, they should always be referred to and followed by engineers in Australia and New Zealand. By spending some time looking around the Standards Australia, Standards New Zealand and ISO websites, you will get a feel for the standards that will govern your future practice in the engineering discipline you've chosen to pursue.

The term 'professional liability' is used to describe the state of being legally obliged and responsible for your work as a professional engineer. There are two types of liability pertinent to the engineer, and they can be loosely termed **tort liability** and **commercial liability** (see figure 4.1). A third form of legal liability — criminal liability — concerns infringement of criminal laws, such as those prohibiting murder or theft, and generally has no place in a discussion of professional and ethical engineering practice.

Tort liabilities, more accurately termed 'civil responsibility liabilities', are those obligations specified by local, national or international laws, or in standards that are specified in local legislation. As shown in figure 4.1, breach of tort law by an engineer comes under the headings of **negligence**, where the engineer fails to act as a 'reasonable person' would, and **malpractice**, where the engineer's failure is of a professional nature and does not measure up to the standard of practice that would be expected of a similarly qualified engineer working in the same location.

FIGURE 4.1 Comparing two bases for legal liability in engineering: commercial and tort

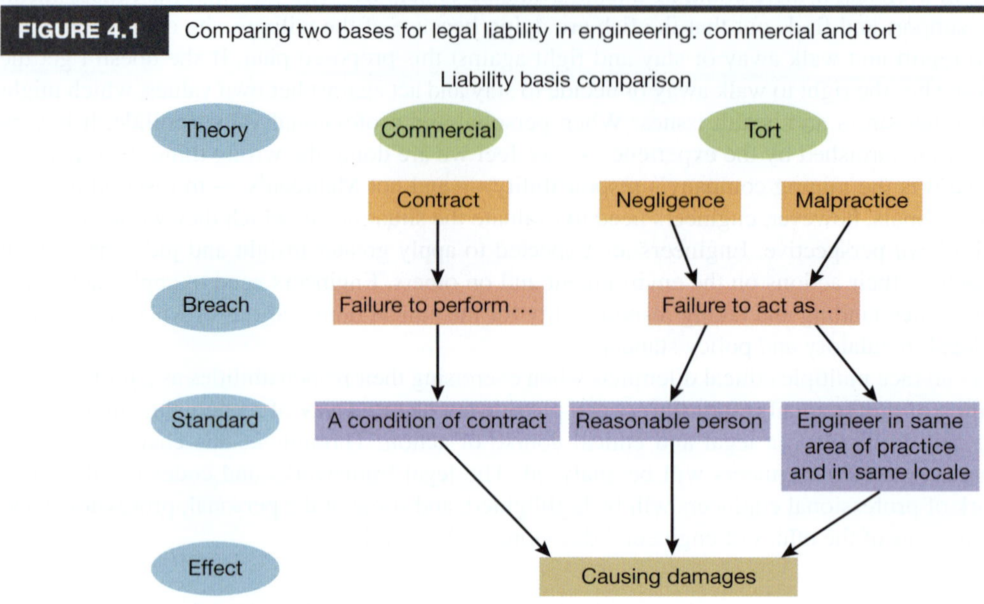

Source: Rubie (2009).

In cases of negligence and malpractice, it must first be proved that the engineer had a 'duty of care' for the affected person or people. A **duty of care** can be difficult to define clearly, and many law firms have built healthy bank balances on the back of protracted legal arguments about its precise definition and application. The Australian National Offshore Petroleum Safety Authority (NOPSA) defines duty of care as 'a legal obligation to have thought or regard for those who may be affected by one's acts or omissions'. In proving an engineer has acted negligently or with malpractice, it must secondly be shown that the engineer breached their duty of care. NOPSA explains how engineers and managers may avoid breaching duty of care: 'Implementing the duty of care principle means planning and taking measures which prevent workplace accidents, injuries and illnesses' (NOPSA 2009). One of the words to note here is *prevent*; this means an engineer can breach their duty of care by failing to think through and mitigate the possible risks and dangers that those in their workplace, or those using their technology (products or processes), might encounter. For example, an electronics engineer designing a solar-powered electric bicycle may have a duty of care to think through the impact if a user attempted to recharge their bike on a cloudy day by plugging into their 240V domestic power supply. The third point that must be proved to show negligence or malpractice is that the affected person experienced negative consequences because of the engineer's breach. An example of this would be where the cloudy-day cyclist found their bicycle seat had become electrically charged and suffered shocking consequences!

As professionals, engineers have a duty of care for the health and safety of others. As with other professions, an engineer's duty of care extends well beyond the people who report directly to them. An engineer in a professional situation has expert knowledge, and thus has an obligation to take action whenever they feel harm is likely to occur. During laboratory classes in undergraduate engineering, you may have a chance to exercise your duty of care for students who are studying with you. Hazards in the lab that should be reported to the lab manager may include faulty electrical equipment, unlabelled or poorly stored reagents, tripping hazards, safety guards that are missing or easily bypassed, and heavy equipment stored at height. You may like to ask the lab manager about the health and safety audit procedures for labs at your university. As a graduate engineer, you will be expected to keep duty of care concerns at the front of your mind as a guide for the work you do.

The following spotlight provides an example of how lawyers evaluate potential cases of tort breach (negligence or malpractice) in the field of engineering. Hembrow (2014) explains that conflicts arise when pathways are shared by pedestrians and cyclists and these complaints:

> are from pedestrians who find the speed of cyclists unacceptable on paths which they use for walking. This is a wholly avoidable problem . . . Where conflict between cyclists and pedestrians occurs, it is almost always due to cyclists being forced to use infrastructure which is not designed for them at all.

SPOTLIGHT

Shared paths and the role of engineers

The following is an extract from a letter of advice written in 2008 by lawyers Slater and Gordon (NSW) in response to a query on the 'civil liability of roads authorities arising from collisions between pedestrians and bicyclists in shared bicycle paths'. The query was sent to Slater and Gordon by Mr Harold Scruby, then-Chairman and CEO of the Pedestrian Council of Australia Limited (Creed 2008). The (abridged) letter shows the process of trying to ascertain which body — the local government or the state government's Roads and Traffic Authority — is responsible for shared footpaths, the role engineers can play as 'expert witnesses' and how lawyers rely on legal precedent (the judgements delivered in previous cases) for establishing whether there are grounds for legal action when a client is claiming engineering malpractice.

> Dear Mr Scruby
>
> . . . in assessing whether a breach of duty of care has occurred, a Court would also consider the existing statutory or regulatory framework applying to the design, construction and use of Shared Bicycle Paths including *Austroads Guide to Traffic Engineering Practice, Australian Road Rules* as well as any offences applying to bicycle use under the *Road Transport Legislation*.
>
> . . . [Precedent was established in] . . . a civil claim brought by Maria Guliano against Leichhardt Municipal Council and the RTA as a result of being struck by a bicyclist in a Shared Bicycle Path on Iron Cove Bridge on 7 March 2002.
>
> . . .
>
> It was contended on the basis of expert engineering evidence that the design of the Shared Bicycle Path in relation to the point of merger of the top landing of the pedestrian stairway and shared path was inconsistent with proper transport engineering standards.
>
> . . .
>
> The expert engineer was also of the opinion that the speed travelled by the bicyclist of 20 km per hour was unsafe for a Shared Bicycle Path.
>
> . . .
>
> I am therefore of the opinion [in response to Mr Scruby's query] that local government road authorities may be found to be in breach of duty of care for failing to impose safe speed limits for bicyclists on Shared Bicycle Paths.
>
> . . .
>
> Allegations of breach of duty of care based upon the design or configuration of Shared Bicycle Paths may also be successful . . .
>
> Yours faithfully,
>
> . . .

CRITICAL THINKING

As a first-year engineering student working on a team project, could you be accused of professional malpractice? In such a project:

- who do you have a duty of care to?
- what breaches or omissions would constitute malpractice (as opposed to negligence)?
- what negative consequences would your team mates suffer?

As is shown in figure 4.1, engineers face two main forms of liability: tort-related and commercial. Commercial liability is the loss that arises from non-compliance with a contract. This form of liability arises in the context of a signed commercial agreement (contract) in which two parties agree that certain activities are to take place within a specified timeframe and some kind of payment is involved. With a commercial breach, the professional engineer fails to comply with a term or condition of the contract, and their failure to comply causes the other contracting party (i.e. the client) to suffer damages. In this definition, the term *damages* means a loss that can be financial or of another nature. Figure 4.2 shows several examples of damages that can arise during an engineering construction project. All of the damages listed need to be thought of prior to the drafting of a contract, and provisions made for preventing or remediating these damages would generally be included in the contract for the body of work that the professional engineer, or engineering company, agreed to undertake.

FIGURE 4.2 Example of damages that can arise from breach of commercial contract

1. Payments (for goods and services, and scheduled payments)
2. Warranties (quality of work completed)
3. Liquidated damages (delays to the schedule)
4. Indemnities (precluding parties sue for commercial damages)
5. Consequential damages (secondary effects of the work)
6. Accidents and pollution (resulting from execution of the work)
7. Insurances (risk management and limitation of financial liability)
8. Jurisdiction (compliance with regulation, legal basis of work)
9. Dispute resolution (economic loss)

Source: Rubie (2009).

From the list in figure 4.2, you can see why engineering contracts are often very long and convoluted documents. Imagine a contract for making a simple cup of black tea, or 'the timely and safe delivery of one insulated receptacle of hot beverage (omitting sweetening products and dairy products, in all their forms)'. The cup of tea contract would run to many pages if all of the risks and potential losses listed above were to be addressed. Further information about contracts and the part they play in engineering work is provided in the chapter on communicating information.

4.2 Work health and safety (WHS) and personal liability

LEARNING OBJECTIVE 4.2 Detail the ways engineers can avoid, manage and remediate WHS risks, discuss how engineers can be held personally liable for their decisions and discuss the advantages and drawbacks of risk assessment frameworks when working globally.

Mistakes, injuries and accidents occur at work. A large proportion of these can be traced back to poor design of equipment or systems. Poor design in this case does not mean technically poor, but poor from a user's perspective. Engineers and engineering managers must be aware of the relationship between technical factors and human factors when considering their decisions related to safety in the workplace. The *Australian workers' compensation statistics 2020–21* records a total of 130 195 serious workers' compensation claims in that year (Safe Work Australia 2022, p. 6). A serious claim is defined as 'an accepted workers' compensation claim for an incapacity that results in a total absence from work of one working week or more' (Safe Work Australia 2022, p. 8). Employees working as health care and social assistance, construction and manufacturing employees represented the highest incidence of all occupation groups: 44 per cent of claims despite being only 29 per cent of the workforce (Safe Work Australia 2022, p. 13). Statistics can be bland, and tend to blunt the impact and gravity of the incidents they report. 'Compensated fatalities' translate to real human costs: leaving families and loved ones bereft. Similarly, serious injury can have lasting effects on the physical comfort, earning capacity and emotional wellbeing of those who suffer them, and ripple effects from the physical and psychological damage can affect the injured person's families, friends and community. For these human reasons alone, **work health and safety (WHS)** (also known as occupational health and safety) is a serious concern to the ethical professional engineer. WHS can be defined as keeping people safe at work while working. The legislation that governs an engineer's responsibilities for WHS differs between Australia and New Zealand. As a graduate engineer, you may choose to live and work overseas. Along with the fun of meeting new workmates and learning to engineer in a new country or culture, comes the responsibility to seek out and familiarise yourself with local WHS legislation and figure out how to fold your WHS responsibilities into every aspect of your new job.

Given the human cost of lapses in WHS and the legislative requirements that engineers need to follow, it becomes important to understand the particular role of engineering design in avoiding WHS problems. Engineers do their best to design processes and technologies to solve problems; however, design deficiencies can inadvertently create accidents and injury. Figure 4.3 details some of the ways in which accidents can result from poorly thought through engineering design.

- *Operator error*. The way a product or process is designed leads the user to operate it in a particular way. If the process is not intuitive or goes against the normal way that the user has operated the process in the past, it increases the likelihood of user error or of the user operating the product or process in the wrong way. This is particularly important in emergency situations.

- *Adverse environment.* If the product is not designed appropriately for the conditions, then accidents can occur. One aspect of this is that choice of material can depend on the environment where the product will be used. For example, aluminium cannot be used in underground coal mines because an incendiary spark can result from impact between some aluminium alloys and steel objects.
- *Design deficiency.* Has the product or process been designed for the user? The designer (and often the client) may not be the 'typical user', and therefore the designer must be aware of who the typical user is, and what their requirements are. For example, a typical user of a prosthetic leg is an amputee. Much retrofitting design that occurs is due to the designer failing to investigate and take account of typical users' requirements.

FIGURE 4.3 Engineers as causers of accidents

Engineering deficiencies that might contribute to or cause accidents

Adverse environment	Poor management	Malfunction and failure	Operator error
Poor maintenance	Miscellaneous other	Lack of knowledge	Design deficiency

Source: Prue Howard (2011).

SPOTLIGHT

Creating recycling facilities in a post-conflict country

The organisation Waste for Life (wasteforlife.org), worked together with the University of Western Australia and several local universities to support the development of local community waste-based businesses in Sri Lanka from 2014 to 2017. The project was funded by the Australian Department of Foreign Affairs and Trade and supported communities in learning how to hot-press plastic waste and reinforcing fibres into composite products to sell on the local market.

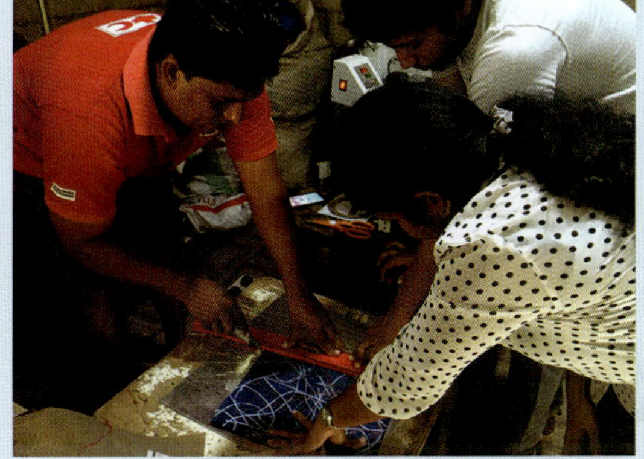

The Waste for Life team in Sri Lanka teaching a community group to make products from waste.

A three-year feasibility study was conducted before the project was attempted and a risk assessment matrix was created at the start of the project. This used a simple risk assessment matrix, whereby the likelihood of the risk is multiplied by the potential consequence. Table 4.1 shows four items from this initial risk assessment. The risk being considered here was 'risk to the success of the project' to determine whether the funders should invest in this project. Some risks considered would be to workers health and safety but others to whether the project might go ahead or be delayed.

At this point, the team was concerned about delivery of equipment and manufacture of the equipment to appropriate standards as well as basic health and safety in its use. There were also references to the potentially volatile political context (the country emerged from a 26-year civil war in 2009).

What is never clear at the beginning of a project are the real risks that might occur; hence, one must predict all potential issues based on past events within the country or similar contexts. When the project went ahead there were many risks that occurred that had not been specifically identified at the beginning. These included flooding on roads, which prevented workers from getting to work, workers whose houses flooded requiring relocation (and even risk of snakes and crocodiles), dengue fever rapidly increasing as a health concern in Sri Lanka (even more than predicted) due to climate change and affecting the health of many team members, the past conflict in the north causing a reluctance of team members to travel across country, and new political disturbances and fuel crises creating delays.

TABLE 4.1 Four risk items from the Waste for Life Sri Lanka risk assessment

End of year outcome/end of activity outcome	Risk no.	Risk: event, source & impact (What can happen (event); how can that happen (source); what will the impact be if it happens?)	Existing controls (What is currently in place?)	Risk rating with existing controls in place			Is risk rating acceptable? Y/N (If no, please propose treatments)
				Consequence	Likelihood	Risk rating	
Composites processing facility implemented in three locations	1	Equipment does not get delivered — caused by issues with source. Equipment does not get manufactured to specification — caused by faulty manufacturing. No adequate space is provided for the processing facility. Facility completion would be delayed while resolving the issues.	Verification of reliable source using advisory panel expert. Expert on advisory panel will oversee local manufacturing for quality assurance. University sends a letter of support committing them to provide space for the facility.	Moderate	Unlikely	Moderate	Y
	2	Project does not get immediate support of the local government, which results in delays to start of project.	Government official has agreed to sign proposal.	Major	Unlikely	Moderate	Y
	3	Travelling to parts of the north of the country may incur some risk.	Transport plan for all staff involved in the program is implemented.	Major	Unlikely	Moderate	Y
	4	There are health and safety issues related to the use of the facilities.	Safeguards will be put in place in all processes and procedures from previous risk assessment of health and safety use of such equipment.	Major	Rare	Moderate	Y

The project was, however, extremely successful despite these delays and concerns, and the project manager became a champion at innovative workarounds. However, it serves as a lesson to those working in sites very different to their own that it is never possible to predict everything using the risk assessment matrix, and it is important to track projects in similar contexts and prepare for the unexpected.

CRITICAL THINKING

During his final year of studies, a student from University of Western Australia went to Sri Lanka to work on the Water for Life project to conduct an internship. After graduation, he returned to work for the project for one year as an engineer, in charge of research and production of the waste-based materials. Imagine you are this student. You have no experience of working outside the laboratory and you are now meeting different people every day who treat you with respect because you are a professional engineer, and they expect you to have all the answers. Think about what you would do if the project is progressing too slowly because your team do not come to work regularly — due to almost weekly public holidays, illnesses, floods and so on — and when they do, they only do exactly what they are asked. You find you have to micromanage all the work, even though you are needed elsewhere to be planning and coordinating. What do you say to your boss, who is the overall project manager? What strategies can you take with your team? What other chapters in this text can help you prepare for such a situation?

One of the key concerns for an engineer in preventing workplace injuries and deaths is eliminating or engineering out hazards. As a professional engineer, you would be expected to use formal analytical frameworks to think through how a WHS risk could or should be designed out of engineered products, technologies or processes. An engineer faced with solving a problem would first employ a risk-assessment framework to:

- identify hazards
- assess risks
- determine control measures
- check whether the new control measures introduce new hazards.

While consideration of hazards is a serious business in professional engineering, you may like to think back to the idea of a contract for making a cup of tea — what are all of the hazards that could befall the tea maker, the tea consumer and any other people in the general vicinity of the kitchen where the tea is made and consumed? If you get creative, you'll be able to think up lots of sensible and silly scenarios for tea-related hazards. Hot water burns are an obvious risk, but what about the risk of slipping on the teabag, drinking the teaspoon or setting off the tea drinker's hot water allergy?

Risk assessment and management are about thinking through the likelihood that one of the hazards you have thought of will occur, and the degree of harm that would happen if it did. Considering tea again, you may like to think through the likelihood and harm of hot water burns. Risk assessment and management are addressed in detail in the chapter on the engineering method — you could read those sections now and think through how risk assessment would look in the case of the hazards identified earlier.

In determining control measures, there is a hierarchy aimed at creating safe places as opposed to safe persons. In other words, good engineering design creates workplaces and equipment that:

- make it difficult for people to have accidents
- don't rely on people having special knowledge to stay safe
- account for the fact that people are creative, distractible and inquisitive and will often bend the rules or take risks where they can see some benefit from doing so.

Table 4.2 shows the hierarchy of control measures.

TABLE 4.2 **Hierarchy of control**

Control measure	Description
Elimination (the most desirable option)	If you eliminate a hazard you completely eliminate the associated risk
Substitution	You can substitute something else — a substance or a process — that has less potential to cause injury
Isolation/engineering	You can make a structural change to the work environment or work process to interrupt the path between the worker and the risk

(continued)

TABLE 4.2 *(continued)*

Control measure	Description
Administrative	You may be able to reduce risk by upgrading training, changing rosters or other administrative actions
Personal protective equipment (the least desirable option)	When you can't reduce the risk of injury in any other way, use personal protective equipment (gloves, goggles etc.) as a last resort

Source: Prue Howard (2011).

The hierarchy of control measures suggests the first and best course of action would be to eliminate the hazard, which means to remove the hazard entirely, but in the case of crocodiles in floodwater or dengue fever personal protective equipment may be the only remaining option.

Product recall

In order to avoid accusations of engineering malpractice and related legal action, companies sometimes launch a product recall. The term *product recall* is most often used in cases in which a manufactured item is found to be unfit for the purpose for which it was manufactured (e.g. headache tablets that have no effect on headaches, an air fryer that does not cook food or a software package with fatal bugs) or constitute a risk to users (e.g. baby cots that fold up with their occupants in them, or chewy dog toys that disintegrate and result in dogs choking to death).

Consider the dilemma for an engineering company that finds it has manufactured a product that is not fit for its purpose or that represents a direct or indirect risk to consumers. It is expensive to recall a product, because customers must be reimbursed for the product and the company may need to bear the cost of freighting the faulty product to safe disposal. It is also expensive in terms of a company's reputation, in that customers may be reluctant to continue to buy products from a company that has previously sold them a faulty item.

Consider the cost to consumers and a company when a product that *should* be recalled is not. In cases in which a company fails to recall faulty product or a product that is creating harm through misuse, the legal fallout and damage to reputation can be phenomenal. One well-documented case of this is the decision by multinational food company Nestlé to promote and distribute infant formula (powdered milk) to mothers in the developing world. It came to public attention that it was often impossible for mothers in the developing world to sterilise bottles and water and, hence, for them to make milk substitutes that were safe for their infants to drink (Newton & Schmidt 2004). In this case, Nestlé experienced massive consumer boycotts throughout the 1970s and 1980s, and were still involved in debate on the topic into 2012. The take-home message from the Nestlé experience is that there can be very real commercial consequences when the public perceives a company is not upholding its responsibility for monitoring and intervening to ensure the safe use of its product.

Several well-documented cases exist in the motor vehicle industry of vehicles that some consumer groups felt were faulty and should have been recalled. A recent case of recall involves the Takata airbag. Fifty million airbags were recalled from 37 million vehicles worldwide because they can explode when deployed and there is a risk of metal fragments and shrapnel being propelled from the bag potentially causing death (National Highway Traffic Safety Administration 2018). The Australian Competition and Consumer Commission (2018a) lists the many car models under recall in Australia and note that they believe 'the Takata alpha airbags have a manufacturing flaw, as they were not produced according to design standards and do not deploy as intended' (Australian Competition and Consumer Commission 2018b).

As a professional engineer, your career path may take many twists and turns, and you may use your high-level analytical ability and creative problem-solving skills to develop solutions for any number of technical and social problems. Through all of this, it will be essential to maintain awareness of the responsibilities that come with the honourable tag of 'professional'. These responsibilities include complying with the legislative conditions, standards and codes set down for practice within your field; and paying close attention to those for whom you have a duty of care — both in your workplace, and in the communities that are affected by the engineering decisions you make. Beyond your legal responsibilities as a professional engineer, you will also have responsibility to practise in a way that is ethical. In the following section, we explore engineering ethics and come to understand how ethical reasoning can help the professional engineer in cases where the correct technical decision may be obvious, but what would constitute a 'good' decision or outcome is much less clear-cut.

4.3 Engineering ethics

LEARNING OBJECTIVE 4.3 Explain the influence of values on ethical decisions.

During discussions with your classmates, you may have realised different individuals hold different ideals, values and morals, and that these ideals, values and morals may suggest different courses of action. An example would be where one student's personal moral code would stop them from cheating on an assessment task, and another student would view cheating as a clever way to save time and effort. It is important for professions to ensure that their members practise in a way that is ethical and that does not bring the profession as a whole into disrepute. It is also important that all members of a profession know and understand what the agreed and shared ethical standards are, and know how to apply these agreed standards in their day-to-day work. For these reasons, professional engineering organisations publish codes of ethics. A commitment to abide by a registering body's code of ethics is part of the process of registration as a professional engineer in Australia or New Zealand. In this section, the Engineers Australia (2022) *Code of Ethics and Guidelines on Professional Conduct* (Engineers Australia Code of Ethics) will be discussed and contrasted with the Engineering New Zealand's (2016) *Code of Ethical Conduct* and the code (2020) that governs the practice of engineers who register with the Institution of Electronics and Electrical Engineers (IEEE).

The Engineers Australia Code of Ethics provides guidance for engineers on deciding how to act in situations where a 'good' course of action is hard to work out, or where there is the potential for unprofessional or illegal conduct. Examples of this include dishonest dealings or corruption, inadequate consideration of financial risk or risks to people, and conflicts of interest — or the potential for an appearance of a conflict of interest. Until 2010, the Engineers Australia Code of Ethics was lengthy, with three Cardinal Principles and nine Tenets (Engineers Australia 2000), but the Code of Ethics, released in 2010, took a different line. It is made up of only four main values and principles (figure 4.4) and, like the Engineering New Zealand *Code of Ethical Conduct*, which is discussed in the following section, includes an expanded section of Guidelines. The Guidelines are not binding, but intended to help professional engineers interpret and apply the values, which are binding but open to interpretation.

> Ethical engineering practice requires judgment, interpretation and balanced decision-making in context. Engineers Australia recognises that, while our ethical values and principles are enduring, standards of acceptable conduct are not permanently fixed. Community standards and the requirements and aspirations of engineering practice and members' behaviour more generally will develop and change over time ... Within limits, what constitutes acceptable conduct may also depend on the nature of individual circumstances (Engineers Australia 2022, p. 3).

The Engineers Australia Code of Ethics makes it very clear a professional engineer is required to behave with integrity, do their job competently, and be a leader in and credit to the profession as well as promote sustainability. Thinking back to the chapter on sustainable engineering, you will recall that sustainability is defined in many different ways; in figure 4.4, the meaning encompasses consultation, human and environmental welfare, and intragenerational equity. Given your own understanding of sustainability, how well does this meaning fit with yours? Will you have any difficulties aligning your own view of sustainability with the requirements of the new Engineers Australia Code of Ethics?

FIGURE 4.4 Engineers Australia Code of Ethics

1. Demonstrate integrity
   - 1.1 Act on the basis of a well-informed conscience
   - 1.2 Be honest and trustworthy
   - 1.3 Respect the dignity of all persons
2. Practise competently
   - 2.1 Maintain and develop knowledge and skills
   - 2.2 Represent areas of competence objectively
   - 2.3 Act on the basis of adequate knowledge
3. Exercise leadership
   - 3.1 Uphold the reputation and trustworthiness of the practice of engineering
   - 3.2 Support and encourage diversity
   - 3.3 Make reasonable efforts to communicate honestly and effectively to all stakeholders, taking into account the reliance of others on engineering expertise

4. Promote sustainability
    4.1 Engage responsibly with the community and other stakeholders
    4.2 Practise engineering to foster the health, safety and wellbeing of the community and the environment
    4.3 Balance the needs of the present with the needs of future generations

Source: Engineers Australia (2022, p. 2).

An interesting point to consider is that the Engineers Australia Code of Ethics does not explicitly discuss the professional engineer's responsibility to further the interests of their clients and employers. In fact, the Code of Ethics appears merely to require honest, competent service. Should a Code of Ethics exhort engineers to excel or to offer their employers innovation or loyalty? Is a Code of Ethics about excellence or minimum acceptable performance? Is a Code of Ethics about the engineer's relationship with their employer or with the wider community? These are important questions because answering them may give you greater insight into what a Code of Ethics is, and is not, for.

KEY POINT

A commitment to abide by the registering body's code of ethics is part of the process of registration as a professional engineer in Australia and New Zealand.

SPOTLIGHT

Design for the dump

The provocative films created by The Story of Stuff Project (2024) take us through the life cycle of a variety of common products such as water bottles and computers, focusing on end of life and waste disposal. The creators of these films are asking us to consider the question of responsibility for this waste and believe that the cost of the recycling or safe disposal of products at end of life should rest with the manufacturers. They believe this would assist with designs that would help create solutions to waste problems rather than the current 'design for the dump' mentality. In *The story of electronics* (2011) we are asked to think about the tangled mess of old computer and camera leads and chargers that we have in our drawer, and of the printer, which requires replacement ink cartridges that cost more than a new printer. We are further asked to think about what happens to all of these at the end of life, once we no longer require them. They are rarely recycled locally but are shipped off to countries in the Global South such as China, where communities are all too happy to retrieve the valuable components, but at a huge cost to their health. Any remaining parts that are unwanted are often simply burned in open dumps.

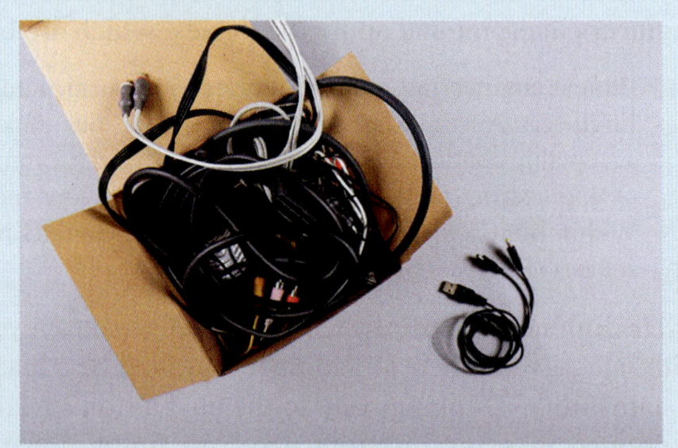

Designers are rewarded for the functionality of the products but not their longevity. We are convinced that buying the next greatest gadget is the most desirable way to live, not keeping hold of our outdated phone and getting laughed at, even if it still works and we are used to it. Such designers rarely need to worry about the end of life of a product, unless they are forced to through regulations such as the Land Fill Tax, which was introduced in the United Kingdom in 1996 and resulted in a lot of research into environmentally friendly materials (Baillie 2004). The film shows us that engineers can change their philosophy from 'design for the dump' to 'design to last'.

CRITICAL THINKING

The lobby against 'design for the dump' would say that engineers have the capability but the companies they work for ask them to design for profit rather than longevity and potential for recycle and re-use. How does this fit with an ethical responsibility of an engineer? Some would say that engineers have a duty to the shareholder to create profit. Others would say that they have a duty to the greater public to create products, which are cheap, regardless of their impact on the environment. Yet others would be more concerned about the conditions of workers who manufacture these low-cost products, often in developing countries with no national laws to protect them. Do you think a company should value workers' health and working conditions, clean air, sustainability of natural resources, access to low-cost commodities or profit for shareholders?

As you can see, this is where the issues of ethics and their relationship to engineering and political positions get muddled. We might have a code of ethics that we all agree with but the interpretations of the code are wide and varying. Whose values are right? What happens if your personal values are in conflict with the professional values of your company?

As indicated in figure 4.4, an ethical engineer not only holds integrity and sustainability foremost, but has a responsibility to be aware of, and work within, the limitations of their knowledge (and also to continue to maintain their knowledge) (Tenet 2). This tenet is enforced by Engineers Australia through the continuing professional development (CPD) requirements for engineers registered under Engineers Australia. You will learn more about CPD in the chapters on self-management and your engineering future.

SPOTLIGHT

Learning from nature's design principles

The critical principle of sustainability is not as easy to fulfil as we might think. Some say that sustainable development is an oxymoron, that development as understood by many as growth in production capacity cannot be sustainable, that we must in fact reduce consumption and produce less. As the Story of Stuff authors suggest, we might therefore consider new ways of designing for sustainability or design for less. Nature is very good at doing this. Trees don't take in more than they need and they produce no waste.

Modern remote spinifex bough shed construction at Dugalunji Camp, 2011. Photo by Tim O'Rouke.

Learning from nature or 'biomimetics' (Biomimicry Institute 2018) has been a key area of science and engineering for about 40 years. First Nations cultures have been learning from nature for tens of thousands of years. In Queensland, a team of engineers, architects and anthropologists from the University of Queensland and a group of traditional landowners (Indjalandji, Wakaya, Bularnu and Alyawarr peoples) collaborated to revitalise the ancient Aboriginal knowledge of spinifex grass, which is embedded in myths and sacred histories (Memmott, Martin & Amiralian 2017). The team are learning from spinifex to develop 'self-reliant, self-healing building technologies that can operate in an environment of energy poverty and water scarcity' (Memmott et al. 2017, p. 189).

CRITICAL THINKING

What engineering practices do you think are unsustainable? What could be done to make these more sustainable and who is responsible to ensure these changes take place? How can nature help inspire us to think sustainably?

The Engineering New Zealand *Code of Ethical Conduct*

Engineers who gain professional registration through Engineering New Zealand are required to adhere to the Engineering New Zealand *Code of Ethical Conduct*. The current code came into force in July 2016 with very specific obligations for its members who are required to report potential adverse consequences — engineers must take action if they observe something of concern (Engineering New Zealand 2016). They are also required to report breaches of the code by other engineers. There is also a specific requirement to keep up to date with professional development and to treat others with respect and courtesy. There are eight significant parts of the code that must be upheld as follows.

1. Take reasonable steps to safeguard health and safety.
2. Have regard to effects on environment.
3. Report adverse consequences.
4. Act competently.

5. Behave appropriately.
6. Inform others of consequences of not following advice.
7. Maintain confidentiality [unless it would necessitate a breach of the ethical code of conduct].
8. Report breach of code (Engineering New Zealand 2016).

It is interesting to compare the Engineers Australia Code of Ethics with the Engineering New Zealand code. The layout and content are different, in addition the emphasis of the Engineering New Zealand code is on the obligation to report adverse consequences and actions of engineers, whereas the Australian code only implies this duty (e.g. being honest, and maintaining the good reputation of the engineering profession).

Some discipline-based professional societies have codes of ethics that are specific to the practice of engineering in that field. Where an engineer registers with two professional bodies — their national body and a professional body specific to their discipline — the engineer must reconcile the ethical standards of both bodies. An example of this would be an electrical engineer working in New Zealand who registered with Engineering New Zealand and with the IEEE.

The *IEEE Code of Ethics*

The IEEE represents over 430 000 members in more than 160 countries. The scope of the IEEE has substantially broadened from its initial focus on electronic and electrical engineering. The body now serves engineers in all engineering disciplines related to electrical or electronic engineering. Can you think of many facets of engineering that do not use these fundamental technologies? As is apparent from the *IEEE Code of Ethics* (IEEE Code) reproduced in figure 4.5, IEEE members have a range of generic ethical responsibilities that do not differ substantially from the codes already discussed. One difference, however, is the ethical duty to educate the public about technology, its use and consequences (Tenet I.2). This is a specific mandate compared with the ethical responsibilities on communicating about the impacts of technology described in the Engineers Australia and Engineering New Zealand codes. Another notable ethic is the explicit rejection of bribery (Tenet I.4). The Engineers Australia and Engineering New Zealand codes require members to act with 'integrity' and 'honesty', which would preclude involvement in bribery, but it is interesting that the IEEE Code is very direct and explicit in naming bribery as unethical. We will return to the ethics of bribery later in the chapter in the section on culture and ethics.

FIGURE 4.5 The *IEEE Code of Ethics*

We, the members of the IEEE, in recognition of the importance of our technologies in affecting the quality of life throughout the world, and in accepting a personal obligation to our profession, its members, and the communities we serve, do hereby commit ourselves to the highest ethical and professional conduct and agree:

I. To uphold the highest standards of integrity, responsible behavior, and ethical conduct in professional activities.
    1. to hold paramount the safety, health, and welfare of the public, to strive to comply with ethical design and sustainable development practices, to protect the privacy of others, and to disclose promptly factors that might endanger the public or the environment;
    2. to improve the understanding by individuals and society of the capabilities and societal implications of conventional and emerging technologies, including intelligent systems;
    3. to avoid real or perceived conflicts of interest whenever possible, and to disclose them to affected parties when they do exist;
    4. to avoid unlawful conduct in professional activities, and to reject bribery in all its forms;
    5. to seek, accept, and offer honest criticism of technical work, to acknowledge and correct errors, to be honest and realistic in stating claims or estimates based on available data, and to credit properly the contributions of others;
    6. to maintain and improve our technical competence and to undertake technological tasks for others only if qualified by training or experience, or after full disclosure of pertinent limitations;

II. To treat all persons fairly and with respect, to not engage in harassment or discrimination, and to avoid injuring others.
    7. to treat all persons fairly and with respect, and to not engage in discrimination based on characteristics such as race, religion, gender, disability, age, national origin, sexual orientation, gender identity, or gender expression;

8. to not engage in harassment of any kind, including sexual harassment or bullying behavior;
9. to avoid injuring others, their property, reputation, or employment by false or malicious actions, rumors or any other verbal or physical abuses;

III. To strive to ensure this code is upheld by colleagues and co-workers.

10. to support colleagues and co-workers in following this code of ethics, to strive to ensure the code is upheld, and to not retaliate against individuals reporting a violation.

Source: IEEE (2020).

The idea of discrimination, oppression or marginalisation of one group of people by another is often an unfamiliar idea to us; even if we have been experiencing it (or doing it), we may not be aware of it.

Interpreting and applying codes of ethics

As discussed in the previous section, part of the process of learning engineering is working out how to *align* your personal moral stance and your values with the professional ethical code you will come under as a graduate engineer. A code of ethics is not negotiable, but it is open to (and requires) interpretation. Aligning your personal moral stance with a code is unlikely to be particularly difficult because most engineering codes of ethics are reasonably generic and commit members to uphold the law, serve the community and maintain professional integrity. Learning to *apply* engineering ethics is the more difficult aspect. In Australia, students are expected to demonstrate a 'commitment to uphold the Engineers Australia — Code of Ethics' (Engineers Australia 2019, p. 6) by the time they graduate, but some would suggest that deep understanding of the relevant code and wrestling with real-life ethical dilemmas as a professional engineer are required in order to really master the application of ethics in practice. Stephan (2001, p. 12) states:

> [Is it] right to claim that graduates are truly prepared for a lifetime of engineering work with only a superficial understanding of engineering ethics? Unlike more technical skills whose half-life may be measured in years or months, the ability to do engineering ethically is something that can improve with years of experience.

Ethical challenges can vary greatly and have multiple dimensions. Sometimes, it is not clear that the discomfort engineers feel in a professional situation is an ethical dilemma, and it may often be difficult determining how to analyse and make a decision in a situation requiring interpretation and operationalisation of a qualitative statement like a code of ethics. Consider the next spotlight about an engineer working in an aluminium plant, facing multiple ethical dilemmas.

SPOTLIGHT

Reporting a leaky pipe

Sally* was an engineer working for a large aluminium producer. Her team was responsible for maintaining two long parallel pipelines carrying a slurry of alkaline sand on its way to the next stage of processing. This sand had a strong eroding effect on the pipelines. To monitor and manage this erosion, the team carried out numerous and frequent non-destructive testing (NDT) on the pipeline. They used the results of this testing to identify sections of the pipeline needing to be replaced due to wear. One year at budget time, the team reported to the plant manager that both pipelines needed changing out due to significant wearing of the pipe. The impact of thin pipeline is an increased likelihood of leakage from the pipe, leading to caustic slurry leaking into the environment. This loss of containment would need to be reported to the government as a breach of the plant operating licence. Sally and her team knew that the pipe had to be replaced or there would be spillage. She also knew that although the line replacement would not have involved production downtime, it would have involved significant cost. When she reported this to her plant manager, he said 'No. Manage it'. As predicted, sometime later there were spills that were reported to the Department for the Environment and ultimately ended up being on the front page of the newspaper.

Incident investigation found that one of the NDT tests was not followed up properly and decided that it was all the fault of the maintenance superintendent, who worked directly for Sally. Sally defended the maintenance superintendent when called to a meeting at Head Office. She told them there were many NDT tests reporting thin pipe that needed to be changed out. They all said the same thing — the pipeline sections needed changing — and this had been reported to the plant manager. An area manager was brought in to take over the role of Sally for this particular issue and to 'fix up the problems'. This new area manager and Sally worked together to change out the pipelines and during the process Sally learned that the plant manager had denied any knowledge that the pipelines were thin and required replacement. When one of the pipes was replaced it was so thin that it crumpled under its own weight while being removed.

Some months later, the company promoted the plant manager. Subsequently, while at a social event, Sally bumped into the former plant manager, who was now in a position of great financial responsibility. He said, 'If you ever need any funds for anything, just let me know'. Understandably, this felt awful for Sally. What did he mean? She pondered this long and hard after the event. Did he really think she had not made a fuss to protect him? Was he saying that so that she would not now say anything about how she had in fact reported the issue to him in the first place and he had told her to 'manage it'?

After ensuring that the team was not going to suffer the consequences, Sally resigned from the company.

*Based on a true incident. A fictitious name has been used to protect the identity of the engineer and company concerned.

CRITICAL THINKING

Sally acted according to her conscience and engineering knowledge in reporting the need to replace the pipelines, in supporting the individual who was inappropriately blamed and in her decision to leave the company. Which tenets of the code of ethics was Sally upholding? What else could she have done? What would you do?

4.4 Ethical theories

LEARNING OBJECTIVE 4.4 Give examples of how moral theories can be applied in ethics.

The study of human behaviour, attitudes and values has a long and interesting history. For as long as there has been human reasoning and the capacity for analysis, people have tried to understand why individuals behave the way they do and how behavioural standards can be improved. As an engineer, developing the capacity to analyse and understand the ethics of your own decisions and behaviour will not only help you in improving your practice and avoiding litigation, it will also assist you in being a role model for good values. Developing the capacity to understand what motivates others to be ethical will help you to be a good manager and a responsive employee or consultant. In this section, we will define important terms in ethics, provide a brief explanation of the main moral theories that underpin ethical decisions, and touch on some tests from the fields of psychology and economics that give insights into why ethical behaviour is usually in an engineer's own best interests.

KEY POINT

Ethical dilemmas vary greatly and occur in the working life of all engineers.

Morals and ethics

In a discussion of engineering ethics, the terms 'moral' and 'ethical' are sometimes used interchangeably. However, the two terms have distinctly different meanings and understanding the difference helps in working out how to apply morals and analyse ethics in real life. A simple definition is that **morals** are the rules that govern how we behave towards other people. Newton and Schmidt (2004) explain that morals are rules or duties that most of us learn as children so that we do not:

- hurt other people
- tell lies
- take more than our fair share.

More formally, morals are, 'principles or habits with respect to right or wrong conduct' (Macquarie Dictionary 2024). To be moral is to pursue or practise 'right conduct' and to understand and respect 'the distinction between right or wrong' (Macquarie Dictionary 2024). A person's moral code develops as a result of their upbringing, culture, life experiences and ongoing evaluation of their own actions and reactions in the world. Morals are often implicit, subconscious or unexamined, and people tend to use them like reflexes. That is, they often use their moral code unconsciously to make decisions that feel good or right.

Ethics is 'a system of moral principles, by which human actions and proposals may be judged good or bad or right or wrong' (Macquarie Dictionary 2024). This suggests that morals can be thought of as a necessary basis or precondition to being ethical. In a sense, morals act as the raw material or ingredients that allow us to think through and 'cook up' ethical behaviours. Ethics can be further understood as the conscious questioning of how people apply morality in a given situation (Baura 2006).

In the workplace, the application of ethics by professionals is an attempt to quality assure or to adapt the application of a personal moral code. Ethics are often decreed or described by a professional body because employees, including engineers, have personal moral codes. So, members of a professional body need an explicit statement of the ethics that they all agree to abide by. Several codes of ethics were discussed earlier in this chapter. These codes attempt to explain and guide how engineers should reconcile ethical dilemmas. Supporting each of these codes is a mixture of moral theories that show different ways of making an ethical decision. These are now discussed and you may like to consider how each of these theories fits with the codes of ethics you read about earlier in this chapter.

Ethical egoism

This moral theory invites a person to think first and foremost of themselves and their own interests. The theory states that an act is intrinsically ethical if a person acts to protect or further their own interests (Holtzapple & Reece 2008). **Ethical egoism** may be used, for example, to justify self-serving or self-preserving acts, such as taking the last life jacket in a sinking ship, killing an axe-wielding maniac in self-defence or eating the last Tim Tam™ in a packet. All of these acts put the safety, health and happiness of the person who is making the decision ahead of others' safety, health and happiness (assuming a Tim Tam™ satisfies happiness in greater measure than it affects health).

The difficulty with ethical egoism as a founding platform for ethical engineering is that most ethical dilemmas in engineering do not concern just one individual. Engineering is a profession in which decisions are made that have an impact on workers, consumers and communities. Also, ethical egoists often find that their interpersonal relationships are compromised because colleagues cannot rely on them to take account of the needs of the team or company when making decisions. While it is tempting to practise ethical egoism, and while engineers may use this moral theory unconsciously at times, it is not a particularly useful platform for professional engineers.

Utilitarianism

Utilitarianism is a moral theory developed by philosophers Jeremy Bentham and John Stuart Mill. It holds that an act can be judged moral or immoral depending on its consequences. The objective of utilitarianism is to maximise human happiness and minimise human suffering. In analysing the ethics of a decision, happiness and suffering are represented by a relative scale. That is, while these things cannot accurately be measured or quantified, estimates can be made of how happy an action might make a group of people and this estimate can be compared with an estimate of likely human suffering. Utilitarians use relative measures of happiness, called *hedons*, and compare them with relative measures of suffering, called *dolors*. Hedons are calculated by estimating the benefit of an action and multiplying it by its estimated importance. Dolors are calculated by estimating the likely harm of an action and multiplying it by its estimated importance.

Utilitarian decision analysis attempts to optimise the happiness objective function. So:

$$\text{Happiness objective function} = \sum(\text{Hedons}) - \sum(\text{Dolors})$$

Referring to Mehreen's dilemma at the start of the chapter about balancing the needs of a community for fishing livelihoods and the aluminium mining venture, a utilitarian approach to deciding how to proceed ethically would suggest that Mehreen could attempt to estimate the social and economic happiness that the mining activity generates for the communities near the mine. She could then compare this with the social and economic significance of fishing for the local community.

Importantly, utilitarians attempt to calculate the overall happiness and suffering a target group will experience, and avoid considering the reality that the happiness and suffering may be unequally distributed within the group. Mehreen would need to work out the happiness objective function for the community, but not for the individual families who may be disproportionately affected (e.g. families who may lose their livelihood in the event of continuing fish kills).

Another shortcoming of utilitarianism is the obvious difficulty with accurately estimating and quantifying happiness and suffering. For example, can you quantify your happiness at passing an exam? If you can do this, would you be able to compare this quantity of happiness with the quantity of happiness or suffering from other events, such as spending the evening before the exam in a gaming marathon, or missing the exam to participate in a hockey team grand final? Another consideration is whether estimates of hedons or dolors are universal (i.e. if individuals' quantity of happiness or suffering is equal for a given event).

Duty ethics

Duty ethics is a moral theory that states an act can be viewed as ethical if it is:

- conducted for the sake of duty
- guided by maxims that could be seen as universally applicable
- a respectful way to treat humanity (Baura 2006).

This position was originally described by Immanuel Kant, who felt that what was morally good should be distinct or detached from self-serving motives. Clearly, Kant would not have been in favour of ethical egoism. Kant felt ethical people should be guided by their duty to others and should use maxims (subjective rules) that could be universally applied (applied to *all* decision making in a way that was consistent, impartial and fair). Unlike utilitarianism, this moral theory suggests that consequences should not be considered, and that the quality, fairness and consistency of the decision process are of greatest importance.

Rights ethics

In some ways, **rights ethics** is similar to duty ethics. Like duty ethics, rights ethics has as its focus the process used for decision making and holds that the decision maker need take little or no responsibility for considering the consequences of an action. In rights ethics, the main foundation for making an ethical decision is upholding the rights of all parties to a decision. Rights vary by country and culture. In Australia and New Zealand, rights that are commonly assumed to exist include the right to life, the right to free speech (although this is not enshrined in Australian law), the right to an education, the right to individual ownership of property and the right to vote (although this is limited to certain age classes and can be withdrawn in the case of a serious criminal offence). Some of these are legal rights and some are rights that are defined for citizens in the Australian and New Zealand constitutions; others are termed *natural rights* (rights assumed to be ordained by God, discoverable in nature or history, or essential to human progress) and are seen as so self-evident that there need be no debate or justification for upholding them.

In most Western democracies, the legal system enshrines, defines and upholds individual rights. However, it is not always possible to uphold the rights of all parties. Consider the construction of a sewerage pipeline or municipal waste treatment facility. It is likely individual residents' property rights and right to live without undue disturbance from odour or noise could be infringed in the case of excavation for sewer lines or the establishment of a landfill and waste treatment centre. In this example, the rights of individuals would conflict with the rights of the community to safe sanitation and waste disposal.

The following spotlight could be viewed as an example of the outcome of a rights ethics position — where the financial profits a company was duty bound to generate for shareholders were placed before their moral responsibility for the impact of their operations on past employees and members of the wider community.

SPOTLIGHT

James Hardie and asbestos-related disease

Consider construction materials manufacturer James Hardie's lengthy battles with victims of asbestos-related disease. Asbestos was once considered a durable, low-flammability, low-cost construction material and was widely used in residential and commercial buildings. Many asbestos-related disease victims worked or lived at James Hardie's asbestos mining and processing sites during the 1950s, 60s and 70s, and later suffered disproportionately high rates of death and disease as a result of asbestos-linked conditions such as mesothelioma. During the 1960s and 70s,

James Hardie manufactured asbestos products including fibro sheeting, blocks and pipe sections at its plant in Western Sydney. One worker at this plant was Mr Bernie Banton.

Bernie Banton worked as a lathe operator in asbestos products manufacturing for James Hardie from 1968 to 1974. In 1998, Banton was diagnosed with asbestos-related pleural disease (Sexton 2007). He chose to fight for compensation and became the public face of the campaign to win compensation for all James Hardie asbestos-related disease victims. The company fought hard to deny liability and limit victim compensation, but ultimately justice prevailed and what would become a $1.8 billion compensation fund was established for victims. Banton had pursued James Hardie through the courts and won, but by August 2007, Banton's illness had developed into deadly mesothelioma and he died in November 2007.

The cost to James Hardie in dollars and time spent contesting the lawsuits brought on by victims of asbestos-related diseases was astronomical, and the costs to the company in terms of reputation and fiscal return were arguably higher. The nature of the court action and James Hardie's response in delivering the compensation awarded to the victims brought strong community condemnation. Ugly scenes ensued, involving shareholders at annual general meetings, union blackbans and hours of adverse media coverage on news and current affairs programs.

Part of the company's response to being found liable for asbestos-related diseases had been to hive off the part of James Hardie responsible for the compensation fund. This decision prompted the Australian Securities and Investments Commission (ASIC) to launch legal action against ten James Hardie directors and executives on the basis that they were aware, or should have been aware, that the compensation fund was hugely underfunded (Klan & Kelly 2008). The legal action against the James Hardie directors was unsuccessful, with ASIC, a taxpayer-funded entity, incurring an estimated $18 million in legal costs (Sales 2012).

In May 2012 the High Court ruled that seven former James Hardie non-executive directors breached corporate law by making a misleading statement about the compensation fund (*ABC News* 2012). However, the New South Wales Court of Appeal reduced the penalties imposed on them in November 2012 (Sales 2012). Asbestos has been a killer all over the world and the material is now used with great caution. The problems mainly occurred during the years when companies were discovering its carcinogenic behaviour but did not wish to disclose this to the public until they were 'convinced' by the data. The many years of silence are considered to be the cause of far too many deaths. Geoffrey Tweedale's book *Magic mineral to killer dust: Turner & Newall and the asbestos hazard* (2001) was reported by the British Asbestos Newsletter (2000) to be:

> [a] thorough and at times harrowing examination of medical files, company reports, memoranda and correspondence which led Tweedale to conclude: 'The company's attitude towards matters of health over so many years may be regarded as strikingly irresponsible. It neglected to implement . . . schemes fully in both the UK and especially overseas; it failed to warn customers; refused frequently to admit financial and moral liability for the consequences of its actions; often paid only token amounts of money for industrial injuries and deaths; tried to browbeat doctors, coroners, and the Medical Board; sought to suppress research linking asbestos and cancer; gave the government inaccurate data about disease amongst its shipyard workers; and disseminated imprecise information about the "safety" of asbestos'.

Despite the now common knowledge that asbestos causes cancer, it is still widely used in construction work in many contexts, especially in the Global South. Communities in post-conflict contexts and in desperate need of housing — for example, in Jaffna within Northern Sri Lanka — are offered houses that can be built cheaply and quickly. For the price tag of just $5000 it is possible to build a house that provides comfortable accommodation for a family of four. Asbestos is still used for roofing materials in many such houses and local labourers are offered no advice or protection related to its use.

CRITICAL THINKING

What would you do if you worked for an aid organisation that selected asbestos for use in the roofing, knowing that it could harm the workers or even the house dwellers themselves?

Virtue ethics

According to the Greek philosopher Aristotle, whether a decision is ethical should not be judged by its consequences or by whether the person taking the decision had the right to do so. Aristotle held that ethical decisions are motivated by *virtue*. Applying this moral theory, a good engineering decision is one in which the engineer applies personal tools of virtue (perseverance, courage, compassion and self-regard) in making a decision. The quality of the decision is judged by **virtue ethics** on whether it strikes a balance between excess and deficiency in terms of its impact on the individual making the decision. In some senses, this moral theory has similarities with ethical egoism, duty ethics and rights ethics, as it promotes consideration of the individual's thinking process and motivation above the outcomes of the decision, or the consequences of the decision on others. In virtue ethics, the outcome is seen as less important than the morality and character demonstrated by the individual in making the decision.

Ethics is an essential part of operating as a responsible professional; however, the philosophy of ethics, codes of ethics and the application of ethics are acts that require the individual to interpret a situation and work out how to balance various moral positions against each other. The moral theories that have been described have strengths and limitations, but they show how ethical decisions and engineering codes of ethics sit upon a set of theoretical assumptions about what is a good way to act. In the face of an ethical dilemma, a professional engineer needs to interpret and operationalise their knowledge of these moral theories to devise a process for making an ethical decision. Any engineering decision needs to fit within a professional code of ethics, comply with the legal and natural rights of parties to the decision, and needs to result in an outcome the individual engineer can live with (from the perspective of their own moral code). At the same time, engineers must also consider and take into account the ethics of the *consequences* of their actions.

You may like to review the different moral theories and consider which aspects or elements of each you could use to inform your decision making as a student and, later, in professional practice.

4.5 Common ethical dilemmas in engineering

LEARNING OBJECTIVE 4.5 Identify the critical importance and the potential limitations of codes of ethics.

While it is always a requirement that engineers adhere to a professional code of ethics, less attention may often be paid to the consideration of ethics compared to the technical, financial or environmental aspects of engineering work. Ethics might not be regularly or explicitly discussed in the tearoom at work or in the coffee shop on campus. Many engineers or engineering companies are not willing to openly discuss times when they have faced and failed an ethical challenge. So how do you know when you are stuck in an ethical dilemma? Seversen (1997) highlights one way you might know. The *conscience* is one potential signalling system that can alert us to decisions that require the application of ethics. The conscience tends to send signals in the form of feelings and emotions, rather than as conscious, rational thoughts. Signals from the conscience include feeling guilty or apprehensive, or being struck by an instinctive and firm feeling about the goodness or wrongness of an action. This means that if your conscience begins to signal, you need to dig into the engineering ethics toolkit and take the time to analyse the moral dimensions of the decision or situation that you feel are wrong. Your conscience might be said to be driven by your morals, as discussed above, but also by your values, or *what you value in the world*, and this in turn is informed by your upbringing, parents, schooling, culture, religion and so on. Problems arise when you are not yet sure of what your values or morals are, or when you have adopted those of others without really thinking them through. The map in figure 4.6 was created by one of the authors of this text together with a group of researchers from Canada who explored the values of engineers and engineering students in both the United States and Canada (Nahar et al. 2010). The engineers surveyed fell into one or more of four quadrants — where they differed in their focus on autonomy versus authority (Y-axis) and their focus on their profession or the broader society (X-axis). Many more engineers fell into the top left quadrant compared with the bottom right. The map was not created to 'type' engineers or suggest that one group was better than another, but to demonstrate that a balance is needed for engineering work to be done in ways that are ethical and with net positive impact on all societies and the environment. For most engineers, the challenge is to learn how to develop their sense of values in that bottom right quadrant.

Herkert (2005) noted that there are two different kinds of ethical dilemmas — **micro ethics** and **macro ethics** — and that it is critical for engineers to understand both. It is easier for an engineer to focus on micro ethics if they are in the top left quadrant than it is for them to consider macro ethics. Furthermore, the kinds of micro ethical dilemmas you are likely to face as a professional engineer have plenty of parallels in student life, so you will have many undergraduate opportunities to develop your ethical conscience. Unfortunately this is less often the case for macro ethics. In the following sections we look at typical examples of micro and macro ethical dilemmas.

KEY POINT

Ethical dilemmas occur when there are discrepancies between the requirements placed on you by your employer, the code of ethics you are committed to uphold, and/or your own personal moral code.

FIGURE 4.6 North American Engineering values map

	Conformity/Authority		
Profession/Engineering	Confidence in big business Confidence in engineering Family motivation for engineering profession Power of money Usefulness of the work	Balance and resolution Comfortable mobility and safety Ethical Governmental regulation National pride Survival Social and moral responsibility Safety and training	Social/Society
	Dream of engineering career Enthusiasm for technological advance Financial and material success Fulfilment through work Honesty and loyalty to work Potential impact of engineering	Awareness of ecological health Belonging to the global village Critical view Empowerment ofindividual Financial security for local people Humanity and social development Introspection and impulse Peace Social learning Sustainable world	
	Ideals/Autonomy		

Source: Nahar et al. (2010).

Micro ethics

Respecting intellectual property

Intellectual property is the creative, original and innovative ideas that others have. Having respect for the intellectual property of others means only using what they are willing and are legally entitled to share, and always acknowledging when an original idea has come from another source. Consider the following example of an ethical dilemma about intellectual property from a practising engineer during the tendering phase of an engineering project.

Trevor turned up to the office as usual on Monday and noticed a few of his colleagues clustered around a report in the conference room. On the desk there was a draft tender document with a prominent 'confidential' watermark and their major competitor's logo on it. The manager of Trevor's firm had arranged for someone to go through their competitor's rubbish bins in an attempt to find information on the tender they were putting in for a job that Trevor's firm was also bidding on. They managed to find draft design drawings, budgets and notes on preferred contractual arrangements. Trevor's firm had effectively stolen the intellectual property of their competitors. Trevor was disgusted at the lack of ethics his firm had shown.

Apart from the lack of dignity and integrity of those involved in stealing another engineer's work, the above case illustrates an important ethical principle — respect for the intellectual property of others (Severson 1997).

One of the first examples of this principle that you will encounter is your university's academic honesty policy (sometimes called an 'anti-plagiarism policy'). Such policies will usually explain that if you use an idea, information or phrases from another source, you must honour the intellectual property of whoever wrote or said it by correctly citing them. As a professional engineer, using the work of others without permission or attribution is severely frowned upon and considered to contravene the high ethical standards expected of a trusted professional in society. Getting into the habit of keeping track of and acknowledging whose ideas you are using as a student will stand you in good stead for your later working life.

Falsification of records

The work of professionals has the potential to affect the health, safety and wellbeing of others, and may be the subject of legal proceedings. This means that the documentation and notes produced by an engineer are legal documents; therefore, an important part of an engineer's legal, professional and ethical duty is to keep accurate documentation. Consider the following account adapted from a passage in Baura (2006, pp. 187–189).

Lyn was a biomedical engineer who was working as a director in a hospital. A significant part of her job was the preventative maintenance (PM) of equipment, including the critical care devices such as those used in intensive care. There were around 7000 pieces of equipment on the PM schedule, and only five technicians on her staff. Despite their best efforts, at that staffing level they just could not keep up with the PM schedule. About 50 per cent of the equipment was routinely off calibration. This was a big problem, as the equipment they were maintaining meant the difference between life and death for a lot of patients.

Lyn reported her staff shortage to upper management, and for the next two years she continued to request additional staff. The additional technicians were never hired, and in the lead-up to the external audit of the hospital there was a suggestion that she should falsify her documentation to cover up the holes in the PM record. The external auditors came through and then listed her department as an area in need of improvement. Lyn was called into her manager's office and asked to resign because of the adverse audit report. Lyn believed she had performed her job well under the circumstances, but in the end took a severance package and resigned. In the case of this biomedical engineer, the falsification of documents about maintenance of equipment in a hospital would have allowed hospital administrators to avoid being accountable if an equipment failure led to the untimely death of a patient.

What are examples of falsification that might tempt undergraduate engineering students? One example might be when a lecturer asks you to keep a weekly logbook or journal and to present it for assessment at the end of semester. Would it be falsification if you wrote up this supposedly 'semester-long' logbook two days before it was due to be handed in? What about lab results? Would it be falsification to copy lab results from a previous year, instead of generating the data yourself? While the main consequences of these forms of falsification are that the student misses out on learning, students need to cultivate the habit during their undergraduate years of keeping the accurate records that are required of practising professionals.

Recognising personal limitations in professional practice

Professional engineers must recognise and stay within the limitations of their own competence. Guideline 2.3 of the Engineers Australia Code of Ethics (2022) states that engineers must 'act on the basis of adequate knowledge'. Applying this idea of practising within personal competence, an electrical engineer, for example, is neither qualified nor certified to offer expertise on the gear–lever ratio of a high-performance car, nor to offer expertise on the risk of residential landslip under conditions of soil saturation.

When you become an engineer, you will need to remain aware of what you are qualified to comment or work on, and maintain and update your expertise so that the quality of advice or service you provide is of the best possible calibre. The maintenance of professional knowledge is an ethical responsibility in engineering. Guideline 2.1 of the Engineers Australia Code of Ethics states that engineers must 'maintain and develop knowledge and skills' (Engineers Australia 2022).

An interesting problem raised in relation to the responsibility for engineers to stay within their competence is what to do about innovation and about extending expertise through new experiences. Mundel (1991, p. 4) asks, 'Is it ethical for [an engineer] to assume they will be able to do a job well and honestly even though they had not done a similar or related job heretofore?' Think back to the engineering masterpieces and innovations you have read about in the preceding chapters. Was it ethical for the engineers involved to move beyond their existing expertise to create these marvellous innovations?

This question suggests that there are many aspects of the technical work of engineers that begin to lead into the arena of ethical dilemmas. Numerous tasks relating to technical engineering work have the potential to create confusion about what is a good (as opposed to the correct) course of action. Some examples are provided in table 4.3.

TABLE 4.3 Routine technical engineering tasks and possible associated ethical problems

Task	Ethical problem
Conceptual design	Engineer's knowledge out of date
	Violating patent or intellectual property
Preliminary analysis	Overly detailed analysis of engineer's main area of expertise and limited analysis in other aspects of the design
Design specification	Unrealistically tight, leading to likely cost and time overruns
	Design changes not incorporated into specification

Tendering	Specification written to favour one vendor or contractor
Fabrication of parts	Variable or poor-quality workmanship, materials or components Design alterations not detected or noted
User safety	Reliance on complex or failure-prone safety devices Too high an expectation of user compliance with safe use instructions
Product implementation or commissioning	Insufficient instruction given to purchaser in use of the design Training subcontracted out without sufficient care
Use and maintenance	Inadequate or overly expensive supply of spare parts Failure to offer adequate ongoing product support

Source: Adapted from Schinzinger and Martin (2000).

KEY POINT

'Claiming' your failures and attempting redress is part of being a responsible and ethical engineer.

Owning up to mistakes

A possibly uncomfortable but very necessary part of being a responsible engineer is 'owning up' to design failures or mistakes. Engineers have a personal responsibility for the quality and consequences of their professional work. As such, the engineer who is aware of faults in their work is obliged to notify those who could be affected and to attempt some form of remedy for the faulty work. Admitting failure is even more important in contexts where the community who the engineering system or product is designed for is vulnerable and voiceless. This is often the case in the international development sphere. There are many stories of development engineers making incorrect decisions and creating more problems than they solve, due to a lack of adequate understanding of context. The following spotlight draws attention to an admirable act of one former Engineers Without Borders member in Canada.

SPOTLIGHT

Admitting failure

Ashley Good returned home from what she described as a 'not very successful' visit with Engineers Without Borders (EWB) Canada, working with subsistence farmers in Northern Ghana. She wanted to find a way to move forward and improve the effectiveness of development engineers. EWB had already started to document its failures with annual Failure Reports, and as Good (2019) attests on her website:

> The Failure Report had become more than simply an internal learning document. EWB was working towards change in the development sector, towards humility, innovation and learning, and the Failure Report was our way of practicing what we preached. This left me wondering how we could spread this change and make it easier for other actors in the development sector to report their failures – and the spark for Admitting Failure was lit.

She went on to work with others to launch her website (see figure 4.7), which asks other similar organisations to also share their failures. Articles about these failures on the site are organised by engineering sector, as well as thematically. Learning from failures is critically important but often, due to concerns about liability or funding, we do not wish to air our issues. Understanding our ethical obligations as engineers helps us realise that in fact the more important thing is to continually improve the way we work and the impact we have. If our failure story can help others in the future from financial, health or safety risks, we have a duty to report on these failures. Ashley went on to start her own consultancy organisation Fail Forward (https://failforward.org), demonstrating how a young engineer can not only be 'good' and do the right thing, but also create new and interesting opportunities for a career. Now that really is *Engineering your future*.

FIGURE 4.7 Articles on the Admitting Failure website

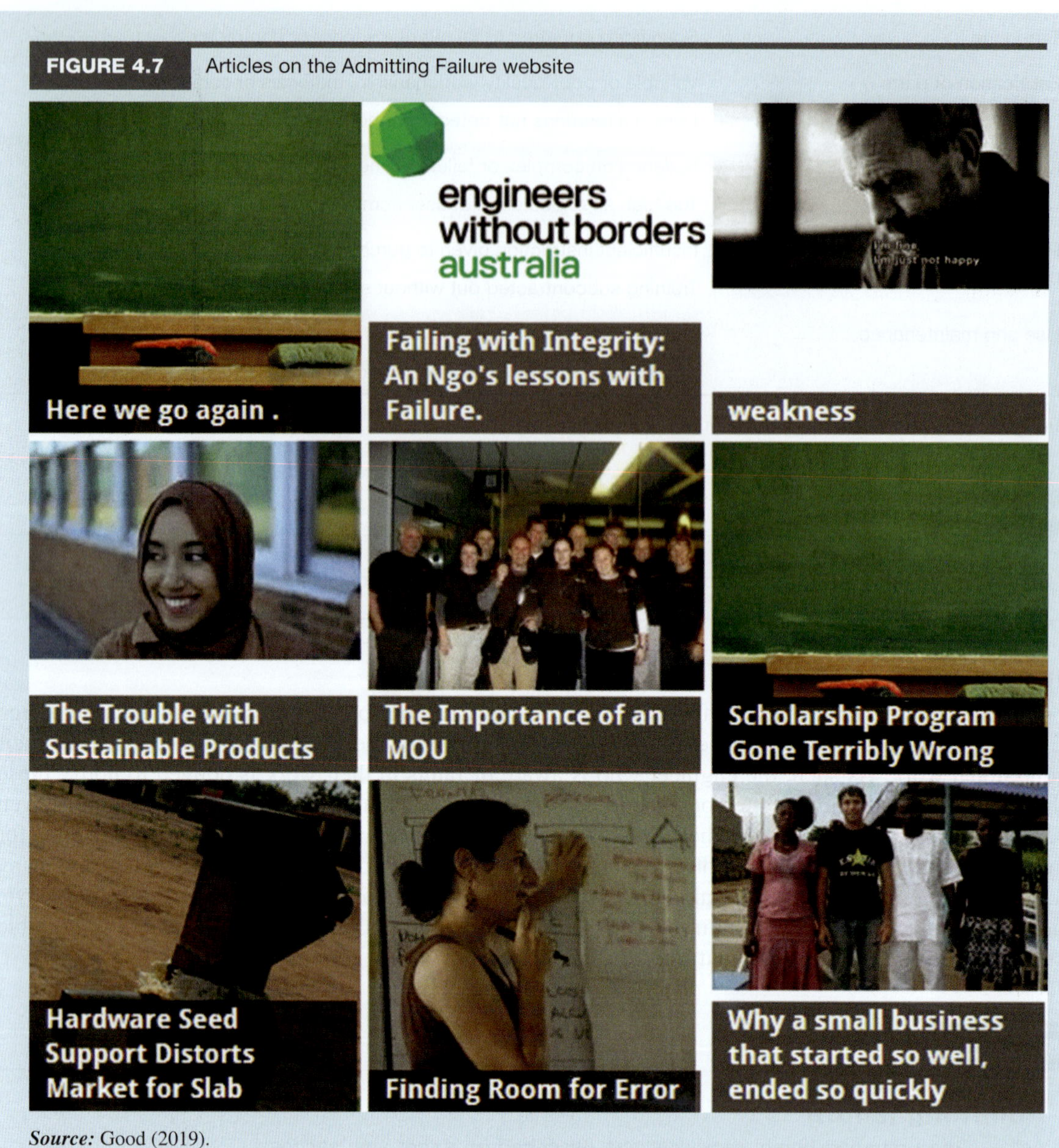

Source: Good (2019).

CRITICAL THINKING

Reflect on a time you have had to admit to being wrong. How did you go about it? How did it feel? How did the person or people you were admitting your error to react? Could you have handled your admission better? By analysing your past experiences of admitting error, it is possible to plan a good approach for future times when you need to eat humble pie and own up to being human.

Balancing conflicting interests

It can sometimes be difficult for an engineer to identify who they owe ethical allegiance to. One of the ways that engineering ethics becomes complicated is when it is not immediately apparent exactly who engineers should be serving in their professional role. When you graduate, you may become strongly committed to an employer. This situation will likely work well if you are able to maintain a separation between your personal life and your professional life and if the company culture is a good fit with your beliefs, responsibilities and allegiances outside work. Difficulties can occur when the requirements placed on you by your employer begin to contravene the code of ethics you are committed to uphold, or your own personal moral code.

KEY POINT

Engineers need to consciously consider ethics and morals in order to decide on a personal and a professional moral code.

Intertwined responsibilities

The term **intertwined responsibilities** is used by Schinzinger and Martin (2000) to describe how engineers exist within many different systems, and have responsibilities and allegiances in each. As a student, you are part of several groups or communities: you are a member of the student body within your faculty; you may participate in a sporting club or hobby group; you may identify with a particular cultural or nationality group; and you may participate in a church, political or online community. Ethics are comparatively easy to negotiate when these groups remain distinct but there may be times when your affiliation with multiple groups throws you into a difficult position and requires a decision that will call on you to interpret, integrate and decide how to apply your personal and professional codes of ethics. Building from the words of Mundel (1991, p. 4), do 'individuals owe a greater loyalty to their boss, themselves, the public', their company, their family or the environment?

Corporate responsibilities and loyalties

There will be instances in your career as an engineer in which you will be required to make a difficult moral or ethical decision or take appropriate action when faced with an ethical dilemma.

People are social by nature. When a person enters a new social grouping, there is a period during which the person comes to understand the new culture. This period is called enculturation. It allows people to fit into and begin to work effectively within a new system and with new people. For engineering students, enculturation has been shown to happen relatively quickly (i.e. within the first month or so for most students) (Godfrey 2004). Enculturation is an important part of a new job and many companies are keen to secure strong allegiance from their employees.

The process of enculturation and social cooperation within the workplace can make it difficult to object when the voice of conscience suggests unethical behaviours are taking place.

In an organisation with clear procedures in place to address breaches of ethics, the process of confronting unethical behaviour is eased by formal reporting mechanisms (e.g. a manager who is nominated to be the company's 'ethics ombudsman', clear policy guidelines for reporting breaches and active support offered to those who attempt to confront wrong behaviour). If you hold a strong view about ethics and the role it plays in engineering work, there are several features that define an ethical corporate climate, which you may like to look for when choosing a company for your holiday work or first engineering job (Schinzinger & Martin 2000). These features are as follows.

1. Ethical values are widely acknowledged and understood by managers and employees and include responsibility to all stakeholders (e.g. customers, clients, shareholders, employees and community).
2. Ethics and ethical language are used as part of daily business and pervade company documentation.
3. Top managers set the moral tone through their words, policy, procedures and, particularly, by example.
4. There are broadly accessible and generally well-known procedures for conflict resolution, and these procedures have been applied to resolve conflict.

Where such structures do not exist, or exist only on paper but not in practice, an engineer concerned about unethical behaviour may consider whistleblowing.

Whistleblowing

Whistleblowing is the act of an employee or insider taking their claims of unethical practice to someone in a position to act on it, but outside the normal channels for resolving grievances. Whistleblowers are described as 'internal' whistleblowers when they take their claims to someone within their organisation (e.g. to the head of another section, to an internal ombudsman or direct to the CEO). 'External' whistleblowers take their grievances into the public arena. Whistleblowers usually take this somewhat drastic action as a last resort when they have been stymied in attempting to stop unethical practices via internal means, or when they feel the system they are working in is so fundamentally corrupt that their chances of stopping the unethical practices within it are severely limited. The following spotlight details the actions of a Japanese whistleblower in the nuclear energy industry.

SPOTLIGHT

Whistleblower slams Japan nuclear regulation

Following the earthquake and tsunami that struck Japan on 11 March 2011 and damaged the Fukushima Daiichi power plant, whistleblower Dr Masashi Goto criticised the Japanese government, Japan's nuclear industry and nuclear regulators over their safety record. Dr Masashi Goto, a nuclear engineer, resigned due to safety concerns from his role as a nuclear engineer at Toshiba Corporation — the supplier of two of the reactors at the Fukushima plant. In Dr Goto's opinion the Fukushima crisis highlighted Japan's poor safety record and showed that Japan had not learned from its mistakes.

At the height of the 2011 crisis, Dr Goto was interviewed by ABC correspondent Eric Campbell and criticised Japan's record on nuclear safety.

The Fukushima nuclear power plant that was damaged by Japan's 11 March 2011 tsunami; only one of the four pale blue reactors shown in this picture looks undamaged.

> 'We have the government commission overseeing nuclear safety standards and in my opinion they are not doing their job'
>
> . . .
>
> 'At Three Mile Island the nuclear fuel melted. Fuel is melting here now,' he said.
>
> 'We have to design reactors to withstand melting fuel rods. Right now the reactor will break down due to the heat generated by the melting rods.'
>
> Dr Goto alleges that in Japan's nuclear industry profits take precedence over safety standards.
>
> 'No-one says it officially or openly. When setting standards for future earthquakes, the thought is of money — how much is it going to cost?' he said.
>
> 'This underlies the government's decision making. They are thinking the costs could have a bad repercussion on the economy.'
>
> Dr Goto says one of his special research interests at Toshiba was how to make containment vessels stronger.
>
> He says Japan's nuclear safety standards have been based on an insufficient acknowledgment of the potential severity of natural disasters.
>
> 'What's wrong with the standards is that the anticipated level of the worst-case-scenario earthquake is not correct,' he said.
>
> 'Seismologists have different opinions and predictions. Some say bigger quakes are coming. Others say a big one is unlikely.
>
> 'Decisions have been made based on the opinion of the more optimistic seismologist and the opinions of the pessimistic ones are ignored.'
>
> The earthquake that shook Japan on March 11 was magnitude 9.0 — the strongest recorded earthquake in Japan, and far stronger than the country's nuclear industry had anticipated.
>
> Despite this, the Tokyo Electric Power Company, which operates the Fukushima plant, boasted in its corporate publicity that its nuclear power stations were 'designed for the largest conceivable earthquake' and that 'all designs provide margins of safety capable of withstanding even natural disasters' (McDermott 2011).

Decommissioning the damaged Fukushima power plant has created 'unprecedented technological challenges' and the Japanese government has estimated US$188 billion costs, which consumers will bear through increased electricity costs and taxes (Suzuki 2017). The costs are approximately one-fifth of the entire general accounting budget of the country and have caused great loss of faith in nuclear power.

CRITICAL THINKING

In Australia there is legislation to protect whistleblowers. Why might whistleblowers need legislative protection?

Based partly on the personal cost it can extract (e.g. loss of employment, legal action, social sanctions), and on the potential reputational damage to corporations, Baura (2006) suggests that whistleblowing should be used only as a last resort. She suggests that an employee who is considering or pursuing whistleblowing should become scrupulous in collecting evidence to support their allegations, and evidence of conversations or correspondence relating to attempts to resolve the grievance. It is a good idea for a whistleblower to attempt to gain support from other engineers, impartial parties outside their company and from their professional body (e.g. Engineers Australia, Engineering New Zealand or IEEE) and seek legal advice prior to taking public action.

4.6 Macro ethics

LEARNING OBJECTIVE 4.6 Describe macro ethics and how it is different from micro ethics.

The previous section considers what is often referred to as 'micro ethics'. Macro ethics, on the other hand, is not as well-defined as micro ethics because it has only recently been recognised as critically important. There is an increasing awareness of the importance of understanding the social and environmental consequences of engineering products and systems. People on both sides of the political fence would argue that it makes sense to concern yourself with the potentially harmful impacts of your work that can end up having legal repercussions — however, there are differences of opinion as to how much engineers are, and can be, responsible for the broader consequences of their work. This is where macro ethics comes into play.

Sustainable development is one area of macro ethics that has been given more attention in recent years and you can read more about this in the chapter on sustainable engineering. Engineers are beginning to see how they are potentially responsible for, and able to contribute to, sustainable development. However, other areas of macro ethics are less clear to some engineers. Can engineers assist with poverty reduction in the world? If they can, is it their responsibility to do so? As we pondered previously, allegiance to whom is more important: an engineer's company, its shareholders or the people in society who they serve, and if so, which parts of that society? There has been much recent activity and attention to the social impacts of engineering as companies are beginning to realise the negative effects of non-action, or of ignoring the issues at stake.

Corporate social responsibility(CSR), **social sustainability** and **social impact assessment** are examples of where we might see macro ethics acting within companies. Many large global corporations have policies or strategies addressing at least one of these, as they relate to their area of work. A mining company might have a CSR policy that relates to the needs of local communities affected by their activities — where and how to support local schools or health care as recompense for granting the right to mine on local land — this in turn will enable them to receive their **social licence to operate**. A pesticide company might have a social and environmental impact assessment to consider the potential harmful effects of pollution on water sources for neighbouring communities. Increasingly, people are thinking about 'social sustainability', although it is hard to find an example of a company with a policy in this area that is meaningfully created and implemented. An example might be an automotive company who wish to guide the development and implementation of a code of conduct for working with factories producing parts in developing countries over which they have little control of working conditions. There is a very ad hoc and piecemeal approach to macro ethics and it is the area that needs urgent attention in engineering.

KEY POINT

Engineers working internationally need to be mindful of different political and social systems, values, cultures and beliefs.

SPOTLIGHT

Free prior and informed consent

Macro ethics is often related to what we might call justice. Baillie and Levine (2013) have considered what it means to be a 'just' engineer, using Rawls (1971) as their guide. As outlined by Baillie and Levine (2013, p. 4), Rawls's original position:

> is a hypothetical situation in which no one knows what particular abilities or knowledge they will have in society or what position (role, job, etc.) they may occupy. Placed in this hypothetical position — a blind spot with regard to any and all personal circumstances including accidents of birth such as innate intelligence and talents — we are then asked to determine which principles of governance are 'just' or fair.

Consider the rights of the traditional owners of the land when a mining company comes knocking at the door. Imagine in Rawls's thought experiment you are that person who has had a relationship to that country for generations: your family and your culture are strongly tied to this place but you are being asked to move to another, very foreign, location, away from the coast to the desert. However, your community is very poor and in need of food, water and basic services. You agree to relocate in exchange for money and jobs. How do you feel? What long-term impact does this have on your people?

One of Australia's largest man-made mining holes, Kalgoorlie, Western Australia

Free prior and informed consent is the process by which mining companies have to operate since 2009 when the Australian Government stated their support of the United Nations *Declaration on the Rights of Indigenous Peoples* (2007). It affirms that 'Indigenous Peoples shall not be forcibly removed or relocated from their lands or territories without their Free, Prior and Informed Consent' (UN Declaration 2007, Article 10).

Baillie and Levine (2013, p. 6) ask us:

> Is their agreement to the development of a new mine on their land (when they do not necessarily understand the implications, even if they are 'told') a free and un-coerced agreement? Is it an 'agreement' at all? A non-coerced, non-manipulative, and non-deceptive agreement may not be possible for those who have been socialized into accepting hand-outs as their traditional means of living. The agency and personal decision making capacity of some Indigenous people may have been gradually eroded by Western urbanization, subjugation and indeed oppression. They are being asked, unjustly on some accounts, to play a game (follow a procedure) which they do not fully understand by means of rules they may not accept.

Sadly, the way that free prior and informed consent is used in many countries does not protect the rights of the people it was intended to. Some countries have recently changed who is deemed 'Indigenous', including those adjacent to mine sites; as free prior and informed consent is only intended to protect 'Indigenous' people, this rule then no longer applies to them. In other contexts there is also a move to distort the meaning of the term 'consent'. Legal precedent within a country often determines whether this means that the local community does have the power of veto or the right to say no, or whether it means they simply have the right to engage with the company.

CRITICAL THINKING

Now imagine you work for the mining company that is asking the community to relocate. How can you be guided by ethics to understand how to think about this? Do any parts of the codes of ethics considered earlier help you in this case?

4.7 Culture and corruption

LEARNING OBJECTIVE 4.7 Explain the role culture and corruption play in ethical decisions and perspectives across the globe.

One exciting aspect of contemporary engineering work is the opportunity to travel. Engineers are in demand in many parts of the world and opportunities abound to undertake professional work in many far-flung corners. In figure 1.4 of the first chapter, countries that are signatories to the Washington Accord were listed — you may like to review that list to see which countries will accept your engineering qualifications on graduation. Regardless of whether an engineer is seeking to work overseas or succeed domestically with international clients, it is important to consider and maintain respect for the cultural differences that exist between countries and among different racial and ethnic groups. As we move to an increasingly globalised world, with multinational companies working in many different countries, it is increasingly necessary to understand cultural differences. As with previous discussions about ethics, there is no right answer. However, when acting across cultures the idea of making decisions based on morals or values that have been developed in *our culture* renders making appropriate decisions much more difficult. What will inform our actions in this case? The reason macro ethics takes on a different and more complex set of challenges is that it brings us face to face with political and social issues that we had no idea related to engineering. In the following sections we will first consider cultural issues at a micro level before touching upon the macro socio-political arena.

International business etiquette

Etiquette is a relatively old-fashioned term that refers to the often unwritten rules for decorum, propriety and good manners. While the term might seem old-fashioned, the function of etiquette remains a highly important one in civil society and in business dealings. Observing the rules of etiquette during social and business interactions helps to build good interpersonal relationships and trust. This is because etiquette is an implicit or behavioural way of expressing respect for and sometimes deference to the person on the receiving end of your good manners. When engineers are dealing with colleagues and clients from their own culture, it is likely the rules of etiquette are reasonably well known to both and there is little chance of one or the other causing unintended offence. The point here is that etiquette can differ substantially in different cultures and what is acceptable behaviour in a boardroom or restaurant in Perth or Wellington may have you losing a contract in another part of the world. Here are a few examples of behaviours that are considered normal in some parts of the world, and very poor form in others.

Touching

- In much of South America men may greet each other with a hearty hug and some women will give or expect a kiss on either cheek in greeting.
- Men hugging men in the Middle East may be acceptable but it is rarely acceptable for a man to touch a woman during a business meeting.

Business cards

- In Australia and New Zealand, exchanging business cards is a fairly casual affair and often the card is received and tossed in a folder or pocket without much of a look.
- The giving and receiving of business cards is a formal and important affair in many parts of Asia, and the polite engineer will give and receive a business card with both hands, and take time to examine and store the card with care.

Drinking alcohol

- Bar hopping and karaoke are often important elements of business dealings in Hong Kong and Japan.
- The consumption of alcohol is illegal in some Middle Eastern countries, and may be offensive to business partners or clients of the Muslim faith.

These are just a few examples of how to observe and respect the values, culture and beliefs of others if you chose to work internationally. You can learn more about cultural etiquette by discussing the accepted social graces and customs of your culture with a classmate from a different culture. A further element of international engineering work, which is much less savoury, is the need to negotiate political and social systems in which corruption and bribery are established and pollute the practice of engineering.

Corruption and bribery

It is an unfortunate fact that in many parts of the world the payment of bribes, incentives and kickbacks are a customary part of how business is done (Czinkota et al. 2009). These practices all fall under the heading of **corruption**, which Transparency International (a group committed to fighting corruption) defines as the abuse of entrusted power for private gain. Corruption is fundamentally unfair to business in that companies do not win work on the basis of their expertise or a competitive bid; rather, the most unethical company with the deepest pockets wins. Corruption undermines civic trust, in that individual citizens cannot rely on their public representatives to look out for the best interests of the people or the country, particularly in the developing world where corruption robs those who are already underprivileged and impoverished. According to Gupta et al. (1998) this happens through:

> direct impacts ([e.g.] increasing the cost of public services, lowering their quality and often all together restricting poor people's access to such essential services as water, health and education) and the indirect impact ([e.g.] diverting public resources away from social sectors and the poor, and through limiting development, growth and poverty reduction).

In the 2023 annual study by Transparency International, levels of perceived public sector corruption were surveyed by country and used to generate a corruption perception index map for the world (figure 4.8). As is apparent from figure 4.8, several of Australia and New Zealand's trading partners and some of the countries that are signatories to the Washington Accord feature as countries where corruption is likely to be a problem. This is a significant challenge for engineers intending to work in countries where bribery and corruption pervade. Such regimes might affect, for example, an engineer's capacity to meet with government officials, the submission and winning of tenders, cooperation of sections of the workforce and the policing (or waiving) of environmental standards. For an engineer faced with paying a bribe or losing a multimillion dollar contract, the temptation to become involved in corruption might be strong. Fortunately, Australia has a strong disincentive for those working abroad to practise bribery. For Australian citizens working overseas, bribery of a public official is a serious criminal offence under Section 70.2 of the *Criminal Code Act 1995* (Cwlth). The penalties for this crime include up to ten years imprisonment. A court can impose a fine instead of, or in addition to, a term of imprisonment.

FIGURE 4.8 Corruption perceptions index (CPI) map 2023. The CPI measures perceived levels of public sector corruption.

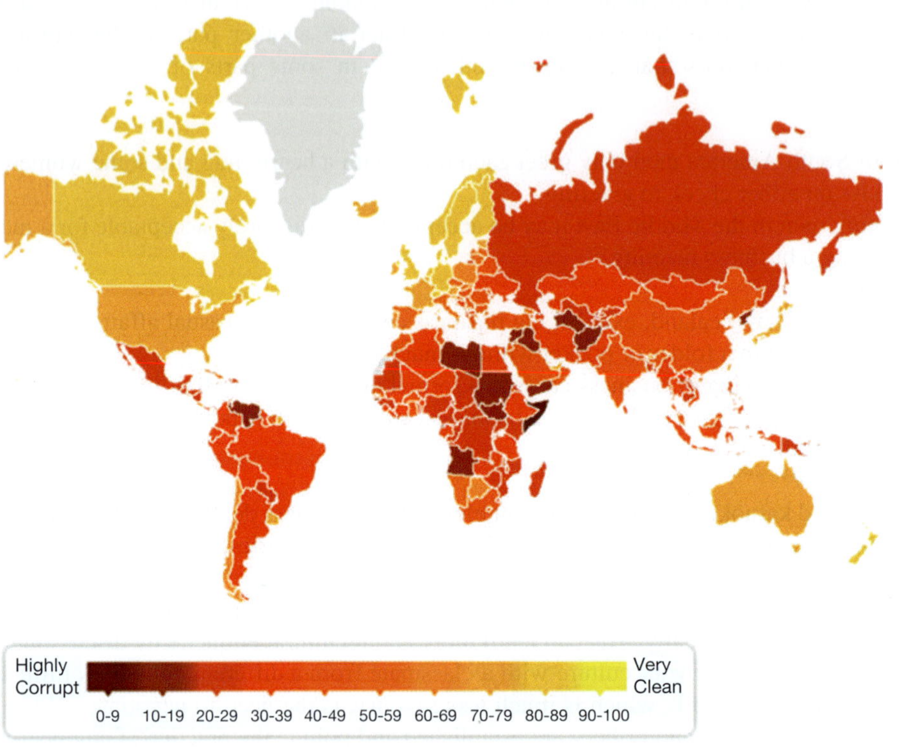

Source: Transparency International (2023).

Corruption is considered by many to be present in all countries. However, in the **Global South** corruption is more explicit. In the **Global North** (or developed countries), kick-backs and bribes may exist, but not on paper and they are not easily traced. Culturally, we have different ways of dealing with corruption. The following spotlight shows how even one of the world's biggest companies in agricultural and genetic engineering can fall into the iniquitous practice of bribing public officials. This is the case of Monsanto's prosecution for attempting to avoid environmental impact evaluation on the use of their genetically modified cotton in Indonesia.

SPOTLIGHT

Monsanto penalised for bribery

In 2005, the United States agrochemical giant Monsanto agreed to pay a US$1.5 million fine for bribing an Indonesian official. The company admitted that the senior official had been paid the bribe by one of its staff members two years earlier in a bid to avoid environmental impact studies being conducted on its cotton. In addition to the fine, Monsanto's business practices were closely monitored by the American authorities for the next three years, with the company referring to the illegal dealings as 'improper activities' (*BBC News* 2005).

At a time when Monsanto was facing roadblocks from activists and farmers who were campaigning against its plans to produce genetically modified cotton in Indonesia, a former senior manager directed a consulting firm to give a US$50 000 bribe to an Indonesian environment ministry official in 2002. The company was told to disguise the bribe as 'consultancy fees', and it was falsely entered into the books as a result. Despite having received the bribe from Monsanto, the official did not waive the environmental study requirement. Monsanto later admitted to paying bribes to other ministry officials between 1997 and 2002 (*BBC News* 2005).

Monsanto said it became aware of irregularities in the books at the Indonesian subsidiary in 2001, subsequently launching an internal investigation before contacting the US Department of Justice and the Securities and Exchange Commission (SEC). As a result of its admission, the firm faced both criminal and civil charges from the Department of Justice and the SEC (*BBC News* 2005).

In addition to paying US$1 million to the Department of Justice, Monsanto agreed to adopt internal compliance measures and cooperate with continuing civil and criminal investigations. In order to settle the bribe charge and other related violations, the firm paid an additional US$500 000 to the SEC. Monsanto stated that it accepted full responsibility for its employees' actions, adding that it had taken steps to reduce the likelihood of a future event of this nature (*BBC News* 2005).

However, issues surrounding Monsanto and environmental and health concerns continue today. In August 2018 in the United States, a landmark legal battle was won by Dewayne Johnson, a 46-year-old groundskeeper with terminal cancer. Johnson was the first person to take Monsanto to trial over concerns that it was their weedkiller Roundup that had caused the cancer. The jury ruled in favour of Dwyane and noted that Monsanto was responsible for negligent failure. Then in 2019, Roundup was found by a Californian (US) District Court to be a 'substantial factor' in causing non-Hodgkin's lymphoma in 70-year-old Edwin Hardeman (Lannin 2019). In 2020, Monsanto's parent company Bayer settled almost 100 000 lawsuits in the US, costing the company US$10.9 billion (Cooper 2024).

Despite the US rulings and settlements, and a call from the Australian Cancer Council in 2018 to review Roundup's use (March 2018), it is still approved for use in Australia, and in 2018, a representative for Monsanto Australia declined to admit their responsibility (Zhou 2018). However, a class action lawsuit currently underway in Australia, involving more than 800 people who claim Roundup is carcinogenic, could lead to regulatory changes restricting its use across the country (Cooper 2024).

This demonstrates that decisions about the approval of products as safe is a complex issue. Based on reports given to them, politicians must decide whether there is a risk to health of their citizens. They weigh the knowledge against the power of voters or corporate donors to their campaigns. As students we often think that engineering and scientific data is black and white, that there must be a clear answer to the question 'is it safe?' However, it comes down to what scientists say is the specific level of allowable toxin in our systems, the accuracy of the measurement system, the variability of the results, the number of trials, and so on. Furthermore, it is possible for a company to prevent publication of any research they have funded, if they do not want the results to become public.

CRITICAL THINKING

Imagine the following scenario: you worked for Monsanto and found out that Roundup is still available in Australia as a result of acts of bribery and manipulation of legal power. What would you do?

SUMMARY

In this chapter, some of your future professional responsibilities and liabilities have been canvassed, and both 'everyday' and 'once-in-a-lifetime' ethical dilemmas faced by practising engineers have been explored. How the undergraduate training of engineers provides opportunities to hone ethical decision-making skills has also been reviewed. The responsibilities attached to life as a professional can seem onerous when decisions of life, death, livelihood and legality may rest in a single pair of hands. Conscious and reflective consideration of personal morality, professional liability and ethical practice is a necessary habit for those who choose to accept these important roles in society. We will now briefly revisit each of the chapter's learning objectives.

4.1 Explain the difference between negligence, malpractice and commercial liability.

The professional engineer has a duty to practice according to local legislation and the standards that govern practice in their discipline. Failure in this duty amounts to professional negligence. A second liability for professional engineers concerns the quality of their work. An engineer must execute work to the standard that would be expected of a similarly qualified engineer working in the same location. Commercial liability relates to the conditions laid down in a signed contract, and breaches of those conditions may lead to action by the client to recover damages.

4.2 Detail the ways engineers can avoid, manage and remediate WHS risks, discuss how engineers can be held personally liable for their decisions and discuss the advantages and drawbacks of risk assessment frameworks when working globally.

Good engineering requires both excellent technical analysis and problem solving and scrupulous attention to the interaction between people and technology. Good engineering aims to create safe workplaces, rather than rely on people to keep themselves safe in the workplace. Risk-assessment and management frameworks and the hierarchy of control are key thinking tools for engineers to meet their workplace WHS responsibilities.

In terms of the duty of care for others and in the case of corruption and bribery, engineers may be held personally liable for the decisions they make. This personal liability compounds the importance of engineers honing their skills in methodical, logical analysis of the legality and ethics of various actions, and the importance of developing the capacity to recognise when a situation or incident feels unethical.

Risk-assessment and management frameworks and the hierarchy of control are key thinking tools for engineers to meet their workplace WHS responsibilities. They are critical to enable an engineer to ensure safe and healthy practices within a tightly controlled environment such as a factory within the home country of the engineer. However, it is difficult to predict all of the potential hazards and consequences when working globally and, in particular, when the health and safety practices of the country are not to the same as Australian standards. Outsourcing of critical components, manufactured at a much lower economic cost to the company, may bring social and environmental costs to the source country workers. Taking care of risks along the whole supply chain needs much more attention to detail and an understanding of local culture and laws.

4.3 Explain the influence of values on ethical decisions.

We are taught so many rules and theories as engineers that we think every decision must come from a chart or calculator. However, when considering the ethics of what we are doing, or not doing, we have to draw on different parts of our experience and knowledge. These are informed by our values. The North American Engineering values map shown in figure 4.6 demonstrates how different engineers value different things in life. Some are focused on doing the right thing for their company, for their country or for the profession of engineering; others are more concerned about human rights and the environment. The values we have developed over the years, and that are highly related to our family, culture and customs, will determine how we interpret the code of ethics. Knowing what we value and why is key. It is also critical to consider the values of those we are engineering for.

4.4 Give examples of how moral theories can be applied in ethics.

The moral theories that can be used to consider and structure a response to an ethical dilemma include ethical egoism, utilitarianism, duty ethics, virtue ethics and rights ethics. Ethical egoism positions the interests of the individual making the decision as having primary importance, and it can result in consequences for the engineer who is interested in maintaining good working relationships. Utilitarianism attempts to optimise human happiness, but it can be based on subjective or erroneous assumptions about what different individuals and groups perceive to be likely to produce happiness. Other moral theories attempt to provide just outcomes by applying set rules or principles and taking no account of consequences.

4.5 Identify the critical importance and the potential limitations of codes of ethics.
Practising engineers in Australia and New Zealand are bound by the code of ethics of their registering professional body. These codes vary in emphasis and offer different levels of guidance to the engineer seeking to interpret and apply them. Engineers are duty bound to comply with their professional body's code of ethics; however, it is important to develop the skills to interpret the relevant code and decide on a clear and rational course of action that is ethical within the constraints of the situation or project. Codes will, however, not offer guidance on what to do and when, or even how. The level of guidance they offer can be interpreted in different ways. They are very much guided by underlying values and also politics. Hence, what one person thinks is the right thing to do will differ from another. Often the dominant population will be making the decisions and the minority voices will not be heard. It is therefore critical to ensure that engineers consider everyone that might be affected by their work, and not only those who have the power to seek liability.

4.6 Describe macro ethics and how it is different from micro ethics.
Micro ethics include dilemmas ranging from pressure to falsify results or records to incentives to overlook safety or quality considerations. Macro ethics address the broader picture and how we as engineers have an impact on society, in particular on marginalised groups within a global context. They are often enacted through social sustainability, corporate social responsibility, social impact assessment or the social licence to operate. However, our understanding of the critical importance of macro ethics is only just being realised.

4.7 Explain the role culture and corruption plays in ethical decisions and perspectives across the globe.
Researching and developing sensitivity to different social norms will allow an engineer to operate effectively on the international stage. An aspect of operating on the international stage is understanding the consequences of different forms of corruption: lack of fairness in international engineering work, reduced confidence of citizens in their public officials, and direct and indirect impacts, which fall hardest on the poor and underprivileged.

KEY TERMS

commercial liability Liability relating to losses that arise from non-compliance with a contract.
corporate social responsibility (CSR) An umbrella term used to assist companies in responding to the growing expectation to understand the social impact of their activities and to take steps to mitigate negative impacts.
corruption The abuse for private gain of the power entrusted to public officials.
duty ethics A moral theory stating an act is ethical if it is conducted out of duty, follows universal maxims and treats people respectfully.
duty of care A legal obligation to have thought or regard for those who may be affected by one's acts or omissions.
ethical egoism The view that an act is intrinsically ethical if a person acts to protect or further their own interests.
ethics Study or analysis of how people apply morals in a given decision or situation.
etiquette The (often unwritten) rules of decorum, propriety and good manners.
free prior and informed consent Whereby communities can offer their un-coerced agreement to systems or projects that will affect them.
Global North A term used to describe industrialised countries.
Global South A term used to describe countries with emerging economies. (*Note:* Australia is designated Global North despite its geographical location.)
intertwined responsibilities Engineers exist within many different systems in which they have responsibilities and allegiances.
macro ethics Broad issues related to the global and social impacts of engineering.
malpractice Improper professional action, either through ignorance or neglect.
micro ethics Local issues related to people and the consequences of their actions.

morals Personal rules of behaviour that people use to distinguish between goodness and badness, and right and wrong.

negligence Failure to exercise that degree of care that would be expected of a 'reasonable person'.

rights ethics A moral theory stating that an ethical decision is one that upholds the rights of all parties to the decision.

social impact assessment An assessment of social issues affected by the engineering system or product and how they will positively and negatively affect particular social groups.

social licence to operate Stakeholder perception of the legitimacy and community acceptance of a project.

social sustainability Aspect of sustainability that relates to impacts on and influences of social issues.

tort liability Liability related to civil responsibilities that are specified by local, national or international laws, or in standards referred to in local legislation.

utilitarianism The belief that an act can be judged moral or immoral depending on its consequences.

virtue ethics A moral theory stating an act is ethical if it is motivated by the virtues of perseverance, courage, compassion and self-regard.

whistleblowing The act of an employee or insider taking their claims of unethical practice to someone in a position to act on it, but outside the normal channels for resolving grievances.

work health and safety (WHS) Keeping people safe at work while working.

EXERCISES

1 How do your values affect your understanding of ethics?

2 Imagine you are a team member for your faculty's solar challenge racing car. The race is in two weeks. With the time so short and the team budget down to a few dollars, the team are starting to argue. You have been responsible for specifying and purchasing the car's solar panels but you think you might have messed up the calculations. When you do the calculations again, it looks as if you might not have the energy you need to get the necessary speed. Do you keep going and hope for the best or do you admit what you have done? What might the consequences be for you and your team if you do or do not own up to the problem?

3 The wording and emphasis can vary between codes of ethics and professional societies. Re-read the Engineers Australia Code of Ethics and the Engineering New Zealand *Code of Ethical Conduct*. Which of the two codes of ethics do you prefer? Which makes more sense to you? Which would provide you with better guidance as an engineer when faced with a dilemma?

4 Revisit the dilemma Mehreen faced at the start of the chapter. Take a few moments to note down or to talk with your classmates about how you would react in this situation. You might be surprised to discover that your classmates have different responses to you. Try to find out why their responses differ from yours.

5 Consider an engineered item or service that you have recently bought or used (for example, a phone, a hairdryer, public transport or yoghurt). Try to trace all the materials and processes involved in its manufacture and transport to where you purchased or used it. Describe three tasks that could have been conducted in an unethical way associated with your item or service. What might the consequences of unethical acts have been on the health, safety of vulnerable people? What does this tell you about the importance of ethics in engineering practice?

6 Consider the risk assessment of a project you or someone you know has been involved with and that went wrong. Would you or they have been able to predict this at the beginning? If not, why not? What does this tell you about what you must do before starting any project?

7 Consider the ideas about macro ethics in relation to an engineering product that you own. What have been the social impacts of that product in your country? In a country of the Global South? How has the product influenced the culture of the people in that country?

8 How does the rule of free prior and informed consent get used in Australia? How does it get used in Peru and Colombia? How can you find out?

PROJECT ACTIVITY

Re-read the Engineers Australia Code of Ethics and consider the following. What does a 'well-informed conscience' mean when it comes to understanding different cultures and customs? How can you move outside of your own cultural bias and values when we have suggested that these are what shape your conscience? Discuss with a group of your peers who are from very different backgrounds.

REFERENCES

ABC News 2012, 'James Hardie directors breached duties: Court', 3 May, https://www.abc.net.au/news/2012-05-03/high-court-rules-in-favour-of-asic-in-hardie-case/3987196

Australian Competition and Consumer Commission 2018a, 'Takata airbag recalls list', https://www.productsafety.gov.au

——— 2018b, 'Takata alpha airbags require immediate replacement', https://www.productsafety.gov.au

Baillie, C 2004, *Green composites*, 1st edn, Woodhead Publishing and CRC Press, Sawston, Boca Raton.

Baillie, C & Levine, M 2013, 'Engineering ethics from a justice perspective: A critical repositioning of what it means to be an engineer', *International Journal Engineering, Social Justice and Peace*, vol. 2, no. 1, pp. 1–11, https://doi.org/10.24908/ijesjp.v2i1.3514

Baura, GD 2006, *Engineering ethics: An industrial perspective*, Elsevier, California.

Bella, T 2023, 'The atomic bombings left Oppenheimer shattered: "I have blood on my hands"', *The Washington Post*, 21 July, https://www.washingtonpost.com/history/2023/07/21/oppenheimer-truman-atomic-bomb-guilt

BBC News 2005, 'Monsanto fined $1.5m for bribery', 7 January, https://www.bbc.com/news

Biomimicry Institute 2018, 'What is biomimicry?', https://biomimicry.org

British Asbestos Newsletter 2000, issue 37, Winter 1999/2000.

Cooper, L 2024, 'Landmark class action seeks to determine if glyphosate is cancer-causing, but how important is it in Australia?', *ABC News*, 18 January, https://www.abc.net.au/news/rural/2024-01-18/glyphosate-trial-determine-if-roundup-causes-cancer-what-happens/103298898

Creed, P 2008, 'Civil liability of roads authorities arising from collisions between pedestrians and bicyclists in shared bicycle paths', *Letter to Pedestrian Council of Australia Limited*, https://www.walk.com.au

Czinkota, MR, Ronkainen, IA, Moffett, MH, Ang, SH, Shanker, D, Ahmad, A & Lok, P 2009, *Fundamentals of international business*, 1st Asia–Pacific edn, John Wiley & Sons, Brisbane.

Einstein, A 1954, *Ideas and opinions: Based on mein weltbild*, Crown Publishers, New York.

Engineering New Zealand 2016, *Code of Ethical Conduct*, https://www.engineeringnz.org/engineer-tools/ethics-rules-standards/code-ethical-conduct

Engineers Australia 2000, *Engineers Australia Code of Ethics*, Engineers Australia, Canberra.

——— 2019, *Stage 1 Competency Standard for Professional Engineers*, Engineers Australia, Canberra, https://www.engineersaustralia.org.au/publications/stage-1-competency-standard-professional-engineers

——— 2022, *Code of Ethics and Guidelines on Professional Conduct*, Engineers Australia, Canberra, https://www.engineersaustralia.org.au/publications/code-ethics

Godfrey, E 2004, *Engineering education: Enculturation, assimilation or just passengers on the bus?*, Proceedings of the 15th conference of the Australasian Association for Engineering Education, Toowoomba, Qld.

Good, A 2019, 'Admitting failure', https://www.admittingfailure.org

Gupta, S, Davoodi, H & Alonso-Terme, R 1998, *Does corruption affect income inequality and poverty?*, working paper, International Monetary Fund, Washington, DC.

Hembrow, D 2014, 'Shared use paths create conflict and cause complaints about "speed"', *A view from the cycle path*, 21 April, https://www.aviewfromthecyclepath.com

Herkert, JR 2005, 'Ways of thinking about and teaching ethical problem solving: Microethics and macroethics in engineering', *Science and Engineering Ethics*, vol. 11, no. 3, pp. 373–385, https://doi.org/10.1007/s11948-005-0006-3

Holtzapple, M & Reece, W 2008, *Concepts in engineering*, 2nd edn, McGraw-Hill, Australia.

Institution of Electronics and Electrical Engineers (IEEE) 2020, *IEEE Code of Ethics*, https://www.ieee.org/content/dam/ieee-org/ieee/web/org/about/corporate/ieee-code-of-ethics.pdf

Klan, A & Kelly, J 2008, 'Hardie case challenges for title of most costly', *The Australian*, 4 October.

Lannin, S 2019, 'Roundup found to be "substantial factor" in causing US man's cancer', *ABC News*, 21 March, https://www.abc.net.au/news/2019-03-20/roundup-again-linked-with-cancer-by-us-court/10920974

March, S 2018, 'Cancer Council calls for Australian review amid Roundup cancer concerns', *Four Corners*, 8 October, https://www.abc.net.au/news/2018-10-08/cancer-council-calls-for-review-amid-roundup-cancer-concerns/10337806

Macquarie Dictionary 2024, *The Macquarie Dictionary online*, Pan Macmillan Australia, https://www.macquariedictionary.com.au

McDermott, Q 2011, 'Whistleblower slams Japan nuclear regulation', *ABC News*, 21 March, https://www.abc.net.au/news/2011-03-21/whistleblower-slams-japan-nuclear-regulation/2651

Memmott, P, Martin, D & Amiralian, N 2017, 'Nanotechnology and the dreamtime knowledge of spinifex grass', in C Baillie & R Jayasinghe (eds), *Green composites: Natural and waste based composites for a sustainable future*, Woodhead Publishing, Duxford, UK, pp. 181–198.

Mundel, AB 1991, *Ethics in quality*, ASQC Quality Press, New York.

Nahar, Y, Baillie, C, Catalano, G & Feinblatt, E 2010, 'Engineering values: An approach to explore values in education and practice', *Developing Professionalism in Higher Education*, SCEPTrE, HEFCE.

National Highway Traffic Safety Administration 2018, 'Takata recall spotlight', https://www.nhtsa.gov
Newton, LH & Schmidt, DP 2004, *Wake-up calls: Classic cases in business ethics*, Thomson South-Western, Canada.
National Offshore Petroleum Safety Authority (NOPSA) 2009, *Health and safety under the Offshore Petroleum and Greenhouse Gas Storage Act: What operators and workers need to know: Understanding duty of care*, NOPSA, Perth.
Rawls, J 1971, *A theory of justice*, Harvard University Press, Cambridge, MA.
Rubie, P 2009, *Professional engineering liability*, lecture for Engg1803 Australian Centre for Innovation Limited, Faculty of Engineering & IT, University of Sydney.
Safe Work Australia 2022, *Australian workers' compensation statistics 2020–21*, Safe Work Australia, Australia, https://www.safeworkaustralia.gov.au/doc/australian-workers-compensation-statistics-2020-2021
Sales, L 2012, 'James Hardie win disappoints asbestos campaigners', *7.30 Report*, https://www.abc.net.au/news/2012-11-12/james-hardie-win-disappoints-asbestos-campaigners/4367900
Schinzinger, R & Martin, MW 2000, *Introduction to engineering ethics*, McGraw-Hill, New York.
Severson, RW 1997, *The principles of information ethics*, ME Sharpe, New York.
Sexton, E 2007, 'Bernie banton 1946–2007', *The Sydney Morning Herald*, 27 November, https://www.smh.com.au/national/bernie-banton-1946-2007-20071127-gdrozn.html
Standards Australia 2018, 'Standards Australia is the nation's peak non government, not-for-profit standards organisation', https://www.standards.org.au
Stephan, KD 2001, 'Is engineering ethics optional?', *Technology and Society Magazine*, IEEE, vol. 20, no. 4, pp. 6–12, https://doi.org/10.1109/44.974502
Story of Stuff 2011, *The story of electronics*, https://www.storyofstuff.org/movies/story-of-electronics
——— 2024, https://www.storyofstuff.org
Suzuki, T 2017, 'Six years after Fukushima, much of Japan has lost faith in nuclear power', *The Conversation*, 10 March, https://theconversation.com/six-years-after-fukushima-much-of-japan-has-lost-faith-in-nuclear-power-73042
Transparency International 2023, 'Corruption perceptions index 2023', https://www.transparency.org/en/cpi/2023
Tweedale, G 2001, *Magic mineral to killer dust: Turner & Newall and the asbestos hazard*, Oxford University Press, Oxford.
Zhou, N 2018, 'Australia urged to restrict Monsanto's Roundup after US court rules it caused cancer', *The Guardian*, 13 August, https://www.theguardian.com/business/2018/aug/13/australia-urged-to-restrict-monsantos-roundup-after-us-court-rules-it-caused-cancer

ACKNOWLEDGEMENTS

Photo: © Caroline Baillie
Photo: © svittlana / Adobe Stock Photo
Photo: © Paul Memmott
Photo: © Gorodenkoff / Shutterstock
Photo: © Bloomberg / Getty Images
Photo: © Gamma-Rapho / Getty Images
Photo: © Hans / Adobe Stock Photo
Photo: © spixel / Shutterstock
Figures 4.1 and 4.2: © Rubie, P 2009, *Professional engineering liability*, lecture for Engg1803 Australian Centre for Innovation Limited, Faculty of Engineering & IT, University of Sydney.
Figure 4.3 and table 4.2: © Prue Howard 2011.
Figure 4.4 and text: © Engineers Australia 2022, *Code of Ethics and Guidelines on Professional Conduct*, https://www.engineersaustralia.org.au/publications/code-ethics
Figure 4.5: © Institution of Electronics and Electrical Engineers (IEEE) 2020, *IEEE Code of Ethics*, https://www.ieee.org/content/dam/ieee-org/ieee/web/org/about/corporate/ieee-code-of-ethics.pdf
Figure 4.6: © Nahar, Y, Baillie, C, Catalano, G & Feinblatt, E 2010, 'Engineering values: An approach to explore values in education and practice', *Developing Professionalism in Higher Education*, SCEPTrE, HEFCE.
Figure 4.7 and text: © Engineers Without Borders.
Figure 4.8: © Transparency International 2023, 'Corruption perceptions index 2023', https://www.transparency.org/en/cpi/2023
Text: © Pedestrian Council of Australia, http://www.walk.com.au
Text: © *British Asbestos Newsletter* 2000, issue 37, Winter 1999/2000.
Text: © McDermott, Q 2011, 'Whistleblower slams Japan nuclear regulation', *ABC News*, 21 March, https://www.abc.net.au/news/2011-03-21/whistleblower-slams-japan-nuclear-regulation/2651
Text: © Baillie, C & Levine, M 2013, 'Engineering ethics from a justice perspective: A critical repositioning of what it means to be an engineer', *International Journal Engineering, Social Justice and Peace*, vol. 2, no. 1, pp. 1–11, https://doi.org/10.24908/ijesjp.v2i1.3514

PART 3

PROFESSIONAL SKILLS

CHAPTER 5

Self-management

'Don't bother just to be better than your contemporaries or predecessors. Try to be better than yourself.'

William Faulkner

LEARNING OBJECTIVES

After studying this chapter, you should be able to:

5.1 identify the personal characteristics and strengths you can use to manage your work

5.2 describe the factors that inspire and motivate you to become an engineer

5.3 develop your engineering knowledge by exploring the engineering in your world

5.4 use self-management skills to plan and manage your work

5.5 use life-long learning skills to plan and manage your learning

5.6 review your performance using reflective practice techniques.

Introduction

University education is designed to provide engineering students with the knowledge and skills required to enter the profession as a graduate engineer. During undergraduate years, work placements, paid holiday work, and seeking interaction with other students and engineers will build your professional network and confidence in 'talking the talk'. Knowledge plus network is a winning combination for landing your first job when you graduate.

The focus of this chapter is self-management. This requires graduates to have 'capabilities for self-organisation, self-review, personal development and life-long learning' (Wright et al. 2010). Graduates need these capabilities to be able to:

- manage their own time effectively by prioritising competing demands to achieve both personal and team goals
- undertake regular reviews of personal performance to inform future practice and identify learning needs
- manage continuing personal and professional development.

All engineers must undertake continuing professional development (CPD) activities to improve their practice, address previously unseen problems, take on roles they have not previously held, assess and implement new technologies, react to changing priorities, respond to new requirements, and operate in new community, economic and environmental contexts. Good engineers know they need to review the way they do things to see if there are better ways to deliver required outcomes. They maintain their creativity to develop innovative and sustainable solutions to emerging engineering problems.

Next time you drop by the local fast-food drive-through, spare a thought for the global packaging engineers who ensure your greasy burger, dripping with hot sauce and melted cheese, behaves itself. Being able to eat your burger one-handed in the car requires packaging that will keep all the sauce, grease and pickle juice from dripping onto your lap. The wrapping needs to be slippery enough that your melted cheese won't stick to it, and also environmentally acceptable. No more plastic straws or cling wrap — get with the program! A convenient, one-handed burger needs compostable packaging to support society's transition to a sustainable, circular economy.

The average Australian packaging engineer has faced quite a few substantial self-management and professional learning challenges to keep up with recent changes in how we expect food, and takeaway food in particular, to be presented.

Fast-food packaging has evolved rapidly in the last few decades. In the 1970s, your parents or grandparents would have torn into Friday night fish and chips drenched in vinegar and wrapped in recycled newspaper. As public sentiment shifted about potentially ingesting newsprint ink, takeaway shops started wrapping fried foods in butcher's paper coated with 'sizing'. The process of sizing involves impregnating dried paper with a solution of starches in order to strengthen the paper, and can also involve the addition of chemicals to improve characteristics like brightness and resistance to water and grease (Glittenberg 2012).

Butcher's paper was fine for a time, but the evolution of takeaway food service meant one-size-fits-all butcher's paper packaging didn't last. Franchise takeaway food outlets popped up all over, and these businesses knew there were financial gains to be made if individual items could be cooked and packed quickly, presented more attractively, kept warm, eaten with one hand and upsold in 'meal deals'.

The 1980s saw the introduction of polystyrene packaging, particularly the revolutionary 'clamshell', which looked groovy, kept burgers hot, and caught all manner of drips and leaks. The clamshell was accompanied by a wide range of convenient, single-use plastic products (plastic straws, lightweight plastic wrapping, polystyrene drink cups and plastic lids) which, after a profligate decade or two, became a public relations and environmental nightmare.

Synthetic plastics are derived from crude oil, natural gas or coal, which now have an inconvenient association with climate change. Litter is also unpopular. Clean Water Action, a powerful American environmental organisation, reported that currently 49 per cent of street litter in the USA comes from fast-food outlets (Samson 2023). Single-use plastics are also terrifically durable, which means landfills are being clogged with these wastes; the half-life of a single-use plastic like dog poop bags is around 20 years, and some so-called biodegradable plastics do not degrade so much as disintegrate into micro- and nano-sized particles, with a range of adverse impacts on the environment.

International marine research has illustrated the problem to sea life of single-use plastics from takeaway food (figure 5.1). Dr Carmen Morales-Caselles, at the University of Cádiz, Spain, led the research and said:

> We were not surprised about plastic being 80% of the litter, but the high proportion of takeaway items did surprise us, which will not just be McDonald's litter, but water bottles, beverage bottles like Coca-Cola, and cans (Carmen Morales-Caselles et al. 2021).

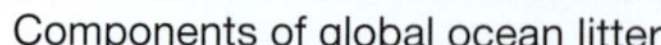

FIGURE 5.1 Components of global ocean litter

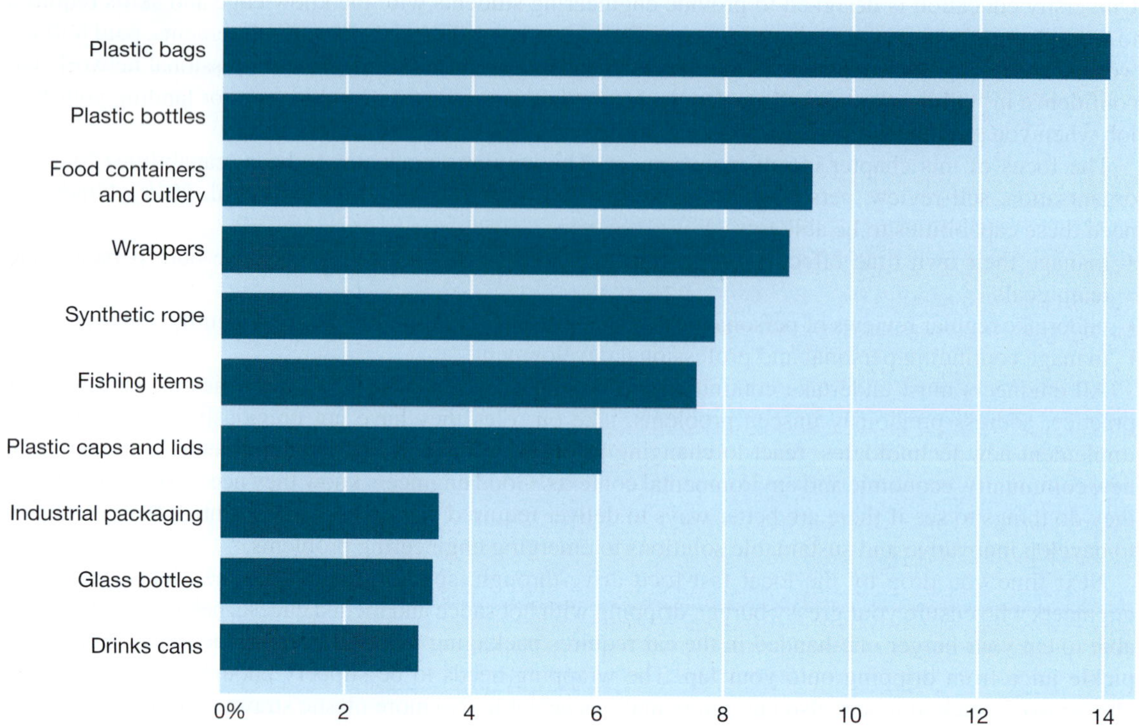

Source: Adapted from Morales-Caselles et al. (2021).

Packaging engineers in Aotearoa New Zealand have needed to upgrade their knowledge and think creatively since 2022 when the Ministry for the Environment made it illegal to provide, sell or manufacture the following plastic products in Aotearoa New Zealand:

- single-use plastic drink stirrers (all plastic types)
- single-use plastic cotton buds (all plastic types)
- plastics with pro-degradant additives (subset of plastic type 7)
- certain PVC food trays and containers (plastic type 3)
- polystyrene takeaway food and beverage packaging (plastic type 6)
- expanded polystyrene food and beverage packaging (plastic type 6) (New Zealand Ministry for the Environment 2023).

Fortunately, a solution was near to hand in the form of compostable takeaway food containers. These items were developed and marketed as green solutions that could replace single-use plastics. They were made from wholesome materials like bamboo, reprocessed sugarcane and paper, and, as the name suggests, they were able to be composted and hence circulated back into the systems we use for food production . . . or not.

To stop foods sticking to compostable takeaway containers, many of them are coated with per- and polyfluoroalkyl substances (PFAs). PFAs have come to be known as 'forever chemicals'. In 2020, the Australian Government released a position statement on the management of PFAs, which reads, in part:

> polyfluoroalkyl substances (PFAS) may cause human and environmental harm. All Australian governments have agreed that further release of PFAS into the environment from ongoing use should be prevented where practicable.

This statement led to bans on compostable packaging in local government food and organics waste collection (FOGO) bins. As such, PFAs coated packaging can no longer be sent to council-run organics composting facilities; the NSW Environment Protection Authority (2023) said:

> Fibre-based food contact materials including baking paper, coffee filters, paper towels, serviettes; fibre-based food containers such as coffee cups, pizza boxes, plates, bowls; paper bags; cardboard packaging; and other compostable plastic bags do not go in the FOGO bin.

While compostable food containers are still available for purchase and use in Australia, an ongoing court action suggests PFAs are unlikely to endure as a component of takeaway food packaging.

> Global chemicals maker 3M will be accused of a decades-long campaign to deceive the public about the risks of its controversial 'forever chemicals' as it faces a series of bombshell legal claims worth up to US$40 billion ($58 billion) that a judge described as an 'existential threat' to multiple defendants' survival (Fellner 2023).

The key message here is that as knowledge, legislation and public expectations continue to evolve, engineers must continually go back to the drawing board (and the technical literature) to keep up to date. Engineers also need the skills to communicate their environmental or social concerns to decision makers and leaders within their workplace.

What a worthy thing, however, to be a lifelong learner and to be the engineer tasked with devising the next environmentally acceptable solution for the challenge of sticky, greasy, drippy, yummy, one-hand-edible burgers and the like.

This chapter discusses learning, self-management and reflective practice, and will equip you with the knowledge and skills that will help you to manage your work effectively and to reflect on, and learn from, your experiences.

5.1 Understanding self

LEARNING OBJECTIVE 5.1 Identify the personal characteristics and strengths you can use to manage your work.

Personal characteristics, capabilities, experiences and preferences play an important role in the way people work, the way they interact with other people and the way they manage their lives. Considering the journey from engineering novice to professional engineer as a continuum, it is likely that each engineering student will start their journey at a different point on that continuum, with personal characteristics influencing progress along the way. For this reason, it is important to understand and capitalise on strengths, as well as understand and address any perceived weaknesses in experience, knowledge and skills.

KEY POINT

Engineers who know their strengths can use them to optimise their performance.

Your personality and attitudes

Psychologists define **personality** as the 'enduring patterns of thought, feeling, motivation and behaviour that are expressed in different circumstances' (Burton, Westen & Kowalski 2023, p. 664). Personality is complex, and it is a distinguishing feature that helps individuals establish their identity. It influences the way people react to places, objects, situations and, therefore, it also affects their attitudes. The Macquarie Dictionary (2024) defines **attitude** as a 'position, disposition, or manner with regard to a person or thing'.

In a 2015 survey of 367 employers of graduates in Australia, the selection criteria rated as most important were interpersonal and communication skills, cultural alignment/values fit and emotional intelligence (including self-awareness, self-regulation, self-motivation) (Graduate Careers Australia 2016, p. 19).

More recent analysis by the Australian National Skills Commission (2020) examined the employability skills employers valued most highly. They found:

> Employers seek staff with the right qualifications and skills, and many require relevant work experience. While all employers are unique and place emphasis on different attributes, they will not compromise on employability skills (National Skills Commission 2020, p. 1).

The National Skills Commission (2020) differentiated between traditional employability skills that are still highly sought after, and emerging needs among employers for professionals who will position businesses to be ready for the challenges of the future (table 5.1).

The National Skills Commission (2020) also found that 75 per cent of employers place at least as much emphasis, if not more, on personal qualities than they do on technical skills when recruiting.

TABLE 5.1 Employability skills

Traditional employability skills	Twenty-first century skills
Reliability and punctuality	Creativity
Customer or client service skills	Problem solving
Positive attitude and motivation	Critical thinking
Ability to work in a team	Digital and financial literacy
Good personal presentation	Presentation skills

Source: National Skills Commission (2020).

The following spotlight describes an Australian research project that identifies the importance of both engineering and personal attributes. The study was undertaken by Dr Sally Male, an electrical engineer and one of the authors of this text.

SPOTLIGHT

The Competencies of Engineering Graduates Project

DR SALLY MALE, UNIVERSITY OF WESTERN AUSTRALIA

An overwhelming finding of the Competencies of Engineering Graduates (CEG) Project was that while technical competencies are necessary, they are not sufficient for performing engineering jobs well. This result, from the first large-scale quantitative study in Australia, is consistent with large-scale studies in Europe, the United States and New Zealand, as well as some small-scale studies in Australia. It emphasises the importance of non-technical competencies in the engineering industry. The CEG Project defined generic engineering competencies as the amalgamations of knowledge, skills, attitudes and dispositions required to respond well to the demands of an engineering job in any discipline.

To answer the question 'What are the generic engineering competencies required by engineers graduating in Australia?', the CEG Project team developed a questionnaire for engineers with 5 to 20 years' experience. This group was chosen following advice from industry advisors that, as employers usually seek graduates who will become engineers who can add value, the competencies should be rated on their importance for performing the job of an 'established engineer' well.

The questionnaire was based on a set of 64 generic engineering competencies (Male et al. 2009) that had been refined from a larger list identified in the literature. Three hundred established engineers rated each of the 64 competencies on their importance for performing their job well (1 = *Not needed*; 5 = *Critical*) (Male et al. 2011). The outcomes of this survey were confirmed in a second survey completed by 250 senior engineers. Just over one-third of the competencies had a mean rating of between four and five on the five-point scale. These competencies are listed as follows, from highest to lowest mean rating. Another way of reporting the results is to identify those competencies that were rated as *critical* by more than 50 per cent of the participants. The eight competencies that met this criterion are marked with an asterisk.

1. Communicating clearly and concisely in writing*
2. Managing own communication*
3. Managing self*
4. Using effective verbal communication*
5. Working in teams*
6. Speaking and writing fluent English*
7. Interacting with people in diverse disciplines, professions and trades*
8. Being committed to doing your best
9. Solving problems*
10. Demonstrating honesty

11. Making decisions within time and knowledge constraints
12. Being positive, enthusiastic and motivated
13. Demonstrating practical engineering knowledge and skills, and familiarity with techniques, tools, materials, devices and systems in your discipline of engineering
14. Sourcing, understanding, evaluating information
15. Acting with exemplary ethical standards
16. Presenting a professional image in demeanour and dress
17. Thinking critically to identify potential possibilities for improvements
18. Using effective graphical communication
19. Being flexible, adaptable, willing to engage with uncertainty or ill-defined problems
20. Thinking laterally, using creativity, initiative and ingenuity
21. Being concerned for the welfare of others in your organisation
22. Negotiating, asserting, and defending approaches and needs
23. Managing information/documents (Male et al. 2009).

The ratings from the first survey were factor analysed to group items with correlated ratings, revealing 11 underlying generic engineering competency factors, with each factor being named after the group of competencies it was based on (Male 2012). They are, in descending order of rating, as follows.

1. Communication
2. Teamwork
3. Self-management
4. Professionalism
5. Ingenuity
6. Management and leadership
7. Engineering business
8. Practical engineering
9. Entrepreneurship
10. Professional responsibilities
11. Applying technical theory.

A focus group of 12 engineers was used to refine the descriptions of the generic engineering competency factors. The top four factors, and the competencies they are based on, were:

1. communication: graphical communication, English, written communication, verbal communication
2. teamwork: interdisciplinary skills, diversity skills, teamwork
3. professionalism: honesty, loyalty, commitment, ethics, demeanour, concern for others
4. self-management: managing development, information-management, self-management, managing communications, action orientation.

CRITICAL THINKING

Use a five-point rating scale (1 = *poor*; 3 = *OK*; 5 = *very good*) to realistically rate yourself against each of the 23 competencies. Consider how you might improve your rating for the three personal attributes that scored the lowest in your self-assessment.

A study that defined the graduate capabilities of mining engineers also highlighted the importance of the following personal attitudes and capabilities (Dowling 2014):

- demonstrates a proactive approach (e.g. volunteers for tasks)
- demonstrates initiative
- demonstrates a capability to be innovative
- considers new ideas with an open mind
- adapts to unfamiliar tasks quickly and effectively
- thinks and acts like an engineer/manager
- identifies gaps in personal capabilities and skills
- owns and takes charge of tasks
- takes responsibility for their work and actions
- uses proper planning to achieve targets and deliver outcomes on time (e.g. a systems approach)
- negotiates personal task deliverables based on effective time management
- develops and executes daily work plans
- allocates time effectively and is punctual
- demonstrates an understanding of the difference between work and activities (i.e. value-add and non-value add)
- reviews personal performance objectively and develops and implements improvement strategies
- maintains a work/life balance

- demonstrates an understanding of the importance of codes of conduct and ethics
- practises in accordance with job description and code of conduct
- practises ethically and with integrity
- respects lines of command.

This text addresses the majority of these competencies so, while you are at university, you should take advantage of the many opportunities that are provided in your program, and in extracurricular activities, to develop these personal attributes and capabilities. For example, the last weeks of each semester will provide you with opportunities to trial different self-management techniques to manage the pressure of assignments and exam preparation. Some of these techniques are discussed in this chapter and others in the chapter on managing engineering projects. You will also learn other techniques in management and project management subjects.

Your strengths

The Gallup Organization has surveyed more than 10 million people worldwide about their thoughts on employee engagement. The interviewees were asked to use a four-point scale to respond to the statement 'At work, I have the opportunity to do what I do best every day', indicating whether they *agree, strongly agree, disagree* or *strongly disagree* with this statement. Only one-third of the interviewees strongly agreed with the statement (Rath 2007). This means that two-thirds of the respondents do not believe that they are working to their strengths. If this is true, then the implication is that these people are under utilised.

Buckingham and Clifton (2002) define strength as a 'consistent near perfect performance in an activity'. Learners may prefer to adopt a less rigorous definition that recognises room for improvement — even in areas considered strengths. From this perspective, a **strength** may be considered as an ability to perform in an activity at a higher level than peers. Peers may also recognise each other's strengths, and their weaknesses! For example, an individual may always be asked to lead a team if leadership is perceived to be one of their strengths.

Buckingham and Clifton (2002) suggest that the signs that point to your strengths are when:

- you like doing a particular task and you are successful
- you look forward to doing that task
- time speeds by when you are doing that task
- you are fulfilled through doing that task.

It is generally accepted that people are the most important resources in an organisation, especially their knowledge, skills and experience. A key management strategy should be to harness these capabilities and ensure that all employees utilise their strengths during the major part of their working day.

Knowing your strengths will enable you to recognise, and seek out, opportunities where you can play to those strengths.

Your personality

As a student, your understanding of personality traits may help you to better manage your performance as an individual and in group or team activities. There are numerous **personality tests** that purport to measure personal characteristics. Some of these tests are widely used by psychologists working in educational and industry organisations to assess individuals' suitability for particular careers. These tests are a useful indicator of a person's mix of traits, but they should never be regarded as concrete evidence when making important decisions. The fact is, there are many different ways to assess personality; therefore, the results of any one test are unlikely to provide a complete view of a person's traits. The results of a test should therefore be treated as useful, but not as a definitive guide to an individual's personality attributes.

Two popular tests are the Myers-Briggs Type Indicator (MBTI) and the Five Factor Model. While participants are normally charged a fee to undertake these tests, many universities provide students with opportunities to take these or similar personality tests at no cost or at a discounted cost. Some versions of the tests are also available online. If you take a personality test, you should find out as much as you can about the meaning of your results so you can use this information to your advantage. For example, traits including agreeableness (how easygoing you are) and extroversion (a position between extrovert and introvert) affect the way people work in team situations; therefore, knowledge about personality traits may help you understand the way you and your fellow team members behave. You may even use this understanding to help you select the members of a team or assign team roles.

Spatial ability

Engineering educators have recognised that spatial ability is an important cognitive ability for engineers (Magin & Churches 1996; Sorby 1999; Sutton et al. 2009a, 2009b). In fact, Ezaki and colleagues have commented that the ever-increasing use of 3D computer design and modelling software systems in the engineering profession requires 'graduates to have developed a greater level of spatial ability than that demanded in previous years' (cited in Magin & Churches 1996, p. 95).

Navigating a car through a strange city can be a daunting task for many people, although a GPS tends to alleviate this problem. Navigation is one example of the use of spatial intelligence or ability. To be successful, the navigator must be able to visualise the information from the map and relate it to the real world in real-time. Another common example of the use of spatial intelligence is when someone asks you to give them directions on how to get to a particular building or address. If you have good spatial intelligence, you should be able to visualise a route you know and give clear instructions about the roads and streets the person can follow to get to their desired destination. You may even be able to picture the whole locality and rotate it in your mind so you can pick the easiest route. The ability of a person to comprehend your instructions may also be affected by their spatial ability.

Burton (1998) found **spatial ability** involves 'the ability to deal with spatial information and, as such, is often recognised as a facility for manipulating visual stimuli, rotating shapes in the mind's eye, and imagining how things would look from another perspective'.

People have different levels of spatial ability and researchers have found that a number of factors influence an individual's spatial intelligence, including age, experience, childhood environment and gender. Many studies have found there are substantial differences in the spatial abilities of men and women, with men generally having higher levels of spatial intelligence (Gorska et al. 1998; Sutton et al. 2009b), although ethnicity may also influence the level of spatial ability. It is likely that this is due to different experiences because spatial ability is successfully improved with learning and practice and indeed this has been found beneficial for engineering studies (Sorby & Baartmans 2000).

Your spatial ability

There are numerous tests that can be used to measure spatial ability, although normally each test only measures one type of spatial ability. For example, **spatial scanning** is the ability to mentally scan a map or object and find a path or connection between two points when speed is a factor. These timed tests include activities such as finding a route through a maze. On the other hand, **spatial relations** has to do with the ability to manipulate or rotate simple visual patterns when speed is a factor. These timed tests include activities such as matching objects in a series of objects that have been rotated in space.

Many engineering schools include spatial ability testing in their first-year graphics course. They may also schedule remedial programs to enhance the spatial ability of the students who were identified as having low spatial abilities. If you find the graphics components of your subjects challenging you should consider ways that you may enhance your skills by seeking help from the staff in your school.

In this section we have discussed a number of personal characteristics that may affect the way you manage yourself, your life, your studies and your work. An understanding of these characteristics can improve your skills to enhance your learning while you are at university and, later, your performance as an engineer. There are many other factors that influence our performance, one of which is motivation.

5.2 What motivates you?

LEARNING OBJECTIVE 5.2 Describe the factors that inspire and motivate you to become an engineer.

What is it that excites you about engineering? What is it that led you to this point where you are starting your degree? Did you choose engineering because you:

- look forward to the challenge of design and creativity
- like applying mathematics and science to solve problems
- want to apply your engineering knowledge to help people and make a better world
- like building things
- like to know how things work
- would like to use your profession as a means to travel
- want to receive the high salaries and quality working conditions many engineers enjoy?

Perhaps it is a combination of some, or even all, of these reasons.

Mature-age students are often motivated by other factors. For example, because they may have worked in the engineering industry and know the career path they wish to pursue. They may also have given up full-time employment to pursue their dream job, and are driven to succeed partly because of the financial cost of their decision. Many students are motivated by the fact that other people such as parents or relatives are making monetary or time-related sacrifices to assist them with their studies.

Motivating factors are positive because they inspire you to press on and succeed. However, while you may be motivated to do well, there are some factors that may have a negative impact on your study. For example, you may:

- feel guilty if you have to take time out from family duties or leisure activities to study
- feel depressed if you find a subject difficult to understand
- want to give up if you fail a major assignment or an exam.

Compare how your reasons for studying an engineering program align with, or differ from, the motivations of the graduates interviewed in the following spotlight.

KEY POINT

Engineers are aware of and inspired by the engineering in their environment.

SPOTLIGHT

1. A civil engineering graduate's perspective

INTERVIEW WITH CAROLINE BUT

After five years' experience working as a site engineer in a construction company, Caroline But talked about her experience as a student and the transition to work.

Why did you choose to study engineering?

As a child I always admired crossing over bridges and staring up at tall buildings and always wondered how they were built. But no one in my family had ever introduced me to the profession of engineering. It was only in high school that my physics teacher opened my eyes to the various STEM career paths I could take, and he encouraged me to pursue my studies in engineering. I thought it was a ridiculous idea — how can a girl study in such a male-dominated environment! But his words stuck with me and I pursued getting into the course.

Caroline But

What fields of engineering interested you when you started at university?

I didn't realise how many fields of engineering there are at university, and it was initially overwhelming trying to figure out what majors I wanted to be in. However, I was naturally drawn to the concepts and theories of civil engineering and it certainly helped that I could physically see it everywhere.

What kept you motivated during your years at university?

I was blessed to have volunteered and taken on committee roles with Engineers Without Borders UWA. The organisation inspired me to view engineering as a profession that can help improve the wellbeing of the most under-served communities. It was motivating to work on projects with like-minded people and it certainly helped me engage with my studies.

Did you get engineering work experience while you were studying?

I got into a cadetship program with Georgiou during my studies and worked with them once or twice a week during the university semester and full-time during the breaks. This continuity of employment gave me practical experience on how projects were managed from start to finish, and it felt great knowing that I knew more about the workplace compared to my other classmates.

How did you find your first full-time engineering job?

I had already been working with Georgiou for three years when I was offered a full-time graduate position with them after I graduated. I didn't even consider applying for other companies as I was eager to learn as much as I could with them.

Was the transition from university to work difficult?

Personally, it wasn't difficult for me as I enjoyed the responsibilities in my role and I loved how I didn't have to worry about assignments and studying after work or during the weekends.

Has working as a structural engineer met your expectations?

My initial expectation after university was that I'd still be doing a lot of learning in the workforce. But what it didn't prepare me for, was how much time I'd spend interacting with people. Everything I do is collaborative and because of that I've met some brilliant people who've inspired me along the way.

What is the greatest challenge you have faced in your job so far?

As a woman in the construction industry I feel I must make more effort to prove my capabilities before my client and colleagues compared to my male counterparts. Most people would initially assume I'm an administrator and are surprised when I inform them of my role.

Have you continued to learn since graduating? If so, why and how?

I've always thought concrete was a pretty dry topic. But since I'm always dealing with it, I've been learning about its various applications and properties during my spare time.

What is the most satisfying aspect of your work?

The most satisfying part of my work is being appreciated by my colleagues and manager for my efforts in the project. I love collaborating in a team and working towards a common goal. The best part is finishing a project and saying, 'We built that together'.

How would you sum up your career to date?

Every year or two I would move onto another project and learn to interact with a new team. It's exciting, challenging and extremely rewarding to see structures that you've helped construct from just a concept.

2. An electrical engineering graduate's perspective

INTERVIEW WITH MILES CATTACH

After two years working as an electrical engineer in a consulting engineering company, Miles Cattach talked about his experience as a student and graduate.

Miles Cattach

Why did you choose to study engineering?

Growing up I spent a lot of time tinkering, building and taking things apart. I have always enjoyed finding out more about how things work and if they could be improved. By the end of high school, physics was my best and favourite subject, in which we learned the way the world worked in an in-depth way that I had not explored before. It was a natural transition from high school physics into tertiary engineering and I have not looked back since.

What fields of engineering interested you when you started at university?

Early on I enjoyed the mechanical engineering units the most, from analysing structures through to mechanisms and machines. From day one I found electrical engineering units at university to be challenging, but the challenging nature of the units made me more curious. The further I progressed with studying electrical engineering, the more I came to realise that I enjoyed the challenge that it is.

What kept you motivated during your years at university?

You always hear about the first engineering lecture, when the lecturer says, 'Look to your left, now look to your right, only one of you will come out of this with an engineering degree'. Of course it's not like that anymore. However, from day one I was determined to be the one out of three. There were many challenges along the way, but with the first engineering lecture in the back of my mind and by surrounding myself with supportive and passionate classmates I found that my motivation levels remained high throughout my studies.

Did you get engineering work experience while you were studying?

I worked on-site at an alumina refinery for my work experience during university. While you learn so much during your time at university, you really learn the most vital information and lessons once you start working. I strongly recommend that all engineering students actively seek work experience early on in your studies as it is never too early to start developing yourself as a professional.

How did you find your first full-time engineering job?

During my studies I was fortunate enough to complete a company-affiliated unit, during which time I created strong professional relationships with key members of the respective company. As the relationship grew and trust was built, a job opportunity developed.

Was the transition from university to work difficult?

The transition into the workplace is a large lifestyle change, but also one of the most exciting times in your professional life, as all your years of hard work come to fruition. The transition was not difficult because the excitement and enthusiasm of the big changes makes the challenges enjoyable.

Has working as an electrical engineer met your expectations?

Working as an electrical engineer has met and gone beyond my initial expectations. The role of an electrical engineer requires a vast diversity of skills, from the technical and theoretical knowledge to allow for the completion of complex modelling and calculations, to managerial skills and critical thinking to understand the challenges that are faced day to day.

What is the greatest challenge you have faced in your job so far?

The greatest challenge has been developing the self-confidence required as an engineer to make decisions and be able to back myself in choices. Looking back, there have been many instances in the beginning of my career when I had the required knowledge and understanding to back my decisions, but did not.

Have you continued to learn since graduating? If so, why and how?

Definitely. If I wasn't learning I would be doing something wrong. Engineers are constantly challenged to learn, whether they are early, middle or late in their careers. The world is forever changing and so is the profession of engineering. Standards, technologies and systems are constantly changing and being updated, requiring engineers to always be learning and up to date. Most of the learning is on the job, with some formal education along the way.

What is the most satisfying aspect of your work?

Working as a team to deliver a high-quality project. Working with a diverse team makes a project more satisfying and rewarding when it comes to completion.

How would you sum up your career to date?

I have been able to work for multiple clients across multiple industries as an electrical engineer. This has involved everything from modelling and calculations through to site work and client interfacing. The learning curve has been steep and overwhelming at times, but the support and understanding of the working team makes it all manageable and truly rewarding.

CRITICAL THINKING

What is your story so far? Write it in approximately 200 words using the first few headings in this spotlight to guide your narrative.

It is important to understand the factors that motivate you to study engineering as they will help keep you motivated throughout your program and that will make managing your life and your studies so much easier. You will find other factors to motivate you while you are at university. Perhaps some of your lecturers will inspire you and encourage you to pursue a particular subject area or career, or perhaps you will have a life-changing work experience. In the meantime, strategies that may help you stay focused include:

- forming friendships, discussing problems with your classmates and encouraging others to do well; celebrating when you are successful
- arranging work experience to identify your preferred field of employment and study
- speaking to a career counsellor (or your course coordinator) about short- and long-term employment options
- joining student societies, or Young Engineers Australia
- joining Engineers Without Borders (EWB), Australia.

Whatever your motivation was to get to this point, you have now started on a life-changing journey. As an engineer, you will have opportunities to pursue your dreams, to be creative, to change your part of the world, and to perhaps also change other parts of the world. Your work as an engineer will be challenging, stimulating, satisfying and rewarding and you will be able to choose to work as an individual, or as part of a team. You may choose to work on large or small projects, in Australia or overseas. Of course, you may have to balance some potential downsides associated with these perks, such as long hours, being away from family and friends for periods of time, or working in remote areas.

An important factor in career planning is to understand what is motivating you to pursue a particular career option and what will continue to motivate you once you begin working in that field. One thing that motivates people to pursue a particular career is inspiration. People can be inspired by another passion or the actions of other people, or by an event, a need or a challenge. They can also be inspired by nature, technology or the built form.

Sources of inspiration

We live in an engineering-rich environment. Residences are connected to a range of engineering services such as water, electricity, wastewater systems, telecommunication networks and the internet. The latest devices, such as mobile phones and music players, are engineered. Household appliances and cars are engineered. In fact, we are probably so accustomed to living in a highly engineered environment that we take the engineering for granted. Yet, the creation of each of these goods or services is the direct result of innovative work performed by engineers. Imagine if you had been a part of the team that invented the latest must-have gadget. Many engineers are inspired by the challenge of creating or implementing new technologies. The benefits of introducing a new technology are often well known because they were the drivers for the development and implementation of the technology.

Engineers are critical not only in providing our basic needs such as shelter and food, and supporting exciting and comfortable enhancements to our quality of life, they are critical also to human survival. Humanity is facing pressing problems such as climate change, pollution, extinctions, demand for water and diseases. Engineers collaborate with others to address these local and global challenges. Students and graduates find passion in opportunities to make significant contributions to communication, food production, renewable energy, recycling, biomedical devices, space exploration, and monitoring and reducing water usage, power demands and pollution.

While some engineers are inspired by the challenge of developing or deploying new technologies, others are inspired by the challenge of improving the lives of people in their community or the lives of people in other countries. The following spotlights introduce two engineers who have put their passions into practice in the fields of electronics engineering and civil engineering.

SPOTLIGHT

Putting passion into practice 1

INTERVIEW WITH MADHU BHASKARAN

Associate Professor Madhu Bhaskaran was awarded the prestigious Batterham Medal by the Australian Academy of Technology and Engineering in 2018. She shared her story.

Tell us about your childhood and why you studied engineering.

I grew up in Chennai, India. Being a daughter to two medical doctors, I was always encouraged to choose a STEMM-based career. I kept my options open to do either medicine, science or engineering until Year 12 as I studied maths, physics, chemistry, botany and zoology all through the high school years. It was the time when India underwent the IT revolution, but my curiosity was in trying to understand what made computers and mobile phones function effectively and how did electronics get miniaturised to slowly become an integral part of a household. Hence, the choice to study electronics engineering.

Where did you study engineering?

I did my Bachelor of Engineering in Electronics and Communication at PSG College of Technology, Coimbatore, India. It was one of the top 10 engineering universities in India with an internationalised and up-to-date curriculum in engineering.

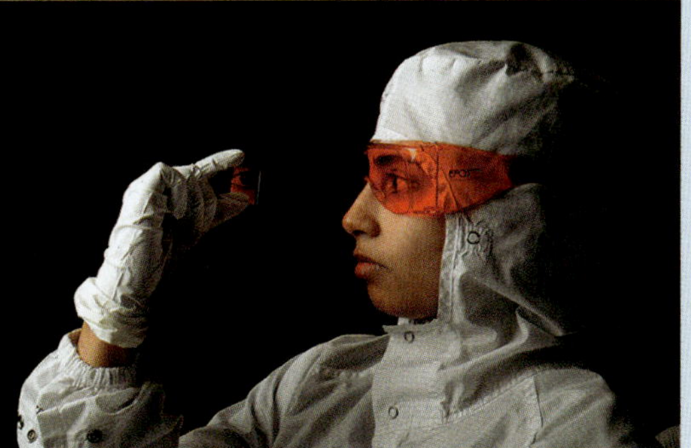

Madhu Bhaskaran

Why did you choose to go into your research field after graduation?

I came to RMIT University to complete a Master of Engineering in Microelectronic Engineering. Going into 'clean' rooms to fabricate my own silicon chips for the first time was an amazing experience. It was also fascinating to work at the micro and nano scales at which materials show new properties and behave differently as opposed to macro scale. I was keen to work more closely with electronic materials and design my own experiments. I was encouraged to undertake a PhD and my research was in electronic materials and design structures to reduce power consumption in silicon chips.

Congratulations on being awarded the Batterham Medal 2018. What was it for, and what benefits will it provide for society?

My PhD in electronic materials engineering equipped me with skills in materials and micro/nano fabrication. For the past five years, my focus has been on developing stretchable oxide materials. Oxides are essentially glass-like materials (which actually feature quite significantly in all of our smart devices) but are incredibly brittle. With the future moving towards stretchable, foldable and unbreakable electronics, my research solved the bottleneck of actually integrating oxides with stretchy rubber (similar to contact lens material) to realise wearable electronics. These could be sensors worn on skin or clothes to sense dangerous gases or monitor our UV exposure on a daily basis or form the building blocks to Harry Potter 'invisibility' cloaks.

What do you enjoy most about your work as an engineer?

I love the fact one can keep turning science fiction into reality. Engineers can come up with ingenious solutions to problems and it is satisfying to work with industry partners to see that some of my research will make its way out of the lab to benefit people.

CRITICAL THINKING

You could be part of an engineering team that overcomes an important challenge or realises a life-changing vision that previously seemed impossible. In 100 words, detail what achievement you would love to contribute to as an engineer.

SPOTLIGHT

Putting passion into practice 2

NIC HAHN, CIVIL ENGINEER

Finding myself responsible for the coordination of Water, Sanitation and Hygiene (WASH) for 180 000 internally displaced ethnic Tamils in Manik Farm, Sri Lanka, in the aftermath of the Civil War in 2006 was the biggest challenge and most shaping experience of my life. For most of the 11 months I was there, Manik Farm was surrounded by barbed wire, was heavily militarised and movement was restricted. Internally displaced people relied completely on assistance to access the most basic human rights of water supply and toilets.

Nic Hahn

Ever since I was in Year 8 and read an article about the valuable impact engineers can have on fixing water supplies to prevent disease I've wanted to be a civil engineer and work in emergency relief. So, after leaving high school I studied civil engineering at Monash University, focusing on water and geo-engineering. That decision took me on an incredible journey, from First Nations communities in outback Australia, to the earthquake and cholera response in Haiti, to the internally displaced persons camps in northern Sri Lanka following the Civil War, to typhoon devastated areas in the Philippines.

As an absolutely privileged Australian — born into a life of safety, free choice, relative wealth, excellent education and opportunity, and the most loving community of family and friends — it can be really difficult to comprehend the challenges that a large majority of the world face on a daily basis. It really is incredible to see people's faces the first time water comes out of a hand pump in their village, when previously they had to walk hours for water every day, or the look of relief on people's faces as they arrive safely into the sanctuary of a neighbouring country, after having to live for weeks, months or years in fear.

RedR Australia is a not-for-profit organisation that provides international emergency relief through the provision of skilled people and training to help communities rebuild and recover in times of crisis. Generally, RedR personnel are seconded to United Nations agencies on the ground to provide additional skills and manpower for three- to six-month stints. Since joining RedR back in 2004, I have been on an incredible journey through nine emergencies. I've worked in lots of different roles: as a water specialist, information manager, site planner, field officer, trainer and cluster coordinator. These emergencies include East Timor IDPs 2006, the Swaziland Drought 2007, Ethiopia/Somali refugees 2009, Sri Lanka IDP Situation 2009–10, Haiti Earthquake and Cholera Response 2010, Liberia/Cote d'Ivoire Refugee Response 2011 and Philippines Typhoons Response 2011–12 and 2013–14.

The most amazing part has been the people I've met along the way, such incredibly resilient human beings — the unsung heroes — who made it through war, natural disasters and their related traumas, but continued to smile, carry on with their lives and yet still give so freely to others. I challenge you to find a way to give to your wider community, whether it be here in Australia or overseas, because I'm absolutely positive that this journey I've chosen has given me and my life way more than I ever thought I could give.

CRITICAL THINKING

Which field of engineering most interests you at this point in your career? Consider how this field of engineering may be used to improve people's lives.

As you become an experienced observer of engineering you will be inundated by the engineering that surrounds you as you go about your daily life. It is likely you will be inspired by what you see; you may be inspired to develop an engineering solution to a problem you see in your community. These inspirational experiences motivate us to learn, and to use our engineering knowledge and skills to improve our world.

5.3 Developing your skills

LEARNING OBJECTIVE 5.3 Develop your engineering knowledge by exploring the engineering in your world.

One important skill that engineers use is problem solving, whether they are designing complex systems, investigating equipment malfunctions or exploring systems. Problem solving involves the balanced application of logic, judgement and decision making to analyse and synthesise information and develop a solution. The first step in this process is to understand the problem and the context. This stage is often called the problem definition stage of a project.

One strategy that is used for problem definition is to seek the answers to six simple questions. The six questions are: *Who? What? When? Where? Why?* and *How?* This approach helps us to understand all of the different aspects of the problem, and may even provide us with a solution. The following sections explain how these questions can be used to develop our critical thinking skills and enhance our understanding of engineering systems.

KEY POINT

Engineers are life-long learners.

Developing an inquiring mind

An inquiring mind is often the motivation that drives the pursuit of knowledge and the need to understand how things work, or why an action leads to a specific outcome. Many engineers and scientists enter the field because they have an inquiring mind — they want to spend their lives searching for answers to technical problems. Of course, there are many other reasons that people study engineering. For example, following a study of more than 300 students in para-professional engineering programs, Dowling (2010) found that the two main influences that led to their decision to study engineering were 'I like finding out how things work' and 'I like building things'. If your inquiring mind motivates your learning, then you are fortunate. If not, you should try to develop the habit of thinking this way, perhaps by using the six simple questions.

Two approaches are described in the following sections. Other modes of enquiry are discussed in the critical thinking section in the chapter on the engineering method and in the chapter on engineering design.

The black box strategy

The term **black box** is used to describe a device, process or system when it is only being studied from an input and output perspective; thus, the study only looks outside the box or the system boundary. For example, most of us treat a computer as a black box because we are satisfied when it produces the required output for the given input. Because we are not concerned with how the computer or software package works, we are treating it as a black box. This strategy is normally used to design large or complex systems that consist of many components and linkages. At the system design level, each component would be treated as a black box with specified inputs and outputs. Once the overall system is defined, then each component would be designed in detail or purchased as a stand-alone item.

The black box approach is often used when a complex software package is being developed. For example, a software package that manages the storage and retrieval of shipping containers at a cargo wharf would consist of a number of components or subsystems that:

- read the code number on a container
- store the location of a container in a database
- instruct a vehicle to locate and retrieve the specified container
- instruct a vehicle to take a container to a specified location and store it
- record and report on all of the completed tasks.

When designing the software system, the software engineer would treat each of these components as a black box. The key information at this stage would be the input data that the component would need to function and the output data that would be returned to the system. Once the overall system had been designed, the software could be developed for each component. It is at that point that the engineers would begin to work within each component to design and develop the software.

Exploring the black box

In this era of rapidly changing technology, it is becoming increasingly difficult to keep up to date, but engineers need to understand how each new technology operates before they use it or recommend it to a client. One way they do this is to conduct some trials with the new technology and then compare its performance against the design specifications or other requirements. Sometimes it can be difficult to access the inner workings of the technology and only the inputs and outputs can be studied. This may prove to be a problem on some projects. The following spotlight demonstrates this dilemma.

SPOTLIGHT

Hydrographic surveys

The maintenance of the advertised depth and width of shipping channels is a critical component of harbour management. Today's supersized vessels have little room to manoeuvre, particularly close to wharves and other docking facilities, whether they are close to shore or built to service offshore oil and gas facilities. Shifting sand and silt can quickly change the geometry of a channel and render it unserviceable. Therefore, shipping channels are regularly measured to monitor clearance and to identify any high or narrow sections and, if necessary, to implement dredging programs to clear any obstructions.

HMAS *Leeuiwn* is a hydrographic survey ship for the Royal Australian Navy.

An example is the upgrade of the channels at Fremantle Port. During 2009, the Western Australian government funded a $250 million project to address a slump in shipping cargo trade through Fremantle Port. The depth of the inner harbour was deepened from 12.7 metres to 14.7 metres so that the new-generation of post-Panamax freight vessels, so called because they are too big to navigate the Panama Canal, can enter the inner harbour. Prior to the upgrade, these vessels could only enter the harbour partly laden, sometimes with less than 80 per cent of their cargo capacity. The size of the project is demonstrated by the fact the dredged material was used as fill to create nearly 27 hectares of reclaimed land behind a kilometre of new seawall constructed at Rous Head (Engineers Australia 2011).

With a navigable depth of 14.7 metres, Fremantle Port is well placed to attract additional trade. Between 2015 and 2016 Fremantle Port attracted 58 cruise ship visits carrying over 150 000 passengers, an increase from 43 visits in the previous year (Western Australia Department of Transport 2016, p. 8).

The measurements of navigable depth are obtained by undertaking a *hydrographic survey* that results in the production of a plan or map of the sea floor. The map also shows the depth of water across the surveyed area at a point in the tide cycle, such as mean sea level. The depths are shown by writing the depths at the map locations where the actual measurements were taken, and/or by a series of contour lines.

If you were learning how to measure a shipping channel 30 or 40 years ago, you would have learned how to use a sextant and an echo sounder and perhaps even a lead line. The position of each depth measurement would have been defined by simultaneous angle measurements made by two sextant observers to a series of known control points. The echo sounder trace would have been marked at the instant the angles were measured and the position and depth of the measurement would be manually plotted on a plan in real-time. This measurement process would involve at least four people on the vessel, which would have travelled up and down the channel until all of the required measurements were observed.

While they were taking measurements, the observers would have relied on a skilled coxswain to steer the vessel along a series of parallel lines within the channel, allowing the measurements to be taken along the channel at fixed intervals on those lines, forming a grid of measurements. This would not have been an easy task. The coxswain would have had to steer along these imaginary lines while simultaneously correcting the vessel's alignment to counter the effects of tides and river currents.

If you were carrying out this hydrographic survey today, you would use an integrated hydrographic surveying system consisting of an autopilot system to steer the vessel, an accurate global positioning system (GPS) to measure the location of the vessel, a digital echo sounder to measure the depth to the seabed or river bed, and a computer-based hydrographic software package to drive the systems and to log and validate the data. The process would be fully automated, with the autopilot system steering the vessel along pre-planned survey lines and the computer logging depth and position measurements at predetermined time intervals. Only one or two people would be involved in the measurement process, and their role would be to monitor progress on the computer screen and to review the computer-generated plan showing the corrected measurements. The measurements would also be plotted in real time on a computer-driven plotter. This would enable the hydrographic surveyor to check the progress of the survey in real time.

A hydrographic surveyor would normally calibrate the equipment before and after a survey to ensure the system (the black box) is working correctly and the measurements meet the specified accuracies for the project. This may be achieved by navigating the vessel along a test line and measuring the location and depth of one or more constructed objects fixed to the sea bed as part of a calibration system.

While the technology used in hydrographic surveying has changed dramatically over time, the fundamentals of the discipline have not changed, including the use of position-fixing geometry, the principles of depth measurement in tidal waters, and the format of hydrographic charts. Only the measurement and processing tools used for these tasks have changed.

CRITICAL THINKING

What can be controversial about the deepening channels to enable larger ships to enter a port? What design specifications and environmental conditions are important to define during the approval process? Has the shipping traffic other than cruise ships increased in line with predictions in the Fremantle harbours? Have cruise ship visits continued to increase? How does the navigable depth of Fremantle Port compare with that of other ports? What factors other than navigable depth influence shipping traffic?

The important principle to learn from this example is that we need to go beyond the technology and learn the underlying theories and principles the technology is based on, and how it can be calibrated. While it is important for you to learn how to use current technologies, it will be your understanding of engineering principles and practices that will enable you to adapt, design and implement new technologies in the future.

If you understand the theory and assumptions a technology is based on, you will also know its limitations and the contexts in which it can be used. This is essential knowledge if the technology is to be used correctly and in a professional manner.

Beyond the black box

Most engineering does not occur in isolation. For this reason, it is important to understand the impact of engineering work on other components in the system, as well as on society and the environment. It is also crucial to understand the impact other components, the environment and society may have on engineering work. If it was as simple as throwing a stone into a pond and seeing the point at which ripples hit the bank (or some other object), this would be easy. Unfortunately, this is not the case. It can often be very difficult to understand the impact an engineering product might have — although this should not stop you from exploring the possibilities. This may mean you have to think broadly, develop your understanding and enquire about the theories and practices of allied disciplines.

As an example, consider the issues related to the installation of telecommunications towers, such as the difficulties of locating and maintaining them and the possibility they may be harmful to our health. Do you think the engineers who designed the first telecommunications towers were concerned about these impacts? Would they have taken these concerns into account in their designs? We may never know the answer to those questions, but we know the engineers who are designing towers today are very aware of these issues. They take them into account during the design process and assess the impact on society and the environment.

5.4 Self-management skills

LEARNING OBJECTIVE 5.4 Use self-management skills to plan and manage your work.

When you are busy, it is easy for your life to be driven by demands rather than by priorities (Motyer 1978). An example you may relate to is that it is easy for students to succumb to the pressure of assignments and follow a study pattern driven by due dates, rather than by knowledge structures or learning sequences. To be able to effectively manage your life, you need to firstly understand how your personal characteristics may contribute to, or impact on, your performance. Then you can establish realistic goals for the different spheres of your life: your personal, career and work lives. Once you have established goals, you can develop strategies to achieve those goals and then establish priorities. Finally, you can use **self-management** skills to help you to achieve your goals. In this sense, the term self-management 'is an individualized approach to using time best according to one's particular need' (Grohar-Murray & DiCroce 2003). We will explore four aspects of self-management, each of which will help you achieve your goals.

Developing goals and strategies

Most successful people have set personal and professional goals and managed their lives to achieve those goals. The Self Management Group (SMG) (2008) states, 'Our 30 years of research with the world's leading organisations has shown that self-management is the number 1 competency of all top performers'. When we establish goals, we should ensure that we have a good balance between work (or study) and other aspects of our lives — this is commonly referred to as work–life balance. We should establish both short-term and long-term goals. From a career perspective a short-term goal could be to complete a specific project on time and budget, or to develop knowledge and skills in a new field. A longer term goal could be to establish our own business. We might set a mid-term goal as well; for example, to obtain overseas experience before we establish a business.

If you have not already done so, you may like to establish some goals. For example, a short-term goal may be to arrange vacation experience in a specific field of engineering, a medium-term goal may be to graduate with honours, and a long-term goal could be to be employed by an international company as a software engineer.

Once you have adopted a set of goals, you should develop some strategies that you believe will help you to achieve them. For example, as a student you may be unsure about which career options will be better for you. A strategy would be to make sure you undertake part-time employment in each of the areas that interest you. For example, you could try working for a small engineering firm; a manufacturing company; and a large, multidiscipline, international company. Then, when you finish university, you will be able to make an informed decision about which jobs to apply for.

Once you have developed your goals and the strategies you plan to use to achieve them, it is time to prioritise them and establish a schedule so that you achieve them in a timely fashion. You should regularly test the effectiveness of your strategies to see if they are helping you to achieve your goals. If they are ineffective, then you should adjust them or develop new strategies. Remember, you can set goals for every sphere of your life so that you achieve in each area and lead a balanced life.

Being responsible

A key self-management skill is the ability to accept responsibility. It is important to be accountable for the actions you take and the decisions you make rather than blaming circumstances, equipment or someone else. Taking responsibility for your actions can be challenging, particularly when you have difficult tasks to perform, or when things go wrong. However, you must be prepared to accept responsibility if you make a mistake, be honest about the circumstances and tell your supervisor (or while you are at university, your teacher).

Being professional

While you are a student, you should learn to act professionally when you work with other students and interact with staff. This is a key skill for engineers when they are dealing with their supervisor, with other members of the engineering team, with clients, and with consultants and contractors. By using a consistent professional approach, engineers build trust with those they work with and their clients. Some examples of acting professionally, both as a student and engineer, include:

- advising people in advance when you cannot meet an agreed deadline
- asking your supervisor when you genuinely need more resources to complete a project
- applying personal standards in all that you do and not compromising your ethical standards or personal values
- seeking advice when you are unsure of how to proceed
- resolving issues before they escalate and cause greater problems.

Take the time to review the twelve personal attributes listed earlier in this chapter. You will see that many of them are required if you are to act professionally and be responsible.

Managing your time effectively

Time management 'is the systematic, priority-based structuring of time allocation and distribution among competing demands' (webFinance 2008). Therefore, it involves setting priorities for the various activities you have planned and then allocating time to each activity. You can employ time-management techniques to help you use your time effectively, meet deadlines and achieve your goals in all spheres of your life.

While some people may spend a lot of time working, their efficiency and effectiveness may be compromised if they have not taken the time to properly plan and schedule their work. Sound time-management techniques will not guarantee success, but they will make the journey towards success easier. Some strategies are described next.

Time-management strategies

1. *Timesheet.* Keep a timesheet that indicates the amount of time spent on the tasks that you routinely undertake daily, weekly or monthly. Then, over time, you will be able to estimate the time you will require to complete one of these tasks or assignments. This understanding is necessary if your time-management planning is to be effective. Once you start scheduling tasks, you can compare the time you scheduled for a task with the actual time you needed to complete the task. Over time, you will improve your ability to estimate the time required to complete projects. This is an important skill that engineers use when they prepare a tender or quotation. Many companies use timesheets and require staff to fill these out in half-hour or even quarter-hour blocks. These sheets are then used by the companies to log (and charge) employees' time against the projects they are working on. In this situation, a key measure of work effectiveness is the ratio of chargeable hours to total hours worked.
2. *Plan.* Develop schedules, or Gantt charts, that list the deadlines for each task or project and estimate the amount of time each task will take. You can then work back from the completion date and schedule the date you should start each task. Initially, you should add extra time to allow for any changes in the task or for any inaccuracy in your estimation of the time required for each task. A variety of computer software packages can be used to automate this process. As a student, you can use this technique to schedule your assignments, laboratory reports and other assessment tasks throughout the semester.
3. *Prioritise.* Prioritise tasks in order of importance. While you are at university, many of your tasks will be allocated by your lecturers and tutors. You will need to prioritise these tasks and other activities that you plan to undertake during a semester. Where there are competing demands, your goals and strategies will help you to prioritise your activities and allow you to make evidence-based decisions.
4. *Organise each day.* At the beginning of each day, you should list all of the tasks you want to achieve. This list may be a subset of a list you prepared at the beginning of the week, or it may form part of a monthly plan. Many organisations provide employees with printed pads that allow them to list in priority order the activities they are undertaking. These simple 'to do' lists generally have a check box to the side of the page that can be ticked when a task is complete. You can also use software for the same purpose, for example, the Tasks function in Microsoft Outlook.

5. *Be responsive.* You must be prepared to be flexible and respond to changing circumstances. For example, you may have underestimated the time it would take to complete an assignment and this means that you will not have time to complete the next assignment. When circumstances change, you should review your priorities and reschedule your activities to ensure you achieve your goals. This may mean you have to work over a holiday weekend, or miss a social event. Alternatively, if you complete tasks ahead of time, you should consider rewarding yourself with some extra time off.
6. *Identify barriers to success.* You should try to identify the things that hinder your work. Do you find it difficult to start projects (or an assignment) or to get going in the morning? Do you find it difficult to make decisions? Procrastination can stop you from making the most of your time. By identifying barriers to productivity you can work out areas for personal improvement. You will need to develop techniques that enable you to break down these barriers. For example, some people find it easier to start a new task late in the day rather than early in the morning. So, their strategy is to try to start new jobs late in the day. Then, at the beginning of the next day, it is much easier for them to start work, as they just continue working on the project.
7. *Reflect.* It is important to reflect on how well you managed your time and the activities you completed while working on a project. This will help you to identify areas for improvement for the next time you undertake a similar project. You may be able to improve the way you study, the way you work, the processes you use, your relationships with other students or employees, your knowledge and skills and your level of satisfaction. This reflection may also help you to review your goals, or the strategies you are using to achieve them. Perhaps they need to be more realistic?

5.5 Life-long learning

LEARNING OBJECTIVE 5.5 Use life-long learning skills to plan and manage your learning.

Your university program is designed to equip you with the knowledge and skills required to begin a career in a field of engineering. However, while you are at university you will also have the opportunity to acquire and practise a range of learning skills and strategies that you can use to manage your learning. Once you graduate, and start working as an engineer, you will be able to use the same skills and strategies to maintain the currency of your knowledge and skills. You will continue to use these skills throughout your career because, like most professionals, you will need to learn throughout your working life to integrate new knowledge, materials, methods, processes and technologies into your practice.

Like many professional organisations, Engineers Australia expects its members to undertake continuing professional development (CPD). Its Chartered Members must record the professional development activities they undertake each year, which may be in their engineering field or in non-engineering fields (such as project management). Engineers Australia undertakes regular audits of the CPD records kept by its Chartered Members.

You have begun a life-long learning journey — a journey that will be made easier if you acquire the skills and strategies that will enable you to become an independent learner and manage your learning. These skills will enable you to evaluate and implement new technologies and, where necessary, review and change practices and processes as part of the implementation strategies.

KEY POINT

Engineers use life-long learning skills to maintain the currency of their knowledge and skills throughout their career.

SPOTLIGHT

Life-long learning as a career develops

KATHERINE NGUYEN

Katherine Nguyen has been working for five years and undertook further study as the interest and need to do so arose. She shares her journey.

I studied a Bachelor of Electrical and Electronics Engineering, which was the obvious choice being academically strong in the math and science subjects. While studying, I spent a number of my university summers interning at the roads authority as one of their scholarship cadets. The Roads Authority offered a diverse learning experience for a young eager electrical engineer. Although the work on traffic signals, streetlights and variable message signs was interesting, I soon sought another level of challenge and excitement. So I moved on to work part time as a Project Coordinator at a resources company. This role was not an internship where you would usually expect to have your hand held to complete tasks. It was a steep learning curve, and it was during my time here that I challenged myself and learned the required industry language and the skill sets necessary to survive in a high-pressure environment where I was expected to deliver.

Katherine Nguyen

Upon finishing university, I wanted something different. I wanted to work in an environment that would provide me with the opportunities to explore possible career paths as an electrical engineer. I took a graduate position at an electrical utility where their graduate program allowed me to rotate across the business. These rotations exposed me to various areas of the business including asset management, engineering design and asset operations, where I came to realise my interest in the commercial aspects of an organisation. During my rotation in asset operations, I was given the opportunity to contract and construction manage various projects that included the delivery and assembly of a transformer and the construction of a new substation.

My interest in contract management has led me to further my commercial knowledge by completing a Masters in Construction Law. Completing the Masters has opened many doors for me including moving interstate and now working as a consultant at a contract and commercial advisory firm specialising in major infrastructure projects. Looking back, I wouldn't have imagined this career path for myself but am excited for what lies ahead in my future.

CRITICAL THINKING

In this chapter, Miles Cattach and Caroline But talked about learning at work and in their spare time; Katherine Nguyen and Madhu Bhaskaran completed additional degrees. What future learning and studies might you undertake during your career?

SPOTLIGHT

A 3D printed car

A car was printed, assembled and driven during the five-day period of the International Manufacturing Technology Show (IMTS), held in Chicago in September 2014. While 3D printers had previously been used to layer print panels and other components for a car, (e.g. the 2010 Urbee), Local Motors claims it was the first time the body of a car had been printed in one piece.

FIGURE 5.2 The car body before routing

The Strati resulted from the 3D Printed Car Design Challenge, conducted using crowd-sourcing technologies by Local Motors, which is based in Phoenix, Arizona. The company believes it is at the forefront of the 3D Industrial Revolution and describes itself as a 'free online and physical workspace' that enables people to turn their vehicle innovations into reality and rewards them for their contribution. The key issues discussed in the brief for the design challenge were as follows.

- Currently, producing a new car from a new design represents a significant investment in tooling to create the many parts that are required to produce the structure of a car (e.g. exterior body panels, trim, internal structure for rigidity, interior panels, dash covers and the seats). The fundamental issue is, what can be done to reduce the initial investment in producing a design, reduce the part count and reduce the follow-up investment that will be required if the design changes?
- Imagine if you could create the major elements of the exterior, the structure and the interior associated with a vehicle in one part. Then think if changing the design, or even taking on an entirely new one, represented no additional cost in tooling. This is what we want investigated in the Direct Digital Manufacturing Project!
- We are focused on investigating the proper use of a hybrid additive/subtractive machine being developed at Oak Ridge National Labs. This machine uses a large diameter extrusion head to 3D print objects at high speed, then on the same head it also uses a router to come back and machine surfaces to a more precise specification where required. This means that we can create car-scale forms very quickly and freely to machined precision, but without the necessity of forming tools, and so on.
- What will the best structure look like?
- What materials should be used and when?
- What is the best way to fasten mechanical and other components to that structure?

The answers to these questions would then be used to achieve objectives to:

- create the major structure of a new vehicle using an additive/subtractive hybrid methodology
- sufficiently define this methodology so an aesthetic study can be undertaken into how the vehicle can be styled
- apply an electric powertrain to this vehicle and have the structure support it
- demonstrate that this methodology can be more economical than existing methods.

FIGURE 5.3 The components that were used to assemble the car

FIGURE 5.4 The completed Strati

The challenge was opened on 12 April 2014 and included detailed specifications and CAD files that could be downloaded from the website. By the due date, 13 May 2014, 216 entries had been received from more than 30 countries. Michele Anoe was announced the winner in June and received the US$5000 cash prize and the honour of seeing his design, the Strati, printed at IMTS.

The manufacturing process for the Strati had three phases. First, the body was printed using a Big Area Additive Manufacturing machine from Cincinnati Inc. & Oak Ridge National Laboratory. This machine used a large diameter extrusion head to 3D print objects at high speed, as shown in figure 5.2.

Then, a router on the same head was used to machine surfaces to a more precise specification where required. Finally, the 48 additional components (see figure 5.3) were attached to the printed body of the vehicle.

The car took 44 hours to print, one day to mill and two days to assemble, making it a five-day build process in total. The completed Strati (see figure 5.4) was fully functional and was driven for the first time on the morning of Saturday 13th September — the last day of the show.

CRITICAL THINKING

What do you think will be the major changes in the car manufacturing process that will result from the adoption of additive/subtractive manufacturing?

The spotlight demonstrates how engineers used a new technology to explore the use of radically different car manufacturing processes. Time will tell whether these new processes become mainstream. For example, will future car buyers go to their local garage and select the make, model, colours and accessories of the car they wish to purchase, with the garage proprietor ordering the components and manufacturing the car on site once a deposit is paid?

3D printing is also challenging engineering practice in other fields. For example, in September 2014 NASA ferried a 3D printer to the International Space Station to test the use of relatively low-temperature plastic feedstock to produce 3D multilayer objects. The study was the first step towards establishing an on-demand machine shop in space (NASA 2014).

NASA is not alone in exploring the use of 3D technologies in space. In 2013 the European Space Agency established a consortium to explore the use of 3D printing to construct a habitable base on the Moon using lunar soil (regolith) as the building material. The aim was to minimise the amount of 'ink' used to bind the regolith, while maintaining the required strength of the structure (Foster + Partners 2013).

The key message from these three examples is that engineers need to keep learning and evaluating their methods if they are to successfully integrate new technologies into their practice. They also have the opportunity to be creative and innovative when adopting new technologies.

Testing the 3D printer in the Microgravity Science Glovebox Engineering Unit at Marshall Space Flight Center

Knowledge frameworks

As you learn about different aspects of engineering it is useful to use an organisational structure to store the different pieces of knowledge — a knowledge framework. As an engineering student, a knowledge framework will help you understand what you know, what you still have to learn, and why you need to learn it. It will help you to manage your learning and to reflect on your learning journey. There are a number of reasons why it is useful to consider your engineering knowledge as a framework, perhaps in the form of a matrix.

- You will gain an understanding of how various strands of knowledge fit together, and how they rely on each other for support.
- You will gain a context for your learning, which will help you understand why you should learn about a particular topic, or acquire a specific skill.

- You will be able to place new pieces of knowledge into the framework and immediately see the connections to other pieces of knowledge, which should give you a sense of progress as, over time, you fill in gaps in the framework.
- You will be able to identify gaps in your knowledge and skills, and plan how you are going to fill those gaps.

You already have some understanding of the work engineers do, and perhaps the work different types of engineers do. If you worked in the industry before coming to university, then your knowledge and understanding of engineering will be greater than that of a student who has come directly from high school. Your current knowledge and understanding is your engineering knowledge framework, although it may lack structure. You should therefore consider using a well-developed framework to organise your knowledge, which will grow rapidly over the next few years.

We will now look at some frameworks, or knowledge structures, that you could use for this purpose.

The program framework

University engineering programs are designed to give students the knowledge, skills and experiences that will enable them to commence practice as an engineer. It is important to realise that while the content and structure of engineering programs at different universities will be similar, they will not be the same. Each program will have its own structure and emphasis. Some universities offer different types of programs, each aimed at equipping students for a particular career in the engineering industry.

- Four- or five-year programs lead to qualifications that enable employment as a graduate engineer.
- Shorter qualifications enable employment as a graduate engineering technologist, or engineering technician (called engineering associates or engineering officers in Australia).

The structure and content of your program will provide you with a useful knowledge framework while you are at university. It will help you to define the extent of your knowledge and skills and the work you are capable of undertaking successfully at a particular point in time, and will change over time as you learn new things, acquire new skills and complete engineering projects. Your knowledge framework will also enable you to identify your limitations — those activities that lie outside your expertise. This is particularly important once you have entered the engineering workforce. When you know your limitations you will know when to seek advice or assistance from your supervisor, more experienced engineers, or practitioners from other disciplines.

Programs accredited by Engineers Australia (2008a, (2008b) or by Engineering New Zealand (2019) provide for criteria-based membership of these institutions at the appropriate level. The documentation developed for your **program** — structured learning that leads to a qualification — will normally contain a set of aims that describe the overall purpose of the program. Your program may also allow you to specialise in a particular field such as electrical engineering or mining engineering. A specialisation is often called a **major**.

A program will also include a set of objectives that provide more detail about its purpose. Detailed learning outcomes are generally defined at the **subject** level. These are often called learning objectives, or objectives. If you collate the learning objectives for the subjects you study, you will in effect be constructing a program knowledge framework.

Engineering schools develop curriculum maps to show how the learning objectives for each subject contribute to the achievement of the overall program aims and objectives. Schools often provide this information to their students in the program handbook or during orientation week. You can use this knowledge framework to understand why you need to learn particular subjects and topics, and this may help to motivate you and enhance your learning.

Most university programs are also designed to give students the opportunity to acquire generic attributes and capabilities — which are defined by the university — by the time they graduate. In these universities, the engineering school is required to show graduates have been given opportunities to learn, practise, enhance and demonstrate their achievement of its nominated attributes. You may like to review your university's graduate attributes.

Engineers Australia's framework

A number of professional organisations define sets of 'graduate attributes' that support graduates to gain professional registration or membership. Examples include Engineers Australia, the Australian Nursing and Midwifery Council (ANMC) and Certified Practising Accountants (CPA) Australia. When an engineering school seeks accreditation from Engineers Australia or Engineering New Zealand, they design their undergraduate programs to ensure graduates have acquired the graduate attributes defined by the relevant professional body.

In 2004, Engineers Australia translated their Generic Graduate Attributes into a detailed set of competencies known as the Stage 1 Competency Standard. Since then, engineering schools seeking accreditation of their programs by Engineers Australia must demonstrate that graduates from the programs have acquired the relevant competencies. The Stage 1 Competency Standard was recently reviewed, and the revised set of Standards was published in 2019 (see https://www.engineersaustralia.org.au). Engineering schools have been using these accreditation standards to develop and quality assure courses for more than a decade. You may decide to use the Stage 1 Competency Standard as your knowledge framework if information about your undergraduate program is not available or suitable for this purpose (Engineers Australia 2019).

In some engineering disciplines the graduate outcomes have been defined in great detail, either as a set of graduate capabilities or as a body of knowledge for the discipline. The following examples illustrate these two approaches. Students in these disciplines would use these as their knowledge frameworks.

Discipline frameworks

During the period 2010–12, the leaders of the Define Your Discipline (DYD) project (funded by the Australian Learning and Teaching Council) collaborated with members of the Environmental Engineering College of Engineers Australia to define a national Graduate Capability Framework for professional level environmental engineering degree programs. The aim of the project was to define the specific knowledge, skills and capabilities an environmental engineering graduate would require to demonstrate achievement of Element 1.3 of the Stage 1 Competency Standard: 'in-depth understanding of specialist bodies of knowledge within an engineering discipline'. The Evidence of Attainment statement for Element 1.3 was: 'proficiently applies advanced technical knowledge and skills in at least one specialist practice domain of the engineering discipline'. The framework was endorsed by the Environmental Engineering College Board and published as a guide for universities in 2013 (Dowling & Hadgraft 2013). It is used as a companion document to the Stage 1 Competency Standard when an Engineers Australia Accreditation Panel reviews an environmental engineering program.

Three types of capabilities were identified (see table 5.2):

- *technical capabilities* — the knowledge and skills of a typical environmental engineering graduate
- *process capabilities* — the processes that environmental engineering practitioners use to apply their knowledge and skills
- *generic capabilities* — the capabilities that graduates from most engineering disciplines would be expected to have. Some of these would also be classed by universities as graduate attributes.

TABLE 5.2 **Environmental engineering graduate capabilities**

Technical	Process	Generic
Water resources and supply	Investigation	Project management
Stormwater management and re-use	Modelling and analysis	Ethics
Water and wastewater treatment	Integrated design and implementation	Communication
Resources and waste management	Assessment of impact, risk and sustainability	Innovation
Soil and geology	Environmental planning and management	Information
Air and noise	Audit, compliance and review	Self-management
Energy systems and management		Teamwork

Source: Dowling and Hadgraft (2013).

Students studying environmental engineering should use this document as their knowledge framework.

Discipline body of knowledge

In 1998, the American Society of Civil Engineers (ASCE) adopted Policy Statement 465, which sets out the academic requirements for licensing and professional practice, including the attainment of a defined body of knowledge (BOK). The first BOK was published in 2004. After receiving input from stakeholders, the BOK was refined, expanded and published in 2008 as the *Civil engineering body of knowledge for the 21st century* (CEBOK). The draft third edition was available for comment until 12 October 2018.

The CEBOK lists 24 outcomes in three categories and the level of achievement required for each of the outcomes at the end of an undergraduate degree (and, for some outcomes, at the end of a masters degree (or equivalent) and/or a period of approved postgraduate experience in the civil engineering field). The outcomes and levels of achievement are shown in figure 5.5.

FIGURE 5.5 CEBOK: outcomes and levels of achievement

	Level of achievement					
	1	*2*	*3*	*4*	*5*	*6*
Outcome number and title	*Knowledge*	*Compre-hension*	*Application*	*Analysis*	*Synthesis*	*Evaluation*
Foundational						
1. Mathematics	B	B	B			
2. Natural sciences	B	B	B			
3. Humanities	B	B	B			
4. Social sciences	B	B	B			
Technical						
5. Materials science	B	B	B			
6. Mechanics	B	B	B	B		
7. Experiments	B	B	B	B	M/30	
8. Problem recognition and solving	B	B	B	M/30		
9. Design	B	B	B	B	B	E
10. Sustainability	B	B	B	E		
11. Contemp. issues & hist. perspectives	B	B	B	E		
12. Risk and uncertainty	B	B	B	E		
13. Project management	B	B	B	E		
14. Breadth in civil engineering areas	B	B	B	B		
15. Technical specialisation	B	M/30	M/30	M/30	M/30	E
Professional						
16. Communication	B	B	B	B	E	
17. Public policy	B	B	E			
18. Business and public administration	B	B	E			
19. Globalisation	B	B	B	E		
20. Leadership	B	B	B	E		
21. Teamwork	B	B	B	E		
22. Attitudes	B	B	E			
23. Lifelong learning	B	B	B	E	E	
24. Professional and ethical responsibility	B	B	B	B	E	E

Notes:
B: portion of the BOK fulfilled through the bachelor's degree
M/30: portion of the BOK fulfilled through the master's degree or equivalent
E: portion of the BOK fulfilled through the licensure experience

Source: American Society of Civil Engineers (2008, p. 6).

The outcomes in the Foundational and Professional categories could be regarded as generic engineering graduate outcomes, as they would be appropriate for graduates from other fields of engineering. Some of the outcomes in the Technical category would also be applicable for some other engineering disciplines, although the emphasis and applications would change. Outcomes 14 (Breadth in civil engineering areas) and 15 (Technical specialisation) are specific to civil engineering, and the details of the expected cognitive achievement levels for these outcomes are reproduced as table 5.3. Note that the level of achievement statements in these tables do not mention any specific fields of civil engineering, such as road design, structural engineering or water infrastructure.

TABLE 5.3 Level of cognitive achievement

	Level of cognitive achievement					
Outcome title	**1**	**2**	**3**	**4**	**5**	**6**
14 Breadth in civil engineering areas	Define key factual information related to at least four technical areas appropriate to civil engineering (B)	Explain key concepts and problem-solving processes in at least four technical areas appropriate to civil engineering (B)	Solve problems in or across at least four technical areas appropriate to civil engineering (B)	Analyse and solve well-defined engineer-ing problems in at least four technical areas appropriate to civil engineering (B)	Create new knowledge that spans more than one technical area appropriate to civil engineering	Evaluate the validity of newly created knowledge that spans more than one technical area appropriate to civil engineering
15 Technical specialisation	Define key aspects of advanced technical specialisation appropriate to civil engineering (B)	Explain key concepts and problem-solving processes in a traditional or emerging specialised technical area appropriate to civil engineering (M/30)	Apply specialised tools, technology or technologies to solve simple problems in a traditional or emerging specialised technical area of civil engineering (M/30)	Analyse complex system processes in a traditional or emerging area appropriate to civil engineering (M/30)	Design a complex system or process or create new knowledge or technologies in a traditional or emerging advanced specialised technical area appropriate to civil engineering (M/30)	Evaluate the design of a complex system or process, or evaluate the validity of newly created knowledge or technologies in a traditional or emerging advanced specialised technical area appropriate to civil engineering (E)

Source: American Society of Civil Engineers (2008, p. 109).

Managing your learning

It is often easy to learn things you really want to learn. For example, how fast did you learn to use your mobile phone? Do you know how all of the functions work and how to set it up so that it suits your needs and preferences? How long did it take for you to set it up? Because you want to use all the capabilities of your phone you are *motivated* to learn how to use all of the functions and apps.

It is the same when you are studying at university. You will find you enjoy studying some subjects in your program more than others. This may be because you understand their relevance to your future career, or because you are interested in these topics. Conversely, it is usually much harder to learn things you are less interested in, and you may therefore find learning about some engineering topics less enjoyable or difficult.

In this section, we will discuss some strategies that will help you manage your learning and overcome these and other learning difficulties. If you apply these strategies, you will learn more efficiently and have a deeper appreciation of the learning experience.

Identifying your learning needs

When people review their performance during a work experience, they may become aware of some factors that have affected their performance. They can use that understanding to identify strategies that will help them develop the knowledge, skills or processes to improve their performance. For example, you may not like working in teams because you believe you work better on your own. By learning about team roles and the factors that influence team success, you can offer to undertake roles that fit your skill set. This may mean that you feel more comfortable and confident in team situations. In time, you may even find that by working in a successful team you can achieve better outcomes than you would if you were working on your own. The key to taking action is to recognise the need for change.

While you are at university you are expected to learn all of the concepts, facts, processes and skills in each of the subjects in your program. The *learning objectives* for a subject define what you will need to learn to successfully complete that subject. Assessment tasks are generally designed to enable students to demonstrate their level of achievement of these objectives. Therefore, for each learning objective, you should identify what you will need to know to demonstrate achievement of the objective. You may find this detail in the materials provided for the subject. If not, you will have to identify this information as you study each topic. Once you have worked out what you need to learn for a subject, you can monitor your learning against that checklist. This will prove invaluable as you get closer to a major assignment or examination.

As you progress to the later years of your program there may be opportunities for you to choose one or more elective subjects. Some students select subjects that interest them, whether they are engineering subjects or subjects from other fields. Other students take the opportunity to specialise in an area that they plan to work in during their career. Thus, they have identified their individual learning needs.

Once you begin work as an engineer, you will find that you will need to acquire additional knowledge and skills, both in your field of engineering and in non-engineering areas, such as risk management and workplace health and safety. You may learn about these topics by attending training courses, short courses, or by enrolling in a postgraduate program. The principles of this life-long learning process are to:

- identify your learning needs
- develop strategies that could deliver that learning
- undertake the chosen learning experience
- embed your new knowledge and skills into your practice
- reflect on, and evaluate the success of, the learning experience.

Strategies for learning

The Chinese philosopher Confucius reportedly said, 'I hear and I forget. I see and I remember. I do and I understand'. Research (Arnold et al. 2001; Laird 1985) supports the general thrust of this statement, reporting that students retain about:

- *twenty* per cent of what they *hear*
- *thirty* per cent of what they *see*
- *fifty* per cent of what they *see and hear*
- *seventy* per cent of what they *see, hear and say*
- *ninety* per cent of what they *see, hear, say* and *do*.

This suggests we should involve as many senses as possible in the learning process. For example, as a student you will learn more effectively if you listen carefully and write notes during a lecture, discuss what you learned with classmates, read about the topic in a textbook, write a summary of your lecture notes and, if appropriate, apply your knowledge by solving problems or completing practical or project work that is related to the topic.

Learning approaches

To be an effective engineer you will need to learn in a way that leads to a genuine understanding, rather than just remembering facts. This is because engineers often need to apply their knowledge to solve problems that are new or complex. Solving problems requires a deeper understanding of the fundamental concepts and principles of engineering and science and the ability to apply them in new and creative ways.

SPOTLIGHT

The law of the pendulum

Ken Davis is an author and professional speaker and in one of his books, *How to speak to youth . . . and keep them awake at the same time* (1996), Davis recounts a presentation he gave as a college student in the United States. The presentation was to be marked on creativity and the ability to make a statement 'in a memorable way'. Davis began his presentation — titled 'The law of the pendulum and belief' — with a thorough breakdown of the physics and workings of a pendulum. That is, that '[a] pendulum can never return to a point higher than the point from which it was released' (Davis 1996, p. 123). This is because friction and gravity slow the pendulum and therefore it will be unable to reach the original release point on its return. Each time the pendulum completes an arc, the height of the arch reduces, until the pendulum comes to a stop.

Davis added a demonstration to his explanation of the pendulum. He attached a string to a weight (in his demonstration he used a spinning top — a children's toy) and attached this string to the top of the class blackboard. He raised it to a starting point, marked this starting point on the blackboard and then released it, recording each peak of the arc on the blackboard with a mark. The marks became lower and lower with each return of the pendulum and thus confirmed the theory of a pendulum.

At this point he asked if the other students in the class believed in the theory he'd demonstrated of the pendulum and they (and the teacher) confirmed they did. It was here that Davis's presentation got interesting and it's here that the connection between his title and his presentation becomes clear.

> Hanging from the steel ceiling beams in the middle of the room was a large, crude but functional pendulum (250 pounds of metal weights tied to four strands of 500-pound-test parachute cord).
>
> I invited the instructor to climb up on a table and sit in a chair with the back of his head against a cement wall. Then I slowly brought the 250 pounds of metal up to his nose. Holding the huge pendulum just a fraction of an inch from his face, I once again explained the law of the pendulum he had applauded only moments before, and I said, 'If the law of the pendulum is true, then when I release this mass of metal, it will swing across the room and return short of the release point. Your nose will be in no danger.'
>
> After that final restatement of this law, I looked him in the eye and asked, 'Sir, do you believe this law is true?'
>
> There was a long pause. Huge beads of sweat formed on his upper lip and then he nodded weakly and whispered, 'Yes.' I could see looks of understanding begin to appear in the audience. The connection between the objective of the talk and this illustration was beginning to come together.
>
> I released the pendulum. It made a swishing sound as it arced across the room. At the far end of its swing, it paused momentarily and started back. I never saw a man move so fast in my life. He literally dived from the table. Deftly stepping around the still-swinging pendulum, I asked the class, 'Does he believe the law of the pendulum?'
>
> The students answered unanimously, 'NO!' (Davis 1996, pp. 124–126).

CRITICAL THINKING

Review the learning strategies you are currently using in your subjects. Identify some examples where you believe deep learning occurred. What learning approaches did you use to achieve these deep learning outcomes?

Did you ever have such an exciting time in your classroom? Can you imagine all of Ken's classmates telling their friends and family how they nearly saw their teacher's head smashed to pulp, and how their teacher was a coward? One thing is for certain, the students will never forget that class; nor will they forget 'The law of the pendulum and belief'. This is an example of deep learning.

Biggs (1987) used the terms deep learning and surface learning to differentiate between reflective (deep) learning and non-reflective (surface) learning. A surface learning approach may be used to memorise the facts required for the task at hand, for example, preparing for a quiz show. A deep learning approach is used by a learner who wishes to understand the material and be able to apply it in current or future situations (Moon 2004).

At the end of the first part of Ken's presentation, most of the students probably had a surface understanding of the law of the pendulum. After the demonstration they had a deeper understanding. Many would now believe the law of the pendulum, with their belief based on a real understanding of the behaviour of the pendulum.

Deep learning

Deep learning occurs when learners look for meaning in a task and try to relate it to their experience, knowledge and understanding.

Biggs (1987, p. 15) suggests that deep learners:

1. are involved in the academic task and derive some enjoyment from it
2. search for the meaning inherent in the task
3. personalise the task based on their own experience and the real world
4. relate the current task to the bigger picture, including previous knowledge
5. try to form hypotheses about the task.

The deep learning approach therefore involves:

- being actively interested in the material by considering the big picture to define your learning needs and relating the topic to a gap in your personal knowledge framework (your existing knowledge base)
- drawing sketches and diagrams to understand spatial locations and relationships
- actively looking for patterns and underlying principles
- critically examining the logic of any arguments
- analysing the evidence provided to see if it supports the arguments, conclusions or recommendations
- discussing the topic with classmates; by explaining key concepts to others, you are reinforcing your own learning.

When you have achieved a depth of learning in a topic you should be able to use your knowledge and understanding to analyse, synthesise and compare data. You should also be able to apply this knowledge in different situations, transferring your skills to another field or domain.

Surface learning

Surface learning occurs when learners memorise facts without understanding.

Biggs (1987, p. 15) suggests that surface learners:

1. consider the educational experience a demand or imposition
2. consider the experience as a discrete activity, unrelated to other tasks
3. worry about the time the educational experience takes
4. avoid considering the meaning that the experience may have
5. rely on memorising facts, words and diagrams in order to reproduce aspects of the material.

Although it is always preferable to take a deep learning approach, a surface learning approach may be required in some situations. You may not have time for deep learning due to a looming assessment task, or because you have to know a lot of factual information. Sometimes you may be forced to adopt a surface learning approach, as you may not understand aspects of a topic and do not have the time to find an explanation. Hopefully this is only a short-term strategy. Some approaches that are useful for surface learning include:

- memorising facts, procedures and processes. For example, creating lists of facts and recording the information on a DVD or media player, and playing the information back repeatedly
- treating facts as separate items of information
- studying material without reflecting on its purpose or meaning.

There are limitations to this learning approach. If you only have a surface-level understanding of a topic, you may not be able to apply your knowledge, particularly in new situations.

It is important to remember that when you only have a surface knowledge of a topic, your understanding is limited. Before you apply that knowledge in an engineering context, you should try to improve your understanding.

Practical techniques

There are some practical techniques you can adopt that will improve your learning. By taking a proactive and strategic approach you will increase the likelihood of achieving good results as you tackle new areas of engineering in your program. It is important you find a balance that enables you to achieve your learning goals without spending an excessive amount of time studying. Your overall strategy should be to maximise your learning and minimise your effort. Here are some recommendations to improve your learning.

1. *Aim high.* From the beginning of your program you should aim for the highest possible grades. Higher grades will increase the likelihood that you will get your dream job. They will certainly be important if you want to graduate with honours or undertake a research qualification, such as a PhD.
2. *Manage your time efficiently.* You may lead a busy life. When you are studying it is important to plan your week carefully to ensure you do not waste time. Start keeping a weekly schedule and fill in all of your obligations, such as classes, work commitments, extracurricular activities and other weekly events. Add in your proposed study times. How much time have you got left for socialising?
3. *Establish good study habits.* You should try to study consistently throughout the semester, rather than in peaks and troughs. You should monitor and reflect on how effective your study sessions are. Do you work better in long sessions or does your effectiveness drop off after an hour or two? You should plan your study sessions accordingly.
4. *Find a suitable study environment.* You are going to spend a lot of time studying, so you should feel comfortable and enjoy being in the study environment you create. You should try to have a place you can call your own — a place where you can leave your books and resources when you take a break. You should also establish whether you work better in a quiet room or with noise (such as music) in the background.
5. *Consider your strengths and weaknesses.* It is important that you use your strengths whenever possible. This enhances your learning efficiency and is particularly important in team situations. You also need to manage weaknesses so they do not affect your learning. This may mean consciously seeking out opportunities for improvement in these areas. For example, you may seek help from your university's learning support centre if you have trouble writing reports.
6. *Balance your life.* It is important you balance your studies with other activities, such as sport, recreation and social activities. Things may start to go wrong if you get too far out of balance. For example, if you do not do enough exercise to keep fit, you may become tired and lethargic, and this will affect your ability to study.
7. *Celebrate good results.* If you celebrate when you do well, you are more likely to stay motivated during tougher times. You should celebrate good results with the people who are supporting you while you are at university.

Specific skills

The following skills can be used to enhance your learning while you are at university. The first two of these are based on studies in engineering units by Male and colleagues (2015) and Male and Leggoe (2017).

- Prepare before your first classes by reading the unit outline, buying any textbook and reading any material on the learning management system. Note your assessments for the semester, especially any that will be due concurrently, or clash with your non-study responsibilities.
- Keep up between classes. Discuss your learning with others, practise applying concepts and be sure to complete any available pre-reading. Try to approach problems in more than one way. If you do not understand a concept, find additional explanations. If you have a problem or project to work on, be sure to work on it before class so that you can make the most of the opportunity to seek help during class.
- Even if your lectures are recorded, keep up with them.
- During a lecture, concentrate on understanding the concepts and the narrative. Write down key points. You should also jot down any comments or thoughts you have about a topic and identify these by writing your initials or another code beside them. Identify any aspects you did not understand and research this area in more depth at a later time.
- Ask questions during lectures and tutorials when you do not understand a topic or cannot solve a problem.

- As soon as possible after lectures, write notes on the topic. Use information from textbooks and other course materials to build your understanding of the key points you gained from the lecture. Using alternative sources can give you different perspectives and improve your understanding.
- Prepare regular summaries of information you have covered in your classes. Reflect on the relationships between the topics and check whether all of the topics will be assessed. Also, establish which aspects of the topics may be assessed, that is, which topics relate to and are aligned with the learning objectives.
- If there is an examination for a subject, remember to review your summaries as you get closer to the exam, ensuring you understand the topics that will be covered. In some subjects, you may also need to review and improve your problem-solving approach.

Other useful learning skills such as note-taking and listening will be discussed in the chapters on communication skills and communicating information.

Learning through assessment

The assessment items you complete in different subjects are designed to enable you to demonstrate your achievement of the learning objectives in these subjects. They are also designed to support your learning. You should therefore use each item of assessment as a means of developing your knowledge and understanding. Here are some strategies that will help you improve your performance in assessments.

1. *Carefully analyse the questions.* Make sure you understand exactly what you are being asked to do. Check with your tutor if you do not understand any part of the assessment or questions.
2. *Check other requirements.* Are there any page or word limits, such as a maximum or minimum number of words or pages for the assessment? What format is required? When is the due date? What are the marking criteria? The answers to these questions will guide the way you respond to the assessment item.
3. *Understand the learning outcomes for which the assessment has been designed.* Your lecturer will have designed your assessment to help you learn as well as to assess you on that learning. Understand the intended learning outcome(s) in order to approach the assessment such that you maximise your learning.
4. *Assess your answers.* Learn to judge whether you have answered questions correctly. You may even like to mark your own answers. This skill will prove invaluable when you are preparing for examinations.
5. *Check before submitting assessment items.* Check the brief or questions for the assessment item to ensure you have met all of the requirements. Carefully proofread your work. You may also ask a friend or family member to proofread your work before you submit it.
6. *Use feedback to enhance your learning.* Learn to use the feedback you receive on assessment items to improve your understanding of the topic. In engineering subjects, it is common for lecturers and tutors to provide students with worked solutions as well as individual comments. You can use this feedback to help you prepare for future assessments.

Once you commence practice as an engineer, you will work under the supervision of a registered engineer until you attain your own independent registration. Under supervision, you may find that you not only have to participate in professional development activities, but also have to foster a culture of life-long learning and skills development in your organisation. Eurobodalla Shire Council, which serves a 110-kilometre stretch of the far-south coast of NSW, provides an example. Like many regional councils, Eurobodalla finds it difficult to compete with other employers when recruiting and retaining skilled staff. In 2005, the council embarked on an ambitious workforce development strategy (Sharpe 2006, 2010). In the following spotlight, Warren Sharpe, Director of Roads and Recreation, outlines the strategy they are using.

SPOTLIGHT

Promoting a culture of life-long learning among engineering staff

WARREN SHARPE, EUROBODALLA SHIRE COUNCIL

We were struggling to get good staff, particularly in supervisory positions. We had a group of staff who were highly skilled in certain areas but lacked self-confidence to push to higher levels. We also had some gaps in skills, which restricted our flexibility and efficiency. My intent was to create the right opportunities for our existing staff, so we could develop the skills of the whole team and offer our people better opportunities to succeed in their careers. At the same time, I reshaped our division to increase our apprenticeship program from 1–2 per cent to 10 per cent of our work team and to create some 'stepping stone' positions to allow people to advance to higher levels.

Unlike past training, we chose to get every single person into a program. A big challenge for sure, but we have great support from our General Manager and the Human Resources group. We had to start with base level training to build confidence and a platform for higher level training. We worked with our partners, Riverina TAFE, to develop three nationally recognised courses that are appropriate for our context and can be delivered in the workplace, which is absolutely critical in regional areas. These three courses are as follows.

- Certificate III — all operational staff complete a trade qualification
- Certificate IV in Civil Construction Supervision
- Diploma in Civil Construction

Brendan Zeigler and other operational staff at their routine tutorial session

We've now trained well over 100 staff in the Certificate III, and graduated 17 with a Certificate IV. Another 15 have now started in the Certificate IV, and another 15 from right across our organisation are doing the Diploma. These qualifications are linked to, and mandatory for appointment to, specific jobs and remuneration levels. We also have some new Civil Construction Trainees who will undertake a Certificate III and then a Certificate IV as part of their four-year traineeship, while getting practical on-the-job experience.

Plant Operator Apprentice Brendan Zeigler outlined what the opportunities mean to him:

> I've completed my Certificate III in Plant Operation already whilst learning to operate graders, backhoes and drive trucks. Now my bosses are supporting me doing a supervisors course through Certificate IV. I'm learning heaps about why we do the things we actually do every day. It's great.

What's really exciting is that we've now formed a partnership with a national skills council, SkillsDMC, for our latest group of staff, and we are working with the University of Canberra and other universities to develop improved articulation pathways into degree courses. Even better is the fact that SkillsDMC has partnered with the NSW Institute of Public Works and Engineering, resulting in another 25 councils seeking funding assistance to run similar programs.

As an engineering manager, there is no greater satisfaction than to see your people succeeding and the consequent improvement in the work we do for our community. Our success has only been possible with the support of my fellow engineers who share my passion and have done a great job to contextualise the course material and offer tutorial support to our staff. We continue to challenge other employers to develop their staff, because we need a unified approach to correct the supply/demand chain and ensure sustainable numbers of people skilled in civil works for the future.

Eurobodalla Shire Council won a National Employer of the Year Award for its comprehensive organisational training programs.

CRITICAL THINKING

Imagine you are the senior engineer in an engineering organisation. What principles would you use from this example to develop a culture of life-long learning in the organisation? What information would you need to develop a program like that implemented by Eurobodalla Council? Where would you find that information?

This example highlights the importance that practising engineers place on life-long learning skills. This is reinforced by a New Zealand study that investigated the importance of 24 competencies to education stakeholders. They found that while the stakeholders reported that all of the competencies were desirable, two of the four groups — the employers and the recent graduates from the science and technology disciplines — gave the highest rating to the competency *ability and willingness to learn* (Coll & Zegwaard 2006). The top five competencies for each group are shown in table 5.4 in descending order.

TABLE 5.4 **Top five competencies for practising engineers**

Rating	Employers	Recent graduates
1	Ability and willingness to learn	Ability and willingness to learn
2	Teamwork and cooperation	Teamwork and cooperation

(continued)

TABLE 5.4 *(continued)*

Rating	Employers	Recent graduates
3	Initiative	Analytical thinking
4	Analytical thinking	Personal planning and organisation
5	Computer literacy	Computer literacy

5.6 Reviewing your performance

LEARNING OBJECTIVE 5.6 Review your performance using reflective practice techniques.

Like most students, you have probably had some paid, part-time jobs while you were at school and during holiday breaks, or perhaps a period of full-time employment. You may also have participated in other life and community experiences, such as sport, voluntary work, committee work, club organisation or being in a school-related leadership role. What have you learned from these experiences? Perhaps you had a great experience. Can you identify what made it perfect for you? On the other hand, you may have had a job you never want to do again. What was bad about that experience, and why?

One of the skills people use to review an experience is known as *reflective practice* — they think carefully about different aspects of an experience, good and bad, and tease out the key points they can learn from that experience. So, consider again one or two of your work experiences. Try to identify what you learned from each experience. Was it knowledge, a new skill, a way to do a particular task, a way to work effectively with a difficult person or the way to handle an ethical issue? How do you think your learnings from these experiences will help you in your career?

Many people reflect without realising it. How many times have you heard someone say, 'I won't do that again!'? They have just demonstrated they have reflected on, and learnt from, an experience. **Reflective practice** can be described as thinking about the past and trying to understand what happened and why. This is just one of the research methodologies that can be used to improve our understanding of our work. Like all skills, people need to practise if they are to become highly skilled at reviewing and learning from their professional experiences, both good and bad. By regularly using these skills, they get into the habit of systematically reflecting on an experience, such as a day's work or an aspect of a project, and then writing down what they learnt from the experience. The following prompts can be used to guide the reflective process.

- What did I do? (describe the situation)
- What worked and what did not work?
- How does this experience relate to similar past experiences?
- How did we work together as a team?
- How did I feel in this situation; what was my response?
- How could I do it better, or be more efficient, the next time I do a similar task?
- What knowledge and skills have I learnt?
- Who should I talk to if I need to rectify any problems?

One way to maximise the benefits from this activity is to write down your answers. Then, when you are faced with a similar experience or task, you can review your previous reflections and learning experiences and apply them to the current situation. People think reflectively all the time, but they often do not write their reflections down. An advantage of writing down your reflections is that you can look back on past reflections and observe how your comments have changed, what you have learnt and how your responses have evolved over time.

Practitioners can use reflective practice to improve their knowledge and skills. For example, some practitioners use reflective practice during the life of a project to monitor their work and methods, and then use that information to change or improve their practice. As part of a personal debriefing about a project, practitioners could compare their *experience* with their *expected experience*, their *learning* with their *planned learning*, and their *feelings* about the completed project with their *initial perceptions* of the project. This knowledge helps them prepare more accurate plans and schedules for their next project.

Practitioners may use reflective practice in other ways. They may look back on a number of completed projects and tease out a list of approaches, principles and processes that worked well on the projects, as well as perhaps highlighting the approaches that did not work. The list of successful approaches may be used to review a set of existing guidelines, or to develop a new set of guidelines that might be used when undertaking similar projects in the future. In this way, reflective practice can lead to innovation in an organisation.

Personal reflections can also be used by a project team as part of the debriefing process, both during and at the end of a project. The following spotlight illustrates this type of reflective practice.

KEY POINT

Engineers learn from their workplace experiences.

SPOTLIGHT

A reflection: working on large projects in isolated areas

Background

Peter worked for three years as a property boundary surveyor in Papua New Guinea: 12 months on the Sepik River and two years in the Highlands. During this time the company he worked for employed at least 12 surveyors, who were normally scattered throughout the country. Although he undertook surveys in towns and smaller settlements, he often worked alone on large projects in isolated areas. When he decided to return to Australia, his boss asked him, 'What essential criteria do you think we should use to select surveyors for large projects in isolated areas of Papua New Guinea?'

Reflection

To answer this question, Peter had to think back on his experiences. He also had to identify the critical attributes and capabilities he believed helped him successfully complete large projects in isolated areas. He thought of a number of questions.

- What personal characteristics and skills are required to operate as a surveyor in Papua New Guinea?
- What project management skills are required?
- What personnel management skills are required?

But the key question he thought of was: what is meant by the phrase 'to successfully complete a project?'

After considerable reflection, Peter came to the conclusion that the following objectives had to be met for a project to be 'successful'.

- The *survey* had to be completed to the standard required by the Papua New Guinea Lands Department, and to ensure that a later visit to the site was not required to attend to requisitions, check measurements or to do additional work (a requisition is a formal request to correct work or undertake additional work).
- The *survey plans* and report had to be completed to a standard accepted by the Lands Department and they had to be carefully checked to minimise requisitions that would require further work.
- The project had to be completed on time and to the satisfaction of the client(s).
- The project had to return a profit of 20 per cent or more to the company.

Recommendations

Peter remembered many of his experiences, and those of other surveyors, before identifying the following attributes and capabilities as being critical skills.

- A high level of surveying knowledge and skills, preferably with bush or jungle experience
- Plan-drawing and report-writing skills
- Project management skills (such as scheduling, costing and job completion skills)
- The ability to work independently in challenging environments for long periods of time (for example, having character attributes such as self-motivation and perseverance)
- Organisational skills (due to requirements to arrange travel and camp accommodation in isolated areas)
- People management and cross-cultural skills to work with, and manage, local employees
- Language and cross-cultural skills to communicate and negotiate with elders from villages and local councils

- The ability to negotiate with clients and produce outcomes that meet client requirements
- First aid skills

Conclusion

After his reflection, Peter realised that the majority of the skills in the list would be required by many other professionals working in isolated areas in Papua New Guinea, such as agricultural scientists, civil engineers, environmental engineers, geologists and mining engineers.

CRITICAL THINKING

Choose a personal or work project you undertook in the last year or so. Think about and then define the criteria you would use to evaluate whether the project was successful. Now, reflect on the activities you undertook during the project, the processes you used to manage the project, the time taken for you to complete the project, and the project outcomes. You should rate these against the criteria you defined for a successful project. Did the ratings surprise you? What would you do differently if you could do that project again? What did you learn from this experience?

Can you see the importance of reflective practice? By adopting the selection criteria and the other recommendations outlined in the above reflection, the company would be able to improve its staff selection and management processes. They could also introduce a training program to ensure their staff have the key skills required to operate in isolated areas. This would minimise the risk to staff who may otherwise find themselves out of their depth in an isolated area. The introduction of the guidelines and training may also improve the company's bottom line.

Engineers can also use reflective practice when they have to apply their knowledge and skills in a new situation that initially appears to be unrelated to any of their previous projects. Through reflection they may be able to identify similarities with past projects, as well as the knowledge, methods and skills that may be able to be transferred and used on the new project. In this case the reflective process would be quite systematic, ensuring the focus remained on identifying approaches that could be used on the new project.

Levels of reflection

We will now explore another aspect of reflective practice. Many writers have proposed reflective models, and many refer to levels of reflection from simple reflection through to deep reflection (Gibbs 1988; Hatton & Smith 1995; Moon 2004). Some of the types of reflection described in these models were synthesised to form the following examples of reflective practice. These examples are listed in order of increasing depth.

- *Descriptive reflection.* A description of the activity and your experience of the activity.
- *Simple reflection.* A brief reflection on the activity and its outcomes.
- *Value-based reflection.* A reflection about your experience and, if others are involved, their experiences (this may include your feelings and any personal challenges you faced).
- *Evaluative reflection.* A reflection about the success of the activity and how it was conducted (this may include technical success and success according to other factors, such as cost-effectiveness and environmental impact).
- *Analytical reflection.* A reflection that seeks to understand what occurred and why (perhaps by comparing outcomes with your previous understanding).
- *Concluding reflection.* A reflection that draws conclusions from your experience of the activity.

A good reflection usually includes aspects of most of these types of reflection.

Kolb's Learning Cycle

In recent years, reflective practice has proven to be a rich field of study for researchers. This has resulted in an increasing body of knowledge that links reflective practice to learning. For our purposes it is only necessary to explore some of the fundamental aspects of reflective practice. These are based on the work of Lewin (1952) and Kolb (1984), which resulted in Kolb's Learning Cycle. The Learning Cycle incorporates reflection as one of four stages in the learning process. The cycle begins with a concrete experience, followed by a reflection about what occurred. The next step involves trying to understand what occurred and why, and this understanding is then tested in the final step in the cycle. These are shown in figure 5.6 and then described in detail.

1. *Concrete experience.* This is when a person is immersed in undertaking an activity and in *the experience*. They may be observing, taking measurements or completing other actions, but these are all part of undertaking the task. For example, preparing a design of a structure, or testing a component of a machine. Questions that should be answered during this stage include: 'What did I do and how?', 'How long did I take?' and 'Why did I do it this way?'.
2. *Reflective observation.* After an activity is finished, consider what occurred and why, how successful the activity was and how the experience affected the participants, and any equipment or machinery used. In order to do this effectively, it may be necessary to step back and look at the experience from a broader perspective and also compare it with other similar experiences. Possible reflective questions include the following.
   - How well did the design process go?
   - How good is the design?
   - What worked?
   - What did not work?
   - What was difficult?
   - What was easy?
   - What external factors impacted on the design?
   - What impact did the design have on external factors?
   - Was the experience enjoyable?
3. *Abstract conceptualisation.* After the period of reflection, the next stage of the cycle is trying to understand what occurred and the relationships between different events. This evaluation may be undertaken against a knowledge base of relevant theories or processes, or it may lead to the development of new ideas or processes. For example, a standard design process may have to be modified to achieve the required outcome. Comparing this with other similar experiences will determine if a new process is required.
4. *Active experimentation.* After the development of a conceptual understanding of the experience, strategies can be planned for improving the activity and for enhancing the experience of the activity. This may mean the development of a new design process or a plan to study some new aspect of the relevant theory.

FIGURE 5.6 Kolb's Learning Cycle suggests there are four stages in the learning process.

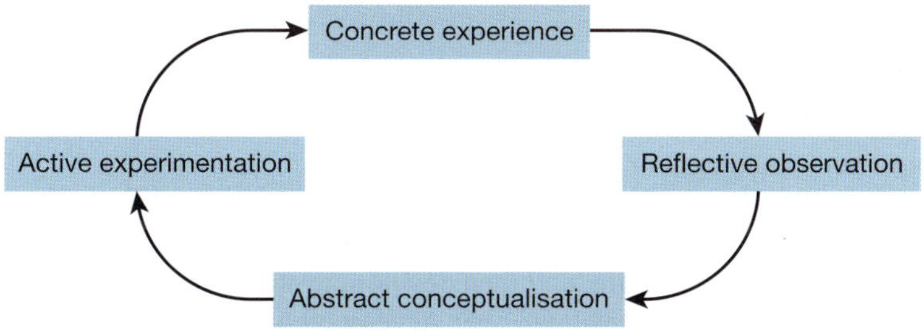

Source: Adapted from Kolb (1984, p. 42).

After stage 4 is complete, the learning cycle begins again at stage 1 when you next undertake the particular activity. In this way, the learning from each experience of the activity is used to improve the activity and the next experience of the activity.

This basic model demonstrates the important role reflection plays in the learning process and the point at which reflection is situated in the learning process. Understanding this process can help you to focus on the way you are learning and how your learning experiences may be enhanced. It also provides you with a tool to manage your learning.

SUMMARY

In this chapter we have explored a range of concepts and skills that will help you to engage in, and manage, your learning while you are at university and during your professional life. Using them will help you develop the deep understanding of engineering principles and practices that engineers must practise. This depth of understanding is required because engineers regularly put their knowledge and skills 'on the line' when an engineering component they have designed is assembled or constructed, such as an aircraft component, a bridge beam, a high-rise building, a chemical process, electricity infrastructure, process-control software or a dam wall. A performance failure in one of these can have catastrophic results, including the loss of human life. You can probably think of many more examples of design errors or material defects that would have similarly serious consequences. This illustrates the importance of managing our learning so that we have the knowledge, skills and capabilities to undertake our role. We will now briefly revisit each of the learning objectives for this chapter.

5.1 Identify the personal characteristics and strengths you can use to manage your work.

Reflect on your personal characteristics, prior learning and experiences, and identify those that you could use to manage your work and your studies. Through an understanding of your personal traits, preferred learning styles and spatial ability, you will be able to recognise and utilise your strengths. This knowledge will help you to use more efficient learning approaches and achieve better learning outcomes. Engineers need to use their strengths in their work and use their understanding of their learning preferences to optimise their learning.

5.2 Describe the factors that inspire and motivate you to become an engineer.

An understanding of the factors that led you to study engineering will help you to set goals and remain focused during your studies. You should also try to identify the factors that motivate you so that you can remind yourself of them when the going gets tough. Although the subjects you study will get harder as you progress through your program, your growing engineering knowledge and understanding will inspire and motivate you to continue your learning journey.

5.3 Develop your engineering knowledge by exploring the engineering in your world.

It is important to have an inquiring mind so that you build your understanding of engineering principles and systems. By observing and exploring the characteristics of the engineering infrastructure in your city and neighbourhood you will gain an appreciation of the complexity and diversity of engineering systems, and the fundamental science and engineering principles that underpin their design and development. You will also become aware of the impact these engineering systems have on other systems, on members of the community and on the environment.

5.4 Use self-management skills to plan and manage your work.

Self-management — the ability to establish personal goals and then adopt strategies to help you to achieve those goals — is a characteristic of many successful people. This involves setting both long- and short-term goals in the different areas of your life, and then developing strategies that will enable you to achieve those goals. Once your goals and strategies have been developed, you can use them to help you prioritise your activities. You can then use time-management skills to allocate time to each of the activities on a daily, weekly or monthly basis. By using a timesheet you can compare the time you spend on different activities with the time you allocated to those activities. Finally, it is important that you monitor your progress towards achieving your goals and, if necessary, change your strategies.

5.5 Use life-long learning skills to plan and manage your learning.

You should take responsibility for managing your learning at university as you will develop key life-long learning skills that you will be able to use throughout your career. A number of tips and techniques were discussed, and these can be used to improve learning and the efficiency of learning. A good understanding of the structure and content of your engineering program, and the embedded graduate attributes, will help you to identify, position, and contextualise your learning as you proceed through your studies. It will also help you recognise the limits of your current knowledge and skills and identify your learning needs. Once you have graduated, these life-long learning skills will help you plan and undertake professional development activities so that you maintain the currency of your knowledge and skills throughout your career.

5.6 Review your performance using reflective practice techniques.

It is also important to engage in reflective practice to learn from your experiences. Reflective practice can be used to improve your performance as an individual, and when you work in teams, both at university and in the workplace. In this sense, a reflective review is a feedback loop that will help you to enhance your work outcomes, and your self-management skills.

KEY TERMS

attitude A position, disposition or manner with regard to a person or thing. Attitudes can develop and change over time.

black box The terminology used to describe a device or process when it is only being studied from an input and output perspective.

deep learning When learners look for meaning in a task and try to relate it to their experience, knowledge and understanding.

major A group of subjects within a program that enable students to specialise in a field (e.g. civil engineering).

personality The enduring patterns of thought, feeling, motivation and behaviour that are expressed in different circumstances.

personality test Typically, a self-report measure of personal characteristics.

program Structured learning that leads to a qualification, usually a bachelor degree, associate degree or advanced diploma. A program consists of a number of individual subjects.

reflective practice Thinking about the past and trying to understand what happened and why.

self-management The ability to establish personal goals and then adopt strategies and use time effectively to achieve those goals.

spatial ability The ability to visualise and manipulate two- and three-dimensional objects in the mind's eye and see them from other perspectives.

spatial relations The ability to manipulate or rotate simple visual patterns when speed is a factor.

spatial scanning The ability to mentally scan a map or object and find a path or connection between two points when speed is a factor.

strength An ability to perform in an activity at a higher level than our peers.

subject A course of study within a program that is assessed and graded.

surface learning When learners memorise facts without understanding.

time management The systematic, priority-based allocation of time to competing demands.

EXERCISES

1. Take a few moments to consider your future and imagine yourself as a professional engineer. What do you think you will be doing in five years' time? Where will you be? Do you think you will be satisfied and highly paid? Write down your career plans and discuss them with your classmates. You should refer to your answers from time to time to remind yourself of your career aspirations.
2. What do you believe your strengths are? You may use Buckingham and Clifton's (2002) suggestions as prompts to help you identify your strengths. Write down any engineering-related tasks or activities that meet these criteria for you. Then consider the engineering program you are enrolled in and list some strategies for applying your strengths to your program (such as choosing practical subjects as electives if you are a practical person rather than a conceptual person).
3. Use Kolb's Learning Cycle to review and reflect on the learning activities that occurred during a practical experiment you have undertaken.
4. Develop a timesheet in a spreadsheet program, such as Microsoft Excel, and use it to record and prepare summaries of the time you spend on specific activities (e.g. studying, attending lectures and tutorials, travelling to and from university, working, participating in leisure activities and so on). You should develop the spreadsheet so it automatically provides you with summaries of how you spend your time. After you have a few weeks of data, you should be able to better plan your time so that it is easier for you to achieve your goals.
5. Engineers Australia has stressed the importance of reflection in engineering practice. A reflective journal will improve your reflective abilities. Start a reflective journal that includes your thoughts and reflections on learning as you progress towards becoming an engineer. You may wish to use this journal to reflect

on other exercises and activities you complete as you work through this text. As a first entry, identify the different levels of reflection reported in the 'Working on large projects in isolated areas' spotlight in the chapter.

6 Prepare a weekly schedule of your activities to help you manage your learning.

PROJECT ACTIVITY

Imagine that you want to undertake some vacation work experience with an engineering firm. The firm has asked you to undertake an audit of your learning skills prior to your participation in a training course that is designed to prepare you to participate in one of their upcoming engineering projects. They would like you to carry out an investigation and prepare a word-processed report on your readiness to learn about engineering.

Guidelines for completing this activity

The following steps are undertaken as part of an engineering investigation. A number of sub-processes are listed for each step, although not all of them would be undertaken in each investigation.

1 *Define the problem.* List, in order, the questions to be investigated, research relevant information, and then scope the problem to ensure all relevant aspects are included in the investigation.

For example, questions that would be relevant to your investigation for this activity are as follows.

- What is your motivation for studying engineering?
- What personal characteristics, experiences and strengths do you bring to your study?
- How will you manage your learning?

2 *Plan the investigation.* Define the overall approach that will be used, define the approach for each question, develop a plan to complete each phase of the investigation, define the methodology to be used for each phase, list the proposed outcomes that are expected from the investigation, and agree on the acceptance criteria with the client (in this case, the engineering firm). For example, the investigation for this activity could be made up of three smaller investigations, each addressing one of the three questions in your problem definition. The general approach for each question could be to:

- gather the information required to gain an understanding of the topic
- undertake a gap analysis by comparing that information with your personal information
- write a response to the question
- develop some recommendations on the actions you should take to enhance your skills.

The outcome of the investigation will be a typed document that reports on the overall findings of the investigation and the findings relating to each question, within a specified time frame. This will be in the format of a technical report and, if the client has agreed to this outcome and the proposed presentation format, this will become the acceptance criteria.

3 *Identify and source resources and capabilities.* List the engineering and other skills that will be required to complete the investigation, as well as the resources that will be required to complete the investigation. Plan how these will be sourced. For example, you should be able to undertake the investigation for this activity by yourself, with access to a:

- notebook for your research findings
- computer with word processing software to prepare the final report.

4 *Carry out the investigation.* Research the problem. Think of some alternative solutions, evaluate the alternatives against the selection criteria, choose a preferred alternative, and check all issues and questions have been addressed.

5 *Draw conclusions and make recommendations.* Analyse and synthesise the information, summarising the findings and developing conclusions and recommendations for action.

6 *Write a report.* Prepare a final report, checking the acceptance criteria are met. Submit and review findings with the client.

REFERENCES

American Society of Civil Engineers 2008, *Civil engineering body of knowledge for the 21st century*, 2nd edn, American Society of Civil Engineers, Virginia.

Arnold, R, Burke, B, James, C, Martin, B & Thomas, D 2001, *Educating for a change*, Doris Marshall Institute for Education and Action, Between the Lines, Toronto.

Australian Government 2020, 'National PFAS position statement — publication and consultation', https://www.pfas.gov.au/news/national-pfas-position-statement-publication-and-consultation-1

Biggs, JB 1987, *Student approaches to learning and studying*, Australian Council for Educational Research, Melbourne.

Buckingham, M & Clifton, DO 2002, *Now discover your strengths: How to develop your talents and those of the people you manage*, Simon and Schuster, London.

Burton, LJ 1998, 'A factorial analysis of visual imagery and spatial abilities', unpublished doctoral dissertation, University of Southern Queensland, Australia.

Burton, LJ, Westen, D & Kowalski, R 2023, *Psychology*, 6th Australian and New Zealand edn, John Wiley & Sons, Brisbane.

Coll, RK & Zegwaard, KE 2006, 'Perceptions of desirable graduate competencies for science and technology new graduates', *Research in Science & Technology Education*, vol. 24, no. 1, pp. 29–58, https://doi.org/10.1080/02635140500485340

Davis, K 1996, *How to speak to youth . . . and keep them awake at the same time*, Zondervan, Grand Rapids, Michigan.

Dowling, D 2010, *The career aspirations and other characteristics of Australian para-professional engineering students*, Australasian Association for Engineering Education Conference, Sydney.

—— 2014, *A graduate capability framework for the MEA mining engineering degree programs*, Mining Education Australia, Carlton.

Dowling, D & Hadgraft, R 2013, *A graduate capability framework for environmental engineering programs: A guide for Australian universities*, Office for Learning and Teaching, https://www.engineersaustralia.org.au/sites/default/files/2022-07/graduate-capability-framework-environmental-engineering-degree-program.pdf

Engineering New Zealand 2019, 'Four-year engineering degrees', https://www.engineeringnz.org/engineer-tools/ethics-rules-standards/accredited-engineering-qualifications/accredited-four-year-engineering-degrees

Engineers Australia 2008a, *Accreditation management system: Education programs at the level of professional engineer, Accreditation Board*, 3rd edn, Engineers Australia, Barton.

—— 2008b, *Review of Australian higher education: Submission in response to June 2008 discussion paper*, Engineers Australia, Barton.

—— 2011, 'Deeper port for bigger ships', *Engineers Australia Magazine*, Civil Edition, March, pp. 50–51.

—— 2019, *Stage 1 Competency Standard for Professional Engineers*, https://www.engineersaustralia.org.au/publications/stage-1-competency-standard-professional-engineers

Fellner, C 2023, '$58 billion day of reckoning looms for 3M over toxic "forever chemicals"', *The Sydney Morning Herald*, 8 February, https://www.smh.com.au/national/58-billion-day-of-reckoning-looms-for-3m-over-toxic-forever-chemicals-20230203-p5chri.html

Foster + Partners 2013, 'Foster + Partners works with European Space Agency to 3D print structures on the moon', https://www.fosterandpartners.com

Gibbs, G 1988, *Learning by doing: A guide to teaching and learning methods*, Further Education Unit, National Institute of Adult Continuing Education, Great Britain.

Glittenberg, D 2012, 'Starch-based biopolymers in paper, corrugating, and other industrial applications', in K Matyjaszewski & M Möller (eds), *Polymer science: A comprehensive reference,* Elsevier, vol. 10, pp. 165–193, https://doi.org/10.1016/B978-0-444-53349-4.00258-2

Gorska, R, Sorby, S & Leopold, C 1998, 'Gender differences in visualization skills — an international perspective', *Engineering Design Graphics Journal*, vol. 62, no. 3, pp. 9–18.

Graduate Careers Australia 2016, 'Graduate outlook 2015: The report of the 2015 graduate outlook survey: Perspectives on graduate recruitment', https://www.graduatecareers.com.au

Grohar-Murray, M & DiCroce, H 2003, *Leadership and management in nursing*, 3rd edn, Prentice Hall, New Jersey.

Hatton, N & Smith, D 1995, 'Reflection in teacher education', *Teacher and Teacher Education*, vol. 11, no. 1, pp. 33–49, https://doi.org/10.1016/0742-051X(94)00012-U

Kolb, D 1984, *Experiential learning: Experience as the source of learning and development*, Prentice Hall, New Jersey.

Laird, D 1985, *Approaches to training and development*, Addison-Wesley, Reading, Massachusetts.

Lewin, K 1952, *Field theory in social science, selected theoretical papers*, Tavistock, London.

Macquarie Dictionary 2024, *The Macquarie Dictionary Online*, Pan Macmillan Australia, https://www.macquariedictionary.com.au

Magin, D & Churches, A 1996, *Gender differences in spatial abilities of entering first year students: What should be done?*, 8th Australasian Association for Engineering Education Convention and Conference, Sydney.

Male, SA 2012, 'Generic engineering competencies required by engineers graduating in Australia: The Competencies of Engineering Graduates (CEG) Project', in M Rasul (ed.), *Developments in engineering education standards: Advanced curriculum innovations*, IGI Global, Hershey, PA, pp. 41–63.

Male, SA, Baillie, C, MacNish, C, Leggoe, J, Hancock, P, Alam, F, Crispin, S, Harte, D & Ranmuthugala, D 2015, *Student experiences of threshold capability development in an engineering unit with intensive mode*, Australasian Association for Engineering Education Conference, Geelong, Victoria.

Male, SA, Bush, MB & Chapman, ES 2009, *Identification of competencies required by engineers graduating in Australia*, 20th Conference of the Australasian Association for Engineering Education: Engineering the Curriculum, ResearchGate, Berlin.

—— 2011, 'An Australian study of generic competencies required by engineers', *European Journal of Engineering Education*, vol. 36, no. 2, p. 151, https://doi.org/10.1080/03043797.2011.569703

Male, SA & Leggoe, J 2017, *Student experiences of threshold capability development in a computational fluid dynamics unit delivered in intensive mode*, 28th Annual Conference of the Australasian Association for Engineering Education, Manly, NSW.

Moon, J 2004, *A handbook of reflective and experiential learning: Theory and practice*, Routledge Falmer, London.

Morales-Caselles, C, Viejo, J, Martí, E, González-Fernández, D, Pragnell-Raasch, H, Ignacio González-Gordillo, J, Montero, E, Arroyo, GM, Hanke, G, Salvo, VS, Basurko, OC, Mallos, N, Lebreton, L, Echevarría, F, van Emmerik, T, Duarte, CM, Gálvez, JA, van Sebille, E, Galgani, F, García, CM, Ross, PS, Bartual, A, Ioakeimidis, C & Cózar, A 2021, 'An inshore–offshore sorting system revealed from global classification of ocean litter', *Nature Sustainability*, vol. 4, pp. 484–493, https://doi.org/10.1038/s41893-021-00720-8

Motyer, J 1978, 'The riches and the enrichment of the church in the place of prayer', *Towards the Mark*, vol. 7, no. 6.
NASA 2014, '3D printing in zero-G technology demonstration', https://ntrs.nasa.gov/api/citations/20140012888/downloads/20140012888.pdf
National Skills Commission 2020, 'How employable are you?', https://labourmarketinsights.gov.au/media/xihfsftd/the-importance-of-your-personal-skills-and-qualities-when-looking-for-a-job.pdf
NSW Environment Protection Authority 2023, 'Fogo information for households', https://www.epa.nsw.gov.au/your-environment/recycling-and-reuse/household-recycling-overview/fogo-information-for-households
New Zealand Ministry for the Environment 2023, 'Guidance on plastic products banned from October 2022', https://environment.govt.nz/publications/plastic-products-banned-from-october-2022
Rath, T 2007, *StrengthsFinder 2.0*, The Gallup Organization, New York.
Samson, E 2023, 'McDonald's not lovin' the PPWR reuse mandates', *Sustainable Plastics*, 21 March, https://www.sustainableplastics.com/news/mcdonalds-not-lovin-ppwr-reuse-mandates
Sharpe, W 2006, *The skills shortage — a bottom up approach*, awarded best paper at the Institute of Public Works Engineers Australia (IPWEA) NSW Division Annual Conference.
Sharpe, W 2010, *Managing the skills shortage — a team game*, Roads Summit, Sydney.
Self Management Group (SMG) 2008, *About us*, https://www.selfmgmt.com/about-us
Sorby, S 1999, 'Developing 3-D spatial visualization skills', *Engineering Design Graphics Journal*, vol. 63, no. 2, pp. 21–32.
Sorby, SA & Baartmans, BJ 2000, 'The development and assessment of a course for enhancing the 3-D spatial visualization skills of first year engineering students', *Journal of Engineering Education*, vol. 89, no. 3, pp. 301–307, https://doi.org/10.1002/j.2168-9830.2000.tb00529.x
Sutton, K, Williams, A & McBride, W 2009a, *Exploring spatial ability and mapping the performance of engineering students*, Australasian Association for Engineering Education Conference, Adelaide.
—— 2009b, *Spatial ability performance of female engineering students*, Australasian Association for Engineering Education Conference, Adelaide.
WebFinance 2008, *Business Dictionary.com.*
Western Australia Department of Transport 2016, *Ports Handbook Western Australia 2016*, Government of Western Australia, Perth.
Wright, S, Hadgraft, R & Cameron, I 2010, *Learning and teaching academic standards for engineering and ICT*, Australian Learning and Teaching Council, Strawberry Hills, NSW.

ACKNOWLEDGEMENTS

The authors wish to acknowledge the following spotlight contributors: Caroline But, Miles Cattach, Madhu Bhaskaran, Nicole Hahn, Katherine Nguyen and Warren Sharpe.
Photo: © shock / Adobe Stock Photos
Photo: © Caroline But
Photo: © Miles Cattach
Photos: © Madhu Bhaskaran
Photo: © Nicole Hahn
Photo: © BNK Maritime Photographer / Shutterstock
Photo: © Katherine Nguyen
Photo: © NASA
Photo: © Fer Gregory / Shutterstock
Photo: © Warren Sharpe
Photo: © mavo / Shutterstock
Figure 5.1: Adapted from Morales-Caselles, C, Viejo, J, Martí, E et al. 2021, 'An inshore–offshore sorting system revealed from global classification of ocean litter', *Nature Sustainability*, vol. 4, pp. 484–493, https://doi.org/10.1038/s41893-021-00720-8
Figure 5.2–5.4: © Local Motors
Figure 5.5 and table 5.3: © American Society of Civil Engineers 2008, *Civil engineering body of knowledge for the 21st century*, 2nd edn, American Society of Civil Engineers. With permission from ASCE.
Table 5.1 and text: © National Skills Commission 2020, 'How employable are you?', https://labourmarketinsights.gov.au/media/xihfsftd/the-importance-of-your-personal-skills-and-qualities-when-looking-for-a-job.pdf
Table 5.2: © Dowling, D & Hadgraft, R 2013, *A graduate capability framework for environmental engineering programs: A guide for Australian universities*, Office for Learning and Teaching, https://www.engineersaustralia.org.au/sites/default/files/2022-07/graduate-capability-framework-environmental-engineering-degree-program.pdf
Text: © Fellner, C 2023, '$58 billion day of reckoning looms for 3M over toxic "forever chemicals"', *The Sydney Morning Herald*, 8 February, https://www.smh.com.au/national/58-billion-day-of-reckoning-looms-for-3m-over-toxic-forever-chemicals-20230203-p5chri.html

CHAPTER 6

Working with people

'A fantastic model of collaboration: thinking partners who aren't echo chambers.'

Margaret Heffernan

LEARNING OBJECTIVES

After studying this chapter, you should be able to:

6.1 work effectively with others

6.2 work effectively in groups and teams

6.3 actively participate and contribute in meetings

6.4 describe negotiation strategies and processes

6.5 understand dispute and conflict resolution strategies.

Introduction

To implement the engineering method (as described in the engineering method chapter) engineers need to collaborate with other members of the engineering team as well as people from other disciplines and professions. These collaborations may be with other individuals, members of project teams, consultants or people from other organisations, government agencies and departments.

A group of six young environmental engineers participated in a consultation process where they were identifying the tasks young graduates undertake in their first two years in the workforce. They were also asked to list the key skills they had used in their work but had not necessarily been taught at university. As they had only been working three to five years since graduation, their post-graduation experiences were fresh in their minds. They identified the following list of skills.

- *Career planning and management.* They all agreed that they would have liked to learn about this in their first year at university so that they could be more systematic in planning their work experiences and selecting elective courses.
- *Writing meeting minutes and notes.* They were regularly asked to accompany their supervisor, or another engineer, to meetings both within their organisation and with clients, government officials and other professionals. Normally their role was to support their supervisor by listening to and observing the people at the meeting and to keep accurate notes of the discussions, decisions and other key points. This was a job they did not always enjoy, and they also found it to be challenging. After the meeting they would prepare draft minutes of a meeting and have them approved by their supervisor prior to circulation. Although they did not always enjoy their role at meetings, they found there were many benefits in undertaking this role. They learned to listen carefully; to read body language; to identify points of agreement as well as sticking points and unresolved issues; to understand the different approaches that may be used to achieve outcomes in meetings; and to identify the key players. They also found these meetings provided good opportunities for networking and learning about other disciplines. Finally, they appreciated the opportunities these meetings provided for them to discuss key issues with their supervisor and other engineers.
- *Gathering background information for a project.* This included reviewing journal articles and preparing literature reviews; identifying relevant standards and codes of practice; reviewing supplier catalogues; coordinating the inputs from other members of multidisciplinary teams; writing document summaries; assembling gathered information to form a draft report; and presenting findings to other members of the team.
- *Reviewing tender proposals and contract documents.* The tasks included analysing documents and identifying key issues; evaluating timelines and financial details; using software to prepare realistic project schedules; preparing cost estimates; and assembling, collating and publishing documents.
- *Collaborating with others.* They were required to liaise with senior staff and consultants, work with and supervise drafters and technical assistants, and work in multidisciplinary teams. They agreed that while some of these experiences were difficult at first, they soon became comfortable in team situations and were able to participate in meetings. They noted the importance of speaking up early if they believed there was an issue that needed to be discussed.

At the end of the discussion the recent graduates emphasised the fact that they had found all these tasks difficult in the beginning — in fact, more difficult than many of the technical tasks they undertook. However, they all said that these experiences gave them plenty of opportunities to network, both within their organisation and with people outside their organisation. These experiences also provided them with opportunities to learn about their organisation and the way it operated, meet the key players and learn about their industry. They agreed that through all of these experiences they had learned how to collaborate effectively with others.

At this point you may like to reflect on how you would feel undertaking the tasks identified by the graduates. Where do you believe you will learn how to work effectively with other people? Some of you may have experience in these tasks, perhaps as a part-time employee. Were your experiences similar to the environmental engineering graduates? In this chapter, you will learn some of the key skills engineers require in order to collaborate effectively with other people. Thus, it builds on the self-management skills discussed in the previous chapter.

The chapter begins with a description of some of the skills needed to work effectively with individuals and in groups and teams. This is followed by a discussion about leading and participating in meetings. The penultimate section of the chapter covers negotiation — a key skill that is used during every phase of an engineering project, from the initial discussions with the client through to the final acceptance of the project outcomes by the client. The final section in the chapter discusses techniques that can be used to resolve workplace disputes.

These skills are important because they enable engineers to work effectively with other members of the engineering team, with professionals from other disciplines, with clients and with members of the public. This chapter provides you with an opportunity to learn about these skills and to practise the techniques while you are at university.

6.1 Collaborating with others

LEARNING OBJECTIVE 6.1 Work effectively with others.

Employers expect graduates to be able to collaborate effectively with other people, both within and outside their organisation (Nepal 2012). For example, Allen et al. (2017, p. 28) state:

> Rather than focusing on individual performance, organisations are more than ever trying to develop a culture where the most valuable employees are those who can collaborate and share information to improve efficiency and achieve organisational goals. The trends of rapid change in markets and technologies, and of multiple generations in the workforce at the same time, trigger the need for collaboration. As organisations become increasingly dynamic and horizontally structured, this need for collaboration impacts all types of roles.

They suggest that the essential skills for collaborating are transparency, communication, teamwork, relationship management, organisational awareness, social/cultural awareness, sociability and teaching others (Allen et al. 2017). These skills are discussed in this and other chapters.

Traditionally, the sets of graduate attributes defined by Australian universities have included attributes relating to teamwork and cross-cultural skills. More recent sets have broadened the teamwork attribute to emphasise the importance of the collaboration skills that are used when working with individuals and in groups; for example, one of the seven RMIT University (2024) graduate attributes is 'global outlook and competence':

> Graduates of RMIT University will have had opportunities to acquire professional and cultural skills that enable them to engage thoughtfully and effectively with the great diversity of people and situations they encounter at work and socially.

RMIT University (2024) states one of the ways this graduate attribute may be evidenced is when graduates '[w]ork in diverse teams to solve complex problems through respectful communication, negotiation and cooperation to effect positive change'.

Another approach is for discipline organisations, such as Engineers Australia, to develop a national set of graduate capabilities for the relevant disciplines, which are then interpreted and contextualised by the relevant school or faculty at each university.

As discussed in the chapter on self-management, the Define Your Discipline (DYD) stakeholder consultation process has been used for this purpose to define detailed graduate capability frameworks for a number of disciplines, including two engineering disciplines: environmental engineering and mining engineering (Dowling & Hadgraft 2013a).

The following spotlight describes the resulting framework and shows that employers from these two disciplines believe collaboration and teamwork are *critical* graduate capabilities for engineers working in their organisations.

SPOTLIGHT

The collaboration capabilities for two engineering disciplines

The collaboration capabilities of environmental engineers

The *Graduate Capability Framework for Environmental Engineering Degree Programs* consists of three sets of capabilities and a set of practice contexts. The authors of the framework listed a set of technical clusters and a set of generic clusters, as these would be aligned with the current curricula emphasis on discipline knowledge and generic graduate attributes. They also included of a set of process clusters, reflecting input from recent graduates and practitioners during the consultation process, as they represent the way environmental engineering practitioners undertake their work.

The interrelationships between the three sets of capabilities can be visualised using the environmental engineering capability cube, shown in figure 6.1, where each set is an axis of the cube. When undertaking a project, a graduate uses *generic* capabilities when applying a *process* in one or more *technical* domains. For example, a graduate may be gathering information (generic capability) to prepare a design (process) for a resource management (technical domain) project at a mine site (context).

FIGURE 6.1 The environmental engineering capability cube

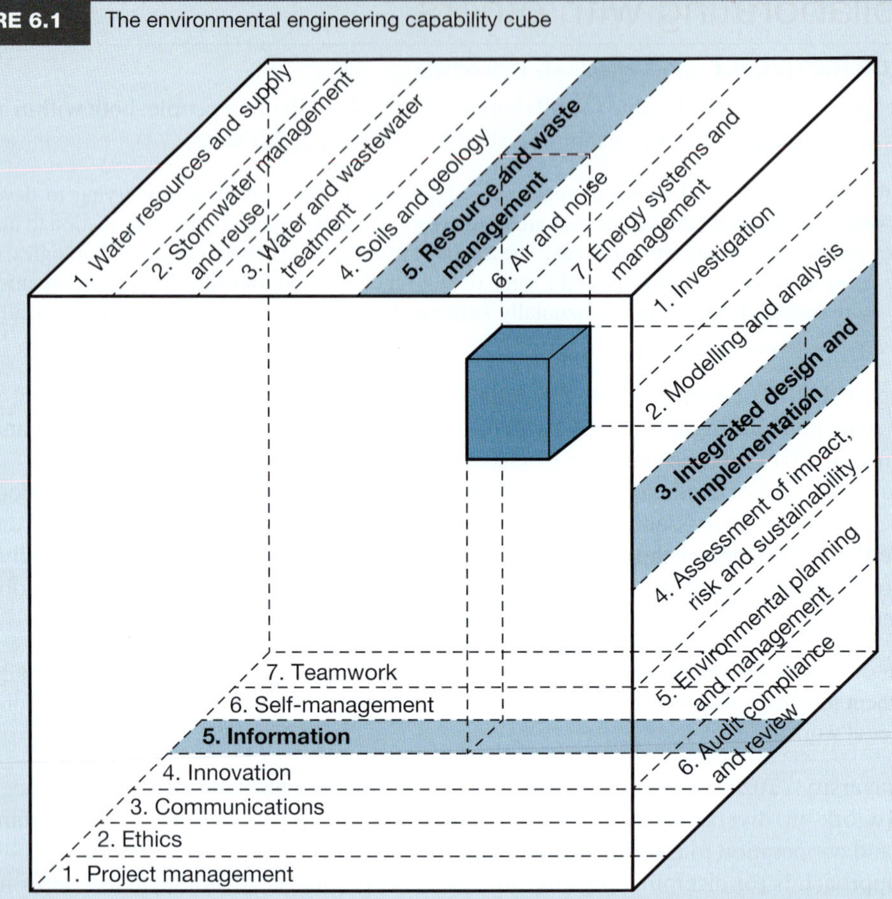

Source: Dowling and Hadgraft (2013b).

The framework included a separate table for each of the domains and processes, which listed the indicative tasks for that domain or steps in that process. A total of 392 indicative tasks were listed for 20 domains. Table 6.1 details the indicative tasks for the teamwork domain.

TABLE 6.1 **Indicative tasks for the teamwork domain**

1.	Uses knowledge of team roles and processes to contribute effectively to project team activities
2.	Works effectively in multidisciplinary project teams
3.	Recognises limits of knowledge and skills and collaborates with team members and others to fill identified gaps in knowledge and experience
4.	Develops leadership and management capabilities to be able to operate in complex multidisciplinary environments

Source: Dowling and Hadgraft (2013b).

A review of these 392 indicative tasks found that a further 20 indicative tasks were related to collaboration capabilities. These are listed in table 6.2.

TABLE 6.2 The collaboration capabilities

1.	Maintains effective and professional relationships with clients, project managers and other members of the project team
2.	Coordinates the information-gathering process for multidisciplinary projects including identifying and contacting key people, and sourcing and requesting information
3.	Compiles and integrates information from a range of sources to prepare initial drafts of tender documents or project reports
4.	Communicates effectively with clients, contractors and members of the engineering team
5.	Engages and consults with other professionals
6.	Communicates effectively with people from other disciplines (e.g. chemical, civil and structural engineers, botanists, chemists, geologists and microbiologists)
7.	Uses appropriate language and techniques to present project information to community groups, business organisations and school groups
8.	Assists with the preparation of reports on multidisciplinary studies
9.	Works with other disciplines to assess the impacts on marine flora and fauna of treated waste from ocean outfalls
10.	Consults relevant regulatory bodies and stakeholders
11.	If appropriate, manages field staff during data collection activities
12.	Presents findings to clients and other stakeholders in meetings and workshops
13.	Plans communication strategies for interactions with stakeholders
14.	During the development of the integrated concept design, seeks advice from specialists where necessary on issues such as architecture, traffic control, WHS operating requirements, noise and air quality
15.	Discusses and reaches agreement with client and relevant approval authorities on the final design concept
16.	Consults with relevant community groups and statutory authorities about the project and the scope of the study
17.	Identifies and engages specialists to undertake measurements, modelling and studies
18.	Coordinates and facilitates the work of specialists from multiple disciplines
19.	Synthesises and integrates information from a range of specialists to form an overall understanding of the impacts
20.	Consults external stakeholders, including regulators and members of the community

Source: Dowling and Hadgraft (2013b).

Table 6.2 shows that environmental engineering graduates are expected to be capable of collaborating with a range of people, including:

- professionals in allied disciplines
- professionals from other disciplines
- field technicians
- clients, consultants, contractors, statutory regulators and business, community and stakeholder groups.

Three of the collaboration capabilities are regarded as *critical* for environmental engineering graduates. These three collaboration capabilities are:

1. works effectively in complex multidisciplinary project teams
2. recognises the limits of their knowledge and skills and collaborates with team members and other professionals to fill gaps in their knowledge and experience
3. coordinates the information-gathering process for multidisciplinary projects, including identifying and contacting key people, sourcing and requesting information; and synthesising and integrating information from a range of specialists (Dowling 2016).

The collaboration capabilities of mining engineers

The *Graduate Capability Framework for the Mining Engineering Degree Programme* (Dowling 2015) also consists of three sets of capabilities:

- Technical capabilities — seven mining engineering technical domains: the mining industry, resources and reserves, mining geomechanics, mining methods, mining equipment and systems, mining services, and mineral production and processing
- Process capabilities — seven mining engineering processes: risk management, mine planning, integrated mine design, scheduling, managing mining operations, monitoring and compliance, and mine rehabilitation and closure
- Generic capabilities — seven generic domains: management, workplace health and safety, communication, sustainability, self-management, working with people, and creative thinking, problem solving and research.

Table 6.3 details the indicative tasks for the generic capability, working with people.

TABLE 6.3 Indicative tasks in the working with people domain

Topics	Indicative tasks
1. Works effectively in teams	Works effectively in crews, groups and teams
	Collaborates effectively with the members of multidisciplinary teams
2. Leads teams	Understands leadership principles and the roles of team leaders and team members
	Secures staff buy-in and ownership for required actions
	Leads small teams
	Leads cross-functional teams to solve problems
	Manages technical teams
	Provides a leadership role model for more junior engineers
	Leads, supervises and provides support to more junior engineers
3. Collaborates with others	Maintains a polite and friendly attitude at all times — within the company and externally
	Engages with all stakeholders — within the company and externally
	Collaborates with other engineers to accomplish the department's goals and targets
	Engages with, and obtains input from, the members of cross-functional teams
	Uses effective questioning techniques to obtain information from operators and people in other departments
	Collaborates well with people from all sections of the mine
	Coordinates supporting disciplines to ensure suitable input to plans and compliance with plans
4. Actively participates in meetings	Participates and contributes to discussions
	Participates and contributes in meetings
5. Mentors and trains people	Identifies training needs of subordinates
	Prepares training programmes for staff and general workforce
	Conducts training activities for staff and general workforce
	Provides training to personnel in relevant management positions
	Trains younger engineers and graduates
	Helps with training undergraduate students on work experience placements
	Mentors junior staff
	Mentors younger engineers and graduates

6. Manages people	Understands and applies HR policies and procedures
	Understands gender, equity and diversity policies and applies best practice
	Understands the legal duties of each person on site
	Understands the job descriptions and roles of managers
	Uses appropriate people management skills to manage people
	Identifies and responds appropriately to different human behaviours
	Contributes to the development of selection criteria for new positions
	Participates in recruitment processes
	Estimates manpower requirements for crews
	Reviews the work of peers and subordinates
	Develops work plans for tasks
	Organises mining crews
	Prepares shift rosters, taking into account health, safety and other requirements
	Allocates tasks
	Supervises a production crew
	Supervises underground drilling and charging processes
	Relieves supervisors for short periods
	Adjudicates workforce discipline issues
	Follows appropriate disciplinary procedure
	Becomes an HR point of contact for graduate recruitment processes
	Prepares vacation training plans

Source: Dowling (2015).

It is important to note that the 'working with people' capability for mining engineers is much broader than the teamwork capability for environmental engineering graduates. Consequently, there are more indicative tasks, which are divided into six topics, including two teamwork topics. A review of the indicative tasks included in the framework revealed an additional 73 collaboration capability tasks, which suggests that mining engineering graduates need a broader set of collaboration capabilities than environmental engineers.

A review of all of these indicative tasks shows that graduates are expected to be able to collaborate with:

- professionals in allied disciplines, such as geologists, geotechnical engineers, metallurgists and surveyors
- professionals from other disciplines, such as accountancy, occupational health and safety, risk management and environmental science
- senior management, such as production supervisors, production personnel, mining crews, operations staff and mine rescue teams
- clients, consultants, contractors, sub-contractors, statutory regulators, equipment suppliers and other suppliers
- business, community, indigenous peoples globally and other stakeholder groups (Dowling 2016).

Five of the collaboration capabilities could be regarded as *critical* for mining engineering graduates. These five collaboration capabilities are:

- collaborates effectively with the members of multidisciplinary teams
- collaborates effectively with people from all sections of the mine, from senior management to operational staff, and with consultants, contractors and suppliers
- understands their role, and where their role and the roles of others in the technical team and production section fit within the organisation, and how they contribute to the mining process and overall business.
- presents data and information in a clear and understandable way to technical teams, management, operational staff and larger audiences.
- recognises the limits of their knowledge and skills and collaborates with team members and other professionals to fill gaps in their knowledge and experience (Dowling 2016).

Thus, mining engineering graduates are expected have the capacity to work effectively with people from their discipline and other disciplines, both within their organisation and from other organisations. They are also expected to work with people at all levels of their organisation including miners, contractors and managers.

CRITICAL THINKING

1. Compare and contrast the lists for the two disciplines to define the common capabilities.
2. (i) If studying environmental or mining engineering, evaluate your capabilities against the relevant list.
   (ii) If studying another engineering discipline, talk to teaching staff, practising engineers and experienced students from your discipline and develop a list of the collaboration capabilities for your discipline. You could work with a small group of students to do this.
3. Review the courses you will study to determine where you will have opportunities to learn and develop these capabilities. You could develop a list of those not covered and then try to develop these capabilities while undertaking work experience or part-time work.

The collaboration capabilities identified in the two disciplines in the spotlight clearly demonstrate the need for graduates to acquire skills that will enable them to collaborate with:

- people working at other levels in their discipline
- people working in other levels in their organisation
- people from other disciplines.
- people from community organisations
- people from other ethnic/cultural groups
- people in government organisations
- members of the public.

This suggests that engineering graduates need to develop a set of discipline-specific, 360-degree collaboration capabilities to be successful practitioners in their field (Dowling 2016).

When communicating with practitioners in the same discipline, people use verbal, non-verbal and visual approaches that are often unique to their discipline, a communication culture that is shared across the discipline or *discourse community* (Sales 2006). When people from two or more disciplines wish to collaborate, each person needs to understand the communication cultures of the other disciplines involved if effective communication is to occur. This includes terminology, ways of thinking, approaches and processes. It is therefore important that graduates have an understanding of the communication cultures used by the practitioners in other disciplines they will collaborate with when they graduate.

Typically, engineering students gain an understanding of the communication cultures used by the common disciplines in the broad field of engineering. They acquire these skills by studying a series of courses that are common across their discipline; for example, all engineering students may study common introductory courses in communication, chemistry, engineering science, mathematics and physics. They will also study some higher-level courses from the other engineering disciplines. For example, an electrical engineering student normally studies a small number of courses from other engineering disciplines, such as mechanical engineering, software engineering and mechatronic engineering. Generally, engineering students will study some courses from other disciplines, such as management.

When students study courses from other specialisations and disciplines they gain an understanding of the knowledge, skills and approaches that practitioners use in those disciplines, as well as the communication cultures of those discourse communities. Thus, these students have laid the foundations for future successful collaborations with practitioners from those fields. Importantly, when students undertake project work in multidisciplinary groups, they are able to further develop those collaboration capabilities.

As we have seen, a capacity to work effectively with other people is high on any employer's list of desirable employee characteristics. This includes working with people from the same department in the organisation, from other departments or from other organisations. Work groups can be established with people from one location, from different locations within the same country or from around the world.

KEY POINT

Engineers should demonstrate 360-degeree collaboration capabilities.

6.2 Working in groups

LEARNING OBJECTIVE 6.2 Work effectively in groups and teams.

If you scan the employment pages of a newspaper or recruitment websites for engineering positions, you will notice many of the advertisements list a capacity to work with other people in groups or teams as an essential requirement. Therefore, being able to *demonstrate* that you have been an effective member in group or team situations will be important when you are seeking employment as a graduate engineer. While it is normal for you to be able to list examples where your group or team has been successful in completing a project, it is also important that you can demonstrate how well the group collaborated in achieving those outcomes.

You will have opportunities to learn and practise these skills while you are an undergraduate student. In this section we will look at a number of characteristics of groups and teams and how they can be managed to effectively achieve their allocated tasks.

KEY POINT

Engineers should be capable of working effectively in groups and teams.

SPOTLIGHT

Team innovation and success: why we should fight at work

REBECCA MITCHELL, UNIVERSITY OF NEWCASTLE

When your staff bicker and compete, your initial response should be to remind them they're part of the same team and encourage them to be friendly, right? Not necessarily; we're now realising that a level of tension and hostility can actually make teams more effective.

Organisations are increasingly bringing together teams of employees from different parts of the organisation to apply a broad range of relevant skills towards complicated tasks. Over time, these teams can suffer from an over-reliance on shared knowledge and fail to share and discuss points of difference.

This is where conflict can help. When team members are asked to be critical and norms of conflict emerge in a group, members are more likely to share their specialised information, which enhances team performance.

Apples and oranges

Interdisciplinary teams are particularly common in health care . . . Teams made up of different professions such as doctors, nurses, dietitians and pharmacists have been shown to improve clinical care, reduce medications per patients and reduce admission to hospital and emergency wards. They're also likely to be more innovative and more effective than homogeneous teams.

Benefits are also seen outside the health-care industry, with research demonstrating that teams of professionals such as engineers, architects and surveyors are able to decrease costs and design more creative products.

Bringing together different professions poses significant challenges including friction and breakdown in communication. This is explained by theories such as the similarity-attraction paradigm: we tend to like and work effectively with people who we perceive as similar to ourselves — at work this is often based on profession — and dislike people who we see as different.

The typical approach to this dilemma has been to charge leaders with minimising negative dynamics and boosting positive aspects of team interaction. But it is very difficult to reduce the sources of conflict that cause hostility, such as differences in professional status and values.

When team members perceive a threat to their profession, such as pressure to compromise on their profession's priorities or change their professional approach, teams are more innovative. When team interaction is characterised by tension and hostility, their work can be more effective.

This seems counter-intuitive, but our findings do not advise a team climate that is overwhelmingly hostile and characterised by threat. Successful leadership requires the ability to create a tension between positive and negative dynamics.

Leadership style

Transformational leaders are known to have high expectations and like to set goals and lead by example. They're also thought to be well-suited to teamwork.

Yet, while these leaders can increase motivation to work across occupational boundaries, this focus on cooperation can lead to premature consensus and conformity. When this occurs, we found that negative mood (reflecting hostility, upset and tension) can provide an effective counter. Negative mood signals to the team that something is wrong, promoting team members to question existing ideas and rely less on assumptions.

In the absence of negative mood, the motivation to work cooperatively reduces effectiveness, but a tension between cooperation and hostility enhances performance.

The same tension between positive and negative dynamics enhances other styles of leadership. Inclusive leaders strive to assure team members that their individual voices and unique perspectives will be valued.

But while creating a participative climate allows team members to express their viewpoints, it may not motivate them to do so. The capacity of inclusive leaders to engender innovation is dependent on whether team members perceive a threat to their profession. Feeling threatened drives members to advocate the distinguishing attributes of their profession's position.

It is the tension between feeling included and feeling threatened that motivates teams to find a solution that incorporates diverse and dissenting viewpoints and increases the likelihood of innovation. When teams appear to move towards compromise at the expense of dissent and critical analysis, negative mood and conflict may introduce a useful tension.

But managers should be cautious about engendering such moods, which have also been linked to team failure. One approach with potential for encouraging conflict within safe parameters is through interventions such as devil's advocacy, which direct dissent to task-related, rather than personal, issues.

Source: Mitchell (2014). Originally published on *The Conversation*.

CRITICAL THINKING

Consider the last team you were involved with (sporting, educational or professional). Did you experience conflict within your team and was it productive or counterproductive? Would your team have been considered interdisciplinary and how did that help/not help your team goal?

Establishing a group or team

In this section we explore the differences between groups and teams. However, it should be noted that many people, including employers, use the terms 'team' and 'teamwork' when they are really talking about a 'group' and 'working in a group'. Thus, for many people, the terms 'group' and 'team' are interchangeable.

'A **group** is a collection of two or more people who work with one another to achieve common goals' (Wood et al. 2025, p. 184). The size of a group can vary from two to more than 20 people. There are two main types of groups:

- *informal* — groups that form spontaneously (e.g. a special interest group)
- *formal* — groups that are established by an organisation to complete one or more tasks (e.g. a software development group in an international company).

Formal groups normally have a leader appointed by the organisation; for example, a manager or supervisor. They may be temporary, such as a task force, or permanent, such as an electricity transmission line design group in an energy provider.

The leader is responsible for ensuring that the group completes the required activities on time and within budget. Thus, the leader is activity focused and would normally delegate tasks to one or more members who are individually responsible for completing the delegated activities and producing the expected artefacts.

'A **team** is a small group of people with complementary skills, who work together as a unit to achieve a common purpose for which they hold themselves collectively accountable' (Wood et al. 2025, p. 217). Teams often share leadership roles and members may spend a lot of time in meetings discussing, brainstorming and problem solving. The members of the team are mutually responsible for the outcomes they produce.

Katzenbach and Smith (2001) suggest that when the tasks allocated to a group require it to produce outcomes rather than undertake activities, then the performance of the group becomes critical. It has to move from being an effective working group to becoming a performance group or team. The decision about whether a group should be a single-leader group or a team, should be based on the task allocated to the group. According to Katzenbach and Smith (2001, p. 13):

> Whenever a small group can deliver performance through the combined sum of individual contributions, then the single-leader discipline is the most effective choice. This choice is fast, efficient, and comfortable, since most organisational units have followed the single-leader model for decades. However, if there must be collective contributions in addition to individual efforts, then the group should apply the team discipline.

Both groups and teams can be high performing and achieve their purposes; however, when a real-time combination of multiple skills, experiences and judgements are required, then a team inevitably gets better results (Katzenbach & Smith 2001). Many organisations have, however, found that developing a group of individuals into a high-performing team is not easy. It is for this reason that an understanding of the principles and processes that lead to effective groups and teams is crucial. Before we analyse this, we need to clarify the meaning of some commonly used terms. In this section the term 'group' will be used because, as we have explained, a team is just a special type of group.

A **virtual group** is a recent phenomenon. Members of a virtual group may actually never meet face to face as they meet and communicate via a range of technologies, such as teleconferencing or videoconferencing (e.g. Jabber and Skype). Virtual group members may be from the same organisation or multiple organisations. The members of a virtual group may be located in one city, in different cities in one country or in cities around the world.

A **teleconference** is a meeting of at least three people who are geographically separated, conducted using a telephone or other audio system. The meeting may include groups of people at some locations and individuals at other locations. A **videoconference** is similar except that both audio and video technologies are used for the conference.

A **groupware**-facilitated meeting is based on a combination of computer hardware and software that together provide *enhanced* videoconference meetings between geographically separated people. Participants in a **groupware conference** use real-time audio, video and text communications and other features to conduct their meeting. While Zoom and Microsoft Teams appear to dominate the groupware market, there are other apps that, while facilitating collaboration, also offer additional tools for use in specialist fields such as project management; examples include Trello and Jira.

Groupware systems are also used by schools and universities to facilitate real-time interactions between teachers and geographically dispersed students during lectures and tutorials. The software enables students to ask questions while a video of the lecture is streamed live over the internet. They can do this directly via a webcam, or by using the keyboard to send an email or text message. The software prompts the lecturer about incoming questions by highlighting these in different sections of the screen.

Many factors influence the way a group is established, and six simple questions can be used to help define the characteristics of a proposed group.

1. *Why* is the group being established? For example, is the group to be part of an organisational structure or is it being established to undertake a specific project?
2. *What* task(s) will the group undertake? The characteristics, size and complexity of a task will greatly influence the size and skill requirements of the group.
3. *Which* skills are required for the group to successfully complete the project? The skill sets required may include technical and professional skills from one or more disciplines, as well as a range of generic skills such as communication and interpersonal skills. For example, a group undertaking a mine development project may include chemical, civil, electrical, mechanical and mining engineers, as well as supply chain professionals and accountants.
4. *Where* will the members of the group be located? Will they all be in one location, or will they be in different locations? This will influence the way the group meets and communicates. For example, the project team for an offshore gas production plant may include on-site construction engineers, on-shore engineering project managers, design engineers in offices around the world, engineers representing the major consultants and engineers representing major component manufacturers.
5. *How* will the group meet and communicate? Traditionally face-to-face meetings have been standard practice, with group members travelling to a central location or head office. Because of the financial costs and travel times, many organisations now use virtual meetings when the members of a group are

geographically separated. The virtual meetings may be interspersed with, for example, a quarterly, an annual or a project milestone face-to-face meeting.

6. *Who* is available to establish and lead the group? It is important that the right person is chosen to establish the group and lead it through the early phases of its life. This is because the selection of the members of a group, and the effectiveness of the first one or two meetings of the group, can be critical to its success.

The answers to these questions will help an organisation to establish a group with an appropriate number of members, the right skill sets and the best possible leader.

In establishing a group, we should remember the key point from the quote at the start of the chapter, reinforced by Oscar Wilde when he said, 'I hate it when people agree with me, because then I know I must certainly be wrong!' Thus, while it is always tempting to establish a homogeneous group, there is evidence to suggest that heterogeneous groups perform better in the longer term (Watson et al. 1993). This is because the members of a heterogeneous group are likely to have a greater diversity of opinions, approaches and experiences than the members of a homogeneous group (Thomas 2014). Traditionally, when companies talked about the diversity of their workforce, their focus was on gender and, to a lesser extent, ethnicity. More recently, many employers have come to value inclusion, equity and diversity leading to the employment of people with a range of abilities in their organisation.

However, to harness the opportunities that diversity and inclusion bring to a group, the leader of the group will need the skills to manage the different opinions that result from such diversity and any conflicts that may arise. This was the focus of the discussion in the previous spotlight.

It is now time to explore the key characteristics of an effective group.

The fundamentals of an effective group

Katzenbach and Smith (2001, p. 3) have found that neither groups nor teams can realise their performance potential unless the following five elements are in place.

1. An understandable charter
2. Good communications
3. Defined member roles
4. Time-efficient processes
5. A sense of accountability

A clear purpose should be added to the list. Each of these elements is now explored in turn.

A clear purpose

All members of a group should have a clear understanding of the purpose of the group. This includes a shared vision and understanding of the tasks allocated to the group, the timeframes for those tasks, the expected outcomes, and how those outcomes may be evaluated by the client. If appropriate, the group should also understand how their performance will be evaluated.

A charter

It is important that a group leader, or team members, establish clear behavioural guidelines at the beginning of group projects. A number of terms are commonly used to describe this type of document; for example, charter, contract or code of conduct. In the workplace, performance expectations may be outlined in formal documents, such as a position description, or clarified in discussions with a supervisor. This is not always the case for student groups and teams.

It is important that a student group or team develops its own charter, and this should be an important part of the discussion during the group's first meeting. The guiding principles that can be used as a starting point for discussions about developing a charter or contract for a student group project at university, include:

- how workloads, task allocations and deadlines will be determined and shared among the group
- the quality of work required (e.g. to achieve a desired mark)
- how often the group will meet
- how the group will communicate
- how any potential disputes will be resolved (e.g. if a group member is not contributing).

A charter must be understandable and be owned by each member of the group if the group is to be effective. Members should sign and date the charter as evidence of this understanding and ownership.

It is also important that the group takes account of the cultures and values of the people in the group. For example, Australians are used to taking an egalitarian approach when working in teams. They expect to be able to state their opinion on the issues being discussed. This is not always comfortable for people from some other cultures where, for example, seniority is respected, and this may inhibit the contributions of younger members of the group.

Communication and coordination strategies

The meeting and communication strategies that will be used by a group need to be clearly stated and should be included in the charter. The strategies would normally include meeting times, length of meetings, meeting protocols, and out-of-meeting communication methods and expectations. In group situations the leader will often develop the communication strategies for the group; in team situations the members of the team should discuss and develop the strategies. It is important that their charter ensures that members will have an equal opportunity to contribute to discussions and that all comments will be valued. This is because a team's performance may be affected by a real, or perceived, imbalance of power. For example, the presence of a manager in a team may restrict the freedom people have to hold a free-ranging discussion about a project and the way it should be managed.

There are many software products that can be used to facilitate collaboration in group or team situations. Among other things, these products enable participants to communicate with each other, hold discussions, develop, modify and store documents, issue meeting agendas, and circulate meeting minutes and action statements. While the software products are generally for PCs, they often include apps for tablets and smartphones so that participants can keep up to date when they are out of the office.

The strategies adopted by virtual groups will be similar to those adopted for face-to-face groups, but they will need to include strategies to ensure all members are included in the meeting and to manage their virtual presence. When representatives from a number of countries are involved in a virtual team, culture, differences in technologies and time zones may have an impact on the quality and effectiveness of communications and meetings. For example: the technologies that are available in one country may not be available in other countries or they may be incompatible with the technologies available in those countries; the working week varies from country to country; public holidays vary from country to country (and, in Australia, from state to state); and differences in time zones may mean that it is difficult to hold meetings when all of the members are at work.

A common solution for these problems is to use **asynchronous communication** methods to communicate information and conduct meetings. This means that people communicate at different times rather than at the same time (i.e. synchronously). Email is an example of an asynchronous method of communication.

As previously noted, virtual groups can conduct synchronous meetings using groupware. They can also meet asynchronously using discussion boards and other web-based technologies. The meetings are held over an extended period to enable people from around the world to participate. For example, a meeting could be conducted over a week. The group leader facilitates the meeting by keeping the discussions on-track, asking new questions, summarising the discussions, drawing conclusions, and developing recommendations for decisions and actions. While other members can participate in the facilitation process, the leader is responsible for ensuring that decisions are reached and agreed within the time limits imposed on the meeting.

Defined member roles

The nature of the tasks being undertaken by the group, and its mode of operation, will determine the skills required to complete those tasks. This includes technical skills, generic skills and group or team skills, with the latter being used to define the roles each member will undertake in the group. Asking prospective members to undertake a skills audit will enable the leader to check that all of the required skills are brought to the group when it is established. There are a number of web-based tools that can be used for this purpose. This knowledge also enables the leader of a group, or the members of a team, to effectively use the skills that are available in the group.

Numerous studies have been undertaken into the factors at play in groups and teams, and how they influence the outcomes of a group. Some researchers have developed models and associated skill audit documentation and many of these kits are available commercially. We will look at two of these models.

John Dew (1998, p. 51) suggested people adopt one of four styles when they work in teams:

1. a team-oriented approach
2. a task-oriented approach
3. a data-oriented approach
4. a process-oriented approach.

Team-oriented people are concerned with how the team is working and with ensuring that all members have an opportunity to contribute and that the team operates by consensus.

Task-oriented people are focused on the task and do not like to be distracted with meetings and discussions about alternative solutions.

Data-oriented people make sure the right data is used and are often happy to work alone to ensure the data is collected, integrated, analysed, managed and used correctly.

Process-oriented people are interested in the process that leads to a solution. They tend to be creative and may stimulate the team to achieve novel solutions. The downside is that once a solution is developed, they may become bored with the tasks undertaken during the implementation phase of the project, or when the team documents and works through the details of the adopted solution. At this point, the other members of the team may perceive them to be disruptive.

Through his research, Meredith Belbin (1993) identified nine different patterns of behaviour, or roles, that individuals exhibit in teams. The nine team roles included:

- three action-oriented roles — shaper, implementer, completer finisher
- three people-oriented roles — co-ordinator, teamworker, resource investigator
- three thinking-oriented roles — plant, monitor evaluator, specialist (Belbin 2011).

Belbin (2011) defined a team role as 'a tendency to behave, contribute and interrelate with others in a particular way'. The Belbin Self-Perception Inventory is a test used to improve understanding of an individual's behavioural type and for allocating people to teams, such as student teams (see https://www.belbin.com for the latest test). The Belbin team role summary sheet is reproduced in figure 6.2, and should provide you with a better understanding of the behaviour of individuals in team situations.

The documents describing team role models normally include discussions about the allocation of roles in groups and teams, including situations where the 'roles' may outnumber the people. In these circumstances people may take on two or more roles. It is also interesting to note that people may swap roles when they change groups, or when new projects are commenced. Finally, it is important to note that, as in all self-completed tests, the outcomes are based on self-perceptions. A more rigorous approach would be to have two or three people who have previously worked with each of the prospective team members complete the test about each team member based on their knowledge of those people. This occurs when the full Belbin test regime is used.

It is also important to note that each member of the team will normally have at least two roles: a technical role aligned with their discipline expertise, and a team role.

Time-efficient processes

Time is money, so individuals working in a group or team need to ensure that they maximise the use of their time. Ten minutes wasted in a meeting can be costly, particularly when a large number of people are involved.

The processes developed for a group, or adopted by a team, need to ensure that the group is focused on the task and on achieving the defined outcomes on time and on budget. This can be particularly difficult in a team situation where, for a variety of reasons, people may have different ideas about how a task should be carried out or a problem solved. At times the group should encourage creative and divergent thinking to find alternative solutions. At other times the group needs to focus on implementing the decisions that have been made. Thus, time needs to be managed carefully to avoid prolonged discussions before a compromise is reached.

Once satisfactory processes have been developed, they should be included in the charter.

Accountability

Members of a group have individual responsibilities generally associated with their skill set and position in an organisation, and these will provide the motivation for their participation in the group. Their formal job descriptions will also define their accountability. Therefore, for small group performance it may be important that the tasks allocated to the small group are aligned with the members' job descriptions.

In team situations there is a shared responsibility for the tasks being carried out by the team. There may also be individual responsibilities as individual members are allocated tasks to complete; thus, their accountability may be both individual and shared. It is important to note here that it is up to the team to define and manage accountability.

FIGURE 6.2 Belbin team role summary sheet

Belbin® Team Role Summary Descriptions

BELBIN

Team Role Summary Descriptions

Team Role	Contribution	Allowable Weakness
Plant 	**Creative, imaginative, unorthodox. Solves difficult problems.**	Ignores incidentals. Too preoccupied to communicate effectively.
Resource Investigator 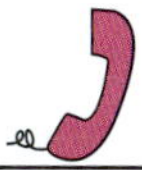	**Extrovert, enthusiastic, communicative. Explores opportunities. Develops contacts.**	Over-optimistic. Loses interest once initial enthusiasm has passed.
Co-ordinator 	**Mature, confident, a good chairperson. Clarifies goals, promotes decision-making, delegates well.**	Can be seen as manipulative. Offloads personal work.
Shaper 	**Challenging, dynamic, thrives on pressure. The drive and courage to overcome obstacles.**	Prone to provocation. Offends people's feelings.
Monitor Evaluator 	**Sober, strategic and discerning. Sees all options. Judges accurately.**	Lacks drive and ability to inspire others.
Teamworker 	**Co-operative, mild, perceptive and diplomatic. Listens, builds, averts friction.**	Indecisive in crunch situations.
Implementer 	**Disciplined, reliable, conservative and efficient. Turns ideas into practical actions.**	Somewhat inflexible. Slow to respond to new possibilities.
Completer Finisher 	**Painstaking, conscientious, anxious. Searches out errors and omissions. Polishes and perfects.**	Inclined to worry unduly. Reluctant to delegate.
Specialist 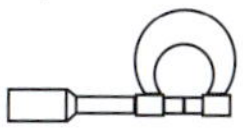	**Single-minded, self-starting, dedicated. Provides knowledge and skills in rare supply.**	Contributes on only a narrow front. Dwells on technicalities.

Source: Belbin (2010).

The life cycle of a team

While proposing a team development model, Tuckman (1965) suggested that a team goes through four developmental phases before becoming effective and achieving its goals: *forming, storming, norming* and *performing*. Following a review of 22 studies, Tuckman and Jensen (1977) revisited the model and added a fifth phase, *adjourning*.

- *Forming.* This phase begins the first time the team meets. It is a transition period and an 'ice-breaker' activity may be used to begin the team bonding process, particularly if some of the members are meeting each other for the first time. The team leader or facilitator will take a leading role in this phase, as tasks, activities and roles are discussed and defined in a charter. The team members are more likely to still have an individual perspective at this time rather than a shared understanding of the tasks and their roles and responsibilities.
- *Storming.* This phase begins when team members begin to put forward different ways of undertaking a task, solving problems or allocating team roles and responsibilities. This is called *storming* because discussions can become heated as the team members jostle to get their ideas and strategies adopted. Team members may also try to establish their position, or level of importance, in the team. Conflict may become a problem during this phase and needs to be resolved quickly so it does not affect the team's performance. The leader needs to ensure all members of the team have equal opportunities to contribute and all contributions are valued. Some team members may not enjoy this phase, particularly if conflict arises.
- *Norming.* In this phase team members begin to work together, recognising and utilising the skills and strengths of each team member. This does not necessarily mean all team issues are resolved and conflict will not arise. Rather, it is likely team members, or a large majority of them, recognise they have a job to do and want to keep the project moving so that they have the time to achieve at a high level. Approaching deadlines may also bring on norming, as members realise that they need to start working together to ensure they complete their allocated task on time.
- *Performing.* Teams that reach this phase are autonomous, perform at a high level and complete tasks on time. By this stage it is likely team members will enjoy working together. They are capable of making decisions, managing differences of opinion and focusing on the team's goal. They achieve synergies from encouraging individuals to work to their strengths and allocating tasks to individuals. They may even delegate responsibilities.
- *Adjourning.* Tuckman and Jensen (1977) called this fifth and final phase *adjourning* because in this phase the team completes its work and disbands. If the team has performed and the experience has been enjoyable, it is likely team members will miss working together. For this reason, teams will often celebrate their success and may even have reunions. The organisation(s) that established the team should ensure the work of successful teams is acknowledged and, if appropriate, rewarded. They should also consider keeping high-performing teams together so they can undertake future projects.

Due to short semesters, and other time pressures, many student teams may not reach their potential and have to adjourn before they reach the *performing* phase. Some student teams may not even reach the *norming* phase! This is unfortunate because they miss the opportunity to experience working in a team that is performing, which can be a special experience.

Improving performance

Each member of a group that performs at a high level will generally have developed the interpersonal skills required to effectively manage their relationships with the other members. These skills include the ability to be assertive; to manage conflict; to accept delegated roles; and to make effective and timely decisions.

- *Assertiveness.* The group should create a culture that encourages members to express their emotions, opinions, ideas and problem-solving approaches. Thus, it allows each member to assert their views without the need to become aggressive.
- *Manage conflict.* Groups that avoid conflict often inhibit the contributions from team members who are contributing ideas and alternative strategies in their roles as plants or shapers (see figure 6.2). Groups that are performing well have generally learned to manage the conflict that can arise when alternative opinions or disruptive questions are asked. Effective collaborations occur when there is a diversity of opinions. Conflict that arises in these situations is often called productive conflict (Schulz-Hardt et al. 2002).

- *Accept delegated roles.* When group members accept a delegated task, they ensure that they understand the task requirements and accept responsibility for completing the task in a timely and effective manner.

SPOTLIGHT

Collaborating and colliding: when alliances go wrong

MARTIJN VAN DER KAMP, UNIVERSITY OF MELBOURNE

Alliances between organisations are a powerful form of collaboration.

But their failure rate of 50% outstrips the average Australian marriage. It's a sobering statistic for anyone involved in sectors where industry partnerships, outsourcing or public private partnerships (PPPs) are the norm, such as banking, IT, construction and government.

Whether we're talking strategic alliances or nuptials, you might assume that a key to success is each party having a clear goal. You'd be wrong: a clear company goal can be very destructive in teams where people from different organisations must collaborate to achieve a common outcome.

My research on conflicts within inter-organisational teams shows that for a venture to succeed, both partners must have clearly defined common goals next to their individual goals for these teams to work with.

The boundary riders

An ambitious project like the Victorian desalination plant is typically too big to be managed by one team; it requires a whole constellation of teams. These are assembled from teams within companies, but also from teams in which people from two or more organisations come as representatives, known as alliance teams.

Subgroups in these teams often form along the organisational boundary lines and are reinforced by the diversity of organisational cultures, personality, gender and functional roles. These lead to splits and conflict in the team, affecting the alliance performance.

There are three types of conflict in these types of teams: people argue about who should do what. They argue about how these things should be done. And finally there is personal conflict.

This sort of problem can be addressed by both motivational and structural solutions. Having strong common goals for the team, next to separate goals for each company, is one motivational solution. Having teams assigned clear tasks and working procedures is a structural solution.

Start sharp-shooting

One European telecommunications giant in the Netherlands outsourced its human resources services. Of course, the HR provider wanted to deliver great HR services — that was the whole purpose of the collaboration. This provider also wanted the deal to be profitable. On the other hand, the telecommunications giant did not want to pay too much for the services it was getting.

So both parties set to work on some ground rules and structures. The collaboration went very well, especially when teams were talking about a clear future that would bring them both value. But in the next meeting when they began discussing finances, they began using 'us' and 'them' language: 'You should pay for this and we don't want to turn up for that.' The companies' common goals were not sharp enough to overcome these conflicts that arose outside their initial meeting.

A common goal holds value for both companies and therefore makes the partnership and its costs easier to sell internally and it makes people realise that they are in this together so they form one team instead of two subgroups.

The importance of good bones

What's especially relevant in this case is to not just have a shared goal, but to have the structures in place to realise these goals — in other words, translating a goal into specific measurements that can be applied on the team level.

It takes some effort to do this. Companies invest a lot of time to do this for their own company, but they should realise it is important in an inter-company situation as well to get it down to the team level. Part of this structure is to think about how to divide future revenues and costs over the partners.

There was one team concerned with daily activities in the telecommunications partnership. This meant if there was an incident or something went wrong with a certain project they would discuss what was required to reach a solution no matter what was involved.

There was a separate team who would deal with the financial part of that solution. This separation of interests and tasks works very well and is a good preventer of conflicts.

It means you have to be very clear about which team does what tasks [and] how they contribute to the overall goal. The more specific you can be about this, the easier it gets to work in these teams.

Corporate bonds

A more motivational solution is to organise social events, say every six months, in which everyone involved in the collaboration, and their immediate work colleagues, comes together for a fun day out.

In most events like this, team members talk about the work they were doing . . . up to a point. The simple act of getting to know each other means they came away better understanding the corporate context that each person works in.

One such successful social event was held at a farm. Over cakes and coffee in the sun, everyone reflected on the past half year — what went well and what went wrong. The day was anchored in a formal presentation.

But the informality of the surroundings — an old cow shed — encouraged involvement, understanding, and open and transparent communication that would not have taken place via email over distance. An after-lunch round of golf in a paddock with grazing cows, trenches, water and bridges cemented the connections that had started to take place at a personal level.

People stopped talking about work and started talking about their families and the schools their children went to, and what they were doing on their holidays. They then took those bonds back into the office.

In-site learning

Another way of fostering empathy between teams is to locate new team members at the supplier's site for several days, so they know what is going on and how the work is done.

In the example above, the telco workers also agreed to have more real-life meetings at each other's sites so they could be more aware of the context in which they were working.

This gives the companies the opportunity to emphasise common goals and to discuss conflicts that had occurred in the past and learn from those conflicts and mistakes.

This 'reflexivity' is an important structural solution that improves working relationships and brings the team closer together. It's easy to put a moment of reflection on the agenda of every meeting: to reflect on collaboration, how tasks should be done and how tasks have been done.

Prior to this type of reflection, teams would set out the whole process of starting a project, from start to finish. Person A must send a work assignment to person B, who has to send it to his or her manager. The manager then sends a formal request to his leadership team asking for approval to begin work, then that approval is handed back down the line again.

In time and through reflection, team members established enough trust and confidence to overcome the need of control. By simplifying the rules and structures they were able to bypass the formal procedures and simply use email or telephone to communicate and confirm a new project.

Colliding becomes collaborating

Organising for effective collaboration between organisations is very different from organising within organisations. Teams fulfil an extraordinarily important role in these settings, managing their performance in relation to the overall goals is of importance to the success of the venture.

Source: van der Kamp (2011). Originally published on *The Conversation*.

CRITICAL THINKING

Investigate the example of the Victorian desalination plant (https://aquasure.com.au/history). How many stakeholders were involved in the project? Find the company's goal (or philosophy). How may it have shaped their interactions with their stakeholders? How did they build community?

Leadership

'Effective teamwork doesn't just happen, it is made to happen' (Caspersz et al. 2006, p. 1). For groups, the leader is responsible for the success of a group and its outcomes (Mayne 2012). In teams, the team members share that responsibility, but generally they will have nominated a leader to facilitate their work and take on the responsibility of guiding the team to success. There are many different styles of leadership including the following.

- *Transformational leadership.* A transformational leader inspires the members of a group or team to perform. These leaders can be visionary and often create synergies, allowing the team to produce more than would be produced if the individuals worked on their own.

- *Transactional leadership.* This is a traditional form of leadership for groups where all of the members of the group acknowledge the leader's position and obey the leader. A team including a senior manager may revert to this mode of operation even if the manager has asked the team members to treat them as an equal participant.
- *Mixed-mode leadership.* A combination of transformational and transactional leadership. This style of leadership can be effective in groups.
- *Shared leadership.* This is the preferred style for teams as the team uses democratic decision-making methods to appoint a member to act as a facilitator. The facilitator guides the team and keeps it on track so it can achieve its goals.

Wood et al. (2025, p. 221) offer the following guidelines for leaders to help them create, and sustain, a high-performing group or team.

- Communicate clear high-performance standards.
- Set the tone at the first meeting.
- Create a sense of urgency; set a compelling context for action.
- Make sure the team members have the right skills.
- Establish clear rules for behaviour by the team.
- As team leader, 'model' the expected behaviours.
- Identify specific objectives that can be achieved to create early 'successes'.
- Continually introduce new facts and information to the team.
- Make sure the team members spend a lot of time together.
- Give positive feedback; reward and recognise high-performance results.

Some more practical suggestions for leaders include the following.

- *Delegate effectively.* The ability to delegate effectively is an important skill for any leader, particularly for engineers. This requires the communication of clear instructions to the person or people who will have the delegated authority to undertake a role or task. The instruction should also clearly define the boundaries and responsibilities of the role being delegated so that the leader will be informed if the work is going beyond those boundaries.
- *Seek input.* Don't be afraid to ask for advice. You are not expected to know everything about the project. Build a team culture where team members are able to provide free and frank advice. This includes providing ample opportunities for team members to give advice and comment on proposals.
- *Listen carefully.* Listen carefully when proposals are being discussed by your team to ensure you do not miss what might prove to be new ideas, important details, alternative approaches, or problems that may delay or even derail a project.

Regardless of the leadership approaches and strategies you use success is not guaranteed as there are many factors that can affect the success of a group or team. Lencioni (2002) suggests that there are five common dysfunctions in a team: an absence of trust, fear of conflict, lack of commitment, avoidance of accountability, and inattention to results. Kimble et al. (2000) identified that trust and identity are key issues in virtual teams. This is because it is not as easy to establish trust and social bonding in virtual teams where members have not met face to face. In fact, a group may struggle with these issues until the members successfully complete some tasks.

While student teams experience these same issues, student complaints about teamwork often revolve around the following four issues.

1. The 'free rider' or 'social loafer' who does little work compared with other team members, but often receives the same mark.
2. The 'know-it-all' who wants to do all of the work.
3. Mismatched expectations within the team about the mark the team should aim to achieve for their work.
4. Complications with allocating tasks, roles and responsibilities and meeting project deadlines.

Team issues are often difficult to resolve, but having open and effective communication strategies can mitigate some of these problems. A good resource to resolve these problems is provided by Oakley et al. (2004, pp. 32–34).

The benefits of working with others

As pointed out in a previous section, many employers, industry groups and engineering professionals consider the ability to work well with others as an essential prerequisite for engineering employment. In addition to having positive effects from an employment perspective, team experiences help undergraduate students to develop their knowledge and skills. Teamwork can also stimulate the development of technical knowledge (e.g. engineering science) and facilitate the application of this knowledge (e.g. engineering design). Beyond technical learning, involvement in teamwork during university can help students develop crucial generic skills, such as negotiation, conflict resolution and time-management strategies.

Positive team interactions can enhance the self-esteem of individual team members, improve their interpersonal communication skills and develop their self-appraisal skills so they can identify their preferred and most effective team roles. When individuals work together, they have the potential to draw inspiration from each other. This *emergence* — or development — of new ideas can improve the performance of the group. In organisations where innovation and creativity are essential for maintaining the provision of high-quality service, and a commercial edge, asking employees to work in multidisciplinary groups may create the ideal conditions for the emergence of outcomes that could not be achieved by individuals. In this situation, the team members (e.g. an engineer, an accountant and a sociologist) are learning from each other and creating synergies that may enable them to perform at a higher level. This type of team project can become an exciting journey — with ideas and outcomes based on the creative amalgamation of information drawn from the different types of experiences, perspectives, knowledge and values that each member brings to the task.

Many variables can affect the quality of a group or team experience and the outcomes they produce. Working well in a team can provide many benefits for the group members. Developing an understanding of how you perform most effectively in a group will help you improve the overall performance of the teams you participate in. Therefore, it is important to understand how you contribute effectively to a team and the responsibilities that are associated with your role as a team member.

6.3 Meetings

LEARNING OBJECTIVE 6.3 Actively participate and contribute in meetings.

As a student you will participate in a range of different types of meetings, from student group meetings through to sporting club meetings. In your working life, you will also attend meetings — meetings with your manager, meetings with company executives, meetings with teams, meetings with clients and meetings with members of the public.

The Macquarie Dictionary (2024) defines a **meeting** as an 'an assembling, as of persons for some purpose; an assembly of people with responsibilities towards an organisation, held to conduct the business of that organisation'. Thus, a meeting is normally called for a specific purpose. Ideally, meetings should be efficient, enjoyable, productive and even exciting. In this section some strategies and skills are discussed that will help you participate in, or oversee, efficient meetings. After looking at the guidelines on how to perform well in a meeting, the discussion moves to how a meeting is organised and the role technology has to play in meetings.

Before a meeting you should establish:

- the purpose
- the style
- the timing
- the participants
- the procedures
- your contribution.

These points are now discussed in detail.

KEY POINT

Meetings should be efficient, productive and lead to agreed outcomes.

The purpose

To be well prepared for a meeting, you need to understand the reason or purpose of the meeting. Sometimes this is obvious — but at other times the purpose of the meeting may not be clear. For example, engineers may understand that a project team meets weekly to update members on the progress of a project, but may need to ask the reason for a meeting between their supervisor and government officials that they have been asked to attend. With experience, it is easy to plan your contribution to a regular meeting of a group you belong to. However, it can be difficult to plan your participation for a one-off meeting, particularly if you do not know the people involved. The following questions will help you to establish the purpose of a meeting.

- Who called the meeting?
- Why did they call the meeting?
- What do they expect to achieve at the meeting?
- Why have I been included in the meeting?
- Who else is participating?
- Is this a one-off meeting or the first meeting of a planned series of meetings?

You may also try to discern if there are any hidden agendas.

For formal meetings, it is normal practice to circulate an agenda prior to the meeting. This would usually contain information that would answer most of these questions and it enables the participants to prepare for the meeting. Figure 6.3 shows an example of a hypothetical action-based agenda that could be used as a template by student teams preparing for a meeting.

The style

The way you prepare for a meeting will depend on the style of meeting you are asked to attend. There are a number of variables to consider, including the formality of the meeting, any legal requirements the meeting may be fulfilling, the format of the meeting and the mode of the meeting. From an engineer's perspective, meetings with colleagues are likely to be informal compared to a company board meeting, a meeting with staff from a government agency or a professional organisation.

One of the factors that can influence the type of meeting is the status of the meeting from a legal perspective. For example, company law requires company boards to hold an annual general meeting (AGM) in accordance with a strict set of guidelines. This is because an AGM is the meeting where a company's financial results are formally considered, changes to its constitution are made and its directors are elected. Generally, it is the only meeting that shareholders are able to attend. Similar regulations apply to registered community, professional and sporting organisations.

It is also important to consider the mode of the meeting. In addition to conventional face-to-face meetings, companies are increasingly using groupware for their meetings. This mode is discussed in more detail later in this section.

The timing

From a timing perspective there are three types of meetings: regular meetings, irregular meetings and one-off meetings. A regular meeting may be scheduled daily, weekly or monthly, for example, a weekly project team meeting. An irregular meeting is normally organised on a needs basis; that is, when the participants need to discuss some issues or some decisions need to be made. For example, a project team leader may organise a meeting with the client when a particular milestone has been achieved or when an issue needs to be discussed and decided. One-off meetings are commonly called to discuss a specific issue and are called on a needs basis.

Meetings can be scheduled at any time of the day or night. Traditionally morning, lunch-time or afternoon meetings have been the most popular times for meetings. However, breakfast and early evening meetings are becoming more popular, and night meetings are often scheduled for meetings with overseas participants. It is important to note that virtual meetings (e.g. Zoom) often have set start and end times.

The length of the meeting

The length of time allocated for a meeting depends on the objectives of the meeting and should be aimed at achieving effective outcomes in an efficient manner. Generally, sufficient time needs to be allocated to enable the participants to give their reports, discuss the scheduled business and make evidence-based decisions. The initial meetings of a project team should allow additional time for team members to undertake team bonding activities.

It is important to recognise that inefficient meetings can affect the productivity of a project team or an organisation. The true cost of a meeting can be calculated by multiplying the sum of the charge-out rates of the attendees by the length of the meeting in minutes and then adding any travel or other costs. The effectiveness of the meeting is then assessed by weighing the outcomes of the meeting against the cost.

FIGURE 6.3 Example of an Annual General Meeting agenda

<table>
<tr><td colspan="8" align="center">Northern Regional Council — Renewable Energy Advisory Committee (REAC)
Annual General Meeting — commencing at 2 pm on Friday 19 July 2024 in the Council Chambers
Agenda</td></tr>
<tr><td>1</td><td>Welcome</td><td colspan="6">Members</td></tr>
<tr><td></td><td>Attended</td><td colspan="6">John Field (President), Marita Coe (past President), Henry Coates (Secretary), Jenny Katseva (Treasurer), Yan Lui (Publicity).
Members of the technical sub-committee: Odette Manu (Convenor), Kylie Meadows, Joel Monk, John West</td></tr>
<tr><td></td><td>Apologies</td><td colspan="6">Ali Sitani</td></tr>
<tr><td></td><td></td><td>Chair</td><td></td><td>Discussion</td><td>Who?</td><td>Action</td><td>Due</td></tr>
<tr><td>2</td><td>Minutes</td><td></td><td></td><td></td><td></td><td></td><td></td></tr>
<tr><td>2.1</td><td>Confirmation of previous minutes</td><td>JF</td><td>Minutes of the Annual General Meeting held on 21 July 2023</td><td>Motion: That the minutes of Annual General Meeting held on 21 July 2023 be adopted as a true and correct record of the meeting.
Moved:
Seconded:
Outcome:</td><td></td><td></td><td></td></tr>
<tr><td>2.2</td><td>Business arising</td><td>HC</td><td>Items not included separately on the agenda</td><td>Renewable Energy Workshops
The secretary is to report on whether the state government will subsidise the costs of providing the workshops in five of the major centres in the region.</td><td></td><td></td><td></td></tr>
<tr><td>3</td><td>Reports</td><td></td><td></td><td></td><td></td><td></td><td></td></tr>
<tr><td>3.1</td><td>President's Report</td><td>JF</td><td>Report attached</td><td>The report highlights the committee's work over the year ending 30 June 2024.</td><td></td><td></td><td></td></tr>
<tr><td>3.2</td><td>Treasurer's Report</td><td>JK</td><td>Report attached</td><td>The report includes two recommendations:
1. The meeting adopt the Treasurer's report for the financial year 2023/24.
Moved:
Seconded:
Outcome:
2. The meeting endorse the appointment of F Monroe to the position of auditor for the financial year 2024/25.
Moved:
Seconded:
Outcome:</td><td></td><td></td><td></td></tr>
</table>

4	**Business** a lot of time can be wasted when attendees						
4.1	Election of committee members	JF	The constitution stipulates that one-third of the nine members must retire each year. The retiring members may stand for re-election (the past president retires)	1. **Motion:** The Committee acknowledges the important roles undertaken by the retiring members and thanks them for their contribution: Joel Monk, Ali Sitani and Marita Coe. Moved: Seconded: Outcome: 2. **Ballot:** A ballot will be held at the meeting as there are seven nominees for the four vacant positions on the Committee.			
4.2	Budget	JK	Budget papers attached	**Motion:** The meeting endorses the budget for the financial year 2024/25. Moved: Seconded: Outcome:			
4.3	Sustainable display house proposal	OM	Papers attached	**Motion:** The technical sub-committee recommends that the Committee ask the Northern Regional Council to prepare a tender for the design of an energy efficient and sustainable house that is powered by renewable energy and suitable for the micro-climates of the major regional centres. Moved: Seconded: Outcome:			
5	**Other business**		(Items to be raised at the meeting)				

If the meetings of a group develop a trend to always go over the allocated time, or finish without completing all of the key items of business on the agenda, then the leader or chair will need to work with the team to implement strategies to meet deadlines. For example, the chair of a committee that often met for 2–3 hours, instead of the scheduled hour and a half, moved the starting time from 2 pm to 3.30 pm. The fact that most participants normally finished work at 5 pm led to more efficient meetings and business being completed on time!

The participants

The meeting may involve two or three people, or it may involve a large number of people. For example, a meeting with a client may involve two or three people and may be informal. A project team meeting may involve 20 people and may be formal. A professional association meeting may involve a large number of people and will likely follow a formal meeting procedure. A public consultation meeting may also involve a large number of people and may be formal or informal, depending on the purpose of the consultation.

The procedures

The procedures for a meeting will vary, depending on the formality of the interaction. A formal meeting usually begins with a welcome, which is delivered by the person who is chairing or facilitating the meeting (normally referred to as the **chair**). If there has been a previous meeting then the *minutes*, or official record of proceedings, may then be discussed, amended and, if required, formally adopted as a true and correct record of the meeting. If the minutes identified that specific actions were to be undertaken then the chair may ask the nominated person, or group, to report on any actions they have taken and any outcomes resulting from those actions. The *items of business* listed on the meeting agenda are then discussed and, where required, decisions are made on any recommendations. This may be followed by a discussion of any other relevant matters. Generally, there will be an opportunity for participants to ask the chair to accept other items of business that are not on the agenda. Once all of the relevant business items have been discussed the chair will formally close the meeting. The meeting secretary will record details of the meeting and then prepare and circulate the minutes prior to the next meeting.

In contrast, a casual meeting is less structured and minutes may not be required, although members may keep personal notes of the meeting outcomes, particularly the actions.

Many organisations are required to formally record the minutes of their meetings; for example, company boards and registered community, professional and sporting organisations. The content, style and format of minutes varies from organisation to organisation. Traditionally, minutes have been written in narrative style and included all of the information relating to a meeting: the participants, any apologies, the items of business, the actions taken and the motions passed. In recent times, there has been a move to more action-oriented minutes, which are more focused and emphasise, or only include, the participants, the issues discussed, the outcomes, actions and accountability. Figure 6.4 provides an example of the minutes that may have been taken from the hypothetical meeting referred to in figure 6.3. In this case, these minutes were added to the agenda template. To illustrate this method, the minutes or notes added during or after the meeting are highlighted in yellow in figure 6.4.

Your contribution

There are strategies you can use to ensure you make a valued contribution to meetings where you are not the chair. Before you attend the meeting, you should make sure you have a good understanding of the matters that will be discussed. You should begin your preparation by reading the agenda for the meeting, the minutes of the previous meeting (if applicable) and any information or papers associated with the listed items of business. You should also check to see if you require any additional information about those items. If, at a previous meeting, you were given actions to complete, try to complete them and then plan how you will report on the progress you have made, any problems encountered or the outcomes of these actions. If you have placed an item of business on the agenda, include sufficient information so that the participants are able to understand the item. If you want the meeting to make decisions or recommend actions relating to your item, then include this in the information you provide. You may also include proposed action statements for the meeting, to be adopted with or without modifications.

You can contribute to the meeting by taking a positive approach and encouraging other team members to adopt a similar attitude.

FIGURE 6.4 Example minutes for an Annual General Meeting

Northern Regional Council — Renewable Energy Advisory Committee (REAC)
Annual General Meeting — 2.00–3.40 pm on Friday 19 July 2024 in the Council Chambers
Minutes

1	**Welcome**		**Members**				
	Attended		John Field (President), Marita Coe (past President), Henry Coates (Secretary), Jenny Katseva (Treasurer), Yan Lui (Publicity). Members of the technical sub-committee: Odette Manu (Convenor), Joel Monk, John West				
	Apologies		Ali Sitani (technical sub-committee), Kylie Meadows (technical sub-committee)				
		Chair		**Discussion**	**Who?**	**Action**	**Due**
2	**Minutes**						
2.1	Confirmation of previous minutes	JF	Minutes of the Annual General Meeting held on 21 July 2023	**Motion**: That the minutes of Annual General Meeting held on 21 July 2023 be adopted as a true and correct record of the meeting. Moved: Coe Seconded: Lui Outcome: Carried			
2.2	Business arising	HC	Items not included separately on the agenda	**1. Renewable Energy Workshops** The secretary reported that the state government will subsidise the costs of providing the workshops in five of the major centres in the region.	**OM**	Technical sub-committee to prepare plan for five workshops	Sept. meeting
3	**Reports**						
3.1	President's Report	JF	Report attached	**The report highlights the committee's work over the year ending 30 June 2024.**			
3.2	Treasurer's Report	JK	Report attached	**The report includes two recommendations:** 1. The meeting adopted the Treasurer's report for the financial year 2023/24. Moved: Katseva Seconded: West Outcome: Carried 2. The meeting endorsed the appointment of F Monroe to the position of auditor for the financial year 2024/25. Moved: Katseva Seconded: West Outcome: Carried	**HC**	Write to F Monroe	ASAP

		Chair		Discussion	Who?	Action	Due
4	**Business**						
4.1	Election of committee members	JF	The constitution stipulates that one-third of the nine members must retire each year. The retiring members may stand for re-election (the past president retires).	1. **Motion:** The Committee acknowledges the important roles undertaken by the retiring members and thanks them for their contribution: Joel Monk, Ali Sitani, Yan Lui and Marita Coe. Moved: Coates Seconded: Manu Outcome: Carried 2. Ballot: Ali Sitani, Yan Lui, Nigel Lewis and Mary Down were elected.	**HC**	Advise new members of meeting times and so on	ASAP
4.2	Budget	JK	Budget papers attached	**Motion:** The meeting endorsed the budget for the financial year 2024/25. Moved: Katseva Seconded: Monk Outcome: Carried			
4.3	Sustainable display house proposal	OM	Papers attached	**Motion:** The technical sub-committee recommends that the Committee ask the Northern Regional Council to prepare a tender for the design of an energy efficient and sustainable house that is powered by renewable energy and suitable for the micro-climates of the major regional centres. Moved: Manu Seconded: West Outcome: Carried	**HC OM**	Write to Northern Regional Council and enclose the proposal	Sept. meeting
5	**Other business**		No matters were raised				

It is easier to direct a meeting if you are the chair as you are then able to keep the meeting focused on the purpose of the meeting and achieve the intended outcomes. You can also ensure that all of the participants have the opportunity to express their point of view. As chair you can also establish a positive tone for the meeting, and moderate the interactions of participants.

Organising a meeting

In engineering organisations, project team leaders and work supervisors organise many of the regular meetings that members of the engineering team attend. If you are asked to organise a meeting, you should ensure participants have the opportunity to check that they can attend at the scheduled time and prepare for the meeting. The process that is used to organise a formal meeting is as follows.

- *Call for items of business.* Participants are notified of the proposed date, time and location of the meeting and are asked to advise the meeting secretary, or organiser, of their ability to attend and any items of business they would like included on the meeting agenda. This enables the organiser to prepare a draft agenda and propose a new date if key people are not able to attend.
- *Establish the meeting agenda.* The organiser of the meeting develops the items of business into an agenda for the meeting. A team of employees, such as an agenda committee or an executive committee, may perform this role for formal meetings.
- *Circulate the meeting agenda.* The meeting agenda is circulated to the meeting participants, together with papers or documents relating to the agenda items. This is the formal notice for the meeting and it should contain the confirmed date, time and location of the meeting, and, for virtual meetings, the URL and key or code. Normally the agenda is sent so that recipients receive it at least a week before the meeting. It is important that meeting organisers include all of the data and information required to enable the attendees to dissect, discuss and make decisions on each agenda item. In some cases, the meeting papers may include drafts of recommended decisions, solutions and motions (and alternatives) for the attendees to consider. This is because a lot of time can be wasted when attendees discuss issues without the relevant information, or develop recommendations during the meeting.

The role of technology in meetings

Earlier in the chapter, it was noted that organisations use a range of technologies to conduct their meetings. Three technologies were discussed: a *teleconference*, a *videoconference* and a *groupware-*mediated conference.

During a teleconference, the participants can only hear what people have to say. Compared to a face-to-face meeting a teleconference is restrictive, because the participants do not see all of the non-verbal cues. Another disadvantage of teleconferences is that the participants can be distracted by emails and other matters if they are sitting at their work desks. The chair of a teleconference should implement strategies that will ensure the participants are engaged in the discussion. For example, the chair can do this by asking members by name for their opinion or recommendation on each item. This should be done in a systematic fashion so that all of the participants are regularly polled for their opinion.

Meetings conducted with videoconferencing technology have the potential to be more inclusive than meetings that use teleconferencing technology; however, the effectiveness of videoconferencing may depend on variables such as the number of participants that can be seen at any one time and the quality of the transmission and available bandwidth. The financial cost of this medium may also be a limiting factor.

Groupware technologies are a relatively recent addition to the list of technologies that can be used for videoconferences. Meetings conducted using groupware have all of the advantages of a videoconference and the added advantage that the participants can also communicate using text messages, which are listed on the computer screen. However, the same technological issues that affect videoconferencing can affect groupware-mediated meetings.

For many years teleconference and videoconference technologies were important components of the communications strategies that many engineers used to establish and develop business relationships. In the early 2020s, COVID-19 lockdowns meant that groupware technologies were more widely adopted by businesses, community organisations, schools and universities to:

- link employers and employees working from home
- facilitate collaboration in groups, teams and organisations
- teach school and university classes
- conduct business meetings and transactions
- maintain contact between family members, friends and community groups.

It should also be noted that the COVID-19 pandemic led to the rapid development and deployment of structural changes and new features in groupware.

While the personal etiquette rules and management techniques for groupware-facilitated meetings have a lot in common with face-to-face meetings, the challenges of virtual meetings mean that different rules and management strategies are required to facilitate cohesive, efficient, inclusive and productive meetings.

In addition to applying the appropriate etiquette rules, each participant should have a good understanding of the features of the groupware package being used, for example: muting the microphone; switching the camera on and off; selecting a background, including a virtual background; using a 'chat' button to make a comment or ask a question; and using a 'raise your hand' button to get the attention of the chair. Additional techniques are required to chair a virtual meeting.

The following spotlight lists many of the tips and techniques currently used for groupware meetings. They have been compiled from the authors' experiences and the tips on etiquette provided by two of the popular apps, Zoom and Teams, as well as those provided by some universities for their students.

SPOTLIGHT

Etiquette tips for virtual meetings

Preparing for the meeting

- Set up a quiet, well-lit space with a neutral background. Be aware of your surroundings as meeting attendees will see and hear what is going on around you. Some apps enable users to blur their background or to select a virtual background.
- Make sure that your back is not to a window or open door, as the light may overwhelm your camera and make your face too dark to be easily seen.
- Dress as you would normally dress for a business meeting or, if you are a student, dress as you would for a class.
- If possible, use a headset with external microphone.
- Try to eliminate any distractions, for example, close any other applications, such as email, and mute or switch off your phone.
- Learn how to use the controls available in the groupware package you will be using, for example, the mute button, the camera on/off switch, the 'chat' button, the 'raise your hand' button, and the volume adjustment slider.
- Do not participate in a virtual meeting while operating a vehicle or other equipment.

During the meeting

- Make sure that you take the time to be prepared and ready to join the meeting a few minutes before the scheduled start time.
- Use the mute function when joining a meeting and remain muted when you are not speaking. When it is time for you to speak, check your mute button is off and then speak.
- The meeting host will probably welcome people as they join the meeting. Be ready to give your name and, if appropriate, why you are attending the meeting and what you hope to gain from the meeting. Make sure that your name is displayed clearly throughout the meeting.
- Be present at the meeting — try not to be distracted by your surroundings, noises or other events, such as pinging emails, ringing phones, typing on a keyboard or by doing other work.
- You should have your camera on as much as you can, as it will make the meeting more engaging for everyone involved. When the participants can see each other, they can assess facial expressions and body language and gauge the reaction of other attendees to the speaker. Having video on reminds people to stay seated and concentrate on the proceedings. You should only turn your camera off if this is permitted, or requested, by the host.
- Be yourself, be patient and respect the other people at the meeting. At all times, use appropriate language and grammar. Respect the contributions, opinions and experiences that are shared by the other attendees.
- Listen carefully to what is being said, particularly before you enter a discussion. Try not to talk over others.

- Try to maintain eye contact with the other participants. To do this you may have to adjust your computer so the camera is at eye level, and then focus on the camera lens during the meeting. This can be difficult if you are trying to watch the other participants and read chat messages.
- Use the 'chat' function to ask questions silently. Remember, in most apps the host can see all of the chat messages — so stay on topic.
- Where appropriate, use the 'reaction' buttons to respond to other contributions.
- Use the procedure suggested by the chair or host when you want to speak in a meeting. This may be by using the 'raise your hand' function or by sending the host a message using the chat function. You should then wait until you are called on before speaking.
- Use simple language to ensure all the participants will understand your contribution to a meeting. Remain on topic and, where appropriate, be factual.
- During the meeting take notes of key points, decisions, agreed actions and your thoughts. This is good practice, even if you can revisit the session recording after the meeting has concluded.
- Resist the temptation to multitask during a virtual meeting. Treat a virtual meeting just like a face-to-face meeting and give it your undivided attention. This is easier said than done, but remember, the other attendees will notice when you are working on another task or checking your email.
- Do not record or share a copy of the meeting without the host's permission.

Hosting a meeting

- Make yourself familiar with the host features of the app you are using; for example, how to welcome participants from a waiting room, how to mute other attendees, how to share a screen, how to set up breakout rooms and how to record the meeting.
- Appoint a co-host, particularly for large meetings, where managing the welcome, chat, mute and other controls is likely to interfere with the host/chair role during a meeting. Preferably appoint the co-host before the start of the meeting so you can agree on procedures and how they will manage administrative tasks. Delegating this role ahead of time will allow you to ensure they understand their role.
- If you plan to share your screen during a meeting to show documents, plans or a video, ensure you understand how this is achieved with your app. This includes how to return to the meeting. Advise the participants before you switch screens so they understand what is happening.
- Regularly advise the meeting where you are up to on the agenda, and poll the participants to check everyone is on the same page and managing to keep up. Using pauses like this will help the participants to maintain focus.

Closing the meeting

- Before closing the meeting, the host should:
  - check there is no further business
  - list the decisions and any allocated tasks, due dates, and so on
  - advise the participants when the next meeting is scheduled
  - thank the participants for their contribution.
- The host should farewell each participant and be the last person to leave the session. This provides an opportunity for people to stay behind and ask questions.

CRITICAL THINKING

Consider the good virtual meetings you have attended and select two of the better ones. Why did you find them so good? How did the host manage the meetings? What techniques and technologies did they use to manage the meetings? What can you learn from those experiences?

The downside of working with others

In an earlier section, we discussed the benefits of working with other people. Now we will look at the other side of the coin: what are the costs of working with others? The potential costs may include monetary costs, time costs, administration costs, meeting costs, communication costs, personal costs and relationship costs. It is important to identify the costs so that a realistic cost–benefit analysis can be undertaken for a proposed collaborative project.

Rob Cross and colleagues (2017) undertook research to identify the most effective collaborators in 28 major organisations they studied in the United States. They found that time demands for collaborative activities had risen by more than 50 per cent in the last decade and that knowledge workers *spent up to 85 per cent of their work time* on emails, in meetings and on the phone. This does not take into account the impact of social media-based collaboration activities after-hours and on weekends.

Clearly, if collaborations are not managed properly, they may have a huge impact on the productivity of the participants and their workplaces. This is particularly true when the participants are not efficient or effective collaborators — that is, they waste their own time as well as the time of those they collaborate with. Cross et al. (2017) listed seven habits of highly *ineffective* collaborators as follows.

1. Trying to accomplish too much through email.
2. Running ineffective meetings.
3. Holding onto a central position in the network rather than delegating in a way that creates clarity and engagement.
4. Being rigid instead of adapting behaviours — theirs or others — to promote the effectiveness of interactions over time.
5. Failing to put structure into work through strategic calendaring and 'to do' lists.
6. Using instant messaging or social media excessively.
7. Holding onto counterproductive beliefs — often driven by fear, identity and power — that result in becoming overwhelmed by network demands.

Having identified these bad habits Cross and his associates went on to develop training courses to help people in those organisations to become more effective and efficient collaborators. It is important to note that these seven habits were identified in a study of the professional staff in large organisations.

While you are a student you have the opportunity to learn from the study by Cross et al. (2017) and develop good collaboration habits so that you become an efficient and effective collaborator when you begin work as a graduate in the engineering industry. As a graduate, it is likely that most of your collaborations will result from tasks allocated by your supervisor. However, from time to time you may be offered the option of participating in other collaborations or colleagues may ask you to carry out tasks for them. When given the opportunity to collaborate you should consider the following questions.

- Has the request come through my supervisor? If not, do I need to inform my supervisor?
- Is this a real opportunity to collaborate, or is it just a request to complete some routine work, or to lighten a colleague's work load?
- Is this request related to my current employment or is it an opportunity to work with people in other sections of my organisation, with people from other disciplines, or with people from outside my organisation?

Regardless of the source of the request for a collaboration, you should consider the following questions as they will help you to fully understand the nature of the proposed collaborative project.

- What is the purpose of the collaboration?
- What are the benefits of this collaboration?
- Will my participation in the project add to my experience and enhance my CV?
- What are the expected outcomes of the collaboration?
- Why was a collaborative approach adopted rather than having the work completed by individuals?
- What role will I be expected to undertake?
- Why was I offered this opportunity?
- What will I be expected to contribute?
- What impact will this collaboration have on my workload?
- Will I have to drop, or reduce my effort on, other projects to undertake this collaboration?
- What do I expect to learn from this experience?
- Will collaboration and/or project management software (e.g. Trello, Project) be used to facilitate the work?
- Will I need to learn to use new software? If so, does the size of my role in the project warrant the time I will spend learning to use new tools?
- Who are the other members of the group?
- What networking opportunities will arise from the proposed collaboration?
- Will there be opportunities for me to share my skills with new technologies, methods and so on?
- What are the risks of participating in this project: personal, professional and organisational?

The answers to these questions will help you determine your role in the project and the level of commitment required. Of course, the answers will also help you decide whether to participate in the project (if it is optional). You can do this by weighing up the potential benefits against the costs of participating.

Having decided to commit to a collaborative project you can use the following management principles and strategies to ensure your contribution to the project is efficient and effective.

1. Managing a collaboration

The self-management skills you learned in the previous chapter and the team-working skills you have learned in this chapter can be applied when you are collaborating with one or more colleagues. Some key strategies are:

- carefully and honestly managing the team's expectations regarding your abilities, availability and contributions
- helping to clearly define the expectations for each member of the group; for example, what is the expected workload for each member (e.g. hours per month); what are the realistic timelines for the project?
- setting aside regular time periods for collaborations in each member's diary or digital calendar
- ensuring that individual tasks are delegated to the most appropriate person, and that clear instructions and boundaries are provided with each task
- acting professionally at all times so that you gain the trust of the other members of the team (e.g. achieving objectives and meeting timelines)
- changing your approach, manner and availability, where appropriate, to fit the preferences of the other group members rather than expecting them to change to suit your preferences.

2. Managing communications for a collaboration

As mentioned, Cross et al. (2017) found that some knowledge workers spent up to 85 per cent of their work time on 'emails, in meetings and on the phone' — that is, communicating. Communication is also a large part of the work undertaken by engineers. It is for this reason that three chapters in this text are devoted to communication. While the many communication principles that underpin effective, and successful, collaborations are outlined in those chapters, four principles are critical.

1. Clearly define the purpose of each communication; that is, the information to be communicated in the message.
2. Carefully select the best method for communicating the message, for example, face-to-face meeting, telephone call, email, virtual meeting or letter.
3. Ensure the information in the message is brief, clear, precise and unambiguous.
4. Ensure that the information is clearly understood by the recipient(s).

3. Managing collaboration meetings

The previous section in this chapter provided a wide-ranging discussion about meetings and offered many principles that can be used to ensure meetings are effective and efficient. Three principles are key.

1. Ensure each meeting has a purpose and expected outcomes. Cancel meetings if there is little or no urgent business.
2. Review the proposed agenda to see if it is necessary for each member to attend. Can a subset of the group achieve the purpose and objectives?
3. Ensure the meeting achieves its objectives efficiently and effectively.

In the end, the aim is to maximise the benefits and minimise the costs of working with others in a collaboration. Successful collaborations are all about good communication, self-management, team management and networking practices. While you are at university you can practise applying these principles and learn to use collaboration and project management software as you work with other students in both formal and informal groups.

6.4 Negotiation

LEARNING OBJECTIVE 6.4 Describe negotiation strategies and processes.

Negotiation is a convergent process that brings the positions of two or more parties to a point of agreement. The agreement may be a salary, a product price, a contract, a design, a share of ownership or the resolution of a dispute. The ability to negotiate is a life skill, a skill people often apply subconsciously; for example, when bargaining with a salesperson over the price of a product. The ability to negotiate is an important skill for engineers, and many engineers use their negotiation skills on a regular basis.

A negotiation process can lead to a variety of outcomes. It may result in a mutually beneficial outcome, an outcome that benefits one party more than another, the process stalling for a period, a party withdrawing from the process or the parties failing to reach an agreement. The negotiation process can be amicable, or it can be unpleasant for one or more of the parties and may result in the need to enter a formal dispute resolution process. This is a process that may be initiated when the parties to a dispute are unable to resolve the dispute.

The process an engineer follows while negotiating an agreement depends on the complexity of the issues being negotiated, the number of parties involved and the value of the matters being negotiated (this is often measured financially). Different approaches can be followed to achieve viable negotiated outcomes. The following sections discuss examples of these approaches.

Fisher and Ury (1999, p. 4) suggest that the following three criteria may be used to judge a negotiation method.

1. It should produce a wise agreement if agreement is possible.
2. It should be efficient.
3. It should improve, or at least not damage, the relationship between the parties.

You should keep these three criteria front and centre while preparing for, undertaking and finalising negotiations.

There are three phases in the negotiation process: preparation, negotiation and agreement (Spegel et al. 1998).

KEY POINT

Effective preparation is vital before entering into a negotiation process.

The preparation process

Good preparation is necessary if you want to negotiate a good outcome. In order to be well prepared, you need to have a good understanding of your position, as well as an understanding of the likely positions of the other parties and the possible outcomes of the negotiations. The amount of preparation you will need to do will depend on what is being negotiated. You may like to use the following action plan to prepare for your next negotiation.

- Define the aspects of the project you are willing to negotiate and the aspects you may be asked by the other parties to negotiate on. There may be technical details you are willing to negotiate (or modify) and other details you are not willing to negotiate (e.g. design standards). You should consider whether you are willing to revise your position on details such as costs, timelines, access and payment regimes.
- Identify any issues you require further information about.
- Define your preferred position on each of the issues being negotiated, as well as the lowest, or weakest, acceptable position for each issue — or the 'walk away zone' (Spegel et al. 1998). The difference between these two positions (i.e. the 'preferred' and the 'weakest' positions) can be considered the negotiation zone for each issue. This negotiation zone is the 'good outcome' zone. The preferred position defines the position that you are comfortable with from a design, financial or risk perspective. The lowest, or weakest, position defines the beginning of the 'no outcome' zone. Beyond this there are no outcomes that you consider acceptable. You should outline the risks if you move below this position, and what the consequences will be if you walk away. The two positions may be the same for some issues and different for other issues.
- Identify the strengths and weaknesses of your preferred positions. This should help you identify any arguments the other parties may use to weaken your case.
- Think about your relationships with the other parties involved. The following questions may help to clarify this for you.
  - Is this the first time you have worked with the parties, or do you have a long-term relationship with them?
  - What do you know about the other parties?
  - Do you want to maintain or build a long-term relationship with them?
  - What are the current characteristics of the relationships?
  - How do you want the relationships to change (e.g. would you like to have a higher level of collaboration or trust with these parties)?
  - Are there any prospects for mutual benefits?

- Undertake scenario planning sessions.
  - What positions are the other parties likely to adopt for each issue and how will you respond to these approaches?
  - Can you predict the likely outcome for the issues being negotiated?
  - How will you react if you do not achieve your desired outcomes?
  - Are you prepared to compromise and accept a poor position for one issue in exchange for a more favourable outcome on another issue?
- Identify when you should withdraw from the negotiation process. For example, you may like to withdraw from negotiations when you do not achieve satisfactory outcomes or if the negotiations have covered issues or solutions you did not foresee as relevant. You may need to discuss this with others, or obtain more information or undertake additional preparation before you are willing to resume the negotiations. You will need to consider the consequences of your actions prior to withdrawing. For example, as a result of your decision you may lose a contract for a project. In some cases, your withdrawal may result in a dispute that may need to be resolved by a third party.
- Are there any objective measures that can be used to help the negotiation process?

Once you have prepared your negotiation action plan you should seek any approvals that you may need from your organisation before you begin the negotiations.

Approaches to negotiation

There are many approaches that can be used during a negotiation process. The most common approaches are the win–win, or principled approach, and the win–lose, or positional approach. The **win–win approach** should mean that both parties benefit from the negotiated outcome, while the **win–lose approach** may result in most of the benefits being given to one party. The negotiation approach you use will depend on variables such as the matter being negotiated, your relationship with the other parties, and your desired outcomes. You will usually decide on the approach you will adopt during the preparation phase.

The win–lose approach

The win–lose approach is the more traditional approach to negotiations. It is an adversarial approach, as the parties try to protect their position while weakening the position of the other party. In the win–lose approach, the parties enter negotiations aggressively, in a similar style to a competition. Each party is attempting to get the best possible outcome for themselves, without any real concern for the other party. The win–lose approach can have catastrophic consequences. For example, one party's 'win' may result in the 'losing' party becoming bankrupt. This approach is often used in situations where parties are negotiating for the first time, or on a one-off basis.

In an engineering context, the **tender process** could be considered to be a highly restricted win–lose negotiation. This process results in companies submitting a **tender** (a detailed quotation) to undertake specified work or provide specified services. The organisation that advertises a tender (the client) establishes its position in the tender documents by clearly stating the required outcomes. All of the organisations that lodge a tender have an equal opportunity to put their positions forward. The client then considers each of the submitted tenders and evaluates them against the requirements in the tender documents. In this process the tenderer has the position of power and will generally win by being able to select the lowest price.

Sometimes, however, selecting the lowest price may not be in the best interest of all parties. For example, the organisation that won the tender may have incorrectly calculated the cost of the project and this could mean that it goes out of business while undertaking the contract. This can lead to long delays and additional costs as one contract has to be terminated and a new contract negotiated. This process may also involve legal proceedings. For this reason, many tenders often state that 'the lowest tender price may not be accepted', and the client carefully assesses the risks associated with each tender before deciding on the winning tender.

John Glenn, the first American astronaut to orbit the earth, is reputed to have said, as he was being launched into space, 'I felt exactly how you would feel if you were getting ready to launch and knew you were sitting on top of two million parts — all built by the lowest bidder on a government contract'.

The win–win approach

The win–win approach grew out of the groundbreaking work undertaken by Mary Parker Follet early last century (Spegel et al. 1998). This approach is usually adopted if you are negotiating with a party you know and trust, and regularly work with,. In this approach, all parties enter negotiations seeking outcomes that will be mutually beneficial; that is, with all parties being content with the outcomes at the end of the process. The parties are more concerned about their interests than their positions. Thus, the parties are willing to change their position as long as their overall interests are protected. This is why this approach is often referred to as being interest-based.

Another key difference is that when this approach is used the parties separate the people from the problem. In this sense they see themselves as problem solvers (Fisher & Ury 1999). Advocates of this approach believe opportunities for growth and other benefits are created by such a strategic relationship. In colloquial terms, the parties may attempt to 'grow the pie' (overall business) so they all get a larger share in the longer term.

Stephen Covey (2004) states that a win–win approach is a character-based code for human interaction and collaboration, and that people who approach conflict with this attitude exhibit character traits such as:

- integrity, by sticking with their true feelings, values and commitments
- maturity, by balancing the courageous expression of their ideas and feelings with consideration for the ideas and feelings of others
- an abundance mentality, by believing there is plenty for everyone (Covey 2004, pp. 217–221).

Table 6.4 shows some of the differences between the win–win and win–lose approaches.

TABLE 6.4 The characteristics of two approaches to negotiations

Factor	Win–win approach	Win–lose approach
The relationship between the parties	Both parties want to establish or build a long-term relationship.	The parties are not concerned about the relationship.
The size of the pie	The parties hope they can work together to grow the pie for their mutual benefit.	The parties compete for a larger share of the pie.
Achieving outcomes	All parties are likely to develop and accept mutually beneficial outcomes that are fair and just.	The aim of one or more of the parties is to maximise their share of the pie.
Being objective	The parties develop criteria that can be used to measure outcomes.	The parties maintain their positions.
Managing risk	A low-risk approach that will ensure all parties are happy with the outcomes and the project is likely to run smoothly.	May be a high-risk approach as one party may walk away from the negotiations, or go out of business.
Negotiating behaviour	The parties are likely to be communicative, relaxed, trusting and willing to compromise.	The parties are likely to be aloof, confident, firm and uncompromising.
Negotiation strategies	A problem-solving approach may be used by the parties to reach agreement.	Each party may begin with an ambit claim and use a range of game-based tactics, such as bluffing.
Using emotions	The relaxed and comfortable negotiating environment may mean the parties are comfortable about showing their emotions.	The parties will try to hold their emotions (such as anger) in check, unless they use them as a tactic.
Willingness to reach a compromise	The parties are likely to be willing to concede so they can achieve mutually beneficial outcomes and their interests are protected (e.g. they may be willing to compromise on one outcome in return for a better outcome in another area).	The parties may not be willing to concede on any issue as it may weaken their position. This may mean that they do not recover from that position and as a result, they are unlikely to achieve their objectives.

Source: Adapted from Fisher and Ury (1999); Spegel et al. (1998).

Another important consideration is the mode of negotiation. Normally negotiations are carried out at a face-to-face meeting; however, they may also be conducted by telephone and by correspondence. Each of these methods has advantages and disadvantages (Spegel et al. 1998).

Outcomes of the negotiation process

Agreement on the outcomes of the negotiation process is the final stage in the process. The following five common outcomes are possible.

1. The parties reach agreement on all of the matters being negotiated.
2. The parties reach agreement on most of the matters being negotiated and agree to resume negotiations on any outstanding matters at a later date.
3. The parties reach agreement on most of the matters being negotiated and recognise they will not reach agreement on the outstanding matters. This outcome is only possible if the outstanding matters can be removed from the agreement.
4. The parties reach agreement on most of the matters being negotiated and seek assistance from a third party to facilitate the negotiations on the outstanding matters. For example, a third party may be used to arbitrate over the fees a consultant should receive for the additional work that has already been completed on a project. At this point the parties have entered a dispute resolution process.
5. The parties may not reach an agreement on the key matters being negotiated and decide to terminate the negotiations.

It is important that the outcomes of negotiations are accurately documented and that all of the parties sign a document that lists the negotiated outcomes. Verbal acceptance of outcomes is also important. For example, a party may demonstrate acceptance by stating, 'We are happy to accept your offer, and we will confirm the details of this agreement in writing in the next seven days. We look forward to working with you to help you successfully complete this project'.

SPOTLIGHT

Negotiating tight spaces

WITH MATTHEW PRESTON, CONNELL WAGNER PTY LTD

The design of commercial buildings is normally undertaken by a team of engineers and other professionals drawn from different companies and authorities. The following example illustrates how the members of a team cooperated to design building services and the spaces they occupy within restricted wall and ceiling cavities.

The Cherrell Hirst Creative Learning Centre is a six-storey educational facility designed to accommodate the music, drama, creative technologies and visual arts faculties at the Brisbane Girls Grammar School. It has a total of 8500 m^2 of floor space, which includes separate large volume performance spaces for orchestras and choirs, computer laboratories, general learning areas, a student hall, a two-level car park and a commercial kitchen.

The general learning area, showing coordinated roof structure, lighting, specialist fabric ductwork, smoke detectors, projector and ceiling fans

Matthew Preston from Connell Wagner Pty Ltd (now Aurecon), the lead multidisciplinary consulting engineering company for the project, managed the services engineering component, and another engineer within Connell Wagner Pty Ltd managed the structural design. The project manager led the total building services design and construction team, which included the following members: client's representative, managing contractor (builder), architect (building surveyor and certifier), quantity surveyor, and a variety of engineers (including electrical, fire, mechanical, hydraulic, vertical transportation, structural, civil and acoustics engineers). The main items that required design team coordination were:

- sustainable development features (e.g. energy efficiency targets and water usage reduction methods)
- building and infrastructure authority requirements
- safe maintenance provisions
- acoustics
- switchrooms, data cupboards, cable risers and audio-visual equipment
- power and lighting
- external cooling plant, air-conditioning water chillers, air-conditioning plant rooms and ductwork
- pipe routes for water, sewerage, stormwater and gas
- digital building management control systems to control building services
- lift design

- beam locations, services penetration positions, riser shaft positions, roof penetration positions, structural wall positions and access walkways
- floor to floor heights, ceiling heights, and wall, door and architectural feature locations
- the client's building fit-out requirements
- budgets
- services design in the limited ceiling space areas.

Early in the project, the team required more general conceptual and space planning meetings than normal, and these were facilitated by the architect and project manager. After this stage, the team moved onto more detailed meetings and the construction documentation phase. These meetings were facilitated by the relevant parties involved and frequently would require small groups to meet and solve design and space coordination issues.

This services-intensive building meant compromises in space allocation were often required. Each member of the team was given equal input during the meetings and then a decision was made in the best interests of the project, rather than to meet a specific engineering or architectural requirement.

The design process commenced in February 2005 and the building was constructed during the period from December 2005 to April 2007. The building was finished three weeks early and on budget. The project achieved a high level of success, with the building receiving a number of industry awards for architectural and engineering design.

CRITICAL THINKING

Consider a typical design/build/test project in your discipline. List the negotiations that may be required during the design and build process, and the disciplines of the people who would be party to those negotiations.

6.5 Dispute resolution

LEARNING OBJECTIVE 6.5 Understand dispute and conflict resolution strategies.

When an issue cannot be resolved between two or more parties it is normally called a **dispute**. One way of resolving a dispute is to involve another party in the process; that is, a third party. We will look at three processes: *mediation, arbitration* and *litigation*. In the first process, mediation, the parties produce the outcome and therefore retain some control over the process. This is not the case with arbitration and litigation, where the outcome is decided by a third party, or parties.

A mediation process may be used when all parties voluntarily agree to abide by the outcomes of the process. A neutral person is appointed to the role of mediator and helps the parties reach an agreement. The mediator may meet with the parties separately or together at different times in the process. Their role is to keep the parties focused on resolving the dispute, and to act as a calming influence.

The second process is arbitration. Normally the parties have to agree to be bound by the arbitrator's decision before the arbitration process begins. The arbitrator, a neutral person, listens to the cases put forward by all of the parties and then makes a decision. One of the disadvantages of arbitration is that the parties may have to accept outcomes that, from their perspectives, are not ideal.

Litigation should be used as a last resort because it means that the dispute will be resolved by a court, and this means that the process can be expensive. Litigation should only be contemplated if there are no other ways of resolving the dispute. Like arbitration, litigation may not lead to a satisfactory outcome for the parties involved.

Conflict resolution

We all encounter conflict in our lives, either directly or as observers. This section discusses some of the principles that may be used to deal with conflict and to help others who are involved in conflict. A **conflict** arises when there is a serious disagreement or argument between two or more parties. Conflicts occur between individuals in the workplace, and in the wider community.

Conflict in the workplace is a serious matter and most organisations have human resource policies and processes to deal with it. It is important to identify conflict as early as possible so these processes can be commenced before the conflict escalates (Conflict Research Consortium 1998).

Some conflict resolution strategies and outcomes are as follows.

- *Constructive engagement.* This process attempts to determine the underlying reasons for the conflict and then seeks ways to resolve them. The answers to the following questions will help the parties understand the reasons for the conflict.
  - Who is the conflict between?
  - What are the symptoms of the conflict?
  - Why is there a conflict?
  - What are the reasons for the conflict; that is, what is the dispute about?
  - Is this the first dispute between the parties?
  - Are there any other stakeholders with an interest in the outcome of the conflict?
- *Compromise.* Are the parties willing to meet in an attempt to resolve the conflict? What is necessary to bring about such a meeting? Is there a need for confidentiality? What sort of resolution are the parties looking for? How will the meeting end if a resolution is achieved?
- *Forcing.* One party may force the other party to accept a solution. This may result in a resolution, but it may prove to be a temporary solution.
- *Smoothing.* One party backs off and lets the other party claim a victory.
- *Avoidance.* The parties avoid each other and hope the conflict will dissipate over time (adapted from Conflict Research Consortium 1998; State Services Authority 2011).

One way of avoiding disputes and conflict in the work environment is for the organisation to adopt transparent and consultative processes. This can be beneficial for its employees and also with its dealings with members of the public.

SUMMARY

In this chapter we have outlined a number of key skills engineers need so that they can work effectively in the workplace. These skills enable engineers to relate to, and work with, people from all sectors of society, including community groups, government organisations, and small and large businesses. On some projects these skills can be just as important as technical skills. We will now briefly revisit the learning objectives from this chapter.

6.1 Work effectively with others.

A capacity to work effectively with other people is high on any employer's list of desirable employee characteristics. The essential skills for collaborating are transparency, communication, teamwork, relationship management, organisational awareness, social/cultural awareness, sociability and teaching others (Allen et al. 2017). Engineering graduates need to collaborate with a variety of people including those working at various levels in their own discipline, those in other levels within their organisation, those from other disciplines and from community organisations. They also need to be able to work with people from a variety of ethnic and cultural groups as well as people from government organisations and members of the public. Therefore, engineering graduates need to develop a set of discipline-specific, 360-degree collaboration capabilities to be successful practitioners in their field.

6.2 Work effectively in groups and teams.

Engineering is a 'team sport' and it is essential that engineers have the skills to be able to work effectively in groups and teams. Group members need to be clear about: why the group has been established; what tasks are to be undertaken; what skills and knowledge the team members possess, or may need; roles; and how the group will operate, meet and communicate. This information should be recorded in a charter. Advances in communications technology enable groups to hold virtual meetings. During your engineering studies you will have many opportunities to develop your team skills, including leadership. It is important to learn from team experiences through reflective practice.

6.3 Actively participate and contribute in meetings.

Meetings with both internal and external stakeholders are an integral part of an engineer's work activities. It is important that the purpose of a meeting is clearly defined and understood by the participants. Meetings should be both effective and efficient. Participants should come to a meeting fully prepared and ready to contribute to achieve quality outcomes. Formal meetings may have set procedures that need to be followed, and these should be clearly understood by the participants. A number of factors may dictate the level of formality required in a meeting, including legal requirements. Finally, it is important that the outcomes of a meeting are recorded.

6.4 Describe negotiation strategies and processes.

Negotiation is a skill that is used throughout the engineering industry to bring the positions of two or more parties to the point of agreement. It is important to carefully prepare before beginning to negotiate. This involves understanding your preferred position, deciding on the approach to be used (e.g. win–win or win–lose) and what an acceptable outcome will be. One outcome may be a dispute that may need to be resolved with the help of a third party.

6.5 Understand dispute and conflict resolution strategies.

Three approaches may be used to resolve disputes: mediation, arbitration and litigation. The process used will depend on the type of dispute, the matters under dispute and the parties involved. The parties to a dispute have some control over the outcome when a mediation process is adopted, but not when arbitration or litigation processes are used.

Disruptive behaviour and conflict in the workplace may affect the people involved, the work environment and productivity. It is important that conflict and disputes are addressed as soon as they become evident. In most organisations there are standard processes to be followed and these are set out in the organisation's human resource policies and procedures.

KEY TERMS

asynchronous communication Communication between parties that occurs at different times. Email is an example of an asynchronous communication method.
chair The person who presides over a meeting. The role of a chair is to ensure a meeting is conducted efficiently and in accordance with the constitution of the organisation.
conflict A state of disharmony between two parties.
dispute An argument between two or more parties who contest an issue.
group A collection of two or more people who work with one another regularly to achieve one or more common goals.
groupware A combination of computer hardware and software that facilitates virtual meetings between geographically separated people.
groupware conference A virtual meeting of geographically separated people who use groupware to facilitate real-time audio, video and text communications.
meeting An assembly of people for a specific purpose.
negotiation A convergent process that brings the positions of two or more parties to a point of agreement.
team A small group of people with complementary skills, who work together as a unit to achieve a common purpose for which they hold themselves collectively accountable.
teleconference A meeting of geographically separated people conducted using a telephone or other audio technology.
tender A notice requesting businesses to submit a quotation for specified goods or services.
tender process The advertising, submission and evaluation of quotations.
videoconference A virtual meeting of geographically separated people conducted using audio and video technologies.
virtual group A group whose members use a range of technologies to meet and communicate.
win–lose approach A situation where the majority of the benefits accrue to one party to the negotiations.
win–win approach A situation where all parties benefit from a negotiated outcome.

EXERCISES

1 Discuss the importance of collaboration in your field of engineering.
2 Think about three or four of your life experiences that could be defined as teamwork. For example, paid jobs where you have worked with a group of other employees over time, experience in a student club or sporting team, or working as a community volunteer. Perhaps you did a substantial group project at high school. For each of your experiences of teamwork, note down:
   - how many people were in the team
   - how long you worked together
   - what you were trying to do (task/s).

   Reflect on your experiences and identify the team skills you learned or developed through each experience.
3 Compare your examples of teamwork with two or three other students in your class and evaluate your collective experiences of teamwork.
   (a) Was your experience with a group or a team? What criteria did you use to select your answer?
   (b) What range of team sizes have members of your group experienced?
   (c) Were the teams long-lived or short-lived?
   (d) Have members of your group been in a team with defined or finite tasks (i.e. tasks that were clear and had a defined end point)?
   (e) Have they worked in a team that had evolving, perpetual or rolling tasks (i.e. tasks that were either not clearly defined or were ongoing)?
   (f) What positive or negative aspects of teamwork have you and your other group members experienced? Why? Are there common themes for your positive and negative team experiences?

4 Reflect on the team or group leaders you have worked with.
   (a) Which ones were successful? Why?
   (b) What were some of the techniques they used that worked? Why?
   (c) What techniques did not work? Why?
   (d) Consider which style of leadership would be suitable for you. What techniques would you use?
5 Use Zoom or similar groupware to hold a virtual meeting of your student project team. Reflect on your experience, and the success of the meeting, by asking the following questions.
   (a) Was the meeting successful?
   (b) Did all of the members contribute to the meeting?
   (c) Were the contributions different from those that you would have expected if a face-to-face meeting had been held? Why?
   (d) Do you think it was easier or harder to hold a meeting this way?
   (e) Do you think a better outcome was achieved?
   (f) What would you do better next time?
6 How would you approach the negotiations about a dispute between your company and a contractor over a large amount of money? What dispute resolution strategies would you use if a satisfactory outcome could not be negotiated?
7 The tendering process can be considered as a form of negotiation. Review and evaluate the tender advertisements online or in a Saturday newspaper.
   (a) What are the common conditions? Why?
   (b) Are there any conditions that surprise you? Why?
   (c) Why do you think conditions are included in the tender advertisements?

PROJECT ACTIVITY

Identify an engineering job you would really like, preferably one that has recently been advertised. Assume that you have been granted an interview. The company has sent you a program of activities that you will undertake as part of the interview process. It includes the following activities.

- 10.00–11.00 am: Tour of the plant, its facilities, offices and other work areas with your potential future supervisor.
- 11.00 am–12.00 pm: Formal interview with the interview panel.
- 12.00–1.00 pm: Lunch with the company CEO, your potential future supervisor, the interview panel members and other short-listed applicants.

During the interview process you will therefore participate in three different types of meetings. Prepare documentation that incorporates all of the information you will need for each meeting. For example:

(a) research and compile as much background information as you can about the organisation and your job
(b) compile evidence to demonstrate your technical abilities as well as your communication, critical thinking and teamwork skills
(c) prepare a list of questions you would like to ask
(d) develop strategies to negotiate satisfactory conditions in the event you are offered the job.

REFERENCES

Allen, R, Teodoro, N & Manley, C 2017, *Future skills and training: A practical resource to help identify future skills and training*, Australian Industry Skills Committee.

Belbin, M 1993, *The Belbin team roles package*, Pfeiffer & Company, California.

——— 2010, 'Team role theory', Belbin Associates, https://www.belbin.com

——— 2011, 'Team roles in a nutshell', Belbin Associates, https://www.belbin.com/media/1288/belbin-for-students.pdf

Caspersz, D, Skene, J & Wu, M 2006, *Managing student teams*, Herdsa Green Guide, Higher Education Research and Development Society of Australasia.

Conflict Research Consortium 1998, *International online training programme on intractable conflict — active listening*, University of Colorado, Boulder.

Covey, S 2004, *The 7 habits of highly effective people: Restoring the character ethic*, rev. edn, Free Press, New York.

Cross, R, Heen, S & Zehner, D 2017, *Reclaiming your day: How successful people manage collaborative overload*, Connected Commons.

Dew, JR 1998, *Managing in a team environment*, Quorum Books, Connecticut.

Dowling, D 2015, *A graduate capability framework for the mining engineering degree programme: A guide for MEA universities*, Research Report for Mining Engineering Australia, Adelaide.
——— 2016, 'Employing STEM graduates: Identifying collaboration capabilities wanted by employers', in G Chapman, *The STEM ecosystem: Building cross-disciplinary leadership capacity in science, technology*, The Writing Bureau, Melbourne, pp. 113–126.
Dowling, D & Hadgraft, R 2013a, *The DYD stakeholder consultation process: A user guide*, Office for Learning and Teaching, https://ltr.edu.au/resources/PP9-1280_Dowling_DYD_Stakeholder%20Process%20Web_2013_0.pdf
——— 2013b, *A graduate capability framework for environmental engineering degree programs: A guide for Australian universities*, Office for Learning and Teaching, https://www.engineersaustralia.org.au/sites/default/files/2022-07/graduate-capability-framework-environmental-engineering-degree-program.pdf
Fisher, R & Ury, W 1999, *Getting to yes: Negotiating an agreement without giving in*, 2nd edn, Random House Business Books, London.
Katzenbach, JR & Smith, DK 2001, *The discipline of teams*, John Wiley & Sons, New York.
Kimble, C, Barlow, A & Li, F 2000, 'Effective virtual teams through communities of practice', *Department of Management Science Research Paper Series*, vol. 2000, no. 9, https://doi.org/10.2139/ssrn.634645
Lencioni, P 2002, *The five dysfunctions of a team: A leadership fable*, Jossey–Bass, New York.
Macquarie Dictionary 2024, *The Macquarie Dictionary Online*, Pan Macmillan Australia, https://www.macquariedictionary.com.au
Mayne, L 2012, 'Reflective writing as a tool for assessing teamwork in bioscience', *Biochemistry and Molecular Biology Review*, vol. 49, no. 4, pp. 234–249, https://doi.org/10.1002/bmb.20621
Mitchell, R 2014, 'Team innovation and success: Why we should fight at work', *The Conversation*, 20 January, https://theconversation.com/team-innovation-and-success-why-we-should-fight-at-work-20651
Nepal, K 2012, 'An approach to assign individual marks from a team mark: The case of Australian grading systems at universities', *Assessment & Evaluation in Higher Education*, vol. 37, no. 5, pp. 555–562, https://doi.org/10.1080/02602938.2011.555815
Oakley, B, Felder, RM, Brent, R & Elhajj, I 2004, 'Turning student groups into effective teams', *Journal of Student Centered Learning*, vol. 2, no. 1, pp. 9–34.
RMIT University 2024, *Graduate attribute: Global in outlook and competence*, https://www.rmit.edu.vn/business-and-industry/graduate-attributes
Sales, H 2006, *Professional communication in engineering*, Palgrave and Macmillan, Basingstoke, UK.
Schulz-Hardt, S, Jochims, M & Frey, D 2002, 'Productive conflict in group decision making: Genuine and contrived dissent as strategies to counteract biased information seeking', *Organizational Behaviour and Human Decision Processes*, vol. 88, no. 2, pp. 563–586, https://doi.org/10.1016/S0749-5978(02)00001-8
Spegel, N, Rogers, B & Buckley, R 1998, *Negotiation: Theory and techniques*, Butterworths, Sydney.
State Services Authority 2011, *Dealing with high conflict behaviours*, State Government of Victoria.
Thomas, TA 2014, 'Developing team skills through a collaborative writing assignment', *Assessment & Evaluation in Higher Education*, vol. 39, no. 4, pp. 479–495, https://doi.org/10.1080/02602938.2013.850587
Tuckman, B 1965, 'Developmental sequence in small groups', *Psychological Bulletin*, vol. 63, no. 6, pp. 384–399, https://psycnet.apa.org/doi/10.1037/h0022100
Tuckman, BW & Jensen, MAC 1977, 'Stages of small group development revisited', *Group and Organizational Studies*, vol. 2, no. 4, pp. 419–427, https://doi.org/10.1177/105960117700200404
van der Kamp, M 2011, 'Collaborating and colliding: When alliances go wrong', *The Conversation*, 18 April, https://theconversation.com/collaborating-and-colliding-when-alliances-go-wrong-807
Watson, WE, Kumar, K & Michaelsen, LK 1993, 'Cultural diversity's impact on interaction process and performance: Comparing homogeneous and diverse task groups', *Academy of Management*, vol. 36, no. 3, pp. 590–602, https://doi.org/10.2307/256593
Wood, J, Wiesner, R, Morrison, RR, Factor, A & McKeown, T 2025, *Organisational behaviour: Core concepts and applications*, 6th Australasian edn, John Wiley & Sons, Brisbane.

ACKNOWLEDGEMENTS

Photo: © Halfpoint / Adobe Stock Photo
Photo: © N Hiraman/peopleimages.com / Adobe Stock Photo
Photo: © nenetus / Adobe Stock Photo
Photo: © Brisbane Girls Grammar School
Figure 6.1 and tables 6.1–6.2: © Dowling, D & Hadgraft, R 2013, *A graduate capability framework for environmental engineering degree programs: A guide for Australian universities*, Office for Learning and Teaching, https://www.engineersaustralia.org.au/sites/default/files/2022-07/graduate-capability-framework-environmental-engineering-degree-program.pdf
Figure 6.2: © Belbin 2010, 'Team role theory', Belbin Associates, https://www.belbin.com
Table 6.3 and text: © Dowling, D 2015, *A graduate capability framework for the mining engineering degree programme: A guide for MEA universities*, Research Report for Mining Engineering Australia, Adelaide.
Text: © Allen, R, Teodoro, N & Manley, C 2017, *Future skills and training: A practical resource to help identify future skills and training*, Australian Industry Skills Committee.

PART 4

COMMUNICATION

CHAPTER 7

Understanding communication

'The single biggest problem in communication is the illusion that it has taken place.'

George Bernard Shaw

LEARNING OBJECTIVES

After studying this chapter, you should be able to:

7.1 explain what communication is and discuss the types of communication skills used by engineers

7.2 discuss the historical development of communication theories and outline the components of the PCR model

7.3 describe how contextual factors can influence the effectiveness of communication

7.4 describe the communication methods commonly used by engineers

7.5 discuss the roles of the creator, gatekeeper and consumer in the communication process.

Introduction

We all communicate! This is because communication is the lifeblood of social interactions. In fact, as Professor Dominic Rowe from Macquarie University stated, 'Communication is life; if you can't communicate, living is almost impossible' (quoted in Wong & Champness 2018). Professor Rowe's opinion was based on his experience as a neurologist working with people suffering from motor neurone disease (MND). People with MND do not have a long life expectancy as they gradually lose their motor functions and thus their ability to speak or to move their limbs. In addition to their physical suffering, they are unable to communicate or manage even the most basic day-to-day activities. In short, they feel isolated as they are unable to interact with their families and friends, or control their lives.

Of course, many other people, such as those with cerebral palsy, locked-in syndrome or spinal muscular atrophy, and those who have suffered a stroke, a spinal cord injury or traumatic brain injuries, experience similar feelings of isolation and lack of control. Those of us with normal communication abilities would find it difficult to understand the impact a lack of ability to communicate would have on our lives.

Over recent years, engineers have worked with researchers from other disciplines to develop a range of hardware and software technologies that facilitate communications between people with disabilities and their families, carers and friends. Generally, these devices rely on the user's ability to move an eye or a part of their body to trigger an electronic signal that enables them to select options on their computer or tablet device. For example, some of the movements that are used to trigger the different devices are a breath, a finger movement or an eye movement.

Generally, two actions are required to select an option on a device, for example, the letter 'P on a QWERTY keyboard. The first action is to *locate* an option and the second is to select it, or switch it. When eye movements are used to make a selection, a camera on the computer or tablet tracks the user's eye movements. Firstly, the user locates the desired letter and then dwells on it for up to a second of time to select it. The user can repeat this process to produce a string of words, sentences or paragraphs. They can then select other keys to send the message as a text or email, or to run it through an app to produce a verbal message.

As this can be a slow and tedious process for all concerned many researchers and biomedical companies are developing new technologies to speed up the process, as well as help other non-verbal groups communicate. The key factors to be addressed are accuracy, flexibility, speed and signal type. The flexibility of a system, and the types of signals used, determine its suitability for users, based on their cognitive and physical abilities. The innovative technologies developed by two Australian biomedical companies will be used to illustrate these principles.

A recent innovation developed by Control Bionics, an Australian company, enables small electronic signals in a patient's damaged muscles to control basic computer functions. In 1989, Sydney entrepreneur and medical technology designer Peter Ford found that a patient's damaged muscles still emitted small electrical signals that could be used to control basic computer functions. The result was the formation of Control Bionics and the development of the technology over recent decades. The invention was tested by Professor Hawking over a five-year period, which provided Control Bionics with invaluable insight into how to improve it. 'I think he was the toughest teacher on the planet to take your homework to' (Ford quoted in Wong & Champness 2018).

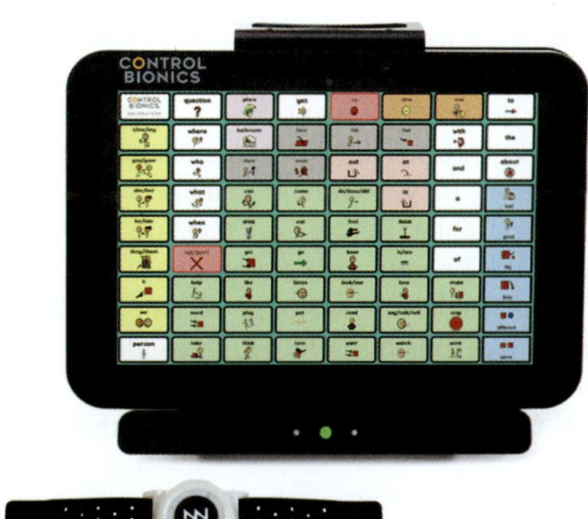

NeuroNode EMG device and Trilogy tablet

Source: Control Bionics (2024).

The first version of the technology was the NeuroNode, a wearable, wireless device that can be connected to computers, phones and tablets. It uses the body's bioelectrical electromyographic (EMG) signals to completely control a computer to generate speech, browse the web, listen to music, and so on (Control Bionics 2018). The electrodes for the NeuroNode were attached above the target muscle using adhesives. Only one effective target muscle was required and it could be on the forehead, arms or legs. The target muscle should respond, at least minimally, to a command to contract it and then it should return to a resting state in a timely manner. Thus, the user has only to think about using a muscle to be able to control computer-based devices (Rob Wong, Control Bionics, pers. comm. 2018).

The latest version of the NeuroNode sensor increases the flexibility of the system and collects an additional signal. It enables the user to control their device either by muscle movements (EMG) or by a 3D spatial movement. When combined with the NeuroNode Trilogy speech-generating device, four access methods can be used: touch control, eye control, EMG or spatial movement from the NeuroNode sensor (Control Bionics 2024). This multimodal access allows the user to increase the speed and ease of communication. For example, eye movement can be used to locate an option on the screen and then an EMG signal or spatial movement can be used to select that option.

A more recent Australian innovation is the brain computer interface (BCI) developed by Synchron. It seeks to track thought patterns to give the power of communication to those who do not have the capacity to send signals from their brain to their muscles to initiate movement. The key innovation is the Stentrode, a 4 cm long metallic stent fitted with a series of electrodes to monitor neural signals (Synchron 2024). The Stentrode was developed by a team led by biomedical engineer Professor Nicholas Opie and Professor Tom Oxley, both from the University of Melbourne, co-heads of the Vascular Bionics Laboratory and co-founders of Synchron. They worked with a large team of researchers drawn from a range of disciplines including neurointerventionists, neurosurgeons, coders, engineers and other allied disciplines (University of Melbourne 2024).

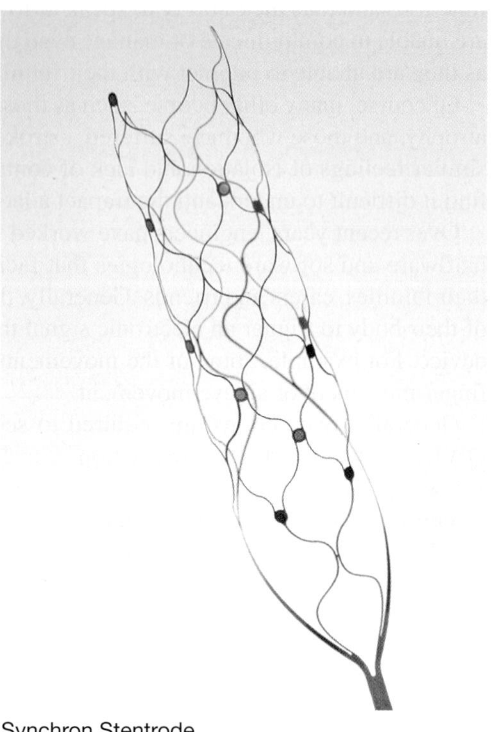

Synchron Stentrode

Source: Synchron (2024).

Typically, BCIs require major open brain surgery to implant them, a risky process that may lead to rejection and other negative outcomes. A major advantage of the Stentrode is that open brain surgery is not required, as the device is implanted via a person's vascular system. It is inserted through the blood vessel network until it reaches the desired location close to the motor cortex of the brain — the movement control centre — where it will remain permanently to record brain signals. The signals are fed via a lead in the vascular network to a transmitter implanted in the patient's chest — much like a pacemaker. The signals are then transmitted via Bluetooth to an external signal processing device which, in turn, sends them on to an app on a smartphone, tablet or other device (Synchron 2024).

Users are provided with regular and personalised training to enable them to learn how to link the thought processes they would use to make a physical movement (e.g. move a foot) to a particular action, or switch on their tablet. Before long they are able to communicate with their family and friends, reconnect with the world and regain their autonomy.

The late Matthew Hodge was an MND sufferer who used the NeuroNode technology to take control of his life despite his debilitating condition. With the NeuroNode attached to his arm, he simply had to think about moving a muscle to control a connected laptop computer, which allowed him to communicate with his family and friends. The value of this was outlined by Wong and Champness (2018) in their interview:

> 'It let us communicate with him and know what he was thinking and what he was feeling,' his wife Joanne said. 'He could use the same technology to do the online shopping, to do banking, to send emails, to send text messages as well as use it to speak.'

One of the first recipients of the Synchron system wrote a book!

Please visit the company websites for the latest information on these devices, or to watch videos of their development and use.

What a wonderful outcome for the recipients of these innovative Australian technologies, which are now beginning to be used across the world. They have been reconnected to their families, friends and the wider world by an engineering innovation. This story illustrates how important the ability to communicate is, and how engineering technology can be used to make positive contributions to people's lives. It should also encourage us to make the most of our communication abilities.

A student's life is communication-rich, with lectures, tutorials, practical sessions, presentations, excursions and social interactions — both face-to-face and through social media — providing opportunities for them to hone their communication skills. Students rely on their communication skills to acquire knowledge, understanding and new skills. They then use their communication skills to complete assessment tasks where they demonstrate their achievement of the learning objectives defined for each subject.

This chapter is the first of three chapters about engineering communication that will help you on the journey to becoming an effective communicator. In this chapter, the focus is on understanding communication: the how, why, when, where, who and what of communication. The aim is to understand why communication can be successful and how communication can break down. The key features of communication theories and models are synthesised to develop a model that will help you understand the communication process, as well as the verbal, non-verbal and visual communication methods that are used in the workplace. The model is then used to plan effective communications. In the chapter on communication skills, the focus will shift to basic interpersonal communication skills, such as reading, writing notes and listening, while the chapter on communicating information describes many of the methods that engineers use to communicate information. The knowledge gained in these chapters should provide you with a sound understanding of the communication skills that you will need to be a successful engineer. You will have many opportunities to develop these skills while you are a student.

7.1 What is communication?

LEARNING OBJECTIVE 7.1 Explain what communication is and discuss the types of communication skills used by engineers.

Communication can be defined as 'the imparting or interchange of thoughts, opinions, or information by speech, writing, or signs' (Macquarie Dictionary 2024). It is important to distinguish between this singular meaning of the word and the plural form, **communications**, which the dictionary defines as 'the science or process of conveying information especially by electronic or mechanical means' (Macquarie Dictionary 2024). This is normally associated with communications technologies, for example, the transmission of data and information by telephone, radio, television and other electronic transmission modes.

The singular use of the word 'communication' is the focus of this chapter.

Communication is one of the most important ways that humans interact and there are numerous reasons why they communicate, including to:

- interact socially
- inform
- conduct business
- create a shared understanding of an event, an object, or a concept
- collaborate
- instruct
- impress
- lead.

Think of some of the types of communication that you regularly use and identify the reason for each one. Are there some additional reasons you might not have readily thought of?

Just as there are many reasons for communicating, there are also many ways that humans communicate; for example, **verbal communication**, which includes written and visual communication, and **non-verbal communication**. Non-verbal methods include facial expressions and hand and arm movements. Like other professionals, engineers communicate for a variety of reasons and they use many different communication methods. A good communicator has a range of communication skills and techniques that can be adapted to suit the purpose of a communication, the needs of the audience and the contexts in which the communication will occur (Mottard & Casteleyn 2008).

The following spotlight highlights the importance of communication in the workplace, both within the engineering profession and across industry and organisational boundaries.

KEY POINT

Engineers use a range of skills and techniques to communicate effectively.

SPOTLIGHT

Catastrophic communication breakdown

On 25 September 1998, an explosion at the Esso gas production and processing plant at Longford, Victoria, killed two people and injured eight others. The accident meant that the gas supply to commercial, domestic and industrial users in Melbourne and other regions of Victoria ceased for two weeks.

The immediate cause of the accident was the reintroduction of warm liquid into a heat exchanger vessel, which fractured because the metal was so cold it had become brittle. The explosion occurred when the escaping gas was ignited by a nearby ignition source (Hopkins 2000, p. 152).

In 1998, the Esso gas plant explosion in Longford, Victoria, killed two people and injured eight others.

A Royal Commission was established to discover the causes of the disaster and it reported in June 1999. Hopkins (2000, pp. 20–21) comments that the report 'made a distinction between "immediate" and "real" causes. *Immediate causes* referred to the sequence of technical events, starting with the process upset, which culminated in the rupture of the heat exchanger and the escape of the inflammable gas. But the report identified that the *real cause* was the inadequate knowledge and training of the operators, which prevented them from taking appropriate preventative action as the accident sequence developed'.

One of the important outcomes from this and other accidents at major-hazard facilities (MHF) is that the reasons for the accident and recommendations for improvement are *communicated* widely in the relevant industries and among the professionals who work in those industries. In the Longford case, the communication process included:

- the findings of the Royal Commission, *Report of the Longford Royal Commission: The Esso Longford Gas Plant Accident* (Dawson & Brooks 1999)
- a book, *Lessons from Longford: The Esso Gas Plant Explosion* (Hopkins 2000); a review of the book was published by Engineers Australia in 1999
- a special issue of the journal of *Occupational Health and Safety Australia and New Zealand*, 'Lessons from Longford: The Trial' (Hopkins 2002)
- a report, 'Have Australia's major hazard facilities learnt from the Longford disaster?' (Nicol 2001). This report was the result of an Engineers Australia sponsored review of the impact the Longford explosion has had, from an engineering perspective, on the policies, procedures and equipment of MHF operators. Following his review, Nicol (2001, p. 23) wrote:

> The recognition of the need for better inter and intra-company communication has now been recognised by many operators. For example, this was one of the key lessons that Shell's Geelong refinery took from Longford. Communication breakdown at all levels occurred at Longford. Shell is concerned to ensure a similar communication breakdown does not occur on their sites. Areas where communication is essential are between management and staff; between engineering and operations; and between shifts.

The Longford case highlights the centrality of communication in the day-to-day work of engineers, as well as the need for 360-degree collaboration skills, discussed in the chapter on working with people.

CRITICAL THINKING

The reporting of reasons for accidents and resulting recommendations for improvement does not happen in some fields of engineering, where this information is often not disclosed due to non-disclosure clauses in the documents that result from litigation or dispute resolution processes. What happens in your intended field of practice? Why?

Key communication skills for engineers

When we reflect on earlier chapters, a common thread when discussing engineering is 'communication'. This is not surprising, as the ability to communicate effectively is a critical skill for engineering graduates. The Graduate Careers Australia (2016, p. 19) survey asked participating Australian employers to nominate which three selection criteria they most used when recruiting graduates. The respondents from all industries, including engineering, ranked '*interpersonal and communication skills (written and oral)*' as the most important selection criterion, which was consistent with the findings from previous years.

The fact that the employers who recruit graduate engineers rate communication skills so highly demonstrates the importance of these skills in engineering practice.

The results of the Australian Competencies of Engineering Graduates (CEG) Project (Male et al. 2009) gave an insight into the importance of a range of competencies, including communication competencies, from the perspective of experienced engineers. (The findings of this study were outlined in the chapter on self-management.)

You will recall that just over one-third of the 64 competencies had a mean rating between four and five on the five-point scale. All *eight* of the communication competencies were in that group of 23 competencies, and six were ranked in the top ten. The eight competencies, and their ranking, are listed in table 7.1. It is also important to note that *five* of the competencies rated as *critical* by more than 50 per cent of the respondents were communication competencies. These are shown in bold in the table.

TABLE 7.1 **The eight communication competencies and their rank**

Communication competency	Rank
Communicating clearly and concisely in writing (e.g. writing technical documents, instructions and specifications)	1
Managing own communication (e.g. keeping up to date and complete, following up)	2
Using effective verbal communication (e.g. giving instructions, asking for information, listening)	4
Working in teams (e.g. working in a manner that is consistent with working in a team, trusting and respecting other team members, managing conflict, building team cohesion)	5
Speaking and writing fluent English	6
Interacting with people in diverse disciplines, professions and trades	7
Using effective graphical communication (e.g. reading drawings)	19
Negotiating, asserting and defending approaches and needs	23

Source: Male et al. (2009).

Are you surprised by the importance of these communication skills? Their importance is reflected in the curriculum of the engineering degree programs offered by Australian universities, particularly those programs that are accredited by Engineers Australia where it is expected that graduates will have satisfied the Stage 1 Competency Standard for Professional Engineers. This standard includes an element of competency (3.2) relating to communication: 'Effective oral and written communication in professional and lay domains' (Engineers Australia 2019). Two indicators of attainment are listed for this element of competency and they describe the expected breadth and depth of the communication skills a graduate should be able to demonstrate.

(a) Is proficient in listening, speaking, reading and writing English, including:
- comprehending critically and fairly the viewpoints of others
- expressing information effectively and succinctly, issuing instruction, engaging in discussion, presenting arguments and justification, debating and negotiating — to technical and non-technical audiences and using textual, diagrammatic, pictorial and graphical media best suited to the context
- representing an engineering position, or the engineering profession at large to the broader community
- appreciating the impact of body language, personal behaviour and other non-verbal communication processes, as well as the fundamentals of human social behaviour and their cross-cultural differences.

(b) Prepares high quality engineering documents such as progress and project reports, reports of investigations and feasibility studies, proposals, specifications, design records, drawings, technical descriptions and presentations pertinent to the engineering discipline (Engineers Australia 2019, p. 6).

The supporting documentation notes that the indicators of attainment should not be interpreted as discrete sub-elements of competency because each element of competency will be assessed in a holistic sense. In fact, additional indicator statements can be used where they complement the competencies listed.

Although you will have many opportunities to acquire and enhance these skills while at university, you should monitor your progress against these indicators of attainment, particularly as you approach graduation. Once you graduate and gain at least three years of engineering experience you may apply for Chartered status.

To become a Chartered Professional Engineer, Engineers Australia requires that the applicant demonstrate competence in 16 elements of competency, including element 8: 'Communication'. To demonstrate this element they must demonstrate that they:

- can communicate in a variety of different ways to collaborate with other people, including accurate listening, reading and comprehension, based on dialogue when appropriate
- can speak and write, taking into account the knowledge, expectations, requirements, interests, terminology and language of the intended audience (Engineers Australia 2012, p. 6).

The indicators of attainment listed for this communication element of competency are:

- respect confidentiality obligations
- build and maintain collaborative relationships with other people, gaining their respect, trust, confidence and willing, conscientious collaboration
- exercise informal leadership in order to coordinate the activities of diverse people who contribute to engineering activities
- collaborate effectively within multi-disciplinary teams including other professions in the workplace
- lead and sustain discussion with others and, where appropriate, integrate their views to improve deliverables
- convey new concepts and ideas to technical and non-technical stakeholders
- deliver clear written and oral presentations on engineering problems and engineering activities in English or in a language appropriate to the engineering work (Engineers Australia 2012, p. 6).

Applicants address as many indicators of attainment as they need to demonstrate the element of competency.

7.2 Communication theories and models

LEARNING OBJECTIVE 7.2 Discuss the historical development of communication theories and outline the components of the PCR model.

Research in the communication field is undertaken by researchers from a number of disciplines, such as psychology, the social sciences and journalism. Many theories have been developed and generally they are associated with one or more of these disciplines. Communication theories may also be categorised by fields of communication, for example, interpersonal communication, mass-media or organisational behaviour. Communication theorist David Foulger (2004a) suggests that the communication field may never be united by a common theory of communication.

Communication theories help to explain why people communicate effectively and why they sometimes fail to communicate or are misunderstood. An understanding of some of the factors that facilitate good communication, as well as those that inhibit effective communication, will help you learn to be a better communicator. This section begins with a review of some well-known theories to distil the key principles for effective communication. In the next section, these principles are used to develop a communication model for engineers, which is then used to guide the discussion in the remaining sections of this chapter and the relevant sections of the following chapters.

KEY POINT

Although communication models and theories have evolved over many years, the fundamental principles remain the same.

The communication process

Writing around 300 BC, Aristotle stated in his treatise on rhetoric that the primary **purpose** of communication is persuasion; that is, creating a desired **response** in listeners (cited in Lane 1932). He believed it was the *listener* who held the 'key' to the success of a speech; that is, whether the purpose of the communication was achieved. Thus, to be effective, a communication must firstly engage and then achieve the desired response from the audience. Maxwell (2010, p. 11) suggests that this only occurs when a connection is made: 'when you connect with others you position yourself to make the most of your talents and skills'.

Aristotle's fundamental communication principle is still valid, not just in speech making, but also in most other forms of communication. In this context, the purpose of a communication is to elicit the desired response from the listener or reader, such as an action, a behaviour or a reply.

Lasswell (1948, p. 37) used a series of questions to suggest that communication is a process.

Who?	Says what?	In what channel?	To whom?	With what effect?

The components of Lasswell's communication process are speaker, message, channel, listener and response. A **message** is 'information conveyed by any means from one person or group to another person or group' (Eunson 2016) and a communication **channel** is the way that a message is conveyed.

Shannon and Weaver (1949) developed a model of the communication process that was based on contemporary telecommunication technology. A schematic of this model is reproduced in figure 7.1. Through the use of the terms *information source* and *destination* rather than *speaker* and *listener*, this model expanded the notion of communication beyond speech. The model also introduced two additional components between the information source and the destination: *transmitter* and *receiver*.

FIGURE 7.1 The Shannon–Weaver model of the communication process

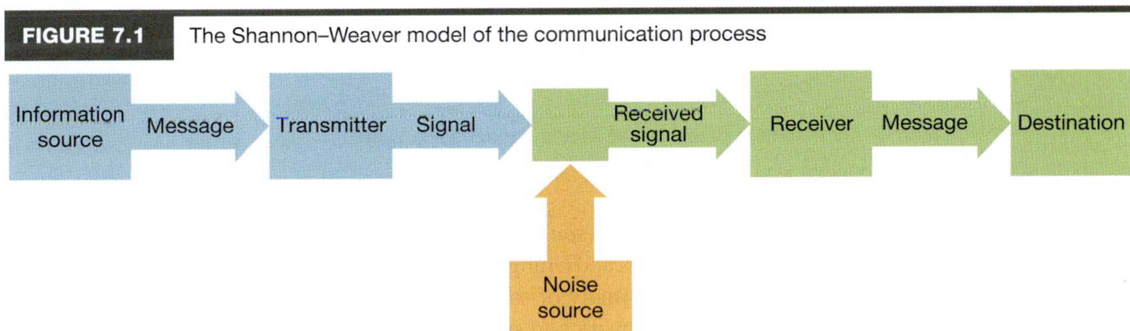

Source: Adapted from Shannon and Weaver (1949, p. 7).

Shannon and Weaver also highlighted the fact that a message can be subject to interference by **noise**, which may mean that the message does not get through, or that it may be corrupted. The concept of noise is easy to understand when a radio analogy is used. In this case noise is static or another physical phenomenon that causes interference to the transmitted signal. In our broader concept of communication, noise is harder to define. One example is the difficulty of having a conversation in a crowded room where the noise from multiple conversations, music and other sources affects the conversation and inhibits communication. Some words may be missed, misinterpreted or misunderstood. A similar effect occurs when a student listens to music during a university lecture.

Lasswell (1948) included the concept of a *message controller* in his model. This concept re-emerged in later models, although the terms *intermediary* or *gatekeeper* were used to describe this role (Foulger 2004a). A **gatekeeper** is located between the *source* and the *destination* and occupies a position of power. The gatekeeper may change or even stop a message before it is transmitted. Gatekeepers include editors, lawyers, reviewers, public relations consultants, and TV directors and producers. For example, during a radio talkback session, a delay of at least four seconds is built into the transmission process; this allows the producer to intervene and terminate a conversation if it goes 'off-topic'.

An engineering manager who has the responsibility of checking and approving the release of documents, including plans, to clients or external organisations is a gatekeeper. This is a critical role in any engineering organisation as the gatekeeper is responsible for ensuring that the information in a message is accurate and correct, and meets the company's legal requirements and quality standards.

Of course, the increasing adoption of AI may mean that the role of the gatekeeper is more complex as it involves checking the source of the artifact. Another possibility is that the role of the gatekeeper may in future be undertaken by an algorithm, particularly in social media contexts.

In the two models we have reviewed, communication was shown as a one-way process when in fact communication is normally a two-way process; for example, a conversation between two people. As the conversation continues, the roles of participants change from source to receiver and back again. This two-way process enables a source to verify if the communication elicited the desired response from the receiver; that is, whether the purpose of the communication was achieved. The second phase of this two-way process is often called a feedback loop.

A contemporary model

Foulger (2004b) captured the key concepts of earlier models, including the feedback loop, in his ecological model of the communication process, reproduced in figure 7.2. The model also incorporated some new concepts.

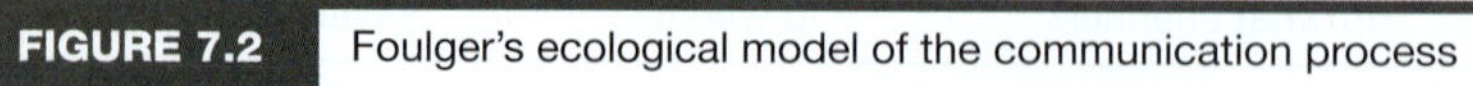

FIGURE 7.2 Foulger's ecological model of the communication process

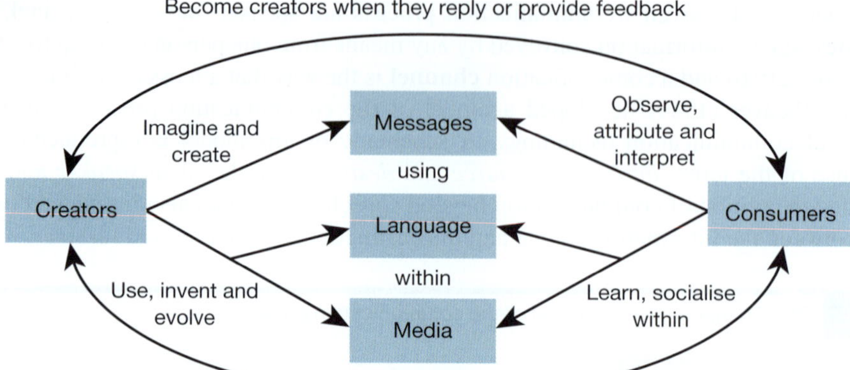

Source: Adapted from Foulger (2004b, p. 7).

First, Foulger uses the term **creator** rather than *sender*, because people use their creativity when they develop a message. They also use different media and a range of technologies to create their messages, often creating new types of messages. For example, many of the messages communicated today are mediated by one or more forms of technology, both hardware and software; for example, email, Facebook, Skype, Snapchat, Teams, X and Zoom.

Second, messages are created and consumed using a **language** that is within the context of media. For example, digital messaging often involves a new language that has evolved with mobile phone technologies. In some ways these new languages could be regarded as codes, as only those who regularly use them are able to understand the content of the text messages.

Third, receivers in a communication context are called **consumers**. This seemingly subtle change in terminology highlights the fact that the receiver can *choose* whether to access a message or not. This is emphasised in figure 7.2 by the direction of the arrows, which go from the consumer to the message, unlike those in the other models we have looked at. This is an important concept as it highlights the *power* that the consumer has in the communication process. It may also explain why some of our messages are not accessed, resulting in unsuccessful communications. Remember, the effectiveness of a communication can only be judged by the impact it has on the consumer (Burke 1969). For a stand-up comedian, the first gag is critical. If it does not engage the audience (the consumers), they may switch off and become passive or disruptive by showing their displeasure — not the desired response!

Fourth, people create and interpret messages within the *contexts* of their perspectives, such as culture, language, relationships and values.

In Foulger's ecological model, the creator's perspectives and relationships influence the way a message is coded and assembled, in the same way the consumer's perspectives, relationships and observations will influence the way a message is interpreted. Thus, it is likely that there are two versions of a message: (1) the message the creator believes was transmitted; and (2) the version that was interpreted by the consumer. For example, a person who creates and sends an email may be in a stressful environment and may be tired, frustrated and even angry. The email that is sent may be influenced by those contexts and may be fired off before the content, format and style are checked. When it is accessed hours, or even days, later, the consumer, who may be in an entirely different context, may misinterpret the message, and even take offence at the language and tone of the email. The creator may, after a period of reflection, regret having sent the email.

It is important to note here that digital records, such as emails, messages, photos and videos, are able to be legally accessed by the police, lawyers and court officials and are routinely used as evidence in court cases, Royal Commissions and other enquiries. Over recent years there have been numerous cases where personal information has been leaked to social media or traditional media, causing much distress to those involved. Authors and creators should therefore carefully consider the possible risks of publishing digital information before pressing upload buttons!

Reflecting on the communication models explored in this section, many of the important concepts included in each model were adopted and, in some cases, adapted in later models. Some of the differences in the models occur because the models show the fundamental concepts of the communication process from the perspective of the researcher's discipline.

It is also interesting to note that in most instances the creators of communication models have relied on both words and drawings to communicate the components of their model, demonstrating the use of more than one channel to convey their message. Whether this was intentional or not, such an approach caters for the different learning styles of the reader (consumer) and encourages them to engage with the document and choose to receive the message package. Before moving to the next section, it is important to note the key concepts and components of the communication process that were defined in this section: purpose, connection, creator, message, gatekeeper, channel, noise, contexts, consumer and response.

Developing a communication model for engineers

In this section, the key concepts and components from the previous models are synthesised to develop a model that can be used by engineers to plan and analyse a communication. The components of the model are briefly discussed while the model is being developed in this section. Once the model has been completely developed each component will be discussed in detail. The first step in developing the model is to create a one-way communication process.

A one-way communication model

Figure 7.3 shows a simple model of a uni-directional communication that includes the key concepts and components. This simple model will help you to consolidate your understanding of the components of a communication process.

FIGURE 7.3 A model of a one-way communication process

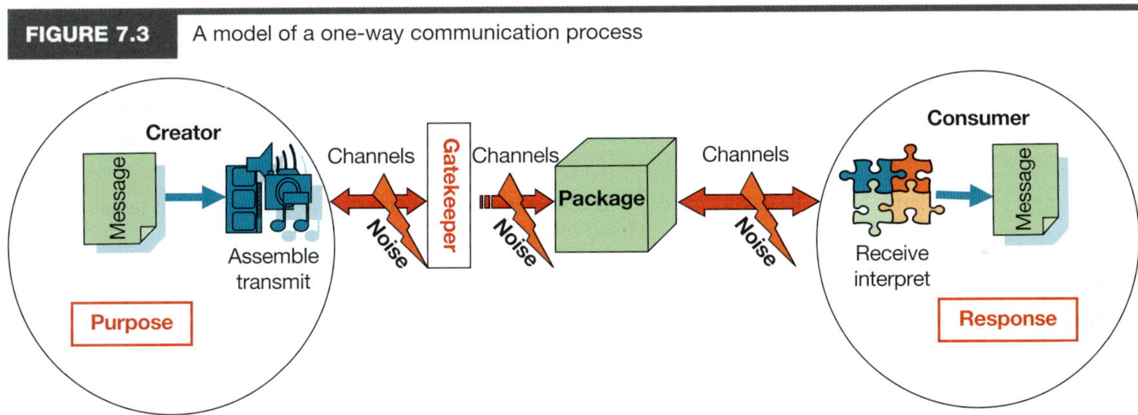

The terms *assemble* and *transmit* have been included to indicate that communication may consist of a number of messages that are assembled to create a **communication package**, which is then transmitted using one or more channels. When someone speaks, they use their voice to transmit words, while the pitch, tone and volume of their voice add other levels of meaning to those words and the message. They may also use facial expressions and body language to transmit other components of the message. Thus, in this case, the communication package consists of the message in the words; the message conveyed by the pitch, tone and volume of their voice; and the non-verbal messages.

The non-verbal components of the message may be assembled either consciously or unconsciously. For example, the person may deliberately frown to convey the seriousness of a message, or they may be unaware that they are frowning.

For the same reason, the words *receive* and *interpret* have been included at the consumer end of the model. This is because a consumer may receive different components of the message from different channels. These components have to be accessed, interpreted and synthesised to recreate the message(s). Therefore, the aim of the creator is to ensure that all of the elements of a message are consistent; that is, when the consumer interprets the non-verbal and other elements they reinforce the main message. The alternative is that the consumer receives mixed messages.

A contemporary example of mixed messages arises when a person demonstrates their interest by logging into a groupware-mediated meeting and participating in the conversation. However, through their online behaviour they may be demonstrating a lack of interest in the meeting and the contributions of the other participants. For example, they may look away from the screen, or be obviously undertaking other tasks such as responding to emails.

In table 7.2 this model is used to study an exchange of information between a mechanical engineer and a chemical engineer who are working on a project to build a new coal seam gas extraction plant in a remote location. Jim, a mechanical engineer, and Kavita, a chemical engineer, work for different organisations.

Jim is responsible for the design of a major filtration vessel. His work is supervised by Bill, who has the gatekeeper role for Jim's work.

TABLE 7.2 Analysis of a one-way communication between engineers

Component	Who	Action
Purpose	Jim	Aim of the message: to obtain technical information (the desired response) from Kavita, the consumer
Creator	Jim	Creates the messages
Message	Jim	Advice required on the corrosive impact of any gas impurities on the materials that seem to be the best options for lining the filtration vessels Include tables showing the properties of the proposed materials Asks about any treatments or other materials that may prolong the life of the linings
Assemble	Jim	Part 1: Text on project letterhead: *Dear Kavita. Please provide advice on the corrosive impact of any gas impurities on the materials we believe may be suitable for lining the vessel (see attached specification sheets). I would like this information by next Friday, 10th July, as we hope to begin the detailed design process the following week. Please contact me when you receive this letter and if you require any other information. Regards, Jim* Part 2: Material specification tables: Downloaded from supplier's electronic catalogue with proposed materials highlighted. Additional comments added in red ink
Channel	Jim	Transmits the communication package by faxing the draft letter and tables to his supervisor Bill, at head office, for approval
Noise		One of the pages is distorted and has to be resent
Gatekeeper	Bill	Accesses the message and suggests an additional material type and then approves the communication
Creator	Jim	Edits the letter and includes the additional material specification sheet
Channel	Jim	Posts the package in the internal mail system
Consumer	Kavita	Receives the package Exercises consumer's choice by opening the package and notes the package is from Jim. Because Kavita regards Jim as a domineering person, she decides not to read it for a couple of days and puts it in her Inward Mail tray
Interpret	Kavita	Reads the letter two days later and reviews the information
Response	Kavita	Kavita is not impressed with the amount of work Jim has given her; decides there is no option but to complete the task

The communication was mediated by a gatekeeper, affected by noise and then delayed because the consumer exercised her right to choose if, and when, to access the communication package. When Kavita did choose to access and respond to the communication, it was out of a sense of duty rather than as a result of being engaged by the communication. Thus, although Jim's communication was received by Kavita, it failed to make a good connection between them because of their poor relationship. Maxwell (2010, p. 44) believes that 'connecting begins when the other person feels valued'.

A simple interactive model

The next stage of the development of the model is the addition of the feedback loop to the one-way model. This is shown in figure 7.4. Note that in the feedback loop the roles of the people are now reversed and that the components are the same as the components in the initial communication. This type of two-way communication model is often called an **interactive communication model**.

FIGURE 7.4 A simple interactive model of the communication process

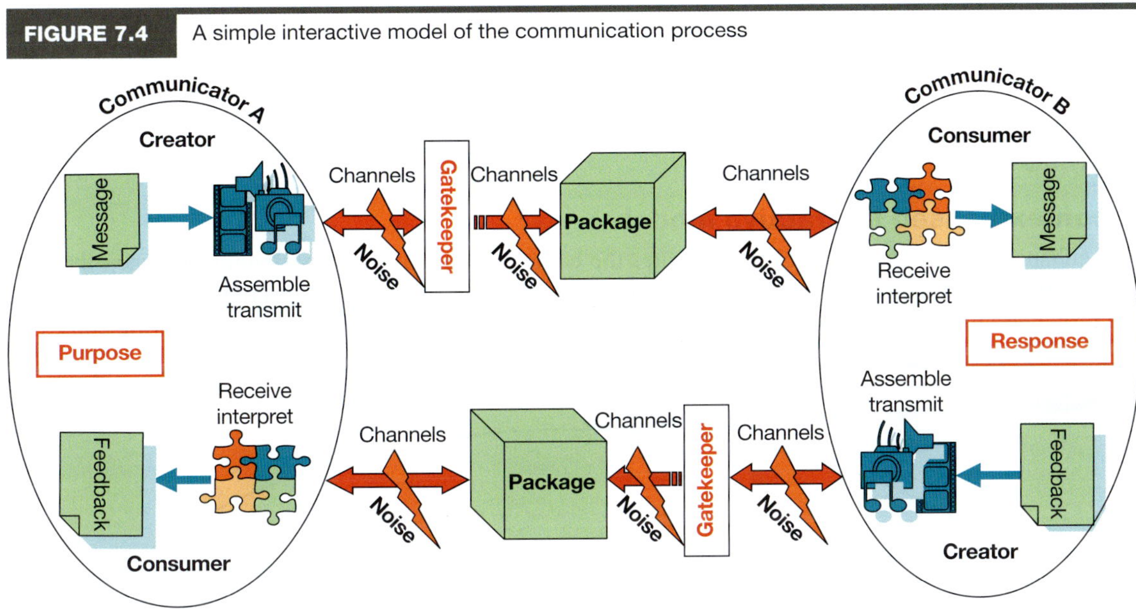

In this model, the people are called Communicator A and Communicator B to avoid confusion. Each person has two roles — creator and consumer. This model is used in table 7.3 to follow the feedback that Kavita gave to Jim.

TABLE 7.3 Analysis of a two-way (interactive) communication between engineers

Component	Who	Action
Purpose	Kavita	To advise Jim she will undertake the task but cannot meet the due date
Creator	Kavita	Message: advise Jim the request has been received, that I will prepare a report and it will be ready a week after his deadline
Assemble	Kavita	Part 1: text: *Jim. I will investigate the impact the gas impurities may have on the materials that you have suggested, and two other materials that I think may be suitable. I have attached information on those materials. My report will include recommendations and other relevant information. I will also investigate any treatments that could be used to extend the life of these materials. Due to other priorities, I will not be able to get the report to you until Friday 17th. Regards, Kavita.* Part 2: PDF: Kavita scanned the specification sheets for the two new materials
Channel	Kavita	Voice: Kavita discussed the proposed response with Helen, her supervisor
Gatekeeper	Helen	Helen approved the proposal by signing the draft
Channel	Kavita	Message emailed. Kavita added her name to the blind copy line so she would have a copy of the email for her records
Consumer	Jim	Accesses the email and, noting it is from Kavita, decides to read it immediately
Interpret	Jim	Reads the text and then prints and files the PDF document. Jim notes that the report will be a week late. He reviews the specifications for the two new materials that Kavita suggested
Response	Jim	Decides he can wait an extra week for the information, particularly as one of the new materials looks promising. He decides to email Kavita to tell her he is OK with her schedule

It was noted earlier that communication occurs in specific contexts and environments (Foulger 2004a). The communication between Jim and Kavita outlined in tables 7.2 and 7.3 occurred in the following contexts and environments.

- The language context was English.
- The media and channel contexts were voice and writing (print, facsimile, post, PDF and email).
- The work context was a large project.
- There were two organisational contexts.
- There were two discipline contexts.
- The social contexts involved relationships, perspectives and gender.
- There were natural, constructed and other environments associated with the remote site.

The final stage of the development of the communication model is the integration of these contexts into the model.

A contextualised interactive model

The enhanced model in figure 7.5 incorporates the contexts that may influence engineers when they are communicating with other people. Thus, the model is not a universal model of the communication process, but provides a useful framework for planning and understanding communication in engineering contexts.

FIGURE 7.5 The PCR model

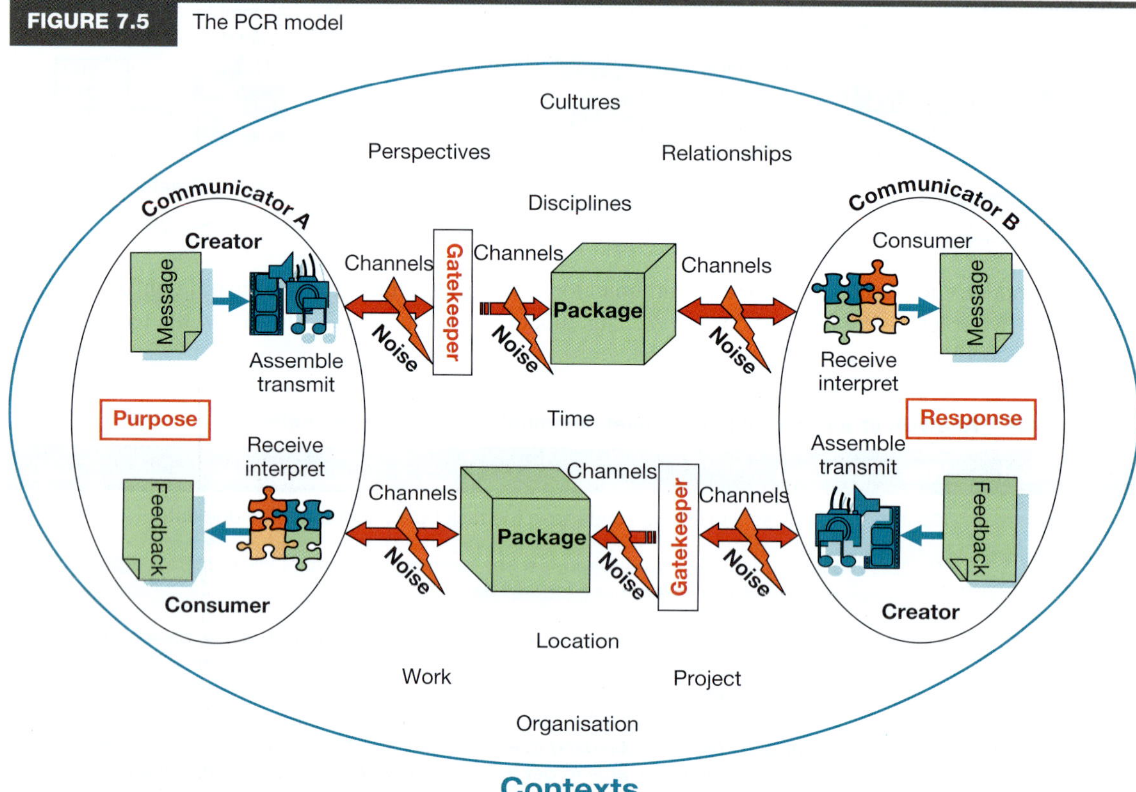

This model is called the PCR model, PCR being an acronym for Purpose — Communication — Response. This name emphasises the fact that the communication links the purpose with the response. There are *ten* components in the PCR model, with the eight components of the communication sandwiched between the purpose and the response. The following list of eight 'C' words will help you remember those eight components: creator, communication package, colleague (gatekeeper), channel, corruption (noise), contexts, connection and consumer. These are represented by the 'C' in PCR.

The PCR model has been developed to help you understand the communication process, and to help you plan communications. A communication can be as simple as a conversation between two people and as complex as the final report for a multimillion dollar project, which may consist of many volumes including text, design plans, databases and multimedia presentations. Therefore, the time required to plan a communication is very much dependent on the size and complexity of that communication.

Although the communication process is often far more complex than any model can depict, a model can be used to gain an insight into the most important components and processes involved in communication. To do that, it is necessary to have a sound understanding of the model.

In this model, a *communicator* may be an individual, a group or an organisation. The different contexts in which an individual, group or organisation may communicate include the following.

1. *Individual to individual.* An example of individual to individual communication is a conversation between two people, whether this is face-to-face or technology mediated. This is a synchronous, interactive communication process. Jim and Kavita's communications are examples of individual to individual, asynchronous communication, because of the time delays between the various messages.
2. *Individual to group.* An example of individual to group communication is when a lecturer gives an assignment to a group of students. If the resulting report is a team report, rather than a set of individual reports, this is a communication between a group and an individual. Another example is the communication between an autonomous underwater vehicle and a swarm of other underwater vehicles.
3. *Individual to organisation.* An example of individual to organisation communication is when an individual writes a letter to an electricity company complaining about a bill. When the electricity company writes back, the communication is written by a person on behalf of the electricity company and so the communication is between an organisation and an individual.
4. *Organisation to organisation.* An example of organisation to organisation communication is when a representative from the government-operated Office of Fair Trade writes to the representative of a petroleum company about price fixing.

It should also be noted that a communicator may seek assistance from others in developing different components of a communication package, for example, a drawing, or a table of statistics. In this case, the communicator would still be regarded as an individual, because of the control they have over the final message. Of course, there may be regular communication between the communicator and the people who are assisting in assembling the various components of the package.

The components of the PCR model, and the communication process, are explored in the following sections. First, the influence that personal and environmental contexts may have on a communication are discussed in the next three sections. Then the penultimate section explores two key components of the PCR model — channels and noise; while the final section looks at the roles people play in the communication process.

7.3 Communication contexts

LEARNING OBJECTIVE 7.3 Describe how contextual factors can influence the effectiveness of communication.

The PCR model depicted in figure 7.5 highlights the fact that engineers communicate within, and are influenced by, a range of **contexts**. Contexts are the personal, organisational and work environments in which a communication occurs. These include: the characteristics of the communicators, such as their perspectives, relationships, cultures and disciplines; time-related factors; and environmental factors, such as their organisation, the characteristics of their workplace and their geographical location. In some circumstances, the people involved in a communication may be situated in similar environments; in other cases, the environments may be quite different. For example, an engineer located in the head office of a mining company in Melbourne is in a very different context to that of the mining engineer she is communicating with at a mine in the Northern Territory.

KEY POINT

Engineers should understand the variety of contexts in which they may need to communicate.

Characteristics of communicators

In a one-way communication model, people may undertake one of three roles during a communication: (1) the creator, (2) the gatekeeper or (3) the consumer. In interactive communication, people are often creators and consumers at the same time, although generally they do not appear to be undertaking these roles at the same time. For example, during a conversation, people continuously swap roles, from speaker to listener, from listener to speaker, and so on. However, while they are listening, in their mind they would normally be creating a response to what they are hearing.

The personal characteristics of the people involved in a communication influence the creation, transmission and interpretation of messages; therefore, these personal characteristics should be considered when a communication is being planned, as they may influence the choice of methods and channels that will be used to communicate the message. Berlo (1960, p. 72) listed five personal characteristics that may influence the creation and interpretation of a message: communication skills, attitudes, knowledge, social system and culture.

The following spotlight highlights the subtle way that one context influenced communication in a company, and the impact that it had on the business.

SPOTLIGHT

Releasing Fletcher Aluminium's invisible handbrake

Communication between management and the engineering and manufacturing teams at Fletcher Aluminium, New Zealand, improved dramatically over the period from 2006 to 2008. This resulted in greater productivity and improved workplace health and safety (Tatham 2008). This outcome was achieved when management acted after recognising the impact the lack of communication skills was having on its business.

Although management knew that English was a second language for 85 percent of its 150 manufacturing staff, the impact this had on its operations was hidden. Human resource manager Warwick Milbank knew that it was difficult to get some members of staff to contribute at meetings, but he was not aware of the poor comprehension skills of many of his staff. He outlined how speaking to someone and having them say 'yes' doesn't mean that they have understood you and how if someone isn't understanding you then they are most likely to switch off (Tatham 2008).

Many of the staff spoke their own languages on the job, which, while good for staff engagement, meant that management and other staff were often unaware of the issues affecting production. These problems surfaced when the company introduced the concept of self-managing teams, where the teams make their own decisions rather than being told what to do. 'We couldn't go down that track unless we had a higher level of communication skills and understanding', Milbank said (Tatham 2008).

A training consultant was engaged to assess the English and comprehension skills of the staff and develop a program. Milbank knew that promoting the program as a literacy skills course ran the risk of insulting those who needed it most, including one employee who had a university degree from his home country but struggled with English. To combat this, the program was presented as a communication course (Tatham 2008).

The program lasted 48 weeks, with staff doing some of the learning in their own time. There were no complaints about the program because the employees recognised that the benefits of the program would flow through to other aspects of their lives. The popularity of the course was demonstrated by the fact that when they called for volunteers for the 20 places in the course, 40 applied (Tatham 2008).

The program began with a goal-setting exercise, including an outline of the company's goals. It then covered communications styles (including assertiveness), listening skills, questioning techniques, clarifying and paraphrasing. Because many of the staff had a fear of computers, the program also covered basic computing skills, such as familiarity with the keyboard, and writing and sending emails. Most of the sessions were conducted one-on-one (Tatham 2008).

Milbank said 'The difference has been massive. People are now challenging the status quo; they are contributing at meetings and they feel more confident at work'. Productivity improved along with workplace health and safety — the company had been accident-free for two years at the time.

Some of the feedback from the participants included how the course had made a huge impact on their lives, that they appreciated the opportunity provided by the company, that the course introduced them to using a computer for the first time and encouraged them to speak up as well as listen properly to others (Tatham 2008).

CRITICAL THINKING

It is likely that there are people in your classes with English as their second language (ESL) — you may be one of them. Take the time to meet with three or four students from different countries and discuss the difficulties ESL students have in understanding the communications they participate in while at university, such as lectures, tutorials, team projects and administration activities. Reflect on the impacts this has on their learning and their ability to participate effectively in class and team activities. What strategies have they used to overcome these impacts? Find out what services are available at your university to assist ESL students.

The spotlight highlights the importance of ensuring you connect and engage with the people you are communicating with, particularly when you are working with people from other cultures and language groups. This is likely to occur when working in the engineering industry as it includes many people who came to Australia as skilled migrants, or on 457/Temporary Work or similar visas, and who work at all levels of the industry. It also reinforces the need for engineers to develop and effectively use 360-degeree collaboration capabilities, which were discussed in the previous chapter.

Communication skills

Sound communication skills are required to effectively prepare and communicate a message, particularly a complex message. The creator should be aware of their own communication abilities, and those of the consumer, so that the most effective method is selected to communicate each key point. During a conversation, communication happens instantaneously because each person is constantly planning their messages during the conversation. However, for more important conversations, the creator should write down and rehearse the words they will use, and how they will be spoken, prior to a meeting or a telephone call. One advantage of verbally communicating a message is that the consumer will generally provide a response (verbal or non-verbal) and, if necessary, the message can be repeated using the same or different words to facilitate the desired response.

Normally, people have more time to plan a written communication, and they should use that time to carefully create the message. This is because message creators may not get a second chance. Remember, to achieve its purpose the written message must engage the consumer and create the desired response. Creators should also develop a strategy to ensure they get feedback from the consumer so they know if the message elicited the desired response.

It is important to note here that, unlike verbal communications (unless a conversation has been recorded), a written message 'lives on' after a communication has occurred. It can be filed and revisited at a later date by both parties. There are important implications for a message creator when their work correspondence is filed. First, any factual errors live on and may, at a later date, be used as evidence in a dispute. This reinforces the importance of carefully checking a communication before it is sent, and highlights the role of the gatekeeper. Second, the document may be used by others for unintended purposes, or leaked to the media. A good example was the unofficial publication (leaking) of some key emails sent between Seqwater staff managing the Wivenhoe Dam in the period leading up to the 2011 Queensland floods. The content of these company emails was printed on the front page of *The Australian* on Friday 21 January 2011, just eight days after the flood. (The chapter on self-management contains more details about the flood.)

Attitudes

The effectiveness of communication is influenced by the attitudes of both the creator and the consumer, including their self-attitude, their relationship, their attitude towards each other and their attitude towards the subject of the communication. Their attitudes may also be influenced by their cultures, perspectives and values.

Attitude can influence speech, facial expression, body language and dress. All of these non-verbal communication methods can send important messages to consumers. For example, people often use facial expressions and gestures to emphasise what they are saying. Other people, such as poker players or professional negotiators, may try to control their non-verbal communication so they do not send any messages, or send a false message!

Knowledge

The creator's knowledge of the subject of the communication, and their understanding of the consumer's knowledge of the subject, are major influences on the content of a message and the way it is assembled. If the creator knows little about a subject, the communication is unlikely to be effective. The same outcome is likely to occur if the consumer does not have sufficient knowledge about the subject to correctly interpret a message. This is a key consideration for engineers and other professionals when they are communicating with members of the public or with people from other professions. For example, telecommunications engineers may be required to hold a community consultation meeting when they plan to build a telecommunication tower. They would need to plan their communication carefully to ensure they use language and visual and non-verbal communication methods that will be understood by members of the public.

Social and cultural contexts

The identification of Australia and New Zealand as multicultural societies has highlighted the notion of cultural difference. Culture pervades what people do, the language they use, the way they talk and the way they relate to others. Cultural differences can have a positive impact in the workplace as people from different backgrounds and cultures bring different educational experiences and engineering expertise to an organisation; however, cultural differences can also have negative impacts if they are not carefully managed. This was highlighted in the Fletcher Aluminium spotlight where a nuanced training solution was used to minimise the impact of language differences in the workplace.

The social and cultural background of the creator can influence the language and format of the message and the ability of the consumer to access and interpret the message. Language is an obvious example of cultural difference, but other cultural differences, such as lifestyle, workplace culture, behaviour and personal values may have less obvious, but just as important, effects on communication.

Cultural differences often cause communication problems when non-verbal communication methods are used, such as nodding the head to communicate agreement; this may have no meaning for one culture, or different meanings in other cultures. Another difference that can cause problems is time. In some countries, such as Australia, time is important and therefore it is considered impolite to arrive late for a business meeting; in fact, it may have an impact on the outcome of the meeting. In other cultures, time is not as important and people are more spontaneous with how they use time. This may be very frustrating for someone from a different culture who 'just wants to get the job done'.

The subtleties of a language can also impact on communication. For example, the grammar of some languages does not include a double negative construct like English. This can mean a person new to the language can badly misinterpret a statement. For example, if you were to say to a member of your team, who was not a native English speaker, 'You did not find the old boundary marker?', they might answer 'Yes', meaning to affirm your words that they did not find it. In standard English the answer would more likely be 'No', meaning no, they did not find it.

From a communication perspective, a key principle to guide cross-cultural communication is that knowledge of the cultural contexts should be used to inform the development of a communication. Care should be taken to ensure that the verbal, non-verbal and visual communication methods used do not lead to a misunderstanding or cause offence. These outcomes can build barriers between the communicators that may lead to a communication breakdown, which in turn may mean a contract is not negotiated, or the specified outcome for a project is not achieved.

Another key principle is that checking the consumer's response is vital to the success of a communication. Every opportunity should be taken to check that the consumer understands the details of the communication and that no cultural barriers are interfering with the communication. The importance of cultural understanding in engineering work was covered in the chapter on professional responsibility and ethics.

Discipline contexts

Over time, each profession develops a knowledge base with a unique language and its own values — in effect, a distinctive culture. Practitioners learn to communicate with their colleagues in the languages and styles of the discipline in which they practise. Therefore, their communications may not be accessible to people from outside the discipline because they are often filled with jargon and acronyms. Thus, discipline cultures can have an impact on the effectiveness of communication with people outside the discipline. In this respect, they are similar to any other type of cross-cultural communication. For example, communication between an engineer and a lawyer crosses a cultural boundary. Great care needs to be taken when communicating across discipline boundaries to ensure that the consumer is able to access and correctly interpret the message.

The previous sections have highlighted the importance of understanding the impact that personal and professional characteristics and competencies can have on the effectiveness of a communication in achieving its purpose. The following spotlight introduces the concepts of personal and professional brands.

SPOTLIGHT

Developing, managing and communicating our brand

The concept of managing a professional profile has been around for many years and is one of the Indicators of Attainment for the Stage 1 Element of Competency 3.5: 'Orderly management of self, and professional conduct' (Engineers Australia 2019, p. 6). The relevant Indicator of Attainment is: 'Presents a professional image in all circumstances, including relations with clients, stakeholders, as well as with other professional and technical colleagues across wide ranging disciplines' (Engineers Australia 2019, p. 6).

Therefore, our image and reputation are important aspects of being a professional engineer, as they may affect the way colleagues, clients and members of the public interact with us. A poor reputation can have a serious impact on an engineering consultant's business, and even the profession as a whole.

Contemporary career advisers and marketing gurus often use the phrase 'professional brand' instead of professional image or reputation. They may also use the phrase 'personal brand' when referring to someone's personal image, which is often based on the way they dress, their behaviour and the way they interact with other people.

In the past, it has been possible to keep these two brands separate; however, with the explosion of social media it is much more difficult to keep our personal and professional profiles separate. This is because it is so easy for people to search the web and find information about other people. For example, employers can do a reality check by searching social networking and other sites to find out if the details provided by a job applicant are genuine. Of the employers who responded to the 2015 Graduate Careers Australia survey over one-third conducted qualification checks (Graduate Careers Australia 2016, p. 16).

For you, this means that as well as planning and managing your career, you will also have to develop, manage and communicate your personal and professional brand. David Ogilvy, founder of Ogilvy & Mather, defined a brand as 'The intangible sum of a product's attributes: its name, packaging, and price; its history and reputation; and the way it is advertised' (in Dvorak 2010, p. 10). In the context of this chapter, it is important to consider our image and how it is affected by the way we act, communicate, dress and interact. You may like to reflect on the following attributes.

- *Name*. Your signature is part of your brand, so you should use a signature that is neat and stylish. Choose an appropriate name as part of your email address — lagerlover@hotmail.com does not portray an appropriate image.
- *Packaging*. Thompson (2011) advises that we only get one chance to make a first impression. 'The way you dress is not just a matter of taste — it plays a role in your workplace survival. Your overall appearance not only affects the way you think, feel and act — it also influences how other people react to you' (Nelson 2011, p. 13).
- *Price*. Your personal attributes and professional capabilities will determine the salary you receive or the fees you can charge.
- *History*. A well-written CV will provide potential employers with a good overview of your history. Hopefully it will lead to an interview where you can tell them more of your story.
- *Reputation*. An old but accurate saying is, 'Your reputation precedes you'. Your reputation is based on your professional conduct and your personal attributes, such as honesty, integrity, initiative and punctuality.
- *Advertising*. Your presence on the internet, particularly on professional and social networking sites, should be managed carefully, including ensuring that appropriate privacy settings are in place. Graduate Careers Australia (2016, p. 16) reported that, overall, 12.5 per cent of the employers who recruited graduates in 2015 indicated that they had viewed an applicant's social media profile as part of the recruitment process, although this varied between disciplines. The main platforms they checked were Facebook (87.0 per cent), LinkedIn (78.3 per cent), Twitter (now known as X) (32.6 per cent) and Instagram (21.7 per cent). The main reasons employers check candidates' social media were to:
  - give more insights into their personality
  - confirm what they have provided in their CV/interview
  - see if they can fit into an organisation
  - gauge their professionalism
  - do a background/reference check (Graduate Careers Australia 2016, p. 17).

> Graduates were asked, assuming recruiters might check their social media profiles, what changes they made to address this possibility. Almost half (47.1 per cent) indicated that they used ways to improve the security or privacy of accounts so that potential employers could not see anything that the graduate did not wish them to see. Another 45.1 per cent chose to be more cautious about what they posted and with whom they publicly associated online (Graduate Careers Australia 2016, p. 38).

CRITICAL THINKING

Reflect on both your personal and emerging professional brands using the points discussed in this spotlight. You may use your role in part-time work, or over a vacation, as your emerging professional brand. How do they compare?

While you are at university, you will have plenty of opportunities to develop your brand. Make sure you take every opportunity to network with staff and employers, particularly when you are undertaking vacation work in your chosen field. You should also consider joining the Young Engineers group of Engineers Australia. This is all part of developing and communicating your brand.

Environments

Engineers communicate within and across a range of environments, all of which may influence the type and structure of messages, as well as the channels and languages they use to communicate. The following spotlight highlights the impact that context and environment can have on engineering activities and the way engineers communicate. Lori Sowa has first-hand experience of the impact contexts and environments can have on communication, as she explains in the following spotlight.

SPOTLIGHT

Environmental engineering in Alaska

LORI SOWA, UNIVERSITY OF ALASKA SOUTHEAST

I grew up and studied environmental engineering in the eastern states of the US. When I moved to Alaska I found out I was from 'Outside', a term Alaskans used to refer to anywhere that is not in Alaska. The move to Alaska also involved cultural and environmental shifts, and I had a lot to learn about the culture and lifestyle of the Alaskan people, which is unique, and still intimately connected to the land, resources and weather.

Water sampling in Alaska

Consider the following facts about my working environment. Alaska is larger than most countries and has over 33 000 miles of coastline, more than three million lakes, over 12 000 rivers, thousands of streams and creeks, and an estimated 100 000 glaciers. The weather in Alaska is extreme — varying from 4.8 inches of rain per year in Barrow on the Northern slopes to 220 inches in Little Port Walter in the south-east, with temperatures that range from temperate to very, very cold. Living and working in Alaska is different from any other place, to say the least. For example, Juneau, the capital city of Alaska, is not connected by road to any other community!

How do these cultural and natural environments affect my work as an environmental engineer? In my studies I learned about how chemicals move through the environment; how important site considerations (such as soil type, hydrology and water chemistry) are to site assessments; and how it is important to understand the environmental regulations that dictate clean up levels and treatment system design. When I began work I realised that actually *applying* these theories to real problems is always the most challenging (and interesting!) part of practising engineering. However, applying them in an Alaskan setting requires not only an understanding of the specific physical differences in the environment, but also the differences in communication, transportation and the available resources.

Many places I visit are only accessible by boat, small aircraft, helicopter or snow machines. Therefore, finding a solution to the water treatment problem involves not just finding the appropriate system, but a treatment system: that can be broken down and transported to the site by small aircraft; that can continue to function at temperatures below freezing for up to half of the year; that can be powered by an off-grid power supply (that I have to select components for); and, finally, that requires little maintenance.

The routine testing of those treatment systems can be another challenge altogether. Take drinking water and wastewater treatment systems as an example. These systems must be tested periodically for various parameters to ensure they are functioning properly, and that human health and the environment are protected. Typically, these samples are collected and, to ensure accuracy of the analysis, quickly taken by car to a laboratory so the analysis can be started within the prescribed holding time. In south-east Alaska, I've learned this is not so simple. With no roads connecting the communities and because the laboratories certified to conduct the testing are only found in the larger towns, transporting samples means putting them on small aircraft. Even this can pose problems when there is only one flight per day. And, during the rainy, windy, foggy Autumn weather, it is not uncommon to be 'weathered out' of an area for a week. Try meeting a six-hour holding time, typical for bacterial enumeration, in those conditions!

So, what is my advice for you when you are faced with the challenge of working in environmental and cultural contexts that are so different from what you know? First of all, sit back and *listen* to the people who live and work there. Even with your completed engineering degree and years of experience, you still have a lot to learn. Don't be afraid to ask questions, be flexible, and don't get stuck thinking about all of the things that can't be done the same way you've done it before. Instead, relish the challenge of finding a new, unique way to make it work.

CRITICAL THINKING

Compare the environments that Lori communicates in with those that a power engineer may encounter when visiting a remote Australian First Nations community to commission a new solar energy power system. Use the PCR model to guide your comparison and to structure your response.

The key lesson from this spotlight is to 'Look, listen and learn' when working in new contexts and environments because the success of a communication may depend on the strategies employed to minimise their impact. In this case the ecological, geographic, organisational and project environments could be described as difficult or even extreme. Lori's story also highlights the fact that the units of measurement engineers use vary from country to country. Although the same language might be used, many of the words and phrases may be different from those used in Australia or New Zealand.

This spotlight and the Fletcher Aluminium spotlight illustrate how subtle the impact of external factors can be, and how easy it is to overlook them. This is why it is important to use a tool like the PCR model to ensure that all of the factors are considered when an important communication is being prepared.

Some environments that may have an impact on communication are discussed in the following sections.

Time

Time is a major factor in the planning and creation of a communication. There is little point in preparing a comprehensive report if project deadlines do not allow sufficient time for the gatekeeper and consumer to properly review and respond to the document. Consideration should be given to the time:

- required to create and assemble the message package
- required for the message package to be reviewed by the gatekeeper
- it will take to forward the message package to the consumer
- it will take for the consumer to unpack, interpret and review the communication
- it will take for the consumer to prepare and forward a considered response.

When all of these aspects have been considered as part of the planning process, the resulting communication is more likely to be effective, particularly in meeting project timelines.

Organisational environments

Engineering organisations normally have communication policies and protocols that define how both internal and external communications should be undertaken, managed and filed. In small organisations the protocols are likely to be relatively simple and flexible, and may represent the preferences of the manager or directors. In large global organisations, the communication policies and practices are likely to be complex and more rigid. This ensures a consistent approach is taken, and employees are able to easily find, access and interpret the documents they require. New employees would normally receive training in an organisation's policies and practices so they can operate efficiently within the organisation and communicate effectively with their colleagues, clients, consultants, contractors and external organisations.

An organisation's communication policies and protocols therefore have a significant influence on the way the employees in the organisation communicate and may also influence the organisation's brand.

Workplace environments

An engineering organisation may have a number of different workplaces at the one location. For example, a power station may have on-site management and administration sections, design and maintenance sections, as well as the power-generating plant. Larger organisations usually have multiple sites. For example, an energy company may have power-generating plants in the regional areas of a state, and offices in cities throughout the state. They may produce and supply one form of energy, such as electrical power, from a single source, or from multiple sources (e.g. coal, wind and solar), or they may produce a range of energy types from different sources.

The environments in each of these workplaces may be quite different. For example, the people working in the workshop and maintenance areas of the power plant will be technicians, trades people, apprentices and labourers; and the cultural environment is likely to be male-dominated. This may mean that bad language and rough behaviour dominate and that there is little respect for bosses, engineers and other professionals — until they earn respect. This would be very different from the cultural environment in the site office where the managers, secretaries, administrators, engineers and other professionals work.

The communication requirements for each of these types of workplaces can be quite different and they may, over time, develop relatively independent systems to suit their purposes. This can pose significant problems when people from one workplace have to communicate with people who work for the same organisation, but at another workplace.

Geographical location

The geographic location of the communicators can influence the way they communicate. Two examples illustrate this point. The communication options at an isolated mine site may be limited, compared to those available in a company office in a capital city, and this may restrict the number of viable channels that can be used. Also, when the creator and consumer are in different countries their communication may be affected by time zone differences, technology interfaces and mail inefficiencies. In this case, the relative locations of the communicators may restrict the number of viable channels, and the channels may be subject to noise; for example, there may be mail delays or even losses.

Project environments

The people required for a project may be employed directly by a project consortium, or seconded from organisations that have formed an alliance. In either case, the project manager needs to quickly establish communication policies and protocols so all members of the team learn to use them, rather than following the procedures they used at their last place of employment. See the spotlight in the previous chapter on when alliances go wrong.

Digital communication

In this digital age, a large percentage of the communication activities that occur in the world are automated communications between digital devices. For example, river gauges and weather stations are regularly interrogated by control systems to access sensed data and status information. Increasingly, AI is being used to initiate and manage automated messaging systems.

A more sinister example is the automated interrogation of computers and mobile phones by third parties who seek information about the user, such as physical location and usage data. For example, the 2013 leaking of secret surveillance documents by Edward Snowden, a contractor with the US National Space Agency, caused embarrassment for the Australian and other governments. These leaks highlighted increasing involvement of governments and commercial enterprises in the gathering, storage and use of personal, commercial and industrial data. Unfortunately, there have been many examples in the 2020s of these databases being hacked, resulting in the sale and publication of key financial, medical and identification documents belonging to millions of people in Australia.

The PCR model can be used to design the communications and protocols that will be used for communication between these devices, although it may be difficult to identify the components in these systems. This is because these systems are developed from a synthesis of communication and telecommunications theories and principles. For example, programmers will embed the standard message set in the system software and develop the protocols that will control the creation, transmission and interpretation of those messages. However, many devices, such as weather stations, will also incorporate measured data in messages in accordance with established protocols. The following spotlight describes the use of automated communication by the global positioning system (GPS).

SPOTLIGHT

Communicating from space

The global positioning system (GPS), which is part of the global navigation satellite system (GNSS), is a good example of a system that uses automated communication. The system has three components: the satellite system, the control system and the user systems. The GPS satellite system consists of a minimum of 24 Navstar satellites that circle the Earth every 12 hours in nearly circular orbits, about 20 200 kilometres above the surface of the Earth. The satellites are continuously tracked by control stations spread around the globe — the control system. The main control station is at Schreiver Air Force Base near Colorado Springs, USA.

There are many different types of user systems, and the list continues to grow as 'navigation' chips become cheaper and smaller; for example, mobile phones, car navigation systems, sports watches, aircraft navigation systems, truck and taxi position monitoring systems, and land surveying systems. The communication between the satellite system and user systems is normally just a one way-communication, that is, satellite to user. Each satellite broadcasts three distinct types of data in a standard format every 30 seconds — a 1500-bit frame that is divided into five 300-bit sub-frames. Together, this information makes up a communication package.

Each sub-frame consists of ten words, each 30 bits long. Each sub-frame begins with a telemetry word (TLM) and a handover word (HOW). This is followed by the relevant data communicated in eight words.

- The first data type is communicated in one 300-bit sub-frame over six seconds. It consists of precise date/time information from the on-board atomic clock.
- The second data type is communicated using two 300-bit sub-frames. It consists of precise positional information about the satellite's orbit at the beginning of the transmission of the frame. A satellite's orbit is known as its ephemeris.
- The third data type is communicated using two 300-bit sub-frames. It consists of different types of system data that, together, require 25 frames to communicate. For example, the fifth sub-frame contains almanac data for all of the satellites currently in the constellation. The almanac data informs the user system about the current and future (less precise) locations of the satellites in the constellation. This enables the user system to search for satellites, and users to plan GPS-surveying activities.

A complete navigation message takes 25 frames to communicate and 12.5 minutes of time. This means that a user system (receiver) should remain in a fixed position for this length of time if the data is to be used for high-precision position fixing purposes. Generally, a user system requires simultaneous signals from at least four satellites to be able to fix a position. Data decoding information is available so that users can locate and use the data they require for their purposes.

All of the satellites use the same carrier frequency, which is generated from the atomic clocks on board each satellite. However, the signal from each satellite is able to be distinguished because they each use unique codes that are modulated onto the carrier wave. The communications between the satellite system and the control system are similar to, but more complex than, those between the satellite system and user systems. This is because it is a two-way communication system.

Many user systems have complex automated communication systems to convey position information in a form that is easily accessible by the relevant user. For example, a car navigation system synthesises this data with map data from a geographic information system (GIS) and communicates it to the user using verbal and graphical communication methods.

Further information on GPS is available from the Navigation Centre (1996) and GPS.gov (2024).

CRITICAL THINKING

Use the names of the components of the PCR model to describe how a car navigation system communicates direction information to the driver. Then, describe the actions typically taken by a car navigation system when it detects that a driver has not followed an instruction, for example, to turn left into a street.

Of course, the technology discussed in this spotlight is old technology because the GPS system was initially developed for defence purposes in the 1970s. Public access was allowed in the late 1980s and the system became fully operational in 1994 when all 24 satellites were deployed.

The increasing number of drones and other automated systems, such as self-guided agricultural systems, which are being deployed means that increasingly complex communication protocols will be required to, first, maintain contact with vehicles and, second, to maintain vehicle-to-vehicle separation and prevent accidents. To date, the majority of these deployments have been undertaken by government agencies but this is changing as more commercial and recreational users are adopting these technologies — for example, the use of drones for commercial and recreational purposes.

As with the introduction of many new technologies in the past, governments are struggling to deal with the commercial, environmental, legal, privacy and safety issues resulting from the introduction of these technologies. Perhaps this will change as governments face pressure to control the use of drones and other automated systems deployed by government agencies for activities such as surveillance, bushfire monitoring and environmental studies, and for commercial activities such as parcel delivery.

A recent innovation is the deployment of swarms of aerial or underwater automated vehicles to undertake mapping and sampling tasks, surveillance, or search and rescue activities. The following spotlight describes a research project where communication and communications theories were combined to develop a communications strategy for a hostile environment. Dr Gunilla Burrowes describes the challenges encountered during the project she completed for her PhD.

SPOTLIGHT

Swarm communication

GUNILLA BURROWES, DIRECTOR, BLUEZONE GROUP

The oceans, which cover over 70 per cent of the Earth's surface, are the new frontiers as countries seek information about the resources in these vast ecosystems, their place in the carbon system and potential impacts on climate change. Around the world there are many projects being undertaken to develop technologies that can unlock the oceans' secrets. The development of small autonomous underwater vehicles (AUVs) has been rapid over the last 20 years and there are many such systems that operate successfully as single units.

The ability to use swarms of these vehicles, like the stylised SeaVision© vehicles shown in figure 7.6, has the potential to dramatically increase data gathering and research capability. However, a robust and reliable communications system is required before a swarm of AUVs can be deployed. This is because the vehicles need to communicate with surface vessels and with each other to avoid collisions between vehicles, or with marine or constructed objects. More importantly, a good communications system will empower the swarm to operate as a team, which will result in emergent behaviours. Two key components needed to enable communications amongst a swarm of AUVs are the optimisation of the acoustic communication channel and an effective communication protocol.

FIGURE 7.6 Autonomous underwater vehicles (AUVs)

Source: Burrowes and Khan (2011).

Most of the AUV development work has concentrated on the vehicles themselves without giving much attention to the development of the swarm architecture that requires a mobile communication networking infrastructure. The aim of the project was to use a short-range acoustic communication channel model and its properties to design and evaluate the MAC (medium access control) and routing protocols for short-range ad hoc sensor networks to support network-enabled AUVs. The communication protocol for an underwater swarm communications system must provide vehicle-to-vehicle and vehicle-to-many-vehicle communications and may include communications with surface vessels.

An underwater swarm communications system uses acoustic instead of radio wave propagation. The model was designed to enable individual AUVs to operate within 10 metres of each other and out to a range of 100 metres.

A schematic of the underwater acoustic environment for a single transmitter–receiver pair is shown in figure 7.7; obviously the environment is far more complex in a swarm of vehicles that is also communicating with surface vehicles.

FIGURE 7.7 Underwater acoustic environment for a single transmitter–receiver pair

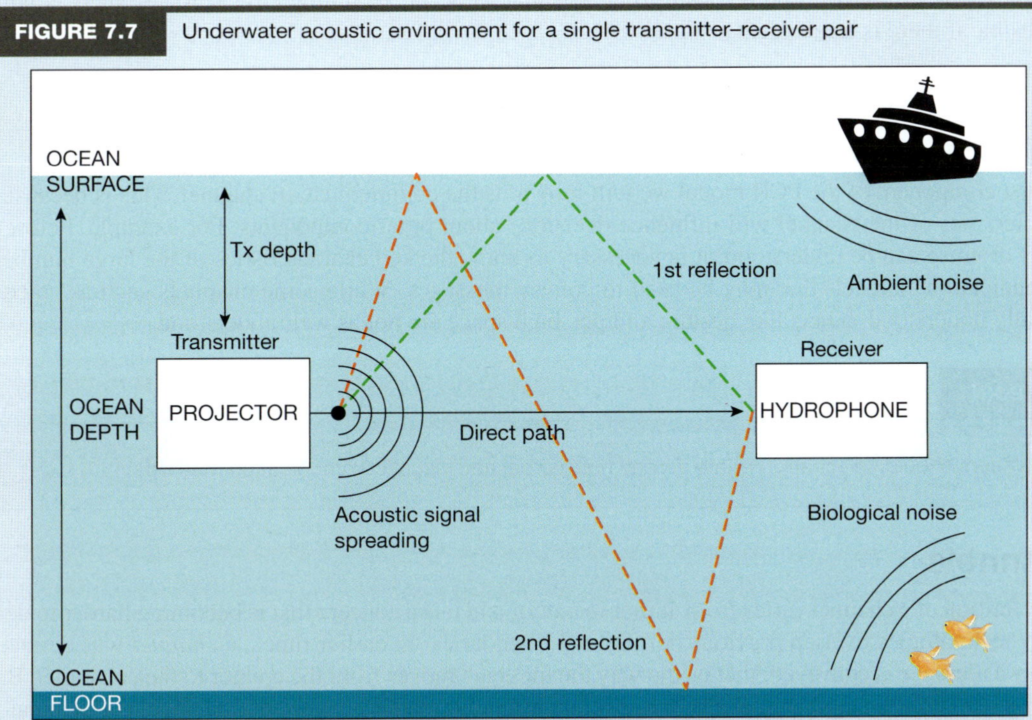

Source: Burrowes and Khan (2011).

It was found that signal strength (path loss) was dependent on a range of factors such as the signal frequency, the salinity, depth and temperature of the water, and the distance between the vehicles. It was also found that the signal frequency and channel bandwidth were limited to low KHz spectrums (30–60 kHz) and that surface wind, which affects wave action, was a significant component of noise.

Some of the challenges of a MAC-based short-range acoustic communication design are:

- the impact of high propagation delays (from the slow and variable propagation speed) require time synchronisation and localisation (mobility) functions
- the low data rates (due to both the latency and limited bandwidths) require bandwidth efficient modulation methods
- high bit error rates are caused by significant multi-path signals, shown as reflections in figure 7.7; these and fading are both range and condition dependent
- the reliance on battery power means careful use of transmitter power and the duty cycle modes may mean vehicles might include sleep cycles to reduce energy consumption.

With autonomous operations, energy consumption is critical, yet transmitter power levels are shown to significantly affect signal-to-noise ratio (SNR) and data rates. Smaller packet size was also shown to improve effective data rate performance.

With the increasing potential of AUV swarm operations there are significant benefits that can be gained by matching the very harsh acoustic communication channel conditions with the operational requirements of the swarm, and in the efficiency of underwater operations when the potential benefits of utilising the emergent properties of a team are realised. To do this the communication methods and, in particular, the design of the MAC and routing layer protocols, are critical in providing the most efficient communication system for the swarm.

CRITICAL THINKING

Carefully review the communications system for the underwater acoustic environment for a single transmitter–receiver pair and then use the ten components in the PCR communication model to identify the role of each of the components of this communication environment.

The successful deployment of automated vehicle technologies by governments, commercial organisations and members of the public will only be achieved if engineers, scientists and other researchers are able to develop appropriate protocols and regimes that will ensure the commercial, communication, environmental, legal, privacy and safety issues are resolved. One of the challenges researchers are currently grappling with is the safe introduction of self-drive cars onto public highways. You may like to reflect on the issues that will need to be resolved to enable that to occur, including ethical issues such as who is responsible if there is an accident.

7.4 Communication methods

LEARNING OBJECTIVE 7.4 Describe the communication methods commonly used by engineers.

The first component of the PCR model we will look at is the communication channel. This is because the characteristics of the channel will influence decisions about other components. For example, before the impact of noise can be understood, it is necessary to know the key characteristics of the large number of communication channels that may be used to convey messages. While some channels such as voice are generally understood, others like gesture and personal space are not as well understood.

KEY POINT

Engineers should select appropriate verbal, non-verbal and visual communication methods.

Channels

The definition of a channel varies from author to author and it is a concept that is becoming harder to define as new technologies overlap previous definitional boundaries. In earlier times, a *channel* was defined as the way a message is conveyed; that is, the way the message moves from the creator to the consumer. Berlo (1960, p. 63) used a cargo ship analogy to explain a much broader concept of a channel. In the analogy, a ship is supported by water as it carries cargo from one dock to another. The boat, the water and the two docks represent three meanings of the word 'channel' in Berlo's analogy.

Two examples show how this concept of channels can apply to a communication. First, when someone talks, their message is carried from their mouth to the listener's ear by sound waves that are supported by the air. The sound waves and the air are the channel. Second, an email is a technology-mediated communication. The message is conveyed from a keyboard to a computer screen and then to the consumer's computer screen by digital coding over the internet. The fact that the email can be printed adds another layer of complexity to this concept of a channel.

Berlo (1960) also highlighted the fact that all five senses can be used to communicate: seeing, hearing, touching, smelling and tasting. While seeing and hearing are the most common forms of accessing messages, the other senses are regularly used, although the consumer may not always be aware of it.

Some examples of the communication methods that are regularly used today are described in the next section to demonstrate the many facets of the channel concept. A number of methods could be used to classify the types of communication that are described; for example, they could be classified by behaviour, by expression, by senses, by channels, or by format. However, three categories will be used in this text to group the communication types generally used by engineers: verbal communication; non-verbal communication; and visual communication. Verbal communication includes both oral and written communication using a variety of communication languages.

Communication languages

The languages that are described in this section may be described as formal languages because they are well-defined and therefore can be formally learned; for example, English, Spanish and Mandarin. Another formal language that engineers use is the language of mathematics. Table 7.4 lists many of the formal languages, message formats and channels that are commonly used to transmit and receive messages in the workplace.

Non-verbal communication

Non-verbal communication is the term that is used to describe methods of communication that do not use words. Depending on the situation, non-verbal communication can be less important, as important, or more important than verbal communication. For example, Knapp (1992, p. 15) reported that:

> In a normal two-person conversation, the verbal components carry less than 35 per cent of the social meaning of the situation; more than 65 per cent of the social meaning is carried in the non-verbal band.

TABLE 7.4 Examples of communication languages

Language	Message format	Channels	Senses
English	Verbal	Person-to-person, meeting, debate, speech	Hearing
	Digital/electronic	Phone, radio, TV, podcast, DVD, internet, voicemail, MP3/MP4	Hearing Seeing
	Written	Paper, email, post, SMS, billboard, web	Seeing
	Coded	Braille, sign language, hand signals (e.g. aircraft docking)	Touching Seeing
Mathematics	Formula	Paper, computer	Seeing
	Graphic	Paper, computer, model	Seeing
Music	Written	Paper, computer	Seeing
	Performance	Live, DVD, radio, TV, computer, phone and MP3/MP4	Seeing Hearing

A message may be totally communicated by non-verbal means, for example a hand gesture, or may be totally communicated by words, for example in an email. The important point to note is that both verbal and non-verbal communication methods should be carefully considered when planning a communication.

Non-verbal methods of communication are not well-defined and they are learned through experience. The variety of meanings that can be conveyed when they are used may mean that people misinterpret these forms of communication. For example, our culture influences the way non-verbal communication methods such as dress, gestures and time are interpreted. It should be noted here that even people with the same cultural background may not interpret non-verbal communications in the same way. For example, people on the autism spectrum may have trouble 'reading' common cues, particularly in social contexts.

Table 7.5 lists many of the non-verbal methods of communication, and examples of the message formats, channels and senses used to transmit and receive messages using these methods.

It is interesting to note that people use only three of the senses to access the languages listed in table 7.4, but may use all of the senses to access the non-verbal communication methods listed in table 7.5. Food tops the list — just imagine a beautifully presented meal arriving at your table during a business lunch, sizzling on a hotplate. It will arouse *all* of your senses.

TABLE 7.5 Examples of non-verbal communication

Type	Expression	Channels	Senses
Behaviour	Facial	Smile, frown	Seeing
	Eyes	Narrowed, closed or averted, or a wink	Seeing
	Head	Nod, shake, tilt	Seeing
	Voice	Tone, pitch, volume, style	Hearing
	Posture	Formal, relaxed, aggressive	Seeing
	Gesture	Folded arms, pointing, hands on hips	Seeing
	Body movement	Facing, turned away, shaking	Seeing
	Touching	Handshake, nudge, pat, embrace	Touching
	Personal space	Distant, professional, close, personal	Seeing
	Clothing	Fashion, power, business, casual, uniform	Seeing
	Grooming	Hair style, make-up, jewellery, perfume, aftershave lotion	Seeing Smelling
	Time	Early, on time, late, planned, unplanned	Seeing
Environment	Garden	Landscape, paths, flowers, shrubs, trees, sculptures, furniture, water	Seeing Touching
	Built	Colour, light, decorations, materials, art, furnishings, textures, space, flowers, plants	Hearing Smelling

(continued)

TABLE 7.5 *(continued)*

Type	Expression	Channels	Senses
Hospitality	Food and drink	Confectionery, cake, biscuit, meal, tea, coffee, soft drink, fruit juice, alcohol	Tasting Smelling Seeing Touching Hearing
	Ambience	Colour, light, scent, music, décor	Seeing Touching Hearing Smelling

Remember, these are just some examples of the many types of non-verbal communication that may be used by people when communicating. Other examples include the many types of laughter that may be used, as well as numerous mannerisms that can be important when interpreting a message.

Visual communication

Visual communication methods are used to convey messages in many professions and industries; however, they are particularly important for engineers. This is because a visual method may be the only way to communicate a design, construction details, or the condition of a component. **Visual communication** methods are listed separately because some of them use both verbal and non-verbal communication methods. For example, a plan incorporates words and symbols as well as non-verbal information. While the term visual has been used to describe this category, other senses may also be used to access the information in some of the communication methods in table 7.6, which lists some of the visual communication methods used by engineers and the channels used to convey messages.

TABLE 7.6 **Examples of visual communication**

Type	Message format	Channel	Senses
Drawing	Sketch	Paper, computer, advertisement	Seeing
	Painting	Paper, canvas, digital	
	Cartoon	Paper, computer	
	Animation	Video, TV, movie, computer	
Object	Sculpture	Bronze, plaster, fibreglass, wood	Seeing Touching
	Model	Cardboard, balsa, timber, plaster, polystyrene, digital, metal, 3D print	Seeing Touching Hearing
	Code	Traffic lights, traffic signal	Seeing Hearing
Photography	Still	Paper, digital, computer, satellite, aerial	Seeing
	Motion	Video, movie, computer, DVD, social media	Seeing Hearing
Technical drawing	Plan	Paper, computer software, PDF, dxf	Seeing
	3D	Paper, computer, PDF, 3D print	

The number of verbal, non-verbal and visual communication methods, and the variety of channels described in the three tables, shows that message creators have a wide choice of options to communicate their message. The tables also highlight the fact that multiple channels can be used to convey the same message, with each designed to attract the attention of different consumers by addressing their personal or learning preferences.

Noise

Noise is anything that interferes with or corrupts the communication package, or part of the package, during transmission through a channel. Some examples of noise have already been discussed, but there are many others. Two other types of noise will be discussed here: (1) technical noise and (2) non-technical noise. Examples of technical noise include electrical interference, loss of telecommunication signals, lost mail and poor-quality printing. Examples of non-technical noise include cultural misunderstandings, language barriers, personal values, incorrect names or addresses, and attitudes.

It may be difficult to eliminate technical noise, but non-technical noise can be minimised by careful planning. The key is to ensure that the words, gestures and languages used do not become a barrier to communication. If the consumer takes offence at something that is said or done, then they may disengage from the communication process. If the consumer does not engage with the communication, then the message has not been communicated effectively, and will not lead to the desired response from the consumer.

The following spotlight shows the careful planning required to engage with, and obtain the desired response from, a delegation from an overseas country during a half-day visit to a manufacturing plant.

SPOTLIGHT

The plant visit

Draft program for visit to DiFlexi Australia's manufacturing plant by a Chinese Government delegation.*

Time	Location	Activity	Personnel	Comments
09.30	Car park	PR manager to meet delegation and escort them through front office foyer to the boardroom	Chinese delegation PR manager	Reserve car parks Arrange for security staff to admit delegation Advise foyer staff about delegation Check names of delegation members — family name in capitals Advise delegation members of the location of toilets and other facilities
09.45	Boardroom	Introductions, formal exchange of business cards and morning tea	Chinese delegation CEO PR manager Chief engineer Sales manager Plant manager	Arrange morning tea — brewed coffee, juice and small biscuits and cakes Check dietary requirements Check business cards are current; arrange for information in Mandarin on reverse Display current product range Hang award pictures Display award trophies Dress: suit and tie
10.00	Boardroom	Official welcome	CEO	Prepare one-page overview of DiFlexi in Mandarin
10.10	Boardroom	Presentation on current range of products, manufacturing facilities and capabilities	Chief engineer	Set up multimedia the day before Rehearse presentation Brief staff about Chinese business culture

Time	Location	Activity	Personnel	Comments
10.30	Boardroom	Discussion, questions	Chinese delegation DiFlexi staff	All staff to attend rehearsal
10.45	Plant	Tour of plant	Chinese delegation Chief engineer Plant manager PR manager	Brief plant staff about the visit Plant manager to brief section managers about the visit so they are prepared to guide the delegation through their section Plant staff to wear DiFlexi logowear Provide clean or new PPE
11.45	Boardroom	Initial discussions about the proposed joint venture	Chinese delegation CEO PR manager Chief engineer Sales manager	Brief staff about the proposal CEO to lead the discussion Arrange for copies of proposal to be printed and distributed in Mandarin and English DiFlexi staff to be prepared to answer questions in their domain
12.30	Boardroom	Lunch	Chinese delegation CEO PR manager Chief engineer Sales manager Plant manager	Light two-course lunch to be served Advise caterers of dietary requirements and the need to have culturally appropriate food
13.15	Boardroom	Continue discussions on proposed joint venture	Chinese delegation CEO PR manager Chief engineer Sales manager	CEO to lead the discussion
14.00	Boardroom	Meeting close Farewell delegation	Chinese delegation CEO PR manager Chief engineer Sales manager Plant manager	Arrange for gifts for each member of the delegation and the interpreter; PR manager and chief engineer to escort the delegation to the car park
14.20	Boardroom	Debrief Plan next phase of the project	CEO PR manager Chief engineer Sales manager Plant manager	CEO to lead discussion

* An Australian Government interpreter will accompany the five-person Chinese delegation.

CRITICAL THINKING

Reflect on an important function you have recently attended, such as a high school formal, a twenty-first birthday party, a wedding or a sports club awards night. Make a list of the responses you think the organisers were trying to garner from those who attended. What messages and channels did they use to garner each of the desired responses? How successful were they in communicating those messages, and did they elicit the desired responses?

7.5 Communication roles

LEARNING OBJECTIVE 7.5 Discuss the roles of the creator, gatekeeper and consumer in the communication process.

The roles people play in the communication process are discussed in this section: the creator, the gatekeeper and the consumer. Most people spend their day swapping between the roles of creator and consumer as they communicate with others at home, at work, while shopping, playing sport or socialising. Fewer people have a gatekeeping role, particularly at their workplace. Understanding these roles will help you to become a more effective communicator.

KEY POINT

Engineers should use their understanding of the gatekeeper and consumer roles to plan a communication.

The creator

The creator's role begins with a decision to communicate. This is a decision based on the reason for the communication — the purpose. This is followed by the planning of the message and the assembly of all the components of the message into a communication package ready for transmission along one or more channels.

Purpose

Carefully defining the purpose of a communication is a critical first step in the communication process. Remember, people communicate to create a desired response in the consumer, such as an answer to a question, some advice, support for a proposal, a sale, an apology or a better relationship. It should be noted that some messages are not aimed at a specific consumer; for example, when a note to file is written about an interview with a client.

The creator should know the reason for the communication, who the consumer is, and the response they desire from the consumer. The answers to these questions define the purpose of the communication. After this, the creator can begin preparing the messages and assembling the communication package. Of course, the time spent on this process will depend on the type and complexity of the communication. For a conversation this process may be spontaneous, while for more complex messages, such as a project report, the purpose of the communication may be written out in detail, and this can be used as a checklist while the report is being assembled.

Creating the message

The first part of the creator's role is to plan the structure of the message. This is followed by establishing the order of the elements of the message, and the level of detail required for each element. Finally, the creator selects the preferred methods, channels and media for the different elements of the communication package.

The creator then fleshes out this structure by preparing the text for the different components of the message. This may include detailed information relating to each key point, the development of any case that needs to be argued, and the identification of any data or other evidence that is required to support the key points or recommendations.

The characteristics of the consumer should be used to inform the selection of the media and mode. For example, a two-dimensional plan that is dense with information about the components and construction details of a microchip will not help to communicate the message if the consumer does not work in the electronics industry or has low spatial ability.

Finally, the creator needs to ensure that the communication package is attractive to the target consumers so that they decide, firstly, to access the package and, secondly, to engage with it. This is because messages can easily be lost in the 'noise' of an information-rich world. In large organisations this work would be overseen by professionals in the marketing field.

Good public speakers and other talented performers realise that in order to engage with their audience positively, they firstly need to establish a rapport with them. They may start with a story, a joke or some other device to attract the attention of the audience. In terms of the PCR model, the speaker has connected with the audience and engaged them in the communication. The consumers are active listeners at this stage, as they have accessed the first components of the communication package. The speaker then gives the main message in a format that keeps the audience engaged so they access and interpret the message. The speaker may then emphasise the message by using an analogy, an example, a picture or a video, and conclude by summing up the message.

A writer uses different techniques to achieve the same outcomes. For example, a catchy title, a good executive summary, pictures, charts and stories can be used to engage the reader. The message only achieves the desired response when the reader has been actively engaged in reading and interpreting the written message.

Some communication packages consist of more than one message; for example, a consultant may include a covering letter and invoice when sending a client the final report on an investigation.

Assembling the communication package

The creator assembles the communication package by preparing and embedding each element of the message into the relevant methods and media. A creator may choose methods, message formats, channels and media, such as those listed in tables 7.4, 7.5 and 7.6. At the end of the assembly process, it is important to check that the communication package delivers a consistent message, a message that is likely to achieve the defined purpose.

While you are a student, you should use this approach to prepare assignments and write answers in examinations. An examination question is written by a lecturer to provide students with an opportunity to demonstrate what they know about the topic. This is the purpose from the lecturer's perspective. A student's response may be to answer the question in great detail, with the purpose being to trigger a response in the lecturer that results in full marks being awarded for the question. Before the student starts writing the answer, the key points would be noted on the question paper, and then the points would be arranged into the order that will best communicate the message.

When a communication package has been assembled by an engineer, it is normally reviewed by a colleague, or supervisor — the gatekeeper — to ensure that it is correct and is likely to achieve its purpose. Once it has been approved, it is transmitted over the appropriate channels.

The communication package

A communication package consists of all of the elements of a message. Consider an example: imagine two people have met for the first time. They will be decoding elements of messages from a number of verbal statements and non-verbal cues, to assess each other, such as the language they speak, the way they talk, their dress, their appearance, their facial expressions, their stance, and their hand and arm movements. Each person will interpret numerous messages from the other person and slowly synthesise them to build a profile of that person. If one of the early messages triggers a negative response, then the person may disengage and switch off, meaning there is little chance of a successful communication.

A consumer may evaluate a written report in the same way. It may be judged on the text, layout, graphics and other elements of the message on the front cover, or in the first few pages. The consumer may only continue reading if they have been engaged by the message. For this reason, it is important to consider how a communication package is assembled and presented to the consumer.

Recently, at the conclusion of a workshop, a group of employers were discussing the methods they used to shortlist applicants for graduate engineer positions in their companies. The accepted practice in most cases was that they continued reading each application until they encountered grammatical or factual errors. They then put aside that application and proceeded to the next one. Thus, only the applications that survived the error test would be considered for shortlisting. The clear message from this group of employers was that an applicant (and their mentor) should check their application carefully before submission to ensure it is error free.

Closing the loop

The final role of the creator is to check whether the communication was successful and elicited the desired response. In some situations you may think you have communicated a message, only to find out it has been lost in transmission, either through poor translation, or through a misunderstanding. It is important to adopt communication strategies that increase the likelihood of the communication package being received and understood by the consumer.

When an important piece of information is to be conveyed to a consumer, the communication should be treated as a transaction and managed in the same way that the transfer of a large sum of money would be handled. The communication package should be in a form that is acceptable to the consumer and the creator should ask the consumer to acknowledge that the package was received and understood. For an engineer, the loop is closed when the feedback message is filed. This is part of the evidence of the transaction, an important part of a quality assurance system. As a student you may be issued with a receipt when you lodge an assignment. For you, this is evidence that the transaction took place, and you can use it as evidence if the assignment is lost or mislaid by the university.

Checking whether a message has been received and whether it has had the desired effect on the consumer is an important aspect of effective communication. In conversations, you can often get immediate feedback. For example, if the person you are talking to frowns at you, this may indicate the message is not getting through. On the other hand, you can be confident the message got through if a person says, 'That's just what I needed to know; thank you for your help'.

Indicators that communication has not been successful include when a person says, 'I don't understand what you mean?', or if you find yourself saying, 'I didn't mean that . . .' or 'I didn't mean it that way'. It is often more difficult to check if a written communication has been successful, because you may not receive any feedback. A good strategy is to email the consumer on the day you mail the package, with a message such as, 'I have posted the three copies of the report to you today. I would appreciate it if you let me know when they arrive, or if you have any queries'.

Another way to check this is to phone the consumer a day or two after you estimate they would have received the communication. You may like to say something such as, 'I am just phoning to check that you have received the package I sent a couple of days ago. I know you have been waiting for our report so I would not want it to go astray'.

Finally, you should heed the warning in George Bernard Shaw's quotation at the beginning of this chapter. You should never *assume* that a communication has been successful. You should always find a way to check if an important communication has been accessed and if it has had the desired effect on the consumer and produced the desired response.

The gatekeeper

A gatekeeper is the person who has responsibility for checking and approving the release of messages. As a student, you may have been in a team in which information supplied by individual team members was modified by the leader during the preparation of the final report. In this case the leader assumed a gatekeeper role. If this role was not handled carefully the team members may have been disappointed if a key section of text or a drawing they prepared did not make it into the final report, particularly if they spent many hours working on it.

In your studies you will find it is good practice to have someone review an assignment or project report before you submit it. This practice has merit even if the person does not understand the content. This is because they can comment on the grammar, spelling, structure and layout of the assignment or project report.

When you begin working in an engineering organisation you may find one or more people have a gatekeeper role in the organisation. For example, your supervisor may carefully review any documents you prepare before they are sent to a client, or before they are used to implement a design. These people are sometimes called document controllers.

Many large organisations have a communication plan that defines the people who are able to speak or act on behalf of the organisation. Generally, all official communications to and from the organisation will pass across the desk of one or more of these people.

If no-one appears to be reviewing your work, you should ask a colleague to take on this role. You can always offer to review their work in return. You will find this peer review process invaluable as it will help you to learn about the formats, channels and media that the organisation uses to communicate within the organisation, and also with the external world.

To be able to review a communication package, a gatekeeper will need to know the purpose of the communication. Only then can they assess the quality of the communication and, from their experience, decide whether it is likely to create the desired response in the consumer.

It would be easy to assume that the concept of a gatekeeper does not apply to verbal communication in formal and informal meetings, but this is not the case. A team member, leader or supervisor can assume the gatekeeper role before a meeting begins, by saying to the other members of the team, 'Let me do the talking' or 'I will lead the discussion, but I will give you "the nod" if I want you to answer a question'. Using this strategy allows them to control one side of the communication flow.

SPOTLIGHT

Who is the gatekeeper?

In his 'President's Message' on 20 February 2014, the then-President of the Queensland Division of Engineers Australia, Blake Harvey, detailed the institution's response to the establishment of an independent Commission of Inquiry into the failure of a bund wall in Gladstone Harbour, which was announced by the federal Minister for the Environment, the Hon. Greg Hunt MP, on 30 January 2014. The bund wall surrounded a reclamation area that had been constructed to contain millions of cubic metres of dredge spoil removed from the harbour during the development of natural gas processing and transportation facilities on Curtis Island (Engineers Australia 2014).

Aerial photo of the Gladstone Harbour bund wall failure

Towards the end of 2011 community members, environmental groups and the fishing industry blamed the dredging activities for the increased turbidity of the water in the harbour and adjoining waterways, which they believed was connected to a rapid rise in disease and death in fish and other marine species. The issue was raised regularly in the media over the next two years, during which time the Gladstone Ports Corporation (GPC) maintained that there was no connection between the dredging activity and the marine health issue. In fact, based on the results of a number of studies, the Corporation reported that the problem was caused by the large flood events in 2010 and 2011.

Following a request from the World Heritage Committee that had conducted a review of protection and management of the Great Barrier Reef World Heritage property, the Australian Government commissioned an independent review of environmental management arrangements and governance of the Port of Gladstone in February 2013. The initial findings were forwarded to the government in July 2013 and, following a public consultation period, a Supplementary Report was delivered to the government on 1 November 2013 (Department of the Environment 2013).

In January 2013 John Broomhead, who had been an environmental manager with the GPC, came forward with fresh evidence and this led to the commissioning of the independent Commission of Inquiry into the failure of a bund wall in Gladstone Harbour, which began in February 2014.

In his President's Message, Harvey reported that Engineers Australia had 'received a request from the inquiry Co-Chairs that we provide a list of engineers whose qualifications and experience would make them suitable to support the inquiry … We will continue to work with the inquiry panel to ensure that the outcomes above can be achieved' (Harvey 2014).

As President of the Queensland Division, Harvey is the only person who can officially speak on behalf of the Division. While other people normally assist in this process by gathering and reviewing information, and by preparing documents and statements, the President is ultimately responsible for all official communications, even those that have been delegated to others. In this respect the President is the gatekeeper for the Division.

The federal Minister accepted the independent commission's report on 9 May 2014. It contained 37 findings and 19 recommendations relating to the assessment, monitoring and management of developments in coastal environments. A key finding was that 'aspects of the design and construction of the bund wall were not consistent with industry best practice. Inadequate restraint of a geotextile liner, piping of water and sediment through paleochannels under the wall and the erosion of mud outside the wall all contributed to changes in turbidity in the vicinity of the bund wall' (Department of the Environment 2014).

CRITICAL THINKING

What role do you think the gatekeeper(s) at the Gladstone Ports Corporation may have played in this saga? Do you have any evidence to support your conclusions?

The consumer

The consumer's role begins when they sense a message is available for them to access. They may use a range of senses to detect messages; for example, they may hear the phone ring, see if there are letters in the mailbox, hear a *ping* when an email has arrived, or they may smell coffee brewing when it is time for morning tea. All of these signals alert them to the fact that a message is available for them to access.

Accessing the message package

The consumer can choose whether to access a message. In an earlier example, Kavita chose not to read the letter from Jim for a couple of days. Similarly, you can choose whether you want to read an email or a report, or answer the phone. You may even choose not to access the message when somebody is talking to you.

Myrna Marofsky, a management consultant and recognised expert on workplace issues, quipped that 'People have remote controls in their heads today. If you don't catch their interest, they just click you off' (Maxwell 2010, p. 179). Have you ever felt a person you were talking to had switched off and was not interested in having a conversation with you? You might have sensed that while they may have heard you, they did not appear to be listening to what you were saying. You might have tried to detect why they were not listening. Were they bored? Were they thinking about a more important issue? Or were they showing their lack of interest by looking around for someone else to talk to? Whatever the reason, such a reaction signals it is time to cut the conversation, adopt a different strategy, or make it more interesting by changing the topic.

If the consumer does not access a message the purpose of the communication will not be realised, and all of the effort of preparing the message will be wasted. It is for this reason that creators often 'dress up' a message to make it more attractive for the consumer. For example, a sales brochure will be far more attractive if it has been properly designed so the layout draws the consumer's eye to the key messages. This 'dressing up' of the brochure should increase the likelihood that the consumer will be engaged by the message and understand the purpose of the communication. Of course, they will have a different response if they find the data, information and presentation in the text does not align with the high expectations elicited by the 'dressing up'.

Interpreting the message

Once the consumer accesses a communication package, the various components are interpreted and the separate messages are synthesised to recreate the message. Hopefully the recreated message is the same as the message the creator intended to communicate, but this may not always be the case. Sometimes people receive mixed messages; that is, two or more messages that appear to be incompatible. For example, an engineering graduate is asked to attend a job interview. While the panel chair says that the graduate's lack of work experience is not an issue, the body-language of the panel members suggests that they believe previous experience is important. The graduate leaves the interview with the impression that while the chair of the panel would offer him a job, the other panel members would not.

Response

You will remember that the purpose of a communication is to create the desired response in the consumer. Thus, the response is the key measure of the success of a communication.

It is important that when you receive a communication, you take the role of 'consumer' seriously and apply the basic principles of communication etiquette in your response. For example:

- A consumer should advise the creator that they have received the message and provide either an immediate response, or an indication of when they will respond to the communication. This is very easy to do by email, text message or telephone. For example, 'I received your request and will be able to get the work to you early next week'.
- A consumer should acknowledge, and where appropriate, value the effort the creator has gone to when they receive a large or important communication package.

Using the PCR model

You will find that the PCR model will prove to be a useful tool to review and prepare communications during your time at university and later when you become a practitioner. In the chapter on communicating information, you will learn how to use the PCR model to plan a communication and the structure and style of the documents and other communications regularly used by engineers.

SUMMARY

This chapter has explored the meaning of communication, the communication process and the contextual factors that may influence the effectiveness of a communication. Engineers need a sound understanding of the communication process as they must be able to communicate *effectively* with members of the engineering team, with people in other professions and, importantly, with members of the community. The PCR model was developed to enhance our understanding of the communication processes and contexts used by engineers. The components of the model were discussed in detail and you were provided with opportunities to use the PCR model to review different aspects of communication. We will now briefly revisit the learning objectives from this chapter.

7.1 Explain what communication is and discuss the types of communication skills used by engineers.
Many forms of communication were described and it was noted that there is a difference between the singular and plural uses of the word communication, with the singular form of the word being used when discussing the interaction between people. The ability to communicate effectively is an essential capability for engineers in the workplace and is highly valued by employers.

7.2 Discuss the historical development of communication theories and outline the components of the PCR model.
A variety of communication theories were developed over many years, with the fundamental principles remaining the same. Various models emphasise different aspects of the communication process and introduce new perspectives. It is important to carefully define the *purpose of a communication*, which is to create the desired *response* in the consumer, and to be aware that people can choose whether or not to access a communication. This will help you to design communications that will encourage the target consumers to access the messages.

Communication can be one-way or interactive. The PCR model highlights the fact that engineering communications are influenced by a range of contexts and workplace environments.

7.3 Describe how contextual factors can influence the effectiveness of communication.
The context in which a communication occurs can impact on the design, transmission and interpretation of a message. A number of the personal characteristics of the message creator, the gatekeeper and the consumer may influence the communication process. Other contextual factors also affect communication, such as: the organisational, project and workplace environments; and the factors associated with the geographic location of the communicators. An understanding of these factors should be used to create messages and select appropriate communication methods and channels to ensure the communication will be effective and achieve the desired outcome.

7.4 Describe the communication methods commonly used by engineers.
The term channel has a broad meaning that covers the transmission, transport and reception of a message. Communication methods can be categorised under the headings of verbal, non-verbal (such as body language) and visual communication methods (such as drawings and photography). Noise, in a communication sense, is anything that affects the message or its transmission along a channel. Engineers should select appropriate verbal, non-verbal and visual communication methods and channels for their communications, and design them so that noise will have a minimal impact.

7.5 Discuss the roles of the creator, gatekeeper and consumer in the communication process.
The three key roles people play in the communication process are that of creator, gatekeeper and consumer. An understanding of the communication process and the characteristics of the consumer may be used to inform the creation of a communication. Engineers (and students) should create messages that can be interpreted by the target consumer, and should not assume communication has occurred. The gatekeeper plays an important role in the communication process, and most engineering organisations appoint people to undertake this role. As consumers, people should respond appropriately when they receive an important communication. Engineers should always seek feedback when they communicate to ensure that the consumer has accessed and correctly interpreted the message, and given the desired response.

KEY TERMS

channel The way a message is conveyed.
communication The imparting or interchange of thoughts, opinions, or information by speech, writing, drawings or signs.
communication package The package of messages that form a communication.
communications The science or process of conveying information, especially by electronic or mechanical means.
consumer A person who may choose to access a message and receive a communication.
contexts The personal, cultural, organisational and work environments in which communication occurs.
creator The person(s) who develops a message.
gatekeeper A person who has responsibility for checking and approving the release of messages.
interactive communication model A model of a two-way communication process.
language The way messages are created and consumed within the context of media.
message Information that is conveyed by any means from one person or group to another person or group.
noise Anything that interferes with, or corrupts a message.
non-verbal communication A form of communication using non-verbal methods such as facial expressions, hand movements or dress.
purpose The aim(s) of communication.
response The effect of communication on a person or group.
verbal communication A form of communication using verbal methods such as speech in face-to-face or virtual conversations, meetings or entertainment.
visual communication A created communication that is designed to be viewed and may include both verbal and non-verbal information.

EXERCISES

1 Think back to the last lecture or talk that you attended. Use the PCR model to reflect on the communication that occurred. What messages were successfully communicated? Why? Were there any messages that did not get through? Why?
2 Review the communication between Jim and Kavita in the chapter (tables 7.2 and 7.3) and write down the contexts in which the communication occurred.
3 Reflect on the impact your attitude about a subject has on your motivation to read and learn about that subject. Then, reflect on the impact a lecturer's attitude about a subject — and towards the students in the class — has on your learning in the subject. Compare and contrast your reflections.
4 Investigate and identify the components of the channel for a text message. Discuss your answer with classmates.
5 Reflect on the following communication between an engineer, Nigel, and his manager, Susan.

   Nigel was in the lift when, suddenly, he saw Susan heading towards the lift. She looked as though she was in a bad mood — she had a determined look on her face and her brow was furrowed. To avoid her gaze, Nigel moved to the back corner of the lift and looked at the floor. As she entered the lift, Susan looked up and saw Nigel. Smiling, she said, 'I liked the report you wrote about the quality of the bearings we use in the plant. Keep up the good work.' The lift doors opened and she walked out, giving Nigel a smile as she turned into the corridor.

   What were the components of the message packages? What channels were used? What messages were conveyed? How do you think Nigel felt after the exchange? Discuss your answers with your classmates.
6 Explain the roles of creator, gatekeeper and consumer, using an example of communication in an engineering context.

PROJECT ACTIVITY

Use the PCR model to review a significant communication that you have previously used to communicate the outcomes of a university project. Your report on this project should include a:

- brief statement on the way you developed the plan for the communication
- discussion about the communication strategy you used, including communication methods and channels, and how they align with the PCR model
- reflection on any problems you had while preparing the communication, and any assumptions you had to make
- brief discussion about whether the communication achieved its purpose and the reasons for your conclusion.

REFERENCES

Berlo, D 1960, *The process of communication: An introduction to theory and practice*, Holt Rinehart and Winston, New York.

Burke, K 1969, *A rhetoric of motives*, University of California Press, Berkley and Los Angeles.

Burrowes, G & Khan, JY 2011, 'Short-range underwater acoustic communication networks', in N Cruz (ed.), *Autonomous underwater vehicles*, InTech, pp. 173–198, https://doi.org/10.5772/24098

Control Bionics 2018, 'Control bionics trilogy product line', https://www.controlbionics.com

——— 2024, 'Control bionics trilogy product line', https://www.controlbionics.com

Dawson, DM & Brooks, BJ 1999, *The Esso Longford Gas Plant accident: Report of the Longford Royal Commission*, Government Printer, Victoria.

Department of the Environment 2013, *Independent review of the Port of Gladstone — report on findings*, Australian Government, Canberra.

——— 2014, *Independent review of the bund wall at the Port of Gladstone — report on findings*, Australian Government, Canberra.

Dvorak, D 2010, *Build your brand*, Pelican Publishing Company, Gretna.

Engineers Australia 2012, *Stage 2 Competency Standard for Professional Engineers*, https://www.engineersaustralia.org.au/publications/stage-2-competency-standard-professional-engineers

——— 2014, 'Inquiry to investigate Gladstone bund wall failure', Civil Engineers, Australia.

——— 2019, *Stage 1 Competency Standard for Professional Engineers*, https://www.engineersaustralia.org.au/publications/stage-1-competency-standard-professional-engineers

Eunson, B 2016, *Communicating in the 21st century*, 4th edn, John Wiley & Sons, Brisbane.

Foulger, D 2004a, 'Models of the communication process', https://davis.foulger.info

——— 2004b, 'An ecological model of the communication process', https://davis.foulger.info

GPS.gov 2024, *Official US Government information about the global positioning system (GPS) and related topics*, United States Coast Guard, Department of Homeland Security, https://www.gps.gov

Graduate Careers Australia 2016, *Graduate outlook 2015: The report of the 2015 graduate outlook survey: Perspectives on graduate recruitment*, https://www.graduatecareers.com.au

Harvey, B 2014, *President's message, Queensland Division*, Engineers Australia, Canberra.

Hopkins, A 2000, *Lessons from Longford: The Esso Gas Plant explosion*, CCH Australia, Sydney.

——— 2002, 'Lessons from Longford: The trial', *Journal of Occupational Health and Safety — Australia and New Zealand*, vol. 18, no. 6, pp. 5–71.

Knapp, ML 1992, *Essentials of nonverbal communication*, Holt, Rinehart and Winston, New York.

Lane, C 1932, *The rhetoric of Aristotle: An expanded translation with supplementary examples for students for composition and public speaking*, Prentice–Hall, New Jersey.

Lasswell, H 1948, 'The structure and function of communication in society', in L Bryson (ed.), *The communication of ideas*, Jewish Theological Seminary of America, Institute for Religious and Social Studies, Harper, New York.

Macquarie Dictionary 2024, *The Macquarie Dictionary online,* Pan Macmillan Australia, https://www.macquariedictionary.com.au

Male, SA, Bush, MB & Chapman, ES 2009, 'Identification of competencies required by engineers graduating in Australia', 20th Conference of the Australasian Association for Engineering Education: Engineering the Curriculum, Adelaide.

Maxwell, J 2010, *Everyone communicates, few connect: What the most effective people do differently*, Thomas Nelson, Nashville.

Mottard, A & Casteleyn, J 2008, 'Visual rhetoric: Enhancing students' ability to communicate effectively', *International Journal of Engineering Education*, vol. 24, no. 6, pp. 1130–1138.

Navigation Centre 1996, *NAVSTAR GPS user equipment introduction*, United States Coast Guard, Department of Homeland Security, Washington.

Nelson, S 2011, 'Mind your manners', *Graduate Grapevine*, no. 20, Summer, Graduate Careers Council, Australia.

Nicol, J 2001, 'Have Australia's major hazard facilities learnt from the Longford disaster?: An evaluation of the impact of the 1998 ESSO Longford explosion on major hazard facilities in 2001', Institution of Engineers, Australia.

Shannon, C & Weaver, W 1949, *The mathematical theory of communication*, University of Illinois Press, Illinois.

Synchron 2024, 'The brain unlocked', https://synchron.com

Tatham, H 2008, 'Releasing Fletcher Aluminium's invisible handbrake', *New Zealand Management*, vol. 55, no. 6, pp. 44–48.

Thompson, K 2011, 'Building and managing your personal brand', *Graduate Grapevine*, no. 20, Summer, Graduate Careers Council, Australia.

University of Melbourne 2024, 'Synchron: Helping people with paralysis regain mobility', https://research.unimelb.edu.au/strengths/updates/news/synchron-helping-people-with-paralysis-regain-mobility

Wong, L & Champness, L 2018, 'Lives affected by motor neurone disease are changed by Australian medical technology', *ABC Radio Sydney*, 19 April, https://www.abc.net.au/news/2018-04-19/australian-technology-motor-neurone-disease-mnd-neuronode/9670726

ACKNOWLEDGEMENTS

Photo: © Control Bionics
Photo: © Synchron
Photo: © STR / Getty Images
Photo: © Kings Access / Adobe Stock Photo
Photo: © Lori Sowa
Photo: © Andrei Armiagov / Shutterstock
Photo: © Australian Marine Conservation Society

Figures 7.6 and 7.7: © Burrowes, G & Khan, JY 2011, 'Short-range underwater acoustic communication networks', in N Cruz (ed.), *Autonomous underwater vehicles*, InTech, pp. 173–198, https://doi.org/10.5772/24098

Table 7.1: © Male, SA, Bush, MB & Chapman, ES 2009, 'Identification of competencies required by engineers graduating in Australia', 20th Conference of the Australasian Association for Engineering Education: Engineering the Curriculum, Adelaide.

Text: © Nicol, J 2001, 'Have Australia's major hazard facilities learnt from the Longford disaster?: An evaluation of the impact of the 1998 ESSO Longford explosion on major hazard facilities in 2001', Institution of Engineers, Australia.

Text: © Engineers Australia 2019, *Stage 1 Competency Standard for Professional Engineers*, https://www.engineersaustralia.org.au/publications/stage-1-competency-standard-professional-engineers

Text: © Engineers Australia 2012, *Stage 2 Competency Standard for Professional Engineers*, https://www.engineersaustralia.org.au/publications/stage-2-competency-standard-professional-engineers

Text: © Graduate Careers Australia 2016, *Graduate outlook 2015: The report of the 2015 graduate outlook survey: Perspectives on graduate recruitment*, https://www.graduatecareers.com.au

Text: © Department of the Environment 2014, *Independent review of the bund wall at the Port of Gladstone — report on findings*, Australian Government, Canberra.

CHAPTER 8

Communication skills

'We are what we repeatedly do. Excellence, then, is not an act but a habit.'

Aristotle

LEARNING OBJECTIVES

After studying this chapter, you should be able to:

8.1 apply the techniques that will assist you to listen actively, read effectively and write accurate notes

8.2 prepare and deliver verbal presentations

8.3 plan and prepare written communications

8.4 use data tables, graphs and charts to communicate information.

Introduction

Many engineers would agree that communication is not one of their strengths, often because they prefer the technical aspects of their role. For example, they may be more excited about conceptualising, designing, testing and implementing a new device than they are about writing a technical manual to accompany it. But, of course, a good technical manual is essential if the device is to be accepted in the marketplace; if it is to be used safely, efficiently and effectively; and if post-purchase enquiries, complaints and repairs are to be minimised.

In the previous chapter, we discussed the importance of understanding the communication process to enable effective communication. The PCR communication model was used to illustrate the process, and to highlight the importance of firstly defining the purpose of a communication and then carefully assembling the communication package so that it achieves the desired response from the consumer.

Communication skills affect both the way a person communicates and the success of their communications; therefore, you should take every opportunity to enhance these skills. There are many tools and techniques you can use to develop your communication skills, not only while you are at university but also as you communicate with your family, friends and colleagues while you undertake social, sport and work activities.

When communicating, it is important to carefully plan the collection, analysis and presentation of data and information, particularly when it is being used as evidence to support designs, recommendations for further work, or project outcomes. The following spotlight highlights the use of a clear structure and a range of charts, tables and text to communicate statistical information.

SPOTLIGHT

Reporting on data breaches

The Office of the Australian Information Commissioner (OAIC) was established to manage the Notifiable Data Breach (NDB) scheme on behalf of the Australian Government. Under the scheme, which has been operational since February 2018, certain agencies and organisations that operate under the Privacy Act are required to notify the OAIC when an eligible data breach occurs, that is, when the data incident is likely to cause serious harm to one or more individuals. The sources of the breaches include human error, theft of paperwork, identity theft, rogue employees and cyber incidents.

The OAIC publishes twice-yearly reports on the breach notifications it has received during each six-month period. The reports use text, graphics, tables and a variety of charts to summarise and highlight the results of its analysis of the NDB data. The following examples were selected from the data and information published in the report for the first half of 2023 (Office of the Australian Information Commissioner [OAIC] 2023). They highlight the strategies used by the OAIC to communicate different aspects of the NDB data to its stakeholders and the general public.

The 40-page report included a table of contents, a snapshot (a dashboard which summarises the key data in six images), an introductory section, an executive summary, a section on the data for all sectors of industry, a section comparing the data for the five industry sectors with the most breaches (the Top Five!) and a glossary. The structure of the report is shown in figure 8.1.

FIGURE 8.1 The structure of the OAIC six-monthly report

Notifiable Data Breaches Quarterly Statistics Report

Contents

About this report
Executive summary
Notifications received January to June 2023 — All sectors
Awareness and impact of data breaches among the community
A maturing regulatory approach
Number of individuals worldwide affected by breaches
Large-scale data breaches
Kinds of personal information involved in breaches
Time taken to identify breaches
Privacy by design to prevent and detect data breaches
Time taken to notify the OAIC of breaches

Conducting reasonable and expeditious assessments by being flexible and adaptive
Preventing risks arising from working in changed environments
Source of breaches
Malicious or criminal attacks
Remaining vigilant to social engineering and impersonation
Assessing breaches with limited or no evidence
Human error
System faults
Data governance to mitigate the effects of data breaches
Effective information governance
Comparison of top 5 sectors
Time taken to identify breaches — Top 5 sectors
Time taken to notify the OAIC of breaches — Top 5 sectors
Source of breaches —Top 5 sectors
Malicious or criminal attack breaches — Top 5 sectors
Cyber incident breaches — Top 5 sectors
Human error breaches — Top 5 sectors
System fault breaches — Top 5 sectors
Glossary

Source: OAIC (2023, p. 1).

The following charts, graphics, tables and text boxes are examples of the presentation methods used in the report. They will also provide you with an overview of the types of notifiable data breaches that are occurring across Australia.

The executive summary includes the following key findings.

- 409 breaches were notified ...
- Malicious or criminal attacks remained the leading cause (70%) of data breaches.
- Human error breaches were the fastest to be identified with 81% identified in 30 days or fewer. Only 57% of system faults were identified in the same timeframe.
- The health and finance sectors remained the top reporters of data breaches. Health reported 63 breaches (15% of all notifications) and finance 54 breaches (13% of all notifications) (OAIC 2023, p. 5).

The key statistics for the period are shown in an image in the executive summary, which is reproduced in figure 8.2.

FIGURE 8.2 The key data in the executive summary

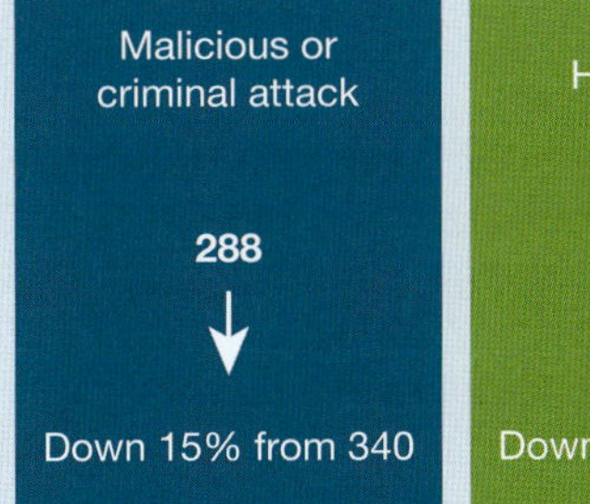

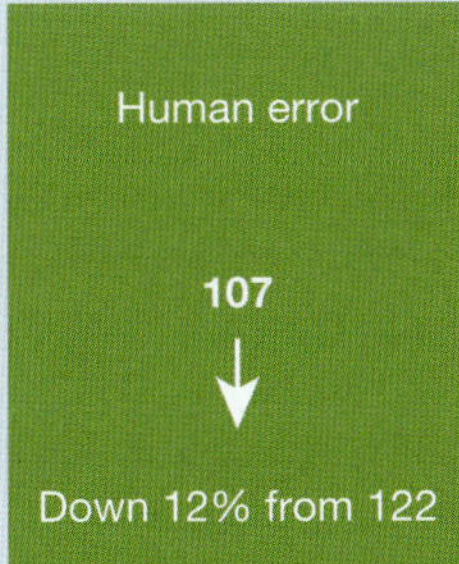

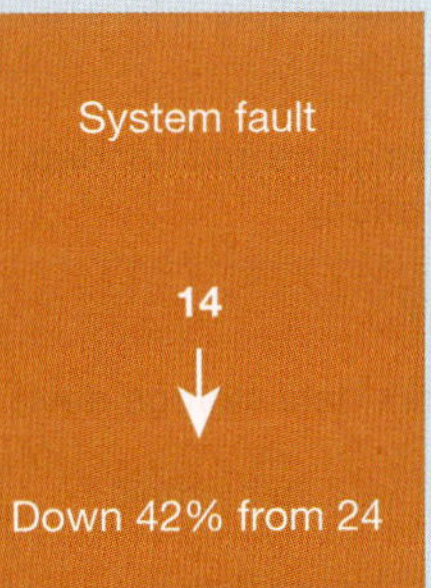

Source: OAIC (2023, p. 5).

It is interesting to note that since the start of the NDB scheme the OAIC has observed a trend where more notifications are received in the second half of each calendar year than in the first half of the year.

Of the 409 breaches reported in the period, most (63%) involved the personal data of less than 100 people. The bar chart in figure 8.3 shows the number of individuals worldwide affected by breaches in all sectors of industry. Twenty-three breaches affected 5000 or more Australians, with two breaches affecting more than one million Australians and one breach affecting more than 10 million, a first for the NDB system.

FIGURE 8.3 The number of individuals worldwide affected by breaches — all sectors

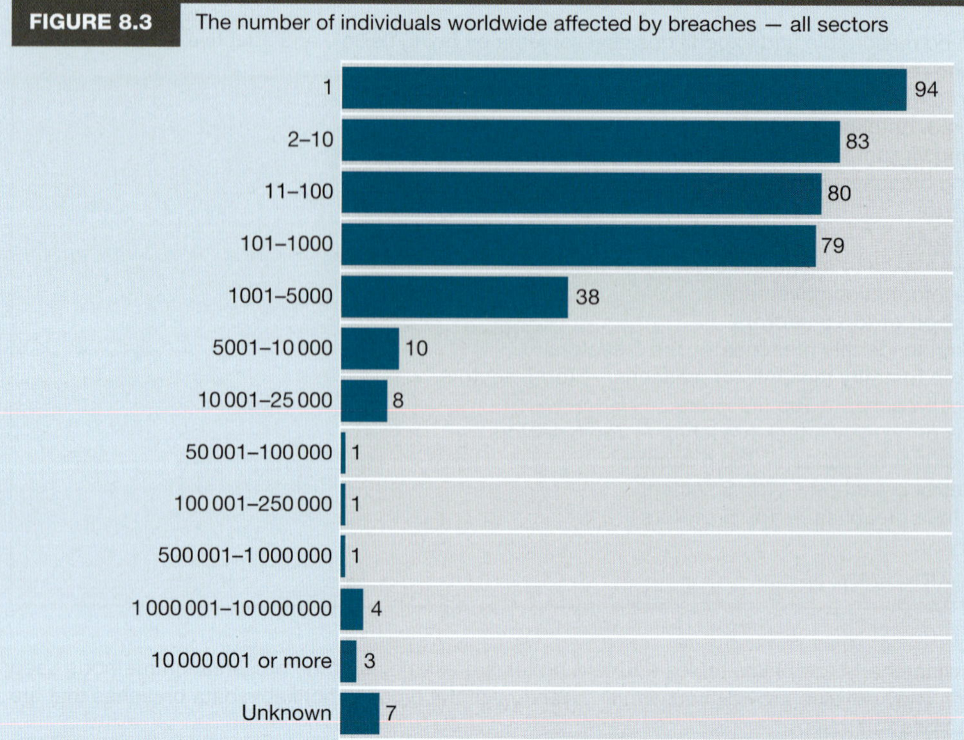

Source: OAIC (2023, p. 10).

Table 8.1 shows the breakdown of the notifications in the 'human error' category of breaches by listing the type of human error, number of NDBs received and the average and median number of affected individuals for each type of error.

TABLE 8.1 The number of NDBs and individuals affected by each type of human error — all sectors

Source of breach	Number of notifications	Average number of affected individuals	Median number of affected individuals
Failure to use BCC [blind carbon copy] when sending email	8	453	185
PI [personal information] sent to wrong recipient (other)	4	276	1
Unauthorised disclosure (unintended release or publication)	19	86	14
Insecure disposal	1	80	80
Loss of paperwork/data storage device	10	69	31
PI sent to wrong recipient (email)	49	38	1
Unauthorised disclosure (failure to redact)	8	3	2
PI sent to wrong recipient (mail)	5	2	1
Unauthorised disclosure (verbal)	3	1	1
Total	**107**	**84**	**2**

Source: OAIC (2023, pp. 25–26).

The bar chart in figure 8.4 shows the kinds of personal information involved in breaches plotted against the total number of notifications from all industry sectors. Note that data breaches may involve more than one kind of personal information.

FIGURE 8.4 The number of notifications for each kind of personal information — all sectors

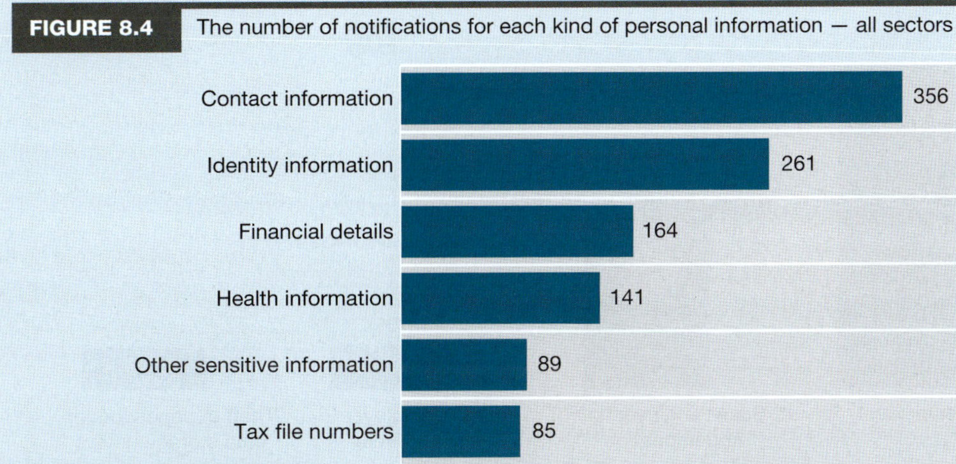

Source: OAIC (2023, p. 12).

The five industry sectors that reported the most notifications in the period were health service providers, finance (including superannuation), recruitment agencies, legal, accounting and management services, and the insurance sector. The chart in figure 8.5 shows the number and types of breaches for each of those sectors.

FIGURE 8.5 The number and type of notifications by industry sector

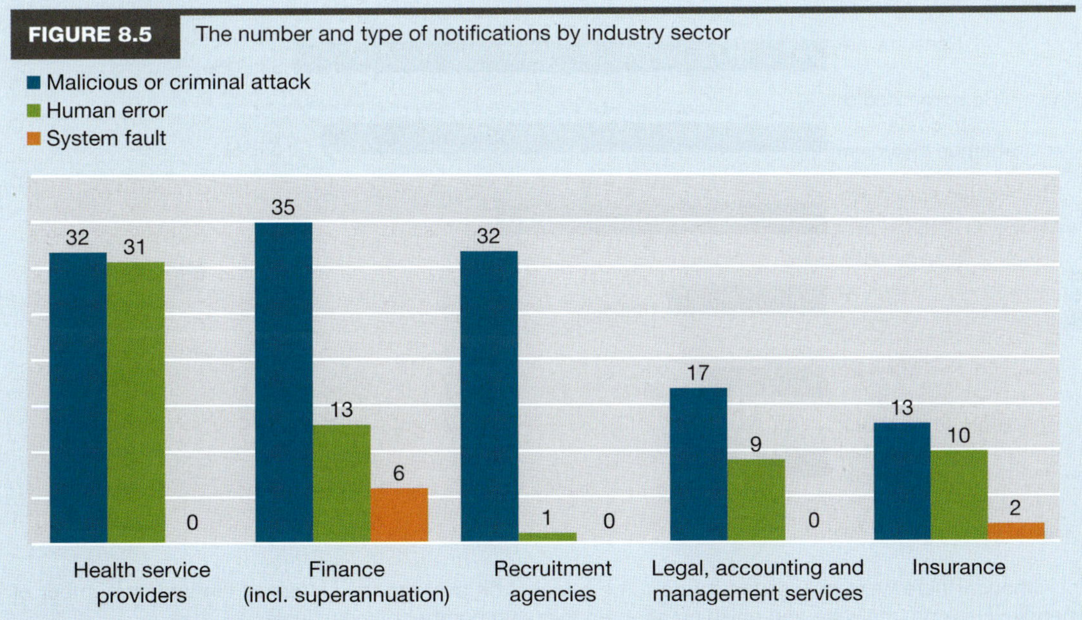

Source: OAIC (2023, p. 32).

The 'malicious or criminal attack' category was the largest source of data breaches, accounting for 70 per cent of the notifications for the period, with 60 per cent of those being cyber incidents. The chart in figure 8.6 shows the number of notifications for each type of breach in this category, for both the current and previous periods.

FIGURE 8.6 The number of notifications by type of malicious or criminal attack

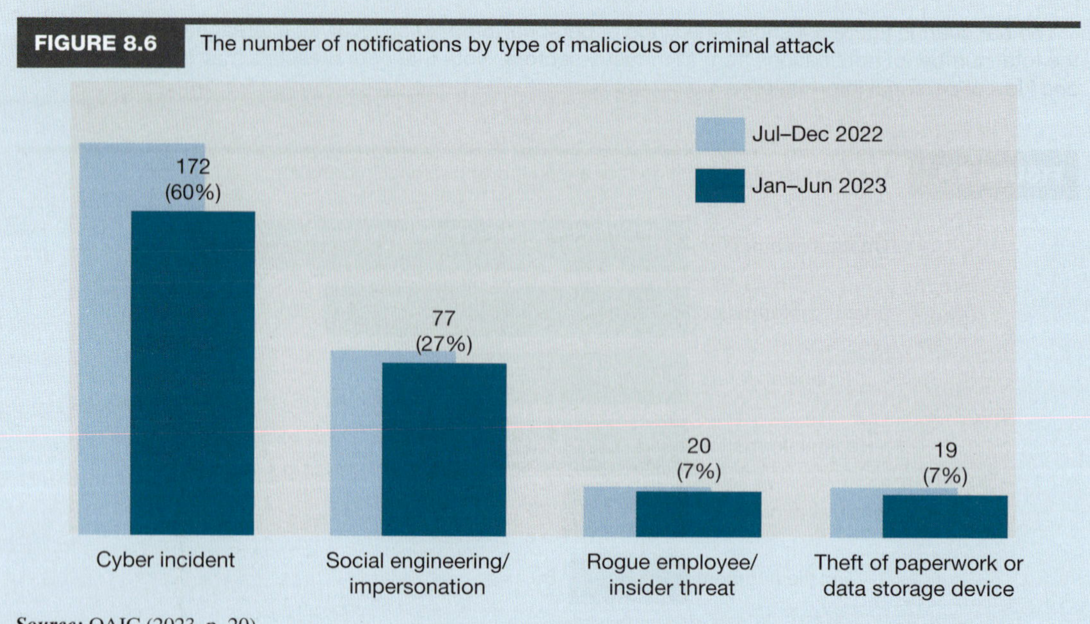

Source: OAIC (2023, p. 20).

The chart in figure 8.7 shows, for all sectors, the percentages for each different type of 'cyber incident' breach for the current and previous periods.

FIGURE 8.7 The number of breaches and percentages for each type of cyber incident

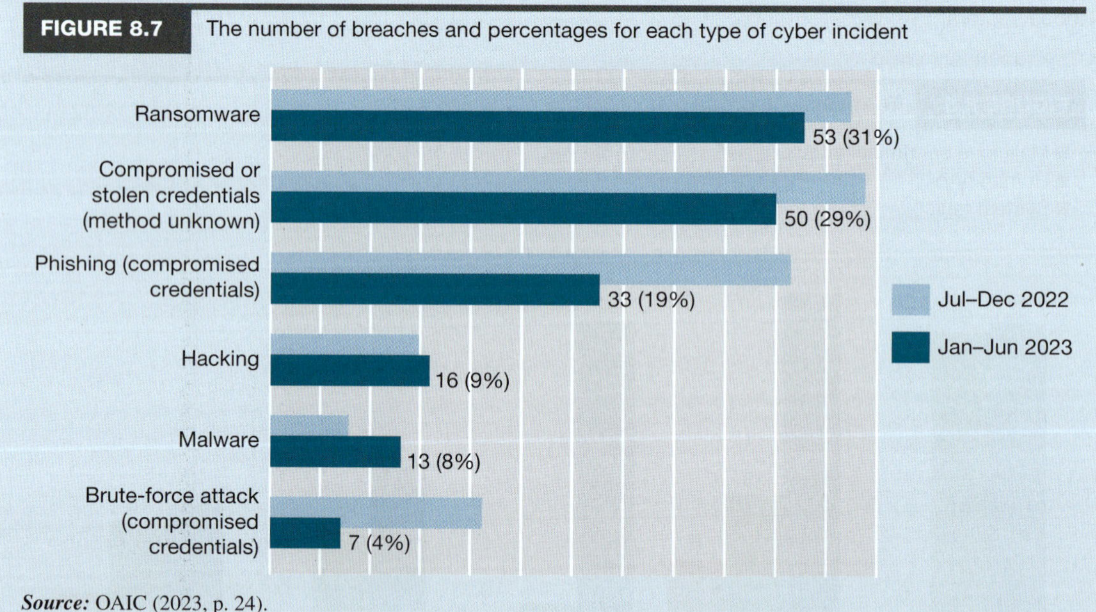

Source: OAIC (2023, p. 24).

Although there were only seven 'brute-force attacks' on average they affected the largest number of people. Hacking, malware and phishing affected the least number of individuals.

CRITICAL THINKING

Carefully review each of the tables and charts and, in each case, consider if there is a better way to present the data so the key messages are highlighted. Download the latest OAIC report and compare the information in that report with the information in this spotlight. What changes have occurred? What are the trends? Has the structure of the report changed? You may like to download more of the reports to gain a better perspective on trends.

In this chapter you will learn about some of the communication skills required to produce engineering reports. Developing these skills will help you to become a more effective communicator, a more effective student and, in time, a more effective engineer. Importantly, you will need these skills to introduce and manage the implementation of AI in your engineering communications. A key part of that role will be to act as a gatekeeper, including by checking the accuracy of the data and information in a document, the style and validity of the presentation of that data and information, and the overall format and quality of the documents.

You will remember that three broad categories of communication methods were briefly described in the chapter on understanding communication, as well as some of the methods in each of those categories. These three categories are:

1. verbal communication — languages, oral communication (speaking and listening) and written communication (reading and writing)
2. visual communication — two-dimensional representations (graphs, charts, drawings, plans, sketches, maps and photos) and three-dimensional representations (models, simulations and video)
3. non-verbal — body language, dress, ambience, hospitality and so on.

In this chapter you will learn more about verbal and visual communication skills.

8.1 Verbal communication skills

LEARNING OBJECTIVE 8.1 Apply the techniques that will assist you to listen actively, read effectively and write accurate notes.

Speaking is the most common form of personal communication and generally people are comfortable when speaking with others, particularly in informal contexts such as office discussions, parties or at a sporting event. Often referred to as oral communication, it is a key skill for all professionals and its importance for engineers was highlighted by the research undertaken by Male et al. (2008), which was discussed in the chapter on understanding communication. The term oral communication is normally used when speaking and listening are the only methods used to convey a message, for example, a conversation. The term verbal communication is used when oral communication methods are used in conjunction with other communication methods, such as plans or a slide presentation, to convey a message.

Oral communication pervades the work of most engineers and, as the examples in table 8.2 show, it occurs in different forms.

TABLE 8.2 Forms of oral communication

Type of oral communication	Examples	
	Informal communication	Formal communication
Listening (mostly one-way communication)	Inviting oral feedback from colleagues and subordinates on how you handled a difficult incident or decision in the workplace	Conducting a community consultation on a range of proposed changes to the municipal waste collection services a local government authority provides
Telling (mostly one-way communication)	Telling a new operator on a coal loader how to clear jams in the conveyor belt system	Presenting a design specification as part of a tender process for a company to construct a new multi-lane overpass on an existing freeway
Discussing (two-way communication)	Brainstorming, debating and bouncing ideas around with a team to devise a creative solution to a difficult design problem	Negotiating safe work practices with union representatives at a workplace. Such negotiations are termed 'formal' because the agreed points are generally written down and form part of an agreement about working conditions at the workplace

Listening

One of the most neglected skills in this fast-paced world is that of listening. Yet it is an essential skill for interacting with others at home, at university, at work and in social situations. Engineers need good listening skills when they talk to clients, colleagues, contractors, tradespeople and professionals from other disciplines; and listening is a critical skill when consulting others about their ideas, opinions or responses to a proposed engineering activity. To be a good listener requires a degree of self-confidence and perhaps some subjugation of the ego.

Listening is not easy. Our minds or eyes often wander during a conversation. This may be because:

- our mind is busy thinking about a problem we have to solve
- we are not interested in what the other person is saying
- we don't have the time to listen properly
- we are looking for an opportunity to interrupt the conversation to say something that we think is far more important than what the other person is saying.

Perhaps this is why Burley-Allen (1995, p. 3) reported that 'On average, people are only about 25 per cent effective as listeners'. Poor listening skills can cause problems because the listener may:

- not concentrate on what is being said and may receive only part of the message and miss information that is critical to understanding that message
- jump to conclusions when they hear a key word or phrase and then miss the real message
- hear only what they want to hear
- kill the conversation if the speaker detects the listener is not really listening.

In an engineering context, any of these outcomes could have an impact on the success of a project, with the result being mistakes, delays, safety issues or the loss of a client. Poor listening may also affect work relationships.

Like any skill, becoming a good listener requires regular and systematic practice. Good listeners concentrate fully when someone is talking to them, particularly when detailed information is being provided. They often use *active listening techniques* (Conflict Research Consortium 1998) to ensure they hear all of the messages being communicated. Piotrowski (2011, p. 69) lists some advantages of active listening as follows.

- You learn more by listening than by talking.
- Actively listening to clients (and others) tells them you are really interested in what they are saying.
- People respond better to those whom they perceive are really listening to them.
- Listening helps you become a more confident person.
- Listening helps you keep an open mind and enhance your critical thinking skills.
- Good listening is very important in team assignments.

The following simple active listening techniques can be used to make sure you have correctly received information that is verbally communicated to you.

1. *Eye contact.* Always make eye contact with the person you are listening to.
2. *Check information.* Once some information has been communicated to you, you can check it by repeating the information back to the other person by saying something like: 'So, if I am hearing you correctly, we need to test the component; is that right?' The most reliable method you can use to check you have correctly understood information, particularly detailed information, is to write down the information as you receive it and then read it back to the other person. You may say, 'Let me check that I have noted that information correctly by reading it back to you'. This is called closed-loop communication.

   Checking methods like these can be used to eliminate one of the most common sources of error that occurs when measurements are being recorded — the transposition of two or more numbers. For example, if the number 36 459 is recorded instead of 36 549, then the numbers 4 and 5 have been transposed. The key to eliminating this error is to read out the number that has been recorded so that the person who made the measurement can verify that the actual measurement has been correctly recorded.
3. *Process information.* You should process information as it is received to identify any gaps in your understanding. You can then ask questions until you have all of the information needed to fill those gaps. For example, 'I am not sure I fully understand that. It would help if you explained what happened again, but this time with a little more detail'.
4. *Write notes.* Wherever possible you should summarise the information as you are listening as this will help you to process that information.

5. *Read all of the messages.* You should 'read' the often subtle verbal and non-verbal messages that are conveyed with a message as they may be crucial to receiving the correct information. Sometimes the words that a person uses, the tone they use, or the way they speak can convey important information about the message they are communicating. Some non-verbal methods or informal languages that can be used to communicate messages include facial expressions, eye contact, hand gestures and the overall 'body language'. For example, a person may prefer to stand during a conversation to emphasise their confidence, strength or power.

   Sometimes you may have trouble understanding a phone conversation because the person is using non-verbal languages to communicate with you, even though you cannot 'read' these messages. Watch someone talking on a mobile phone and observe how they use facial expressions and even hand gestures to emphasise what they are saying. This is despite the fact that the listener is not able to access these messages!
6. *Attitude.* In some instances, your attitude or approach may mean that you do not hear a message you should be hearing. For example, if someone is concerned about how you will react to a piece of information, then they may skirt around the subject while they try to read what your response will be. If they suspect that you will get angry or abuse them, then they may simply choose not to tell you the information, and that may cause you, or them, even greater problems. Many leaders unknowingly encourage this behaviour when they develop a reputation for wanting to only hear 'good' news.

While you are a student you are in an ideal environment to practise your listening skills. Lectures, tutorials and practical sessions are all good environments for you to develop active listening skills. You can then practise using these skills when you work in groups or teams.

In team situations it is important to create an environment where team members are comfortable enough to speak about how they feel about the team, other team members and the project they are working on. This ensures that all team members have the opportunity to hear the messages they should be hearing if the team is to be successful.

But, remember, you have to consciously choose to be an active listener — it will not happen by accident. 'All of us, not just the overtalkers, stand to gain by speaking less, listening more, and communicating with intention' (Lyons 2023). This particularly applies to team leaders.

People who actively listen to what others say gain insights into profoundly different world views. This is particularly so when communicating with people who have English as their second language — remember the spotlight on Fletcher Aluminium in the previous chapter.

Asking probing questions, choosing to accept criticism or opposing ideas, and then *reflecting* on what another person says, and means, are methods we can use to enhance this powerful form of communication. Interpreted and applied in this way, active listening offers the potential to build strong linkages that are based on interpersonal trust and respect. Good listeners have the potential to develop solutions that are founded on an understanding of the values, priorities, feelings and beliefs of others.

KEY POINT

Listening actively is an important skill for students and practising engineers.

Telling

Telling is mostly one-way communication where, for example, a person tells a story to another person, or a group of people. In an engineering context an example of telling would be when an engineer gives instructions to a design team about the design of a component. Telling is a quick and efficient approach and it allows the 'teller' to maintain a position of power, or to act as the expert. The downside is that this approach provides little opportunity for the teller to check that the message was understood, or to find out if the audience agrees with the contents of the message, or if they have a better way of doing the task.

Discussing

Discussing is a two-way form of communication in which the engineer and the person or people they are communicating with get an opportunity to contribute their ideas, thoughts and feelings. While discussions can take longer and be more challenging than telling, due to a perceived loss of authority or control of the meeting, authentic discussions can generate feelings of mutual respect and openness, and create opportunities for the generation of shared ideas, plans and understandings.

A better approach would be to use all three approaches to ensure the message was received correctly. For example, and engineer may give a contractor a formal instruction by *telling* and then *listen* carefully to any questions or responses. Finally, the engineer would discuss the details of the instruction with the contractor to ensure that the instruction was fully understood.

Informal verbal communication

The information in table 8.2 makes a distinction between informal and formal communication. The difference is that informal communication tends to happen frequently and with little preparation, while formal communication is planned and normally occurs less frequently. Informal communication is one of the most frequently used skills for professional engineers.

Some of the key informal communication activities regularly undertaken by engineers include face-to-face conversations, or meetings, and telephone or video-mediated conversations and meetings. A technique commonly used by managers to identify issues and gauge the morale of their staff is to walk through their workplace and talk to some or all of the staff in their offices, laboratories or workshops. These informal conversations can give a manager an insight into what is really happening in the workplace, much of which may not be communicated through normal reporting channels. This approach also builds relationships.

Conversations

Because conversations with clients, colleagues, contractors, suppliers and members of the public may take up most of an engineer's working day it is important to develop the skills required to be able to converse with people from all sectors of the engineering industry, from all sectors of society and from other cultures. While inclusive language should be used at all times, an engineer may need to change the language, manner and style of a conversation to ensure their communication is professional, the information is relevant and that it is communicated effectively.

The telephone has always been an important form of communication and mobile technology has extended its capabilities. The mobile phone is ubiquitous in most engineering organisations because it enables engineers to seek instant answers to questions — whether they are on-site, in the office or on the other side of the world. Mobile phones can save time and resources on a project, because they can be used to resolve problems, as and when they occur. They can also be used to take and transmit photographs or videos of the matter being discussed, such as a component, a structure or a site. One downside, however, is that an engineer may learn to rely on the instant access capability of the technology and hence not properly prepare for their work or ensure they have all of the information and resources they are likely to need for a project.

Engineers often have to ask hard questions in the workplace, such as 'Why aren't you meeting project deadlines?', 'Why did you begin work on the plant before you completed a risk assessment and management plan?' or 'Why did you allow the concrete pour to go ahead when you knew the structural engineer had to check and sign-off on the steel reinforcement before the pour began?'

Asking hard questions is a challenge, especially when you are new to an organisation. For this reason many people avoid confrontation by not asking the questions that should be asked. This often makes matters worse! Good preparation helps as it is easier to ask such questions if you have thought through the different ways you could ask the questions and the way the person may react or respond. There are numerous questioning techniques that can be used to elicit the required information without offending the person being questioned. For example, you could be upfront and ask a question in a straightforward manner, or you could use conversation to create a relaxed environment and then ask the question. If you have previously considered the likely answers to the question, you will be ready for all possible answers, including 'left-field' and unusual answers. For example, how would you feel, and what would you do, if the foreman in charge of the concrete pour replied, 'It is your fault because you forgot to tell the structural engineer to check the mesh'?

One of the most difficult aspects of communication is communicating bad news. It is one of the challenges that we will encounter in both our personal and professional lives. The following spotlight considers communication in the medical profession, providing a history of the evolution of doctor–patient communication and examples of how good communication during everyday consultations, specialist consultations and when delivering bad news can have profound effects.

SPOTLIGHT

Communicating good and bad news

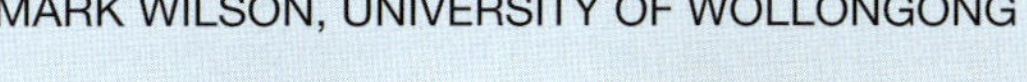

MARK WILSON, UNIVERSITY OF WOLLONGONG

Up until around 30 years ago, in the days of paper patient files, doctors were on pedestals and patients generally did as they were told, in a more patriarchal model of patient care. Over 2400 years ago, Hippocrates wrote:

> The physician ought also to be confidential, very chaste, sober, not a winebibber, and he ought to be fastidious in everything, for this is what the profession demands.

Since Hippocrates, bedside manners have varied greatly, from low points during early 18th century England, when a motley crew of physicians, barber surgeons, apothecaries and quacks 'treated' patients with dirty hands and instruments, selling a range of useless powders and potions, to the more austere and conventional late Victorian age, as medicine progressed towards being one of the most universally respected professions.

Before the first wide-scale use of effective antibiotics from the 1940s, it has been said that medical history was for the most part the history of placebos. The art of medicine, incorporating bedside manner, was often all that either doctor or patient could rely on.

Today, medical professionals are technocrats, consulting with the help of software packages linked to powerful databases and search engines. Patients rightly expect doctors to listen to their health concerns, do appropriate physical examinations and tests, and provide up to date, expert treatment advice.

We recognise a successful consultation with our doctor when we leave the surgery with a practical solution to our problem and a vibe that we have been an equal and respected partner in making health decisions.

Likewise, we know what happens when it all goes terribly wrong: the doctor didn't listen, you were rushed through the consultation, or you were given some medication but little in the way of advice.

Teaching doctor communication

Medical students are taught a lot about communication during their studies. At my university for example, from year one of the graduate medical program, students are given regular feedback on ideal 'doctor–patient' interactions from clinical teachers and from 'standardised patients', the hundreds of community volunteers from all walks of life who play the role of patients at the medical school.

Medical students are taught the skills with which to put their patient at ease, to take a comprehensive history relevant to the problem, and how to explore the biological, psychological and social aspects of a particular health issue for that person. Beginning from when the patient walks in, students are taught to employ open body language, polite and friendly introductions to establish rapport, to reassure regarding confidentiality, and to structure their questions in a logical and sensitive manner.

Even in their final year clinical examinations, 10% of the marks given to each student are from standardised patients, based on how likely they felt it was they would want to see the student again as their doctor.

There's also evidence good bedside manner can benefit the patient, either from care that aims to build positive expectations, or from emotionally supportive care.

A study of patients with high blood pressure told to expect a higher repeat blood pressure reading revealed higher levels than those of similar patients told their second reading would be the same or reduced.

Studies of patients with breast cancer have identified a number of beneficial effects of doctors' positive communication style. Providing treatment choice, involving patients in decision making, or touching the patient's hand and providing support, may reduce long term anxiety and depression, contribute to higher quality of life, or even improve physical functioning.

Later in the medical course, students observe their clinical teachers to get a better grasp on patient communication. It's vital students have excellent modelling of care for a patient in distress, or the considered, compassionate breaking of bad news to a patient.

How much empathy is too much?

This role modelling can be a double-edged sword. Many experienced doctors have evolved superb knowledge, skill and empathy towards their patients, and provide wonderful role models. However, as with any profession, there can also be negative role modelling, which can have a powerful impact on professional development, including bedside manner.

▶

There is some evidence that as medical students evolve into medical professionals, they may experience an erosion of empathy.

The quandary for medical educators, and individual doctors, is how to develop the best balance of empathy with clinical competence. An ability to temporarily suppress empathy may assist the doctor's brain to perform high level cognitive tasks and to emotionally detach while managing complex medical problems.

The practical ability of a surgeon operating on a brain aneurysm is obviously paramount, if they were thinking about their empathy for the patient in this moment they would not be able to perform in surgery. Another part of the puzzle is that, over time, risk of burnout and negative mental health effects may be greater for more empathetic doctors.

In future, a growing number of therapeutic and medical investigation options, the plethora of information driving patient and doctors' choices, and growth in the alternative medicine industry, are all likely to test doctors' bedside manner. Medical educators need to be good role models for incoming doctors, to ensure our future doctors feel happy and our patients feel cared for.

Source: Wilson (2018). Originally published on *The Conversation.*

CRITICAL THINKING

Think of a time when you experienced good personal communication from a medical practitioner. What made this communication good? Have you ever experienced poor communication from a medical practitioner? How was it poor?

Meetings

For many engineers, meetings are an important part of their work. Engineers hold meetings with clients, consultants, contractors, technicians, government personnel, colleagues, business advisers and component suppliers. These meetings may be formal (with minutes being recorded) or informal. Meetings can be held anywhere; for example, they may be in an office, on a construction site, or in a restaurant over breakfast or lunch. Meetings are used to brief clients and consultants, to present progress reports, to discuss and find solutions to problems, to issue instructions, to seek information, and to train staff in new techniques and technologies. See the working with people chapter for more information on meetings.

Recording oral communications

For legal and quality assurance purposes it is important to keep a record of oral communication. Normally, a written record is prepared, unless the conversation or meeting has been recorded electronically. Engineers generally record the key matters raised during a business communication, even if it is only for their personal records. The communication can be captured by summarising the key points, such as the participants in the communication, the time and date of the communication, the information exchanged in the communication and the outcomes of the communication. The key points may be written as a note to file, or may be circulated between the parties to ensure they all agree with the record of the meeting. Informal records of oral communication may prove crucial if a dispute arises between parties. While you are a student you could practise this technique by keeping notes of the team meetings you have while undertaking projects.

8.2 Formal verbal communication

LEARNING OBJECTIVE 8.2 Prepare and deliver verbal presentations.

Formal verbal communications tend to be pre-planned. They are often accompanied by supporting documents such as technical reports, plans, meeting minutes, a running order or printed PowerPoint slides. Delivering formal presentations is an important skill that engineers use to present the outcomes of their work, such as designs, progress reports or investigations. As a student, you should endeavour to learn and become proficient in using formal verbal communication skills, as they usually form part of the assessable tasks in undergraduate engineering programs.

Media releases, interviews and publicity

For large and/or important projects, engineers often use the media to communicate information to the public and to other stakeholders. For example, a project engineer may wish to advise the public that, due to road works, there will be delays on a stretch of road over a two-week period. In this case the engineer chooses which media to use for issuing a media release, for example, giving an interview, organising a press conference, publishing an advertisement or using social media.

The opposite situation can also occur; that is, the media can seek out an engineer to find information about a project or to obtain the opinion of a person working in the relevant field. In some cases, an engineer may face scrutiny over an aspect of a project that has entered the public domain; for example, the design and construction of the bund wall in Gladstone Harbour (see the chapter on understanding communication).

Great care should be exercised when speaking to journalists. Experts often find themselves in difficult situations because they are not able to answer questions for commercial or ethical reasons. If they say the wrong thing, or cannot answer what appears to be a simple question, then they may appear to be out of their depth, hiding something, or even unprofessional. Of course, the expert may have a perfectly valid reason for not answering the question; for example; if a thorough investigation is being undertaken to determine the cause of an incident or if the reason for a failure is technically complex and difficult to explain in simple terms without distorting the facts or attributing blame.

It is for this reason that many engineering organisations have a communication section that handles public relations issues or employ a communication consultant to advise them about these issues. If you decide to be involved with the media, you may benefit from undertaking a training course so you are well prepared and confident.

Presentations

Engineers are often asked to present a report they have prepared at project team meetings, company meetings or public meetings. For example, they may present a report on the design of a new computer system or a new renewable energy technology. Careful preparation is required to ensure the right information is communicated within the given time limits, and that any likely questions can be answered. Careful thought should also be given to the type and content of any media that are used as part of a presentation — such as plans, photographs, posters, videos or a slide presentation.

One way to improve your presentation skills is to study the techniques used by politicians and other speakers who catch your attention and keep you listening until the very end of their presentation. What is it about their presentation that makes them so good? Thinking about the following elements of a presentation may help you determine the factors that made a presentation effective.

- *Topic.* What was the topic? Why was it of interest? Did the presenter do or say anything that reinforced the relevance of the presentation to you? Was a complicated issue explained in relatively simple terms?
- *Pitch and pace.* Did the presenter linger on ideas that were particularly interesting or difficult to explain? Were there any parts of the presentation where you lost interest? If so, what do you think caused you to lose interest?
- *Intonation and language.* Reflect on the sound of the presenter's voice, their use of expression and different intonations. Did they use interesting turns of phrase to make the presentation engaging and easy on the ear?
- *Emotion.* How did the presentation make you feel? Did the presentation make you smile or laugh, or was it sad? Did anything that was said surprise you? Did it change your mind about an issue?

Reflecting on the different elements of formal presentations will help you become highly skilled at developing and delivering high quality and effective oral presentations during your engineering career.

Defining the purpose

The first step in developing a presentation is to define the purpose, that is, the key messages that are to be communicated to the audience. This is necessary because it focuses the presenter on communicating those messages. When doing this it is important to resist the temptation to include additional messages that, while they may be important, may overshadow the key messages and prevent them from being effectively communicated.

Once the purpose of the presentation has been defined, it may be expressed as a series of objectives. The presentation can then be developed to achieve the objectives and, of course, the purpose. For example, a student may initially list the following objectives for a ten-minute presentation. I want to:

- improve my confidence in public speaking
- engage the audience
- present a summary of progress on my project to date
- get an excellent mark for my presentation
- dazzle my lecturer and classmates with my knowledge of the topic
- help my classmates to learn about this complex topic

- get feedback from classmates on my proposed design
- generate excitement, enthusiasm or interest in my topic.

Students gain confidence in public speaking through practice. For the other objectives listed above, however, there are different approaches that can be used to develop the content and plan the delivery. For example, a student who wanted to dazzle classmates would probably pitch the technical or knowledge content of the presentation fairly high by including more complex ideas or theories, and then take the time to explain them to ensure that the audience understands most of what is said.

If, on the other hand, the student wanted to generate enthusiasm and excitement, the approach might be to present less technical detail and provide more examples of the relevance of the topic to everyday life, or examples showing how the topic may have an impact on life.

If the student wanted feedback on a conceptual design or a prototype, then the presentation could end with specific questions for the audience. In this case plenty of time would be allowed for the audience to discuss and answer the questions.

From this discussion it can be seen that it is unlikely that the student is going to achieve all of these objectives in a single presentation. This is because the style of the presentation would need to be different to achieve each of the objectives. Therefore, the student would have to review the objectives and perhaps select those that were more important, or those that could be achieved using the same style of presentation. Once the objectives were decided, the student would plan how they could be achieved by carefully selecting the information to be presented and the appropriate communication methods, formats and media (channels).

The second step in the process is to find out as much as possible about the audience. This ensures that the content, structure and communication strategies used in the presentation are appropriate for the audience and will engage them in a way that helps the key messages to be communicated.

A case study will be used to illustrate how a presentation may be planned and delivered.

CASE STUDY

Planning a presentation: Top that! Metal moguls cap an engineering icon

Elka has been asked by her local Engineers Australia group to prepare a ten-minute presentation for the next quarterly dinner meeting. The topic is 'An Australian engineering icon'. After some research, Elka decided that she would do the talk on the redevelopment of the Spencer Street Railway Station, which transformed the station into the renamed Southern Cross Station. She also decided the purpose of the talk would be to highlight the engineering in the design and to show how beautiful 'raw' engineering can be.

Elka expected between 30 and 40 engineers would attend and that they would be from a range of disciplines. In addition, because partners may attend a dinner meeting, Elka estimated that there would be between 60 or 70 people present for the talk. Elka also noted that the presence of non-engineers meant that the technical content of the talk would have to be limited.

Southern Cross Station is one of the most important rail terminals and public transport interchanges in Victoria.

Working to presentation time limits

The third step in the process is to identify the timing and other requirements for the presentation. Formal presentations vary, but it is reasonably unusual to be invited to speak for long periods of time unless you are widely recognised as having in-depth expertise in a particular area. The key information required from the organiser is, firstly, the time allocated for the presentation and, secondly, the expected time to be allocated for questions and whether this is included in the presentation time.

Generally, speakers are allocated between ten and twenty minutes, with an additional five to ten minutes for questions. The following rules of thumb can be used to manage the allocated time.

- How much information can be presented in ten minutes? It is helpful if speakers know approximately how many typed pages of information they can present in, say, ten minutes. For example, a speaker may present between four and five pages of information in ten minutes. While this will vary from presentation to presentation, and depend on the other media used, this information will prove to be useful when a speaker is developing a presentation.
- What style should be used for the presentation? For example, the speaker could give a speech, talk to a set of PowerPoint slides, or use a conversational approach and allow members of the audience to ask questions during the talk. More information is likely to be presented in a speech than if slides or a conversational approach is used. In fact, if a conversational approach is used, it may mean the speaker is unable to present all of the information because of the time spent answering questions. If a slide presentation is used, then approximately one minute should be allowed for each slide. This means that a ten-minute presentation would normally consist of about ten slides. If a slide includes a picture or photo that requires little explanation, only 30 seconds should be allowed for the slide. If a slide has a table, graph or a complex set of equations that need to be thoroughly explained, then two or three minutes should be allowed for the explanation.

CASE STUDY

Working to time limits: Top that! Metal moguls cap an engineering icon

The meeting organisers had advised Elka that the total length of time available was 20 minutes. Elka decided that the talk would run for about ten minutes and this would be followed by ten minutes for questions and a discussion about the beauty of 'raw' engineering. She had also downloaded a video that could be played later in the evening if there was sufficient interest.

Elka then prepared a rough plan of the presentation.

Introduction: two minutes, two slides

- Fascinating story about the early days of Melbourne's Spencer Street railway station
- Introduction to the presentation

Body: six minutes, six slides

Address six questions about the redevelopment.

- Why redevelop?
- What were the aims?
- What were the design imperatives?
- Who were the people involved in the design and construction?
- What were the construction details?
- What were the outcomes?

Conclusion: two minutes, two slides

- Why is it an icon?
- The raw engineering is there for all to see . . . Isn't it beautiful!

The presentation contexts

Step four in the process is to consider the contexts in which the presentation will occur. The information required includes details about the venue and equipment that may influence the planning and delivery of the presentation.

- What is the size and shape of the room?
- What media will be available in the room (e.g. projector, computer and video projection facilities)?
- Will a computer be provided at the venue? If so, will it be available prior to the function so that any media can be loaded and tested?
- Will a microphone be available?
- Will a laser pointer and slide changer be available?
- Will internet access be available?

CASE STUDY

Presentation context: Top that! Metal moguls cap an engineering icon

The monthly dinner meetings are held in a conference room at a local hotel. The talk will be given between the main course and dessert. A laptop computer, portable projector and screen, lectern, microphone and amplifier will all be provided at the venue. The internet cannot be accessed from the room except by using a mobile broadband device.

Content and structure

The fifth step in the process is to identify and write drafts of the key messages that are to be communicated. The aim is to select the most significant and interesting aspects of the story and to tell them well. There are a number of ways that the content can be selected. For example, the key points can be identified first and then the information gathered to support and present these key points. An alternative method is to gather the information, prepare a full draft of the talk, and then select the key points from the draft talk.

The following points should be considered when deciding what should be included in a presentation.

- What information are members of the audience likely to know already?
- What information will help achieve the objectives of the presentation?
- What information is required as evidence to support the key messages?
- What information is required to make an interesting and cohesive story?

It is very tempting for a speaker to want to tell the audience everything they know about the subject, particularly if the project they worked on was exciting, challenging, innovative or complex. However, this usually means the talk will be too long. One way to check this is to do a timed rehearsal. If the talk goes over time, or if it has to be rushed to fit the time, then the content will have to be reduced.

While there are many different structures for a formal presentation, the basic idea is to use the introduction to tell the audience what the talk is about and, if appropriate, to introduce the key messages. The body of the presentation is then used to tell them about it and the conclusion is used to summarise the information that has been presented and to reinforce the key messages. Following this approach, a general plan for a talk about a project would be as follows.

1. *Introduction.* Introduce the project, topic or research question (including the purpose of the presentation). This should include a clear description of the topic — including the history of the issue — and the main points the presentation will address.
2. *Body.* Describe how the research questions have been answered, and the results that have been obtained.
3. *Conclusion.* Summarise the main points of the presentation and explain the general conclusions of the research.

Remember, in most situations an audience will have decided to attend because they are interested in the project, so the speaker should provide only essential background information. If necessary, additional information can be provided when answering questions. Generally, an introduction should take about 15 per cent of the allocated time for the presentation and the body of the presentation should use about 60 per cent of the time available, leaving 25 per cent for a conclusion. So, for a ten-minute presentation, there should be around a minute-and-a-half introduction (two slides), a six-minute body (six slides) and about a two-and-a-half-minute conclusion (two slides).

Remember, the allocated time to speak is likely to be only part of the time allocated for the presentation. It is customary to leave at least five minutes at the end of a short formal presentation for questions and comments from the audience.

CASE STUDY

Content and structure: Top that! Metal moguls cap an engineering icon

Elka planned the content of the talk, noting the source of each piece of information.

Introduction: two minutes, two slides

- Start with the story of the development of the Batman's Hill Railway Station in 1859 — the first railway station on the site — and how it was nearly removed in the 1880s to make way for the expansion of the city.
- Introduce topic: The 2002–06 redevelopment of the station.

Body: six minutes, six slides

- *Why redevelop?* The existing station and its facilities were not able to support the growing transport needs of the city, the planned Dockland developments adjacent to the site, or act as an efficient transport hub for regional Victoria.
- *What were the aims of the project?* To transform the drab 1960s architecture and facilities of the existing station into a world-class station and transport interchange, with supporting shopping and commercial facilities.
- *What were the design imperatives?* The design aimed to provide:
  - modern and fully sheltered train and coach platforms and enclosed waiting areas
  - efficient entry and exit points and interchange facilities
  - aesthetic and efficient links to the city and the new Docklands precincts
  - a shopping and commercial precinct
  - car drop-off zones and an 800-space car park.
- *Who were the people involved in the design and construction?*
  - Public–private partners: Victorian government and the Civic Nexus consortium
  - Architects: Grimshaw Jackson
  - Structural engineers: Winward Structures
  - Construction: Leightons Contractors
  - Design award: The people involved received the 2007 British Architects' Lubetkin Prize for the most outstanding building outside the European Union. (Elka noted that she should check if the project had received any other awards.)
- *What were the construction details?* The redevelopment occurred during the period 2002–06. There were major construction issues related to keeping the station operating while removing the old structure and installing the new structure in among the power lines that drive the suburban train system. Leightons claimed they had trouble getting enough time every evening to make this happen and it was one of the reasons stated for completion delays, as well as the cost over-runs and a subsequent 'variations' claim.
- *What were the outcomes?* The $700 million project has transformed the station precinct to create an aesthetically stimulating transport facility that is light, open and safe, and provides passengers with a twenty-first century travel experience. The engineering structures that support the architectural design features are highly visible, particularly the structures that support the long external glass walls and the rolling moguls that form the 37 000 square metres of roof over the terminal complex and the platforms. The sustainability features built into the design, such as the light pillows in the valleys of the roof, and fume extraction systems at the peak of the moguls.

Conclusion: two minutes, two slides

- Highlight the engineering in the design. The 'raw' engineering is there for all to see — beautiful!

Elka noted the sources for the information she planned to use in her talk: Das (2005a, 2005b), Department of Transport (2009a, 2009b), Rollo (2006), V/Line (2009).

Selecting and preparing media

The sixth step in the process is to select the most appropriate communication method, channel and media for each message. Presenters can use a range of media to enhance their presentation, including demonstrations, slides, video clips, maps, models, plans, posters, and examples of components and equipment. The media should be carefully selected to ensure they help achieve the objectives by supporting the communication of the message. Only one or two types of media should be used so that the audience is not distracted, and to avoid wasting time while switching between media.

Preparing slide presentations

Think about a good slide presentation you have seen lately. What made it so good? Was it the colour scheme or the layout of the information on each slide? Or was it good because the presenter included a mix of pictures, text and diagrams? Think of a bad slide presentation you have seen lately. What made it so bad? Why did the presentation fall into the 'Death by PowerPoint' category?

The following tips can be used to prepare clear and engaging slide presentations. You can build on these tips and modify them to your own style as you become more accustomed to preparing and delivering formal oral presentations.

- *Lettering.* The lettering should be at least 24 point, have a clear typeface, be printed in a strong colour, and be well spaced. Different typefaces and colours should be tried to achieve a strong contrast between the lettering and the background. This should be tested by standing well back from the computer — if the words are difficult to see, then a different font or colour should be used. The presentation should also be trialled in the room that is going to be used for the actual presentation. This is because some colours tend to wash out over longer distances. The effectiveness of the slides may also be affected by the quality of the screen.

- *Information.* The information should be presented in short and simple sentences, each with a clear meaning. The spelling should be carefully checked, as mistakes are distracting and look unprofessional. If bullet points are used, the content should be abbreviated. Preferably only one main heading or theme should be allocated to each slide, and no more than 30 words per slide. The most powerful slides often just contain a simple statement and a relevant photo or graphic.
- *Tables and diagrams.* The information in tables should be at least 16 point and include horizontal lines for ease of reading. Avoid excessive amounts of detail in tables, graphs and diagrams. A lot of detail will confuse the audience and might obscure the main message being communicated.
- *Pictures.* Include graphics such as photos and cartoons as these will add interest to a presentation. However, pictures should only be used when they assist in telling the story. Cute or funny cartoons can make a presentation appear lightweight or silly and distract the audience. Make sure images are clear and visible from the back of the room.
- *Motion.* The motion capability of a slide presentation can be used to capture attention. This feature should be used selectively, as it can become repetitive and lose its effect. Having words and images zooming, dematerialising and bouncing all over the slide may also irritate some members of the audience. The use of motion may also decrease flexibility and cause problems with timing, particularly if multiple clicks are required to bring up the lines of text on a page.
- *Humour.* Starting with a carefully selected and, relevant, humorous story, picture or cartoon can engage the audience from the outset. A poorly chosen feature may have the opposite effect.

The key is to keep the slides simple so that the main points are clearly communicated to the audience. We will revisit this topic in the chapter on communicating information, where the focus will be on the use of slide presentations to communicate engineering information.

CASE STUDY

Preparing media: Top that! Metal moguls cap an engineering icon

Elka developed the content for each of the slides.

Introduction

- Slide 1: Title and an early picture of Spencer Street railway station
- Slide 2: Introductory statement and picture of Spencer Street railway station in 1960

Body

- Slide 3: The reasons why the station was redeveloped and project timelines
- Slide 4: The aims of the project
- Slide 5: The design imperatives
- Slide 6: List of the companies involved in the design and construction
- Slide 7: Construction challenges — picture of suburban train, platform and overhead power cables
- Slide 8: List of key engineering and sustainability outcomes

Conclusion

- Slide 9: Why it is an icon? Picture of the structural members supporting the roof — the beauty of raw engineering fully exposed
- Slide 10: Conclusion

Elka also prepared a one-page handout summarising the main points of the talk and the references.

Assembling and rehearsing the presentation

The seventh step in the process is to assemble the communication package and, in this case, test it by rehearsing the presentation. Rehearsing helps to reduce anxiety in the lead-up to the presentation. It is important to run through a presentation to check that the information is communicated clearly, effectively and within the allocated time. The speaker also becomes comfortable with using the selected media and is able to identify when best to change to a new slide or to give a demonstration.

Generally people tend to speak faster when they are nervous, so it is important to practise speaking slowly. Some speakers print a notice on the first page of their notes reminding them to SLOW DOWN!

Another reason for practising is that with practice the speaker should become less reliant on reading from the notes and slides. The potential flow-on benefit from this will be increased eye contact with the audience, which should help engage the audience and make them feel included and want to be actively involved with the presentation.

Finally, a speaker can arrange for a presentation to be recorded on video. This can be done prior to the presentation and the speaker can then review it to identify any omissions or weaknesses. This can be an illuminating experience, as the speaker will see any annoying mannerisms and hear the sound of their voice. The actual presentation may also be recorded by the organisers, perhaps for uploading to the organisation's website. The speaker may review and critique the presentation a day or two after the event and identify ways in which it could be improved; this information can then be used to improve future presentations.

As a student, you may consider rehearsing in front of one or two classmates, perhaps with the other members of a project team. They should provide valuable feedback and may ask you some good questions that will help you to refine or even change the content. If that is too challenging, you may like to video and then review your presentation.

CASE STUDY

Assembling and rehearsing the presentation: Top that! Metal moguls cap an engineering icon

Elka wrote out a draft of her talk and rehearsed it with the slides to check timing. As it was three minutes too long she removed some sections and rehearsed it again. This time it was just under ten minutes.

Elka then thought about the questions that the audience may ask and researched answers for those questions.

Is approval required?

When a talk is being presented on behalf of an organisation or is about a project being undertaken by an organisation, the speaker may need the approval of that organisation for both the content and the presentation. This is the eighth step in the process, where the communication package is forwarded to the gatekeeper for checking and approval.

Presenting through words and action

The ninth step in the process is delivering the presentation. During the presentation it is important to remember that the members of the audience will be receiving verbal, non-verbal and visual messages. It is clearly important for a speaker to consider the use of non-verbal methods such as hand movements. Speakers can use the following methods to manage their voice and actions to enhance their presentation.

Voice

- Speakers should talk slowly, clearly and a little more loudly than normal.
- Speakers should memorise opening lines so they can deliver them calmly, get into the swing of their talk, and establish eye contact with the audience.
- Speakers who have a quiet voice, or who tend to mumble, should try to open their mouths wider, breathe deeply and make their tone deeper than usual.
- Speakers should speak from their diaphragm as it uses less energy than speaking from the throat or nose; it is also less likely to strain vocal chords. They can practise this by putting a hand on their lower chest, breathing in deeply and attempting to make their hand vibrate with their words.
- Speakers should use a tone of voice that shows they are enthusiastic about their topic.
- Speakers should vary the speed and loudness of their voice to maintain audience attention and to signal important points in the talk. They should use pauses to get the attention of the audience, and not be afraid of silence.

Actions

- Speakers should make eye contact with people in different parts of the audience and overcome nervousness by finding a couple of people in the audience who look friendly and glance at them.
- Speakers should be aware of distracting habits and control them, habits such as hair flicking, fiddling with clothes, hand gestures, and rocking backwards and forwards.

- Speakers should use exaggerated gestures and facial expressions when speaking to large audiences; this makes it easier for people in the back row to get enthused about the presentation.
- Speakers should always face the audience and feel free to walk around the stage or floor area, using hand movements to emphasise important points or to help describe something.
- When standing, speakers should stand up straight, maintain a comfortable stance and avoid slouching.

Answering questions

The tenth and final step in the process is to provide an opportunity for members of the audience to respond to the presentation. For some speakers, delivering a formal presentation is a 'walk in the park' compared with the apprehension they feel about answering questions from an audience. While question time can be a great opportunity for a speaker to parade their knowledge, a common concern is, 'What if they ask me something I just can't answer?' The prospect of being 'dumbstruck' in front of an audience is profoundly unsettling. There are several ways to avoid this and to deal with it if it happens.

The first strategy for answering difficult questions is to prepare by thinking in advance about the questions an audience might ask and planning one or two sentence replies. When a talk is being prepared the speaker should try to think of any questions they really hope nobody will ask, and prepare to respond to them (even if it means having to concede a gap in their thinking, research or presentation).

If a speaker is stuck on a question a strategy is to stall the answer and 'buy' some time to think by repeating or rephrasing the question or by asking the questioner to clarify the question. A useful phrase is, 'If I am right, you are asking . . .' or 'I didn't quite follow that question, would you mind repeating it?'

Questions often have a few parts, so a useful strategy for the speaker is to answer the part of the question that they know, and then comment in general terms on what they know about the other parts of the question. If a hard question appears to be outside the scope of the work or topic being presented, one strategy is for the speaker to politely explain to the questioner why their question is beyond the topic or set of ideas that are being presented. A final strategy is for the speaker to concede that they do not know the answer to the question but will follow up and get back to the questioner with an answer at a later time. This strategy should only be used as a last resort because a speaker who has thoroughly researched a topic will generally be able to answer the majority of questions about that topic.

Speaking at a formal level is daunting for most people. As a budding engineer, you will need to learn how to speak in a confident and relaxed manner in formal gatherings. Like most other things in life, practice will help you improve your performance. Everyone can learn to speak effectively, but most people have to recognise their need for practice and training before they are willing to make the effort to improve their skills. Your engineering program will provide you with teaching and practical activities that will help you to enhance your oral communication skills. You may also like to consider joining a public speaking organisation, such as Toastmasters, to help you develop your skills.

Finally, our confidence in speaking about a topic also depends on our knowledge about the topic and, where appropriate, our experience working in that field.

8.3 Written communication skills

LEARNING OBJECTIVE 8.3 Plan and prepare written communications.

Engineers and engineering students use a wide range of written communication skills and these are discussed in the following sections. The first skill is reading.

KEY POINT

Reading effectively is an important skill for students and practising engineers.

Reading

As a student in an engineering program you will have plenty of opportunities to practise and enhance your reading skills. The key is to focus on your purpose for reading a document. One way to define your purpose is to ask some questions before you begin reading. The questions you ask will, of course, depend on the type of document you are reading, and the context.

For example, as a student, you could consider the following questions when reading textbooks or other course materials.

- Why am I reading this document?
- What information am I looking for?
- Is this information from a reliable source?
- What are the key pieces of information in the document?
- Are the outcomes supported by valid evidence?
- Do I fully understand the information and how to use it?
- Is this information outside the scope of my project?
- How does this information fit with the lectures, tutorials or practical sessions I have attended on this topic?
- Where does this information fit on my engineering knowledge framework?
- What else do I need to know about this topic?
- Where can I find out more about this topic?
- How can I easily remember this information?

In addition to these questions, engineering practitioners are likely to consider the following questions when reading articles, practice notes, technical reports or research papers.

- Do I agree with the findings?
- Does this fit with my experience?
- How can I use this information in my work?

A key concept to note here is that information often comes with **metadata**, which can be described as information, or data, about data. For example, when you take a measurement in a laboratory you would normally record the following metadata: the name of the people who took the measurement; the time and date; the manufacturer, model and serial number of each piece of equipment used to take the measurement; the reason the measurement was taken; and any assumptions you made when taking the measurement. Thus, metadata consists of information about the measurement, which is the context, the time and date, the equipment and the personnel. Then, when the data is used, it is important that any relevant metadata is considered as part of that process. This will ensure that the data is used correctly. For example, a report on an operational incident in a power plant would normally include the time, date, production levels and other relevant statistics. This metadata may help an engineer identify and understand the reason why the incident occurred.

One strategy that can be used to gather information efficiently is to scan documents to quickly determine their relevance. This is called skimming and the aim is to locate any of the key words or phrases you are interested in. You should initially look at the title, abstract, introduction, table of contents, headings and conclusion. If any of the key words are found, then the document may contain information that is relevant to your needs, in which case it warrants reading more carefully. It may be necessary to read the whole article, the rest of a chapter and any other relevant chapters to gain the information you require. When reading reports or journal articles you should avoid the temptation to read only the executive summary or the abstract and then simply use or cite that information in your work.

The key point is that, before using any information, you should ensure all of the relevant sections of the document have been read so that you have a good understanding of the information, the context and the metadata. This topic will be addressed in the chapter on understanding the problem, which also includes a section on analysing information and evaluating sources of information.

When engineers read technical documents they normally take notes and summarise all of the relevant information and the sources of that information. You should do the same.

Writing

The preparation and use of many different types of written documents is a major part of an engineer's work. In fact, just over 65 per cent of the UK engineers that responded to a survey reported that they spent at least 30 per cent of their time writing at work, with 15 per cent of them reporting they spent more than 60 per cent of their time writing (Sales 2006, p. 7).

Engineers write many types of documents and each type of document has a particular purpose and is associated with particular conventions relating to the type of information that should be included, the structure of the information, and the format and layout of the document. These conventions enable regular users of these documents to efficiently navigate their way to the information they require for their purposes.

Informal writing

Writing notes

In this section, we focus on learning some techniques that you can use to write notes while you are a student. These notes will enable you to keep accurate records of the key pieces of information that you hear in lectures, meetings or other gatherings. There are many note-taking techniques, so people select the techniques that suit their individual preferences or a particular context. Common techniques include:

- noting key concepts, principles and issues
- writing as much as possible to summarise a whole lecture
- writing short sentences, or dot points, from which more detailed notes can be written later
- annotating the printed handouts from a slide PowerPoint presentation
- typing notes directly into a laptop computer or other electronic device, or even handwriting notes on a tablet style computer that automatically converts the notes to a standard electronic text document
- sketching concept maps to create the structure of a lecture. These maps can be further enhanced by using coloured pens to define different parts of the structure.

Regardless of the note-taking technique used, it is important to record any metadata associated with the information that is recorded. For example, when writing notes about a lecture, the notes should include the lecturer's name, and the date and time of the lecture. This not only defines the source of this information, but also enables you to follow up with the relevant person if you need to clarify something at a later date. Any references that a lecturer refers to during a lecture should also be recorded, as these sources can be used to gather additional information.

As noted earlier, when you are reading you should write notes about the key ideas, the details of the source, publication details, page numbers, URLs and dates of access. This ensures that the information can be easily sourced again at a later date, and also facilitates correct citing and referencing in assignments and reports. Information sourcing and referencing will be covered in more detail in the chapter on understanding the problem.

Note-taking can also be an effective way of recording the important points discussed during telephone conversations. Once again, the metadata about the conversation, such as date and time, is noted so that the information can be cited at a later date if required.

When you are deeply engaged in reading a document or listening to a presentation, you may find your mind buzzes with ideas about the topic. You may think of other examples related to the topic or issues that you think have not been addressed, or how you could use the idea, technique or information in your work or study. You may even have a flash of brilliance and solve a problem, or think of a way to extend or adapt the concept being described. You should capture these thoughts before you lose them, noting that they are your ideas or reflections. You can do this by simply writing your initials beside these notes.

It is easy to evaluate the effectiveness of your note-taking techniques by reflecting on the following five questions.

1. Are your notes easy to understand when you re-read them at a later date; for example, when you are preparing for an examination?
2. Which note-taking technique provides you with the right amount of information for your needs?
3. What information do you consistently forget to note?
4. What sort of information do you find difficult to understand?
5. How easy is it to find all of the notes you have made on a topic?

The answers to these questions will highlight any weaknesses in your note writing capabilities and you can then change your approach to improve your techniques.

Finally, it is important that you file your notes so that they are easily accessible. This may be in an electronic database, or in a project file.

KEY POINT

Writing accurate notes about conversations and meetings is an important skill for students and practising engineers.

Formal writing

A formal communication should be written in a style that is engaging, factual and straightforward; it should be devoid of the gestures and accents of normal conversation. Wherever possible, simple English should be used and significant terms should be defined so that the reader has the same understanding as the writer. Some of the other key points are as follows.

- Clichés, colloquialisms, euphemisms, doublespeak and slang should not be used, and wordiness — using unnecessary phrases — should also be avoided.
- Jargon (discipline-specific and specialised terms) should not be used unless the creator knows the consumer will understand the jargon.
- The full spelling of words like minimum or maximum should be used rather than shortened versions such as 'min' or 'max'.
- The full text of acronyms and abbreviations should be given when they are initially used, along with the acronym or abbreviation shown in brackets; for example, the Office of the Australian Information Commissioner (OAIC). The abbreviation can then be used in the remainder of the text. Abbreviations commonly used when texting mobile phone messages should *not* be used in professional emails or other work-related forms of communication.
- Inclusive and non-discriminatory language should be used so that the communication does not offend, abuse or harass the reader. Most organisations have human resource policies that are designed to ensure communication is conducted in a manner that does not offend, abuse, discriminate or harass. These matters are also regulated by federal and state legislation; for example, work health and safety (WHS) legislation (which was discussed in the chapter on professional responsibility and ethics).

When a type of document is regularly used in an organisation, such as a formal instruction, a template is developed so that only the information that is project- or task-specific has to be added or amended to create a new document. In some cases it is easier to recycle an existing document than to start afresh. In this case, care should be taken to ensure that there are no copyright issues, and that the existing document does not contain any sensitive information. Once these issues have been resolved, a copy should be made of the existing document and amended to create the new document. This needs to be done carefully to avoid the retention of information that is not relevant or appropriate for the new document. One way to avoid this error is to change the colour of the text in the copied document to another colour, for example, blue. Then, each section will remain blue until it has been reviewed and either rewritten or deleted.

Essays

As a student you are probably familiar with this style of writing and know that essays are used to present and argue ideas. While essays are not a common form of assessment in engineering schools, they are widely used in other disciplines, such as those in the arts and business faculties. Engineering students need to be able to use this genre because they are likely to study one or more subjects offered by those faculties. In most cases the structure of an essay is the same: introduction, body and conclusion.

The *introduction* is used to introduce the reader to the topic. The aim is to entice the reader to engage with the ideas presented in the body of the essay. The introduction also sets out the author's response to a topic and the approach that is taken in the essay. The introduction should be written boldly, so readers are encouraged to read on and engage with the author's arguments.

The *body* of the essay consists of a series of paragraphs, with each paragraph addressing a specific topic or argument. The arguments are normally supported by evidence and are arranged so that they build the author's case in a logical manner. The paragraphs may therefore be gathered into a number of sections, each addressing a different topic or argument.

The *conclusion* should draw the main ideas together and demonstrate that the aims of the essay have been achieved. The arguments may be summarised and key points may be highlighted, but no new material is presented in the conclusion.

There is some variation in the academic styles and formats used in different professions and even within engineering disciplines. Therefore, it is important that you check the style and format required for each assignment. For example, the style and format requirements for an essay on contract law in a subject offered by a law school are likely to be different from those in a subject offered by an engineering school. The same variation can occur within engineering, with some disciplines using different formats and referencing styles.

Reports

Reporting is a regular activity for most engineers. There are many different types of written reports, including project reports, and the structure of each type varies depending on the purpose. Generally, however, reports contain all of the following sections, although the order of the sections may be varied to suit the contents of the report. For example, the table of contents may be placed immediately after the title page.

Part 1: Front matter
- Title/title page
- Purpose
- Acknowledgements
- Executive summary
- Table of contents (including, where appropriate, a list of tables and figures)

Part 2: Body
- Introduction
- Information — main section and subsections
- Discussion
- Conclusion
- Recommendations

Part 3: End matter
- References/bibliography
- Index
- Glossary
- Appendices

In small reports the parts may not be separated, and some sections may be merged and others deleted. In large reports the structure may include some additional sections. The same structure is often used for student project reports, although this will depend on the degree program being studied and the educational institution. Generally, the structure and formatting details are provided with the instructions for the project or in a project guide that is used throughout the program. For example, the *Report writing guide for mining engineers* (Hagan & Mort 2018), and the accompanying video of a discussion on report writing, can be downloaded from the UNSW Mining Engineering website. Final-year student project reports will normally include additional sections such as acknowledgments, statement of originality, and so on.

There are many types of reports and some of the more important ones are described in the following sections.

A progress report

This is a status report on a project, or a report on a person's role in a project. Progress reports usually contain a description of progress against targets, a discussion about any incidents that should be reported, a list of changes or variations that have occurred, a detailed list of the work proposed to be undertaken over the next period, and any recommendations for future phases of a project. The *progress against targets* section may include progress against defined work targets, against the project budget and/or against other performance criteria. Progress reports may be required periodically (e.g. once a month) or when project milestones are reached.

A final report

A final report is written when a project has been completed. It should describe and comment on all aspects of the project and provide technical details, costs and schedules, as well as an evaluation of the project outcomes against the specifications and other contract requirements. A final report should also provide detailed information about any variations to the contract, plans or specifications.

A technical report

A technical report is used to document the findings of an engineering investigation. Technical reports usually contain a description of the problem being investigated, as well as details such as the methods used, the results, an analysis of the results, issues to be resolved, recommendations and a conclusion. Technical reports normally list and discuss the impact of any assumptions that were made while using the technology, developing a process, or calculating statistical data, as well as the results of any measurements that were made, or the source and quality of any existing data that were used.

Delivering a report

Once completed, a report would normally be sent to the client with a covering letter, often called a letter of transmittal. The sample covering letter presented in figure 8.8 highlights the standard information that may be included in a covering letter. The optional information is shown by dotted lines.

FIGURE 8.8 A sample covering letter for a technical report

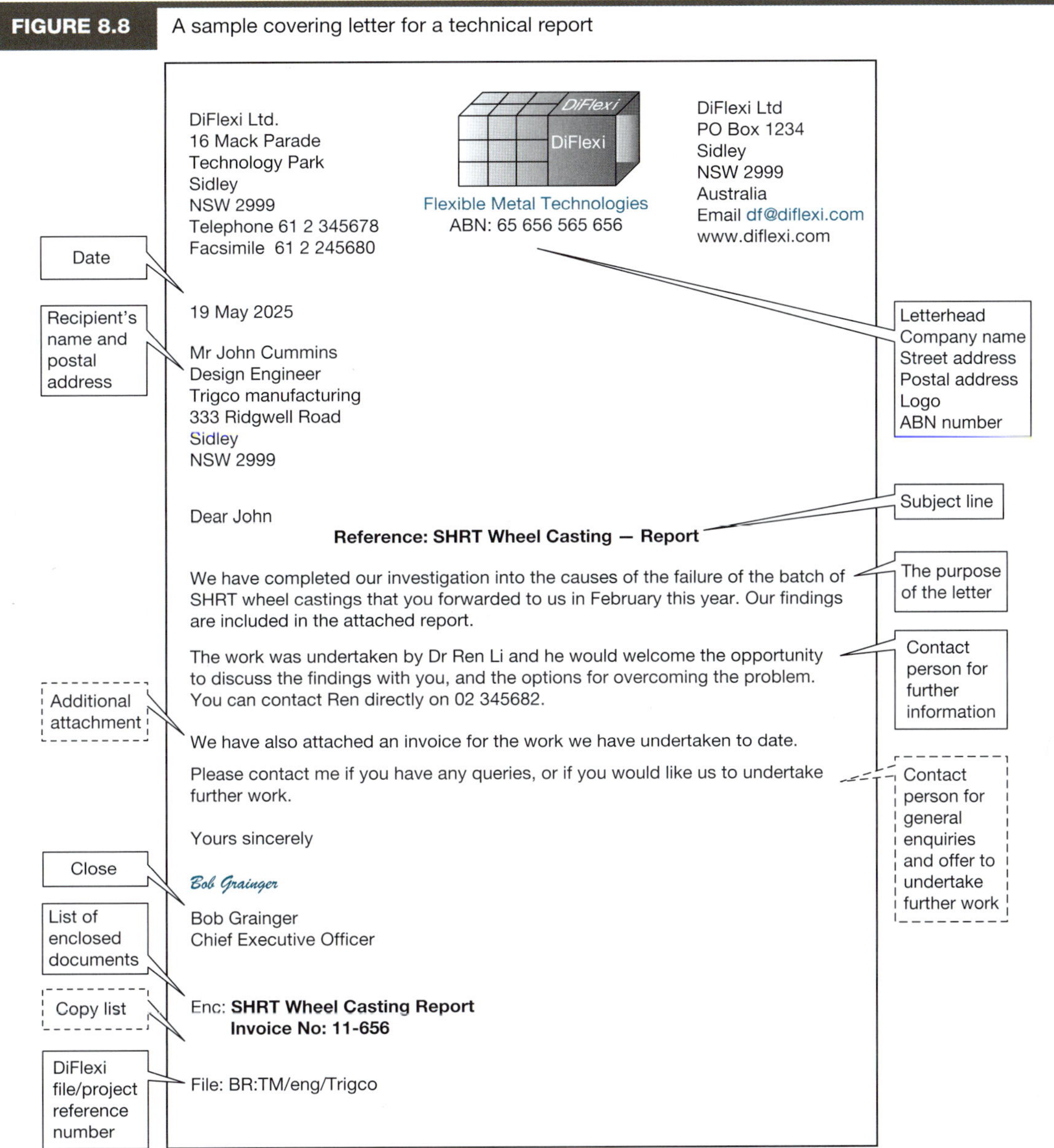
DiFlexi Ltd.
16 Mack Parade
Technology Park
Sidley
NSW 2999
Telephone 61 2 345678
Facsimile 61 2 245680

DiFlexi
Flexible Metal Technologies
ABN: 65 656 565 656

DiFlexi Ltd
PO Box 1234
Sidley
NSW 2999
Australia
Email df@diflexi.com
www.diflexi.com

19 May 2025

Mr John Cummins
Design Engineer
Trigco manufacturing
333 Ridgwell Road
Sidley
NSW 2999

Dear John

Reference: SHRT Wheel Casting — Report

We have completed our investigation into the causes of the failure of the batch of SHRT wheel castings that you forwarded to us in February this year. Our findings are included in the attached report.

The work was undertaken by Dr Ren Li and he would welcome the opportunity to discuss the findings with you, and the options for overcoming the problem. You can contact Ren directly on 02 345682.

We have also attached an invoice for the work we have undertaken to date.

Please contact me if you have any queries, or if you would like us to undertake further work.

Yours sincerely

Bob Grainger

Bob Grainger
Chief Executive Officer

Enc: **SHRT Wheel Casting Report**
Invoice No: 11-656

File: BR:TM/eng/Trigco

Research and practice articles

Engineers write conference papers or journal articles to report on an innovative practice they have implemented, the findings of an investigation or a research project. The articles contained in a refereed journal are peer reviewed to verify that the research process and results are valid and supported by evidence and by earlier articles. The reviewers are normally drawn from a panel of experts in the relevant field. Generally, authors do not know the names of the people who have reviewed their work as the copies of the reviewers' comments, suggestions and so on are anonymised. This opaque review process is normally referred to as a 'blind' review. Conference papers may also be refereed.

Often researchers and practitioners present their work or initial research outcomes at a conference to receive feedback from their peers about their work. They may then incorporate any new ideas or approaches into their work, which, when completed, may be published as a journal article.

Research papers normally begin with a description of the **hypothesis** that was explored during the research. According to the Macquarie Dictionary (2024), a hypothesis is:

> a proposition (or set of propositions) proposed as an explanation for the occurrence of some specified group of phenomena, either asserted merely as a provisional conjecture to guide investigation (a working hypothesis), or accepted as highly probable in the light of established facts.

A journal article generally includes the following sections.

- *Abstract.* This is a brief summary of the topic that is addressed in the paper, the hypothesis, the approach taken and the findings. An abstract should entice the reader to read the rest of the paper.
- *Introduction.* This section provides the reader with background information about the topic. The author normally cites and discusses recent research and aspects of current practice that relate to the hypothesis.
- *Method.* The methods used to conduct the research are described in sufficient detail so others may replicate the work.
- *Results.* The results of any experiments, surveys or trials are given in sufficient detail so the reader is able to see how they support the outcomes deduced by the researcher.
- *Discussion.* The author discusses the results in relation to the hypothesis — noting whether the results support the hypothesis, whether there are any unexpected outcomes, and if there are any issues that may have impacted on the results.
- *Conclusion.* This is a summary of the research and the findings. The conclusion may contain recommendations for further research.

The research context

While you are at university, you will likely do some academic writing, for example, your final-year project report. Academic writing is used by lecturers, tutors, students and practitioners. University academic staff use academic writing when they prepare an article for a conference or journal; particularly if the article is to be refereed by their peers. This genre becomes the main form of written communication for people who pursue a career in research and development, or in academia.

The aim of academic writing is to influence thinking on a subject, or to change practice, by presenting ideas or research outcomes to practitioners and to the wider academic community. Most research papers are peer reviewed before they are accepted for presentation at a conference or for publication in a journal. In the PCR model, the peer reviewers undertake the role of gatekeepers because they decide whether an article will be published. If they accept an article for publication they are endorsing the ideas, innovations or research outcomes contained in the article. This does not mean that the ideas or research outcomes are faultless; rather, that they have been judged worthy of consideration by the wider community.

The academic publishing system relies on referees to detect and report on papers that lack rigour, containing, for example, poor methodology, poor writing, fake data or plagiarism. There is growing concern that the peer review process may not be as rigorous as it is commonly believed to be. A number of factors may be influencing the integrity of the system, including:

- the pressure on academic staff to publish papers in high-ranking journals
- the fact that the peer review process is generally not funded and relies on senior researchers to volunteer their time to review papers, a task that requires time and expertise if it is to be rigorous
- the fact that both the number of papers being submitted and the number of journals being published have multiplied in recent years, making it difficult for researchers to keep up to date in their field.

This concern has led to the establishment of a number of organisations who monitor and check research papers, for example Retraction Watch (2024), which was set up by two scientific journalists in 2010. An indication of the size of the problem is that more than 10 000 articles were reportedly retracted in 2023.

If there is a lack of confidence in the process then it will undermine the integrity of academic scholarship.

There are many indicators that can be used to assess the impact of a research outcome (such as if the idea is used by others, especially if it becomes part of common practice). Thus, the communication of research outcomes is an important part of the research and development process. In recent years, research funding organisations and academics have adopted a relatively crude measure to assess the impact of their research. This measure is the number of times a published article has been cited by other authors, and it is used as an indicator of the acceptance of the research. Of course, this does not guarantee that the research has changed common practice.

File sharing

For many years wikis were the only tool that students could use to collaborate on the preparation of a written report. A **wiki** is a web page that allows a group of people to access and communally develop and edit text. Unlike standard web pages, contributors do not have to use a special language such as HTML.

In recent years wikis have been replaced by software packages such as SharePoint and Teams, which are powerful tools for collaborative work and this is why they are often used by student teams to prepare project reports. They are also widely used in the engineering industry as secure cloud-storage systems have become available.

Web pages

Most organisations have created websites to let the world know about their organisation, what they do and the services they may offer. While many achieve their purpose, others fail because their website is poorly designed, or because it does not get listed near the top of the search results provided by a search engine. There are many books and websites about the creation of 'good' websites and there are many individuals and organisations who consult in this field by creating websites for other organisations. The following basic principles should be considered when creating a website.

The first decision to make is to carefully define the purpose, or purposes, of the website. Once this has been agreed, the site should be carefully *planned* so that it provides the appropriate information to the target consumers in a structure and format that they will be able to negotiate easily. An understanding of what information the target consumers will require and how they may read a website should be used to inform the design of the content, the structure and layout of the pages, the structure of the site, and the way the pages are linked. Navigating the site should be intuitive for consumers and a consistent layout and style should be used for each page.

The content of each page should be *simple* so that download times are minimal and the reader can quickly scan the page, particularly the text, to see if it is relevant for their purpose. The layout of each page should be designed to focus the reader's attention on the key pieces of information on the page.

The content should be kept *up to date* so that readers are encouraged to keep returning to the site. The creation date, or the date the page was updated, should be listed at the bottom of each page, along with the contact details of the website manager, usually called the webmaster.

The cost of maintaining the site should be carefully considered during the planning stages. It is important for an organisation to be realistic about its ability to allocate the staff and resources required to update and maintain the site on a regular basis. The cost of this will, in part, depend on the currency of information to be placed on the site. The cost of maintaining a site hosting information about products or services that only change every year or two is going to be much less than one where the information is changing on a daily, weekly or monthly basis.

The proposed web pages should be carefully *checked* before they are published. The content should be checked for accuracy and someone other than the creator should proofread the text. All of the links should be checked to ensure that they correctly navigate to the target pages, and that it is easy to return to previous pages.

Finally, it is critical to ensure that an ongoing element of website design is protection of the site from cybersecurity threats, which are becoming more sophisticated and ubiquitous each year.

A good way to learn about the key principles of web design is to evaluate the web pages and websites you visit. Make a record of the ones that are easy to navigate, those that are efficient, those in which the information you wanted is in a format that is easy to read and access, and those that are easy to find. Then, try to assess what made these sites user friendly. You can then use your research to design web pages.

8.4 Visual communication

LEARNING OBJECTIVE 8.4 Use data tables, graphs and charts to communicate information.

Since ancient times, humans have used drawings to illustrate stories, communicate information and explain concepts. In today's world, people are bombarded with graphical images and computer-generated models; for example, on billboards and in movies, computer games, newspapers and magazines. The power of visual communication is illustrated by its importance in advertising.

Graphs, charts and tables are standard business communication tools and there are many tools and techniques that can be used to prepare and present information using these methods. While tables are used to provide data, graphs and charts are used to communicate important information *about* data, such as contrasts, similarities, trends and forecasts. When they are used correctly, graphs, charts and tables can convey critical information in a way that will engage and inform stakeholders. Used incorrectly, they can confuse or mislead stakeholders.

Data

In the introduction to this chapter it was noted that careful planning is required at the beginning of a project to ensure that by the end of the project the data required to support the findings and recommendations have been collected. Too often you hear people say, 'I wish we had collected some more data on the process' or 'I wish we had measured the deflection in that beam before it collapsed completely'. Careful planning and preparation is required, along with a degree of flexibility to collect data about unexpected events.

Once the data have been collected it can be analysed using formulae and statistics to calculate results, and the accuracy and precision of the data. Graphs and charts can be used to learn more about the data, to discern trends and to detect errors and inconsistencies in the data. Graphs and charts are also used to present supporting data and information in oral presentations, and in written and electronic documents.

Graphs

Traditionally engineers have used graphs to plot observed data and to mathematically *analyse* data; for example, to determine lines or curves of best fit. The data on these graphs is normally accurate (although this depends on scale) and therefore users can measure values on these graphs. A range of different types of graph paper was used to plot information, for example, square graph paper and log-log graph paper. These days there are a number of specialist software packages that can be used for this purpose, such as Excel, Mathcad, MATLAB and Mathematica. You will learn how to prepare graphs, and to analyse data using graphs, in your mathematics and discipline-based technical subjects.

Tables

Engineers often use tables to report data, and the data in these tables is generally accurate, particularly when it is compared to the data in a chart that has been produced from this data. Therefore, when a chart is used in a document, the data it is based on should also be provided in the document. Small tables can be in-text, while large tables are normally separated from the text and placed on a separate page or in an appendix. Table 8.3 is an example of a small table; it shows a set of monthly rainfall data recorded at the Sydney Observatory Hill Station, which was opened in 1858 and has an elevation of 39 metres. This set of data will be used to illustrate how tables and charts can effectively highlight the significant features of a data set.

TABLE 8.3 Monthly rainfall recorded at Sydney Observatory Hill Station, 2012–16

	Jan	Feb	Mar	Apr	May	Jun	Jul	Aug	Sep	Oct	Nov	Dec	Total
2012	*138.8*	*111.0*	*269.8*	*187.0*	*37.2*	*244.2*	*56.2*	*19.0*	*23.8*	*29.4*	*52.0*	*45.2*	*1213.6*
2013	*137.8*	*165.4*	*65.6*	*199.8*	*110.2*	*316.4*	*32.6*	*14.8*	*35.8*	*42.2*	*192.8*	*31.0*	*1344.4*
2014	*17.4*	*58.2*	*102.6*	*121.0*	*27.4*	*68.0*	*16.4*	*215.2*	*50.4*	*86.6*	*16.0*	*118.0*	*897.2*
2015	*164.8*	*59.0*	*65.2*	*366.8*	*109.8*	*110.6*	*47.0*	*71.8*	*80.0*	*43.0*	*122.6*	*96.6*	*1337.2*
2016	*249.8*	*25.8*	*193.2*	*155.0*	*7.2*	*305.0*	*104.6*	*151.4*	*70.0*	*31.4*	*27.2*	*65.0*	*1385.6*
Five-year mean	*141.7*	*83.9*	*139.3*	*205.9*	*58.4*	*208.8*	*51.4*	*94.4*	*52.0*	*46.5*	*82.1*	*71.2*	*1235.6*
All years mean	101.7	117.5	130.8	127.9	118.0	133.2	96.0	80.3	67.8	77.0	84.2	77.5	1215.7
All years highest daily	191.0	243.6	280.7	191.0	212.3	150.6	198.1	327.6	144.5	161.8	234.6	126.0	

Notes:
1. Rainfall data in millimetres — data in italics yet to be quality controlled.
2. Mean monthly rainfall data based on measurements in Sydney for the period 1858–2016.

Source: Bureau of Meteorology (2018).

A careful analysis of the data is required in order to make statements about the significance of the differences and any trends that may be occurring. An inspection of the data in the table shows that the yearly rainfall for three of the five years in the period 2012–16 was equal to or above the long-term mean (i.e. the average over the period 1858–2016).

Once the data is in a spreadsheet, the software can be used to manipulate and analyse it. For example, the yearly data could be converted to a percentage, based on the long-term mean data. As shown in table 8.4, colour can also be used to highlight information to detect trends, for example, drought years. In this table the cells where the monthly rainfall was less than the mean rainfall are shown in pink and those with higher than average rainfall are shown in white. The use of colour highlights trends in the data. For example, the chart clearly shows that all of the May readings were below the average for that month and the rainfall in the months in the second half of the year were mostly below average.

TABLE 8.4 Monthly rainfall data in Sydney Observatory Hill as a percentage of mean, 2012–16

	Jan	Feb	Mar	Apr	May	Jun	Jul	Aug	Sep	Oct	Nov	Dec	Total
2012	136.5%	94.5%	206.3%	146.2%	31.5%	183.3%	58.5%	23.7%	35.1%	38.2%	61.8%	58.3%	99.8%
2013	135.5%	140.8%	50.2%	156.2%	93.4%	237.5%	34.0%	18.4%	52.8%	54.8%	229.0%	40.0%	110.6%
2014	17.1%	49.5%	78.4%	94.6%	23.2%	51.1%	17.1%	268.0%	74.3%	112.5%	19.0%	152.3%	73.8%
2015	162.0%	50.2%	49.8%	286.8%	93.1%	83.0%	49.0%	89.4%	118.0%	55.8%	145.6%	124.6%	110.0%
2016	245.6%	22.0%	147.7%	121.2%	6.1%	229.0%	109.0%	188.5%	103.2%	40.8%	32.3%	83.9%	114.0%
Mean	101.1	118.0	129.7	127.1	119.9	132.4	97.9	79.8	68.4	76.9	84.3	77.3	1212.8

Note: Mean monthly rainfall data in millimetres and based on measurements at Sydney Observatory Hill for the period 1858–2016.
Source: Adapted from data from Bureau of Meteorology (2018).

We will now look at some other data sets while we explore graphing and charting techniques.

Charts

Like other professionals in the business world, engineers use charts to *communicate* the results of their work. Business charts are normally produced using the 'chart' function in spreadsheet software. While the data these types of graphs are based upon is usually accurate, the charts themselves should be regarded as indicative or representative, rather than numerically accurate. There are many different types of charts, with the most common being bar charts, column charts, pie charts, line plot charts and scatter plot charts.

Some key aspects of preparing charts include the following.

- *Use of data tables.* Charting software develops a chart from data in a 2D matrix. The way the data and labels are entered into the rows and columns influences the way the data will be represented in a chart. When only part of the data in a table is used, the way the data is selected will also affect the way it is charted.
- *Chart selection.* The type of chart selected depends on the type of data and the messages being communicated.
- *Chart verification.* Because charts are produced automatically by the software, they need to be carefully checked to ensure that the data has been correctly represented. Problems are generally caused by the selection of the wrong type of chart for the data, or the way the data is set out in the data table.
- *Chart distortion.* The proportional dimensions of a chart are normally defined by the software when a chart is produced; however, it is easy to distort a chart when it is being stretched, copied or pasted. Distortion may mean that the chart provides a misleading representation of the data.

The following examples were produced using the chart function in Excel.

Scatter plot charts

Scatter plot charts are used to plot individual points and can be used to show *variation* and *trends*. Lines and curves can be added to the plot by selecting an appropriate curve fitting option. Figure 8.9 is a scatter plot of the monthly rainfall totals for the year 2016. The software automatically allocates a numerical scale to the X-axis, and this was adjusted to show the 12 months. A polynomial trend line was added to demonstrate this function.

Line plot charts

Line plot charts are generally used for *chronological* data and can be used to show trends and to compare sets of chronological data. Figure 8.10 shows a line plot of the monthly rainfall data for the Sydney Observatory Hill Station over the five-year period 2012–16 and the mean monthly rainfall for all years the weather has been recorded at the station. Note that the software has used the month data from the table for the X-axis labels. The plot shows that while the rainfall trends are consistent, the monthly rainfall can vary greatly from year to year.

Two other data sets will be used to illustrate the use of line plots to discern trends. They are extracted from the results of the annual Quality Indicators for Learning and Teaching (QILT) Graduate Outcomes Survey (GOS), which is administered by the Social Research Centre on behalf of the Australian Government Department of Education and Training. The GOS is completed by graduates of Australian higher education institutions approximately four to six months after completion of their programs. Thus, the 2022 data shows the employment rate for graduates who completed their degree in 2021. The GOS reports also provide information on the labour market outcomes and further study activities of graduates.

FIGURE 8.9 A scatter plot of the monthly rainfall at Sydney Observatory Hill in 2016

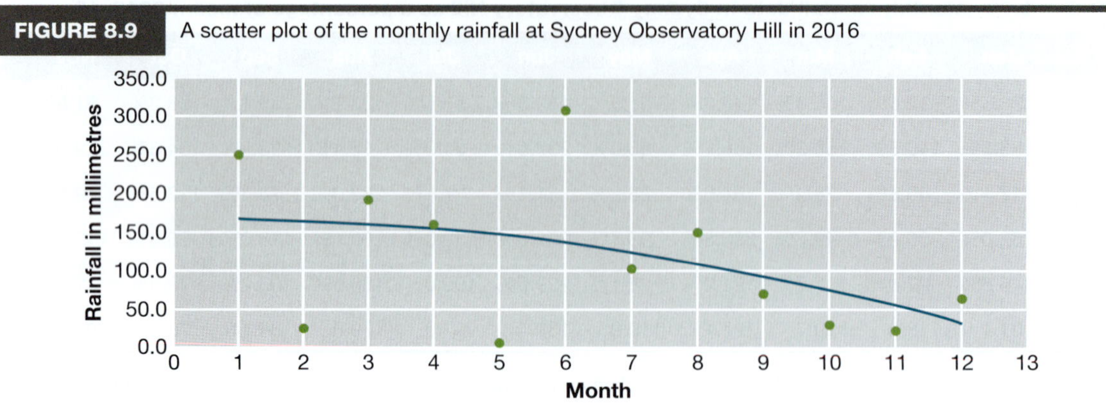

Source: Adapted from data from Bureau of Meteorology (2018).

FIGURE 8.10 Line plot of Sydney Observatory Hill rainfall data, 2012–16

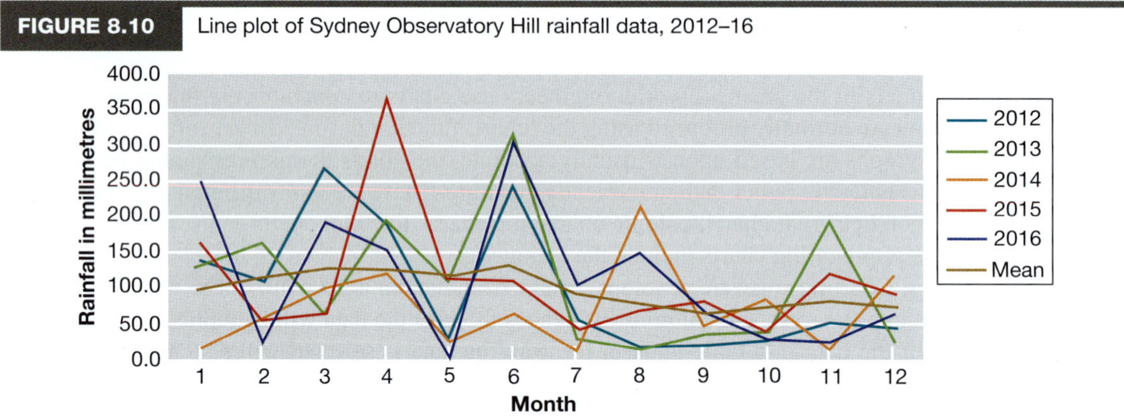

Source: Adapted from data from Bureau of Meteorology (2018).

Table 8.5 shows the percentage of graduates who were in full-time employment, by discipline, for ten of the 21 study areas included in GOS reports. The study areas are based on the ASCED 'Fields of education' — see the 'Working with rubbery figures' spotlight later in this chapter.

TABLE 8.5 Percentage of graduates in full-time employment, by selected discipline

	2015	2016	2017	2018	2019	2020	2021	2022
Architecture and built environment	75.4	75.2	75.2	77.7	74.5	67.7	65.2	78.8
Business and management	72.7	75.5	76.5	77.9	76.6	74.3	72.8	84.2
Computing and information systems	67.0	72.5	73.3	73.2	75.9	72.1	67.9	76.6
Engineering	73.9	76.4	79.4	83.1	84.8	83.0	80.3	87.5
Humanities, culture and social sciences	59.3	61.8	62.2	64.3	64.3	60.9	57.9	72.9
Law and paralegal studies	73.0	72.6	74.8	77.2	77.3	75.7	72.5	80.2
Medicine	96.3	98.2	95.9	94.9	91.1	86.7	90.2	93.0
Nursing	78.7	82.5	79.3	78.7	76.3	72.7	74.2	82.6
Science and mathematics	49.5	61.0	59.0	64.6	63.4	59.1	61.1	72.5
Teacher education	71.7	80.3	81.7	83.3	80.8	80.6	79.1	86.7

Source: Adapted from GOS data from The Social Research Centre (2016, 2018, 2019, 2020, 2021, 2022, 2023).

The data for the ten disciplines is reproduced in figure 8.11, which shows consistent trends for all ten disciplines, although the employment rates for each discipline vary. There was a drop in the employment rates in all disciplines during the 2020–21 COVID-19 pandemic, and a sharp recovery in 2022. The employment rates for engineering have generally been higher than most of the other disciplines, while the rates for science and the humanities have been consistently lower. The nursing discipline exhibited more variation from year to year compared to the other nine disciplines. You might like to gather more recent data and continue to compare the trends.

FIGURE 8.11 Line plot of annual graduate employment rates, by discipline, 2015–22

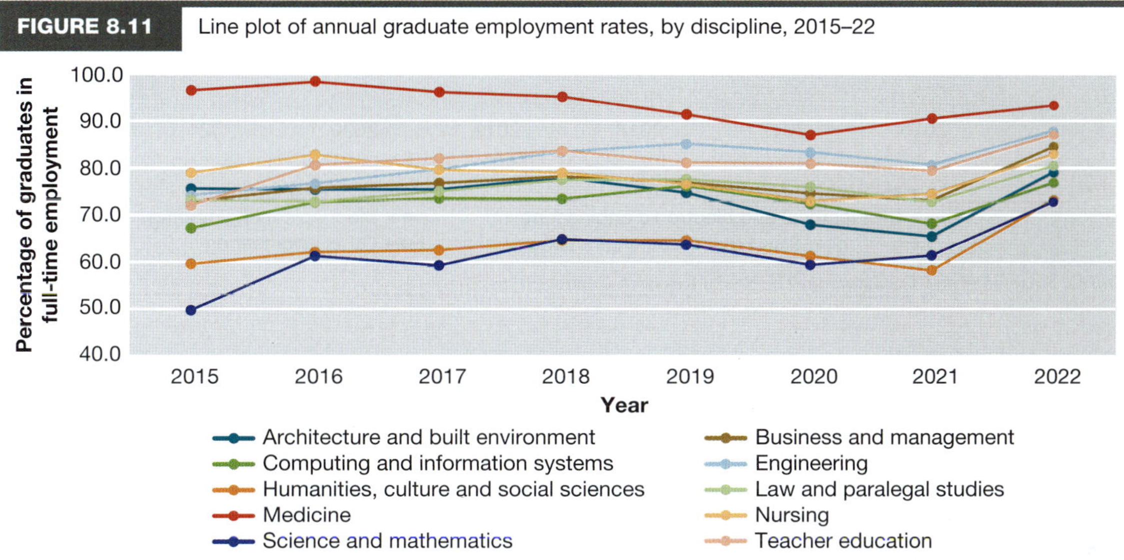

Source: Adapted from GOS data from The Social Research Centre (2016–22).

It is important to note here that employment rate varies between the individual disciplines within each GOS study area (for example, between disciplines in the 'Science and mathematics' study area such as biology, physics and mathematics).

For comparison purposes we will now look at the graduate employment data for some of the engineering disciplines. Table 8.6 shows the annual percentage of the graduates in full-time employment for five specific engineering disciplines and the 'Other engineering' category, over the period 2016–22, the latest available GOS data. The data are reproduced in figure 8.12, which shows that while the employment rates for all engineering disciplines were affected by COVID-19, the impact was greatest in the aerospace and mechanical engineering disciplines. While the graduate employment rate for aeronautical engineers was close to the other disciplines in 2016, it steadily declined during the period, while the rates for the other disciplines rose during the period. While the rate for aerospace engineers recovered to its 2016 level in 2022, it was more than 15 per cent less than the rate for the other engineering disciplines.

TABLE 8.6 **Percentage of engineering graduates in full-time employment, by engineering discipline**

	2016	2017	2018	2019	2020	2021	2022
Aerospace engineering	68.5	70.1	70.2	67.8	64.4	59.0	69.4
Civil engineering	81.7	84.3	88.2	89.0	85.6	86.0	91.1
Electrical and electronic engineering	75.4	76.1	85.5	85.7	82.8	78.9	87.8
Mechanical engineering	72.3	76.5	78.4	82.9	79.9	76.1	85.7
Process and resources engineering	69.6	74.4	80.6	82.3	79.9	78.7	84.4
Other engineering	79.2	82.8	85.4	86.7	86.8	82.7	90.3

Source: Adapted from GOS data from The Social Research Centre (2016–22).

FIGURE 8.12 Line plot of employment rates for engineering graduates, by discipline, 2016–22

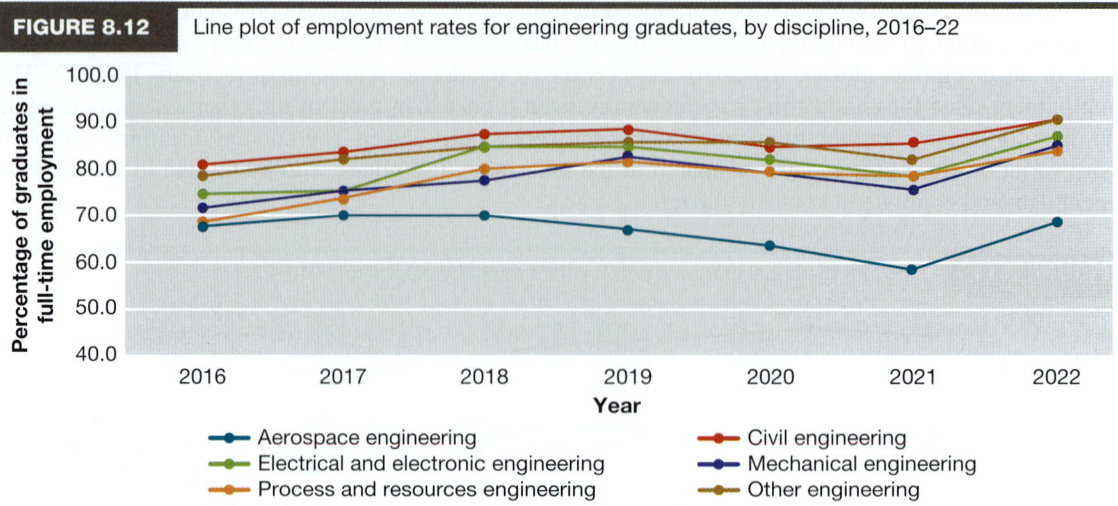

Source: Adapted from GOS data from The Social Research Centre (2016–22).

Column and bar charts

Column and bar charts are used to show the magnitude of data and to compare and contrast data. A bar chart is a horizontal column chart and needs to be carefully prepared as people may misinterpret the data because of their expectation that the size of data is normally represented on the Y-axis and time represented on the X-axis. Column and bar charts can be 2D, 3D or stacked. These are illustrated in the following figures.

Figure 8.13 is a bar chart of the rainfall data at the Sydney Observatory Hill Station from table 8.3. The chart highlights the problem of interpreting bar charts, particularly when there is too much information.

FIGURE 8.13 Bar chart of monthly Sydney Observatory Hill rainfall, 2012–16 and overall mean

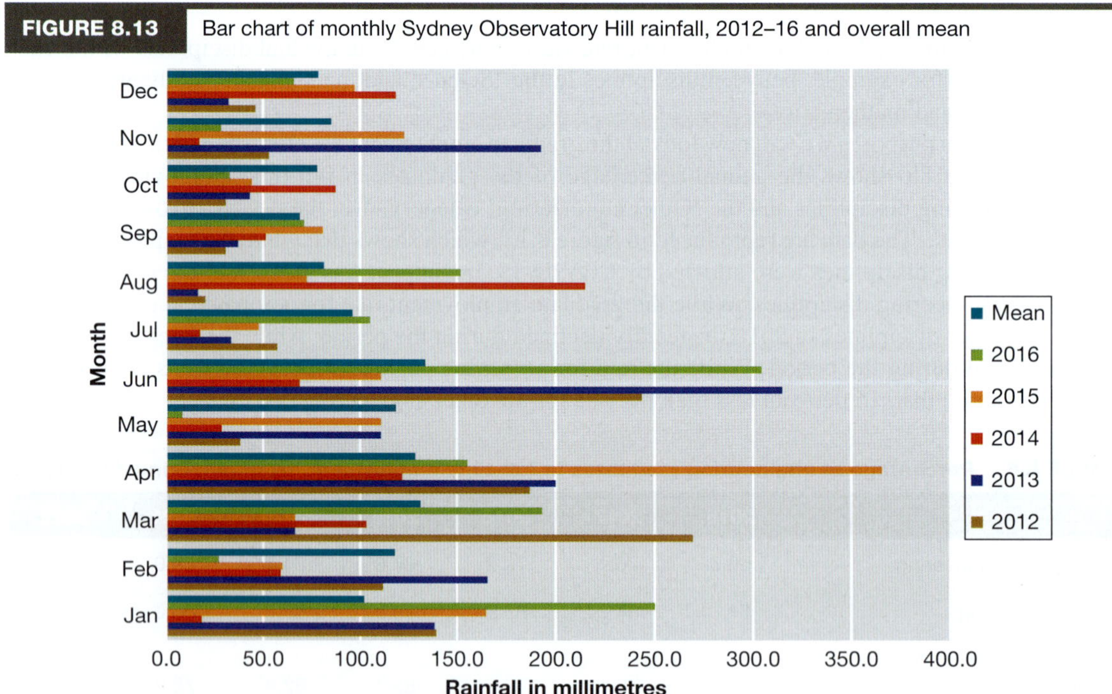

Source: Adapted from data from Bureau of Meteorology (2018).

Figure 8.14 is a column (vertical) chart of the Sydney Observatory Hill rainfall data. Although the data in this chart appears easier to interpret, the amount of data in the chart still makes it difficult to identify trends and other significant features.

FIGURE 8.14 Column of monthly Sydney Observatory Hill rainfall, 2012–16 and overall mean

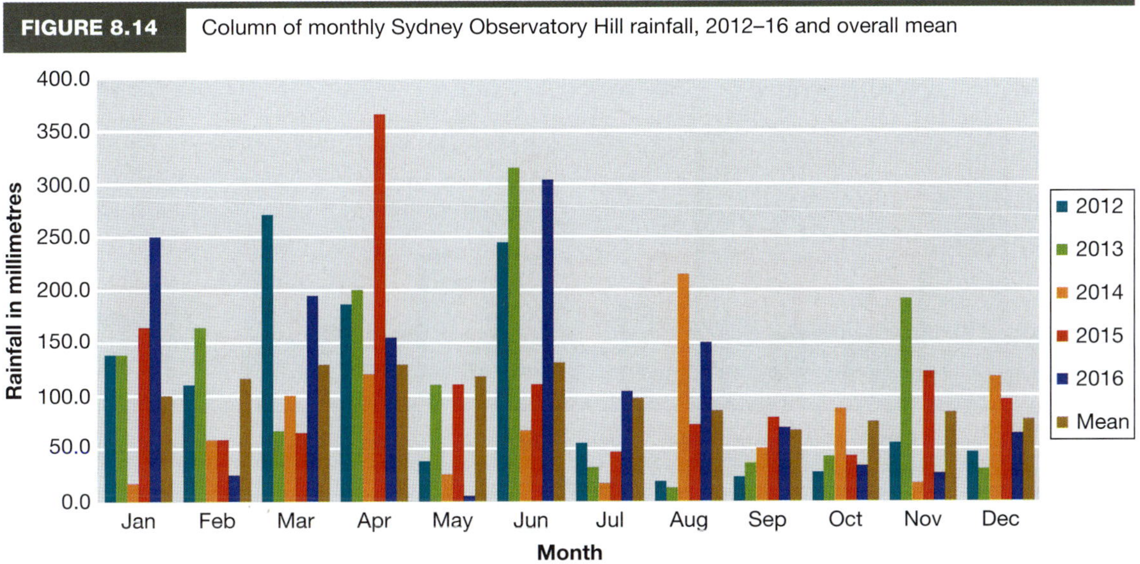

Source: Adapted from data from Bureau of Meteorology (2018).

For comparison purposes the graduate employment data in table 8.6 is reproduced in figure 8.15. This representation clearly shows the trend for each of the disciplines and enables comparisons between disciplines to be made. Note that data labels have been included in this chart.

FIGURE 8.15 Column chart of percentage of engineering graduates in full-time employment, by discipline, 2016–22

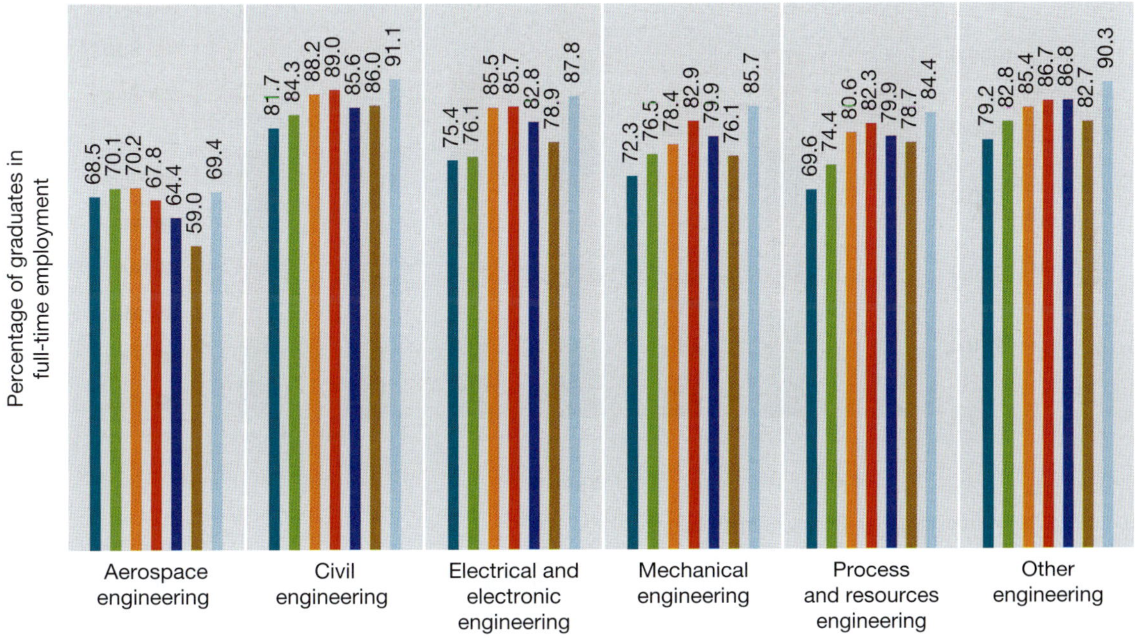

Source: Adapted from GOS data from The Social Research Centre (2016–22).

A 3D column chart of the rainfall data is shown in figure 8.16. This representation of the data is easier to interpret, although some data is hidden. This chart clearly shows the trends over the three-year period and contrasts those with the mean values, which are at the back of this view. The order of the data can be changed by adjusting the order of the rows in the table.

FIGURE 8.16 A 3D column chart of Sydney Observatory Hill rainfall data, 2014–16 and mean

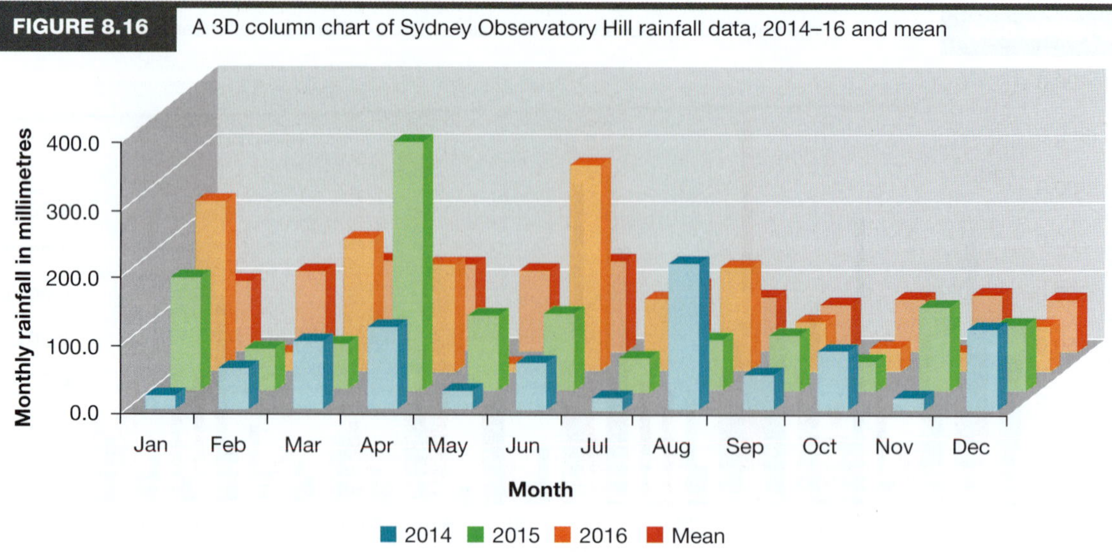

Source: Adapted from data from Bureau of Meteorology (2018).

Figure 8.17 shows a stacked column chart of this data with the five-year mean included for comparison purposes.

FIGURE 8.17 A stacked column chart showing the monthly rainfall data for the years 2012–16 and the means

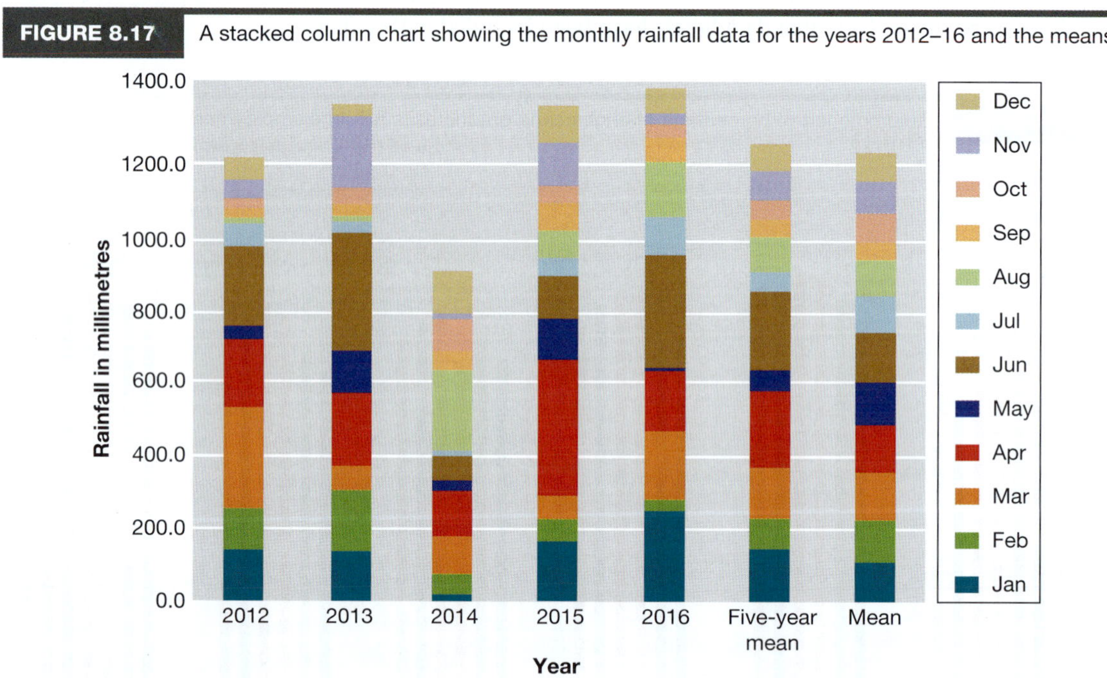

Source: Adapted from data from Bureau of Meteorology (2018).

Figure 8.18 is a stacked column chart of the commencing student data in table 8.7 in the spotlight at the end of this chapter. The stacked column chart visually reinforces the overall trends in the data and the trends in the individual sets of data. They also highlight any year-to-year variation in the individual data sets. These variations are much greater in the rainfall data in figure 8.17 than they are in the commencing student data in figure 8.18.

Pie charts

Pie charts are used to show the *proportion* of each of the categories that make up the total of the data. Thus, a pie chart is generally used to show percentages. The segmented pie chart in figure 8.19 shows the proportion of each type of breach in the 'malicious or criminal attack' category of the cyber incidents reported in the first half of 2023. The data is the same as that used in the chart in figure 8.7 in the OAIC spotlight at the beginning of the chapter.

Engineers often use data from a wide range of sources — data that was collected and reported by third parties. Normally this is because it would not be possible, or economically feasible, to collect the required data themselves. However, using third party data can be frustrating because it may not exactly meet the requirements or specifications of the project. The following spotlight provides some examples of this and the mechanisms used by authors to ensure readers are aware of the assumptions and issues faced when using the data.

FIGURE 8.18 A stacked column chart showing the commencing students by program type, 2012–20

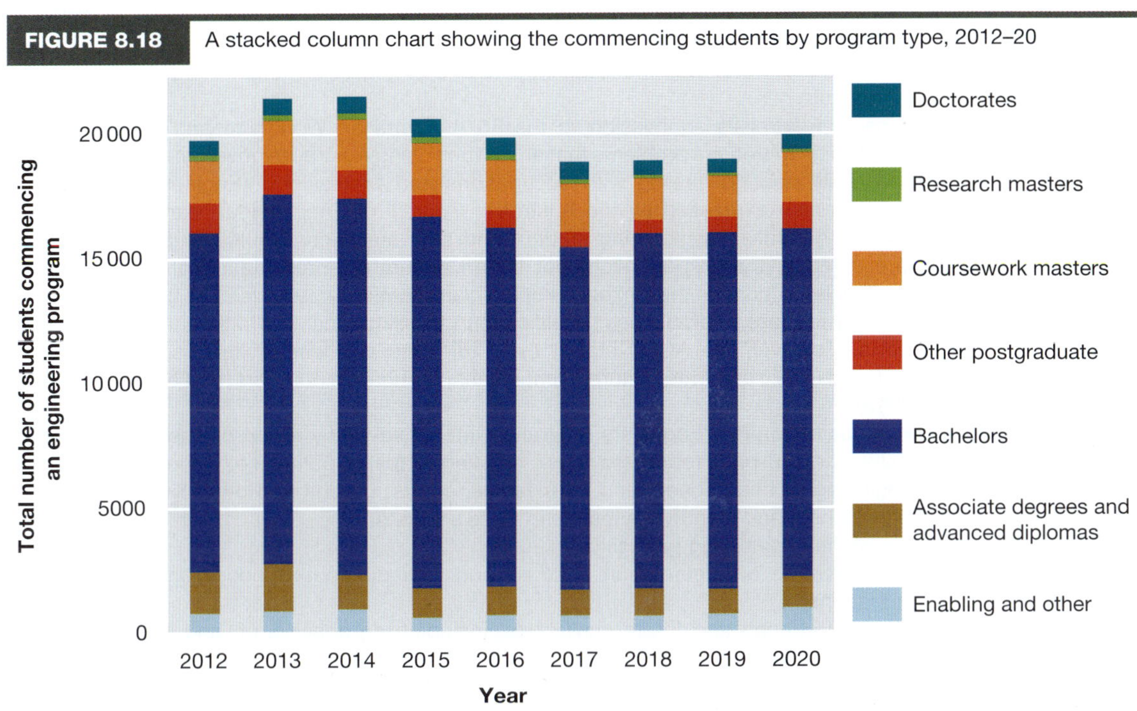

Source: Adapted from data from King (2022, p. 35).

FIGURE 8.19 Proportions of types of cyber security breach — first half 2023

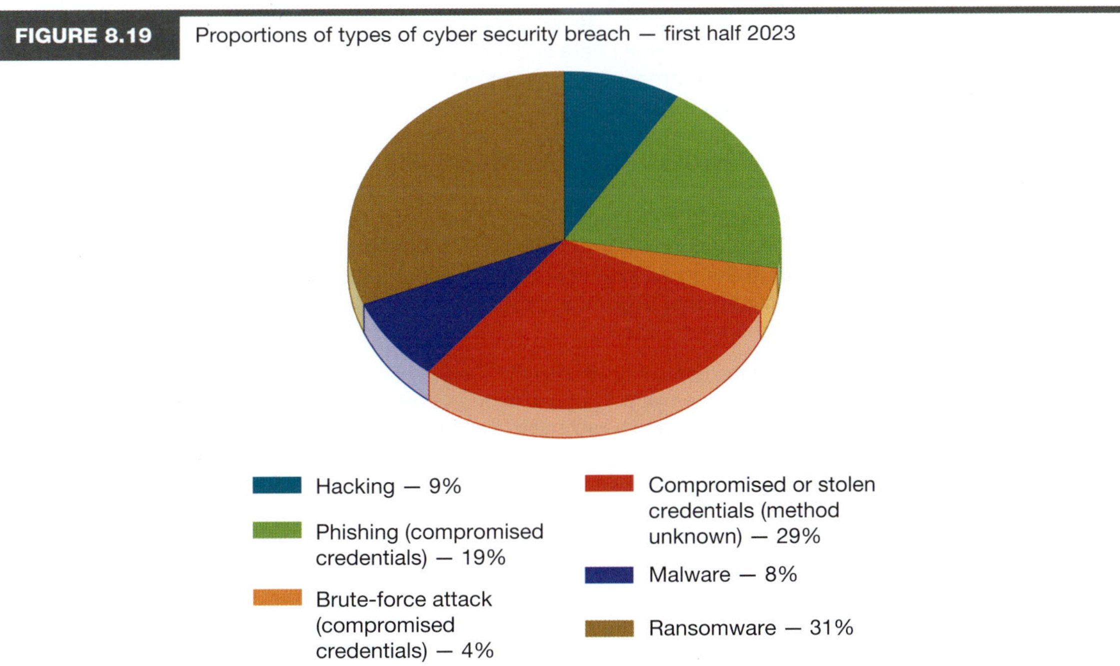

Source: Adapted from OAIC (2023, p. 3).

SPOTLIGHT

Working with rubbery figures

Engineers Australia regularly prepares submissions that address important issues being considered by the federal or state governments, industry sectors and the engineering profession. In its submissions, Engineers Australia presents data, information, insights and recommendations, all from an engineering perspective and it does so on behalf of its members and the profession, and for the public good. Generally, it relies on data published by third parties, such as government agencies, although it sometimes has to purchase specific data sets that better suit the requirements of a project. Even then, the data does not always meet the requirements.

For example, Engineers Australia's submission to the 2014 Australian Workforce and Productivity Agency's workforce study highlighted the problems that arise when using Australian government statistics: 'official data are notoriously poor in dealing with occupations and professions' (Engineers Australia 2014). This sentiment was also expressed by the Australian Council of Engineering Deans (ACED) in its 2013 annual report on Australian engineering student and staff statistics. The ACED also felt that some of the data presented in their report might not be an accurate representation of engineering education in Australia. This is because of the limitations of the classification system used by the Australian Bureau of Statistics (ABS) to classify disciplines, qualification classifications, and differences between the universities' reports. However, the ACED stated that, '[d]espite all these limitations, these data compilations provide a reasonably accurate picture of engineering education numbers, and enable trends to be seen' (ACED 2013).

The problems persist, as is noted in Engineers Australia's latest five-yearly report on the engineering profession (Briggs 2023). The 130-page report provides a detailed snapshot of the engineering profession in Australia at the beginning of this decade, and uses tables, charts, graphs and other illustrations to compare data, highlight trends, and identify key points in three sections.

- Section 1: The state of the engineering profession in 2023
- Section 2: The characteristics of engineers in Australia and their distribution geographically, within industries and occupations
- Section 3: The nature of change and future challenges, including an analysis of, and insights into, engineering education, graduate outcomes, industries and occupations

The report is mainly based on the relevant components of the 2021 Census data, which were released by the ABS in October 2022. For the first time the ABS data is accessible in the report via an interactive data dashboard implemented using Microsoft Power BI. This enables the reader to delve into the data that most interests them. Data from other sources is also used in the report, such as:

- student data from the Australian Government Department of Education
- migrant data from the Australian Government Department of Home Affairs
- student experience data, graduate outcomes data and employer satisfaction data from QILT
- student PISA scores, Australian Government Department of Industry, Science and Resources
- engineering student data, Australian Council of Engineering Deans
- participation rates in high school mathematics — Australian Mathematical Sciences Institute.

The report states that the definitions of terms for fields of education, industries, occupations and other parameters were generally derived from the ABS or, where applicable, from the entity responsible for the relevant data. The three main classification systems used for analysing the Census data are:

- the Australian and New Zealand Standard Classification of Occupations (ANZSCO)
- the Australian and New Zealand Standard Industrial Classification (ANZSIC)
- the Australian Standard Classification of Education (ASCED).

The authors described the methodology used:

> We take segments (or slices) of the population at different levels of detail in the classification schemes with other census parameters. We provide the population size of engineers that these segments represent, as not all respondents are able to be encoded at all levels of detail within ANZSIC, ANZSCO and ASCED. This means that the entire population of engineers and those working in engineering are unable to be represented at lower levels of the classification scheme. Also, at the highest levels of the classification schemes, there are significant proportions of responses which are 'inadequately described' or 'not stated' (Briggs 2023, p. 5).

Importantly, the report notes that:

> All statistical systems involve some compromise and over time can be overtaken by social and labour market developments. For example, in engineering, the discipline of mechatronics is playing an increasingly important role in both society and industry. However, mechatronics is not currently included in either the official education or occupational classification systems in its own right. Offsetting this disadvantage is that ABS classification is the mainstay of statistics used in official advice to ministers and governments (Briggs 2023, p. 2).

The limitations of using some of the other datasets were also highlighted in the report, and included the following.

- Census data is self-reported by the respondent, which means that it relies on the understanding and judgement of the respondents to answer the Census survey honestly and accurately (Briggs 2023, p. 5)
- The term 'engineer' is 'not a legally protected title within Australia (except where statutory registration schemes require particular types of engineering work to be undertaken by those with appropriate qualifications and experience)' (Briggs 2023, p. 5).
- The Census data classifies engineers who have a higher qualification in a non-engineering field in that non-engineering field (Briggs 2023, p. 74).
- The fact that the three categories in the engineering workforce (Engineering Associates, Engineering Technologists and Professional Engineers) align with a separate level of undergraduate qualification is not always acknowledged in the education and occupational classification systems (Briggs 2023, p. 3).
- 'The ANZSCO and ASCED schemes used in classifying Census data do not fully capture the breadth or depth of what engineers do or what engineering can entail. This includes emerging fields and disciplines' (Briggs 2023, p. 110).
- The way universities use the ASCED scheme to report student data to the Department of Education and Training is inconsistent, both between universities and from year to year. The lack of consistency makes it difficult to accurately monitor and report on Australia-wide trends in engineering education (Kaspura 2017).

We will now explore the impact of some of these limitations in more detail using graphs and charts to highlight those limitations in specific sets of data.

Matching student data with workforce categories

A table published by the Australian Council of Engineering Deans (King 2022) shows the data for commencing domestic students in Australian undergraduate and postgraduate engineering programs for the period 2012–20 (table 8.7). The table is based on data provided by the federal Department of Education and Training (DET), which normally only provides one set of data for bachelor degrees in engineering — in this case the total number of students commencing in three-, four- and five-year engineering degree programs in a calendar year. This means that the data for the three-year degree, which leads to the Engineers Australia Engineering Technologist membership category, is combined with the data for four- and five-year degree programs (some of which are Masters programs), which lead to the Professional Engineer category of membership — that is, Engineers Australia Stage 1 accredited programs. This can be misleading if this limitation is not clearly reported when the data is published.

TABLE 8.7 **Commencing domestic students in Engineering and Related Technologies programs, 2012–20**

Level	2012	2013	2014	2015	2016	2017	2018	2019	2020
Doctorates	601	662	673	718	701	712	589	562	597
Research masters	231	234	258	253	214	176	128	120	140
Coursework masters	1 690	1 780	2 043	2 091	2 023	1 931	1 671	1 646	2 003
Other postgraduate	1 186	1 167	1 118	844	682	594	519	609	1 040
Bachelors	13 595	14 817	15 085	14 896	14 390	13 736	14 238	14 291	13 938
Associate degrees and advanced diplomas	1 659	1 890	1 370	1 178	1 136	1 031	1 095	995	1 239
Enabling and other	748	836	909	564	655	631	616	687	943
Total commencing students	**19 710**	**21 386**	**21 456**	**20 544**	**19 801**	**18 811**	**18 856**	**18 910**	**19 900**

Source: Adapted from King (2022, p. 35).

The following line plot is based on the data in the table and it shows that the overall number of students commencing an engineering program increased until 2014 and then steadily declined through to 2017, after which it was relatively steady until it rose again in 2020. Figure 8.20 also shows that the majority of engineering students commence their studies in a bachelor degree program.

FIGURE 8.20 Commencing student trends, 2012–20

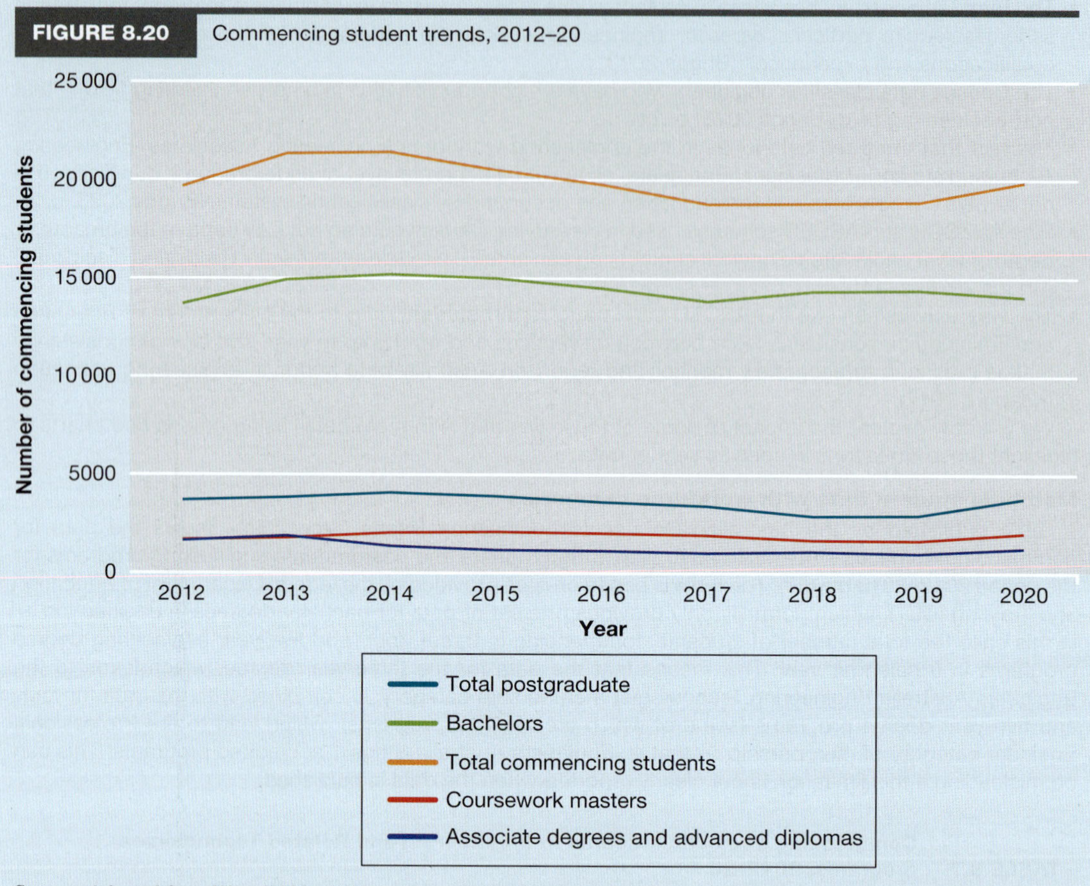

Source: Adapted from King (2022, p. 35).

The figure also shows there was an increase in commencing students in the associate degree and advanced diploma category until 2013, with a decline in the following years until 2019, and then a rise in 2020. Graduates from these programs are eligible for the Engineers Australia Engineering Associate membership category. One of the factors contributing to the rise and fall in enrolments in these programs was the shortage of professional engineers during the period 2005–13. During that period many employers recruited high school graduates as trainees, who then worked full time while studying part time in one of these programs, often by distance education.

The figure also shows the steady increase in enrolments in coursework masters programs until 2016 and then a steady decline until 2019 before a sharp rise in 2020. Many of these programs have been accredited by Engineers Australia and this means the graduates are eligible for membership in the Professional Engineer category. This adds to the complexity of determining the precise number of students who commenced in Engineers Australia Stage 1 accredited programs.

The Australian Standard Classification of Education system

The second problem with some official statistics results from the use of the Australian Standard Classification of Education (ASCED) system, which was the subject of the earlier ACED statement.

The Field of Education (FOE) codes in the ASCED system are designed to include an education code for all the occupations taught in the tertiary education sector in Australia (i.e. both the higher education and vocational education and training (VET) sectors). Note, the GOS Study Areas are based on the ASCED Fields of Study.

The two-digit ASCED FOE code for engineering is 03 Engineering and Related Technologies; it includes nine four-digit sub-categories (e.g. 0307 Mechanical and Industrial Engineering and Technologies) and a set of discipline codes within each sub-category, each with a six-digit code (e.g. 030701 Mechanical Engineering).

Table 8.8 shows the full list of four-digit sub-codes and names and the six-digit codes and names of the disciplines that may be studied in university degree programs. To allow for new disciplines, a code for 'Other' is included in the list of four-digit sub-categories (e.g. 0399 Other Engineering & Related Technologies — n.e.c. (not elsewhere classified)) as well as in each list of six-digit discipline codes (e.g. 030999 Civil Engineering — n.e.c.).

TABLE 8.8 ASCED four-digit sub-codes and the six-digit discipline codes for degree programs

03 Engineering and Related Technologies Four-digit subcodes and selected six-digit discipline codes	
0301	**Manufacturing Engineering & Technologies** 030101 Manufacturing Engineering 030199 Manufacturing Engineering & Technology, n.e.c.
0303	**Process & Resources Engineering** 030301 Chemical Engineering 030303 Mining Engineering 030305 Materials Engineering 030307 Food Processing Technology 030399 Process and Resources Engineering, n.e.c.
0305	**Automotive Engineering & Technologies** 030501 Automotive Engineering 030599 Automotive Engineering and Technology, n.e.c.
0307	**Mechanical & Industrial Engineering & Technologies** 030701 Mechanical Engineering 030703 Industrial Engineering 030799 Mechanical & Industrial Engineering & Technology, n.e.c.
0309	**Civil Engineering** 030901 Construction Engineering 030903 Structural Engineering 030905 Building Services Engineering 030907 Water and Sanitary Engineering 030909 Transport Engineering 030911 Geotechnical Engineering 030913 Ocean Engineering 030999 Civil Engineering, n.e.c.
0311	**Geomatic Engineering & Technologies & Related Technologies** 031101 Surveying 031103 Mapping Science 031199 Geomatic Engineering, n.e.c.
0313	**Electrical & Electronic Engineering & Technologies** 031301 Electrical Engineering 031303 Electronic Engineering 031305 Computer Engineering 031399 Electrical & Electronic Engineering & Technology, n.e.c.
0315	**Aerospace Engineering & Technologies** 031501 Aerospace Engineering 031599 Aerospace Engineering and Technology, n.e.c.
0317	**Maritime Engineering & Related Technologies** 031701 Maritime Engineering 031799 Maritime Engineering and Technology, n.e.c.
0399	**Other Engineering and Related Technologies** 039901 Environmental Engineering 039903 Biomedical Engineering 039905 Fire Technology 039999 Engineering and Related Technologies, n.e.c.

Source: Australian Bureau of Statistics (2001, pp. 75–76).

As noted previously, table 8.8 only shows the disciplines that may be studied in university degree programs. It does not include most of the six-digit codes for the disciplines normally offered by Vocational Education and Training (VET) institutions. For example, the set of subdisciplines for the 0307 FOE Mechanical and Industrial Engineering and Technologies includes 030705 Toolmaking and 030707 Metal Fitting, Turning and Machining.

It should be noted that some of the codes indicate that two different levels of employment have the same name (e.g. Mechanical Engineering), one achieved after undertaking a VET diploma or advanced diploma program and the other after completing a degree.

Some of you will find that the discipline you are studying does not have a four- or six-digit code, and it may not be listed in the 'Other' category (under the 0399 code). This highlights the problem that both Engineers Australia and ACED grapple with each year: the engineering disciplines in the ASCED system do not accurately reflect the engineering disciplines currently being practised in Australia. The key issues are as follows.

- The relevant two-digit engineering and related technologies category includes a number of disciplines, such as air traffic control and fire technology that are not normally associated with engineering. Geomatic engineering (spatial science, surveying) is now included in Engineers Australia data because it has accredited some programs. However, the majority of professionals in that field would not regard themselves as engineers.
- Some current disciplines, such as mechatronic engineering, power engineering and software engineering, are not listed.
- While the majority of engineering specialisations (e.g. structural engineering, environmental engineering) have six-digit codes, others have four-digit codes (e.g. civil engineering and electrical and electronic engineering).
- To further complicate matters, there are degree programs at the four-digit code level and at the six-digit code level in the same field, for example, civil engineering and the disciplines in that field such as structural engineering and transport engineering.

These factors make it difficult to report engineering data at the specialisation level unless the data are collected at both the four- and six-digit code levels, which is often not the case. For example, some Australian universities use the general code 03 Engineering and Related Technologies to report their data to the federal government, while other universities use the 0399 Other Engineering & Related Technologies code. The Engineers Australia report (Kaspura 2017, p. 57) highlighted this issue of universities choosing between the four- versus six-digit code levels, outlining how university reporting protocols to the Department of Education and Training are:

> inconsistent year on year with large numbers of completions allocated to 'general' categories in some years but not in others. This limits the degree of disaggregation that is possible.

The lack of consistency makes it difficult to accurately monitor and report on Australia-wide trends in engineering education and to undertake workforce planning. This is highlighted in data supplied by the Department of Education and Training for the year 2017. A statistical analysis of the data shows that the usefulness of the data is severely compromised by the fact that, together, the graduates in the general '0300 Engineering and Related Technologies' category and the '0399 Other Engineering & Related Technologies' categories, accounted for nearly half (45.5%) of the graduates in 2017.

To address their continuing frustration with inadequate data ACED identified a list of thirteen engineering branch groups, and an 'Other' category, primarily based on the titles of degrees (King 2021). Except for 'Electrical and Electronic', dual-title degrees were assigned to the more specialised branch. For example, Civil and Environmental degrees were classified as Environmental Engineering. It will be interesting to see if this classification system persists and leads to changes in the ASCED and other classification systems.

The thirteen Engineering branches are listed in table 8.9, which also shows the number of domestic and international on-shore graduates at Australian universities for the years 2016–19. The data includes graduates from accredited Bachelor (Hons) degree programs, that is, accredited programs that lead to Stage 1 Competence at the Professional Engineer level.

It is important to note that the number of graduates in the 'Other' category is only a very small percentage of the total number of graduates.

This data is reproduced in the column chart in figure 8.21, which clearly shows the majority of students still graduate from the four traditional disciplines: civil engineering, chemical engineering, electrical engineering and mechanical engineering. The chart also shows the growth in some of the newer disciplines. While there has been some growth in the number of domestic graduates over the period, there has been a larger growth in international graduates who studied on-shore at an Australian university.

TABLE 8.9 **On-shore domestic and international graduates in Accredited Bachelor (Hons) degrees, 2016–19**

Engineering branches covered		2016		2017		2018		2019	
Number	Branch	Dom	Int	Dom	Int	Dom	Int	Dom	Int
1	Civil, structural, construction, infrastructure	2 060	667	2 075	792	2 186	867	2 094	1 053
2	Environmental, civil and environmental	158	44	182	60	195	61	159	77
3	Chemical, materials	407	239	538	259	552	367	516	313
4	Mining, petroleum, metallurgy	244	146	283	186	220	140	141	103
5	Electrical, electrical and electronic, energy	630	347	687	370	831	458	896	446
6	Electronics, computer systems, telecommunications	187	149	255	177	299	170	278	147
7	Software, IOT	119	23	159	21	156	45	245	66
8	Biomedical	71	10	110	23	149	31	207	39
9	Mechanical, manufacturing, industrial, production	1 027	414	1 188	523	1 296	609	1 186	608
10	Aeronautical, aerospace	125	58	179	74	204	77	220	87
11	Mechatronics, robotics	272	92	391	107	447	110	485	183
12	Naval architecture, maritime	32	27	31	26	31	28	33	26
13	Geospatial, surveying	3	1	15	1	15	2	13	4
14	Other	19	1	31	1	38	4	40	2
	Total graduates by category	**5 354**	**2 218**	**6 124**	**2 620**	**6 619**	**2 969**	**6 513**	**3 154**
	Total graduates by year	**7 572**		**8 744**		**9 588**		**9 667**	

Source: King (2021, p. 4)

FIGURE 8.21 Column chart showing graduates in Accredited Bachelor (Hons) degrees, 2016–19

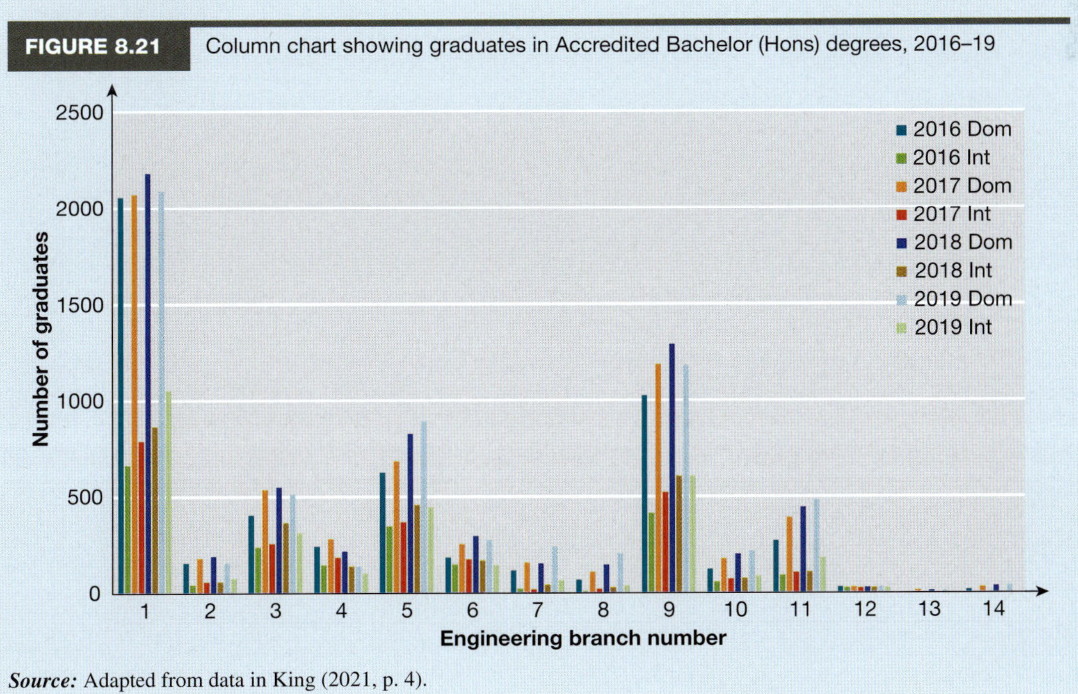

Source: Adapted from data in King (2021, p. 4).

CRITICAL THINKING

What other trends can you see in the data? How does your university report its data?

Download a copy of the latest Statistical Overview of the Engineering Profession from the Engineers Australia website, and the latest ACED report, and review the trends in STEM education, university enrolment and graduation data, and employment trends by engineering discipline.

The spotlight highlighted the importance of using valid data and authentic presentation techniques to support the key messages in a communication. This requires careful planning during the collection, analysis and synthesis phases of a project, and during the development of tables, graphs and charts used to present the data and information. It also highlights the need to clearly state the limitations of the data used, and any assumptions made when reporting that data.

These topics are discussed in more detail in the following chapter.

SUMMARY

Communicating information is a critical skill for engineers, and this chapter discussed a number of the key skills that are used to communicate data and information. The examples and spotlights in the chapter highlighted different aspects of the communication process, and what happens when communication does not occur. We will now briefly revisit each of the chapter's learning objectives.

8.1 Apply the techniques that will assist you to listen actively, read effectively and write accurate notes.

Three key communication skills are listening, reading and writing notes. As a student you should strive to continually hone these skills as they will help you with your studies and be beneficial for your career. Listening is not easy. This is because it is difficult to concentrate on what others are saying while accurately hearing and noting the key messages. Active listening is a critical skill for students and engineers and you should use active listening techniques to ensure that you hear and understand the messages being communicated. This is particularly important when technical data and information are being communicated. Active listening involves listening carefully to what a speaker is saying, observing any non-verbal messages being conveyed and checking with the speaker that you have accurately received information.

When reading, it is important to have a clear focus, normally based on your purpose; that is, you need to clearly understand why you are reading a document and the information you are seeking. As a student and an engineer, you will need to write in a way that is accurate, focused, simple and to the point. Effective note-taking is a simple way of recording activities undertaken and information exchanged in important conversations, discussions and meetings.

8.2 Prepare and deliver verbal presentations.

Oral communication, including discussing and telling, is a key aspect of most engineers' working lives. Adequately preparing for oral communications, even informal communication, is important in a professional environment. A variety of tips and techniques can be used to enhance formal presentations. Preparing and delivering formal presentations is a routine activity for many professional engineers, and these presentations should be carefully planned so that they achieve their purpose.

8.3 Plan and prepare written communications.

Engineers can spend a lot of their time at work writing or responding to written documents. This includes correspondence, project initiation documents, financial documents, human resource documents, workplace health and safety documents and project reports. Each of these documents has a purpose, and most have a defined format. Engineers need to consider the legal and regulatory contexts of all of the documents they prepare, not just contracts. Engineers should always use an appropriate format, structure and style for a written communication so that the information is accessible by the audience and achieves its purpose. The gatekeeper role and the control of the document distribution processes (especially electronic documents) are important as they help to ensure documents do not reach an unintended consumer.

8.4 Use data tables, graphs and charts to communicate information.

Visual communication methods have always been an important part of engineering. In this chapter we discussed the use of data, tables, graphs and charts to highlight the important messages we want to convey, messages that provide the evidence for the statements or recommendations we make at the end of a project. In the chapter on communicating information we will discuss other forms of visual communication, such as the plans and models used to communicate design information to clients, users and the other members of the engineering team.

KEY TERMS

hypothesis A proposition that is suggested to explain a group of phenomena.
metadata Information or data about data.
wiki A web page that allows a group of people to access and communally develop and edit text.

EXERCISES

1. Compare the writing styles and approaches used in the articles in an engineering magazine with those used in the advertisements in the same magazine. Are they different? If so, why?
2. Write a formal covering letter that you could attach to your next major assignment.
3. Review the report submitted by a winning team in the annual Engineers Without Borders (EWB) contest (https://ewb.org.au), or another undergraduate engineering student competition. What is the structure of the document? What are the key messages the document conveys? What methods, approaches, formats and media are used to convey those messages? What was your response?

REFERENCES

Australian Bureau of Statistics (ABS) 2001, *Australian Standard Classification of Education (ASCED)*, ABS, Canberra.

Australian Council of Engineering Deans (ACED) 2013, *Australian engineering education student and staff statistics from national data collections*, November, ACED.

Briggs, P 2023, *The engineering profession: A statistical overview*, 15th edn, Engineers Australia, Canberra, https://www.engineersaustralia.org.au/publications/engineering-profession-statistical-overview-15th-edition

Bureau of Meteorology 2018, 'Climate statistics for Australian locations. Monthly climate statistics: Sydney — Observatory Hill', http://www.bom.gov.au

Burley-Allen, M 1995, *Listening: The forgotten skill*, 2nd edn, John Wiley & Sons, New York.

Conflict Research Consortium 1998, 'International online training programme on intractable conflict — active listening', University of Colorado, Boulder.

Das, S 2005a, 'All change at Spencer Street', *The Age*, 9 July, pp. 1–2.

——— 2005b, 'Contractor denies it is sacrificing quality finishes to save money', *The Age*, 9 July, p. 2.

Department of Transport 2009a, 'Southern Cross Station project', Victorian Department of Transport, Victorian Government, Victoria.

——— 2009b, 'Early history of Southern Cross Station', Victorian Department of Transport, Victorian Government, Victoria.

Engineers Australia 2014, *The engineering workforce: Engineers Australia's response to the AWPA paper*, Engineers Australia, Canberra.

Hagan, P & Mort, P 2018, *Report writing guide for mining engineers*, 10th edn, Mining Education Australia, Adelaide, https://www.unsw.edu.au/engineering/sites/default/files/documents/Report_Writing_Guide_for_Engineers_2018ed.pdf

Kaspura, A 2017, *The engineering profession: A statistical overview*, 13th edn, February, Engineers Australia, Barton, ACT.

King, R 2021, *Professional engineering graduates by branch of engineering*, 21 May, Australian Council of Engineering Deans, https://www.aced.edu.au/downloads/ACED%20Graduates%20by%20Branch%20of%20Engineering%20May%202021%20-%20RKing.pdf

——— 2022, *Australian engineering higher education statistics 2010–20*, April, Australian Council of Engineering Deans, https://www.aced.edu.au/downloads/ACED%20Engineering%20Statistics%20April%202022.pdf

Lyons, D 2023, *STFU: The power of keeping your mouth shut in an endlessly noisy world*, Henry Holt and Co, New York.

Macquarie Dictionary 2024, *The Macquarie Dictionary online*, Pan Macmillan Australia, https://www.macquariedictionary.com.au

Male, S, Chapman, E & Bush, M 2008, 'Generic competencies required by engineers graduating in Australia', unpublished paper, School of Mechanical Engineering, The University of Western Australia, Perth.

Office of the Australian Information Commissioner (OAIC) 2023, *Notifiable data breaches report: January to June 2023*, 5 September, https://www.oaic.gov.au/__data/assets/pdf_file/0020/83702/OAIC-Notifiable-data-breaches-report-January-to-June-2023-final.pdf

Piotrowski, C 2011, *Problem solving and critical thinking for designers*, John Wiley & Sons, Hoboken.

Retraction Watch 2024, https://retractionwatch.com

Rollo, J 2006, 'Under the Big Top', *The Weekend Australian Magazine*, 1–2 April, p. 16.

Sales, H 2006, *Professional communication in engineering*, Palgrave and Macmillan, Basingstoke, UK.

The Social Research Centre 2016, *2016 Graduate Outcomes Survey (GOS): National report*, Quality Indicators for Learning and Teaching, Victoria.

——— 2018, *2017 Graduate Outcomes Survey (GOS): National report*, Quality Indicators for Learning and Teaching, Victoria.

——— 2019, *2018 Graduate Outcomes Survey (GOS): National report*, Quality Indicators for Learning and Teaching, Victoria.

——— 2020, *2019 Graduate Outcomes Survey (GOS): National report*, Quality Indicators for Learning and Teaching, Victoria.

——— 2021, *2020 Graduate Outcomes Survey (GOS): National report*, Quality Indicators for Learning and Teaching, Victoria.

——— 2022, *2021 Graduate Outcomes Survey (GOS): National report*, Quality Indicators for Learning and Teaching, Victoria.

——— 2023, *2022 Graduate Outcomes Survey (GOS): National report*, Quality Indicators for Learning and Teaching, Victoria.

V/Line 2009, 'Southern Cross Station', https://www.vline.com.au

Wilson, M 2018, 'Patriarchs on pedestals? How doctors are taught to improve their bedside manner', *The Conversation*, 19 June, https://theconversation.com/patriarchs-on-pedestals-how-doctors-are-taught-to-improve-their-bedside-manner-95863

ACKNOWLEDGEMENTS

Photo: © Monkey Business Images / Shutterstock

Photo: © Neale Cousland / Shutterstock

Figures 8.1–8.7, table 8.1 and text: © Office of the Australian Information Commissioner (OAIC) 2023, *Notifiable data breaches report: January to June 2023*, 5 September, https://www.oaic.gov.au/__data/assets/pdf_file/0020/83702/OAIC-Notifiable-data-breaches-report-January-to-June-2023-final.pdf

Table 8.3: © Bureau of Meteorology 2018, 'Climate statistics for Australian locations. Monthly climate statistics: Sydney — Observatory Hill', http://www.bom.gov.au

Table 8.8: © Australian Bureau of Statistics (ABS) 2001, *Australian Standard Classification of Education (ASCED)*, ABS, Canberra.

Table 8.9: © King, R 2021, *Professional engineering graduates by branch of engineering*, 21 May, Australian Council of Engineering Deans, https://www.aced.edu.au/downloads/ACED%20Graduates%20by%20Branch%20of%20Engineering%20May%202021%20-%20RKing.pdf

Text: © Wilson, M 2018, 'Patriarchs on pedestals? How doctors are taught to improve their bedside manner', *The Conversation*, 19 June, https://theconversation.com/patriarchs-on-pedestals-how-doctors-are-taught-to-improve-their-bedside-manner-95863

Text: © Briggs, P 2023, *The engineering profession: A statistical overview*, 15th edn, Engineers Australia, Canberra, https://www.engineersaustralia.org.au/publications/engineering-profession-statistical-overview-15th-edition

PART 5

APPLYING THE ENGINEERING METHOD

CHAPTER 9

Information skills

'A wealth of information creates a poverty of attention.'

Herbert Simon, Nobel Laureate

LEARNING OBJECTIVES

After studying this chapter, you should be able to:

9.1 differentiate between data, information and knowledge
9.2 identify the information needs for a project
9.3 locate and retrieve high-quality information
9.4 evaluate the quality and reliability of information and information sources
9.5 integrate, manage and use information
9.6 cite and reference information.

Introduction

Previous chapters highlighted the fact that a major part of an engineer's work involves the use of information. An early chapter described the engineering method, with the second step in the method being to gather data, information and knowledge (research) — and, in the process, clearly define the problem and the performance criteria. This chapter describes a range of information skills that are necessary to source data and information, and to use data, information and knowledge successfully. Engineers source, analyse, synthesise and manage data to assemble the new information required for a project. They also use existing information to create new designs, and they interpret, massage and reformat information to make it accessible to their clients, other professionals and members of the public. The information they produce and use may be in the form of words, mathematical symbols, tables, graphs, charts, computer programs, plans or models.

Many engineering students undertake the Engineers Without Borders (EWB) Australia Challenge in their first year at university. Each year, since the program began in 2007, EWB sets a new challenge in collaboration with a local non-government organisation (NGO) or a local community partner. Through workshops and discussions, EWB and the partner organisation develop a series of community-identified issues and opportunities that would benefit from appropriate and sustainable technical and engineering solutions. Over the last few years these have included the following EWB Challenges.

- *2019.* EWB partnered with WaterAid in Timor-Leste:

  > The 2019 EWB Challenge design brief focused on WaterAid's work with communities across Suco Holarua, in the Manufahi District of Timor-Leste. Suco Holarua, along with much of Timor-Leste, has recently seen significant improvements in areas such as energy access, water supply and road networks. While coverage is not yet comprehensive, these initial infrastructure improvements are enabling an increase in community opportunity, sustainability and wellbeing [as well as the inclusion of all individuals in project planning and outcomes] (EWB Australia 2019).

- *2020.* EWB partnered with the Centre for Appropriate Technology (CfAT) in Australia. 'The EWB Challenge design brief considered appropriate technology to support traditional owners living and thriving on homelands and outstations, with a focus on CfAT's work with communities in the Cape York region of Far North Queensland' (EWB Australia 2024a). Specific partnership activities include:

  - access to energy — exploring and developing self-reliant models including community enterprise and impact investing
  - appropriate technology development — appropriating technology to make it more suitable for Indigenous communities for sustainable livelihoods on Country
  - land-use planning — supporting Ranger programs and the Healthy Country Planning process with appropriate enabling infrastructure
  - specific community infrastructure support projects (EWB Australia 2020).

- *2021.* For the second year running EWB partnered with the Centre for Appropriate Technology (CfAT), delivering projects in the same contexts as those used in the 2020 Challenge.
- *2022.* EWB partnered with Dawul Wuru Aboriginal Corporation (DWAC). Dawul Wuru is a community organisation owned, managed and governed by Aboriginal traditional owners that exists to 'protect, secure, support and promote the rights and interests of local Aboriginal Traditional owners and custodians' (Dawul Wuru Aboriginal Corporation 2024). For this EWB Challenge, 'the project briefs support Yirrganydji people to sustain their wellbeing and culture, and care for their rainforest and coastal land and sea between Cairns and Port Douglas, in Far North Queensland in Australia, for the benefit of current and future generations' (EWB Australia 2022a).
- *2023.* This was the first year that the EWB Challenge ran two contexts in an academic year, one focusing on an international community and the other within Australia's First Nations communities. Universities undertaking the Challenge could choose one of the following contexts.
  - *Context 1.* EWB partnered with EWB Australia's Engineering team in Cambodia. The Challenge was delivered through live collaboration with the team with a focus on Pu Ngaol village in Mondulkiri province. The Project briefs were developed using EWB's Technology Development Approach, which involved community workshops and interviews with Pu Ngaol village residents. Priority issues and aspirations of the community included themes around water supply, agriculture, education and

sanitation. Student design ideas would support EWB's Engineering team and their work on the ground in Mondulkiri province, which aims to positively contribute to the lives of Pu Ngaol residents (EWB Australia 2022b).
  - *Context 2.* Due to university demand, EWB again partnered with the Dawul Wuru Aboriginal Corporation. This time the focus was on DWAC's Ranger programs.
- *2024.* Once again universities undertaking the EWB Challenge could choose between two contexts.
  - *Context 1.* EWB partnered with the Torres Strait Island Regional Council (TSIRC) on Saibai Island, marking the first collaboration with a local government council. This new partnership added a unique and valuable context to the EWB program as it brought an enriched dimension to its projects, and an awareness of the Torres Strait islands. It was aligned with EWB's commitment to making a positive impact in diverse communities. Key challenges at this location were waste management and climate change response, given the fast-rising sea levels in the area. This context required students to explore the circular life of design, starting from conception until disposal, as it is very difficult to dispose of waste in these remote islands (EWB Australia 2024b).
  - *Context 2.* EWB again partnered with EWB Australia's Engineering team in Cambodia. Like the 2023 Challenge, the 2024 Challenge was focused on Pu Ngaol village in Mondulkiri.

We will use the 2018 EWB Challenge as an example of the information resources needed for a project. In that year EWB partnered with the Cambodian Rural Development Team (CRDT), 'a local NGO focused on community development and natural resource management throughout the north-east of Cambodia. The EWB Challenge design brief [given to the participating student teams] focused on CRDT's work with three adjacent communities along the Mekong River in the Sambo District of Kratie Province: [Ksach Leav, Koh Khnear and Puntha Chea]' (EWB Australia 2024a).

The Challenge focused on seven separate design areas, developed through workshops and interviews with community partner staff and community representatives, which would contribute to the sustainable development of the communities. The student teams were asked to develop a design concept that addresses one or more of the projects listed in the design brief. The design areas were:

- *water supply* — for example, optimising the rainwater collection systems, or the quality of the water
- *sanitation and hygiene* — for example, designing appropriate and affordable toilets or sanitation education and awareness programs
- *energy* — for example, designing low-cost and low-energy refrigeration systems or optimising cookstoves
- *agricultural systems* — for example, designing crop irrigation systems or greenhouses for vegetable production
- *waste management* — for example, designing a solid waste management system or increasing the opportunities for recycling
- *information and communication technologies* — for example, designing systems for water monitoring or to connect producers to markets
- *conservation and livelihoods* — for example, designing an appropriate and viable bamboo processing system (EWB Australia 2018).

The information provided for the 2018 EWB Challenge gives an insight into the types of information that engineers use in their work. Examples of projects underway in the partner organisation or wider community are one source of information that can help students build an understanding of the context. For example, figure 9.1 shows one view of the CRDT 'model farm' on Koh Dambang in Cambodia, which was captured as part of the 2018 EWB Challenge resources.

The following information resources were provided by EWB Australia (2018) to assist the student teams to conceptualise the problems and develop a deep understanding of the project context through exploring a variety of different perspectives.

- Background information about the work undertaken by the Cambodian Rural Development Team.
- CRDT 'project snapshots' — short summaries of previous projects, including what led to success or failure.
- A design brief that provided detailed information about each design area.
- Contextual information and materials about Ksach Leav, Koh Khnear and Puntha Chea.
- Maps of the Ksach Leav, Koh Khnear and Puntha Chea areas, maps on the Open Development Cambodia website and a map of the villages in the Kratie province.
- Interviews with previous guests of CRDT's ecotourism enterprise.

- A guide on what makes a technology appropriate.
- Costing data — examples from Kratie province.
- Videos — Transport around Kratie province, Understanding Kratie town, Overhead view of the Koh Dambang model farm, Overhead view of rice farms, A walk through the Koh Dambang model farm, Rice farms and Farm homes.
- The Rural Water Supply, Sanitation and Hygiene Strategy.
- The results of water tests — Stung Treng.
- Water supply and sanitation sector assessment, strategy and road map.
- Links to other sources of information provided in the 'About Cambodia' and 'About CRDT' pages of the EWB website, including examples of cultural/governance information.
- Reports from the previous EWB Challenge finalists.

FIGURE 9.1 A view of the CRDT model farm on Koh Dambang in Cambodia

This list of resources included a wide range of information and media, some of which were developed by EWB and CRDT for the purposes of the 2018 EWB Challenge, and some of which were compiled from existing sources such as government reports. The information in this list can be separated into five broad categories, which are indicative of the types of information that engineers use to inform their work, as follows.

1. Information about the client(s) — CRDT.
2. Information about the task(s) — the design areas.
3. Information about the users and other stakeholders — the people of the Kratie province who will use the products or services, their culture, characteristics, lifestyles and requirements.
4. Information about the context(s) — towns, villages, existing rural industries; housing; waste management; water test results; energy systems; infrastructure; transport; governance structures; and site characteristics such as climate, topography and vegetation.
5. Information from previous projects and practitioners.

During your first year of studies, you should acquire some of the key skills an engineer uses to access and use the types of information relevant to your discipline. You will enhance these skills and acquire new skills as you progress through your program. These skills are set out in Engineers Australia's Stage 1 Competency 3.4: Professional use and management of information. Engineers Australia's (2019) stated evidence for demonstrating attainment of this Element of Competency are that an engineer:

(a) is proficient in locating and utilising information — including accessing, systematically searching, analysing, evaluating and referencing relevant published works and data; is proficient in the use of indexes, bibliographic databases and other search facilities
(b) critically assesses the accuracy, reliability and authenticity of information
(c) is aware of common document identification, tracking and control procedures.

Another key personal skill you will require is the ability to recognise gaps in your knowledge and then develop and implement strategies to access and integrate the information required to fill those gaps.

Beginning with a discussion about information, this chapter then focuses on the six competencies in the information process: identifying information needs, locating and retrieving information, evaluating the quality of information, integrating information into existing knowledge frameworks, managing information, and citing and referencing information sources.

9.1 Data, information and knowledge

LEARNING OBJECTIVE 9.1 Differentiate between data, information and knowledge.

The terms 'data' and 'information' are often used interchangeably and the meaning of these terms can vary between professions — and even between different engineering disciplines. Understanding the meaning of terms such as data, information and knowledge is an important starting point for the successful acquisition and use of information. The effectiveness of communications between people depends on a common understanding of these terms.

A number of models have been proposed to define the relationship between data, information and other concepts such as knowledge and wisdom (e.g. Cleveland 1982; Sharma 2008; Zeleny 1987). Although these models are generally referred to as models of the Data Information Knowledge Wisdom (DIKW) hierarchy, some models include additional concepts such as understanding.

KEY POINT

Engineers use data, information and knowledge in their practice.

Data

Davenport and Prusak (1998, p. 2) defined **data** as 'a set of discrete, objective facts about events' and a **fact** is defined by Macquarie Dictionary (2024) as 'something known to have happened; a truth known by actual experience or observation'. While some of the data used by engineers is measured automatically (e.g. the electricity consumed by a household is measured by an electricity meter), most of it is **empirical** — that is, 'derived from or guided by experience or experiment' (Macquarie Dictionary 2024). Put simply, empirical data results from verifiable and objective measurements, observations and experience.

Empirical data is to some degree subjective because of human involvement in its creation. The degree of subjectivity depends on the level of human involvement. When a student observes a measurement on a scale there is normally a low level of subjectivity. A higher degree of subjectivity is involved when a mechanical engineer uses sight and touch to gather data about the metal fatigue characteristics in a failed component. This empirical data could be verified by checking the fatigue characteristics of other failed components to see if there is a pattern, or by conducting tests on the component.

The singular form of the word *data* is **datum**. This term is used in some fields of engineering to describe a reference point to which other measurements are related. For example, the heights of all of the features of a building, such as ceilings, floors, arches and roof structures, may be related to a particular point on the floor of the building (for example, the entrance step) or a survey nail in the roadside kerb adjacent to the building. This point is known as the *datum* for the heights or levels. The official datum for the heights of topographic features in Australia is the Australian Height Datum (AHD), which was adopted in 1971 (Geoscience Australia 2023). It was derived from mean sea level data measured by tide gauges at 30 locations around the coastline, between the years 1966 and 1968. The Australian Height Datum (Tasmania) was adopted in 1983 (AHD-TAS83) and is based on 1972 mean sea levels at tides gauges in Hobart and Burnie (Geoscience Australia 2023).

New Zealand Vertical Datum 2016 (NZVD2016) is the equivalent for New Zealand and its offshore islands (Land Information New Zealand 2024).

It is important to note that the term *data* may have a different meaning when it is used by engineers working in an *information technology* field such as computer systems engineering or software engineering. In this context McDonald (2002, p. 63) suggests '*Data* is a kind of statement with a specific structure. It is a statement that identifies the value of an attribute of a phenomenon'. On the other hand, Pressman (2005, p. 896) defines *data* in this context as being 'raw information — collections of facts that must be processed to be meaningful'.

Finally, it is important to note that data is usually accompanied by metadata, which was described in the chapter on communication skills. In most cases, the metadata defines how the data was obtained, how it should be used and its limitations. Some examples of the metadata for measured data are accuracy, precision, the equipment used to measure the data, the date the data was measured and who measured the data. The *accuracy* of a measurement indicates the difference between the measurement and the correct or true value of the phenomena being measured. Thus, an error-free measurement would be accurate.

Unfortunately, the true value of the phenomena being measured is not normally known, so other methods are used to evaluate accuracy. First, key measurement systems are normally calibrated against a standard. The results of the calibration process define the differences between the measurement system and the standard. This benchmarking process is a requirement for most engineering testing equipment and must be undertaken at specified intervals to retain certification. When calibrated equipment is used correctly, there is a high degree of confidence that the measurements are accurate.

Second, measurements are often repeated to gain an indication of the *precision* of the measurements. Precision is therefore a measure of the *repeatability* of the measurement using the same measurement system. Often, to enhance confidence, measurements may be repeated using different equipment and personnel. This assesses the reproducibility of measurements. The precision of a set of measurements is an indication of the magnitude of the random errors in the measurement sample.

It is common practice to use the bullseye analogy to illustrate the difference between accuracy and precision (see figure 9.2). The shots fired by shooter (a) are both accurate and precise, as the seven shots are clustered in the bullseye. The shots fired by shooter (b) are precise but lack accuracy, as the shots are tightly clustered but away from the bullseye. This indicates the presence of systematic error in the rifle or the shooter's technique. The shots fired by shooter (c) are accurate as they are clustered around the bullseye, but lack precision as they are scattered. The shots fired by shooter (d) lack both accuracy and precision.

Statistical methods are normally used to evaluate the accuracy and precision of a sample, or a number of samples of data. Where there is evidence of a systematic error, or a bias, in a measurement (for example, as the result of a calibration process), then it may be possible to correct the data. On the other hand, accidental and non-identifiable errors, normally referred to as random errors, cannot be corrected.

The results of a measurement should be reported so that only the significant digits are included, that is, the digits that may be used with confidence. Scientific notation can be used to communicate this for large numbers, as this format highlights the significant figures of the data. Measurement data is normally accompanied by statistical data.

We all use smart technologies such as phones, watches and other devices for work, recreation and social activities. In fact, many of us rely on them to help us through the day (and night). While they are pre-filled with lots of amazing features, we can import apps to add new features and to personalise them. The accuracy and precision of the measurements made by these technologies and the associated apps continue to be studied, and the examples in the following spotlight highlight some of the inconsistencies found by ordinary users.

FIGURE 9.2 An illustration of accuracy and precision

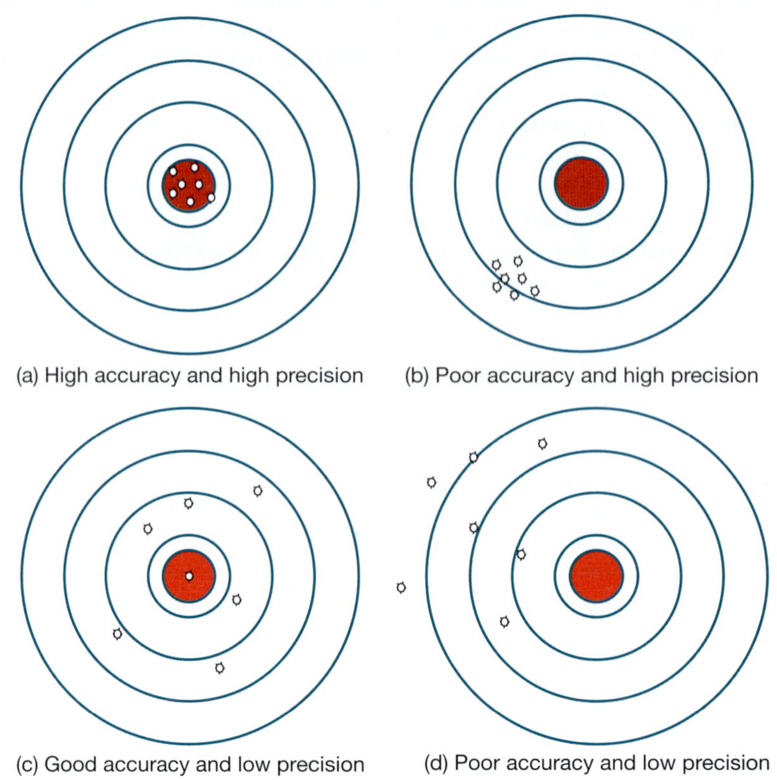

SPOTLIGHT

Smart technologies: how accurate are their measurements?

Example 1

Jess was surprised at the number of steps her Fitbit had counted on a regular school day. The number was way higher than on other days, yet she could not remember doing any physical activity that could account for the extra steps. Then she remembered her half-hour session playing a marimba percussion instrument. Had her new Fitbit counted her more intense hand movements as steps? Jess searched the Fitbit website and found that her device uses a finely tuned algorithm to interpret the data from the device's 3-axis accelerometer. The algorithm assesses the type, size and intensity of the motion patterns to determine if they are indicative of walking or running. If they meet the threshold then they are counted as a step. Jess was most interested to read that 'when working at a desk, cooking or doing other arm movements, a device on your wrist may pick up some extra steps if it thinks you're walking . . . For the vast majority of customers, the number of stray steps accumulated is negligible when compared to the entire day' (Fitbit 2018). Jess concluded that this explanation probably accounted for her extra steps, which in her case made up a large part of her step total. To test this, Jess timed a practice session on the marimba and found she did 1200 steps in 15 minutes, while virtually standing still!

Example 2

Early one morning, while on a 10-kilometre walk, David was alarmed to see that his TomTom Adventurer Sportswatch indicated his heart rate was 180 beats per minute (BPM) — a dangerous rate. From experience, he knew this was wrong as his usual BPM when walking on a slight incline (and not puffing) was between 90 and 100. He noticed that the watch had slipped down and was near his wrist bone so he repositioned it a couple of centimetres up his arm and tightened the strap. Within a few seconds his BPM was back in the expected range, at 95. To confirm his action, David reread the TomTom instructions, which stated that the watch 'needs to be worn quite tightly (almost to the point of uncomfortable) and . . . much higher up the wrist than a regular watch' to ensure that no light seeps under the band and causes interference (TomTom 2017). He also researched the accuracy of the technologies used in TomTom cardio watches and found that TomTom uses LifeQ technology to measure heart rate and that the watch uses an algorithm to 'combine motion information and sensor readings to better estimate your heart rate whilst you are doing activities' (TomTom 2017). After extensive testing, LifeQ reported that 'during athletic activity the LifeQ solution shows very high degrees of accuracy, with the vast majority of heart rate measurements within a 5% error of a chest strap electrocardiogram' (LifeQ 2018).

Example 3

Ellie and the other two walkers were baffled when they received their smartphone voice prompts that indicated they had completed the first kilometre of their proposed 15-kilometre trail walk through hilly and, at times, steep bushy terrain. It was their first team training session for a 60-kilometre Oxfam Trail Walk they planned to complete later in the year. They were all logging their journey using the popular Strava App on their smartphones. It provides them with real-time information such as a topographical or satellite map showing the path followed, their walking speed, the metres climbed, the overall time and the moving time (see example in figure 9.3). They had all pressed the 'start button' at the same time when they set off at dawn on the forest trail and had switched on the 'audio cues' feature so that they would receive a voice prompt advising them of the total time and time for last kilometre, as they completed each kilometre.

FIGURE 9.3 Close-up of a smartphone displaying the Strava app, which shows navigation and pace information for a cyclist

The reason they were baffled was that instead of being within a few metres of each other, the three audio prompts were spread over at least 150 metres. They wondered which phone was right and why they were all so different. To figure this out, they reviewed the Strava website. After reading comments in a Strava blog attached to the 'Bad GPS Data' article, they learned they were not alone; many others had similar experiences (Strava 2018a). They gleaned the following points from the information in two articles on the Strava website (Strava 2018a, 2018b).

- No recorded GPS data is perfect.
- Different devices have different qualities of GPS hardware and software.
- The number of satellite signals received and the location of those satellites affect the accuracy of the measurements. A minimum of five satellites are required, although seven or eight are normally needed to calculate the position to within ten metres.
- Environmental factors, such as dense trees, steep hillsides, tall buildings and heavy cloud, can have an impact on, or even interrupt, the signal between the satellite and the device.

The inaccuracy of phone-based measurements was confirmed when one of the walkers completed a number of five-kilometre parkruns (see https://www.parkrun.com.au) and found the distance recorded using Strava on their phone was always at least 3 per cent longer than the actual course distance of five kilometres.

Example 4

To improve his understanding of the accuracy of GPS tracking, and variations found when different devices are used, David recently used a TomTom Adventurer sports watch and the Strava App in an iPhone to record the route of a long walk through bushland parks and urban streets. Following his usual practice, he shared his TomTom results with friends by transferring the TomTom data file to Strava using the preferred file format. The results were as follows.

	Distance (km)	Difference from TomTom	Elevation (m)	Difference from TomTom
TomTom Adventurer	12.73		210	
Strava – through iPhone	13.66	+7.3%	242	+15.2%
Strava – from TomTom data	12.73	0%	234	+11.4%

David was surprised at the magnitude of the differences in the data, particularly the elevation data, which refers to the total number of metres climbed during the walk. The differences are due to the different measurement technologies and the processing software used to determine distance and changes in elevation.

The TomTom Adventurer uses both the GPS measurements and the measurements from the built-in barometer to determine the metres climbed. The Strava App for mobile devices only collects GPS elevation data, which is then cross-referenced with Strava's Elevation Basemap to determine the metres climbed. Strava advise that 'In some areas the underlying basemap data is poor resulting in inflated elevation totals' (Strava 2018c). Strava also uses an algorithm and thresholds to smooth elevation data and remove noise. The climb thresholds are as follows.

- For non-barometric based data, the climb must be consistent over at least ten metres before it is added to elevation gain.
- For barometric based data, the climb must be consistent over at least two metres before it is added to elevation gain.

David also found that differences resulting from the use of algorithmic smoothing processes can be observed when the TomTom Hike activity program is used to record details of a walk (or run) that starts and ends at the same point. When this program is used, both the ascent metres and the descent metres are reported. Generally, they differ by a few metres rather than being the same, although the results depend on the length of the trail and the topography traversed.

It is important to note:

- the accuracy and precision issues discussed here are not unique to the brands and technologies described, as similar issues are reported by the users of the other brands and technologies
- the examples discussed only cover some of the common types of measurements made by 'smart' devices.

The accuracy and precision of 'smart' devices continue to improve as new technologies are integrated into their hardware and software systems, and into their spatial information support systems.

CRITICAL THINKING

Do the results in these examples surprise you? Test the accuracy and precision of your personal devices by using them to record activities over known courses.

From the examples in the spotlight we can see that the *accuracy* and *precision* of the physical activity related measurements reported by personal smart technology devices depend on many factors including the measurement technologies, the algorithms used to process and smooth the data, the location of the device on the person, the weather patterns at the time of the activity, the vegetation and the topography of the area where the activity occurred including both natural and man-made features. Another issue is that the format of the measurement data, and the number of digits reported for each measurement type, often leads the user to believe the accuracy of a device is greater than it actually is.

These conclusions suggest that the measurement data provided by smart devices should be treated with caution and only regarded as an estimate of the actual measurements. They also suggest that engineers need to carefully test, evaluate and calibrate the hardware and software of new technologies that they introduce into their practice. They should review any assumptions that were used to develop the technologies and any limitations that the developers placed on its use.

As we saw in the spotlight, many smart devices rely on GPS and other spatial location technologies to determine the spatial location of the device. These technologies are also the key components of the navigation systems used in cars, trucks, trains, planes and ships. Scientists use them to track animals on land, marine creatures in the sea and birds in the sky. As they are ubiquitous it is important that you have some understanding of GPS technology and the expected accuracy of the measurements made by common devices because you are likely to use these devices in your engineering work. The following spotlight provides an overview of accuracies of the commonly used spatial location devices.

SPOTLIGHT

Exploring the accuracy of spatial location devices

WITH CHRIS MCALISTER, PROFESSIONAL FELLOW, UniSQ

There are a multitude of positioning technologies being used in Australasia, from mobile phone apps for *navigation* to high-precision positioning systems used for *surveying*. In the less sophisticated systems, like phone apps and car navigation systems, the measurement process is fully automated and requires no involvement from the user. In more precise positioning systems, the operator normally must follow a defined measurement process to realise the potential accuracy and precision of the equipment.

Most positioning devices use one of the available global navigation satellite system (GNSS) positioning systems, such as the US global positioning system (GPS), the Chinese system BeiDou or the Russian system GLONASS. All GNSS use similar systems and technology to allow users to determine their position on the Earth via radio signals. These signals are called carrier phase and carry specific 'codes' that tell the receiver about where the satellite is positioned in its orbit, what time its clocks are at and a variety of other information. Some codes are available for civilian use, some are for military use only, while others are used for 'safety of life' — allowing emergency services to locate lost or injured people.

GNSS signals are freely available to everyone and are available 24 hours per day, seven days per week, anywhere on earth and in all weather. All that is required is some form of technology with a GNSS chip, called a receiver (McAlister 2023), and a signal from at least four satellites.

In some countries, the preferred satellite system, such as GPS, has been augmented by other technology, both space and ground based. Communication satellites or ground stations monitor satellite data and can broadcast signal corrections so that enabled receivers can apply the corrections and provide more accurate positional data.

An example of an augmented system is the satellite-based augmentation system (SBAS) in Australia, called SouthPAN. It aims to make GPS accurate to around 1 metre on your average mobile phone. The opportunity for applications of this kind of technology now extend far beyond the basic navigation functions we have previously used GPS for.

The accuracy of a position calculated from satellite measurements depends on many factors, the four most important being:

- the number and location of the satellites (i.e. the satellite geometry) used to calculate the position
- the quality of the satellite receivers used
- the measurement process
- the measurement processing software.

Some of the other factors that affect accuracy are atmospheric conditions, orbital errors, signal multipath and receiver clock errors.

So, how accurate are the position measurements made by some of the more commonly used technologies?

Mobile phone location apps generally use a combination of GNSS signals, wifi systems and mobile phone towers to determine their approximate location. In fact, these systems are designed to provide position data in places where satellite data is not available (e.g. inside shopping centres). Generally, the accuracy of a position determined by a mobile phone is about 10 metres. This means that the true position of the phone is likely to lie within a 10-metre radius circle centred on the calculated position. Depending on the chipset technology embedded in a phone, it may be able to receive more of the types of radio signals used, meaning it will be even more accurate.

The positional information determined by car, boat and other handheld satellite navigation devices is, on average, accurate to about ten metres. If the satellite data is augmented by satellite or ground station data, then positional accuracy may improve up to the point where it is better than one metre accuracy.

A differential GPS (DGPS) system may be used to enhance the simple GNSS positioning techniques. By placing one GNSS receiver over a mark with a known coordinate (called a base station), and then having another GNSS receiver (called a Rover) collecting the data, a comparison can be made between the base's known coordinate and the coordinate the GNSS receiver is producing, and this difference can be applied as a correction to the data the Rover is collecting. This can be done in real time, or later back in the office. This then produces a much improved accuracy of between 0.5 and 1.0 metre relative to the known station. It is important to note that the accuracy of height measurements is generally 1.5 to 3 times less than the accuracy of horizontal measurements.

Surveyors use high-precision satellite surveying equipment to accurately define the location of survey control marks to within a centimetre. While they use a similar process to DGPS, they tend to use a technique called real-time kinematic (RTK) positioning or network RTK (NRTK) positioning so that they can achieve 1–3 centimetres accuracy in real time. These systems use elements of the carrier phase signal, rather than the code part of the signal, to determine the measured distance between a satellite and the ground station. Often the raw data is still recorded, and this can later be put through post-processing software in the office, which uses the actual position of the satellites at the time of data collection rather than the predicted positions of the satellites that were broadcasting at the time. Post processing the data generally achieves sub-centimetre accuracy.

For high-precision surveys, the accuracy statement for the horizontal position of a measured station is called uncertainty, and has two components: (a) the error in mm and (b) the confidence level (normally 95 per cent). Thus, if the accuracy (uncertainty) of a survey mark is said to be ±7 mm with a confidence level of 95 per cent and if the position was measured another 100 times, 95 of the calculated positions would lie inside a 7 mm radius circle centred on the initial position.

CRITICAL THINKING

Review the handbook, or specifications, of a GNSS device to find the specified accuracy and precision of the device. Develop a plan to test the accuracy and precision of the GNSS measurement system used by the device. If possible, use your plan to test the device and compare your results with the manufacturer's specifications.

Information

Macquarie Dictionary (2024) defines **information** as 'knowledge communicated or received concerning some fact or circumstance; news'. In this definition one person's knowledge is communicated to another as information. Thus, information is 'a *message*, usually in the form of a document or an audible or visible communication' (Davenport & Prusak 1998, p. 3). From a DIKW perspective, data becomes information when it is given meaning and/or context by its creator. For example, a household water meter was read on 10 December 2023 and the measurement was recorded as 28 381 689. This data is meaningless until it is compared with the previous reading so the household's consumption of water over the period can be calculated. This gives the data meaning and it is now information. It is important to note that computers, as well as people, can convert data into information. In this case the water authority's computer would use the amount of water consumed to calculate the customer's water bill and then prepare an invoice.

Once data has been transformed into information, that information cannot normally be deconstructed to create the original data (Zeleny 2005). For example, the rainfall for a particular year cannot be determined from the average annual rainfall for a region.

Knowledge

Macquarie Dictionary (2024) provides a simple definition of **knowledge**: 'acquaintance with facts, truths, or principles, as from study or investigation'. But knowledge is a complex concept and Davenport and Prusak (1998, p. 5) provide a more useful explanation:

> . . . knowledge is a fluid mix of framed experience, values, contextual information, and expert insight that provides a framework for evaluating and incorporating new experiences and information.

It originates and is applied in the minds of knowers. In organisations it often becomes embedded, not only in documents or repositories, but also in organisational routines, processes and norms.

When new information is integrated in the mind, learning occurs and new knowledge is created. This knowledge represents the learner's understanding of the topic at that point in time. When surface learning occurs, the resulting knowledge may be based on memorisation rather than understanding (see the self-management chapter). This form of knowledge may be tacit; that is, it may not be retained. Deep learning, however, will be integrated into the learner's existing knowledge framework and will enhance understanding. Most importantly, Zeleny (2005, p. 5) contends that:

> ... knowledge is the primary form of capital. All other forms are dependent and derived ... Without knowledge, money is just a pile of paper, machines just a concoction of metals, buildings just a heap of bricks and concrete, and raw materials remain just that: raw materials. Knowledge gives life to all.

Many writers incorporate wisdom into their definition of knowledge; however, it is worthy of separate mention because engineers often seek wise counsel — wisdom from a colleague when they are faced with a difficult decision. Wisdom is defined by Macquarie Dictionary (2024) as 'knowledge of what is true or right coupled with just judgement as to action; sagacity, prudence, or common sense'. Wisdom, therefore, has to do with using knowledge in a manner that takes account of values, justice and experience. You will develop your engineering wisdom as you gain experience in the engineering industry and as you apply the principles associated with professional responsibility and ethics.

Differentiating between data, information and knowledge

McDonald (2002) used Karl Popper's three worlds theory (Popper 1979) to develop a framework for the integration of the different theories and practices developed by those working in the information professions. This theoretical framework provides an 'Information Age' perspective into the nature of information, how it is created and how it is used.

Popper's theory divides reality into three separate worlds. World 1 is the physical world; World 2 is the world of our conscious experiences; and World 3 is the world of the logical *contents* of books, libraries, computer memories, works of art, music and feats of engineering (Popper 1979, p. 74).

McDonald represented the interactions between these three worlds diagrammatically, as shown in figure 9.4. In this representation, a person interacts with World 1 through actions and perceptions and with World 3 through writing and reading. For example, an interaction between World 2 and World 1 occurs when an engineer observes the real world and perceives a relationship between two objects or events. After reading what others have written on the subject, which is an interaction between World 2 and World 3, and careful consideration (which enhances her knowledge), the engineer may develop a concept, test it and then write an article that is published in a professional journal. This is another interaction between World 2 and World 3, but this time it results in an addition to the information in World 3. It should be noted that while the conceptual component of the article (the information) resides in World 3, the physical component of the article, the words on paper or in electronic storage, resides in World 1. In the diagram, the arrow and the word 'denotation' indicate that the information in the article describes a real-world phenomenon.

Figure 9.5 is an adaption of McDonald's representation and illustrates how this conceptual framework can be used to show the differences between data, empirical data, information and knowledge.

- Data that is automatically (electronically or mechanically) detected in World 1 may be automatically converted into information and made available in World 3.
- Empirical data that is observed, or experienced, in World 1 may be made available in World 3 via World 2. This normally includes measured data because the results are observed and processed by the mind before being documented.
- Knowledge is created from observations and experiences from the physical world and from information in World 3. Knowledge is created and stored in the mind, World 2. When knowledge is documented it becomes information in World 3 and is available for others to encounter and use.
- Information resides in World 3, in physical and/or digital formats.

To research a topic, an engineer may use her knowledge and judgement to analyse and synthesise data, observations and personal experience from the physical world to produce a new engineering idea or principle. This process is represented in figure 9.5 as an idea created in World 2 and documented in World 3. The process is shown in more detail in figure 9.6, which shows that the engineer selects information from a range of information sources from World 3 and then, in World 2, the mind analyses it and compares it with personal experience (from World 1) before creating and then documenting new information in World 3.

FIGURE 9.4 Interaction between the three worlds

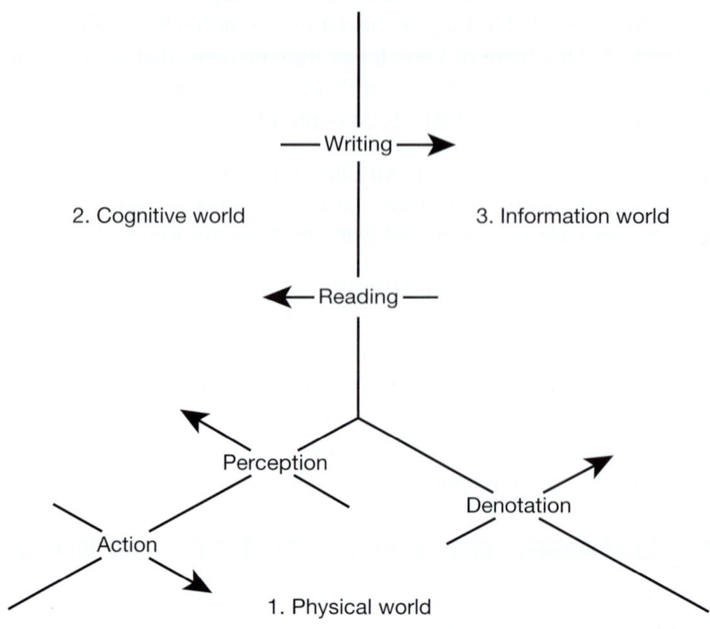

Source: McDonald (2002, p. 60).

FIGURE 9.5 The relationships between data, knowledge and information

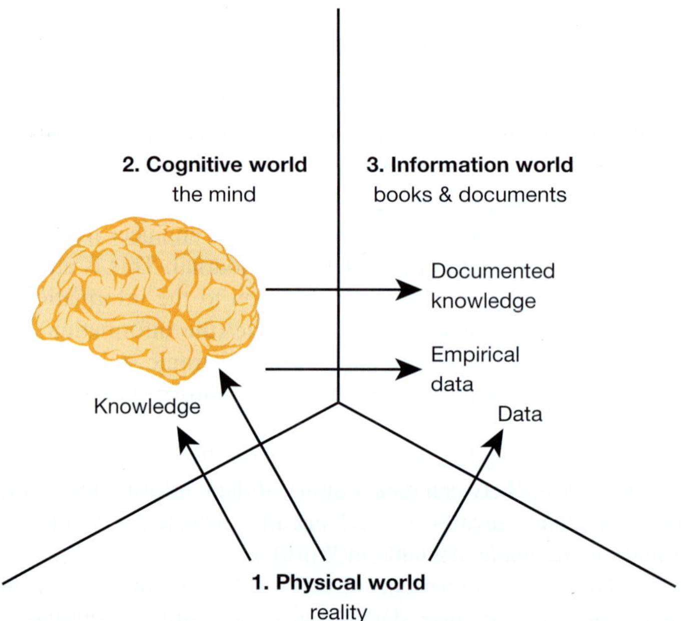

Source: Adapted from McDonald (2002, p. 60).

McDonald (2002) highlights the problems of applying definitions for terms such as data, information and knowledge across all of the disciplines of engineering, particularly in the computer systems and software engineering disciplines.

'The terms data and knowledge are controversial in Informatics, but Popper's three worlds model allows a clear definition of them to be made' (McDonald 2002, p. 63).

Zeleny (2005) highlights another interesting difference between information and knowledge. While people often talk about there being *too much information* and *information overload*, they are unlikely to say *there is too much knowledge*. He suggests that this is because compared to data and information, knowledge is best.

FIGURE 9.6 Creating information. The symbols in the 'Information world' represent different information sources and media

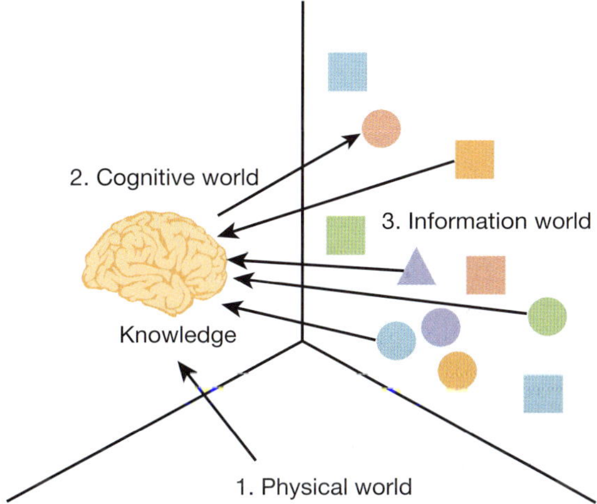

Source: Adapted from McDonald (2002, p. 60).

9.2 Identifying information needs

LEARNING OBJECTIVE 9.2 Identify the information needs for a project.

The first phase of the information gathering process is to identify the information that will be required to begin or undertake a project. This can be a difficult task, particularly when you have not worked on similar projects before. The aim should be to identify *all* of the information needs at the beginning of a project, although this is not always possible.

The time spent defining the information needs for a project is good preparation for the next phase of the process, the search. Proper preparation will save time and money as the searching process will not have to be repeated to find additional information. Good preparation can also provide a better understanding of the project itself, the areas that may cause problems or create risk, and the issues that are yet to be resolved. Information should be analysed as it is retrieved, as this may help to identify additional information needs and sources. Thus, the process of identifying and accessing information can be cyclic and may continue through several stages of a project.

KEY POINT

Engineers need to carefully define the information required for a project.

Investigative questioning

The second phase of the engineering method is research. Critical thinking skills are used to answer two questions that will help to better define the problem.

- What do I know about the problem?
- What do I need to find out so that I can solve the problem?

The answers to the first question are noted and any relevant information located. The answers to the second question can often be much harder to define.

A number of problem-solving methods are cyclic — for example, the evolutionary, incremental and spiral models that are used for software development. When cyclic problem-solving methods are used the information acquisition process may also be cyclic, with additional information being acquired after the development, and perhaps implementation, of each prototype, or version, of a product.

The most common way to define the information needs of a project is to generate a series of questions that identify knowledge gaps. These may be developed by applying the intellectual standards and elements of reasoning associated with critical thinking (see the chapter on the engineering method). Figure 9.7 lists some generic question stems that can be used as a starting point or checklist for developing relevant questions for an information search.

FIGURE 9.7 Question checklist

Who . . .?
Why is . . . important?
Explain **why** . . .
Explain **how** . . .
How are . . . and . . . related?
How is . . . similar to . . .?
How is . . . different from . . .?
How does . . . affect . . .?
How can . . . be used to . . .?
What are the benefits of . . .?
What are the costs of . . .?
What is the purpose of . . .?
What can you conclude about . . .?
What if . . .?
What does . . . mean?

Source: Adapted from King (1992).

As a student you can use these question stems and critical thinking skills to help you understand a topic, identify the outcomes of an experiment, or to prepare for an examination. You should modify and add to these questions so they suit your particular problem or project. A checklist of question stems helps you to define the questions that need to be answered.

The following spotlight illustrates how the question stem approach can be used to define the information needs of a project.

SPOTLIGHT

Flood protection for a mine tailings slurry system

When Julie started working for a mining company in north-west Australia she was the only on-site mechanical engineer. Although she had been working as an engineer for more than five years, this was the first time she had worked in the mining industry.

One of her first jobs was to design a slurry pump and pipe network to remove waste product to a new tailings dam. Julie's supervisor Nate, a mining engineer, gave her the following advice.

> Take your time to get your head around slurry pumps and pipe systems because we have to get this right. The environmental regulations are really strict and we must avoid spills at all costs. More importantly, if we get it right, this is an opportunity for the company to demonstrate its 'good citizen' credentials by protecting the environment. But, at the end of the day it will come down to a costing exercise — we will have to assess the benefits of protecting the environment; that is, the costs associated with minimising the risk against the costs of the potential losses. In this case, the losses may include environmental damage, repair costs, down-time, fines and adverse publicity for the company if the environment is damaged.
>
> Let me know when you are on top of it and we will talk to Kale, who is the civil engineering consultant who will design the earthworks and other structures.

From her knowledge of the site, Julie knew it would be subject to flooding if a cyclone hit the area. She realised one option would be to design the pump station and even the pipe network so they were constructed above the highest known flood level, but a preliminary feasibility study showed this would double the cost of the project. Julie decided to investigate other options, such as protecting the pump against flooding or using a submersible pump system, even though she thought this would be an unlikely solution.

Julie reviewed books and articles in the company's reference collection. *The Pump Handbook*, edited by Karassick, had some useful information, but the edition was printed in the late 1980s so Julie realised she would need to check if there was a later edition of the book, or another book, to see what new technologies and systems were available. Julie wrote a list of preliminary questions she would need to answer before beginning the design.

- What would happen if a standard slurry pump was submerged for a short period during a flood? Would spillage occur and, if so, what damage would it cause?
- What would it cost to repair any damage to the pump?
- How long would it take to repair?
- How would a standard slurry pump operate if it was enclosed by waterproof materials to protect it from a flood? Would the efficiency of the pump be affected because of an increase in operating temperature? Would it overheat and self-destruct?
- How do submersible slurry pumps compare with standard slurry pumps on performance and cost?
- What is the estimated depth and flow rate of recent cyclonic floods? How long do floods generally last? Should the design be based on the height of a 20-year or a 100-year flood?
- What systems can be used to anchor the pipe so that it does not get damaged by a flood?
- Are there any other flood protection measures to consider?
- What is the estimated cost of these flood protection measures?

Julie then decided to categorise her questions and to identify any links between them.

CRITICAL THINKING

Can you think of any other questions Julie should research? Prioritise the list of questions, and your new questions, so that the information search is systematic.

The first step in the problem-solving process is determining the information needs for a project. The problem is 'pulled apart', divided into manageable components and looked at from different perspectives. Once the list of questions has been developed, the searching process begins. As each piece of information is located and evaluated, it is likely that new questions will be identified. This is to be expected because as our knowledge of a subject grows, new issues, problems and opportunities are identified. These are analysed and may result in new questions being added to the list of questions to be resolved for the project.

The question stem approach is particularly useful for identifying the information needs for engineering projects that are open-ended or groundbreaking. In these projects, it is important to explore alternatives and stimulate creative thinking. In particular, the 'what if' questions should be explored, as this type of thinking often leads to innovative solutions. If possible, the problem should be discussed with colleagues in the relevant engineering and allied disciplines, to brainstorm other solutions, and to check and test currency of engineering principles and processes used in proposals.

Categories of information

The investigative questioning approach can be used in conjunction with another technique in large or complex projects. This involves defining the information needs under a series of categories so that the links between pieces of information are more easily defined. This approach can also be used to allocate different aspects of the information gathering process to different members of the project team. The categories should be suited to the characteristics of the project. For example, the categories could just relate to the technical aspects of the project or, as with the EWB Challenge, they could include other aspects of the project such as the information required to manage the project.

The following list would be a useful starting point for many projects, but it is not exhaustive and would need to be adapted for each project.

- *Information about the client.* The client's organisation, the requirements such as timelines and budget, the communication protocols to be used, the key people, including gatekeepers, and a business check.
- *Information about the project.* The specifications, existing products and components that could be adapted or purchased, codes and standards, and similar projects undertaken by the company.
- *Information about the other stakeholders.* User requirements, personal characteristics, culture and skills.
- *Information about the context.* Location, site characteristics, infrastructure — including transport — and the availability of construction materials and technologies.
- *Information from other professionals.* Senior colleagues in the same discipline, colleagues from other relevant disciplines, staff in government departments and agencies, product manufacturers and agents, and university academics who are researching in the field.
- *Information about risk.* Risk identification, assessment and management, including workplace health and safety.

Organising information needs

In some engineering fields, projects may be completed using relatively routine processes. Therefore, an engineer's knowledge of these processes, and the steps in the processes, may be sufficient to define the informational needs for a project. In fact, the organisation may have documented that knowledge and created a template that defines the types of information that will be needed for each stage of a project. It may, of course, be necessary to adapt the template to suit the contextual requirements of each project, but the majority of the questions will be applicable to each new project. For example, while each housing estate project brings its own unique challenges and complexities, engineers can often use the same routine design and development processes for future estates.

An information needs template identifies the information required for the particular type of project and helps to ensure that nothing is missed. More importantly, it ensures that relevant project information is available when each stage of the project begins. Table 9.1 is an example of an information needs template that may be used for a land development project, and includes data from a fictitious project, Project 19/124, to illustrate how the template is used. The categories of information are listed in the first column and the standard types of information required for these kinds of projects are listed in the second column. The third column lists the normal source for each piece of information.

TABLE 9.1 **Extract from the 'Information Needs' template for Project 19/124. Start date: March 2019**

Category	Type	Source	File	Who	When
Site details	Registered owner	Certificate of Title (CT)	DEV/19/124/1	SF	22/03/19
	Title information	CT	DEV/19/124/1	SF	22/03/19
	Area	CT	DEV/19/124/1	SF	22/03/19
	Encumbrances	CT/owner	DEV/19/124/1	SF	22/03/19
Town planning	Allowable land uses	Council town plan	DEV/19/124/1	DK	End March
	Requirements	Council town plan	DEV/19/124/1	DK	End March
	Restrictions	Council town plan	DEV/19/124/1	DK	End March
	Other	Council staff	DEV/19/124/1	DK	5/04/19
Location	Street address	Owner	DEV/19/124/1	SF	22/03/19
	Location plan	Surveyor	DEV/19/124/2	SF	End April
	Facilities plan	Surveyor	DEV/19/124/2	SF	End April
	Infrastructure	Surveyor	DEV/19/124/2	SF	End April
Use	Current	Town planner	DEV/19/124/2	SF	End April
	Future	Real estate agent/valuer	DEV/19/124/2	SF	End April
Site detail plan	Features	Surveyor	DEV/19/124/2	SF	End May
	Vegetation	Botanist	DEV/19/124/2	SF	End May
	Soils	Engineer	DEV/19/124/2	SF	End May
	Hydrology	Engineer	DEV/19/124/2	SF	End May
Design criteria	Client	Owner	DEV/19/124/2	SF	End May
	Lot dimensions	Surveyor	DEV/19/124/2	SF	End May
	Street	Engineer	DEV/19/124/2	SF	End May
	Stormwater	Engineer	DEV/19/124/2	SF	End May
	Sewerage	Engineer	DEV/19/124/2	SF	End May
	Power	Engineer	DEV/19/124/2	SF	End May
	Water	Engineer	DEV/19/124/2	SF	End May
	Communications	Engineer	DEV/19/124/2	SF	End May
	Gas	Engineer	DEV/19/124/2	SF	End May

At the beginning of the information stage of the project, the template would be adapted for the project with any additional information needs added, and any unwanted ones deleted. The 'Who' and 'When' columns are then completed to indicate who is to retrieve each piece of information, and the target date. As information is found and retrieved, the table would be updated to indicate the source for each piece of information, the date it was retrieved and where it is filed.

Another useful tool for organising information is a mind map, particularly for non-standard projects. A mind map can help the project team to understand the complexity of a project or organise the information needs to suit the stages of a project. A mind map can be easily adjusted if there are changes to information needs or the way they are organised. For example, an 'information need' can be dragged from one location to a new location. The mind map can also be adjusted if new links are found between pieces of information.

Figure 9.8 shows a mind map of the questions that Julie developed for her flooded slurry pump project. The mind map provides a graphical representation of the problem and indicates the many facets of the problem to be resolved. From this perspective, the mind map may be more useful than a list of the questions, even if they are categorised.

9.3 Locating and retrieving information

LEARNING OBJECTIVE 9.3 Locate and retrieve high-quality information.

We are living in an information age. Today's engineers face different information-related problems to those faced by the engineers who were practising 20 or 30 years ago. Then, two problems they often encountered were: (1) a lack of information and (2) an inability to quickly and efficiently access information. Today, engineers are likely to face two different problems: (1) having too much information and (2) being able to efficiently and effectively identify and source relevant and reliable information from the long lists of sources provided by the internet and database search engines. The challenge is to identify the key sources and refine the lists to manageable proportions without losing any vital information. A more recent problem is to identify, and cull, 'fake' information.

KEY POINT

Engineers must be able to efficiently locate accurate, relevant and reliable information.

Typical sources of engineering information

There are many strategies that can be used to source engineering information, and some of the commonly used strategies are discussed in this section.

The first step is to correctly identify the most likely sources of the required information. Experienced engineers will normally have established a personal library of the most important sources of information that they use in their practice. These collections normally include hardcopies of textbooks, codes of practice, standards, handbooks, reports on completed projects, tables and journals. Engineers are also likely to have a list of bookmarked internet sites that they regularly use to locate information. For example, the member's area of Engineers Australia's (2024) website (https://www.engineersaustralia.org.au) provides access to the online library and its services, including:

- Engineers Australia conference proceedings
- the seven Engineers Australia technical peer-reviewed journals
- journal articles and historical transactions
- the Ei *Compendex*
- Australian technical bibliographic databases
- Engineering handbooks
- the *Knovel* database containing formulae, interactive graphs and tables
- the *Emerald* database covering management resources
- Engineers Australia webcasts.

Many of the larger organisation that employ engineers have their own technical libraries, or have a commercial arrangement with a public library that provides similar services for a fee. The holdings in these technical libraries are generally quite specialised as they are aligned with the types of engineering work undertaken by the organisation.

FIGURE 9.8 Mind map of slurry pump project information needs

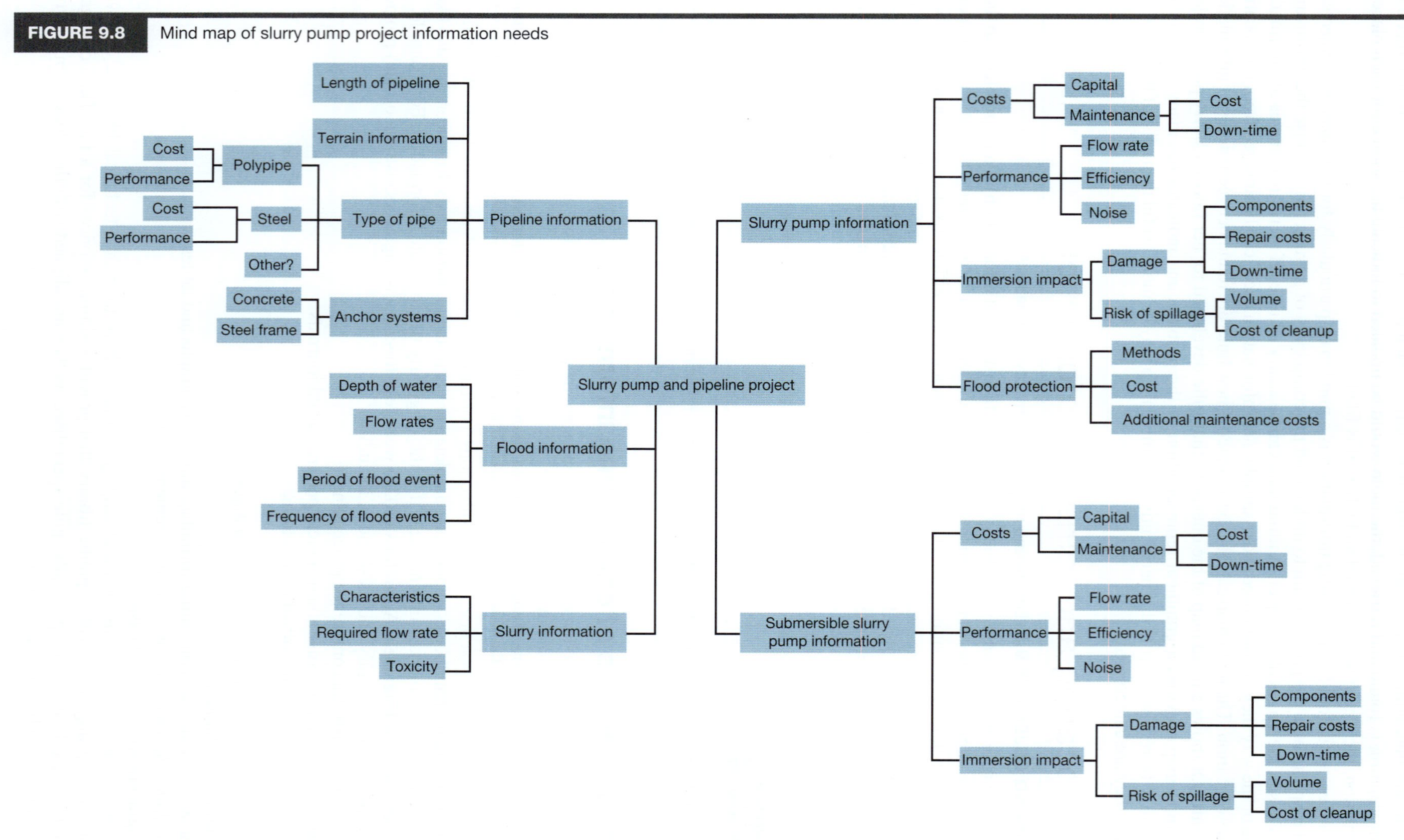

Many technical databases have AI search tools, for example Scopus AI, a new generative AI tool that searches the Scopus database (Elsevier 2024).

Experienced engineers also have a network of colleagues in both their own organisation and other organisations who they contact for specialist advice or to discuss problems. Thus, when engineers need information, they can generally locate it using their personal resources or they know where to look or who to ask in their professional network. This is not the case for students or graduates, who have only just begun to establish their own library and network of colleagues. Therefore, they will have to initiate a search when they require information, beginning with the identification of the most likely and reliable sources. One strategy they may use is to develop a list of sources for the different types of information they would normally require for a project. For example, they could base their list of sources on the six categories of information described in the previous section. Some typical sources of information for each category are:

- *sources of information about the client* — client brief, company literature, company website, communication with client and credit rating check
- *sources of information about the project* — project specifications, books, codes, communication with colleagues, engineering handbooks, internet, journals, legislation, manufacturers handbooks, practice notes, reports on similar projects, standards and technical manuals
- *information about the other stakeholders* — client brief, books, journals, websites and direct communication with users and other stakeholders
- *information about the context* — client brief, building plans, communication with government and local government authorities, geographic information systems, the internet, maps and plans
- *information from other professionals* — communication with colleagues and professionals from other disciplines
- *information about risk* — client brief, company policies, environmental impact reports, financial information, hazard reports, project specifications, standards and workplace health and safety legislation.

The information needs, and the mix of sources, will vary from project to project and between the many engineering disciplines. For example, in some disciplines the end users of a proposed product are the key sources of information for the project. The sources listed in the six categories are discussed in the following sections.

Documents

Many types of documents are used by engineers and the most common ones are listed in table 9.2. This table also includes typical examples of each type of document, their primary uses and where they can usually be found.

TABLE 9.2 Engineering information source documents

Source category	Example	Uses	Sources
Codes of practice	National Construction Code (Building Code of Australia)	Design	Personal library, public library, internet
Component catalogues	JPR Electronics online catalogue	Design, specification	Manufacturers and suppliers
Conference papers	The Institute of Electrical and Electronic Engineers (IEEE) Conference Proceedings	Design, practice	Personal library, internet
Design guides and templates	Your Home design guide	Design	Personal library
Equipment brochures	Warman SHW submersible slurry pumps	Design	Manufacturers and suppliers
Industry magazines	*create* — Engineers Australia Member magazine	Practice	Personal library
Government legislation	Workplace Health and Safety legislation	Practice, design, employment conditions	Library, internet

(continued)

TABLE 9.2 *(continued)*

Source category	Example	Uses	Sources
Handbooks	ASM Handbook: Materials Selection and Design	Theory, design, practice	Personal library, company library
Local government policies	City of Perth local laws and policies Local Environment Plans (LEP) State Environment Planning Policies (SEPP)	Design	Personal library, internet
Patents	IP Australia, Intellectual Property Office of New Zealand	Design	Library, internet
Practice notes	Association of Consulting Engineers Australia	Design, practice	Personal library
Professional organisations	Engineers Australia, Institution of Chemical Engineers (IChemE.org), Engineering New Zealand	Design, practice	Conferences, seminars, internet
Refereed journals	Engineers Australia Transactions	Research, practice	Personal library, library, internet
Reports	MKSEA GHD Engineering Report — City of Gosnells	Design, practice	Internet, request
Standards	Australian Standards, New Zealand Standards, ASTM Testing Standards	Design	Personal library, library
Textbooks	Student textbooks, practice books	Theory, design, practice	Personal library, company library
Online tutorials	AutoCAD (computer-aided design) Learning Path, YouTube	Theory, practice	Internet

Peer review

One measure of the reliability of the information in a document is whether it has been subject to a peer review process. For example, codes of practice are normally prepared by a group of people selected for their expertise in the field. This is followed by a consultation process where one or more drafts are circulated widely in the industry for comment. This process continues until there is general agreement that the document is ready to be published.

Another type of peer review is the process commonly used to review papers submitted for publication in journals and conference proceedings. Before a paper is accepted for publication, it is reviewed, often anonymously, by at least two people selected for their expertise in the field. The reviewers advise the journal editor to accept the article, to reject it or to accept it subject to a list of specified changes being made prior to publication. Journals that have been peer reviewed are normally called refereed journals.

The peer review process is an important part of the quality assurance process of a profession, as it helps to ensure that additions or modifications to its body of knowledge (BOK) are accurate and authentic. This process provides practitioners with a level of confidence to use the information, although they should still evaluate any information before using it. Several information sources are described next, highlighting differences in accessibility and quality.

- *Books.* These include codes, handbooks and textbooks. Codes of practice and handbooks have generally been peer reviewed. Books may have been reviewed. These documents may be borrowed from a library.
- *Periodicals.* These include journals, magazines and other items that are published periodically, either on a regular or irregular basis. Journals may be peer reviewed, particularly academic and research journals.

Hardcopies of these documents normally have to be read in the library as they may not be available for loan. However, as most current periodicals are available as electronic journals they can be read online by library members.

- *References.* Traditionally a reference collection consists of important books and journals, such as government yearbooks, statistics and reports. Reference collections are shrinking in size as more and more of this information is being made available on the internet. Reference collection resources are normally not able to be borrowed from a library.
- *Media.* Most libraries hold collections of newspapers, magazines, films, videos, DVDs, computer games and other multimedia resources. Some of these items may be available for borrowing.
- *E-books.* Libraries incorporate many e-books in their collections. These can be read online when the reader is a member of the library, or downloaded and read on electronic readers such as the Kindle or iPad.
- *Electronic journals.* Some peer reviewed journals are now only available online and the publishers of many previously hardcopy journals are now publishing electronic copies of earlier issues of their journals that were previously only available in print format. These are usually only available through paid subscription services or through a library.

It should be noted that the integrity of the peer review process is being questioned by some academics and industry-based researchers. This is because there have been a number of high-profile cases where papers previously published in prestigious journals have had to be withdrawn by the publishers. In these cases, the papers were withdrawn because they were found to be based on poor research methodologies, biased or selective sampling, or involved incorrect interpretation or falsification of experimental results. A high-quality peer review process would normally be expected to identify these inadequacies and reject the papers, meaning they would not be published. This is because it is expected that the reviewer will have the expertise and time to be able to carefully interrogate the methodology, the logic, the algorithms, the data, the mathematics, the results and the conclusions, and then place the work in the context of current knowledge and best practice.

However, the burgeoning number of papers being submitted for publication, and the increasing number of journals being published, has meant that, except for the top-ranked journals, it is becoming more difficult to recruit expert reviewers for this normally unpaid work. This, and the time pressures on busy professionals, may mean that the peer review process may lack the rigour that the editor and readers expect.

This is another reason why we must always have a questioning mind when we read information, even from the most trusted sources.

Colleagues

A series of articles published in *Fortune* magazine in 1994 encouraged companies to recognise the importance of the intellectual capital (knowledge) of their employees (Davenport & Prusak 1998). The editor of *Fortune* magazine has since published on the subject (Stewart 1997), and many books and articles have been published on the theme since. One way for an organisation to capture knowledge is to ensure that its employees share their knowledge with their colleagues. This can be done through a mentoring program where more experienced staff share their knowledge and experience with graduates and less experienced staff. This information exchange can also be achieved through staff seminars, workshops or formal training sessions. Finally, an organisation can encourage its experienced employees to document their knowledge so that the information is available for other employees to integrate into their practice. To do this an organisation may establish an online knowledge management system. The development of a knowledge management system is an example of the application of the World 2 to World 3 processes depicted earlier in the chapter in figures 9.5 and 9.6.

An organisation that has developed formal processes to document and learn from the experiences, knowledge, practices, processes and wisdom of its employees is often called a learning organisation.

Davenport and Prusak (1998, p. 89) wrote that 'Spontaneous, unstructured knowledge transfer is vital to a firm's success'. These informal exchanges include discussing a problem over morning tea, asking a colleague at the next desk for advice, or talking to a supervisor about a project. According to Fidel and Green (2004, p. 564):

> Studies have shown repeatedly that engineers rely most heavily on internal sources for information, mainly on interpersonal communication with colleagues. Further, the accessibility of an information source is the most prominent factor affecting its use.

Fidel and Green studied the factors that engineers considered when deciding which information source to use. They found the three most common accessibility factors were 'sources I know', 'saves time' and 'is physically close' (2004, p. 572).

By seeking advice from colleagues, engineers can fast-track the process of finding the sources of information. The keyword here is 'advice'. After receiving advice, engineers should always follow up and verify any information given to them by colleagues. They must satisfy themselves that the advice is correct and appropriate for a project. This approach avoids the possibility of using old, incorrect or incomplete information.

Over time engineers build up a network of colleagues and other professionals who share information and provide advice. These networks are important for all professionals and could be viewed as informal subsets of the more formal interactions that occur in professional organisations, such as Engineers Australia. Learned societies, such as Engineers Australia, were formed to maintain, extend and renew the engineering body of knowledge. They do this by providing forums where members share information about their practice and by publishing codes of practice, journals, magazines and other documents. As a student, you can generally join these organisations and the membership fees are often substantially reduced or waived.

Stakeholders

Most engineering projects have one or more stakeholders — people or organisations that have a pecuniary or other interest in the project or the outcomes of the project — for example, product end users, local government authorities, funding organisations, landholders, Aboriginal Land Councils and community groups. Information provided by end users and other stakeholders can be vital to the success of projects that result in a product or a process.

Sound consultation processes and active listening techniques should be used to ensure that the needs and wants of stakeholders are identified and understood. This may lead to an innovative design rather than a design that embodies a traditional approach, or an engineer's preconceived understanding of user requirements.

The following list of principles (Pressman 2005, pp. 133–134) can be used to guide the stakeholder consultation process, even though it was designed to gather information from the customer, or end user, of the product of a software project.

1. Prepare adequately.
2. Choose face-to-face communication when possible.
3. Engage a skilled facilitator when more than a few people are involved.
4. Listen.
5. Maintain focus, breaking the discussion into discrete parts or modules if necessary.
6. Be willing to use pictures and props if they will be more effective than words.
7. Collaborate.
8. Take notes and document decisions.
9. Keep discussions moving; if they become bogged down on a topic, move on and return to it later or at another agreed time.
10. Strive for win–win outcomes — negotiation is not a contest.

These principles may be adapted to develop suitable user information sessions in other fields of engineering. For example, they could be used when community consultation meetings are planned for a new route for a highway or transmission line.

The stakeholder consultation processes used for a project should be genuine and transparent. Poor consultation processes may mean that disaffected stakeholders take legal action or embark on a media campaign based on the premise that they were not properly consulted. This may have serious impacts on project outcomes, for example, time delays, budget overruns or costly redesigns. Of course, even after a genuine and transparent consultation process there may be disaffected stakeholders who do not accept the outcomes of the process and undertake legal action or initiate a media campaign.

There are many examples of wicked consultation problems playing out across the eastern states of Australia during the 2020s involving routes for the increasing number of electricity transmission lines required to connect geographically dispersed renewable energy projects to the cities that need the energy they will produce. There are many dimensions to these problems, and numerous stakeholders in each location. Some of the dimensions are:

- hasty decision making and lack of planning where short-term solutions are selected over long-term solutions
- a free-for-all over where companies can establish solar and wind farms

- climate activists calling for urgent action
- environmentalists concerned about biodiversity, landscapes and viewscapes
- city dwellers demanding cheaper electricity
- country communities concerned about their living environments
- farmers and other landholders concerned about the impact on their land and income
- engineering dimensions; optimal engineering designs (e.g. efficient routes and infrastructure choices) may not be pursued due to hasty decisions and consultation processes
- taxpayers and energy users concerned about funding poor solutions, and decisions based on commercial factors rather than social or environmental factors.

We will see the results of the implemented projects and the answers to these questions over the coming decades. These examples highlight the need to identify, consult and negotiate with the stakeholders who may be affected by a project, either during the construction phase of the project or the operational phase of the infrastructure that results from the project. This includes those who live, or conduct business, on or adjacent to the land where the infrastructure will be constructed. Many of these people will expect to be compensated for the impact the project has on their lives, their buildings and their land.

The rights to compensation for stakeholders affected by government actions or proposed actions are generally prescribed in town planning and land acquisition legislation, although this varies across the Commonwealth, state and territory jurisdictions in Australia. The following definitions will help you to understand some of the key principles that underpin the compensation provisions in those statutes.

Blight

'Blight occurs when land becomes unsaleable, or devalued, because of a belief that it will be required for, or affected by, some public project' (Australian Law Reform Commission 1980, p. 38). For example, when a government plan or town plan defines the corridor for a proposed freeway, the land within or close to the proposed corridor can be blighted because no development is allowed to occur on that land. The value of the land may therefore decrease and the land may be unsaleable until such time as the plans for the freeway are finalised, or even until the freeway is constructed. It should be noted that blight is called 'worsement' in some jurisdictions.

Betterment

Betterment, which is the opposite to blight, occurs when a government decision or project results in a benefit accruing to a parcel of land. For example, when a new town plan, or an amendment to an existing town plan, changes the land use zoning of a parcel of rural land so that it may be developed as a residential precinct. The value of the rural land will increase as a result of the rezoning decision.

Injurious affection

Essentially, injurious affection involves any nuisance, damage or decrease in value of land caused by a planning scheme or construction works undertaken by government. Normally a person cannot seek compensation for injurious affection from a statutory authority acting within its powers. Thus, an individual must suffer a loss for the benefit of the community.

> The only exception to this rule is where part of the claimant's land is taken for the work which causes the adverse, or injurious, effect. In such a case the compensation payable to the claimant for the loss of the part taken may include an allowance for injurious effects on the remainder ... The present law distinguishes between landowners who have suffered a loss of value by reason of a public work not on the basis of the extent of the damage but on the basis of which of them happened to lose land for the work (Australian Law Reform Commission 1980, p. 41).

It should be noted here that where private development causes, or threatens to cause, such a loss, the landholder may take common law action.

Thus, in most jurisdictions, compensation for injurious affection can only be claimed by the owners of land where part of that land has been acquired for the work. The key criticism of this legal concept flows from the inequity between neighbouring landowners that occurs because only those who have had part of their land acquired may claim compensation (Giskes 2004, p. 13). The unfairness of this position is highlighted when those neighbouring landowners, who have not had any land taken, do not receive compensation even though may have been more seriously affected by the work.

Giskes (2004, p. 4) states that injurious affection includes any physical damage to the retained land, any limitations on the use or activities that can be undertaken on the retained land, any interference with the amenity or character of the retained land, any things that may deter purchasers from buying the retained land or any things that increase the expense of using the retained land. The injurious affection may occur during and/or after construction, for example, due to dust, noise, vibration from machinery, night lights, loss of views, cracking of buildings, damage to landscaping, loss of business, loss of access and loss of value.

The legal provisions in New South Wales will be used to illustrate how these generic compensation processes may be implemented by the state government, particularly those provisions that apply to infrastructure projects. This discussion will also aid your understanding of the next spotlight, which is a case study from New South Wales.

The importance of major infrastructure corridor planning is highlighted in the NSW Department of Planning and Environment publication 'Planning guideline for major infrastructure corridors' (NSW Department of Planning and Environment 2016). Planning for major infrastructure such as the WestConnex freeway network is a complex and lengthy process. While the corridors may be defined in the early stages of the planning process the final alignment, mode and type of infrastructure may not be decided for many years. Over that time, many factors can influence the final decision, for example, the economy, new technologies, and changes in urban growth patterns and consumer behaviour.

Under the *NSW Roads Act 1993*, the Department of Roads and Maritime Services (NSWRMS) has the authority to acquire either a part or the whole of a parcel of land if it is 'directly affected' by a project and is needed to deliver that project. The owner's entitlement to compensation is established in the *Land Acquisition (Just Terms Compensation) Act 1991* and the processes are detailed in the NSWRMS land acquisition information guide (NSWRMS 2014). The owners of a parcel of land that was partially acquired are able to claim injurious affection for any impact on the remaining part of the land parcel.

Where appropriate, the Department, or its agent, consults with the owners of neighbouring properties before and during a project to address any concerns those owners may have about potential impacts.

> This can result in changes to the design of the project to avoid or minimise impacts, including specific or targeted mitigation measures. In almost all cases, impacts on properties are temporary, or can be appropriately managed through mitigation measures in the design or as part of the conditions of approval. These measures may include improvements such as noise insulation of homes (for example installing double glazing windows, or noise walls) and landscaping for visual screening (NSWRMS 2016).

Under the Roads Act, the NSWRMS has the authority to compulsory acquire subsurface land for tunnels, easements and pipelines. This power is used to acquire the subsurface land of the properties that lie directly above the tunnel (NSWRMS 2015). The provisions of the Land Acquisition Act mean that compensation is not payable for the subsurface land. However, under the specific circumstances detailed in the Act, compensation is payable for actual damage to the buildings, or the land, and for injurious affection caused by the tunnelling and construction works (NSWRMS 2014). For example the sinkhole that opened up under a commercial building on 1 March 2024 during M6 tunnelling operations in Rockdale, Sydney (Chenery & Daniel 2024).

The following spotlight discusses some of the impacts that people claim to have suffered during the construction phase of the major WestConnex project in Sydney.

SPOTLIGHT

WestConnex stakeholders seek compensation

The WestConnex project involves the construction of a 33-kilometre freeway scheme in Sydney, including approximately 16 kilometres of tunnels. It is a joint project funded by the New South Wales and Australian governments. As construction activities begin on a new stage or along a new section of the project, that area is likely to become a 24-hour/day construction zone and this has led to some compensation claims.

WestConnex tunnel construction underway in Sydney

WestConnex has developed and implemented a range of stakeholder engagement and community participation processes, distributing over one million updates to homes and businesses, hosting over 55 information sessions that allowed locals to speak to project experts and organising 510 briefings, including briefings with council staff and with property owners and other key stakeholders, by the start of 2019.

WestConnex (2018a) strives to involve stakeholders and the community in their projects at every opportunity and to:

- arrange engagement activities at times and places that are convenient for the community and stakeholders, and provide online options
- respond to reasonable requests for additional engagement activities and information
- acknowledge and understand diverse views on the project
- use feedback to positively influence the project.

So, what do the stakeholders think? We will look at just some of the many cases of compensation claimants discussed in the media.

1. The Inner West project

For this stage of the project, WestConnex acquired all the homes they needed for the road corridor. However, the noise, dust and 24-hours-a-day construction schedule has caused problems for the residents in the homes adjacent to what will be a four-lane freeway. Park (2018) reports that residents near the controversial WestConnex freeway in Sydney's Inner West claim the noise and air pollution from the construction activities have caused depression, as well as breathing problems such as asthma. Park (2018) reported on the case of Kate Cotis, who has been a St Peters resident for the past 13 years. She has lived alongside the above-ground construction zone for months and gave evidence at the NSW parliamentary inquiry into the impact of WestConnex (see WestCONnex Action Group 2018 and Cotis & Bryson 2018).

> 'It's dirty, filthy, noisy ... We need to be properly consulted about measures offered to ameliorate noise and pollution. We were told this is what you're going to get and this is how you're going to live your life living next to WestConnex.' To date, the NSW Roads and Maritime Service has only offered mechanical air filtration on her house by way of compensation — a measure Ms Cotis says is 'inadequate' ... 'What WestConnex have done is bought all the homes they need to build the road, with no consideration to homes right next to this road. And that's us,' Ms Cotis said. 'It means our homes are unliveable because of it' (Park 2018).

This is an example of injurious affection on land not acquired for a project.

Park (2018) also reported that Pauline Lockie, who was an Inner West councillor and the co-founder of the WestCONnex Action Group, had said the project has ruined Sydney's Inner West.

> 'You walk through St Peters now and it looks like a warzone,' she said ... 'People's houses, they can't sell them. Or if they sell them now it'll be for hundreds of thousands of dollars less' (Park 2018).

WestConnex is planning to use unfiltered ventilation stacks for its tunnels. The Brittlif family, whose home in Ashfield is 150 metres from one such planned stack, said their:

> concerns over air quality won't end when the temporary construction phase finishes. 'I agree that infrastructure projects are necessary and I accept that it's in my backyard,' Ms Brittlif said. 'What I think the most important thing is, when you are doing this in someone's backyard, you treat those people with respect and dignity, and you genuinely consult with them' (Park 2018).

Park (2018) sought a response from Ken Kanofski, the chief executive of NSWRMS who:

> defended air quality standards for WestConnex. 'The air quality standards for the WestConnex project are equivalent to the best in the world ... They are set independently, subject to expert advice ... and those very strict air quality requirements will be adhered too [sic]' (Park 2018).

Under the Department's Exceptional Hardship acquisition provisions, it may, at its absolute discretion and in exceptional circumstances, purchase a property where the owners are experiencing real and disproportionate hardship (NSWRMS 2016). This may be the outcome if the mitigation measures do not provide a solution for the hardship being experienced by a landowner who is further affected by the road project. In this case, the property owner may request the Department to purchase the property under the Exceptional Hardship acquisition provisions.

2. The tunnel projects

The depth of the proposed WestConnex tunnels meant that it was highly unlikely there would be any impact on the properties above those tunnels (WestConnex 2018b). Therefore, in most cases the consortium only acquired the subsurface land under properties immediately above the proposed tunnels. However, recognising that there was still a small risk of land and buildings being affected during the tunnelling operations, WestConnex identified any properties that may be affected. A corridor was defined on the surface with boundaries located 50 metres from the outer edge of the underground tunnels. Then, WestConnex offered all of the owners of the properties within that corridor the opportunity to have a property condition survey undertaken before construction began. Then, another condition survey would be undertaken at the end of the construction period if the owners believed that their property had been affected. This process ensures there is a clear record of the property's prior condition and any changes that occurred. Any damage attributed to the project would then be repaired at no cost to the property owner (WestConnex 2018b).

Mayers (2018) reported the impact on Umberto Galasso:

> [his] house sits in North Strathfield, Western Sydney — 26 metres above the M4 East WestConnex tunnel. Construction on the project began two years ago, and since that time Mr Galasso claims huge cracks have started appearing in the walls, columns, tiling and doors of his home. 'I fear that my house could collapse or the cracks get worse,' he said. 'I need something done as soon as possible.'

Residents may request that WestConnex implements measures to mitigate dust and noise problems, and they will be able to claim compensation for the damage to their properties. However, they still have a number of issues about the processes, including that they:

- are in a state of limbo as no compensation will be paid until the project is completed
- would prefer that independent engineers assessed the damage rather than WestConnex or their contractors.

The NSW Shadow Transport Minister has also called for an independent property assessment panel to be formed, similar to one the government implemented for the M4–M5 link (Mayers 2018).

The success of WestConnex's stakeholder consultation, engagement and negotiation processes will not be known until the project is completed.

CRITICAL THINKING

In the cases discussed in the spotlight, the issues mainly arose because the landowners are not able to claim for injurious affection. Why is that remedy not available to them? What measures are available to them and on what basis can they claim compensation?

Following a study of the impact of the land acquisition phase and site assembly of land for large scale infrastructure road projects in Australia and their impact on property owners, Mangioni (2018, p. 1) found that the phase that was most important in a large project was 'the planning and consultation phase, which includes most importantly the way in which [affected] owners are informed, assisted and compensated. It is concluded that the processes engaged in by acquiring authorities rather than the statutory provisions available, will determine the success of the land acquisition phase and perceptions of the project'.

In this section we have highlighted the importance of consulting two of the many groups of stakeholders in the early stages of a project, the people who may be affected by the project and the people who will use the products resulting from the project. For many projects it will also be important that these stakeholders are consulted during the life of the project and at the end of the project.

Geographic information systems

An increasingly important source of information, particularly for civil and environmental engineers, is a geographic information system (GIS). A geographic information system is a special type of database that enables users to store, retrieve, analyse and present data that is spatially referenced. Spatial features are located on a digital map system using real-world coordinates. The data associated with each feature is then entered into a data table that is part of the database. The data can be represented in a series of thematic layers, with each layer containing a set of data such as contours, vegetation, roads or gas pipes. The user is able to select the layers to be viewed or analysed by switching each layer on or off as required. This concept is shown in figure 9.9. Linear information such as a road or electricity transmission line is stored in vector format, while other types of information are stored using a raster format by allocating it to a cell on a grid.

A simple example of GIS is the system used by the Bureau of Meteorology (BOM) on its website to show radar images of rain clouds. Users can select the background layers they want to view, for example, locations, lakes and major rivers, roads and topography.

FIGURE 9.9 The concept of layers in a GIS

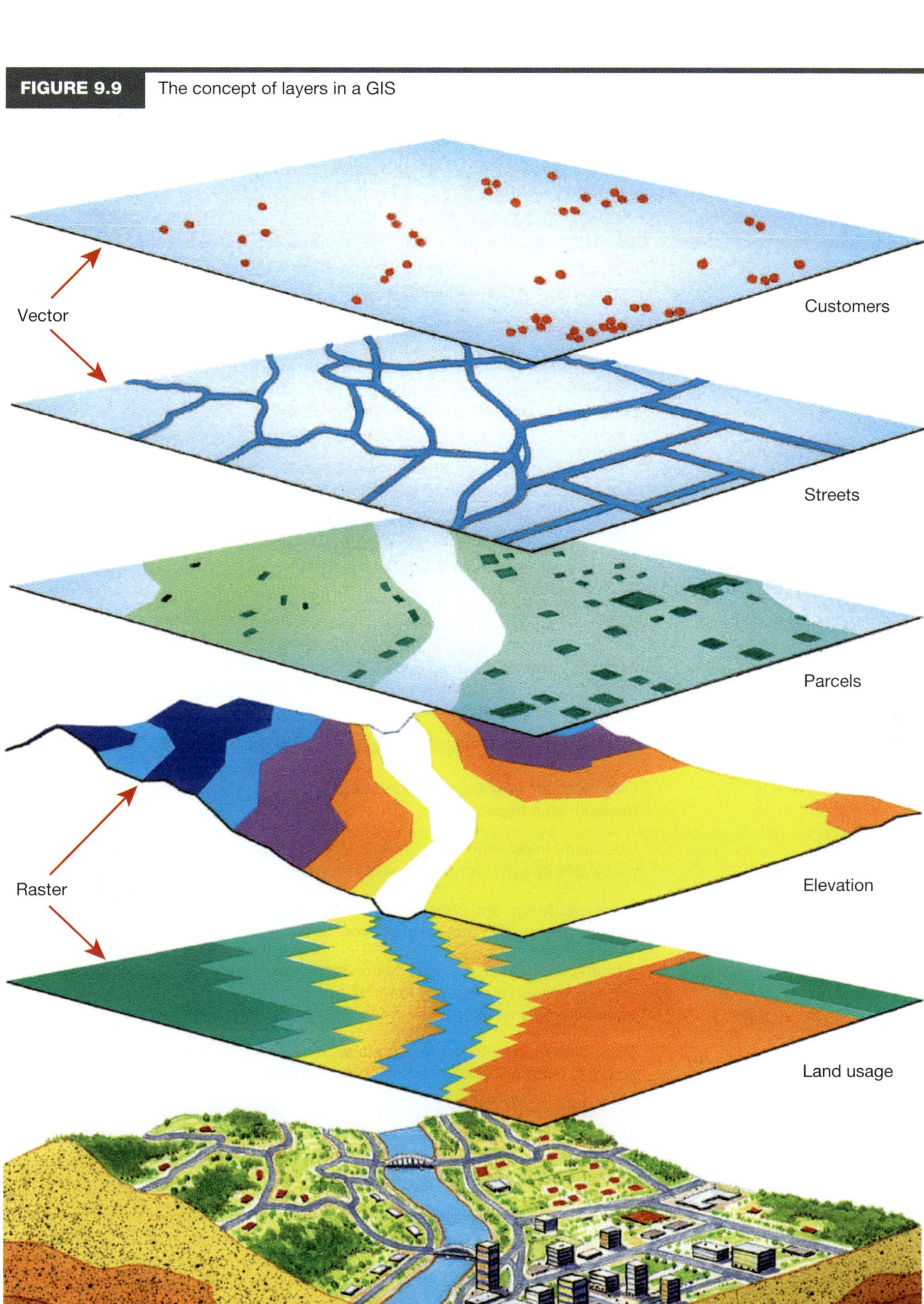

Source: https://www.esri.com

A GIS is an important tool for many professionals, and GIS systems are used for an increasing range of applications. For example, on its website, ESRI (2024) lists current applications in a range of industries (see table 9.3).

TABLE 9.3 GIS applications by industry sector

Industry sector	Current applications
Architecture, engineering and construction	Environmental management, design and engineering, construction management and asset management
Business	Financial services, high-tech companies, insurance, logistics and distribution, manufacturing, real estate, retail
Conservation	Conservation overview, conservation land management, landscape conservation, community-based conservation
Defence and intelligence	Defence, intelligence
Education	Higher education, kindergarten to Year 12 education, students
Energy utilities	Local energy, electric, gas
Global development	Sustainable Development Goals
Health and human services	Humans in crisis, public health preparedness, health equity, access to health care and services, strategic planning
National government	Aviation, defence, earth sciences, elections, humanitarian assistance, First Nations communities, intelligence, national mapping, national maritime and hydrography, official statistics
Natural resources	Agriculture, forestry, mining, petroleum, renewables
Nonprofit and NGOs	Communicate your cause, understand your community, act on your mission, measure your impact
Petroleum	Upstream, midstream, downstream, health, safety and environment, renewable energy, technology and innovation
Pipeline	Planning, design and construction, asset management, integrity management, operations, health, safety and environment
Public Safety	Security operations, emergency communications, emergency management, fire, rescue, and emergency medical services, homeland security, humanitarian assistance, law enforcement, bush fire
Science	Open science, weather and climate science, ocean science, solid earth science, geographic information science, social science
State and Local Government	Economic development, environmental and natural resources agencies, health and human services, housing and homelessness, First Nations communities, land administration and land records, public works and engineering, urban and community planning
Telecommunications	Network operations and maintenance, sales and marketing, customer service, planning and engineering, information technology and location services
Transportation	Airports and aviation, ports, public transport, rail, roads and highways
Water	Water resources, water utilities

Source: Adapted from ESRI (2024).

Google Maps is an example of a large but simple GIS. It enables the user to select different information types, such as Traffic, Photos, Weather or Terrain, with a Map or Satellite image background. When the 'Photos' option is selected, a thumbnail appears for each of the images available in the selected area. These thumbnails are spatially located pieces of information. Finally, the 'Directions' tool is an example of an analytical tool. The user selects two locations and the tool calculates the best route between those locations (based on software algorithms) and lists a set of instructions for the driver, including drive times.

Engineers from a range of disciplines use GIS to manage assets such as energy transmission systems, pipelines, telecommunication systems, roads, railways and building services. The location of each asset is mapped and the associated information, often called an attribute, is entered into the database and linked to the asset.

The database can then be used to inform maintenance and construction activities. For example, if a water main burst, once the street location of the break has been identified, the repair staff can quickly ascertain the exact location and depth of the pipe, the size of the pipe, the manufacturer, the pipe materials and the date the pipe was laid. This enables the repair crew to select the correct replacement fittings before they leave their base to travel to the site.

Most local government authorities have a GIS system that incorporates a large range of information about their region, including buildings, utilities, parks, streets, town plans and pet animals. These systems can be an important source of information for many engineering projects. Search the website of your local government authority to see if there is an online GIS for your area.

Library search tools

During your studies and future career, you may use city, company, government and university libraries. Standard search techniques can be used at most libraries, although the software tools vary from library to library. Libraries normally provide information about the search technologies that can be used to search for and access information in the library, and may provide face-to-face classes or online tutorials to help users acquire information searching skills using their systems. Once search principles have been learned using one tool, they can be applied and adapted when using other search tools.

There are many electronic search tools that can be used to find information stored in libraries and databases. For example, tools that can be used for general searches of the books, journals and other media held in a library collection, and specialist tools to search databases, a term that is normally used to describe a collection of resources held by other organisations. Library search tools generally fall into the following categories.

- *Library catalogue.* These tools enable the collection to be searched using keywords in a number of categories, such as: title, author, subject heading, publisher, call (shelf) number and journal title. These tools generally return an alphabetical list of titles or author's names. Most systems also allow a keyword search using Boolean logic (e.g. and/not/or).
- *Electronic databases.* Users can choose to search a single database, or search databases in a particular field such as mechanical engineering. Some examples of engineering electronic databases include: Compendex — Engineering Village 2, ASM Handbooks, American Society of Mechanical Engineers (ASME) Journals, Scopus, Web of Science, SAI Global (Australian and NZ Standards).
- *Electronic journals.* A search can be conducted using the journal title; once the site is located, it can be searched for particular articles.

Most Australian libraries allow their catalogue to be searched via the internet. Thus, if a resource is not available at one library, an online search of other library catalogues can be conducted until the resource is located. While it may not be possible to download or borrow the resource, it may be reviewed at the selected library, as most libraries allow public access to their collections. Alternatively, it may be possible to arrange an interlibrary loan. Generally, because of licence conditions, only the members of a library are able to search the electronic databases at that library.

As previously stated, AI tools are being integrated with these search tools to enhance the search process.

Internet search tools

There are a number of internet search engines that can be used to find information. Currently, the battle for supremacy is being won by Google, although Yahoo and Bing refuse to die. Other search engines, for example, Dogpile and Webcrawler, search using a group of other internet search engines and then summarise the results. Some of the more useful search engines for finding technical information are:

- Google Scholar — restricts a search to scholarly material such as journal papers and research theses
- WorldCat — an open access repository developed by the University of Michigan that can find libraries near you
- Bing
- Google Books — restricts a search to books and includes previews and extracts from many titles.

Internet search tools generally provide information from a wider selection of sources than subscription electronic databases, but greater care is required in assessing the reliability of the information. For example, while the internet is a good source of information about manufacturers and their products, it cannot be assumed that an internet search will find all of the products, or the most appropriate products for a project. Another reason to be cautious is that some search engines allow sponsored sites to be displayed, normally at the top of the listed sites following a search.

Free online dictionaries and encyclopaedias often appear in the first few entries following a search. The information provided by these sources should be treated with great care as some are open access and allow entries to be submitted by members of the public. Wikipedia is an example of an open access site. Information from these sites should not be used for academic or professional purposes, although they may provide links to more scholarly articles. Remember, we are living in the 'fake news' era and there are many fake or dodgy technical documents being published as well.

Enterprise sites such as Amazon can also be used to find information on books and other sources.

Recent articles in newspapers and journals have discussed the impact on search results of search engine optimisation processes and the move to personalise search results. Some commentators suggest that the increasing inclusion of filters, such as the location of the searcher, may introduce biases into the content of the search results and the order they are listed. Their concern is that the move to personalise web searches may mean that search results lack diversity. Some simple tests may be used to check bias or lack of diversity in search results. First, the same key words can be used with different search engines. Second, two people with different personal and professional profiles can check whether they receive the same research results when they enter the same keywords into the same search engine.

Developing a search strategy

Although there are standard processes that can be followed when searching for information, the starting point for each search is different. For example, one researcher may begin with the name of an author who is recognised as an expert in the field. Another researcher may know less about the subject and begin with some keywords. If there are a number of questions to be answered, then a basic process such as that shown in figure 9.10 can be followed. The process begins once all of the questions have been defined. If keywords are to be used, then the selection of the keywords can be critical to the success of a search. For example, Chanson (2007) found that the quality of results of Google Scholar searches was closely linked to input; that is, the keywords or phrases used for the search. The steps in the process shown in figure 9.10 are as follows.

- *Step 1.* Check if the next question on the list of questions is in your field. If so, begin a search for the information. If not, consult an appropriate colleague.
- *Step 2.* Begin the search.
- *Step 3.* Once some relevant information is found, check that it is applicable for the project and context. If so, note the source details. Try to verify it from another source, particularly if the quality of the first source is unknown.
- *Step 4.* Check the information covers all aspects of the question. If not, begin a search for additional information.
- *Step 5.* Check if the information raises any new questions or changes the direction of a proposed solution. If it does, add the question to the list, or adjust the questions to follow the new direction.
- *Step 6.* Move to the next question on the list and start the process again from step 1.

The actual search process begins at step 2. One key to a successful search is that the searcher must be flexible, while remaining focused on the specific information need. Promising leads should be followed even if they appear to be going beyond answering the question, or even the boundaries of the subject. However, care is needed, because following numerous 'promising' leads can also waste time. The guidelines in figure 9.11 illustrate a simple search strategy you can use for university projects.

If you know the name of a journal or other reference but cannot access it in your library, it is worth searching the internet to see if a copy is available at another location, or if you are able to download a free copy from the journal's website. Sometimes you can download one article from each edition free of charge. If you find you have to pay a fee to access an article, then you may find the same article is freely available at another site. For example, many university libraries and research institutions freely publish an author copy of conference or journal articles in an electronic format and some academics and researchers publish free versions of their papers on their websites.

FIGURE 9.10 Flowchart of information search process

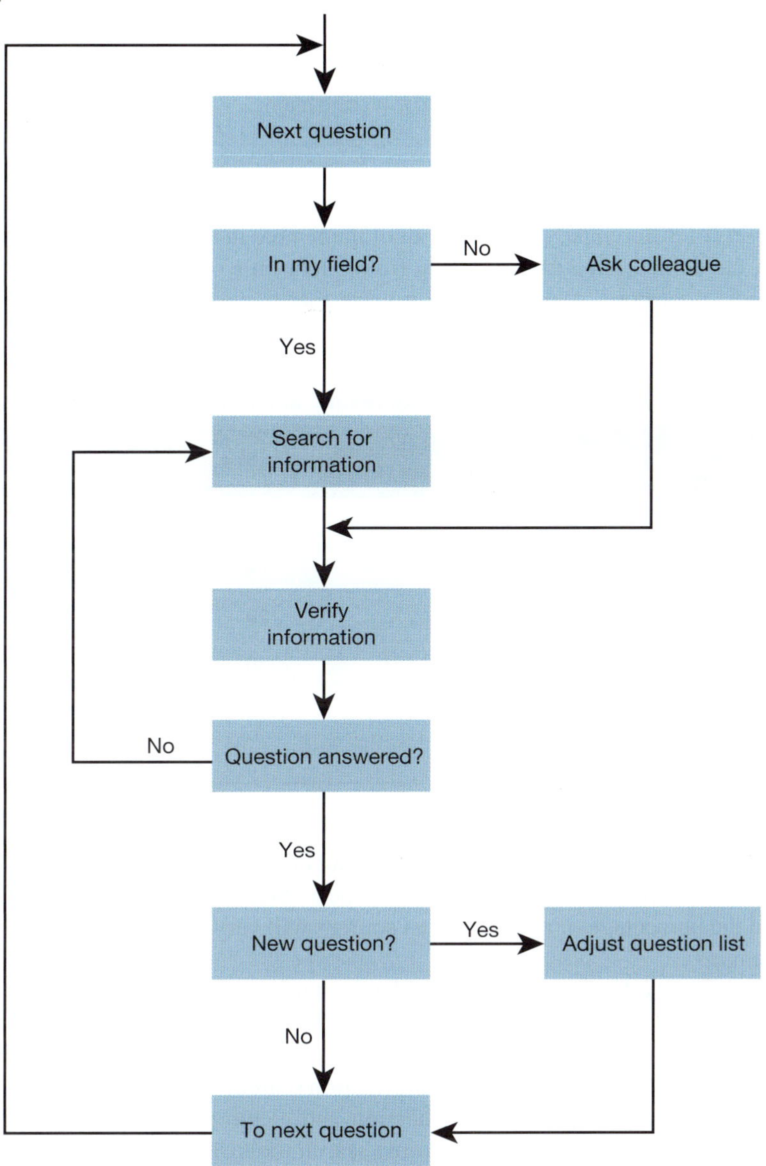

FIGURE 9.11 A simple search strategy

1. Select the most likely sources for the information you are seeking to answer a question. Your study materials may help you with some sources that you can use to start your search. If there is more than one source, try to prioritise them. If you are unsure where to find the information, ask your tutor or a librarian.
2. Select the most appropriate method for accessing the source and retrieving the information. If you know details of the source (e.g. the title of a code of practice) you may start by searching a library collection.
3. Search for and select a book for a more general search. You may decide to browse or to do a keyword search.
4. Begin the search and, if necessary:
   - widen the search (e.g. by removing one or more of the keywords, or broadening the publication date range)
   - narrow the search (e.g. by adding one or more keywords, searching within the list from a previous search, or decreasing the publication date range)
   - change the search method by using Boolean logic to link keywords (e.g. 'slurry and pump').

5. Identify and scan promising documents by reviewing abstracts, content tables, introductions and conclusions. If the source looks promising and you want to review and analyse it, arrange to borrow a hard copy or access an electronic copy.
6. Review the list of references at the end of articles to find other relevant articles.
7. Search for later articles by the same author(s) or organisations.

Chanson (2007) compared the results from a number of searches he conducted using different search tools. He used six sets of different keywords, but all of them were in the same general topic area. Table 9.4 is an extract from the comparative data resulting from Chanson's research. It shows the results for just one of the keywords: tidal bore. A tidal bore is a wave that travels up a narrow river or bay when the tide changes.

TABLE 9.4 **Internet search results showing total results and percentage peer reviewed**

Search tool	Total number of results	Percentage of peer reviewed works
Google Scholar	170	50%
El Compendix (period: 1884–2006)	78	90%
ISI Web of Science (period: 1945–2006)	29	100%
Scirus (now defunct) — total results	1036	90%
Scirus (now defunct) — journal results	8	90%
OAIster (now WorldCat)	13	92%

Source: Chanson (2007, p. 265).

The key point to note is the variation between the outputs from the different search engines, both in the total results and the percentage of those results that are peer reviewed sources. It is also important to note that the results and the percentage of peer reviewed sources will vary when different keywords, topics and search engines are used.

Chanson also searched the University of Queensland catalogue and found books and other sources *not* included in any of the internet searches and concluded that the results clearly show that 'traditional libraries cannot be replaced simply by an internet search' (Chanson 2007, p. 265).

While you are a student you will probably use information searching strategies that focus on the facilities at your university; for example, the library collection and electronic databases. If you are a distance education student, you may focus more on electronic search tools. The main resources you will use may include books, codes of practice, handbooks, legislation, standards, conference papers and journal articles. These may either be electronic or printed resources.

Practising engineers have to adopt different search strategies if they have difficulty accessing a library with up-to-date engineering information. While larger organisations may have extensive libraries, many smaller organisations rely on individual staff collections of texts, codes of practice, handbooks and journals. An engineer's search strategy may therefore be different to that of a student. For example, an engineer may use the following strategy.

1. Search their personal reference collection, which is likely to consist of textbooks, handbooks, codes of practice and journals.
2. Ask a colleague.
3. Search their company library, or in larger organisations, ask a librarian to undertake the search.
4. Search the resources of professional organisations, such as Engineers Australia.
5. Contact equipment manufacturers, agents and suppliers.
6. Search the internet.
7. Search a public, state or university library.
8. Employ an expert consultant if there are aspects outside their expertise.

To be successful and efficient at searching information, it is important to understand the strengths and weaknesses of the information technologies that are being used to access information. It is easy to get a false sense of security from the array of information technologies that are available in libraries and on the internet, and also from the large amounts of information provided at the conclusion of a search. Expertise comes through a variety of experiences, but only if the user reflects on those experiences to identify what worked, and why, and what did not work, and why.

You will use your information skills throughout your professional life, and you will need to maintain the currency of those skills throughout your career. This can be done by attending professional development activities that enable you to update your knowledge and learn new skills and technologies.

SPOTLIGHT

Surf, dive or scan?

In just a few short years you will be beginning your professional life and then you will take on responsibilities that test your professional knowledge base. You will have to extend that knowledge base to fulfil your responsibilities, and to extend your knowledge base you will need to find information.

Finding information in a pressured professional environment is very different from finding information for a university assignment. At university, students often 'make do' with information that they can find quickly but to perform well as a student you need to adopt a professional attitude. As a professional engineer, you must always find the right information to enhance your understanding and the right information for the job. In the professional world, 'making do' leads to problems, such as durability, risk or safety issues with a product, system or service.

Many students 'make do' because learning to find information efficiently is a skill they don't have yet; developing this skill takes time and effort — time and effort that many students feel is not worth spending. In the information-saturated twenty-first century, being an efficient searcher for the right information is important and learning to be an efficient searcher is an investment in your professional future.

There are several things that you can do to develop your information-seeking skills: attend library sessions organised by your lecturers, go to the library and ask for one-on-one help and develop an awareness of yourself as a searcher. Research has identified three types of searchers. Heinstrom (2005) has named these types 'fast surfers', 'deep divers' and 'broad scanners'. You may identify as one of these, or a combination of them depending on the situation.

Fast surfers apply surface thinking to their information seeking. They do not really understand what information they need and do not develop efficient search strategies. They pick the first result that looks good. Inefficient search strategies lead to the retrieval of much irrelevant information and little relevant information (Heinstrom 2005).

Broad scanners are interested in a lot of things and have trouble restricting their information needs analysis and searching to the topic at hand. This lack of focus also influences search strategy efficiency and, as with fast surfers, broad scanners retrieve a lot less relevant information than irrelevant information (Heinstrom 2005).

If you think that you could be a fast surfer or a broad scanner, you can help yourself. Take the time to analyse the problem or question fully, identify your information needs, and develop and record possible search strategies. When you actually search, work from your record of search strategies. You will find much more relevant information and, through practice, you will also find that information more quickly (Heinstrom 2005).

Deep divers are analytical searchers. They analyse their information needs thoroughly and develop effective search strategies. Their search results are generally smaller in number than fast surfers or broad scanners, but that smaller number is characterised by a high proportion of relevant information. Deep divers report they can sometimes be so focused on finding particular information that they do not recognise alternative valuable information and ideas. Heinstrom's advice: stay analytical but keep an open mind to other possibilities (Heinstrom 2005).

CRITICAL THINKING

Use a search engine to find information on a particular topic. Review the first 30 search results, and note the relevance or each. Now, consider what search terms (or keywords) are more likely to improve your results, and then use them in the same search engine. Again, look at your first 30 search results. Are they different from the first set and, if so, how?

Recording data about information sources

There are some good reasons for keeping an accurate record of the details of a search and the sources of the information you find, even if you end up not using them at that time. By keeping a search log, you can easily return to the references or sites you have visited and check the details or, if required, obtain additional information from those sources. It can be really frustrating if you cannot find the source of a piece of information again, particularly if it is a key piece of information for a project. Having to redo the search can waste valuable time — time that may be better spent on other tasks — particularly when a project deadline is approaching. It is for these reasons that you should train yourself to record this information while you are undertaking a search.

When you access and source information, you should develop the habit of recording as much detail about the source as possible; including the name of the author(s), the title, the publisher, the year, the edition and the place of publication. For edited publications, you should include the name of the editor(s) and the title of the publication. For quotations, the page number(s) should also be included in your log. For internet sites, you should include the URL and the date it was viewed. You can use one of the many software referencing (bibliographic) tools, for example, Endnote, to record this information. Once recorded the information is then easily converted into a list of references that can be included in a report or paper. You may have an opportunity to learn how to use one of these tools while you are at university, or you could use an online tutorial, for example the RMIT one. You may also save key URLs using an internet search engine's 'save' or 'bookmark' function.

Engineers may be called upon to justify their work, and a log of the searches they undertake can be used to support their design, investigation or report. This information may also prove to be important if they are asked to give evidence in a court, although this is a relatively rare occurrence for most engineers. It is for these reasons that some companies require their staff to keep a journal of their activities and a log of the information searches they undertake as part of their work.

9.4 Evaluating information and information sources

LEARNING OBJECTIVE 9.4 Evaluate the quality and reliability of information and information sources.

Once the search process has been completed and the available information has been retrieved, both the information and the sources should be evaluated. This is not an easy task, and it takes time and experience to develop these skills. The key principle is that it is wise to be cautious in the use of information, particularly when the information will be used for a design, or for a risk-prone project.

KEY POINT

Engineers need to carefully analyse information before applying it.

Evaluating information sources

One measure of the quality of information is the reliability of a source. For example, information contained in handbooks or codes of practice that have been published by government or industry organisations will normally have more credibility than information published in magazine articles. Wright (2002, p. 144) describes the CARDS method for evaluating information sources. CARDS is an acronym for 'credibility, accuracy, relevance, date and source'.

- *Credibility.* Is this a site or resource known for its credible information? Is the author's name and affiliation listed? Is the author respected in the industry? Has the document been peer reviewed? Have you previously found the source to be reliable? Have you used information from this author before?
- *Accuracy.* Can the information be verified by other sources? Are there any obvious errors in the information? Is the document well written? Has the work been cited by other authors? Is there evidence of bias in the publication or the source? Does the information fit with your understanding of the topic and your experience?
- *Relevance.* Who was the target audience for the information and did the document achieve its purpose? What is the purpose of the website? Is the website relevant for your project?

- *Date.* What is the publication date of the book or document? Is the website up-to-date? Is there a note on the web page saying when the website was last updated?
- *Source.* Was the website, or publication, published by a government or professional organisation? Are the details of the website owner published on the homepage? Does the website link, or refer to, other print or electronic resources? Does the website have a bibliography?

Another important consideration is whether the source is the **primary source** for the information or a **secondary source**. A primary source is a document where information is published for the first time. A secondary source is a document that uses, adapts or refers to information in a primary source. Every effort should be made to obtain information from the primary source. This is to ensure that the information from the primary source was accurately reported, interpreted and properly referenced in the secondary source. Where appropriate, the primary source should also be carefully reviewed to identify the contexts of the reported work, any metadata associated with reported data, any assumptions the author has made, and any limitations placed on the findings or recommendations.

For example, the information contained in table 5.4 (see the chapter on self-management) was originally read in a secondary source. When the primary source (Coll & Zegwaard 2006) was reviewed, it was found that the following information was not reported or was incorrectly reported in the secondary source (published in Australia).

- The secondary document did not indicate that the participants were associated with a single country (New Zealand). Readers may have assumed the participants were from Australia.
- The number of participants was not included in the article.
- The secondary source stated the participants rated 22 competencies when, in fact, they rated the importance of 24 competencies.
- The fifth competency in the employer list was listed incorrectly.

Fidel and Green (2004, p. 574) found that the three most important source quality factors for engineers were: 'can give data that meets the needs of the project', 'is most likely to have the information needed' and 'is reliable'.

Evaluating information

All of the information gathered for a project should be critically evaluated before it is used on the project. This is to ensure the information is accurate, reliable and appropriate for the project. Information should not be used merely because it was found during the search process. Simply because it has been published does not make it incontrovertible. The importance of critical thinking was discussed in the chapter on the engineering method. It is a key skill used to assess the quality of information. Burton et al. (2023, p. 221) list three key principles that underpin critical thinking as (1) scepticism, (2) objectivity and (3) open-mindedness. When this approach is applied in the engineering context, the following questions can be used to evaluate information.

- *Scepticism.* Are the results, conclusions or recommendations in the document supported by evidence in the document or by evidence in other documents? Have the author's ideas or concepts been developed and argued in a logical manner? Is the technical information correct?
- *Objectivity.* What questions have not been answered in the article? Is the information appropriate for your purposes? Is it practical? Is the information applicable for your project contexts; that is, will the applicability of the information be affected by changes in one or more of the following contexts: the technical context, the geographic context; the workplace health and safety context; the legal context; the socio-cultural context; or the environmental context? What limitations does the author place on using the technique?
- *Open-mindedness.* Does the information suggest an alternative approach that should be considered for the project? Is the information incompatible with the other information gathered for the project?

Some additional questions that can be used to evaluate information are as follows.

- What issues need further exploration before the information can be used for the project?
- Has the author, or another author, updated or extended the information, or applied it in a new context?
- Has the document been cited by other authors?
- Are there any other information sources in the document that should be reviewed, such as the references listed in the bibliography at the end of the document?

The analysis of the information in a document should be continued until you are satisfied that you have understood the key concepts, metadata, results and/or recommendations. Only then will you be able to successfully integrate the information with the other pieces of information that have been gathered for a project. It may help to discuss the information and any concerns you have with a colleague. When you discuss information with your colleagues, they may present other viewpoints that may help to clarify your understanding of a topic. This is because you engage in higher order thinking when you discuss or debate a topic. Sometimes you can come to a new understanding just by putting your case to another person.

When reading textbooks, journal articles and other similar documents, it is sometimes useful to engage at the same level; that is, as if you were discussing the topic with the author. One way of doing this is to annotate the document to highlight important sections or to note your thoughts on particular sections. For example, you may underline key sentences or technical information, or highlight it using a marker pen. You may also write succinct comments in the margins explaining why the information is important to you; for example, you could write 'A good idea, use this in assignment 2', 'Try this in the lab', 'Check what the textbook says', 'I don't agree with this', or 'Use in assignment 1'. You should not, of course, annotate books or other documents you have borrowed from colleagues or a library; in this case you should use a notebook.

The importance you place on the quality of information may depend on how you plan to use the information. For example, you might place a much higher value on ascertaining the quality of the information if you are using it to design the cooling system for a new electricity generating plant, rather than using it to design the cooling system for a garden shed. Over time, you will become a good judge of the quality of information you access and the reliability of the sources you use in your practice. The reliability of a source is an important information characteristic that should be included in a personal bibliography.

A final note on evaluating information: in a world where AI-generated information, and fake information, are becoming ubiquitous it is critical that engineers ensure that they carefully evaluate the information they use in their practice.

Refining information needs

When information is accessed, it often provides unexpected answers to the search questions. It may be incompatible with another piece of information that has been gathered, or with the project team's existing understanding of the problem. This may mean that the questions have to be reviewed, or that new questions have to be defined.

The key principle in this situation is that the issue must be resolved. This is the professional approach when information from different sources is incompatible. The aim is to find the correct information, the best available information and the information that is most appropriate for the project. The information search should continue until the issue is resolved, normally by locating compatible information from at least two independent sources.

Sometimes resolving incompatibility is not possible or feasible. When this occurs, engineers should report any inconsistencies to their client, explaining why they adopted a particular approach. This is best practice, and demonstrates an ethical approach that engineers should use throughout their careers. As a student, you should also learn to use this approach.

SPOTLIGHT

The ethics of data management

At the 1985 national Tax Summit the Hawke Federal Government announced a proposal to introduce a national identification card — the Australia Card — to address growing problems with tax avoidance, and health and welfare fraud. Despite the proposal provoking a fierce public and political backlash, legislation for the card was introduced into parliament in 1986, though it failed to pass through the senate. The government eventually abandoned the proposal in 1987.

In the following years, there were a number of other proposals to introduce a national identity card. For example, the final report of the 2014 Financial System Inquiry recommended that the government '[d]evelop a national identity strategy based on a federated-style system in which public and private sector identity providers compete to supply trusted digital identities to individuals and businesses' (Australian Government 2014, p. 156). However, the report also warned that '[m]any Australians may object to this option on the basis of privacy concerns. It could be viewed as a digital version of the unpopular Australia Card initiative, which was rejected in 1987, or the Access Card, which was terminated in 2007' (Australian Government 2014, p. 159).

Today, it is hard to understand how these early proposals for a simple identity card caused such a public backlash, particularly as we live in a world where personal and private data and information is collected, stored, managed and traded at an alarming rate, often without the owner's consent or knowledge. For example, researchers at Cambridge University showed that the easily accessible records of 'Facebook Likes can be used to automatically and accurately predict a range of highly sensitive personal attributes including: sexual orientation, ethnicity, religious and political views, personality traits, intelligence, happiness, use of addictive substances, parental separation, age, and gender' (Kosinski et al. 2013, p. 5802).

Our personal and professional data is being regularly harvested as we go about our everyday activities: shopping loyalty cards, business activities, interactions with government agencies, consulting with health professionals, travel, holidays, attending entertainment shows and sporting matches, and interactions with community and professional organisations are some examples. Much of this data is highly valuable to criminal groups in the hacking business. A recent example of a data breach was the 2023 Optus breach where hackers stole personal data belonging to more than 10 million customers, including customer names, dates of birth, bank account information, residential addresses, email addresses, driver licence numbers, Medicare ID numbers and passport numbers. This data can be used for identity theft purposes.

There is growing evidence that the Australian Government and its agencies are taking action to protect the privacy of Australian citizens on a number of fronts, including via cybersecurity education, by strengthening privacy laws, by strengthening cybersecurity and breach reporting systems and by prosecuting cases. Two recent cases demonstrate this.

1. On 3 November 2021:

> Australian Information Commissioner and Privacy Commissioner Angelene Falk found that Clearview AI, Inc. breached Australians' privacy by scraping their biometric information from the web and disclosing it through a facial recognition tool. The determination follows a joint investigation by the Office of the Australian Information Commissioner (OAIC) and the UK's Information Commissioner's Office (ICO). Commissioner Falk found that Clearview AI breached the Australian *Privacy Act 1988* by:
>
> - collecting Australians' sensitive information without consent
> - collecting personal information by unfair means
> - not taking reasonable steps to notify individuals of the collection of personal information
> - not taking reasonable steps to ensure that personal information it disclosed was accurate, having regard to the purpose of disclosure
> - not taking reasonable steps to implement practices, procedures and systems to ensure compliance with the Australian Privacy Principles.
>
> The determination orders Clearview AI to cease collecting facial images and biometric templates from individuals in Australia, and to destroy existing images and templates collected from Australia (OAIC 2021).

2. On 11 October 2022:

> The Australian Communications and Media Authority (ACMA) ... advised Singtel Optus Pty Limited (Optus) that it [had] commenced a formal investigation in response to the September 2022 Optus data breach. The ACMA will investigate the data breach in regard to Optus' obligations as a telecommunications service provider. These include obligations relating to the acquisition, authentication, retention, disposal and protection of personal information, and requirements to provide fraud mitigation protections (ACMA 2022).

These and other actions taken by government agencies have sent a clear signal to corporations, business, commercial organisations, consultants, community organisations and, of course, all levels of government, that organisations such as OAIC and ACMA will swiftly take action when data breaches occur and privacy is compromised. This issue was emphasised in an article in the *Harvard Business Review* where Segalla and Rouzies (2023, p. 88) stated that, in our current data-driven environment, '[a]s companies jockey for competitive advantage ... they are increasingly penalized for the abuse of data.' They suggest that companies planning to use data (whether provided directly from customers/clients or from existing databases) need to consider five critical issues: 'the *provenance* of the data, the *purpose* for which it will be used, how it is *protected*, how the *privacy* of the data providers is ensured, and how the data is *prepared* for use' (Segalla & Rouzies 2023, p. 88).

Segalla and Rouzies (2023, p. 93) call these issues the *five Ps of ethical data handling*, and suggest questions are asked to address each of the Ps before they commence a project. Some examples of questions to ask could include the following.

- *Provenance.* What is the source of the data? Who owns the data? Was permission granted to use the data? Was the data acquired legally?
- *Purpose.* What was the original purpose of the data? Is the data being used for a different purpose? If so, can it be used for this different purpose (i.e. would the original source approve use for this different purpose/would additional permission be required)? If using dark data — data that 'organizations collect, process and store during regular business activities, but generally fail to use for other purposes' (Gartner 2024) — will it be used within the bounds of the original authority to collect the information?
- *Protection.* What protections are in place to secure the data? Who is responsible for the storing, dissemination and destruction of the data? How long will the data be accessible for this use?
- *Privacy.* Will access to the data be limited to select persons? Who will be able to access the data? Will data be anonymised (to protect individual responses)?
- *Preparation.* How has the data been prepared/cleaned? Has the data been verified as accurate and how was this verification achieved? Has the data been improved and, if so, how? Has anonymity been retained after collation? Has missing data been managed and, if so, how?

Engineers should use these questions to develop a checklist of questions to consider for their practice, for both staff and client data. Other questions should be added to the list, questions that are appropriate to their field of practice. For example, what data should be collected about a new client? What data should be retained long term?

CRITICAL THINKING

Review the data devices and apps you use to determine who has access to the data and what data is currently being harvested by third parties. Who is harvesting the data? What can you do to limit the data being harvested? Examples include smartwatches, phones, notepads, home 'assistants', construction machinery, electronic devices and instruments, and cars.

9.5 Integrating, managing and using information

LEARNING OBJECTIVE 9.5 Integrate, manage and use information.

In the chapter on the engineering method you learned that when information is gathered for a project, it is important that it is organised in a systematic manner so that people can access it efficiently. There are obviously many ways to organise information; for example, by topic and subtopic, by building location, by author or by date retrieved. For small projects, a simple information index or even a mind map can be used to file and locate information.

A common method of organising information is by topic and subtopic, with the information for each topic being filed in a separate section of a folder for a small project, or a separate folder or series of folders for larger projects. This filing system, which may be electronic, could be supported by a small matrix that lists the articles by topic, author and keywords.

For example, a team of civil engineering students may have to design the services for a small residential subdivision for an assignment. Before commencing the design, they would gather and separately file information about aspects of the design: allotment size, shape and orientation; road, pedestrian and cycle path specifications; specifications for electrical and communications systems; and the specifications for sewerage, stormwater and water supply systems.

KEY POINT

The ability to organise and manage information is an important professional skill for engineers. Do not ignore inconvenient truths.

Integrating information

After gathering information from a range of sources, it should be integrated into one or more summary documents. The information then becomes more useable as all of the information relating to a topic will be located in the one document. Finally, the integrating process will help you find any gaps in the information or your understanding, and this will prompt you to search for the missing information. This process should also improve your understanding of the information as you locate each piece so that it drops into place in the same way a jigsaw piece does.

As you are integrating the information, you may be able to add value to the document by incorporating your thoughts, ideas or experience — that is, your intellectual property (IP). For example, you may add a description of your experience, develop and argue a case, or build on the existing information to create a new concept, model or design.

A simple way to develop a draft summary is to write a series of dot points, acknowledging the source of each idea. The summary may take a number of forms, often depending on the type of information that is being synthesised. For example, technical information about a machine or a circuit board may be gathered into a table of specifications.

Publishing information

Many informal summaries developed by engineers contain information that may be useful to other practitioners, either within their organisation, or in the wider profession. If this is the case, then the summary should be published as a report and circulated within the company. The author may receive comments and suggestions from other engineers in the company, and the document can then be amended and re-published, perhaps as an official company document.

The same information may be made available to the wider profession by having it published as an article in an industry magazine or through a paper presentation at a conference. This may mean that your work is questioned or is the subject of debate among interested practitioners. This should be considered a positive outcome because once the debate has finished both you and the participants will have a better understanding of the topic.

Be cautious about using information from published case studies because there is an inherent risk of bias in published articles as they tend to favour projects that were successful. It is less likely that an engineer will publish an article about a project that was less successful than expected. This is mainly because there may by liability issues and the potential for compensation claims from clients. The legal requirements in the contract may also restrict what can be publicly disclosed.

A literature review

A more formal summary is often called a literature review, particularly when the information being synthesised is drawn from refereed journals. In this case, the summary can be viewed as a written debate about ideas and previous research findings. The author guides the reader through the debate, clarifies issues and evaluates evidence if there is disagreement about the accuracy or quality of the information. A literature review can bring readers up to date on a topic, and often ends by drawing attention to any issues that are yet to be resolved. Practitioners can use literature reviews to highlight flaws in existing systems or areas requiring further development or research. Researchers conduct literature reviews before they commence their research in a particular field. The literature review provides them with a 'jumping off point' so that they are able to build on, rather than repeat, the work of other researchers. It helps them to develop their research questions.

An information management system

An engineer may manage a range of information types, including tender documents, contracts, progress reports, client instructions, letters, emails, budgets, personnel reports and, on larger projects, publicity material. Managing information is an important skill for all professionals, whether the information relates to a particular project, or builds on the body of knowledge required to practise in their field. Because gathering information can often be a time-consuming and difficult process, it is important that the information is filed in an easily accessible manual filing system or, preferably, an electronic document management system. From a personal perspective, an engineer will want to be able to easily access information when it is required for a project. A robust and up-to-date catalogue will enable them to easily and efficiently access information.

From a project perspective, the management of information may be critical to the successful completion of a project. For example, one hundred or more plans may be required to document the details of a design for a large manufacturing plant. During the design and construction phase of such a project, it is likely most of the plans will be amended; in some cases, several times. Therefore, it is critical that all of the people working on the project use the latest versions of the plans they require to complete their part of the project.

The key principles of information management systems are:

- information must be readily accessible to those who require access
- an index or cataloguing system should be used to manage the storage and retrieval of information
- the document management system must include a robust versioning system that facilitates the issue of new versions of documents and plans to relevant stakeholders
- relevant records must be amended when a new version of a document or plan is adopted. This includes the metadata, such as the changes, the date of the changes, and the names of the people who made and approved the changes.

These principles apply to all types of information and media, whether in a print or electronic format. A well-designed information management system helps people cope with a constantly increasing amount of information in their field of practice. However, the development and maintenance of a system requires time and resources, and is normally facilitated at the organisation level rather than the project level. An organisation-wide information management system is used in organisations so that all staff can efficiently and accurately prepare, edit, file, manage and retrieve documents. The same approach should be adopted on projects, particularly on large projects involving a number of people and organisations.

The adoption of a single information system and associated processes should be negotiated at the beginning of a project, particularly when it involves staff from a number of organisations. An important part of the project establishment process is the provision of training for all relevant staff in the use of the project information management system.

Another important aspect of an information management system is archiving or disposing of information that is no longer required, or that may be out of date. There are a number of questions to be considered.

- Is the information required for business or legal purposes? If so, it will have to be archived for a specified period.
- Could the information be useful as background information for future projects? If so, it should be archived in a manner that enables it to be easily accessed.
- Does the information have a use-by date?
- What is the risk to the business if the information is destroyed?

Finally, along with the other components of an information management system, the systematic archiving and disposal of information requires the allocation of resources, space and time.

SPOTLIGHT

3D buildings in a 3D world

The Dr Chau Chak Wing building at the University of Technology, Sydney's Ultimo campus, is an architecturally iconic structure. Designed by the world-renowned architect, Frank Gehry, the UTS addition to their Business School challenged engineers and builders to realise the 14-storey structure inspired by tree houses and the shape of a crumpled paper bag.

Source: Gehry Partners LLP.

Communicating the unstructured geometry to structural designers, services engineers and then to foremen and bricklayers was a formidable task. According to Alison Mirams, General Manager of lead contractor, Lend Lease, 'on a scale of difficulty: it's 10 out of 10' (Gilmore 2014). A typical building of this size requires around 2000 drawings. For the Gehry project, 100 000 drawings were needed (Gilmore 2014).

The façade is mostly hand-laid brickwork (Arup 2014). Austral Bricks custom made over 30 prototype bricks in six different factories around Australia before the final product was selected (UTS 2014). Some 320 000 bricks were made in the Austral Brick factory in Bowral NSW for the project (Gilmore 2014; UTS 2014). Five different brick shapes were used to make the complicated curved folds of the façade (UTS 2014). Because the surface did not translate well onto standard 2D plans, a 3D building information model (BIM) was used to document every aspect of the design. Young engineers familiar with BIM were teamed up with the experienced old-school foremen to navigate the model and instruct the bricklayers via a live link to surveyors using GPS coordinates (Gilmore 2014). This was a marriage of the high-tech skills of engineers and surveyors with the artisan skills of master bricklayers.

The supporting structure inside the building was designed by engineers from Arup's Sydney office in constant coordination with the architectural team. Sloping concrete columns were repositioned many times to ensure feasible load paths that did not detract from the architectural vision. The engineers used the same integrated BIM, a program called Digital Project, developed by Gehry Technologies (Arup 2014). This proved vital for managing the 100 000 drawings for the design and construction (Gilmore 2014). Digital Project was used across all disciplines ensuring an integrated model that helped to overcome the geometrical complexity (Arup 2014). Most of the drawings never made it to paper — the building went straight from computer to construction.

The engineers had the task of translating the geometry from Gehry's sculpture to a set of plans that could be used to make a 14-storey building.

Source: Andrew Worssam.

As can be seen from table 9.5, many companies and organisations were part of the design and construction team. Within the contracting team there were dozens more subcontractors (Arup 2014; UTS 2014). Coordination of information is crucial on projects as complex as the UTS building. BIM and Cloud computing helped to ensure all contributors were using the latest information (Arup 2014). The designers were able to see each other's work and avoid clashes between services and structure. Decisions, and the reasons for them, could be documented and catalogued so that unnecessary revisions were minimised (Arup 2014).

TABLE 9.5 The Chau Chak Wing building design and construction team

University of Technology, Sydney — client	Arup — structural engineer, transport, traffic
Gehry Partners — design architect	RPS — statutory planner
Daryl Jackson Robin Dyke — executive architect	Casey & Lowe — archaeological consultant
Lend Lease — main works contractor	Godden Mackay Logan — heritage assessment
AW Edwards — early works contractor	Morris Goding Access Consulting — accessibility
AECOM — services engineer, ecologically sustainable development	Wind Tech Consulting — wind assessment
Australian Museum Business Services — archaeological investigation and excavation	Dominic Steele Consulting Archaeology — Aboriginal archaeological investigation

Source: Adapted from Arup (2014) and UTS (2014).

What makes this project stand out is the way an innovative thinker like Frank Gehry can challenge everyone to do the extraordinary. The building is iconic and its creation stretched the ingenuity of Australia's best engineers and builders. At the time of construction it was also considered a $180 million investment in the training of our future business leaders and an investment in the practice of the innovative engineers of today (Gilmore 2014). As Mirams told *The Sydney Morning Herald*, it was 'the job of a lifetime' (Gilmore 2014).

CRITICAL THINKING

How have the student teams you worked with managed their information, documents and drawings? What changes could be made to improve those methods for your next team project?

Controlling access to your information

Engineering information management is important, but there are other important issues that need to be addressed relating to communication within an organisation and with external parties. There are many reasons why an organisation adopts protocols to control information and documents. For example, a document may contain information that is confidential, information that includes significant IP, or sensitive commercial information such as the budget for a project. All documents are prepared for a specific consumer, or group of consumers, and they are controlled so that other people do not have access to, or receive copies of those documents.

The use of electronic communication methods has greatly increased not only the risk of documents reaching an unintended audience, but also the size and geographic spread of an unintended audience. This is another driver for the changes that are occurring in the systems used to manage engineering information.

Finally, when developing and managing an information management system it is important to recognise the growing threat of people hacking into the system and copying files to gain company or personal information and credentials, or for industrial espionage purposes. The following spotlight highlights the importance of implementing a range of cybersecurity strategies to protect your information.

SPOTLIGHT

Do you need better cybersecurity?

Cybersecurity is continuing to be a problem for governments, businesses and organisations in Australia. Back in 2018, the Australian Crime Commission (2018) reported that a cybersecurity review, led by the Department of the Prime Minister and Cabinet, found that:

> [c]ybercrime is costing the Australian economy up to $1 billion annually in direct costs alone. In addition to loss of money, cybercrime causes other damage including:
>
> - damage to personal identity and reputation
> - loss of business or employment opportunities
> - impact on emotional and psychological wellbeing.

The problem has grown since then, despite the best efforts of the government to implement mitigation strategies and to educate government departments, businesses, community organisations and the general public about cybersecurity. According to the OAIC Australian Community Attitudes to Privacy Survey (OAIC 2023b, pp. 8, 10), undertaken in March 2023:

> - Three-quarters (74%) of Australians feel data breaches are one of the biggest privacy risks they face today.
> - Almost half (47%) of Australians said they had been informed by an organisation that their personal information was involved in a data breach in the 12 months prior to the survey.
> - Three-quarters (76%) said they experienced harm as a direct result [of a data breach].

While there are many security breach events, not all of them lead to an incident where information is accessed by a third party. In Australia, the Office of the Australian Information Commissioner (OAIC) defines a data breach as occurring 'when personal information an organisation or agency holds is lost or subjected to unauthorised access or disclosure' (OAIC 2024a). Some examples of a data breach include when:

> - a device with a customer's personal information is lost or stolen
> - a database with personal information is hacked
> - personal information is mistakenly given to the wrong person (OAIC 2024a).

The OAIC manages the Notifiable Data Breach (NDB) scheme on behalf of the Australian Government. Under the scheme, certain agencies and organisations that operate under the Privacy Act are required to notify the OAIC when an eligible data breach occurs — that is, when the incident is likely to cause serious harm to an individual. According to OAIC (2024b), 'Australian Government agencies (and the Norfolk Island administration) and organisations with an annual turnover more than $3 million have responsibilities under the Privacy Act, subject to some exceptions'.

The NDB scheme has been operating since February 2018 and initially the OAIC published quarterly reports. However, since the middle of 2019, half-yearly reports have been published.

Table 9.6 compares the key statistics for the first half of 2020 with those for the first half of 2023.

TABLE 9.6 A comparison of NRB key statistics between the first halves of 2020 and 2023

	Total number of breaches	Percentage human error	Percentage malicious or criminal attacks	Percentage system faults
1 January – 30 June 2020	518	34.0%	61.2%	4.8%
1 January – 30 June 2023	409	26.2%	70.4%	3.4%

Source: Adapted from OAIC (2020, 2023a).

Over the four-year period the total number of breaches has declined, along with the percentage of human errors and system faults. Conversely, during the same period the percentage of malicious attacks has increased.

CISCO (2019) defines cybersecurity as 'the practice of protecting systems, networks, and programs from digital attacks ... these [attacks] are usually aimed at accessing, changing, or destroying sensitive information; extorting money from users; or interrupting normal business processes'. To have a secure system, three types of technology must be protected: user devices such as computers, notebooks, tablets, phones and routers; networks; and the cloud.

Users of digital technologies are regularly advised to comply with a number of basic data security principles such as:

- implementing multi-factor authentication on their devices
- choosing strong passwords, and a different password for each device and software system
- regularly backing up computer files and data
- being wary of opening or saving email attachments, particularly those from unknown sources.

The Australian Cyber Security Centre (ACSC), which is part of the Australian Signals Directorate (ASD), developed strategies to assist organisations mitigate cybersecurity incidents and protect their systems against cyber threats. According to the ACSC (2017), these threats include:

> - targeted cyber intrusions (advanced persistent threats) and other external adversaries who steal data
> - ransomware ... and external adversaries who destroy data and prevent computers/networks from functioning
> - malicious insiders who steal data ...
> - malicious insiders who destroy data and prevent computers/networks from functioning.

Since 2017 the ACSC has recommended that organisations implement the Essential Eight mitigation strategies, which will make it much harder for their systems to be compromised. The Essential Eight strategies are regularly updated. The current strategies are (ACSC 2023):

1. patching applications — use the latest versions of applications and apply patches immediately to mitigate against known security vulnerabilities
2. patching operating systems — use the latest operating system versions and apply patches immediately to remediate known security vulnerabilities
3. multi-factor authentication — strong user authentication makes it harder for adversaries to access sensitive information and systems
4. restricting administrative privileges — to limit powerful access to systems
5. application control — only listed applications are executed to avoid the execution of unauthorised software
6. restrict Microsoft Office macros — to block untrusted macros
7. user application hardening — disable unneeded features
8. regular backups — to maintain the availability of critical data.

CRITICAL THINKING

What strategies do you currently use to protect your personal and professional information? What additional strategies can you implement to improve your cybersecurity?

Of course, cybersecurity is a fast-changing field and you should try to keep up to date with the latest advice on how to secure your data and information and avoid hacking.

9.6 Citing and referencing

LEARNING OBJECTIVE 9.6 Cite and reference information.

It is important to acknowledge sources when you use information and ideas published by other authors, as it is a demonstration of ethical practice. The inclusion of previously published information as evidence for your work will support your case in written work, particularly if the information is from a reputable source. It will also add authority to your idea, concept, proposal or design. Citations allow readers to go back to source documents to review that work, and to increase their understanding of a topic. Readers are then able to easily identify your work, innovation, creativity and achievement; that is, referencing enables people to recognise your IP.

Engineers may strengthen the case for the acceptance of a project if they acknowledge they have used information from reliable sources to develop the proposal. This is because information from reputable sources may enhance the credibility of the project proposal.

If you do not cite the source of the information you use in an article or other piece of work, you have plagiarised another author's work; that is, you have not acknowledged the use of their IP. This is a serious matter as it means you have misled readers, as they may assume the information or ideas are yours. Acknowledging the IP of other people is a legal requirement in Australia and many other countries. If, as a student, you plagiarise in an assignment or project report, you may be penalised and lose marks.

A publisher's version of the Harvard referencing system has been used throughout this text to cite the work of other authors. For example, 'Fidel and Green (2004, p. 564)' was used to cite the source of a quotation in the text of this chapter. Note that the page number is included in the citation when a quotation from another document is used. The full reference for this document is included in the list of references at the end of the chapter:

Fidel, R & Green, M 2004, 'The many faces of accessibility: Engineers' perception of information sources', *Information Processing and Management*, vol. 40, no. 3, pp. 563–581, https://doi.org/10.1016/S0306-4573(03)00003-7.

There is sufficient information in this reference for readers to be able to locate the original article.

There are many referencing systems; some are used across a number of disciplines, while other systems are used by a single discipline. Two examples of commonly used systems are the Harvard AGPS (or author–date) style and the American Psychological Association (APA) style. Two examples of discipline-specific systems are the Institute of Electrical and Electronics Engineers (IEEE) style and the Australian Guide to Legal Citations (AGLC).

The editors of a publication will nominate the referencing style to be used in that publication. In most cases the same style will be used for all of the publications produced by an organisation. Three styles commonly used in engineering publications will be used to illustrate the common features of referencing systems, and the differences between them: APA, Harvard and IEEE.

KEY POINT

Engineers should acknowledge the source of information they use.

Listing and citing print references

In general terms, a referencing system has two characteristics: (1) it ensures ample information is included in the reference so that readers can readily find and access the original source of the information; and (2) the information is coded and formatted in a systematic manner to enhance readability and minimise the length of the reference. Combinations of the following components are normally included in the reference for a printed document: author(s), year of publication, title of work, name of publication, edition of journal, month of publication, page numbers and print location. The references used in an article or report are normally listed at the end of the document.

The DOI

Many publishers have begun to allocate a digital object identifier (DOI) to each article they publish online, which is then used in citations as it is a persisting URL (see https://www.doi.org). The DOI system is used for managing the identification of content over digital networks and the key features of the system include persistence, network accessibility and interoperability with other identifiers. For example, ISBNs have been used for more than 30 years to identify each unique publication for physical books and related eBooks, software and mixed media publications. The ISBN-A (Actionable ISBN) system enables the ISBN 13-digit identification number to be incorporated into the DOI system. For example, a publication is cited as follows.

Mangioni, V 2018, 'Evaluating the impact of the land acquisition phase on property owners in megaprojects', *International Journal of Managing Projects in Business*, vol. 11, no. 1, pp. 158–173, https://doi.org/10.1108/IJMPB-08-2017-0090.

The following examples show how the Fidel and Green (2004) journal article is cited and referenced using the three different styles.

Harvard AGPS

The Harvard AGPS referencing system was originally developed by the Australian Government Publishing Service (1970–1997) and the latest version is the sixth edition (Snooks & Co. 2002). In this author–date system, the references are listed in alphabetical order by author. The basic format for a periodical reference is:

Author, A, Author, B & Author, C Year, 'Article title', *Journal Title*, vol. xx, no. xx, pp. xx–xx, DOI.

The reference for our example article is:

Fidel, R & Green, M 2004, 'The many faces of acccssibility: Engineers' perception of information sources', *Information Processing and Management*, vol. 40, no. 3, pp. 563–581, https://doi.org/10.1016/S0306-4573(03)00003-7.

If there were two papers by the same author, both of which were published in the same year, then the first would be listed as Brown, J 2005a and the second paper as Brown, J 2005b.

There are a number of ways that a document can be cited in the text of an article or report. For example: 'Brown (2005b) said that . . .' or 'Engineers believe accessibility is just as important as quality when they are searching for information (Fidel & Green 2004)'.

APA

The seventh edition of the APA referencing guide was published in 2020 (American Psychological Association 2020). This author–date system also lists the references in alphabetical order by author. The basic format for a periodical reference is:

Author, A. A., Author, B. B., & Author, C. C. (year). Article title. *Journal Title, volume number*(issue number), page numbers. DOI

For example:

Fidel, R., & Green, M. (2004). The many faces of accessibility: Engineers' perception of information sources. *Information Processing and Management, 40*(3), 563–581. https://doi.org/10.1016/S0306-4573(03)00003-7

An example of an intext citation is: 'Fidel and Green (2004) reported that . . .'

IEEE

The IEEE website includes a section about its publications including an 'Author digital tool box', which contains information, templates and tools to help authors prepare and submit articles. The IEEE Style Manual provides information about citing and referencing articles (IEEE 2018). The references are listed in numerical order, with the number [1] being allocated to the first reference that is cited in the text, the number [2] to the second reference cited and so on. The same number is used if a reference is cited again in the text; that is, there is one number per reference.

The basic format for a periodical reference is:

[1] A. Author, B. Author, and C. Author, "Name of paper," *Abbrev. Title of Periodical*, vol. x, no. x, pp. xxx–xxx, Abbrev. Month, year, doi: xxx.

The references are listed as:

[1] R. Fidel and M. Green, "The many faces of accessibility: Engineers' perception of information sources," *Inform. Process. Manag.*, vol. 40, no. 3, pp. 563–581, May, 2004, doi: 10.1016/S0306-4573(03)00003-7.

[2] J. Brown, 'Citing using the IEEE style . . .'

Although citation in text is optional there are two methods: 'Fidel and Green [4] stated that . . .' or 'According to [1] and [4] the design . . .'

Listing and citing online references

The vast amount of information that is increasingly available online has heightened the need for listing and citing online references.

Examples of some of the different styles used for referencing online documents are provided next.

Harvard AGPS

These are listed in alphabetical order with the other references. The basic format for an online news article is:

Author, A Year, 'Article title', *News Title*, day Month, viewed day Month year, URL.

An example is:

Vaughan, A 2018, 'BP's Deepwater Horizon bill tops $65bn', *The Guardian*, 16 January, viewed 4 March 2024, https://www.theguardian.com.

Online references are cited in the same way as print sources; for example, 'The IEEE (2018) has published a guide for authors . . .'

APA

These are listed in alphabetical order with the other references. The basic format for an online periodical reference is:

Author, A. A., Author, B. B., & Author, C. C. (year, Month day). Article title. *News Title*. URL

An example is:

Vaughan, A. (2018, January 16). BP's Deepwater Horizon bill tops $65bn. *The Guardian*. https://www.theguardian.com/business/2018/jan/16/bps-deepwater-horizon-bill-tops-65bn

Online references are cited in the same way as print sources; for example, 'The IEEE (2018) has published a guide for authors . . .'

IEEE

The online references are listed with the other references. The basic format is:

[1] A. Author, B. Author and C. Author. "Article title." Website Title. Accessed: Abbrev. Month day, year. [Online.] Available: URL.

An example is:

[1] A. Vaughan "BP's Deepwater Horizon bill tops $65bn". theguardian.com. Accessed: Mar. 4, 2024. [Online.] Available: https://www.theguardian.com/business/2018/jan/16/bps-deepwater-horizon-bill-tops-65bn.

Further information about referencing styles

In each of these systems, specific formats and protocols need to be followed when referencing information from different sources. There are many variations on the standards of these sources, such as the one used in this text. These guidelines are generally available on journal websites or through university libraries. You should learn to use them while you are a student because it is important that you correctly use protocols and styles. The system you use will be defined by your school or your discipline. Your lecturer may specify the style for your subject in the subject documentation. You will be provided with a referencing guide for the school's preferred system and may receive some tutorial assistance in one of your first-year subjects or as part of a library orientation program.

The key to success in citing and referencing is to record the required information at the time you access information so you do not have to return to the source to find the date of publication, publisher or other details. You can ensure you have all of the information by writing the reference in the format you have to use for your assignments as you retrieve each piece of information. If you use a software package like Endnote you will have a digital record of those references. By doing this, you will be gaining valuable practice in using the system, as well as staying on track with assignments.

Finally, you should note referencing styles are 'living systems' and so some structural or formatting details may change from time to time. Therefore, you should always check for the latest version of the guidelines before writing an assignment or article. If you are using software such as Endnote, you should ensure you are using the latest version.

SUMMARY

In this chapter, we have looked at how engineers find, use and manage information. The information age is upon us and it is crucial that you are able to navigate your way through it in an efficient and professional manner. Finding, analysing, evaluating, integrating, managing and using information are key skills for students and for engineers practising in the modern professional environment. We will now briefly revisit each of the chapter's learning objectives.

9.1 Differentiate between data, information and knowledge.
The relationships between data, information and knowledge were explored to illustrate the meaning of these concepts and to highlight the differences between them. A model was used to show how data becomes information and how information can be integrated into a person's mind to create knowledge. The importance of knowledge was emphasised, as well as the need to share our knowledge with colleagues. When knowledge is shared it is communicated as information.

9.2 Identify the information needs for a project.
Students (and engineers) need to carefully define the information requirements for a project. This process normally begins with the development of a list of questions that define the problems to be solved, and the information that may be required to understand and solve those problems. On larger, or more complex projects, the information needs are often organised into categories to improve the project team's understanding, and to enable the searching and retrieval process to be split among the team members. The time spent defining the information needs for a project is vital preparation for the next phase of the process — that of searching for the information.

9.3 Locate and retrieve high-quality information.
Information searches are focused on finding answers to defined questions. The key sources of information for engineering students and practitioners include documents, colleagues, clients, stakeholders and GIS systems. 'Asking a colleague' is an important information search strategy used by engineers, often because colleagues may provide information more quickly than using other search strategies. Students can also use this approach, by asking peers, tutors and lecturers for advice. For some projects, it may be important to stage the information gathering process, because the information gathered in one stage may change the direction of the project. It is important to record details of the search and the sources of the information gathered for a project, even if it is not used at that time.

9.4 Evaluate the quality and reliability of information and information sources.
The quality of information sources can vary widely. It is crucial to evaluate the quality and relevance of information used for a project, and the sources used to access the information. The CARDS technique for evaluating sources addresses the key questions of credibility, accuracy, relevance, date and source. Critical thinking skills can be used to evaluate information, including scepticism, objectivity and open-mindedness.

9.5 Integrating, managing and using information.
Ethical and secure processes need to be used to manage the quality and integration of the information gathered and produced during the life of a project. Summarising key pieces of information from different sources into a series of dot points is one way of integrating information. The dot points can then be joined to develop a narrative that summarises the information in that field. This approach can be used by students to prepare assignment reports, literature reviews and project documents.

An information management system must enable users to efficiently locate and retrieve the information they require for their work. A critical component of a quality process is a versioning system for plans and other documents. This is required to ensure all of the people on a project team are using the latest version of project documents. Some other key elements of a secure information management system are effective document access, archiving, cybersecurity and disposal processes.

9.6 Cite and reference information.
For both ethical and legal reasons it is important for engineers and students to acknowledge the source of the information they use. Citing the sources of information can also enhance the credibility of work. Two examples of commonly used referencing systems are the Harvard style and APA style. Two examples of discipline-specific systems are the IEEE style and the Australian Guide to Legal Citations (AGLC).

KEY TERMS

data A set of discrete, objective facts about events.
datum A reference point to which other measurements are related.
empirical data Derived from experience or experiment.
fact Something known to have happened; a truth known by actual experience or observation.
information Knowledge communicated or received concerning some fact or circumstance; news.
knowledge The fluid mix of framed experiences, values, contextual information, and expert insight in the mind of the knower.
primary source A document where information is published for the first time.
secondary source A document that uses, adapts or refers to information in a primary source.

EXERCISES

1. Explain the difference between data and information, giving an example of each for your chosen engineering discipline. Determine the accuracy and precision of some of the data you have collected or measured in your studies.
2. Use the list of question stems in figure 9.7 to define the questions you will use to find the information you need for an upcoming assignment.
3. Develop a search strategy for the questions you developed in exercise 2. Your strategy should show the questions in the order in which you will seek information and, for each question, the sources you plan to use to look for the information and the search strategy.
4. Use the CARDS method to evaluate the information you have gathered for a project.
5. Outline the key principles of information management.
6. What referencing style will you have to use for your assignments while you are at university? Obtain a copy of the guidelines for the style and practise using the style so you are ready for your next assignment. You should also use this referencing style when you develop your catalogue as part of the following project activity.

PROJECT ACTIVITY

Develop a catalogue of the information sources commonly used in the engineering discipline you are studying or intending to study; for example, chemical engineering. The catalogue will be in the form of a searchable electronic database you can continue to build and maintain as a student and throughout your career. The engineering information catalogue you prepare should (for each information source), as a minimum, contain:

- a description of the source of the information; for example, a journal name or URL
- the access details using the referencing style recommended by your school or department
- the type of source
- an indication of the reliability of the information from the source
- keywords that describe the type of information normally obtained from the source
- comments relating to the source.

A template is provided online for a table that you could use to begin this project. An example of a partially completed table is shown in the following table. You may modify this table to suit your discipline, or your needs. You may also prefer to use a software package such as Endnote, to build and maintain your catalogue.

Your project report should include a one-page report on the process you used to develop your catalogue, a mind map showing the structure of your catalogue and a digital or electronic copy of your catalogue.

Sample engineering information catalogue

Category	Type	Field	Source	Access information	Reliability	Comments
Technical	People	Pumps	John Liu	jl@realgoodpumps.com	3	Good advice, but verify from other sources
		Fire systems	Bill Knight	billk@firesystemsoz.com.au	4	Excellent
	Australian Standards		XYZ Standard	www. . . .	1	Good for disabled access
	International standards				2	
	Codes				2	
	Component specifications				3	
	Handbooks				2	
	Material specifications				3	
	Sales brochures				5	
	Other				5	
Theory	People					
	Textbooks					
	Journals — refereed				2	
	Journals — non-refereed				3	
	Industry journals				3	
	Conference proceedings				3	
Practice	People					
	Industry magazines				4	
	Industry journals				3	
	Conference proceedings				4	
	Online tutorials	CAD	AutoCAD			
	Professional information sites			www.civilengineering.com		

REFERENCES

Australian Communications and Media Authority (ACMA) 2022, 'ACMA investigation into Optus data breach', 11 October, https://www.acma.gov.au/articles/2022-10/acma-investigation-optus-data-breach

American Psychological Association 2020, *Publication manual of the American Psychological Association*, 7th edn, American Psychological Association, Washington, DC, https://doi.org/10.1037/0000165-000

Arup 2014, 'Dr Chau Chak Wing Building', https://www.arup.com/projects/dr_chau_chak_wing_building.aspx

Australian Crime Commission 2018, 'Cybercrime', https://www.acic.gov.au

Australian Cyber Security Centre (ACSC) 2017, 'Strategies to mitigate cyber security incidents', 1 February, https://www.cyber.gov.au/resources-business-and-government/essential-cyber-security/strategies-mitigate-cyber-security-incidents/strategies-mitigate-cyber-security-incidents

——— 2023, 'Essential Eight explained', 27 November, https://www.cyber.gov.au/resources-business-and-government/essential-cyber-security/essential-eight/essential-eight-explained

Australian Government 2014, *Financial system inquiry: Final report*, December, https://treasury.gov.au/sites/default/files/2019-03/p2014-FSI-01Final-Report.pdf

Australian Law Reform Commission 1980, *Lands acquisition and compensation (ALRC Report 14)*, Australian Government Publishing Service, Canberra, https://www.alrc.gov.au/report-14

Burton, LJ, Westen, D & Kowalski, R 2023, *Psychology*, 6th Australian and New Zealand edn, John Wiley & Sons, Brisbane.

Chanson, H 2007, 'Impact of commercial search engines and international databases on engineering teaching and research', *European Journal of Engineering Education*, vol. 32, no. 3, pp. 261–269, https://doi.org/10.1080/03043790701423815

Chenery, E & Daniel, S 2024, 'Sinkhole opens in industrial building car park at Rockdale adjacent to M6 motorway as tunnelling work stopped', *The Australian*, 1 March, https://www.abc.net.au/news/2024-03-01/sinkhole-rockdale-carpark-no-injuries-m6-tunnelling-stopped/103531958

CISCO 2019, 'What is cybersecurity?', https://www.cisco.com/c/en/us/products/security/what-is-cybersecurity.html

Cleveland, H 1982, 'Information as resource', *The Futurist*, December, pp. 34–39.

Coll, RK & Zegwaard, KE 2006, 'Perceptions of desirable graduate competencies for science and technology new graduates', *Research in Science & Technology Education*, vol. 24, no. 1, pp. 29–58, https://doi.org/10.1080/02635140500485340

Cotis, K & Bryson, S 2018, 'Submission to the inquiry into impact of the WestConnex Project', *WestConnex*, 30 August.

Davenport, TH & Prusak, L 1998, *Working knowledge: How organisations manage what they know*, Harvard Business School Press, Boston.

Dawul Wuru Aboriginal Corporation 2024, 'About us', https://dawulwuru.com.au/about

Elsevier 2024, 'Scopus AI: Trusted content. Powered by responsible AI', https://www.elsevier.com/en-au/products/scopus

Engineers Australia 2019, *Stage 1 Competency Standard for Professional Engineers*, https://www.engineersaustralia.org.au/publications/stage-1-competency-standard-professional-engineers

——— 2024, 'Resources', https://www.engineersaustralia.org.au/resources

Engineers Without Borders (EWB) Australia 2018, 'Cambodian rural development team (CRDT)', https://ewbchallenge.org/challenge/cambodian-rural-development-team

——— 2019, 'Scoping for best-practice', 12 December, https://ewb.org.au/blog/2019/12/12/scoping-for-best-practice

——— 2020, 'The evolution of a partnership that's delivering engineering solutions for Indigenous communities', 22 September, https://ewb.org.au/blog/2020/09/22/cfat-and-ewb-partnership

——— 2022a, 'EWB Challenge', https://ewb.org.au/project/ewb-challenge

——— 2022b, 'Announcing our 2023 EWB Challenge community partner', *Engineer without Borders Australia*, 18 October, https://ewb.org.au/blog/2022/10/18/announcing-our-2023-ewb-challenge-community-partner

——— 2024a, 'Previous top reports', https://ewbchallenge.org/participant-resources/previous-top-reports

——— 2024b, 'Announcing our 2024 EWB Challenge community partner — Torres Strait Island Regional Council', 30 January, https://ewb.org.au/blog/2024/01/30/announcing-our-2024-ewb-challenge-community-partner-torres-strait-island-regional-council

ESRI 2024, 'Industries', https://esriaustralia.com.au.

Fidel, R & Green, M 2004, 'The many faces of accessibility: Engineers' perception of information sources', *Information Processing and Management*, vol. 40, no. 3, pp. 563–581, https://doi.org/10.1016/S0306-4573(03)00003-7

Fitbit 2018, 'How accurate are Fitbit devices? Article 1136 and 1141', https://help.fitbit.com/articles

Gartner 2024, 'Dark data', https://www.gartner.com/en/information-technology/glossary/dark-data

Geoscience Australia 2023, 'Australian Height Datum', 7 June, https://www.ga.gov.au/scientific-topics/positioning-navigation/geodesy/ahdgm/ahd

Gilmore, H 2014, 'Frank Gehry's Sydney building sculpture revealed', *The Sydney Morning Herald*, 30 August, https://www.smh.com.au/national/nsw/frank-gehrys-sydney-building-sculpture-revealed-20140829-109vfe.html

Giskes, R 2004, 'Injurious affection: The position in Queensland after *Marshall* and under the *Integrated Planning Act 1977* (QLD). Research brief 2004/07', Queensland Parliamentary Library, Brisbane.

Heinstrom, J 2005, 'Fast surfing, broad scanning and deep diving: The influence of personality and study approach on students' information-seeking behaviour', *Journal of Documentation*, vol. 61, no. 2, pp. 228–247, https://doi.org/10.1108/00220410510585205

IEEE 2018, *IEEE Reference Guide*, https://ieeeauthorcenter.ieee.org

King, A 1992, 'Comparison of self-questioning, summarizing, and note taking-review as strategies for learning from lectures', *American Educational Research Journal*, vol. 29, no. 2, pp. 303–323, https://doi.org/10.3102/00028312029002303

Kosinski, M, Stillwell, D & Graepel, T 2013, 'Private traits and attributes are predictable from digital records of human behavior', *National Academy of Science*, vol. 110, no. 15, pp. 5802–5805, https://doi.org/10.1073/pnas.1218772110

Land Information New Zealand 2024, 'New Zealand Vertical Datum 2016 (NZVD2016)', 5 April, https://www.linz.govt.nz/guidance/geodetic-system/coordinate-systems-used-new-zealand/vertical-datums/new-zealand-vertical-datum-2016-nzvd2016
LifeQ 2018, *Validation of the LifeQ solution for the measurement of heart rate*, version 2, https://www.lifeq.com
Macquarie Dictionary 2024, *The Macquarie Dictionary online*, Pan Macmillan Australia, https://www.macquariedictionary.com.au
Mangioni, V 2018, 'Evaluating the impact of the land acquisition phase on property owners in megaprojects', *International Journal of Managing Projects in Business*, vol. 11, no. 1, pp. 158–173, https://doi.org/10.1108/IJMPB-08-2017-0090
Mayers, L 2018, 'WestConnex tunnel construction causing cracks in walls, local residents claim', *ABC News*, 22 November, https://www.abc.net.au/news/2018-11-22/sydney-residents-say-westconnex-causing-cracked-walls/10521220
McAlister, C 2023, *Lost without it*, https://usq.pressbooks.pub/gpsandgnss
McDonald, C 2002, 'Information systems foundations — Karl Popper's Third World', *Australasian Journal of Information Systems*, special issue, vol. 10, no. 1, pp. 59–69, https://doi.org/10.3127/ajis.v10i1.446
NSW Department of Planning and Environment 2016, 'Planning guideline for major infrastructure corridors', https://www.planning.nsw.gov.au
NSWRMS 2014, 'Roads and Maritime Services land acquisition information guide. June 2014', https://www.transport.nsw.gov.au/operations/roads-and-waterways
——— 2015, 'Factsheet: Property acquisition of subsurface lands. January 2015', https://www.transport.nsw.gov.au/operations/roads-and-waterways
——— 2016, 'Exceptional hardship land purchase guideline', https://www.transport.nsw.gov.au/operations/roads-and-waterways
Office of the Australian Information Commissioner (OAIC) 2020, *Notifiable Data Breaches Report: January–June 2020*, 31 July, https://www.oaic.gov.au/privacy/notifiable-data-breaches/notifiable-data-breaches-publications/notifiable-data-breaches-report-januaryjune-2020
——— 2021, 'Clearview AI breached Australians' privacy', 3 November, https://www.oaic.gov.au/newsroom/clearview-ai-breached-australians-privacy
——— 2023a, *Notifiable Data Breaches Report: January–June 2023*, 5 September, https://www.oaic.gov.au/privacy/notifiable-data-breaches/notifiable-data-breaches-publications/notifiable-data-breaches-report-january-to-june-2023
——— 2023b, *Australian Community Attitudes to Privacy Survey*, August, https://www.oaic.gov.au/__data/assets/pdf_file/0025/74482/OAIC-Australian-Community-Attitudes-to-Privacy-Survey-2023.pdf
——— 2024a, 'About the Notifiable Data Breaches scheme', https://www.oaic.gov.au/privacy/notifiable-data-breaches/about-the-notifiable-data-breaches-scheme
——— 2024b, 'Rights and responsibilities', https://www.oaic.gov.au/privacy/privacy-legislation/the-privacy-act/rights-and-responsibilities
Park, A 2018, 'WestConnex construction causing depression and asthma, residents claim', *7.30 Report, ABC News*, 15 November, https://www.abc.net.au/news/2018-11-15/westconnex-construction-causing-depression-asthma-residents-say/10479480
Popper, KR 1979, *Objective knowledge: An evolutionary approach*, rev. edn, Clarendon Press, Oxford, London.
Pressman, R 2005, *Software engineering: A practitioner's approach*, 6th international edn, McGraw–Hill, Singapore.
Segalla, M & Rouzies, D 2023, 'The ethics of managing people's data', *Harvard Business Review*, vol. 101, no. 4, pp. 88–94.
Sharma, N 2008, 'The origin of the Data Information Knowledge Wisdom (DIKW) hierarchy', https://www.researchgate.net/publication/292335202_The_Origin_of_Data_Information_Knowledge_Wisdom_DIKW_Hierarchy
Snooks & Co. 2002, *Style manual for authors, editors and printers*, 6th edn, John Wiley & Sons, Brisbane, Australia.
Stewart, TA 1997, *Intellectual capital*, Doubleday Currency, New York.
Strava 2018a, 'Bad GPS data: What/why/how', https://support.strava.com
——— 2018b, 'Why is GPS data sometimes inaccurate?', https://support.strava.com
——— 2018c, 'Elevation on Strava FAQs', https://support.strava.com
TomTom 2017, 'How accurate is the built-in heart rate monitor?', https://au.support.tomtom.com
UTS 2014, 'Dr Chau Chak Wing Building', https://www.uts.edu.au
WestConnex 2018a, 'Our commitment to engage', https://www.westconnex.com.au/community
——— 2018b, 'Tunnelling', https://www.westconnex.com.au
WestCONnex Action Group 2018, 'Submission no. 436. Inquiry into the impact of the WestConnex Project', 10 September, https://www.parliament.nsw.gov.au
Wright, PH 2002, *Introduction to engineering*, 3rd edn, John Wiley & Sons, New Jersey.
Zeleny, M 1987, 'Management support systems: Towards integrated knowledge management', *Human Systems Management*, vol. 7, no. 1, pp. 59–70, https://doi.org/10.3233/HSM-1987-7108
——— 2005, *Human systems management: Integrating knowledge, management and systems*, World Scientific Publishing Co, Singapore.

ACKNOWLEDGEMENTS

Photo: © ifeelstock / Adobe Stock Photo
Photo: © Rose Marinelli / Shutterstock
Photo: © Thaakirah/peopleimages.com / Adobe Stock Photo
Photo: © Seventyfour / Adobe Stock Photo
Photo: © Gehry Partners LLP
Photo: © Andrew Worssam
Figure 9.1: © Engineers Without Borders Australia

Figure 9.3: © Anthony Brown / Adobe Stock Photo

Figure 9.4: © McDonald, C 2002, 'Information systems foundations — Karl Popper's Third World', *Australasian Journal of Information Systems*, special issue, vol. 10, no. 1, https://doi.org/10.3127/ajis.v10i1.446

Figure 9.9: © ESRI. All rights reserved. Used by permission. https://www.esri.com

Table 9.4: © Chanson, H 2007, 'Impact of commercial search engines and international databases on engineering teaching and research', *European Journal of Engineering Education*, vol. 32, no. 3, pp. 261–269, https://doi.org/10.1080/03043790701423815

Text: © Engineers Without Borders Australia 2019, 'Scoping for best-practice', 12 December, https://ewb.org.au/blog/2019/12/12/scoping-for-best-practice

Text: © Engineers Without Borders Australia 2024, 'Previous top reports', https://ewbchallenge.org/participant-resources/previous-top-reports

Text: © Engineers Without Borders Australia 2020, 'The evolution of a partnership that's delivering engineering solutions for Indigenous communities', 22 September, https://ewb.org.au/blog/2020/09/22/cfat-and-ewb-partnership

Text: © Engineers Without Borders Australia 2022, 'EWB Challenge', https://ewb.org.au/project/ewb-challenge

Text: © Engineers Australia 2019, *Stage 1 Competency Standard for Professional Engineers*, https://www.engineersaustralia.org.au/publications/stage-1-competency-standard-professional-engineers

Text: © Australian Law Reform Commission 1980, *Lands Acquisition and Compensation (ALRC Report 14)*, Australian Government Publishing Service, Canberra, https://www.alrc.gov.au/report-14

Text: © Park, A 2018, 'WestConnex construction causing depression and asthma, residents claim', *7.30 Report, ABC News*, 15 November, https://www.abc.net.au/news/2018-11-15/westconnex-construction-causing-depression-asthma-residents-say/10479480

Text: © Mayers, L 2018, 'WestConnex tunnel construction causing cracks in walls, local residents claim', *ABC News*, 22 November, https://www.abc.net.au/news/2018-11-22/sydney-residents-say-westconnex-causing-cracked-walls/10521220

Text: © Australian Government 2014, *Financial system inquiry: Final report*, December, https://treasury.gov.au/sites/default/files/2019-03/p2014-FSI-01Final-Report.pdf

Text: © Office of the Australian Information Commissioner (OAIC) 2021, 'Clearview AI breached Australians' privacy', 3 November, https://www.oaic.gov.au/newsroom/clearview-ai-breached-australians-privacy

Text: © Australian Communications and Media Authority (ACMA) 2022, 'ACMA investigation into Optus data breach', 11 October, https://www.acma.gov.au/articles/2022-10/acma-investigation-optus-data-breach

Text: © Office of the Australian Information Commissioner (OAIC) 2023, *Australian Community Attitudes to Privacy Survey*, August, https://www.oaic.gov.au/__data/assets/pdf_file/0025/74482/OAIC-Australian-Community-Attitudes-to-Privacy-Survey-2023.pdf

Text: © Office of the Australian Information Commissioner (OAIC) 2024, 'About the Notifiable Data Breaches scheme', https://www.oaic.gov.au/privacy/notifiable-data-breaches/about-the-notifiable-data-breaches-scheme

Text: © Office of the Australian Information Commissioner (OAIC) 2024, 'Rights and responsibilities', https://www.oaic.gov.au/privacy/privacy-legislation/the-privacy-act/rights-and-responsibilities

CHAPTER 10

Engineering design

'Design is not just what it looks like and feels like. Design is how it works.'

Steve Jobs

LEARNING OBJECTIVES

After studying this chapter, you should be able to:

10.1 use a basic design approach to identify opportunities, needs and problems and to generate solutions in a systematic manner

10.2 apply whole system design principles to seek optimal solutions to engineering problems

10.3 generate design solutions, using research, consultation, creativity and biomimicry principles.

Introduction

In an early chapter we introduced the engineering method, a generalised method for proceeding through complex problem solving. In the previous chapter, research and information skills were introduced, which are critical for success in working your way through any new problem. It can get a bit discouraging if we think we are surrounded only by problems that need solving so we will use the word 'problem' to include the identification of a need or an opportunity for improvement. Much of engineering design is concerned with improving existing systems — systems that function and satisfy a critical need but have the capability of being better. We are also always on the lookout for new opportunities.

Fortunately, we are surrounded by a rich pool of information to assist us, accessible to us from our computers and mobile devices. This information can help us understand a new and complex situation. We also have access to centuries of ingenuity, inventions and analysis methods.

In this chapter, we will consider the processes of understanding the problem and finding or generating solutions through research, consultation and creativity. We will look at systems thinking (making sure you are answering the right question), and we will also look at how we can begin to evaluate proposed solutions against the design criteria and constraints. This will lead eventually to a final decision, checking and reviewing the resulting recommendations. Let's look at an example in which engineers would have proceeded through all these stages.

The Falkirk Wheel is a rotating boat lift that links the Forth and Clyde Canal with the Union Canal in Scotland. The wheel is a modern example of creative problem solving in practice. It replaces locks that previously connected the two canals, separated by 30 metres of elevation. The wheel uses two balanced chambers, with barges entering the top and bottom chambers. The structure, which is the height of an 8-storey building, rotates 180 degrees in four minutes, driven by a 22.5-kilowatt (kW) motor (Falkirk Wheel 2024). This requires only 1.5 kWh of energy.

The Falkirk Wheel

The two chambers are inherently balanced — even if there is no boat in one of the chambers. This is because, according to Archimedes' Principle, a boat displaces an amount of water exactly equal to its weight. If the height of the water is the same in each chamber, they contain the same weight.

The Falkirk Wheel is an example of different engineering disciplines being used to achieve a desired result. Civil engineers, mechanical engineers, and electrical and software engineers were all involved in this project. The engineers involved in the project also took aesthetics into account, with the design of the arms based on a Celtic double-headed axe.

This innovative solution is an example of how creativity and systems thinking can be used to apply engineering and science fundamentals to solve problems — in this case, replacing a series of conventional lock chambers that use large amounts of energy and water with a design that uses sound sustainability principles and minimises energy and water use. In conventional canal locks, water is lost from the upper canal to the lower one. With the Falkirk Wheel there is no net flow of water between the two canals. This giant machine is a good example of how interdisciplinary engineering creates new, efficient solutions to old problems (moving barges between canals).

The rest of this chapter will explore some of the techniques the engineers in this project may have used.

10.1 Design = problem solving

LEARNING OBJECTIVE 10.1 Use a basic design approach to identify opportunities, needs and problems and to generate solutions in a systematic manner.

Are design and problem solving the same thing? A lot of design is about asking questions — asking the right questions, in fact. We start out by asking ourselves: what problem are we solving? Answering this question can take quite some time and it's the first major step in the design/problem-solving process.

The next steps involve studying the causes of the problem and researching the science of these causes. Then we ask: what solutions are available for this problem? Most of the solutions are well documented and will be revealed in the research phase. Research skills were discussed in detail in the chapter on understanding the problem.

We need to think about the budget for the task. What will it cost? Is this reasonable? Will the client pay this much? This budget will rely on a feasible solution. Can we *improve* our design and deliver the desired outcome at a lower cost? This might result in a larger profit or a lower bid price to win the job. When we finish the process and deliver the solution, we then examine what we have learned and think how we might do it better next time.

Key ideas in the design process

There are different starting points for the design process and *different types of design effort*:

- needs-based design (identifying a need), which leads to a new product
- improving/competitive — outperforming competitors by adding new features or optimising existing designs
- fixing a problem — repair and modify to avoid repeated failures (this is similar to the previous point).

You need to *understand* the problem to solve it, and you need to spend more time on this than any other phase. This is why engineering consulting companies exist. The problem needs to be solved in virtual space before it is solved in physical space. This reduces the *time* required in physical space, and it reduces the *risks* by anticipating setbacks and constraints. This never ends because design is *iterative; re-calibration and checking* are required constantly as new information emerges.

Expect *emergence* of the problem type or form. Use analogy and search. Expect problems and solutions to be like other ones you've solved before; however, be on the lookout for *new and innovative ways* of solving familiar, well-understood problems.

Developing a *problem statement* is critical. This should define the problem to be solved in as much detail as required.

In brief, remember that:

- there may be no optimal solutions in real life — there are good ones and better ones
- solutions are constrained by cost and time, among other things
- rough approximations get you there quickly, and most of the time are close enough.

CASE STUDY

Building a house extension

Imagine that you've been chatting with friends who are building an extension on the back of their house. The location for the extension is 50 metres from the front gate and is inaccessible except by foot up an inclined narrow path. They are complaining that it is difficult to carry bricks from the footpath to the back of the house and are wondering if there's something that would help them do this more effectively. What do you do? How can design thinking help in this situation?

Well, you'd need to ask a few more questions and gather some more information. *Understand the problem*. What really is the problem? What are the *criteria* and *constraints* to this problem? You also need to know what has already been decided and cannot be changed. Let us assume that the bricks have been ordered and their arrival is in the near future.

For example, if your friends want the problem resolved within a few hours and the job is not too large, you might consider going to the shop and buying or hiring a wheelbarrow. But then you notice that there's quite a steep incline from the front yard to the back, which might make pushing a fully loaded wheelbarrow quite difficult (if not impossible). Besides, just how many bricks are involved?

You might then suggest getting some friends around to form a human chain. How much are your friends willing to pay? If they say $200, then your problem is to determine how many people you need, and how much money each one would want to form a part of this chain. Can you still make a profit for your design skills? Probably not.

You tell your friends that it's not going to be that easy, then sit down with them to consider the options. You decide to go with your first solution; you buy a wheelbarrow, bring it back and start helping your friend move the bricks. But, as you suspected, the incline is quite steep and by now both of you are suffering from severe shoulder and back stress.

Is this any way to run an engineering consultancy? What should you have done?

This is an example of what is at the core of engineering as a discipline. Engineers use a *structured approach* to these types of open-ended and ill-structured problems together with their technical knowledge and problem-solving abilities to avoid wasting everyone's time and money.

The type of problem your friends asked you to solve was a logistics design problem, requiring you to synthesise a solution from a given situation. What was needed was a more detailed description of the *problem state*. To this end, you could have spent more time interviewing your friends to develop a better understanding of the problem by gathering more details. You might have considered asking the following types of questions.

- What is the total volume (and weight) of material to be moved?
- What types of materials need to be moved (size, density, type of substance, fragility)?
- What is the volume (m^3) per trip?
- What is the distance (m) per trip (horizontal and vertical)?
- How many trips will be needed?
- What is the duration (speed) of each trip, and the total duration of all trips combined?
- What is the source of power for doing this work?
- Are there any obstacles between the two ends of the trip?
- What is the budget?
- How much time is allowed to find a solution?

We call this step *generating design criteria* because we need to identify and list all of the *requirements* (i.e. things that the design must be able to do) and the *constraints* (i.e. things that bound the solution choices) that will form the basis for our search for feasible solutions to the problem.

CASE STUDY

Quoting for the job

Having collected all the data from your friends, you decide to press on with the job by quoting in your professional capacity. Say, for instance, that your friends tell you that they are willing to pay up to $5000 for the problem to be solved (as they have discovered that the problem is bigger than they first suggested). Let's assume you think you can solve this problem within five days and that you're happy to make $1000 for those five days to do the work. That leaves $4000 to design and implement a solution. What do you do?

Feasibility

What types of solutions exist so far for this type of problem? Before committing yourself to deliver a solution, you obviously need to know that there is *at least one solution* that fits the budget. The function that you are addressing is moving or conveying the bricks to a new location. It might be useful to search a dictionary for verbs that help define this function to give you other ideas.

Make a *sketch* such as the one in figure 10.1, which will allow the problem to be suitably simplified, and critical issues will, hopefully, begin to emerge. For example, the incline may not be as gradual as shown. There might be a flat section and then a very steep incline. There might be a narrow gate to negotiate or a dog that needs to be kept locked in the backyard. These are further constraints on the solution.

FIGURE 10.1 Sketch of the incline from street level to back of house

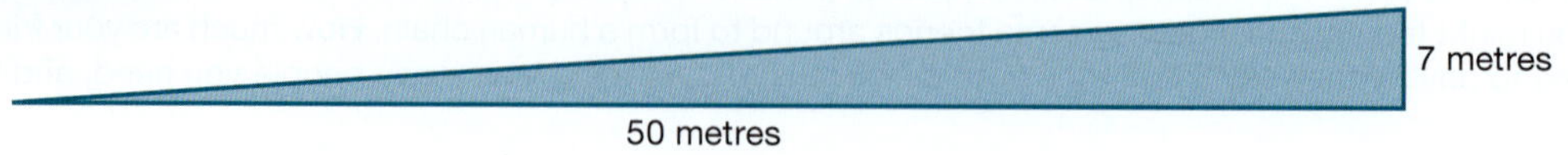

7 metres

50 metres

CASE STUDY

Assessing the data

Let's assume that the following data has been gathered.

- All bricks need to be moved within 24 hours (in 3 × 8-hour shifts) because the bricklayer is coming on-site the day after next and the bricks are currently on a public road necessitating traffic control.
- The bricks weigh 2.5 kilograms each; in Australia, the typical house brick has the dimensions 230 × 110 × 76 mm.
- The extension is 20 metres long by 10 metres wide and 4 metres tall. Ignoring the inclusion of doors and windows, the total number of bricks you need to move is:
  - perimeter length of walls: 60 000/230 = 261 bricks
  - height of wall: 4000/76 = 53 bricks
  - the four walls amount to: 261 × 53 = 13 833 bricks. Modifying for waste and openings, 13 700 bricks are needed.

Let's revisit the wheelbarrow idea. How many bricks could be carried in each load? Let's assume that a safe lifting mass is around 25 kilograms for a person, and that the wheelbarrow will allow you to carry up to 75 kilograms per load if you could stack the bricks in a stable arrangement so that the lifting load is 25 kilograms. This would require most of the weight to be placed on the front wheel.

A brick has a mass of 2.5 kilograms, which suggests that you can carry a maximum of 30 bricks at a time. Considering the size of a typical wheelbarrow (say, around 500 mm × 500 mm), you should be able to place six bricks stacked five high. This arrangement would imply that you'd need to move 13 700/30 = 457 loads in total, or 153 loads in each 8-hour shift (19 loads per hour). If one person is doing the moving, they would have to travel at (19 × 100 m)/1 hr = 1.9 km/hr, which is feasible.

However, you must also consider how long it takes to load and unload the wheelbarrow. Let's assume that it takes two minutes to put the bricks in and two minutes to take them out. That means for 153 loads we'd spend 612 minutes (more than 10 hours) simply loading and unloading the bricks, which means you can't accomplish the task in an 8-hour shift no matter how fast you move the bricks between the two points. What about using more than one wheelbarrow at a time? From the number 612/480 = 1.3, you will need more than two wheelbarrows just to break even on the time required to load and unload. So, let's guess that you need four. That means you are moving 120 bricks at a time, or 38 loads per shift. At 4 minutes to load and unload, this would require 152 minutes. This leaves 328 minutes to move the loads — 8.6 minutes per trip of 100 metres, or 12 m/min. This is feasible, even with a heavily loaded wheelbarrow, given that a moderate walking speed is about 4 km/hr = 67 m/min.

If you pay four people around $30/hr for the 3 shifts, it will cost 30 × 24 × 4 = $2880 for labour. Hiring the wheelbarrows at $25/day × 4 = $100; a total cost of $2880 + 100 = $2980 — way under the budget. You get to make $2120 from this deal!

This is a feasible solution, but is it the best one? We still haven't factored the incline into the solution, although we have acted fairly conservatively, using four wheelbarrows. You might like to check whether three wheelbarrows will do the job. Perhaps we can earn more and still give our friends a solution under budget?

By now, you should be starting to appreciate the iterative nature of design; moving forward with potential solutions, we've had to revisit the requirements several times already, and still only arrived at one feasible solution. Furthermore, this solution will still require more work if we are to maximise our profit and assure ourselves that the labourers can navigate the steep incline in the yard — 38 × 4 = 152 times with 75 kilograms each trip. How much work can someone do in an eight-hour shift? How much energy is involved in raising 38 × 4 × 75 kilograms by 7 metres? That's 79 800 joules or almost 80 kilojoules. This might explain why it is better to opt for four wheelbarrows instead of three.

Safety

There are workplace health and safety regulations on manual lifting of no more than 20 kilograms. By taking this into account and making further calculations on a standard wheelbarrow, you will find that the total load will need to be reduced to 50 kilograms instead of 75. Alternatively, it may be an option to design extension handles to be fitted to the standard wheelbarrows.

CASE STUDY

Further solutions

The more you think about the incline, the more concerned you are that someone will get injured in the process. Can we bring some mechanical effort to bear on the problem, instead of relying on human power, with the potential for back strains, leg injuries or worse? A quick check of the local hire shop shows a range of equipment available for moving this type of material. The most likely candidate is a *tracked dumper*, which is a track-based, motorised wheelbarrow. It costs about $200 per 24 hours to hire, plus $100 in delivery charges.

This barrow can carry 500 kilograms and it moves at 5 kilometres per hour, or 1.2 minutes for the 100-metre round trip (let's assume 1 minute each way). If it takes 4 seconds to load and unload each brick, it will take 800 seconds (13.3 minutes) at each end and 28.6 minutes per trip (let's round that up to 30 minutes). In 8 hours, that means 16 trips to move 3200 bricks (note that most of the job is now loading and unloading the bricks). So, with two of these devices, you could get the whole job done in less than 24 hours with just two labourers. Just to be sure, you decide to hire three labourers, with the idea that there will be one person loading, one person unloading and one person driving the machines and helping with the loading/unloading. This will reduce the loading and unloading time.

Let's say it's 10 minutes now (roughly 30 minutes of work spread across 3 people) — this makes the trip time about 20 minutes, which increases the number of the trips to 24 in a shift (or 4800 bricks × 2 barrows). The whole job will now get done in less than 12 hours (8 × 13 700/(4800 × 2)) — much better than working all night!

So, what's the new cost?

- Two tracked dumpers at $200/day = $400
- Three labourers: 3 × $30 × 12 = $1080
- Delivery, fuel and other expenses: say, $200
- Total: $1680

This is a saving of $1300 on our first solution. You could even afford to give the workers a bonus. The workers are also much less likely to suffer a serious injury, although there's still a lot of manual handling involved. At least you've removed the most serious danger of pushing the bricks up the incline.

Are there other mechanical interventions that you could consider to ease the process of manually loading and unloading the bricks?

So, what have we learned? We started by scoping the problem — understanding the *criteria* and the *constraints* — and assessed what needed to be done and by when. We established a *budget* (the resources) and quickly determined that there was at least one *feasible solution* that would get the job done. Then we considered whether there were *better* solutions. This is the process of *optimisation*. Considerations of safety helped us to consider more mechanical intervention, which resolved the serious safety issues and also sped up the whole job. Ultimately, we were able to employ fewer people for less time.

In this problem we have arrived on the scene after a number of key design decisions have already been made. If you were starting from scratch, what would you advise your friends to think about from the outset? Are there alternative solutions to building an extension out of brick? Could they move to a larger house rather than build?

10.2 Systems thinking

LEARNING OBJECTIVE 10.2 Apply whole system design principles to seek optimal solutions to engineering problems.

The chapter on the engineering method introduced the idea of systems thinking. **Systems thinking** is the process of identifying all the elements of a problem. It is a tool for making sure the right problem is being solved. System diagrams are valuable as they provide groups with a clear understanding of the problem description. Figure 10.2 shows the elements of a system.

FIGURE 10.2 Schematic of a system

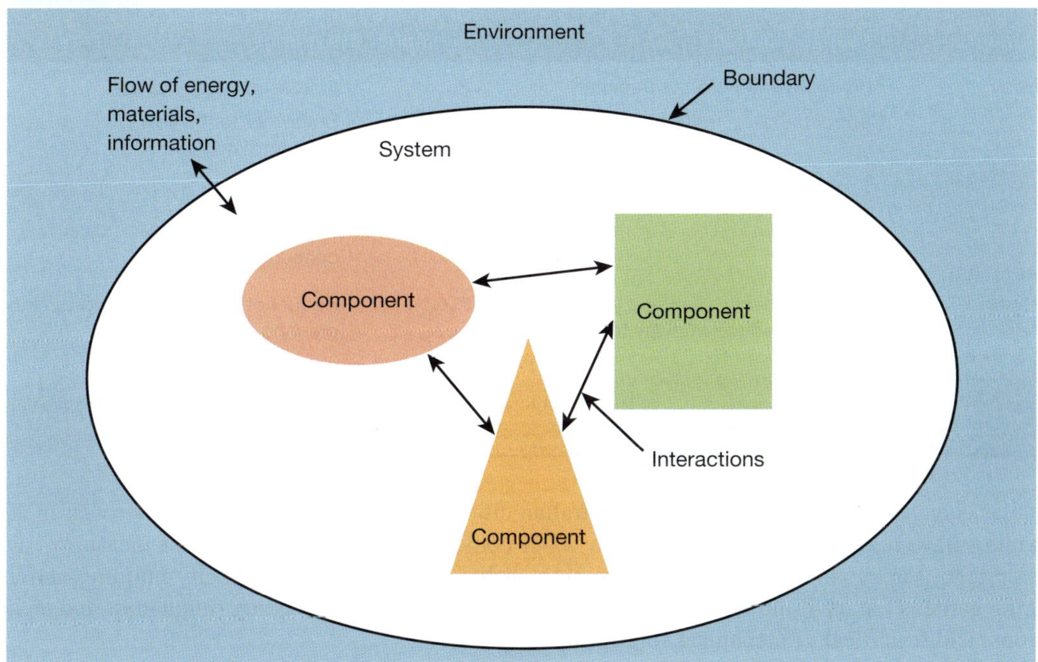

The chapter on the engineering method used the example of a congested freeway to explain systems thinking. We expanded and contracted the system boundary as a means of considering alternative problem definitions. In doing so, it became apparent that a congested freeway could be interpreted as a road issue, a transport issue or a workplace issue (as a result of working from home policies). Expanding the system definition meant alternative solutions to the problem could be considered.

Taking a broader approach to problem identification allows the degree of complexity of problems to be ascertained. Typically, problems require an **interdisciplinary approach**, with professionals from multiple disciplines working together to address a range of interconnected issues.

As the system definition is expanded, other issues may emerge that were previously hidden or just not accounted for (called **externalities**). For example, infrastructure and building projects are now required to consider a variety of environmental issues that were previously *external* to design processes, including long-term energy, water and material consumption, greenhouse gas emissions, and the health of natural systems such as waterways, ecosystems and the atmosphere. Resilience against the effects of climate change is an additional factor for most construction projects. The same kinds of decisions must also be made in terms of internet infrastructure, for example, futureproofing, and the energy consumption of a network switch that will operate 24/7 for several years. Clearly, a low-energy consumption unit is preferable.

Table 10.1 shows how broader system definitions for the congested freeway problem could lead to greater gains in energy efficiency, mostly by relocating people's work and reducing travel requirements. However, these changes are also much more complex to implement and it may take ten years or more to see results. They fall within the scope of urban planning documents such as 'Delivering Melbourne's newest sustainable communities' (Department of Transport and Planning 2010).

It is likely that all of the recommendations in table 10.1 may be implemented. The problem of urban transport is so complex and so large that a multi-faceted solution is required. There may be some quick returns through increasing public transport, for instance, and some longer term solutions through reducing travel via smarter planning guidelines.

As the system boundary is expanded, the range of disciplines involved also increases. Adding a lane to the freeway is mostly a problem for civil engineers, in addition to environmental engineers who manage the impacts of construction. Adding traffic signals to on-ramps and electronic tolling on roads would require electrical engineers, as well as electronic and software engineers. Improving public transport would likely involve mechanical engineers, environmental engineers, urban planners and possibly architects for new stations. Changing work practices would likely require industrial relations experts, change managers and human resources managers; thus, as the system boundary is expanded, the range of solutions available also becomes increasingly interdisciplinary.

TABLE 10.1 The impact of a systems definition on solutions: the congested freeway

System statement	Possible solutions	Green credentials (the greater number of ticks the higher the rating)	
Freeway	Add another lane to the freeway	✓	Increased energy usage
	Control on-ramp traffic to the freeway	✓✓	Smoother traffic flow = lower fuel consumption
	Add tolls to the freeway	✓✓	Reduce traffic
	Improve exit conditions from the freeway	✓✓	Smoother traffic flow = lower fuel consumption
Transport system	Increase capacity of public transport system (to reduce freeway traffic)	✓✓✓	Reduction in fuel consumption (and greenhouse gases)
City	Relocate work via telecommuting or moving businesses out of the central business district	✓✓✓✓	Moving work will reduce travel requirements, fuel consumption and greenhouse gases

Similar changes with water supplies in Australian cities show an interdisciplinary approach in action. Most cities have now reduced their per capita water consumption through a range of measures, such as higher pricing (an economic incentive), education (behaviour change and capacity-building activities), changing gardens (an environmental response), mandatory rainwater harvesting (regulatory and technical solutions) and fewer leaks (a technical solution).

KEY POINT

Systems thinking is a necessary tool in understanding complex problems.

Stakeholders

The people and organisations that have an interest in a problem are defined as **stakeholders**. There are usually many stakeholders associated with a problem. Although it can be tempting for an individual engineer to assume that they have identified the solution to a problem, engaging with others will likely broaden their perspective. A variety of other solutions to the problem can be discovered (as demonstrated with the congested freeway example). There may be more than one appropriate solution to a problem, and it may be appropriate to allocate resources to multiple recommendations; therefore, it is important for engineers to learn how to engage with stakeholders and consider alternative viewpoints.

This process is sometimes called **conversation mapping** and could be done at the start of a project by researching the attitudes of the stakeholders. For example, how do you think all the stakeholders would view the problem of the congested freeway? The challenge is to be able to patiently investigate their different views.

As discussed in the chapter on communication skills, *effective listening* requires the listener to actively attend to what the speaker is saying. Often people spend their time thinking about what they will say next, rather than listening carefully to others. When a speaker finishes, an active and effective listener should be able to summarise what the speaker has said. For example, 'What I heard you say was that you are very concerned about . . .'

A whiteboard or flipchart are tools that can be used for conversation mapping. To begin, you might like to write the problem or the issue on the whiteboard or flipchart. You can build a mind map around this, either by yourself or as part of a group. Figure 10.3 is an example of a possible mind map for the congested freeway problem.

If you are working as part of a group, you may like to start this process by asking other team members the factors they believe have contributed to the problem. These can be shown as 'branches', or sections extending from the identified problem. On the major branches, you can add contributing factors or possible solutions as they are identified by team members. You would continue this process until the team members cannot think of any more ideas, lack enthusiasm or are out of time. You may like to format this information using mind mapping software. A team member may email a final version of the work done to other group members as a starting point for future discussions.

FIGURE 10.3 Mind map of congested freeway

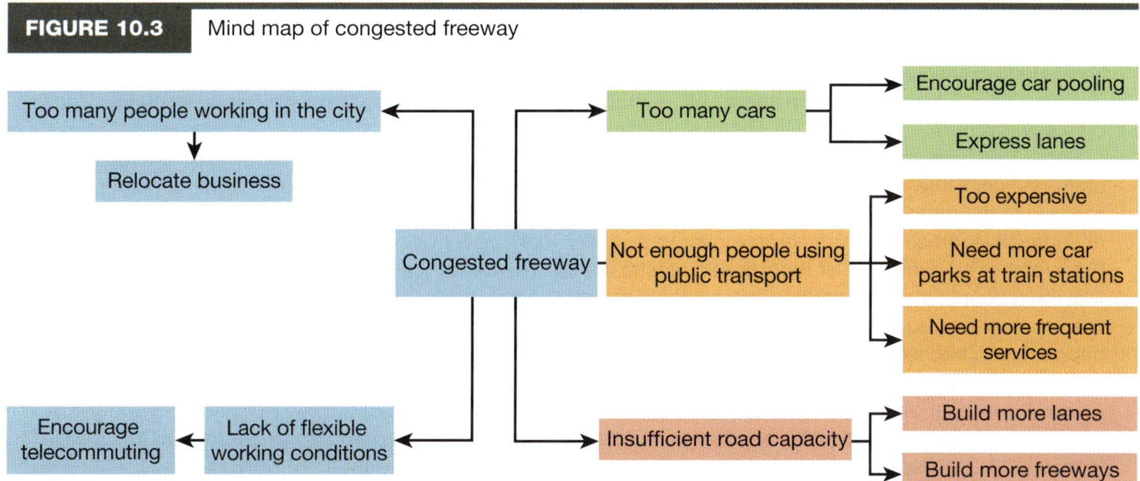

It would also be advantageous for the group to have access to an interactive website. The group could continue to contribute ideas via a discussion forum, a blog or a wiki. A good example of public engagement is *Participate Melbourne* (City of Melbourne 2024a). It provides an easily accessible space for public comment on the future directions for the city. Twenty years ago, public meetings and focus groups would have been the way to gather public input on such an issue. Now, a range of technological solutions allow people to contribute to such a discussion from the comfort of their own homes.

SPOTLIGHT

The good news about plastic waste

DEIRDRE MCKAY, KEELE UNIVERSITY

Waste plastics contaminate our food, water and air. Many are calling for a global ban on single-use plastics because throwing them 'away' often means into our river systems and then into the world's oceans.

Take the UK's single-use plastic bottles: it's estimated that 35 m are used — and discarded — each day, but only 19 m are recycled. The 16 m bottles that aren't recycled go to incinerators, landfill or the environment, even though, being PET (polyethylene terephthalate), they are easily reprocessed. Even those bottles that are placed in the recycle stream may be shipped to Asia, in a global market for waste plastics that is itself leaky.

Blasting cap wrapper backpack being modelled by its producer, a former mine worker

It's suspected that much of the 'recycling' shipped to Asia may be joining local waste in the Great Pacific Garbage Patch. This soupy collection of plastic debris is trapped in place by ocean currents, slowly breaking into ever-smaller pieces, but never breaking down. Covered by bacterial plaques, they are mistaken for food by fish. Ingested, they contaminate the food chain and, potentially, may even be disrupting the biophysical systems that keep our oceans stable, thus contributing to climate change.

So we need to use far less plastic, re-use what we can, and dispose of what we must far more wisely. In facing this challenge, developed countries can learn from innovations in the less-developed world. People, globally, are innovating, creating new processes to use waste plastics and making new objects and art forms.

Filipino plastics

The Philippines, for example, is the world's third biggest plastic polluter. Waste material, even from municipal collections, makes its way into the river systems and, from there, out into the open Pacific, contributing to its 'plastic soup'. But it's largely what happens on land that determines the load the oceans must bear.

Industrial recycling isn't accessible to most people in the Philippines. Even if recycling were available, shipping their waste to China, long the centre of the global plastics recycling industry, is no longer a viable option: the industry is now the target of regulation by the Chinese government, eager to clean up its own environment.

Some waste in the Philippines is reprocessed locally. But people living in remote, rural areas have a stark choice. Either they bury their plastic waste locally, burn it, or come up with innovative solutions to repurpose the material instead. Given the unpalatable nature of the first two options, the country now has an innovative, artisanal plastic craft movement. A few years ago, I worked with craftspeople and artists to put together an exhibition of repurposed plastic waste. The items that really stood out were those things that local people made according to traditional patterns. These were items people used for cultural events that marked their identity as tribal Filipinos. Much of the work of experimentation with the materials was happening in kitchens and workplaces as people shared techniques and tips with each other.

One of our contributing craftspeople, Ikkay, lives in a remote tribal village in Kalinga Province. She makes strands of plastic beads out of bits of waste plastic, using old CD cases and fast food spoons — anything with a bit of gloss. Her beads are replicas of traditional tribal designs. These beads are used in local cultural performances and shipped all over the world for demonstrations of Filipino dance.

Another example of creative re-purposing came from the nearby gold mines. There, mine workers weave the yellow, red and pink plastic wrappers of blasting cap detonators into traditional basket forms. They find the high-quality wrappers just as good as the rattan they'd originally used. Burly mineworkers walking the roads with dainty pink-and-yellow plastic backpacks have become a frequent site in and around mining communities.

In both these examples — beads and backpacks — people had figured out innovative ways to re-purpose waste plastic. They took material that would have been garbage and turned it into an item that expressed important cultural values, making something cool, fun, and desirable.

The material itself — waste plastic — has something to do with this. In these Filipino cases, it made tribal displays of beads extra impressive. The makers and wearers of these items were not just subverting ideas about waste, but about social hierarchies and the social power to innovate and set trends. People who admired the plastic craft started to see plastic no longer as just "stuff" on its way to being garbage. Instead, it became imbued with the potential to become something new and different and a way of asserting local ingenuity and identity in a global world.

Flip-flops and whales

These examples from the Philippines are not isolated ones. Around the world, there are many small-scale groups doing very similar kinds of projects with re-purposed plastic waste. Rehash Trash in Phnom Penh, Cambodia, makes baskets out of used plastic bags. Ocean Sole in Kenya works with discarded plastic flip-flops to make art. Other initiatives are networks, like that created by Precious Plastic a 3D printing initiative from Amsterdam. They have groups all over the world building 3D printers to create re-purposed plastic items for local markets.

In the UK, people have used fine art to communicate the urgency of the issue to the wider public, whether it's a giant plastic whale touring the UK or the artist Stuart Haygarth's collection of plastic waste from UK beaches, hanging in University College Hospital London. But pointing out the problem is only the first step in providing comprehensive solutions. We need to reduce our plastics footprint, seek out items made from recycled materials, and — most importantly — learn, hands on, about new ways to reuse what is now a ubiquitous class of waste materials.

People are meeting the challenges plastic waste poses in creative ways. While the end goal will be to phase out the materials creating the bulk of the pollution, in the meantime we must improve the capture of recycling systems globally and make locally recycled and re-purposed materials more desirable and acceptable. Learning to love plastic — wisely — means taking on the responsibility for our own discarded items.

We could take up the examples of these local innovation workshops and 3D printing groups to get making in communities worldwide. While making new items won't halt the consumption of plastics altogether, it will divert material from the waste stream while helping people to see its potential. Offering the public a chance to co-create part of the solution should make the inevitable regulatory responses — deposits, disposal taxes and more rigorous waste sorting — more acceptable.

After all, waste plastic is still plastic and amazing stuff — you can make it into pretty much anything you can imagine.

Source: McKay (2017). Originally published on *The Conversation*.

CRITICAL THINKING

In 2018, China banned the importation of Australian waste for recycling. This was followed by India banning plastic waste imports in 2019. What new strategies should we put in place to better manage our use and re-use of plastics in Australia and New Zealand?

Socio–ecological thinking

Increasingly engineers and other professionals are expected to take a broader view of the problems they face. Whereas engineers once worried mostly about technical (can it be made to work?) and economic (is it affordable?) issues, more recently environmental issues (e.g. what are the ecosystem effects?) have become just as important. Conversation mapping is a tool that helps open up the complex ways in which problems are seen within our society; it supports **socio–ecological thinking** as an approach to problem solving.

A socio–ecological approach recognises human and social systems as embedded in natural (ecological) systems. Consider that when we produce goods and services for society, there are negative environmental and social impacts attached to these activities — such as water, soil and air pollution — which it would be ideal to reduce or eliminate altogether! This is increasingly being referred to as **decoupling** economic growth from negative environmental pressures, as shown in figure 10.4 (Smith, Hargroves & Desha 2010). The term comes from economics, but engineers can appreciate it for its technical analogy too — think of a car travelling up a hill, hauling a trailer that is *coupled* to the car via the tow bar. If we decouple the trailer, the car will accelerate up the hill, unhindered by the drag from the unintended side effects (the trailer).

FIGURE 10.4 Diagram showing the curve for decoupling economic growth from negative environmental pressure

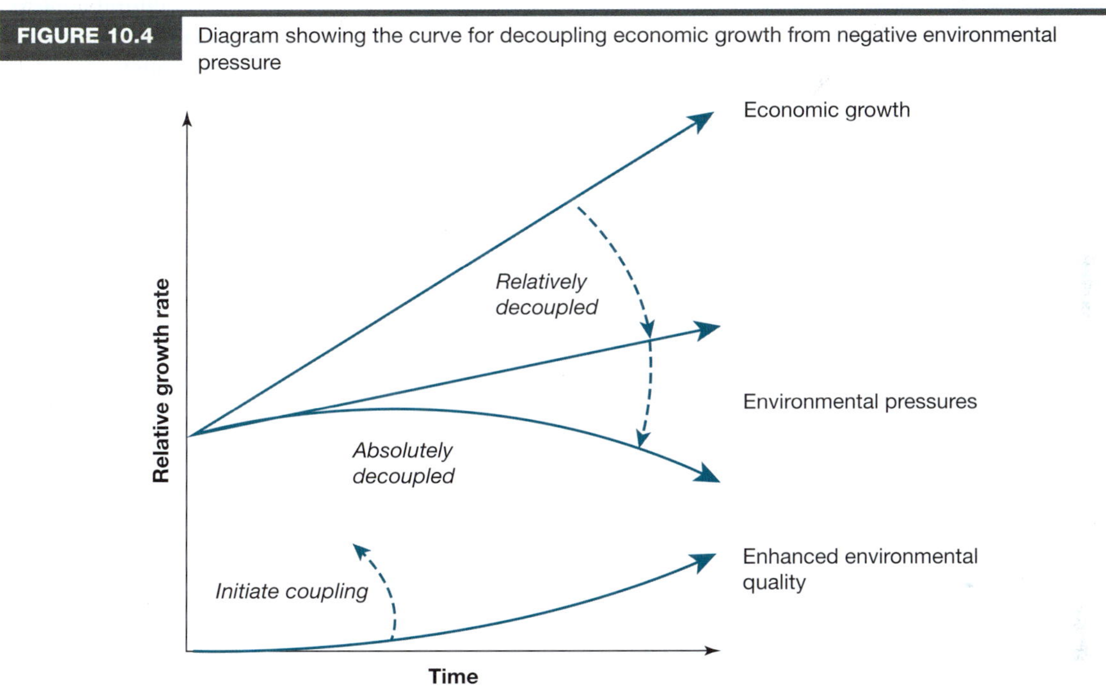

Source: Smith, Hargroves and Desha (2010).

Imagine that the trailer represents all of the negative impacts (like air, soil and water pollution) being decoupled from the car, which represents economic growth. If this could happen — if we decouple the negative pressures from producing goods and services — we should be left with economic growth that progresses our quality of life while not harming the planet in the process.

Examples of this kind of decoupling can already be seen around the world, including the removal of ozone-depleting CFCs from aerosols and lead from petrol in the 1990s (two of Thomas Midgley's damaging inventions — see the spotlight in the chapter on sustainable engineering). Similarly, the building material asbestos was widely used until the 1980s but was not banned until 2003. Clearly, each of these 'decoupling' examples has been successful, without the economy suffering collapse, although we are still paying a price for the problems caused by these materials.

However, as we discussed in the chapter on sustainable engineering, several of our planetary systems are already quite degraded, with unprecedented concentrations of greenhouse gases in the atmosphere, polluted and acidic ocean waters, and depleted biodiversity. Even if we manage to decouple all of the negative environmental pressures from our development in the near future, we will still be living on a planet where things are significantly out of balance. Unless we can also remediate and replenish the impacts of the past, society will continue to experience the effects of deteriorating systems. By adopting *restorative* solutions, we can begin to get the planet back to a stable state.

Thinking more broadly about development, if we can 'recouple' positive environmental and social aspects to producing goods and services, we can contribute to restoring natural systems and communities, many of which have been substantially degraded over the last two centuries. For example, if we use permeable pavements instead of concrete driveways in a residential subdivision, we can increase the amount of rainfall runoff being returned to underground aquifers, helping to restore groundwater reserves and reduce local flooding. We might also use a roads project to increase pedestrian safety in a previously dangerous pedestrian zone.

The Community Health Research Unit (2009), at the University of Ottawa, recommends useful questions for getting started with this approach to considering environmental and social issues.

- What factors are contributing to or causing the problem?
- How are these factors linked?
- How do different stakeholders see the issue?
- What are the local priorities, needs, resources and structures for implementation?
- What system levels, partners, sectors and jurisdictions should be considered and/or involved in planning?

By using a socio–ecological perspective, engineers can become aware of the interdependence and interrelationships existing between themselves, their work, others and the natural environment. This allows engineers to understand how broader physical, social, political, economic, ethical, and cultural contexts and histories influence the ways in which they (and other people) make meaning out of their observations and experiences. Some useful questions that support a socio–ecological approach are as follows.

- What are some of the major trends that are affecting this problem?
- What are the effects of these trends in the problem (system)?
- How is the system currently responding to some of these trends?
- What does this mean for the people in the system? (What is the effect on them? How are they responding?)
- What does this mean for younger people such as university students?
- What do young people need to do, or can they do, about this issue?

The following strategy has been adapted from a document published by the New Zealand Ministry of Education (2023). Originally written in a health education context, the five points help engineers realign their priorities from technology towards the wellbeing of members of society in the global environmental context. With a socio–ecological approach, engineers:

1. research and analyse the wellbeing needs of groups, communities and society
2. identify ways to meet these needs
3. plan appropriate action
4. take action, either individually or collectively
5. reflect on the actions they have taken and evaluate the effectiveness of these actions.

Whole system design goals

An important aim is to define the goals of the system. Later on, these will be used as selection criteria for judging possible solutions. For the freeway problem, possible goals may include having reduced stress, travel time, air pollution, energy consumption and noise.

Generally, a socio–ecological perspective shifts engineers' thinking from making technology work, to managing complex societal and environmental problems in a 'whole system' approach. Here, the goals of the system shift from making the technology work to the societal problem that is being addressed, within the global environmental context. Typical systemic problems addressed by engineers include providing access to water, food and shelter; reducing demand through efficiency measures; and providing solutions for the supply of energy, transport, communication and resources.

In the process, there are opportunities for 'step-change' rather than incremental improvements. For example, rather than 1–2 per cent improvements in energy efficiency, a whole system approach might achieve 80 per cent or **factor 5** improvements, where 5 times less energy is used for the same purpose, or 5 times more can be done with the same energy (Stasinopoulos et al. 2008).

Consider planet Earth, as seen from space. The whole of the Earth is a closed and whole system, bounded by space. Input to this system is primarily energy from the sun. This underlines the importance of protecting the viability of the Earth given its isolation and our current lack of an alternative planet to inhabit! It is our most important 'whole system' — the one upon which up to 10 billion human beings will depend by mid-century. It is also clear that engineers have had a profound effect on the Earth, particularly in the last two hundred years, starting with the Industrial Revolution in the late 1700s. The evolution of human development since this time can be neatly summarised as a series of 'waves of innovation' (see figure 10.5). We might even add an emerging seventh wave of artificial intelligence!

FIGURE 10.5 Waves of innovation since the Industrial Revolution

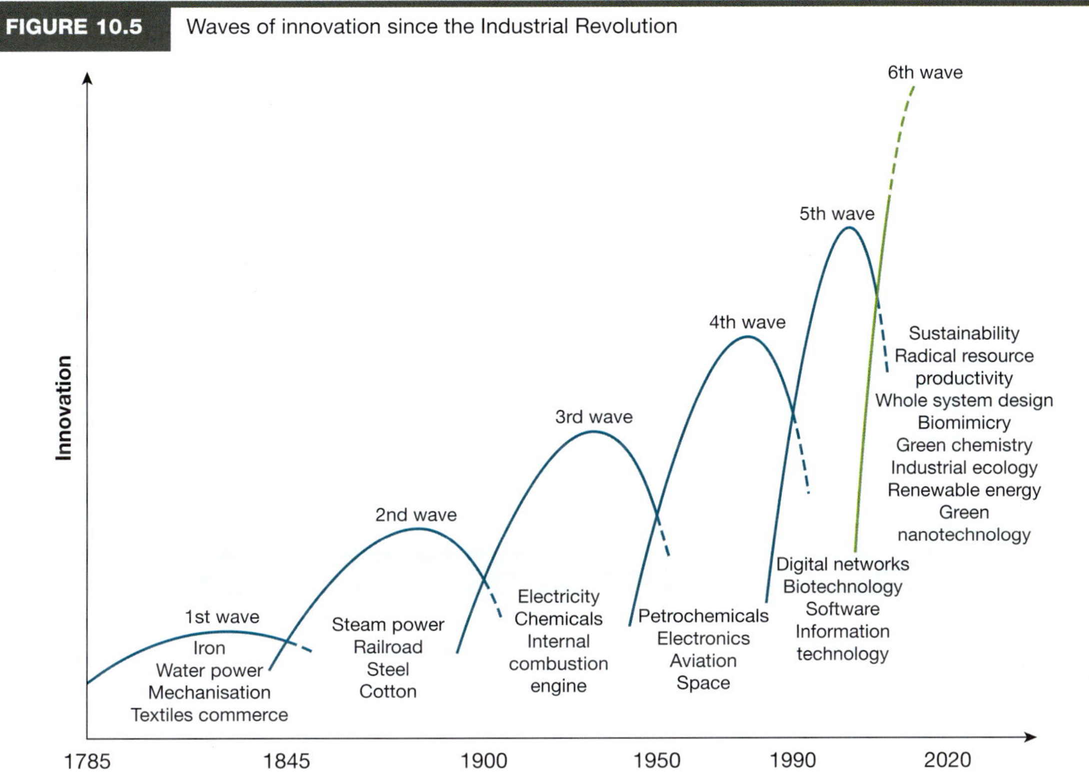

Source: Hargroves and Smith (2005, p. 17).

When viewed in this way, it is easier to see the history of engineering as serving the wellbeing needs of society. Early engineers grappled with housing, water, security, energy and transport; they harnessed machine labour through water and steam power through to petrochemicals. Our current state is riding the wave of digital technology — the bringing together of computer and communication technology, providing a raft of new entertainment and educational experiences, with an ability to connect globally like never before through social networking sites. Artificial intelligence is the latest digital technology to emerge.

Reflecting on the last century in particular, the former president of the American Society of Civil Engineers, Dr Dan Turner, outlined the contribution of engineering to the extension of the US lifespan, which has been extended by 40 years over the past 100 years. Dr Turner reported that of these 40 years, 37 years were due to the advances made by engineering, with the remaining 3 years due to advances in medicine. Dr Turner (in Engineers Australia 2005) outlines that:

> [t]hese advances to public wellbeing through good engineering were achieved by the provision of clean water (civil engineering), the removal of wastes (civil and resource engineering), the upgrading of human living space (civil, mechanical and electrical engineering) and the upgrading of food supplies (agricultural, chemical and electrical engineering).

However, many of the achievements of the twentieth century have resulted in harmful impacts that have recently become evident to us, including holes in the ozone layer, climate change, desertification and ocean acidification. In the twentieth century, concepts such as 'waste' and 'pollution' were taken for granted as consequences of development. What makes the sixth wave unique is our current challenge to continue to progress development while both reducing the environmental impacts of our current actions and addressing the negative environmental and social pressures incurred from our past actions.

This will most certainly require a whole-of-system approach to design solutions to achieve the decoupling of negative pressures and the recoupling of positive aspects discussed earlier in this chapter. Perhaps the ultimate design challenge is to ensure that the global population (currently 8 billion and growing) can live in symbiosis (a kind of steady-state dynamic equilibrium, for those of us who enjoy thermodynamics!) with other systems on this planet, now and forever. The sixth wave is thus emerging as being related to opportunities surrounding sustainability, as businesses are growing in areas such as renewable energy, industrial ecology, biotechnology and lifestyle services.

The following spotlight looks at an emerging sustainable building rating system. It sets tough challenges that will lead to ethical, healthy and more beautiful environments for us to live and work in.

SPOTLIGHT

Living Building Challenge

The Living Building Challenge (LBC) is a building certification program that looks at the whole system when designing and constructing a building. In contrast to Green Star and the US equivalent LEED, LBC considers not just the building design and construction but also the interaction between people, the building and the Earth. It is a holistic philosophy that divides the system into seven petals that together make up the whole flower. Each petal represents a performance category or imperative.

As can be seen from table 10.2, the LBC combines technological challenges for sustainable buildings, such as energy efficiency and renewable energy production, with societal and economic benefits, such as encouraging local industry. Its holistic approach is indeed a challenge for the building designer and developer. The aim is to make buildings regenerative rather than merely sustainable. The Living Future Institute Australia was established in 2012 and the first building went through certification beginning in 2015 (Sustainable Buildings Research Centre 2024).

TABLE 10.2 Living Building Challenge imperatives

Petals	Imperatives
Place	The building must not use greenfield sites. It should promote urban agriculture and habitat exchange, protect wetlands and encourage car free living.
Water	The building must harvest rainwater so as to be Net Positive Water, delivering 105 per cent of its needs.
Energy	An LBC building must be Net Positive Energy. It must reduce energy consumption and generate enough renewable energy to cover 105 per cent of its needs.
Health and happiness	The building must promote wellbeing and provide a civilised environment with healthy air and encourage biophilia (the innate affinity for nature).
Materials	LBC has a Red List of banned materials including lead, PVC, formaldehyde and most ecotoxins. The design and material procurement must minimise the embodied carbon footprint, encourage conservation and reuse, and through appropriate sourcing (source locally) deliver responsible industry. The building must meet the Net Positive Waste imperative.
Equity	The development must be at a human scale and create human places. Social justice and democracy must be evident and the rights to nature must be preserved.
Beauty	Beauty and spirit combine with inspiration and education to create an aesthetically wonderful place.

Source: Adapted from International Living Future Institute (2022).

CRITICAL THINKING

Of these living building imperatives, which one would you consider the most difficult to achieve? How would you go about delivering it?

Whole system design elements

The advantage of systems thinking is that it encourages us to expand our view to recognise the real problem we are trying to solve. This is the basis of **whole system design**. The engineering challenges discussed in the LBC spotlight provide a provocative framework for all engineering disciplines. Each of these complex problems needs to be understood in a system-oriented way, which demonstrates that if we are to make improvement to such problems, an interdisciplinary approach is required. This is discussed in more detail in the chapter on sustainable engineering.

There are essentially ten key elements of whole system design, which build on what has already been described earlier in this chapter.

1. Ask the right questions.
2. Benchmark against the optimal system.
3. Design and optimise the whole system.
4. Account for all measurable impacts.
5. Design and optimise the subsystems in the right sequence.
6. Design and optimise subsystems to achieve compounding resource savings.
7. Review the system for potential improvements.
8. Model the system.
9. Track technological innovation.
10. Design to create future options.

In whole system design, other key language that you will see in the literature includes 'whole systems thinking', 'solution multiplier effect', 'win–win–win' solutions — basically, a shift in thinking from reducing 'bad stuff' to doing 'good stuff'. The idea is to aim for solutions that provide factors of improvement in system performance and that provide a win for all stakeholders. Some additional resources include The Natural Edge Project (Stasinopoulos et al. 2008), Factor 5 (von Weizsäcker et al. 2009) and the Factor 10 Engineering Initiative (Rocky Mountain Institute 2019). With this in mind, let's consider the ten elements in a little more detail.

1. Ask the right questions

Earlier in the chapter, the example of moving bricks showed the value of asking the right questions at the start. By understanding the objectives and constraints on a problem, we can more quickly understand what the problem *is*. A systems view emerges that allows us to concentrate on what we need to pay attention to. So, a good designer knows how to ask good questions.

Asking the right questions at the right time is the primary strategy for acquiring a deep understanding of the system and eliminating costly late modifications. Three primary questions are critical for embedding a sustainability focus into the design specifications.

What is the required service?

A service is an intangible version of a good. Service provision engages goods and resources but does not necessarily consume any resource other than time. It is important to consider the number of times that the service is required during the system's operating life. Designing for too few services means the system could fail prematurely, and designing for too many services means that more resources than necessary are engaged and lost. Thinking in terms of service (pull) rather than product (push) attenuates the temptation to engage resources unnecessarily.

An example of this is SafeChem, which provides solvents through a closed-loop leasing system, wherein all solvents delivered to the customer are recovered, recycled and reused. The company's calculations suggest that the enhanced model can reduce solvent consumption in already low-emission plants by 40 to 93 per cent (SafeChem 2019).

What is the optimal service?

A single service can be provided by any number of systems. The optimal service delivers the most benefits for the least cost, while providing the required service. Throughout its lifecycle, a theoretically optimal system will minimise the degradation of resources, maximise the restoration of resources and the environment, optimise the service life, be cost-effective and provide social fulfilment.

What are the system's operating conditions?

There are two key considerations of a system's operational life. First, a system always operates within a larger system, so it is important to account for external interactions. Second, most systems are required to operate at several different loads and in several different environments, so it is important to optimise the system for the most common operating conditions, while still designing the system to be reliable at maximum load.

The second consideration is often overlooked, for example, in many air-conditioners. To ensure reliability, air-conditioners are designed and optimised to operate at full load on extremely hot days; however, air-conditioners are more often operated at part load on warm days.

2. Benchmark against the optimal system

Most design activity is intended to improve an existing system or is an example of existing design. This means that there are many examples from which to draw inspiration. It is common practice to seek out exemplary solutions as benchmarks, note their key design parameters and determine ways to improve their designs. By taking this approach, however, we may unintentionally build many of the limitations of those benchmarks into our design. An alternative approach is to benchmark against what the laws of physics will allow.

Benchmark targets are embedded in the system specifications. It is important to benchmark against the optimal system, not merely against the best existing system. Most existing systems are sub-optimal and, thus, benchmarking against them can introduce arbitrary constraints that restrict the solution space. In addition, setting ambitious targets encourages breaking away from cultural norms to explore new opportunities for service provision and system development. Benchmark targets are generally developed in a two-stage process (although minor updates are usually made throughout the process as the understanding of the system improves).

1. Initial benchmark targets are developed as part of the design specification. These targets reflect *theoretical optimal service provision* as determined by Element 1: *Ask the right questions* and are largely qualitative.
2. Once the system's function is better understood, the benchmark targets are reviewed. At this stage, *practical constraints* are superimposed onto the theoretical targets to determine the practical targets, many of which are quantitative.

An example of this is the theoretically optimal layout for a pipe system that connects two fluid reservoirs is a single, straight pipe (since there aren't any bends or components that introduce head (pressure) loss through friction). However, this layout is usually impractical for several reasons.

- There may be solid obstructions in the path directly between the reservoirs.
- Some pipe locations may be a safety hazard and inconvenience.
- The pipe may not be available in the required length made from the appropriate material.
- Some type of valve is often required to control fluid flow.
- Using long straight pipes can result in the fluid flow being excessively noisy.
- Placing the pipeline above ground will facilitate maintenance but may be visually unattractive.

3. Design and optimise the whole system

Having identified an optimal system as a starting point, think about the whole system that you are designing. Consider an engineering process that produces low-grade waste heat at some point in the process — such heat is often vented to the atmosphere, sometimes using large amounts of cooling water, which is subsequently lost.

In the past, when water was seen as a limitless resource of very little value, this was considered the optimal solution — or, at least, optimal in cost terms. These days, it is necessary to consider a wider view of what is optimal. This waste heat may be useable by another process as preheating; alternatively, it might be sold to another company to use for heating. In Europe, it is common to pipe such waste heat around the town, providing homes with low-cost space heating. This is the essence of industrial ecology, where the waste from one company becomes the resource for another (discussed in the chapter on sustainable engineering).

Many systems and technologies are believed to be so complicated and refined that only incremental improvements are possible. However, a systems approach to design reveals that the more complicated a technology is, the more opportunities there are for improvements. These opportunities can be captured by considering three opportunities.

1. *Clean sheet design.* The system's development is facilitated by the process of developing a system from only a set of requirements and a 'clean sheet of paper'. This enables the system to be designed and optimised as a whole and, consequently, compromise can be minimised.
2. *Designing the system as a whole.* Placing a system-level emphasis on selecting and integrating subsystems, allowing synergies between subsystems to be identified and optimised. System-level design is assisted by interdisciplinary development teams that have expertise in a broad range of relevant technology and design concepts.
3. *Optimising the system as a whole.* Comparing subsystem modifications against changes in both system service provision and subsystem functionality (usually an iterative process). Comparing against the system and subsystem levels ensures that both the interactions and subsystems are optimised for the

benefit of the system. In the case of a contradiction, optimising system service provision is more important than optimising the subsystem functionality.

4. Account for all measurable impacts

Industrial ecology is a good example of being careful to account for all impacts. It is easy to take the resources of the planet for granted, particularly the atmosphere and the ocean. Engineers in the past have assumed that these are infinite sinks into which we can discharge pollutants of various kinds, but it has become increasingly clear over the last 50 years that this cannot continue. We need to account for all the impacts of our activities.

A defining feature of a system is the fact that making a single modification will create at least one other impact beyond that modification. This feature can be leveraged to assist in developing an optimal system in a short period of time; this leverage is created by designing and optimising for the most positive impacts. The aim is to optimise as much of the system as possible, not just a particular subsystem, with each decision. There are two general types of impacts to consider in determining a decision's net value.

Impacts through synergies

These occur internally on subsystems other than the subsystem that was acted on. These impacts can manifest throughout the life of the system, but not always automatically; thus, further action on the subsystems of impact may be required to optimise the full impact of the original action.

An example of this is RLX's blade server, invented by Chris Hipp and David Kerkeby in 2001. It was an energy efficient stripped down computer processor that generated relatively little heat. Therefore, it did not require auxiliary internal cooling through heat sinks and cooling fans, nor did it require as much external cooling though air-conditioning as did conventional servers. Fewer cooling components made the server smaller and lighter than conventional servers, so more servers could be housed in a single rack. It formed the basis for supercomputing in small spaces (Hipp & Kirkeby 2002; Los Alamos 2001).

Hidden impacts

Hidden impacts occur externally as a result of transforming resources into the state that they are used to create the system, including transportation. These impacts are relevant to input resources for both production and transportation. Hidden impacts can be measured using appropriate ecological indicators, such as ecological footprint (Rees 1996), material input per unit service (MIPS) (Schmidt-Bleek 1993, 1999), and *emergy*, which quantifies the equivalence of different forms of energy (Odum 1995).

EnviroGLAS® create hard surfaces such as floors, countertops and landscaping materials that contain about 75 per cent recycled glass, which consumes 4.7 times less abiotic material and 1.9 times less water than the production of virgin glass (EnviroGLAS 2024). EnviroGLAS materials are sourced from local recycling programs, so fuel consumption, gas emissions and human labour costs are low.

Another example is how, in mild climates, low-energy LED lights have more of an impact on increased demand for heating than on the amount of energy saved on lighting. And dark coloured finishes can cause more lighting to be installed, which causes more cooling load and/or more ventilation load, requiring increased fan power or duct sizing. This leads to suspended ceilings being perhaps a little lower or columns a little longer and buildings a little higher (N Howard 2014, pers. comm.).

Designing for multiple services

Often, the most positive impacts can be achieved by designing for multiple services, which reduces the demand for resources by enabling a single system to perform the services of multiple systems (hence, making those other systems redundant). In the cases where more costly advanced technologies are required to provide multiple services, the additional capital economic cost can be at least partly offset by the reduced demand for other systems and resources. Designing for multiple services is applicable at both the system and subsystem level.

An example of this at the system-level is building integrated photovoltaics (BIPV). H.Glass, now marketed as Sun Station (Freshape 2024), is a building material that acts as a coloured glazed window and solar panel. It uses a dye-sensitised solar cell incorporating vegetable dyes. The advent of translucent and transparent photovoltaics creates enormous potential for electricity generation from building façades.

Another promising technology uses halide perovskites to generate electricity from the non-visible wavelengths of sunlight, allowing the visible light through (Liu et al. 2018).

Tractile is an Australian company that has developed BIPV in the form of a roof tile that incorporates the PV panel and a solar hot water system. The sun on the tiles heats the water, which in turn cools the

PV panel (Tractile 2024). The Tesla solar roof (2024) is another example of integrated solar panel plus tile. Rather than having to retrofit solar panels to a roof, they can be integrated into the actual roof.

An example at the subsystem level is Fisher and Paykel's SmartDrive top-load washing machine, which uses a brushless direct current motor. This eliminates the need for belts, brakes, pulleys, clutches and gearboxes, making the machine lighter, more reliable and more energy efficient than brush-motor washers (Fisher & Paykel 2019).

5. Design and optimise the subsystems in the right sequence

It is important to recognise those parts of the system that have the greatest impact on the design. Which part of the system has the largest impact on the consumption of resources? This is the subsystem to optimise first. In a car, the engine consumes all of the fuel, but in order to reduce the car's fuel consumption, it is more effective to begin by reducing the weight of the body and reducing the drag of the shell and undercarriage.

While a systems approach emphasises system-level design and optimisation, there is a clear role for subsystem-level design and optimisation. The management of subsystem-level design and optimisation is determined by the system synergies. Consistent with *Element 4: Account for all measurable impacts*, the decision to design or optimise a certain subsystem at a given time in the development process depends on its potential impact. Decisions that have the most positive impact are favoured. An extension of this logic is that there is a single subsystem (or small number of subsystems) that are best designed or optimised first, and there is also a logical sequence (or set of parallel sequences) for designing or optimising subsystems that will yield the optimal system with minimal effort and human resource cost. The sequence is generally non-linear, and usually iterative.

The cost of a rooftop solar photovoltaic (PV) system to power a home refrigerator can be minimised by firstly making users aware of simple practices that reduce cooling load, such as not having doors open unnecessarily and not obstructing the internal air ducts. Then, the refrigerator's energy efficiency can be improved by cleaning and repairing components, or even replacing the refrigerator with a smaller and more efficient model. Finally, the PV system should be sized correctly.

6. Design and optimise subsystems to achieve compounding resource savings

It is likely that the interrelationships between subsystems mean that there are gains to be made by optimising subsystems together. Many systems have subsystem synergies that resemble a distinct 'path' originating at a single or small number of subsystems. As discussed in *Element 5: Design and optimise subsystems in the right sequence*, these subsystems usually have the most positive impact and thus are best designed and optimised first. An important observation is that the sequences resulting from applying Element 5 are generally counter to the actual resource transmissions. That is, the subsystem design and optimisation sequences are a set of integrated, general downstream to upstream sequences. The impacts of subsystems in series compound, rather than sum. Thus, it is important to design and optimise subsystems such that the compounded impact is optimised. Compounding impacts can be leveraged to turn several small improvements at the subsystem level into a large positive impact on the system.

A rooftop photovoltaic thermal (PVT) system containing thin film solar cells consumes less energy and fewer material resources than a conventional solar PV system; it also captures thermal energy. In a PVT system, air is channelled beneath the solar panel so that it cools the cell thereby increasing the PV efficiency on hot days. The hot air can then be pumped through a heat exchanger to store heat. In the Solar Decathlon house Illawarra Flame (see the spotlight in the chapter on sustainable engineering), the air is passed over a phase change material that can store the thermal energy for later release in the house. A phase change material is one that melts or evaporates at operating temperatures. The latent heat of the phase change can be used for cooling. The phase change material in Illawarra Flame utilises almost 800 kilograms of a special salt that melts at 22 °C. One sunny winter's day can create enough heat in the thermal store for three days heating of the house (Sohel et al. 2014; University of Wollongong 2014).

7. Review the system for potential improvements

Having made a first pass through the design and optimisation process, consider the whole system again. Are there further gains to be made? Design is fundamentally an iterative and complex process. Once one design has been produced, it becomes easier to see new opportunities. This often happens when consulting with clients and stakeholders — once they see the nature of the design, they suggest other changes and improvements.

It is rare for an optimal system to be developed solely because of the original design decisions. Typically, additional review and investigation, especially into the finer decisions, uncovers opportunities for improving the system towards the benchmark targets and assists in identifying performance inadequacies that were originally overlooked. Reviewing the system can be performed at both the system-level and the subsystem-level.

System level

Review at the system level can be performed by monitoring the system output and comparing the results to the benchmark targets, with discrepancies indicating opportunities for improvement. Monitoring involves any activity that accurately informs of the system's state, real-time function, performance and environmental impact.

The Melbourne Central office building provides a good example of review at the system level (Pears 2004). The building was considered by its engineer to be relatively energy efficient at a rating of 3.5 to 4 stars, but an energy assessment rated it at only 2 stars. Monitoring and analysis of the building's electricity demand showed that its after hours energy consumption was abnormally high, and subsequent investigations and repairs have lifted the building's rating to that expected.

Subsystem level

While review at the subsystem level identifies typically smaller opportunities for improvements than at the system level, there are multiple subsystems, so the total opportunity can be substantial. Furthermore, as discussed in *Element 6: Design and optimise subsystems for compounding resource efficiency*, several improvements applied in the right sequence will compound through synergies (rather than sum) into a large improvement.

Lee Eng Lock is an engineer and technical director at Trane, Singapore. In 2012 he received the Champion of Energy Efficiency Award from the American Council for an Energy-Efficient Economy (ACEEE) for his work that resulted in an 80 per cent reduction in the overall energy consumption of Trane heating ventilation and cooling equipment. He achieved this by fine-tuning the most energy intensive subsystem, the chiller, while considering many of the less intensive subsystems such as supply fans, pumps and cooling towers. Through synergies in the system, improvements in one subsystem led to reduced demand on other subsystems, thereby compounding the benefits (American Council for an Energy-Efficient Economy 2024).

8. Model the system

Modelling is a fundamental aspect of engineering. At university, much of your time is spent learning new ways of modelling systems. This begins with subjects such as engineering mechanics, electrical networks, fluid mechanics, thermodynamics and computer programming. In later years, it involves control theory and other complex ideas. The chapter on evaluating options also considers economic modelling, which is an essential element of optimal design.

Models are tools that assist in understanding a system's function and behaviour. They are especially valuable for complex systems, incorporating a variety of mathematical, computer and physical options. Improved understanding from modelling can assist in both benchmarking and optimisation and can also be used to verify the system's function when physical verification is destructive or cost-prohibitive.

- *Benchmarking.* In developing benchmark targets, models can be used to determine a system's theoretically optimal environmental impact and its expected environmental impact after practical constraints are superimposed.
- *Optimisation.* In optimising a system, models can be used to inform decisions, particularly at the subsystem level where impacts may appear insignificant when considered individually, but can have a large impact collectively when compounded.

An example of this is the rotated arc mixer (RAM), patented by CSIRO's Guy Metcalfe and Murray Rudman in 2001. The RAM relies on very chaotic mixing to mix highly viscous fluids that were previously considered unmixable (Metcalfe & Rudman 2001). The success of mixing is a function of flow rate, rotation rate and flow aperture location, and these parameters are optimised using mathematical modelling. When the parameters are optimised, the RAM mixes twice as well as a conventional mixer, while consuming only one-fifth of the energy previously used.

9. Track technological innovation

Any engineering design field has new ideas emerging continuously. All engineers need to be abreast of the current developments in their field. This is both easier and harder at the current time — easier because we have access to information technology that connects us with innovation all over the world, and harder because there is now so much information — so we need efficient filtering tools to find the information that matters. Technical societies such as Engineers Australia and Engineering New Zealand continue to provide efficient means of remaining in touch with state-of-the-art developments in our fields of practice.

The rate of innovation in science and technology is rapid and increasing. Thus, it is important to conduct research into the latest technological innovations (such as those in appropriate technologies) and design approaches (such as biomimicry and green chemistry) that can effectively fulfil the service specifications.

Appropriate technologies

Technologies that suit the environmental, social and economic states of the intended operating environment, usually with regards to developing countries, are referred to as appropriate technologies. Innovations in appropriate technologies are enabling developing countries to cost-effectively skip intermediate technologies to use advanced technologies that are more environmentally sustainable.

The Appropriate Technology Collaborative is teaching Indigenous Mayan women in Guatemala about electricity, circuits and solar power. The women are also being trained to create their own businesses making and selling solar-powered LED lights to replace inefficient and polluting kerosene lights. Small scale solar home-energy systems provide better light at a lower cost to kerosene or candles for homes without grid electricity (Appropriate Technology Collaborative 2022).

Biomimicry

Biomimicry applies to systems that mimic the forms, processes and behaviours of biological organisms and systems (Benyus 1997). Innovations through biomimicry tend to consume considerably fewer resources and produce less waste in creatively delivering comparable or better services than conventional systems. Many of these innovations are also environmentally benign or restorative.

FIGURE 10.6 Council House 2 south facade water showers

At the system level, waste treatment systems are available that mimic wetland and soil ecosystems by using a diverse combination of plants, animals, fungi, bacteria, algae and other microbes, as well as rocks and other minerals, in specific ratios in order to remove contaminants from water and air.

At the subsystem level, paints and surface sprays are available that mimic the Lotus-Effect®, the self-cleaning process used by the lotus leaf wherein the leaf's microscopically rough surface limits the contact area for water (Brownell 2006). As a result, the water accumulates into a ball and runs off the surface, carrying dirt with it.

Council House 2 in Melbourne was the first 6-star green building in Australia. Its cooling system is based on the way a termite mound cools itself through thermal mass absorbing the heat of the day and using vents to purge that heat at night, making the mound cool for the start of the next day (City of Melbourne 2024b). Council House 2 also employs the physics of latent heat. The south façade of the building, seen in figure 10.6, uses natural cooling by dropping water 15 metres through 'showers'. The naturally cooled water is then passed over a phase change material that 'freezes' at 15 °C. The freezing extracts more heat from the water, which is then pumped through chill beams to create cool air within the building (Adams 2008).

Green chemistry

Minimal toxic substances are required and generated throughout the green chemistry system life cycle (Anastas & Warner 1998). Innovations through green chemistry tend to exhibit similar benefits as innovations through biomimicry.

At the system level, Yield10 (formerly Metabolix), an agriculture bioscience company, developed technology for the co-production of energy, plastics and chemicals from renewable energy crops (Yield10 2024). Their first commercial product was Mirel (DiGregorio 2009), a family of high-performing, biodegradable and greenhouse gas-neutral plastics that are produced by fermenting natural sugars and oils using microbial bio-factories.

At the subsystem level, Lilly Research Laboratories altered the synthetic process of one of its drugs by using a biocatalyst, which prevented chromium waste and the need for solvents and increased the per cent yield by more than 3.4 times (United States Environmental Protection Agency 2018).

10. Design to create future options

Where possible, consider your design in the framework of the life of the product. Is it possible to make the design adaptable to changes in future conditions? In the design of a mineral-processing plant, will there be changes to the composition of the ore in that time? Does that require more flexible process parameters and alter the equipment design? How will the plant account for increases or decreases in throughput of the plant?

Long-term planning for future adaptations can pay off. The Civic Tower in Sydney had two distinct phases in its existence. The first seven storeys were designed and constructed in 1974 by the architect Joseland Gilling for the Sydney Masonic Society. The Society realised that their property was a prime site in the CBD. They did not have the resources to construct more than the seven storeys. However, they instructed their architect and engineers to design the base to be strong enough to carry a tower with an extra 25 storeys.

Sydney Civic Tower

Thirty years later, architects PTW with new developer, Grocon, and engineers, Connell Mott McDonald, finally realised Gilling's master plan and the tower was completed. This solution avoided the demolition of the original building. In fact, the adaptation gave the seven-storey base its true architectural purpose (McCarthy, Sheikh & Gardner 2008).

Adaptive systems will find multiple uses during their lifetime. Systems that can be modified to prevent obsolescence, or from which resources can be easily recovered with their integrity maintained to then provide other services will have increased longevity. Developing systems with these attributes is facilitated using tools such as backcasting and through an emphasis on design for end-of-life processing.

Backcasting involves designing a 'future system', a system for an envisioned future, by considering desired technological and political states, and then working backwards to develop a system that most closely matches that future system with technologies and policies that are currently available (Holmberg & Robèrt 2000).

Design for end-of-life processing has a large influence on a system's legacy and salvage value. Design and optimisation decisions determine the technical potential and, hence, cost of end-of-life processing for a system. Designing for end-of-life processing involves considering ease of disassembly, ease of cleaning, ease of inspection, ease of component replacement, ease of reassembly, reusable components, modular components, diversity of fasteners and diversity of interfaces (Amezquita et al. 1995). End-of-life processing is also facilitated by physically separating different materials in the system, thereby making the reclamation process simpler. Separating materials, especially biodegradable and non-biodegradable materials, prevents cross-contamination and, hence, complications during end-or-life processing.

An example of this is Hewlett-Packard's office equipment, which is designed for recyclability by:

- using modular designs so components can be easily removed, upgraded or replaced
- eliminating glues and adhesives by using snap-in features
- marking plastic parts to aid materials identification during recycling
- reducing the number and types of materials used

- using single-plastic polymers
- using moulded-in colours and finishes instead of paint, coatings, or plating
- using a particular platform with common parts for several products
- providing a product return and recycling service in 74 countries (Hewlett-Packard 2024).

Can you think of systems where this end-of life processing has been well planned for and systems where it has not?

SPOTLIGHT

Formula E — high-performance electric cars

The first Grand Prix for all-electric cars was held in September 2014. Ten teams took part driving the high-performance Spark-Renault SRT_01E on the streets of Beijing. Designed and built by French engineer, Frédéric Vasseur and his team at Spark Racing Technology, the cars reached a top speed of 170 to 180 kilometres per hour and looked every bit the high-performance racer. The car (known as Gen1) was replaced for the 2018–19 season by the SRT05e, also called the Gen2 (Fédération Internationale de l'Automobile [FIA] 2024) and Gen3 was released in 2022 (FIA 2022).

Electric Gen3 cars at the 2023 São Paulo, Brazil ePrix

According to Vasseur, the biggest design challenge was the battery. The maximum power permitted during practice and qualifying was 200 kW, later revised to 250 kW, and then 350 kW for a top speed of 322 km/h (FIA 2022). Charging the battery is not permitted during qualifying or during the race, although this is under review. The race itself lasts approximately one hour and originally involved one mandatory pitstop when drivers switched to a second car. The introduction of Gen2 eliminated the need for two cars and the mandatory pitstop. Gen3 cars are now all wheel drive, with 250 kW of power available to the front axle and 350 kW at the rear, doubling the power available in Gen2 (FIA 2022).

An interesting feature of the electric safety car used on Grand Prix days is that it recharges its battery wirelessly using a system developed by Qualcomm's Halo Division. This uses inductive charging that allows the car to recharge while standing on a powered pad instead of having to be plugged in. In the same way that you can now charge your smartphone wirelessly you will be able to recharge your car's battery as you drive using the power from the roadway (Huang, Toshiyuki & Hori 2014). Indeed, Professor Yoichi Hori of University of Tokyo's Trans-disciplinary Sciences Division, believes that we may even be able to get rid of the need for batteries by receiving all the power from the road as we drive (Hori 2004).

CRITICAL THINKING

What technologies do you see finding their way from Formula E into mainstream automotive manufacture? What are the implications for infrastructure of powering vehicles from induction loops in the road?

10.3 Generating alternative solutions

LEARNING OBJECTIVE 10.3 Generate design solutions, using research, consultation, creativity and biomimicry principles.

So far this chapter has considered how systems thinking can be used with a socio–ecological perspective to increase the understanding of problems. It must be recognised that problems need to be treated in the context of their environment. How a system is defined depends on how much of the environment is included within its system boundary. As the boundary expands, more solutions become available. This section will discuss methods for finding and creating alternative solutions to problems. A variety of techniques can be used to generate solutions. Deciding on which technique to use will largely depend on the type and size of a particular engineering project.

Consider some of these current engineering challenges: solar energy generation, energy storage, mobile phones, iPads and other tablets, energy use of the internet, web services such as Google maps, Bluetooth, wifi, cloud computing services, the National Broadband Network, hybrid cars, electric cars, digital cameras, the bionic eye. What are some of the design innovations required to address each of these issues? What techniques are available to engineers to help them find solutions to problems or ways of innovating for new products?

KEY POINT

Research may be able to uncover standard solutions to problems. A range of creative thinking techniques may be required to solve more complex or unusual problems. Blue sky research might help solve problems we didn't realise we had.

Five Ws and an H

As outlined in earlier chapters, a good place to start when understanding a problem is to ask relevant questions about the nature of the problem. Six basic questions to start with are *Why? What? When? Where? Who? How?* These questions are known as the 5 Ws and an H.

We can apply the 5 Ws and an H to the problem of the congested freeway by asking the following questions.

- **Why** is the freeway congested? (This will depend on our identification of the system boundary.)
- **What** makes the freeway congested? (Too many cars, accidents, breakdowns, rain, snow, distractions, scheduled maintenance, and so on.)
- **How** does the freeway get congested? How do people drive on it; that is, what are their driving habits? How might this situation be changed; that is, what solutions are available?
- **When** does it get congested? (Time of day, days of week, weeks of the year, seasons.)
- **Where** does it get congested? (Perhaps isolated 'bottlenecks' or congested sections will need to be fixed.)
- **Who** makes it congested? Who can help relieve the congestion? (This identifies stakeholders for consultation. Who is affected by the congestion?)

After asking these questions, research and brainstorming will help pinpoint solutions more clearly.

Research

Research is an important stage in problem solving. We live in an age where information gathering has never been easier. Search engines provide fast access to billions of documents. Researching on the internet and via traditional resources such as books and journals can help us find out what others have done with problems such as ours, as discussed in the previous chapter.

The introduction of generative AI tools, such as ChatGPT, has made research even easier. Instead of getting a set of links from a search engine, these tools can provide a concise summary of the key points or weaknesses to be investigated. For example, you might ask ChatGPT: *what are the key factors in integrating a socio–ecological approach into engineering design?* Just be careful to check its advice!

As we have seen above, the point of this research is to find excellent solutions from *credible* sources. Not all problems have to be solved from first principles. An increasingly common form of research is *crowd sourcing* on social media, for example, posting a Facebook status with a question you need answered. This is good when others have already solved the problem such as finding a good restaurant in a new town. However, the advice needs to be validated. Crowd sourcing is not so good for finding solutions to new or complex problems.

As well as searching online and using print sources, an important research skill for an engineer is to tap into their personal network, particularly within their own organisation. In any engineering organisation, there is much expertise, and tapping into this expertise is important. However, it is also important for an engineer to have done some research first so that time is not wasted asking trivial questions of colleagues. One thing that an engineering colleague can often tell, which is not available in the literature or online, is what the particular organisation they are working for is likely to do; that is, what the preferred technologies or approaches to solving a particular type of problem are likely to be.

SPOTLIGHT

Rethinking timber: a story of long life

Timber is one of the most sustainable building materials, provided it is harvested from a forest that is managed sustainably. The main certification schemes include the Forest Stewardship Council (FSC), the Program for the Endorsement of Forest Certification (PEFC) and the Australian Forestry Standard (AFS). There is also an Australian Standard, AS4708-2013 Sustainable Forest Management. These organisations and the Standard administer the certification of forests, timber and paper products, ensuring that all the principles outlined in the chapter on sustainable engineering are applied from plantation to window frame.

Certified sustainable wood products ensure that tree planting equals, or exceeds, the rate of logging and that the forest ecosystem is preserved. Forests are a vital resource in the battle against build-up of CO_2 in the atmosphere. Timber products are effectively a carbon sink as they lock up the CO_2 absorbed during the life of the tree giving the product a negative carbon footprint.

You would think then that all certified sustainable timber is equal. Or would you? Accsys Technologies is a chemical technology group specialising in the acetylation of wood (Accoya 2021). At their facility in the Netherlands, Accsys treat radiata pine, a wood that grows abundantly in Australia, New Zealand and Chile, using their acetylation process. This uses a vacuum to replace the moisture in the pine with, effectively, a natural vinegar. By increasing the level of 'acetyl' molecules in the wood, the cell structure changes making it as strong as a hardwood. As well as increasing its strength, the process greatly increases the durability of the wood from a Class 4 softwood to an effective Class 1, according to AS5604-2005.

A Class 1 timber has a life expectancy of over 40 years, while a Class 4 will last up to seven years in an exposed above-ground environment. The acetylated wood, called Accoya®, is also termite and fungus resistant. The bugs do not recognise it as food. Indeed, window frames made from Accoya are guaranteed for 50 years. So why isn't everyone using this wonder product? That might come down to availability. The only treatment plant is in the Netherlands. Accsys Technologies import pine from New Zealand and Chile, treat it and sell most of the product in Europe. To get it here in Australia and New Zealand we must import it back from Europe. This doesn't help its carbon footprint with a round-the-world voyage. This also makes it about twice the cost of its main competitor, red cedar.

Team UOW used Accoya for the window frames and doors in the Illawarra Flame house.

Now, let us compare Accoya to the red cedar. Red cedar is also harvested from FSC and PEFC certified forests, mainly in British Columbia, Canada. It has a long journey to make but not nearly as long as Accoya. The cedar is also termite resistant and does not require artificial preservatives. Red cedar results in a high-quality product at an affordable price. Typically, the window frames are guaranteed for a service life of 25 years. For a 50-year building life, you might be replacing the windows after 25 years. Then there is the life of the tree. A red cedar typically takes 250 years to mature into the graceful king of the British Columbian forest. It then has a service life of only 25 years. In terms of land use, in the time taken to grow one cedar tree you can grow ten radiata pine trees. Another way to look at the comparison is that the pine is 20 times more efficient than the cedar because it grows in one-tenth of the time and with acetylation lasts twice as long in the finished product (R Chapman 2014, pers. comm.).

CRITICAL THINKING

What parameters would you consider when choosing a material for window frames in a house? What other materials would you investigate?

Brainstorming

Brainstorming is a simple, group-based technique for creating a list of possible solutions to a problem. The basic rule is to concentrate on naming as many solutions as you can think of without evaluating or criticising them. In this process, do not spend time elaborating or explaining a solution, otherwise the whole process becomes very slow, as it may take many minutes to explain and elaborate on each possible solution. That can be done later. The basic process is to 'name it and write it down', then to move quickly to the next idea. The technique is not the natural way for an engineer to think and so we must consciously stick to its method. With brainstorming, it is better to be creative — encouraging impractical or fanciful solutions can help with generating innovative solutions. The value of whacky ideas is not so much the idea itself but the thought process that idea might trigger in the other participants. This is called synectics or serendipity and is further discussed a little later.

With the freeway scenario, possible solutions include:

- flying cars, such as ski lifts
- multi-level freeways
- covering the freeway with a roof (to avoid weather problems and also collect water)
- 24-hour businesses (to avoid peak hour)
- flexible working hours
- flexible workplaces
- more people in each car
- teleportation devices
- additional freeways
- putting cars on trains into the city
- more bicycles
- increased public transport
- reduce population
- pod-like public transport
- public transport that comes on the minute
- more compact and dense housing
- housing closer to the city
- tunnels
- charging vehicles for entering the city.

There are categories of solutions within this list. For example, there are freeway-specific solutions and solutions involving the vehicles used, alternative vehicles and alternative work definitions. After identifying these categories, attention can be turned towards finding category-specific solutions, such as *social* solutions (e.g. carpooling), *economic* solutions (e.g. tolls or taxes as disincentives), *technological* solutions (e.g. electronic tolling, flying cars, matter transfer beams) and *environmental* solutions (e.g. encouraging bicycle usage and planting more trees to absorb carbon dioxide).

Technical solutions are often more expensive in economic terms and in environmental costs, such as embodied energy. Social solutions, such as getting businesses to encourage telecommuting — even one day per week — are often cheaper; however, they can also be more complex to realise. Nevertheless, such social campaigns have been quite successful with other issues, such as reducing water usage in cities through television marketing and education.

Getting started with brainstorming

It can be quite useful to start the brainstorming session by spending a few minutes of quiet time as each person writes down as many solutions as they can think of. The chair of the session could then ask for one idea from each person in turn. This encourages contributions from all group members and makes everyone feel part of the process.

It is also useful to use a whiteboard so that everyone can see all the ideas at any given time. This helps to trigger new ideas. Record the idea in any way you wish, either as a simple, long list, or as a series of lists in categories.

A variation on this approach is to mind map the ideas on a whiteboard (or on a communal computer screen). This breaks the rule of no evaluation, because it is necessary to think about the idea and where it fits the overall schema of ideas. The advantage is that it can be quite a lightweight form of evaluation and it helps the group to see classes of solutions emerging, which can be further explored with other new ideas and new categories of solutions.

Keep going until the group has exhausted their ideas. You can then use evaluative thinking to eliminate some of the more fanciful ideas, or move them to a special list. Perhaps give an award to the craziest idea.

An important aspect of brainstorming is to explicitly separate expansionary thinking from contractive thinking. The purpose of brainstorming is to provide as many solutions as possible to consider expanding both the thinking and the list of solutions. Later, the list of solutions can be contracted by using logical and critical thinking to evaluate them against goals. These two stages are often described as right-brain thinking and left-brain thinking, as the right side of the brain is associated with creative thinking and the left side with analytical thinking.

You will usually find in any group situation that some members of the group are better at right-brain thinking and others at left-brain thinking. This is another dimension to consider when playing to your strengths. The challenge in effective brainstorming is to quieten the left side of your brain (and the left-brain dominant group members) while the right side gets creative. This is easier said than done. Many engineers are naturally left-brain thinkers and feel embarrassed or uncomfortable when engaged in activities that challenge their nature. The session chair can help get over such reticence.

A method of giving brainstorming a new twist is to reverse the problem. For example, how could the freeway be made more congested? Thinking of answers to this problem, where the situation is reversed, potentially enables new solutions to the original problem to be found.

Lateral thinking, parallel thinking and the six thinking hats

Edward de Bono has written many books on thinking: lateral thinking, parallel thinking, thinking hats, tactics (de Bono 1985, 1994). In fact, the term lateral thinking has become a part of the English language (de Bono n.d.). Lateral means sideways, so lateral thinking suggests thinking sideways, to think differently about a problem. Brainstorming is one way of doing this. By trying to name or invent a range of possible solutions, the mind is triggered into thinking differently — that is, thinking sideways — about a problem.

In parallel thinking, de Bono addresses some issues of real-world problem solving, such as those that are discussed in this chapter (de Bono 1994). According to de Bono, much thinking has been based on the 'gang of three' — Socrates, Plato and Aristotle — whose collective thinking style emphasised the search for truth.

However, much real-world problem solving is about seeking answers to complex problems. In these situations, rather than debate or argue, de Bono suggests that all stakeholders be encouraged to think in parallel with each other. For example, the whole group may be seeking solutions to a contemporary problem, such as climate change. Each stakeholder will likely have different views, different priorities and different solutions; however, they are all working towards a similar goal.

Contradictions and clashes can be named and documented in the conversation-mapping process and worked on through the problem-solving process. The aim is to have all stakeholders working constructively in parallel, benefiting from the diverse set of viewpoints that are assembled.

Another de Bono technique that encourages parallel thinking, is the six thinking hats (de Bono 1985). The metaphor of a hat relates hats to roles as in the expression, 'now, with my other hat on . . .' Each of the hats in the de Bono six thinking hats technique has its own colour representative of its function, as outlined in figure 10.7.

These six thinking hats can be applied to problem solving and design. For a new problem, information should be gathered about the problem (white hat). The next phase is to think about what solutions are available (green hat). When evaluating these solutions, the positive aspects of each solution can be identified (yellow hat), as well as feelings about them (red hat) and any cautions or negative aspects (black hat). At key stages, the progress of the whole process should be monitored (blue hat).

FIGURE 10.7 Edward de Bono's six hats of creative thinking

White represents white paper for gathering information

Hunches, feelings and emotions are symbolised by red, the heart colour

Caution and evaluation — black being the colour of a judge's robes

Optimism, positive aspects — yellow is for sunny optimism

Creating ideas and alternatives — green is the colour of life and creation

Controlling the sequence of thinking — blue is for blue sky — being high above the situation; to have the bird's eye view

SPOTLIGHT

Qantas Q Bag Tag

Have you ever stood at the airport behind a hundred other people to check in your bag (likely been running late) and thought 'There must be a better way!'? Qantas (2019a) introduced a solution: a permanent bag tag. No more paper tags will be required. Once you're checked in, just drop your bag on a belt and go! Your bag will magically find its way to your destination.

The Q Bag Tag is a permanent baggage tag that lets passengers check their baggage at domestic airports without attaching temporary baggage tags. Designed by Marc Newson, each tag contains world-first technology that synchronises details on a boarding pass or Qantas card with baggage. In addition, passengers can check in online or with their mobile prior to arriving at the airport, allowing them to go directly to the bag drop (Qantas 2019a). In 2019, the technology was used in major airports across Australia and many regional centres (Qantas 2019a).

Each tag contains radio frequency identification (RFID) that pairs the passenger's boarding pass and baggage at the time of check-in. No personal information is stored, which means that people can lend tags to family members or friends (Qantas 2019b). Scanners around the luggage belt read the tag, link it to check-in details and then route the bag to the right aircraft.

In order to function, an RFID system has three parts (Technovelgy n.d.), consisting of a:

- scanning antenna
- transceiver with a decoder to interpret the data
- transponder — the RFID tag — that has been programmed with information.

The scanning antenna puts out relatively short-range radio-frequency (RF) signals. This RF radiation does two things.

- It provides a means of communicating with the transponder (the RFID tag).
- It provides the RFID tag with the energy to communicate (in the case of passive RFID tags) (Technovelgy n.d.).

Because of the energy provided by the scanning antenna, passive RFID tags do not need batteries and can be used for very long periods of time (maybe decades). Passive RFID tags are 'awoken' by an activation signal when they pass through the field of a scanning antenna (whether this antenna is permanently affixed or portable/handheld). They then send back the information on the chip in the passive RFID tag to the scanning antenna (Technovelgy n.d.).

In addition to passive RFID tags, there are active RFID tags. These active RFID tags require a battery, which allows them to be read from a further distance but means a shorter lifespan than the passive RFID tags (Technovelgy n.d.).

RFID tags have many advantages to other 'read technologies', for example, barcodes, which have an accuracy of only 80–95 per cent (Negroni 2016). Tags can be read in more circumstances and the information can be processed in less than 100 milliseconds. Large numbers of tags can be read at once and the tag can be concealed inside an object, which protects it from wear (Technovelgy n.d.).

CRITICAL THINKING

The Q Bag Tag has been available since 2011 but has had mixed reviews, despite more than a million being in circulation. Check some online travel websites for reviews by users. What extra steps need to be taken to fully establish this technology? Can you think of other applications in your discipline that would benefit from RFID tags?

Synectics

Synectics means 'bringing forth together' and is a way of systematically generating new ideas (Roukes 1988). It relies on the mind's ability to connect quite dissimilar things — an idea articulated by Buckminster Fuller — and requires brainstorming for creative connections. The technique is based on connecting seemingly disparate things to trigger new ideas and, in doing so, mobilising both sides of the brain — the creative right side and the reasoning left side (Gordon 1961). Various trigger words are used to stimulate these new ideas, such as:

- subtract, repeat
- combine, add, transfer
- empathise, animate, superimpose, change, scale
- substitute, fragment, isolate, distort, disguise
- contradict, parody, prevaricate, analogise
- hybridise, metamorphose, symbolise
- mythologise, fantasise.

Consider how these trigger words could be used to think about the congested freeway problem (see table 10.3).

TABLE 10.3 **Synectic keyword prompts for creativity**

Keyword	Action	Example
Subtract	Take something away. How does this change the problem?	Take cars off the freeway. Remove emergency lanes. Restrict to certain types of vehicles (e.g. no trucks).
Add	Make the problem bigger — expand the system boundary.	Add other transport modes such as public transport.
Repeat	Repeat any combination of steps.	Consider repeating any of the steps above and below.
Combine	Combine steps for new arrangements.	Combine any of the examples above and below to find new angles on the problem.
Transfer	Transfer the problem into a new context: geographically, historically.	How do freeways operate in other countries? Are there different design models?
Empathise	Become the subject (problem). How does it feel?	The freeway gets hot and bothered when it is congested. What would it do?
Animate	Turn the problem into a living creature.	If the freeway could move, what could be achieved? Reconfigure lanes perhaps?

Superimpose	Superimpose other views and images, including time.	How does the freeway behave at other times, other days, other seasons?
Change scale	Change the scale to bigger, smaller in space and time.	What if the freeway could only be used by small vehicles or large ones?
Substitute	Substitute for another plan or approach.	What is another use for the freeway?
Fragment	Break the problem into parts.	Think of the freeway as on-and off-ramps, road lanes, signals. How can each component be optimised?
Isolate	Focus on one part at a time.	Consider just the off-ramps.
Distort	Distort into a different shape.	Stretch the freeway or make it much shorter. Do these new views lead to new solutions?
Disguise	Camouflage the problem so it cannot be seen.	Hidden from view, are there new solutions?
Contradict	Contradict the problem's rules, laws and behaviours.	What if there were 'no limits' on the freeway?
Parody	Turn the problem into a joke, parody or cartoon.	The freeway is often referred as the 'south-east carpark'. It is a joke really.
Analogise	Draw comparisons with other similar problems.	If the freeway is like a pipe flowing, what can be done to increase the flow rate or decrease the flow resistance?
Hybridise	Cross-fertilise with another system.	If a freeway could be crossed with a conveyor belt, what would be the result?
Metamorphose	Transform in physical shape.	If the freeway was made of rubber, how could it be reshaped?
Symbolise	Symbols have special meaning — both public and private.	What symbol would be used for the freeway?
Mythologise	Build a myth around the subject.	In ancient times, freeways . . .
Fantasise	Think of preposterous things about the subject.	If only freeways could . . .

The following spotlight features some creative approaches to problem solving and design. It is easy to imagine that many of the techniques outlined for generating alternative solutions in this chapter so far may have been used in the bridge designs profiled next.

SPOTLIGHT

The Gateshead Millennium Bridge

The conventional approach to bridges is to build a straight link across a river — the shortest path between two points. If the section of river requires tall boats to pass, then the bridge must be opened in some way. The most common solution is to raise the central span in one or two parts, such as a drawbridge pivoted at deck level. A counterweight is usually added to reduce the amount of energy required. These mechanisms require the effective input of several kinds of engineers. The easy part is the basic structure. The lifting motors, the power supply, the software control and the safety signals all require input from mechanical, electrical and software engineers.

The Gateshead Millennium Bridge when open, allowing tall boats to pass

The Gateshead Millennium Bridge when closed

The engineers who designed the Gateshead Millennium Bridge in the United Kingdom took a different approach to enable boats to pass by. Instead of splitting in two, with the two parts of the span lifting clear of the shipping channel, the Gateshead Millennium Bridge pivots. The arch acts as a counterweight to the bridge deck, in the same way that a seesaw works (Gateshead Council 2019).

According to the Gateshead Council, the Gateshead Millennium Bridge is designed in the tradition of the great bridges on the River Tyne, with high visual impact and tourist appeal.

This bridge is an example of the synectic principle of *distort* — changing the shape of the bridge from an assumed straight line (by convention), to a graceful curve. This is matched with a *repeat* process, where the supporting arch repeats the shape of the deck as structural support and counterweight. The bridge is also *animated* — it moves to let ships through — and its motion is described as similar to a blinking eye, so this is an example of *analogise*.

The bridge was designed by WilkinsonEyre, who have designed a series of innovative bridges. Another interesting WilkinsonEyre bridge is the Twin Sails Bridge, Poole Harbour's second crossing, which has quite a different approach to an opening bridge, where the deck splits in two diagonally to form a graceful, sculptural display, mimicking duelling racing sailboats. This solution illustrates the synectic principles of *fragment* (split diagonally) and *analogise* (as sailboats). Unfortunately, there have been some continuing problems with the operation of the bridge (Heren 2023).

Twin Sails Bridge, Poole Harbour's Second Crossing Bridge, United Kingdom

CRITICAL THINKING

It is interesting to compare the Gateshead Millennium Bridge and the Twin Sails Bridge designs. What do you see as the advantages and disadvantages of each design? Which do you prefer and why? Research 'bascule' bridges and compare the solutions with the Gateshead and Pool Harbour opening bridges.

TRIZ

TRIZ is an English transliteration of the Russian acronym ТРИЗ for Теория Решения ИзобретателЬских Задач (Theory of Inventive Problem Solving). TRIZ is a generic name for a family of heuristics (tools) for problem analysis, problem reframing, failure analysis and creative problem solving that was conceived in Russia in the 1950s.

TRIZ is grounded in the analysis of thousands of patents that revealed important trends in the development of products. The original six tools of 'classical TRIZ' were developed (by the mid-1980s) specifically for engineering design practice. Today, the TRIZ family contains over 20 tools (including software) ranging from weak to strong heuristics. TRIZ tools are widely used by engineers at multinational corporations like Intel, GE, Philips, Siemens, Bosch, Boeing, Samsung, LG and Cochlear.

40 Innovative Principles

The 40 Innovative Principles of TRIZ (see https://www.triz40.com/aff_Principles_TRIZ.php) are 'solution recipes' that have been applied successfully in thousands of patents in these and other companies. To derive the 40 Innovative Principles, more than 20 000 patents were analysed. The 40 Principles can be used separately, but they yield more focused solutions when used in combination with the Contradiction Table, which lists 39 typical engineering product parameters, such as weight, length, speed, strength, energy consumed, reliability, ease of repair and convenience of use.

Say you are designing an electric scooter and you want to reduce its weight. However, you realise that doing so could compromise the strength of the scooter. So, you are looking to resolve the contradiction between weight and strength.

Using https://www.triz40.com, input the two contradictions: (i) weight of a moving object and (ii) strength. The suggested innovation principles, from the 40 Principles list, are:

1. mechanics substitution
2. cheap short-living objects
3. mechanical vibration
4. composite materials.

Using the idea of *cheap short-living objects*, you might decide to use lightweight materials for the wheels. Unfortunately, these will wear faster, requiring more frequent replacement. This seems not environmentally friendly, so you discount this option.

The last of these seems more promising: *composite materials*. The deck of the scooter might be made from a carbon fibre composite, rather than an aluminium alloy. You may be able to think of other solutions using these four principles.

In essence, the 40 Innovative Principles comprise an idea-generation heuristic. They help a user to search their knowledge and practical experience for analogies that are suggested by a specific principle and use these analogies to solve an engineering design problem.

There are many introductory videos for TRIZ on YouTube, including: https://www.youtube.com/user/triz4u.

Transforming design through biomimetic thinking and design

Building on from knowledge gathered over centuries of harvesting and harnessing nature, engineers and designers are now exploring the exciting field of emulating nature's successes to assist sustainable development. Many of our solutions from the last 300 years have been poorly adapted (or maladapted) to natural ecosystems. In fact, many of these 'solutions' have led to significant global challenges such as those caused by the creation and dispersion of pollution, including greenhouse gases, toxic chemicals and other hazardous substances. Faced with the need to address these challenges, engineers and designers will be tempted to emulate the way humans have problem solved, rather than asking nature's advice. Consider that Russian researchers working on a global database for patents (TRIZ) uncovered an overlap of a mere 10–12 per cent between man-made patents and natural systems! As Janine Benyus, creator of the biomimicry field puts it, 'when we look to nature, 90 per cent of the time we will be surprised!' (Benyus 1997).

Biomimicry is, quite simply, 'the art of asking nature for advice' to assist in creating more sustainable ways of living. Essentially, if we are to achieve harmony between development and nature on a global scale, we need to combine our engineering knowledge with the knowledge contained in natural systems, rather than just extracting resources from it, to deliver solutions that are well-adapted to our global environment — essentially, innovation inspired by nature.

As will be discussed further in the chapter on your engineering future, in engineering terms biomimicry describes the inquiry-based process of studying and mimicking the design and behaviour of nature to inform the development of solutions that meet the needs of society while being in harmony with the planet's natural systems. It is the crossover between natural systems and human systems — using the knowledge of nature and a method of inquiry to inform the built environment. In almost every field of endeavour, innovators are mimicking nature's design elegance to create sustainable solutions. Key principles of biomimicry include the following.

- Nature runs on sunlight.
- Nature uses only the energy it needs.
- Nature fits form to function.
- Nature recycles *everything*.
- Nature rewards cooperation.
- Nature banks on diversity.
- Nature demands local expertise.
- Nature curbs excesses from within.
- Nature taps the power of limits.

Understanding the relationship between natural and human systems

It is apparent on a global level that many current practices harnessing nature's resources are unsustainable — we need to rediscover nature's knowledge. Consider that for most of our time on the planet as a species, we have been hunters and gatherers. As hunters and gatherers (harvesting nature) and then as agrarians through pre-industrial times (harnessing nature), we paid a great deal of attention to natural systems as a source of knowledge, naturally mimicking the organisms that we admired.

As our knowledge of natural systems increased, we began to harness those organisms that we needed, then to process nature's raw materials to produce products and services (for example, through agricultural practices, and steel and plastics manufacturing). Once we realised that we could make value-added products from nature's raw resources, we began paying less attention to natural systems, seeing them more as a source of inputs for our products and services. As we transitioned from organism domestication to mass-production and industrialisation, 'transgenic engineering' emerged, with the mindset of 'animal as factory'.

Today, when we try to solve problems (such as filtration, adhesion, desalination and energy harvesting), we nearly always study the way humans have problem solved in the past, rather than how nature has done the very same thing. However, combining our knowledge of processes with our knowledge of natural systems, we can now build products and services that are in harmony with natural systems — 'biomimetic' solutions.

The relationship between humans and nature (figure 10.8) represents the transition in application of knowledge — from *harvesting* or *using nature*, to innovations that are *inspired by nature*.

- *Harvest* [take] refers to using materials provided by nature — plant or animal — with no human intervention in the production process itself. This includes using rainforest timber, or fishing for seafood. This is also known as 'bio-utilisation'. However, rather than simply living off nature, if we instead saw it as a source of ideas and inspiration ('nature as mentor'), we could seek to live in balance with nature.
- *Harness* [adapt] refers to domesticating the producer — domesticating organisms to assist us in product development. This includes, for example, agricultural practices using 'beasts of burden', and using bacteria for the production of insulin. This is also known as 'bio-assistance'.
- *Harmony* [copy] is the art of asking nature for advice, to assist in creating more sustainable ways of living. As Benyus (1997) explains, 'This includes studying nature's best ideas, designs and strategies and then emulating them so that we might live more gracefully on the planet'. This might include, for example, designing the front of a train like the beak of a bird, or an adhesive tape like the pads of a gecko's feet. This is also known as 'bio-inspired' or 'bio-mimetic' design.

Innovation from nature can be drawn from a number of areas, such as:

- the structure, or *form*, of nature ('nature as model') — aerodynamic shapes, non-chemical adhesive methods and structural finishes and colour
- the *process* of nature — cooling systems, nutrient cycling, filtration, desalination and energy supply
- nature's *ecosystem* — feedback loops, diversity, organism niches and interactions, symbiotic relationships, food webs, energy and material flows, resilience and the role of redundancy.

Besides providing the model, nature can also provide the measure ('nature as measure'). We can look to nature as a standard against which to judge the 'rightness' of our innovations. *Are they life promoting? Do they fit in? Will they last as long as is needed, and no longer?* A well-adapted product or service would address all three innovation categories, whereas maladapted products or services may focus on one or two of the categories to the detriment of the others.

As Benyus (1997) explains, we could manufacture the way animals and plants do, using sun and simple compounds to produce totally biodegradable fibres, ceramics, plastics and chemicals. Our farms, modelled on prairies, could be self-fertilising and pest resistant. To find new drugs or crops, we could consult animals and insects that have used plants for millions of years to keep themselves healthy and nourished. Even computing could take its cue from nature, with software that 'evolves' solutions and hardware that uses the lock-and-key paradigm to compute by touch.

FIGURE 10.8 Natural systems understanding map

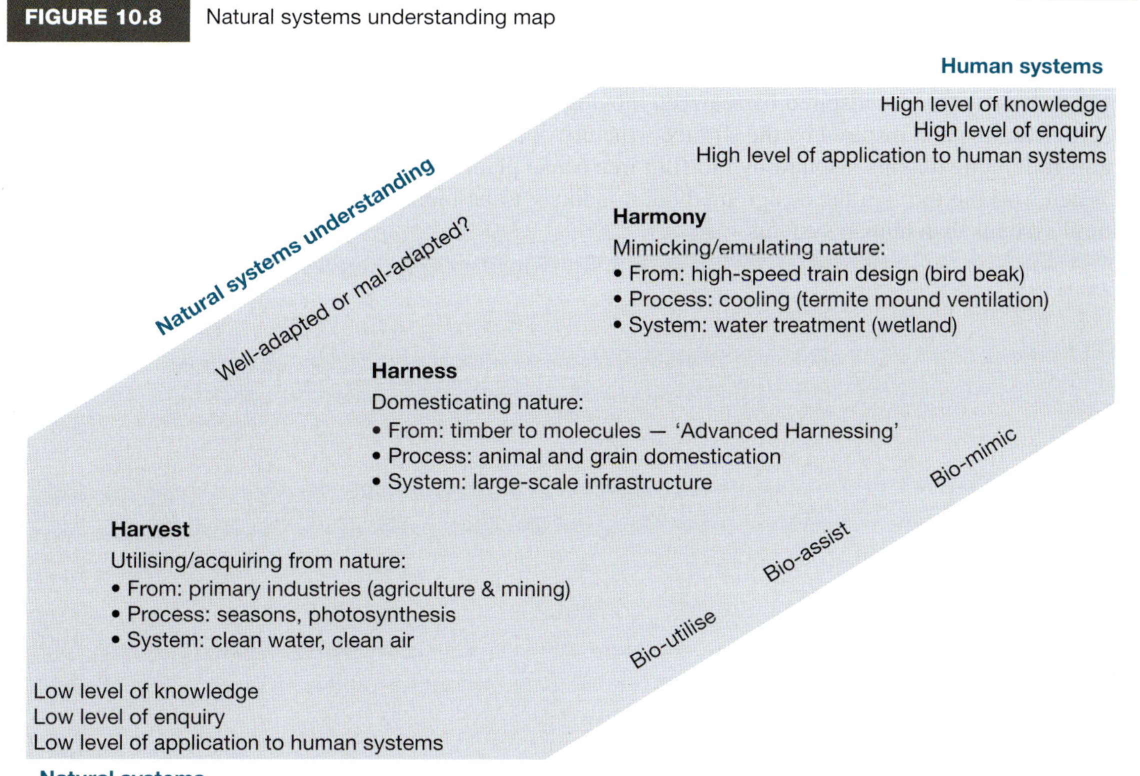

Source: The Natural Edge Project, Biomimicry Guild (2006).

In each case, nature can provide the models: solar cells copied from leaves; steel fibres woven spider-style; shatterproof ceramics drawn from mother-of-pearl; perennial grains inspired by tall grass; computers that signal like cells; and a closed-loop economy that takes its lessons from redwoods, coral reefs and oak-hickory forests. Most of nature's products and services are *biotic*; their processes are carried out in ambient temperature, low pressure and low toxicity conditions (with the exception of a volcano, tidal wave, hurricane or bush fire, which are considered *abiotic*).

Elements of biomimetic design

As we look to nature for advice in design, it is critical to manage business pressures (time and resources) in the process of design innovation. We need to be sure that we have a clear design method, and that we ask the right questions at the right time. In the field of biomimicry, engineers and designers can ask questions of nature, to help integrate natural systems knowledge with our knowledge of human systems as part of the design process.

How can engineers design cities to be as sustainable as a mature forest ecosystem? How would nature clean water? How does nature store energy? The design process is enhanced by the opportunity to look to nature for organisms with a similar problem and context to see what they do, and then to translate the useful forms, processes, and systems within the design context. As this is an emerging field, the challenge is for professions to understand their role within their context.

Based on an evolving methodology developed by the Biomimicry Guild (Benyus 1997), the following list proposes how a biomimicry methodology would work.

- Step 1. Identify the real challenge.
- Step 2. Translate the challenge into biology language.
- Step 3. Define the habitat parameters/conditions.
- Step 4. Re-ask 'how does nature do that function here, in these conditions?'
- Step 5. Find the best natural models (literal and metaphorical).
- Step 6. Mimic the natural model.

- Step 7. Evaluate the solution — 'nature as measure'.
- Step 8. Pay respect to the inspiration.

Each of these steps has a sub-set of questions and tasks to help the engineer and designer focus in on the design solution. Further to the common examples like Velcro® and Gecko Tape®, additional examples of commercialised biomimetic design outcomes that have followed some or all these steps include:

- energy conversion inspired by the swaying motion of sea plants in waves (BioWAVE)
- molecular-sized light sponges inspired by leaves (Dyesol)
- efficient motor blades inspired by seaweed moving in ocean currents (PAX Impeller)
- self-cleaning paint inspired by the surface structure of a lotus leaf (Lotusan)
- anti-fouling treatments inspired by bio-film free ocean plants (Biosignal).

When working through the design method, it is likely to be more difficult to locate information about natural systems than human systems.

SUMMARY

This chapter has explored the process of engineering problem solving and design in some detail, elaborating the process that was introduced in the chapter on the engineering method. This chapter has concentrated on demonstrating design and problem solving as fundamentally the same process. Designing the process for moving a large number of bricks requires the same kind of thinking as designing a new mobile phone: first, understand the problem through a process of questioning and research; consider a range of solutions; and, finally, evaluate the solutions and choose the best or better ones and communicate this to the client.

Systems thinking and whole system design have been demonstrated as key frameworks for the design process in addition to the potential for transforming design through biomimicry — innovation inspired by nature. Methods for generating alternative solutions have also been discussed. The chapter on evaluating options considers evaluation of solutions in more detail. We will now revisit each of the learning objectives.

10.1 Use a basic design approach to identify opportunities, needs and problems and to generate solutions in a systematic manner.

An essential part of problem solving is making sure that the right problem is solved. Using a systems approach provides a structured approach to examining different definitions of the problem by expanding or contracting the system boundary. Depending on the definition of the problem chosen, there will be a corresponding set of stakeholders who need to be consulted. The aim is to use a socio–ecological approach — to think about how this project or design contributes to improving social wellbeing. This generates a set of goals for the problem solution, which will be used in the evaluation of potential solutions at a later stage.

10.2 Apply whole system design principles to seek optimal solutions to engineering problems.

We have learned about the concept of 'whole system design', using ten key principles. These begin with asking the right questions; followed by benchmarking against an optimal system; and designing and optimising the whole system. This requires the designer to account for all measurable impacts, and to design and optimise the subsystems in the right sequence to achieve compounding resource savings. It is then necessary to review the system for potential improvements, model the system, track technological innovation and design to create future options.

10.3 Generate design solutions through research, consultation, creativity and biomimicry principles.

When the problem is suitably defined, a range of solutions can be identified through research and consultation (What solutions already exist?), and through innovation (where a new solution needs to be created). The methods available for creating new solutions include brainstorming, lateral thinking, parallel thinking, synectics and TRIZ. Each of these methods uses different ways of triggering new solutions, often through collaborative activity.

We have learned about the potential for nature to provide design guidance using the concept of biomimicry, which refers to 'innovation inspired by nature'. This field is based on several principles, including that nature runs on sunlight, uses only the energy it needs, fits form to function, recycles everything, rewards cooperation, banks on diversity, demands local expertise, curbs excesses from within and taps the power of limits.

Considering these principles, we can apply a biomimetic design approach where we first identify the real challenge and translate this into biology language. We then consider the habitat or conditions in nature where this problem could be addressed, asking how nature might respond to the problem. We then find the best natural models and mimic these, evaluating our design approach against 'nature as measure'.

KEY TERMS

biomimicry Drawing inspiration for engineered solutions from the forms and functioning of biological systems.

brainstorming A group-based technique to quickly identify a range of solutions to a problem. It relies on not evaluating the solutions until later. Any solution can be included, no matter how impractical.

conversation mapping A process of recording and organising the points of view of the stakeholders.

decoupling The process of separating negative environmental effects, including resource consumption, from positive economic growth.

externality A side effect or consequence of an industrial or commercial activity that affects other parties without this being reflected in the cost of the goods or services involved.

factor 5 A factor of 5 in performance achieves the same outcome at one-fifth the energy consumption (e.g. compact fluorescent lightbulbs compared to incandescent bulbs).

interdisciplinary approach Professionals from multiple disciplines work together on a complex problem.

socio–ecological thinking A socio–ecological approach recognises human and social systems embedded in natural (ecological) systems.

stakeholders People and organisations that have an interest in a problem.

synectics A way of systematically generating new ideas; relying on the mind's ability to connect quite dissimilar things.

systems thinking The process of identifying all of the elements of a problem. It requires explicit attention to the problem boundary and the components of the problem and an understanding of the interactions across the boundary and between the components within the system.

TRIZ The Russian acronym for the Theory of Inventive Problem Solving.

whole system design The process of considering the whole system that will be affected by a design, including benchmarking against the optimal design.

EXERCISES

1 Describe how a systems approach is useful in problem identification. You may wish to do this through an appropriate example from your favourite discipline.

2 Explain and give an example of how the socio–ecological perspective has expanded engineering thinking to include social and environmental impacts, as well as technical and economic efficiency in your discipline.

3 Explain the concept of decoupling economic growth from negative environmental pressure and how it applies to design in your discipline, including why it is important to also consider recoupling opportunities.

4 Consider your ecological footprint.
- Follow the instructions on the Global Footprint Network website (https://www.footprintcalculator.org) to complete an ecological footprint assessment for yourself.
- In a project group, discuss each of your results, and then try varying your inputs to see how different lifestyle decisions could affect your footprint. What lifestyles changes might you make as a result of this activity?

5 Find examples of responses to freeway congestion in various cities of the world. What ranges of solutions have been attempted? Can you think of others from the perspective of your discipline?

6 Create a mind map that shows the systemic interconnections between water, energy and materials use in anything that we do within the engineered environment, with a focus on your discipline. You may use any freeware mind map tool (e.g. https://bubbl.us). Alternatively, you can draw the mind map by hand.

7 Outline the main aspects of each of the following techniques for generating alternative conceptual solutions.
- Brainstorming
- Synectics
- TRIZ (access the TRIZ website, https://www.triz40.com, and apply it to one example from your discipline).

8 Obtain information about the design and operation of Council House 2 in Melbourne and identify the different examples of biomimicry that have been employed.

PROJECT ACTIVITY

Basic human needs include food and water, shelter, security, energy, communication and transport. To these, health services, community services, mining, manufacturing and entertainment can be added. You may be able to think of others.

Choose an engineering opportunity in one of these areas. Some opportunities for the future include better means of growing food, new water sources, renewable energy, energy storage, new communication devices or services, electric cars, public transport improvements, a bionic eye, online communities, new mining technologies, environmentally sensitive manufacturing and new entertainment media and services.

Your group's challenge is to investigate the problem and document a range of solutions to the problem. Your proposal should demonstrate your capability across the range of techniques discussed in this chapter.

Here are some reminders to help you approach this activity.

1 Start your design file. If you are working in a group, you should also request access to an online group space, either on your learning management system or at an external site.
2 Use a mind map, or another suitable tool, to develop your first thoughts about the scope of this project. What is in-scope and what is out-of-scope? This is your first attempt at defining the system that matches your proposal. For example, if you are designing an electric car, do you also need to think through how and where people will recharge it?
3 Engage as many other stakeholders as you can and expand the mind map with their ideas, design criteria and constraints about your project. In this process, you will change roles from *idea generator* to *recorder* of other people's views. Try not to impose your views on others or to spend time justifying your own views or position. Your role is to collect as many different points of view as possible. You will need good active listening skills and you may want to use the brainstorming technique with the stakeholders.
4 Identify key components within your system. Consider several different system definitions. How does each system definition lead to a new range of possible solutions? Visit the website of the Biomimicry Institute (https://biomimicry.org) and take the challenge (https://biomimicry.org/globaldesignchallenge).
5 Research how others have approached the same problem. What solutions are already available? What new solutions can you think of? How do these solutions match the design criteria and constraints from step 3? Considering one or more elements of the whole system design approach, how do these solutions match the design criteria and constraints from step 3? Is there a possible solution derived from a biomimetic approach?

REFERENCES

Accoya 2021, *The Accoya Specification Guide*, https://www.accoya.com/app/uploads/2021/09/Accoya_Specification-Guide_UK.pdf

Adams, R 2008, 'Council House 2 — a video tour', *City of Melbourne*, 24 August, https://www.youtube.com/watch?v=vJV0wnbAZ6M

American Council for an Energy-Efficient Economy 2024, https://www.aceee.org

Amezquita, T, Hammond, R, Salazar, M & Bras, B 1995, 'Characterising the remanufacturability of engineering systems', *Proceedings 1995 ASME Advances in Design Automation Conference*, 7–20 September, Massachusetts, Boston, vol. 82, pp. 271–278.

Anastas, PT & Warner, JC 1998, *Green chemistry: Theory and practice*, Oxford University Press, New York.

Appropriate Technology Collaborative 2022, 'Mayan power and light', https://apptechcollaborative.org/mayan-power-and-light

Benyus, J 1997, *Biomimicry: Innovation inspired by nature*, HarperCollins, New York.

Brownell, B 2006, *Transmaterial: A catalog of materials that redefine our physical environment*, Princeton Architectural Press, New York.

City of Melbourne 2024a, *Participate Melbourne*, https://participate.melbourne.vic.gov.au

——— 2024b, 'Council House 2', https://www.melbourne.vic.gov.au/building-and-development/sustainable-building/council-house-2/Pages/council-house-2.aspx

Community Health Research Unit 2009, *MIP toolkit: Module 1: Conduct a socio–ecological assessment*, MIP Toolkit.

de Bono, E 1985, *Six thinking hats*, Penguin, London.

——— 1994, *Parallel thinking — from Socratic to de Bono thinking*, Penguin, London.

——— n.d., 'Lateral thinking', https://www.debonogroup.com/services/core-programs/lateral-thinking

Department of Transport and Planning 2010, 'Delivering Melbourne's newest sustainable communities', State Government of Victoria.

DiGregorio, BE 2009, 'Biobased performance bioplastic: Mirel', *Chemistry & Biology Innovations*, vol. 16, no. 1, pp. 1–2, https://doi.org/10.1016/j.chembiol.2009.01.001

Engineers Australia 2005, 'Future demand for engineering skills', https://studylib.net/doc/18054886/future-demand-for-engineering-skills

EnviroGLAS 2024, 'Turning trash into terrazzo', https://enviroglasproducts.com

Falkirk Wheel 2024, 'The Falkirk Wheel', https://www.scottishcanals.co.uk/visit/canals/visit-the-forth-clyde-canal/attractions/the-falkirk-wheel

Fédération Internationale de l'Automobile (FIA) 2022, 'Formula E and FIA reveal all-electric Gen3 race car in Monaco', 28 April, https://www.fiaformulae.com/en/news/2456/formula-e-and-fia-reveal-all-electric-gen3-race-car-in-monaco

——— 2024, 'Welcome to the story of the ABB FIA Formula E World Championship', https://www.fiaformulae.com/en/championship/history

Fisher & Paykel 2019, 'SmartDrive™ technology', https://www.fisherpaykel.com/au
Freshape 2024, 'Sunlight collector', https://www.freshape.com/vision/?pop=sun-router
Gateshead Council 2019, 'The Gateshead Millennium Bridge', https://www.gateshead.gov.uk/article/4594/The-Gateshead-Millennium-Bridge
Gordon, WJJ 1961, *Synectics*, Harper & Row, New York.
Hargroves, K & Smith, M 2005, *The natural advantage of nations: Business opportunities, innovation and governance for the 21st century*, Earthscan Press, London.
Heren, K 2023, 'New faults found on troubled WilkinsonEyre bridge after mast snaps', *Architects' Journal*, 24 January, https://www.architectsjournal.co.uk/news/new-faults-found-on-troubled-wilkinsoneyre-bridge-after-mast-snaps
Hewlett-Packard 2024, 'Sustainable impact', https://www.hp.com/us-en/sustainable-impact.html
Hipp, C & Kirkeby, D 2002, *High density web server chassis system and method*, US Patent 6411506 (B1).
Holmberg, J & Robèrt, K-H 2000, 'Backcasting — a framework for strategic planning', *International Journal of Sustainable Development and World Ecology*, vol. 7, no. 4, pp. 291–308, https://doi.org/10.1080/13504500009470049
Hori, Y 2004, 'Future vehicle driven by electricity and control-research on four-wheel-motored "UOT electric march II"', *IEEE Transactions on Industrial Electronics*, vol. 51, no. 5, pp. 954–962, https://doi.org/10.1109/TIE.2004.834944
Huang, X, Toshiyuki, H & Hori, Y 2014, *Optimized topology and converter control for supercapacitor based energy storage system of electric vehicles*, presented at the EVTeC and APE Japan, 23 May.
International Living Future Institute 2022, https://living-future.org
Liu, D, Yang, C & Lunt, RR 2018, 'Halide perovskites for selective ultraviolet-harvesting transparent photovoltaics', *Joule*, vol. 2, no. 9, pp. 1827–1837, https://doi.org/10.1016/j.joule.2018.06.004
Los Alamos National Laboratory 2001, 'Supercomputing in small spaces' https://public.lanl.gov/radiant/research/potpourri/scss.html
McCarthy, T, Sheikh, N & Gardner, A 2008, *Encapsulating sustainability principles for structural design of buildings*, paper 383, 25th International Conference on Passive and Low Energy Architecture, PLEA 2008, Dublin, 22–24 October.
McKay, D 2017, 'The good news about plastic waste', *The Conversation*, 15 November, https://theconversation.com/the-good-news-about-plastic-waste-83742
Metcalfe, G & Rudman, M 2001, *Fluid mixer,* Patent WO2002020144A1.
Negroni, C 2016 'Never lose your luggage again as RFID technology tracks all bags at all times', *Australian Financial Review*, 29 August, https://www.afr.com/companies/transport/never-lose-your-luggage-again-as-rfid-technology-tracks-all-bags-at-all-times-20160829-gr39y5
New Zealand Ministry of Education 2023, 'The socio-ecological perspective', *Health and physical education in the curriculum*, https://hpe.tki.org.nz/health-and-physical-education-in-the-curriculum/underlying-concepts/the-socio-ecological-perspective
Odum, HT 1995, *Environmental accounting: Emergy and environmental decision making*, Wiley, Hoboken.
Pears, A 2004, 'Energy efficiency — its potential: Some perspectives and experiences', background paper for International Energy Agency Energy Efficiency Workshop, Paris, p. 10.
Qantas 2019a, 'Permanent bag tags — Q Bag Tags', https://www.qantas.com/au/en/travel-info/baggage/permanent-bag-tags.html
——— 2019b, 'Qantas marketplace', https://marketplace.qantas.com/au/p/qantas-q-bag-tag-1-pack/AU100020244
Rees, WE 1996, 'Revisiting carrying capacity: Area-based indicators of sustainability', *Population and Environment: A Journal of Interdisciplinary Studies*, vol. 17, no. 3, p. 10, https://doi.org/10.1007/BF02208489
Rocky Mountain Institute 2019, '10xE: Factor ten engineering', https://rmi.org/our-work/areas-of-innovation/office-chief-scientist/10xe-factor-ten-engineering
Roukes, N 1988, *Design synectics: Stimulating creativity in design*, Davis Publications, Worcester.
SafeChem 2019, 'COMPLEASE chemical leasing', https://safechem.com/en/metal-cleaning/complease
Schmidt-Bleek, FB 1993, *The fossil makers*, Factor 10 Institute, France, p. 73.
——— 1999, 'Making sustainability accountable putting resource productivity into practice', in *Factor 10*, Factor 10 Club, p. 41.
Smith, M, Hargroves, K & Desha, C 2010, *Cents and sustainability: Securing our common future by decoupling economic growth from environmental pressures*, in The Natural Edge Project, Earthscan, London.
Sohel, MI, Ma, Z, Cooper, P, Adams, J & Scott, R 2014, 'A dynamics model for air-based photo voltaic thermal systems working under real operating conditions', *Journal of Applied Energy*, vol. 132, pp. 216–225, https://doi.org/10.1016/j.apenergy.2014.07.010
Stasinopoulos, P, Smith, M, Hargroves, K & Desha, C 2008, *Whole system design: An integrated approach to sustainable engineering,* The Natural Edge Project, Earthscan, London.
Sustainable Buildings Research Centre 2024, 'About: Living Building Challenge', https://www.uow.edu.au/sbrc/about
Technovelgy n.d., 'How RFID works', http://www.technovelgy.com/ct/Technology-Article.asp?ArtNum=2
Tesla 2024, 'Tesla solar roof', https://www.tesla.com/en_au/solarroof
The Natural Edge Project, Biomimicry Guild 2006, in M Smith, K Hargroves, C Paten & N Palousis 2007, *Engineering sustainable solutions program: Critical literacies portfolio — principles and practices in sustainable development for the engineering and built environment professions*, The Natural Edge Project, Australia.
Tractile 2024, 'The designer solar roof', https://tractile.com.au
United States Environmental Protection Agency 2018, 'Green chemistry', https://www.epa.gov/greenchemistry
University of Wollongong 2014, 'Illawarra Flame', https://www.illawarraflame.com.au
von Weizsäcker, E, Hargroves, K, Smith, M, Desha, C & Stasinopoulos, P 2009, *Factor 5: Transforming the global economy through 80% increase in resource productivity*, Routledge, London.
Yield10 2024, 'Camelina Platform', https://www.yield10bio.com/crop-science/camelina-platform

ACKNOWLEDGEMENTS

The authors wish to acknowledge Carl Reidsema, who proposed and developed the 'moving the bricks' case study in the chapter, Iouri Belski, who contributed to the section on TRIZ, and Cheryl Desha, Peter Stasinopoulos and Charlie Hargroves, The Natural Edge Project (https://www.iau-hesd.net/organisation/natural-edge-project-tnep), who co-authored a section of this chapter.

Photo: © Alan Crawford / Shutterstock

Photo: © Makinex

Photo: © *The Conversation*

Photo: © Tim McCarthy

Photo: © SPP Sport Press Photo / Alamy Stock Photo

Photo: © Tim McCarthy

Photo: © Qantas

Photo: © Traceyaphotos2 / Shutterstock

Photo: © Stu Collier / Shutterstock

Photo: © Tessa Bishop / Shutterstock

Figure 10.4: © Smith, M, Hargroves, K & Desha, C 2010, *Cents and sustainability: Securing our common future by decoupling economic growth from environmental pressures*, in The Natural Edge Project, Earthscan, London.

Figurc 10.5: © Hargroves, K & Smith, M 2005, *The natural advantage of nations: Business opportunities, innovation and governance for the 21st century*, Earthscan Press, London.

Figure 10.6: © Tim McCarthy

Figure 10.8: © The Natural Edge Project, Biomimicry Guild 2006, in M Smith, K Hargroves, C Paten & N Palousis 2007, *Engineering sustainable solutions program: Critical literacies portfolio — principles and practices in sustainable development for the engineering and built environment professions*, The Natural Edge Project, Australia.

Text: © McKay, D 2017, 'The good news about plastic waste', *The Conversation*, 15 November, https://theconversation.com/the-good-news-about-plastic-waste-83742

CHAPTER 11

Evaluating options

'We can't solve problems by using the same kind of thinking we used when we created them.'

Albert Einstein

LEARNING OBJECTIVES

After studying this chapter, you should be able to:

11.1 evaluate solutions in terms of basic economic principles and consider non-economic criteria in the selection of project options

11.2 consider the technical feasibility of engineering projects

11.3 apply principles of mathematical modelling to conceptual and detailed design.

Introduction

The previous chapter described methods for generating alternative solutions through research, consultation and creativity, and showed simple methods to quantify aspects of the engineering problem being addressed. As that process proceeds, the engineer will be constantly comparing alternative solutions with the original design specification, criteria and constraints. It will be assumed in this chapter that each alternative generated in the design process (see the previous chapter) is technically feasible and that appropriate mathematical modelling has been used in this process.

In any design process, various alternative solutions will be generated but only one solution will be used in the final design. In the evaluation of the various alternatives or options, a set of evaluation criteria will need to be developed. These may be different for different projects but will include economic, environmental, social, institutional and system reliability. Note that safety is a crucial consideration in the technical design process. Projects will vary in complexity and the emphasis on each evaluation criterion will vary from project to project. These criteria can be grouped into economic and non-economic criteria.

Economic criteria are based on capital costs (to construct the project) and life cycle costs (for example, operation, maintenance and replacement costs in future years during the life of the project). Projects require investment capital and so an important consideration is the return on that investment. This can be evaluated as an internal rate of return or a cost-benefit or a payback period.

Non-economic criteria include considerations of the environment (regulations), social implications of the project, the reliability of the system and implications of unreliability as well as institutional factors such as funding capability and the ability to meet time, safety, and sustainability constraints.

To make a final decision on the selected option, usually all of the above criteria are applied to each alternative/option. For each technique, there are several assumptions that need to be made. The final selection may be quite dependent on the actual assumptions made during the analysis.

Figure 11.1 is a framework that can be used in the evaluation of the different options generated to satisfy the requirements of a particular project. In most large and complex projects, a combination of both quantitative and qualitative methods are used to arrive at the final option. Technical and various economic methods can be used as quantitative techniques and usually a range of experts are used to perform qualitative analyses. Fewer of these techniques would be used for smaller and less complex projects but the choice of technique is very important. This chapter provides some examples of the use of some of these techniques.

11.1 Evaluating solutions — economics

LEARNING OBJECTIVE 11.1 Evaluate solutions in terms of basic economic principles and consider non-economic criteria in the selection of project options.

Many engineering tasks start with some basic economic analysis. This was obvious in the chapter on engineering design, where a decision had to be made about whether to accept a job (in that case, moving a large number of bricks) before the technical solution could be refined. An initial feasibility check was done to ensure that the job was possible within the resources available. Billy Koen (2003) illustrated this concept when he stated 'The engineering method is the use of heuristics to cause the best change in a poorly understood situation within the available resources'.

As a list of possible solutions to a problem is developed, the solutions need to be evaluated. An initial evaluation of the list of possible solutions generated may develop a shorter, more probable list. As part of the evaluation, an analysis of both the economic and the technical design feasibility of each of the proposed solutions is likely to be necessary. In the chapter on the engineering method, conceptual design and detailed design were discussed. In conceptual design, the focus is on identifying the preferred design and configuration without having to do a detailed design of every component. Of course, you may have to do some detailed design to get a handle on likely costs. In detailed design, the focus is on preparing prototypes and/or **working drawings** (detailed drawings used for construction or manufacturing) ready for manufacturers or contractors to build the product.

KEY POINT

Economic analysis can be used to compare options with different lifetime costs, while technical analysis is required to perform detailed design of engineering components.

FIGURE 11.1 A framework for project identification and evaluation

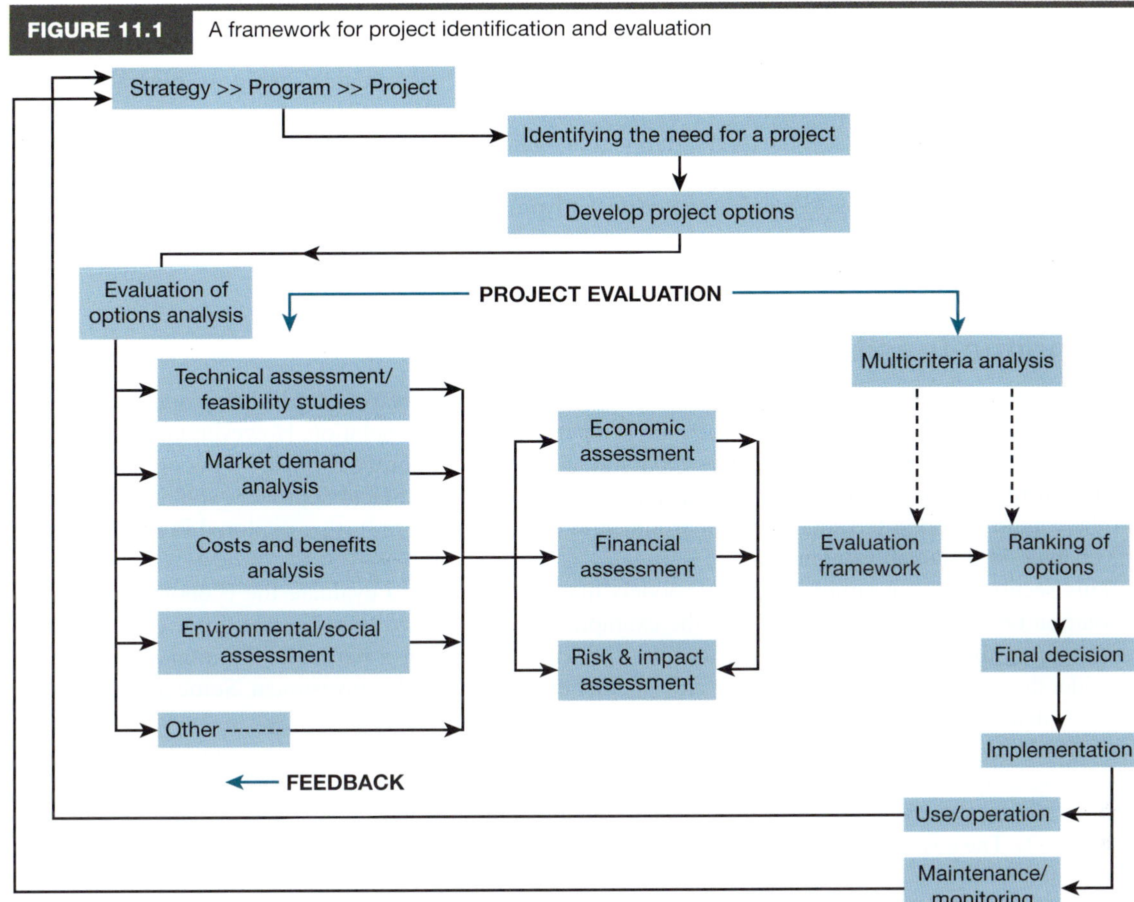

Source: Martland (2006).

Non-economic criteria

Non-economic criteria include (but are not limited to) environmental, social, institutional and system reliability factors.

- *Environmental.* These could include environmental compliance, cultural resources, visual resources, flora and fauna, threatened and endangered species, air and water quality, and noise pollution. Environmental compliance with various levels of government regulations can be significant in the evaluation of various projects.
- *Social.* These might include private land, conservation easements, public acceptability, recreation, environmental justice and native titles. Again, this criterion could be significant in the evaluation of a project.
- *Institutional.* These might include funding capability, ability to meet time constraints, agreements with project partners, government bodies and others, compatibility with growth needs in other areas and development risk.
- *System reliability.* This rating is based on project integrity and system redundancy. In a water supply project, system redundancy would be enhanced by constructing multiple sources of water to meet needs, as opposed to only one source or pipeline.

Qualitative approaches to project evaluation

While quantitative methods provide an important toolset for project proposal evaluation and selection, there is also a growing sense of frustration — especially among managers of complex and technologically advanced undertakings — that reliance on strictly quantitative methods does not always produce the most useful or reliable inputs for decision making and that not all methods are equally suited to all situations (Thamhain 2013). So, it is not surprising that for project evaluations involving complex sets of business criteria, narrowly focused quantitative methods are often supplemented with broad scanning, intuitive processes, and collective, multifunctional decision making such as Delphi, nominal group technology, brainstorming, focus groups, sensitivity analysis and benchmarking. Each of these techniques can be either used by itself to determine the best and most successful or valuable option or integrated into a comprehensive analytical framework for collective multifunctional decision making.

These processes rely on subject experts from various functional areas to collectively define and evaluate broad project success criteria, employing both quantitative and qualitative methods. The first step is to define the specific organisational areas critical to project success and to assign expert evaluators. Ideally, these evaluators should be members of the core team who are ultimately responsible for project implementation.

Economic feasibility

Here, we will consider conceptual design from the economic perspective of whether proposed solutions are financially viable and look at this economic perspective largely in isolation. However, in the chapter on sustainable engineering, we consider this economic dimension in a broader sense — in conjunction with environmental and social criteria. If you have followed a socio–ecological approach, you are likely to be on the way to covering the social and environmental bases, which are further explored in more detail in the chapter on sustainable engineering.

This section looks at simple financial models that can be used to evaluate the economic aspects of a solution or alternative solutions using the example of a wind farm. Imagine that a power company is investigating the feasibility of placing a cluster of wind turbines at a particular location. The job is to consider the economics of this wind farm and whether this is a reasonable investment. Some of the technical issues will also be discussed.

A wind turbine is a complex piece of machinery. Wind turbines come in a range of sizes; the most popular sizes used by wind farms being from 2.0 MW up to 15 MW (Vestas Wind Systems A/S 2024). These turbines are available from a number of manufacturers, such as GE, Vestas and Acciona (see figure 11.2). The cost of the units, installed, is typically $2.2–2.5 million/MW, which includes the turbines, towers and basic site infrastructure. Other aspects of the whole project, such as planning approvals and site access are a fixed cost of roughly an additional $2 million (Blewett 2023).

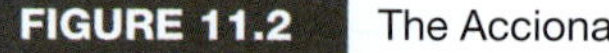

FIGURE 11.2 The Acciona 1.5 MW turbine

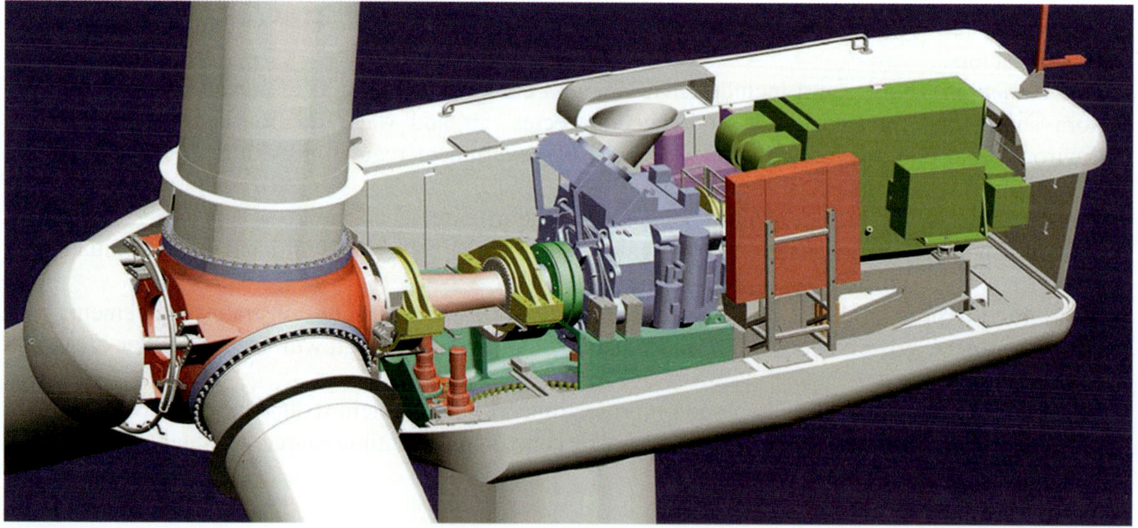

Source: Acciona (n.d.).

Calculating return on investment

There are several commonly used approaches for financial analysis of projects such as the net present value (NPV) approach, the internal rate of return method (IRR), return on investment (ROI), cost–benefit ratio and payback period. We will consider the NPV method first and use that data in the other techniques.

Net present value (NPV)

The **net present value (NPV)** approach reduces all costs and incomes (over the lifetime of the project) to a value in the present (hence the name). This works by recognising that the value of an amount of money in one year's time is less than the same dollar value today (due to, for example, inflation) or in general terms:

$$PV = \frac{FV}{1 + r}$$

Where PV = present value, FV = future value and r = discount rate. The discount rate combines several effects: (1) the cost of borrowing money; (2) the inflation rate; and (3) the risk of the future income or cost.

One way to think about this relationship is to imagine that you need to pay $1000 in 12 months' time. How much money, PV, would you need now to pay that future debt? Suppose you were able to keep your money in a bank account at say, 6 per cent; in 12 months' time, you could have 1.06 times PV. Therefore, if this future value is $1000, the present value PV = (1000/1.06) = $943.

For a value i years in the future, the formula becomes:

$$PV = \frac{FV}{(1 + r)^i} \tag{1}$$

So, for a series of payments over, say, five years, the present value is:

$$PV = \frac{FV1}{(1 + r)^1} + \frac{FV2}{(1 + r)^2} + \frac{FV3}{(1 + r)^3} + \frac{FV4}{(1 + r)^4} + \frac{FV5}{(1 + r)^5} \tag{2}$$

The present value of an *infinite* series of payments equal to FV is even simpler:

$$PV = \frac{FV}{r} \tag{3}$$

Suppose a business has $1 million at its disposal. There are two options to be considered here: invest the money in some project/business where it is expected to return $250 000 each year for five years or deposit it in a bank account that will accrue interest each year.

If both the discount rate and the bank interest are 7 per cent, which option is a better choice for the business?

For the first option, we calculate the present value of the future returns (over five years). If this is a positive number, then it is a good investment. That is, the sum of the future returns, suitably discounted, must exceed the (negative) investment (that is the initial $1 million). Using the basic formula (1) above, we can tabulate the present value (table 11.1), which shows a net present value of $25 049 — indicating that this could be a suitable investment. Note that the NPV value is the sum of the PV column.

TABLE 11.1 Basic net present value example

	Discount rate (7%)		
Year	**Future value**	**PV factor***	**PV**
0	−1 000 000	1	−1 000 000
1	250 000	0.935	233 644
2	250 000	0.873	218 359
3	250 000	0.816	204 074
4	250 000	0.763	190 723
5	250 000	0.713	178 246
		NPV	**25 049**

*The PV factor is $(1.0/1.07)^i$, where i = number of years.

Although this investment shows a positive NPV, perhaps we could do better. What would be the NPV if the investment was taken over ten years? What would the NPV be if the discount rate was 7.5 per cent?

Consider the situation where the business deposited the $1 million into a bank at 7 per cent (table 11.2). Would that be a better investment?

TABLE 11.2 Basic net present value of a bank investment

	Discount rate (7%)		
Year	**Future value**	**PV factor***	**PV**
0	−1 000 000	1	−1 000 000
1	70 000	0.935	65 420
2	70 000	0.873	61 140
3	70 000	0.816	57 140
4	70 000	0.763	53 402
5	1 070 000	0.713	762 895
		NPV (rounded)	**0**

This shows an NPV of zero, which is less than the $25 049 available from the investment scenario (table 11.1). So, keeping the money in the bank at the discount rate of 7 per cent preserves its value over the five years. Note that in year 5, we take the $1 million out of the bank (it is a 5-year term deposit), which accounts for the larger future value in that year.

Note that we have not considered the impact of taxation on the investment decision; nor have we included the effect of inflation (typically around 3 per cent). What if, for instance, in table 11.1, the yearly payments/returns were not fixed at $250 000 but were indexed to inflation at 3 per cent? The NPV of the return on investment will jump to $84 085 (table 11.3).

TABLE 11.3 Basic net present value including inflation (but not taxation)

	Inflation rate (3%)	Discount rate (7%)	
Year	**Future value**	**PV factor**	**PV**
0	1 000 000	1	−1 000 000
1	250 000	0.935	233 645
2	257 500	0.873	224 910
3	265 225	0.816	216 503
4	273 182	0.763	208 409
5	281 377	0.713	200 618
		NPV	**84 085**

Internal rate of return (IRR)

Consider now the internal rate of return (IRR) of the investment proposition from table 11.1. The **internal rate of return (IRR)** is the discount rate for which NPV of the investment is zero. This is, in general, an iterative calculation, equivalent to solving for x in the equation $f(x) = 0$. There are many available algorithms such as Newton–Raphson or the Secant method that can be used to solve this equation. Most of us, however, will simply use a tool such as Goal Seek in Excel.

Recalculating the data in table 11.1 to give a zero NPV, using Goal Seek provides an IRR of 7.93 per cent for this investment, as indicated in the following screenshot of the Excel spreadsheet. This 7.93 per cent is the bank term deposit interest rate needed in order to provide an equivalent return to the proposed table 11.1 investment. The bank is only offering 7 per cent in the table 11.2 example; therefore, the table 11.1 investment proposition is more lucrative than simply investing in the bank term deposit.

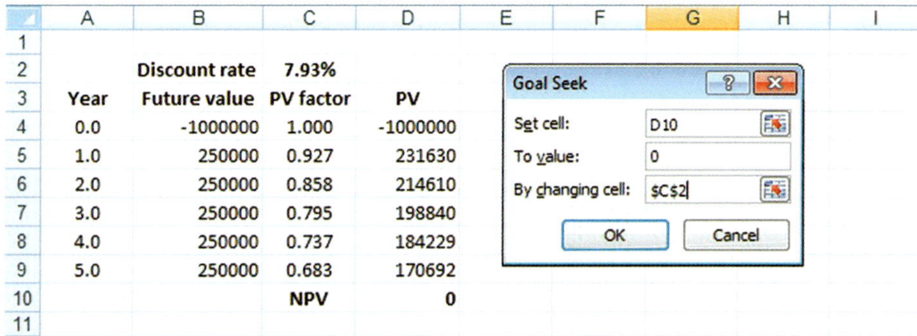

	A	B	C	D
1				
2		Discount rate	7.93%	
3	Year	Future value	PV factor	PV
4	0.0	-1000000	1.000	-1000000
5	1.0	250000	0.927	231630
6	2.0	250000	0.858	214610
7	3.0	250000	0.795	198840
8	4.0	250000	0.737	184229
9	5.0	250000	0.683	170692
10			NPV	0
11				

Return on investment (ROI)

Perhaps one of the most popular measures for project evaluation is the **return on investment (ROI)**:

$$\text{ROI} = (\text{Revenue } (R) - \text{Cost } (C))/\text{Investment } (I)$$

ROI calculates the ratio of net revenue over investment. In its simplest form, the stream of cash flow is not discounted. As a performance measure, ROI is used to evaluate the efficiency of an investment or to compare the efficiencies of several different investments. One can look at the revenue on a year-by-year basis, relative to the initial investment.

In a somewhat more sophisticated way, we can calculate the average ROI per year over a given revenue cycle:

$$\text{ROI} = \frac{(Rn - Cn)/In}{N}$$

Where $n = 1, 2 \ldots$ and N is the number of years.

Although this is a popular measure, it does not permit a meaningful comparative analysis of alternatives with fluctuating costs and revenue streams. Furthermore, it does not consider the time-value of money.

Cost–benefit (CB)

We can calculate the net present value of the total ROI over the project life cycle. This measure, known as **cost–benefit (CB)**, is calculated as the present value stream of net revenues divided by the present value stream of investments. It may be used to compare potential project investments or estimate the value against costs of a single decision, project, or policy. It is an effective measure for comparing project options with fluctuating cash flows.

Payback period (PBP)

Another popular measure for comparing project alternatives is **payback period (PBP)**. It indicates the time period of net revenues required to return the capital investment made on the project. For simplicity, undiscounted cash flows are often used to calculate a quick figure for comparison. This is quite meaningful if we deal with an initial investment and a steady stream of net revenue. However, for fluctuating revenue and/or cost streams, the net present value must be calculated for each period individually, and cumulatively added up to the break-even point in time. The PBP is the time taken for the net present value of the revenue to equal the investment.

Wind farm economics

For the wind farm example, the investor would need to borrow an amount of money up front to pay for the purchase and installation of the turbines. Over the next 25 years, there will be income from the sale of electricity as well as costs for maintenance and operations. Based on the following assumptions, it is possible to tabulate the NPV calculations (see tables 11.4 and 11.5).

CASE STUDY

Wind farm example

Turbine production capacity

- Expected life of the wind farm: 25 years.
- The maximum power-generating capacity of the turbine is 1.5 megawatts (MW), but due to the variability of wind, the average output is estimated to be 0.45 MW (30 per cent of maximum capacity). This could be confirmed through more detailed modelling based on recorded wind data, which will be considered later.
- Total hours per year = 8766 h (365.25 days × 24 hours).
- Total MW·hours generated at average output of 0.45 MW = 8766 x 0.45 = 3945 (MW·h).

Income

- The income is \$80 per megawatt hour (MW·h) = \$0.08/kW·h.
- The annual income at average output = 3945 MW·h × \$80/(MW·h) = \$315 600.
- This income is assumed to grow by 5 per cent per year due to climate change factors (i.e. faster than inflation).

$$\text{annual income}_{i+1} = \text{annual income}_i \times (1.05)$$

Costs

- Capital plus installation cost is \$2.5 million (m) per MW capacity (1.5 × \$2.5m = \$3 750 000).
- The discount rate for this capital is estimated at 9 per cent.
- Annual operation and maintenance is \$20/(MW·h) = \$78 900.
- The inflation rate for costs is estimated at 3 per cent, so:

$$\text{annual cost}_{i+1} = \text{annual cost}_i \times (1.03)$$

Note that the assumed income of \$80/(MW·h) is made up of \$40/(MW·h) as electricity sale and \$40/(MW·h) as the sale of renewable energy certificates (RECs). All non-renewable energy producers are required to buy a certain number of RECs to offset their CO_2 production (Clean Energy Regulator 2024).

Consider the basic NPV analysis. We need to discount future costs and income to present value. The annual cost is assumed to be \$20/(MW·h) of electricity generated and the annual income is based on assumptions made about MW·h generated and the average sale price of the electricity. Both of these are somewhat uncertain. We have assumed that the sale price of the electricity will increase faster than the annual costs, which is assumed to increase only with inflation. A useful aim of the analysis is to determine the break-even point at which the electricity price makes the project pay. Alternatively, the electricity price is known, and we need to determine the interest rate at which the project becomes viable.

Computing tools such as Microsoft Excel make these calculations quite straightforward. Excel's NPV function calculates the net present value of a series of payments or costs:

$$= NPV\,(\text{rate, values})$$

Table 11.4 provides a summary of the data required to perform the NPV calculations for this wind farm example.

TABLE 11.4 Summary of data for the wind farm example

			Factors	NPV
Capacity	1.5	MW		
Average production	0.45	MW	30%	
Hours/year	8766	h		

Total production	3 944.7	MW·h		
Electricity price	80.0	$/(MW·h)		
Interest rate	9	%		
Inflation rate	3	%		
Energy inflation	5	%		
Operating cost	20	$/(MW·h)		
Installed cost	3 750 000	$		−3 750 000
Annual cost	78 894	$	5%	−95 628
Total income	315 576	$		4 791 168
Total				**45 540**

In table 11.5, the NPV is calculated by summing the total of the PV column, which was calculated for each year by multiplying its respective net income by the discount factor.

TABLE 11.5 Net present value analysis for a single wind turbine (dollars)

Year	Annual cost	Annual income	Net income	Discount factor	PV
0	−3 750 000			1.000	−3 750 000
1	−78 894*	315 576*	236 682	0.917**	217 139***
2	−81 261	331 355	250 094	0.842	210 499
3	−83 699	347 923	264 224	0.772	204 029
4	−86 210	365 319	279 109	0.708	197 728
5	−88 796	383 585	294 789	0.650	191 592
6	−91 460	402 764	311 304	0.596	185 620
7	−94 204	422 902	328 698	0.547	179 809
8	−97 030	444 047	347 017	0.502	174 156
9	−99 941	466 249	366 309	0.460	168 659
10	−102 939	489 562	386 623	0.422	163 314
11	−106 027	514 040	408 013	0.388	158 118
12	−109 208	539 742	430 534	0.356	153 070
13	−112 484	566 729	454 245	0.326	148 165
14	−115 858	595 066	479 207	0.299	143 401
15	−119 334	624 819	505 485	0.275	138 775
16	−122 914	656 060	533 146	0.252	134 283
17	−126 602	688 863	562 261	0.231	129 923
18	−130 400	723 306	592 906	0.212	125 692
19	−134 312	759 471	625 160	0.194	121 587
20	−138 341	797 445	659 104	0.178	117 604
21	−142 491	837 317	694 826	0.164	113 742
22	−146 766	879 183	732 417	0.150	109 996

(continued)

TABLE 11.5 *(continued)*

Year	Annual cost	Annual income	Net income	Discount factor	PV
23	−151 169	923 142	771 973	0.138	106 364
24	−155 704	969 299	813 595	0.126	102 842
25	−160 375	1 017 764	857 389	0.116	99 430
				NPV	**45 540**

*These numbers come from table 11.4.
**The discount factor = (1.0/1.09).
***PV = Net income × Discount factor.

The annual costs and incomes shown in table 11.5 yield the highlighted values shown in table 11.4 (−95 628 and 4 791 168, respectively). Summing these values with the initial investment of $3 750 000 provides an NPV of $45 540. Note that this is the same value we achieved from first principles in table 11.5 by carefully discounting each annual cost and income and determining the annual PV before summing the PV column to get the NPV of $45 540.

The net present value for this case suggests that the project is a suitable investment. Before looking at the sensitivity of these calculations to the various assumptions, consider table 11.5, which sets out the calculations in detail as a way of checking that they are correct.

For each year, subtracting the operating cost from the gross income provides the net income. This value is then discounted at the borrowing rate (the cost of capital, which is the assumed discount rate mentioned earlier of 9 per cent) to provide the present value (PV) for each year. Summing the PV column (the *discounted* income over 25 years) provides a *present* value of all the future income the project is expected to derive. This amount, less the initial capital borrowing cost ($3.75m), provides a net present value NPV of $45 540, as expected.

Based on this analysis, and with these assumptions, the wind farm appears to be a worthwhile investment *if the income is at least $80/MW·h*. Sensitivity analysis will be considered shortly. For example, what is the break-even income above which the wind farm is a worthwhile investment?

Internal rate of return

The internal rate of return (IRR) is the interest rate at which the NPV is zero. Recall that NPV is the current value of future costs and revenue. So, it is an estimate of the upper limit on a viable interest rate. If you can borrow money at a lower cost, then NPV > 0 and the investment is worth taking. Above that value, NPV is negative, so it is not such a good investment.

Table 11.6 demonstrates that substituting some alternative interest rate values leads to a very rapid estimate of the IRR for the wind turbine example (an interest rate of 9.12 per cent would yield an NPV of zero), assuming an automatic iterate function is not available.

TABLE 11.6 **Iteration to estimate IRR**

Interest rate (%)	NPV ($k)
9.0	45
9.1	6
9.12	0

Sensitivity analysis

The above calculations suggest the wind farm is a good investment because NPV is greater than zero and IRR is just over 9 per cent. However, at what point does this become an unattractive investment? This process of systematically examining assumptions and parameters is called **sensitivity analysis**.

All the parameters would need to be checked in a full sensitivity analysis. For our purposes, we will look at a couple of these for explanatory reasons. A likely sensitive parameter is the income, assumed to be $40/(MW·h) from selling electricity *plus* $40/(MW·h) from selling RECs (Clean Energy Regulator 2023).

However, at what electricity price would the project become just viable (profitable)? Re-computing via the spreadsheet model provides an answer of \$79.20/(MW·h), which is very close to the figures used, \$40 + \$40 = \$80/(MW·h), so there is little margin for profit. An investor might decide that this wind farm is a suitable investment, particularly if the likely trend in interest rates is downwards rather than upwards and the trend in wholesale electricity prices is upwards. The Australian Energy Market Commission (AEMC) has released several reports over the years investigating price trends, especially for residential energy use. The 2023 report indicated price increases from 12–28 per cent in the Australian Capital Territory, New South Wales, Queensland, South Australia and Tasmania, and 5–10 per cent in Victoria (Australian Energy Regulator 2023).

What other factors might influence the investment decision? The sensitivity to the price of the turbines and their installation and operations and maintenance costs could be analysed. For example, increasing the capital cost by 10 per cent and maintaining 9 per cent as the base interest rate, changes the break-even income to \$85.5/(MW·h) — a change of 8 per cent.

Another likely sensitive parameter is the average amount of power that a wind turbine produces, which had been set at 30 per cent. Capacity factors vary between 25 per cent and 45 per cent for wind farms in Australia (NSW Government 2010).

How sensitive are the calculations to this capacity factor? Substituting an average capacity factor of 31.2 per cent increases the internal rate of return to 9.5 per cent — a 4 per cent increase corresponding to a 4 per cent increase in the capacity factor.

Some of the other parameters are financial or cost factors, which need to be checked with the suppliers or with the banks. In the next section, we will build a more detailed model of the cash flow from this wind farm project. This will reinforce the importance of time-value of money concepts and provide some insight into the financial aspects of such a project.

Is there any benefit in using NPV versus IRR? Certainly, NPV is simpler. It doesn't require any iterative calculation and a spreadsheet makes the calculations quite transparent. NPV can also be used with variable discount rates, which we have not considered above. Over a 25-year project, the discount rate will certainly vary, as a comparison of central bank interest rates over a 25-year period would confirm. However, such variations are impossible to predict, so an average value over the life of the project is usually assumed.

So, why use IRR at all? The main reason is simplicity of reporting. The project can be summed up with a single, break-even interest rate. It is important, however, to remember all the simplifying assumptions that went into its calculation.

A more detailed economic model

A more detailed economic model can also be built in a spreadsheet program. For the wind turbine example, assume the capital cost of the project is to be 70 per cent borrowed from a bank plus 30 per cent from a private investor, and that this money is to be paid back over the life of the infrastructure, namely 25 years. The bank must be paid back on a regular schedule. The private investor is prepared to accept less income initially for a higher rate of return later.

Computing the annual repayment required to the bank is complex mathematically, because each repayment amount must be discounted by $(1 + r)^n$ where r is the interest rate and n is the nth year of the loan. This is such a common calculation, however, that you will find it as a function in spreadsheet programs.

For the wind turbine with a purchase and installation cost of \$3.75 million and a discount rate of 9 per cent, the annual repayment would be \$267 241 over 25 years if 70 per cent was borrowed from a bank or other lender. However, would this annual repayment be affordable, particularly in the first few years of the project? How much would the private investor receive each year?

Table 11.7 shows that (based on the original project assumptions), the expected net income in the first year is \$236 682, which is less than the annual loan repayment required of \$267 241. In this year, the private investor would need to increase their investment by the difference, which is \$30 559.

This would be the case for the first three years (i.e. net income would be less than the necessary loan repayment figure). By year 4, the profit is \$11 868, and it continues to increase until it reaches \$590 147 in the 25th year of the project. Again, we can use the net present value method to decide whether this is a suitable investment. Discounting these future incomes to present value gives a net present value of \$1 170 540, compared to the original investment of \$1 125 000 (30 per cent of \$3.75 million), which suggests a favourable investment.

TABLE 11.7 Cash flow for a wind turbine (dollars)

Year	Bank debt at start of year	Interest	Operating & maintenance costs	Gross income	Net income	Bank debt at end of year	Discount factor	Profit/loss	Investor debt	PV
0						−2 625 000			−1 125 000	1 170 540
1	−2 625 000	−236 250	−78 894	315 576	236 682	−2 594 009	0.917	−30 559	−1 155 559	−28 036
2	−2 594 009	−233 461	−81 261	331 355	250 094	−2 560 228	0.842	−17 147	−1 172 707	−14 433
3	−2 560 228	−230 421	−83 699	347 923	264 224	−2 523 407	0.772	−3 018	−1 175 724	−2 330
4	−2 523 407	−227 107	−86 210	365 319	279 109	−2 483 272	0.708	11 868	−1 163 857	8 407
5	−2 483 272	−223 495	−88 796	383 585	294 789	−2 439 525	0.650	27 547	−1 136 309	17 904
6	−2 439 525	−219 557	−91 460	402 764	311 304	−2 391 841	0.596	44 063	−1 092 247	26 273
7	−2 391 841	−215 266	−94 204	422 902	328 698	−2 339 866	0.547	61 457	−1 030 790	33 619
8	−2 339 866	−210 588	−97 030	444 047	347 017	−2 283 212	0.502	79 776	−951 014	40 037
9	−2 283 212	−205 489	−99 941	466 249	366 309	−2 221 460	0.460	99 068	−851 946	45 613
10	−2 221 460	−199 931	−102 939	489 562	386 623	−2 154 150	0.422	119 382	−732 564	50 428
11	−2 154 150	−193 873	−106 027	514 040	408 013	−2 080 782	0.388	140 772	−591 793	54 554
12	−2 080 782	−187 270	−109 208	539 742	430 534	−2 000 811	0.356	163 293	−428 500	58 056
13	−2 000 811	−180 073	−112 484	566 729	454 245	−1 913 642	0.326	187 004	−241 496	60 997
14	−1 913 642	−172 228	−115 858	595 066	479 207	−1 818 629	0.299	211 966	−29 530	63 430
15	−1 818 629	−163 677	−119 334	624 819	505 485	−1 715 064	0.275	238 243	208 713	65 407
16	−1 715 064	−154 356	−122 914	656 060	533 146	−1 602 178	0.252	265 904	474 617	66 973
17	−1 602 178	−144 196	−126 602	688 863	562 261	−1 479 133	0.231	295 020	769 637	68 171
18	−1 479 133	−133 122	−130 400	723 306	592 906	−1 345 013	0.212	325 665	1 095 302	69 039
19	−1 345 013	−121 051	−134 312	759 471	625 160	−1 198 823	0.194	357 918	1 453 220	69 611
20	−1 198 823	−107 894	−138 341	797 445	659 104	−1 039 476	0.178	391 862	1 845 082	69 920
21	−1 039 476	−93 553	−142 491	837 317	694 826	−865 787	0.164	427 584	2 272 666	69 995
22	−865 787	−77 921	−146 766	879 183	732 417	−676 467	0.150	465 175	2 737 842	69 861
23	−676 467	−60 882	−151 169	923 142	771 973	−470 107	0.138	504 732	3 242 573	69 543
24	−470 107	−42 310	−155 704	969 299	813 595	−245 176	0.126	546 354	3 788 927	69 062
25	−245 176	−22 066	−160 375	1 017 764	857 389	0	0.116	590 147	4 379 075	68 438

SPOTLIGHT

Upgrade of concentrate thickeners

ANDREW SCHRODER, LEAD PROCESS ENGINEER, BECHTEL MINING & METALS

A copper concentrator was considering the upgrade of its existing concentrate thickeners to support an increase in production rate of approximately 50 per cent. The concentrator was already equipped with two concentrate thickeners, which were old and suffering from structural damage and leaks. An engineering team was formed to examine options and identify the preferred scope to a pre-feasibility level of detail. Three options were identified as follows.

- Refurbish the two existing thickeners.
- Replace the two existing thickeners with new units in the same locations.
- Install a single new thickener and then decommission the existing thickeners.

The engineering team then investigated each of these options. The existing thickeners were inspected and potential sites for a new thickener were proposed. Data were gathered from the maintenance department on the condition of the existing thickeners. Budget pricing was sought from thickener suppliers for key components (such as thickener rakes, drives and feedwells).

After a short investigation, the engineering team arrived at the following conclusions.

Option	Assessment
Refurbish the two existing thickeners.	• The existing thickeners were generally in poor condition and the scope of refurbishment was not well defined. Any refurbishment was at risk of high cost and schedule overrun. • It would not be possible to process the ~50 per cent concentrate without greatly reducing the amount of thickening achieved. Hence, the thickened concentrate will contain more water and so the concentrator water demand would increase, possibly beyond the limits of the water supply systems.
Replace the two existing thickeners with new units in the same locations.	• This option would involve the demolition of each thickener, followed by the construction of a new thickener in the same location. This would take many months, during which the concentrator would need to operate with a single thickener. Production would be reduced during this period, with a substantial loss in production revenue. • The thickeners would need to be replaced with larger diameter units to suit the increased production. Larger diameter units would not fit in the existing location and would not suit the existing foundations. • This option would involve construction adjacent to the operating plant. It is likely that construction in this location would be less efficient, as operations requirements would be given precedence. This would increase the construction labour costs.
Install a single new thickener, and then decommission the existing thickeners.	• A single larger thickener would be a lower equipment purchase cost than two smaller thickeners. • A single new thickener could be constructed and commissioned without interruption to production. • A 'greenfield' location was available where the thickener could be constructed efficiently, and installation would not be affected by operations. • A new thickener would be properly sized for the increased production and so would not affect the water balance.

The three options had different profiles for capital cost, project risk, production interruption, interaction with operations and safety risks.

An option was then selected using a weighted scorecard approach, which considered non-financial factors such as risk and safety. This approach was qualitative, as full financial analysis was not required for the pre-feasibility study. Weighting factors were selected to reflect the relative importance of each factor. The engineering team and site then agreed on scores for each aspect for each of the three options. The weighted scorecard results were as follows.

		Score (out of 10)		
	Weighting	Option 1 Refurbish existing	Option 2 Replace existing	Option 3 New thickener
Capital cost	0.25	6	4	7
Operating cost	0.1	5	5	5
Schedule	0.1	3	1	8
Safety	0.15	2	4	8
Production impact	0.25	2	2	10
Project risk	0.15	2	4	8
Total		**3.4**	**3.3**	**7.95**

Outcome

This analysis was performed using a qualitative system for evaluating the three options. This technique requires subjective judgements by using various experts in each category.

In this case, option 3 (install a new thickener) achieved the highest score and so was selected as the preferred option for the upgrade.

CRITICAL THINKING

Consider the sensitivity of the final decision to the various criteria and weighting factors. For example, will the outcome be greatly affected by not considering the operating cost for each of the three options? What would change if the weighting for production impact was doubled?

11.2 Technical feasibility

LEARNING OBJECTIVE 11.2 Consider the technical feasibility of engineering projects.

Now that we have looked in more detail at the cash flows for the wind turbine project, we can consider some of the technical aspects. The only technical issue that has been considered so far is the average amount of power that the turbine will produce. We will return to this issue later in this chapter, in the section on mathematical modelling. During the design of this piece of infrastructure, sustainability should be the prime consideration (Hayes & Hargreaves 2021).

For the wind turbine itself, it has been assumed that specialised contractors will be able to design and construct it in the required location. It has been assumed that the towers will be manufactured locally, and the turbines and blades imported. Local conditions would also need to be considered. Some of the key questions that would need to be answered are as follows.

- Is the soil and/or rock suitable for constructing a series of wind towers?
- What size towers will be required?
- Where will the towers be manufactured and how will they be transported to the site?
- Will they require on-site assembly? How will they be erected (some towers are 100 metres tall)?
- How and where will access roads be constructed? What disturbance to the natural environment will be required (and allowed)?
- What are the right of entry conditions that have been negotiated over private and/or governmental lands?
- How will First Nations lands in Australia, indigenous artefacts globally or significant sites be identified?
- Are there any threatened species?
- What are the expected weather conditions?

- Can the whole project be completed within 12 months, as has been assumed in the economic/financial analysis?
- How will the towers be connected into the electricity grid? Has this cost been included in the earlier financial calculations?
- What size electricity cables will be required?
- What maintenance schedule will be required?
- What approvals are required?
- What safety issues need planning?
- Consider the materials of construction especially in terms of sustainability. Turbine blades, for example, are generally made of carbon or glass fibre-reinforced epoxy plastic composites that are almost impossible to recycle.

These questions cover issues of structural engineering (the towers); foundation engineering (tower foundations, access roads); mechanical engineering (turbines and gear boxes); electrical engineering (generators and connection to the grid); environmental engineering (impact on the natural environment; impact on First Nations peoples in Australia or indigenous peoples globally or their lands); project management (scheduling, safety) and so on. It is easy to see why engineers work in teams. It is impossible to be an expert in all these areas; most engineers are trained and specialise in one discipline; however, they need to work with many other disciplines (including surveying) in the delivery of each project.

A structural engineer will design and check the towers. This is likely to be based on a standard design. Companies will often use a basic design with small modifications on different jobs. Wind turbines are usually placed on top of a tapering steel column of a standard size. The design may change depending on the site conditions. Why? The tower needs to resist the force from the wind. At windier sites, this force can be significant. Do you design all towers across Australia for the windiest site imaginable, or construct towers with slight changes of wall thickness for different sites? This is a trade-off between more complex (and more expensive) manufacturing and cost-saving in reducing material usage for some sites.

The wind force on the tower is also transmitted to the ground, where the foundation must resist the force and overturning moment (figure 11.3). How this is achieved depends on the local foundation conditions, which must be assessed by a geotechnical engineer. The towers could be situated on soil or rock. The soil could be clay, sand or gravel; the rock could be strong and unfractured or fractured and heavily weathered. All these variations affect the foundation design, which would normally be a round concrete slab with some reinforcement. Figure 11.3 shows a rough sketch for the tower loads; it does not include the weight of the whole structure, nor the wind loads on the tower.

The geotechnical engineer will be able to specify the diameter of the slab that will distribute the stresses across the ground, which will ensure the tower is safe under all expected wind conditions. A structural engineer then needs to consider how thick the slab needs to be and how it will be reinforced. The slab is a structure itself and it needs to be designed within acceptable stress limits. As this more detailed design proceeds, the basic assumptions for the economic analysis should be revisited. Detailed design should help to determine whether the foundation and all erection costs will be covered by this amount.

Even before the towers can be designed, the location of the wind farm must be determined. This will be a set of trade-offs dependent on the terrain and foundation conditions, and the distribution of the towers across the landscape, ensuring each gets as much of the wind as possible. If the towers are too close together, some may shadow others, resulting in less energy being produced. The issue of which turbines to use also needs to be analysed. That is, should a small number of large turbines be installed, or a larger number of smaller ones? The smaller ones may be better suited to locations with lower average wind speeds.

Similar considerations would determine the design of the mechanical and electrical work. For example, the location of the towers would determine the cabling connections to the electricity grid. Each turbine is a complex mechanical device with a large rotor driven by the wind connected through a gear box to a generator. All these issues would need to be analysed as part of the technical evaluation of the wind turbine project.

Similar design considerations are required for the design of jet boats. In effect, the water jet is the reverse of the wind turbine, where energy is extracted from the wind flow to create energy via a generator. In the pump, a motor is used (the reverse of a generator) to add kinetic energy to the water as it flows through the pump system and is expelled at the back of the boat.

FIGURE 11.3 Sketch for tower loads

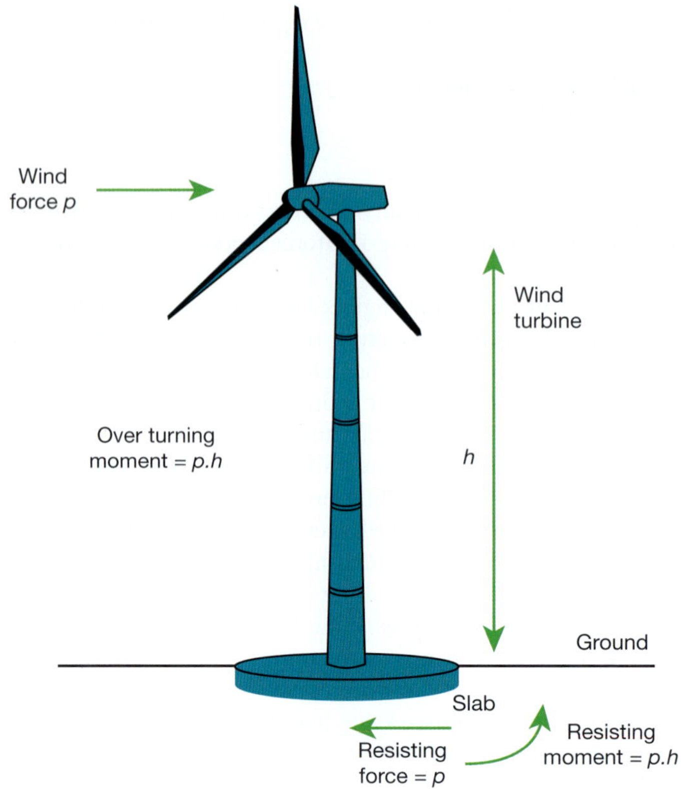

SPOTLIGHT

Water jet propulsion — HamiltonJet

HamiltonJet is a New Zealand company that has pioneered water jet propulsion in boats. This technology is most familiar in jet skis and in tourism advertisements for New Zealand that show jet boats skipping across whitewater in fast-flowing rivers. This commonplace technology is the result of more than 50 years of research and development by the company (HamiltonJet 2024).

Water jet propulsion relies on Newton's third law — to every action, there is an opposite and equal reaction. A pump onboard the boat extracts water from the river and accelerates it to a high velocity and expels it to the rear of the boat. Various pump types have been used over the years by HamiltonJet. Each subsequent development has provided greater power and/or efficiency (HamiltonJet 2024). The pump produces a flow of water at velocity *v*. Now, Bernoulli's principle (see any book on fluid mechanics) states that for an incompressible fluid such as water:

$$\frac{\rho v^2}{2} + \rho g h + p = \text{constant}$$

In this equation, ρ is the density of water (about 1000 kg/m^3), v = velocity of water, g = gravitational acceleration = 9.81 m/s^2, h = measure of elevation (height) above some reference point and p = pressure.

By considering two cross-sections in the water propulsion system (at the discharge of the pump and at exit of the jet), and assuming that there is no significant elevation difference between the two and that the pressure at the back of the boat is zero (atmospheric), then the propulsion equation becomes:

$$p_1 = \frac{\rho(v_2^2 - v_1^2)}{2}$$

In this equation, p_1 is the pressure at the discharge of the pump where there is a velocity, v_1. The high velocity, v_2 occurs within the exit jet where the pressure is atmospheric. Obviously, the larger the difference between the two velocities, the higher the pressure and the larger the thrust on the boat, which is equal to:

$$T = p_1 A, \text{ where } A = \text{cross-section at point 1 in the pump}$$

Note that neither v_1 nor v_2 are the boat velocity. If you wish to determine the boat velocity, you need to start with the thrust force and calculate the friction losses between the boat and the water (this is dependent on the boat velocity and the area of the boat in contact with the water).

This simple analysis shows how basic mathematical modelling can provide technical insight into the creation of an innovation even before any prototypes are built. HamiltonJet has become successful by applying the same basic principle to a range of watercraft, from small pleasure boats to large passenger ferries (HamiltonJet 2024).

CRITICAL THINKING

Consider the design of a personal watercraft, such as a jet ski. How much thrust is required? How much energy is consumed? How efficient are the small petrol engines used in such craft? What is a reasonable design range for such a watercraft? What are the design considerations for an electric jet ski? (See Taiga Motors 2024.)

11.3 Mathematical modelling in design

LEARNING OBJECTIVE 11.3 Apply principles of mathematical modelling to conceptual and detailed design.

As with the financial model previously discussed, most engineering situations require the use of a model to allow the understanding and prediction of system performance. Many of these models are based on 'laws' of physics, mechanics, chemistry and so on. Frequently, these 'laws' are not enough. (These laws are approximations to how the universe really works.) Equations derived from material behaviour, fluid flow, or even human behaviour, are needed to predict the performance of complex systems. These are always *approximations* to real system behaviour, but they can be accurate enough for engineering purposes. These are examples of the heuristics that were discussed in the chapter on the engineering method.

The basic engineering sciences include statics, dynamics, solid mechanics, fluid mechanics, thermodynamics, control (of systems), and circuit theory. These are usually covered in the first two years of any engineering program, although not all engineers will cover all these basics. These basic engineering sciences then underpin the more complex modelling of applications (such as wind turbines).

There are many opportunities to use mathematical modelling within both conceptual design and detailed design. During your first years of undergraduate study, your non-engineering specific subjects and your mathematics and sciences classes you will learn some of the necessary skills in calculus and differential equations, numerical methods, probability and statistics to name a few.

Fortunately, there is mathematical software to help with these tasks — software such as Excel and MATLAB for numerical computation and Maple and Mathematica for symbolic mathematics. Your challenge is to develop skills in using such tools to analyse data and build predictive models of system behaviour.

For the wind farm example, models would be needed for the towers, for soil and/or rock foundations, for the mechanical systems such as the rotor and blades as well as the gearbox, for electrical systems such as the generator, and for the communication and control systems. Models of project management would also be needed to enable the construction to be completed. Data would have to be collected (wind data, topographical data, soil and rock data, and so on) in order to better understand and predict the system to be designed. Engineers use modelling to make sense of these data and to predict system behaviour.

KEY POINT

Mathematical modelling links the physical understanding of the world with data collected about it.

Power output

Consider modelling the power output from the wind turbine example. The power output was just one issue identified from the economic modelling that had a significant impact on the viability of the whole project. How much data would be required to estimate generated power? The local meteorological organisation could be contacted to source all wind data for the nearest recording site. This may be up to tens or even hundreds of kilometres from the intended wind farm site. If an analysis of the metadata associated with the wind data revealed that it was collected at 10 metres elevation, whereas the proposed wind turbine needs to be 80–100 metres high, would this matter? Yes, it would, because wind gets slowed by passing across the ground, so wind speed at 90 metres above ground is usually greater than wind speed measured at 10 metres. This is called the boundary layer effect, a phenomenon found in all fluid flow.

A model could be used in these circumstances to extrapolate from one to the other. More details of such a model are given next. Most wind companies, however, will install instruments at a proposed site to collect data at 80–100 metres. This provides real data on which to base the technical as well as economic modelling.

Any wind data will have large variations that can occur from minute to minute, hour to hour, day to day and season to season. The data therefore needs to be sliced into short time periods and the power output from the turbine in each time period calculated, assuming that it can adjust quickly enough to different wind speeds. This should be a straightforward calculation for an engineer with spreadsheet skills.

Power can be calculated by using the principles of physics. Physics defines power as energy use per unit of time. The SI unit for power is the watt (W) and for energy it is the joule (J). One watt equals one joule per second. So, if an energy flow rate can be calculated, so can an estimate of power. The kinetic energy of a mass m, moving at velocity v, is $\frac{1}{2}mv^2$. For a moving body of air passing through a wind turbine, an equivalent term for it (the mass) needs to be used, since the wind velocity is the data set.

Mass can be estimated by first estimating the volume and then multiplying by the density. If the air is moving at velocity v m/s, the rotor has a swept cross-sectional area of A, and the density of air is ρ, then the mass flow rate is $(\mathrm{A}\cdot v\cdot \rho)$. So, the total power available is:

$$P_0 = \frac{1}{2}(\mathrm{A} \times v \times \rho)v^2$$

In this formula, the flow rate (the volume of air that passes the wind turbine every second) is A times v, and its mass is this volume times its density ρ.

The formula for P_0 is a formula for power (rather than energy) because it measures the amount of energy *per second*. Note that the density of air (about 1.2 kg/m^3) is temperature dependent. Air gets denser as it gets cooler.

Note that the formula for P_0 depends on the velocity to the third power:

$$P_0 = \frac{1}{2}(\mathrm{A} \times \rho)v^3$$

So, when the wind speed doubles, the potential power output increases by a factor of eight. This explains why it is possible to generate a large amount of energy from the wind. We will see later in the chapter that turbines are designed to harvest less than the maximum amount of energy available.

This formula estimates the total power that can be generated by the wind. In fact, it is not possible to extract more than 59 per cent of this power. This physical limit was identified by German physicist Albert Betz in 1919 and published in 1926 (Betz 1926). By a very simple analysis, he showed that the maximum energy can be extracted when the wind speed is reduced in speed to one-third of its initial value as it passes through the turbine blades. This creates an amount of power equal to 16/27 or 59 per cent of the total power available, P_0. So:

$$P_{max} = \frac{16}{27} \times \frac{1}{2} \times \mathrm{A} \times \rho \times v^3 = 0.59P_0 = (\mathrm{A} \times \rho)\left(\frac{2v}{3}\right)^3$$

Now that a simple model has been created for the maximum amount of power that can be generated out of a wind turbine, how could the actual amount of power input be calculated if the wind speed is constantly changing? One alternative would be downloading hourly wind data from the meteorological service, calculating P_{max} for every hour of record, and adding it all up. However, it would not be likely that the turbine is producing the maximum power at every moment. It is more likely to produce some fraction: $\mathrm{k} \cdot P_{max}$, where $\mathrm{k} < 1$ and k is the efficiency of the turbine — the ratio between the actual power output and the maximum theoretically possible. How could the value of k be determined? Most likely, the only people who will know that are the turbine manufacturers. They will have done wind tunnel tests and extensive field tests and will know the response of the turbine under various conditions.

Consider the Vestas 3.3 MW wind turbine. Its power curve is outlined in figure 11.4. Reading this curve suggests that at 11 m/s (40 kph) the turbine reaches maximum power output at 3.3 MW. At higher speeds, the power output is constant. The turbine blades (which have a swept radius of 54.65 m) actually tilt themselves into the wind to stop the rotor from spinning out of control. Wind turbines are built for a long life rather than maximum power output. Now, substituting in the earlier formula for P_{max}:

$$P_{max} = (A \times \rho)\left(\frac{2v}{3}\right)^3 = (\pi 54.65^2)(1.2)\left(\frac{2 \times 11}{3}\right)^3 = 4.4\,\text{MW}$$

FIGURE 11.4 Power curve for the Vestas V112–3.3 MW® wind turbine

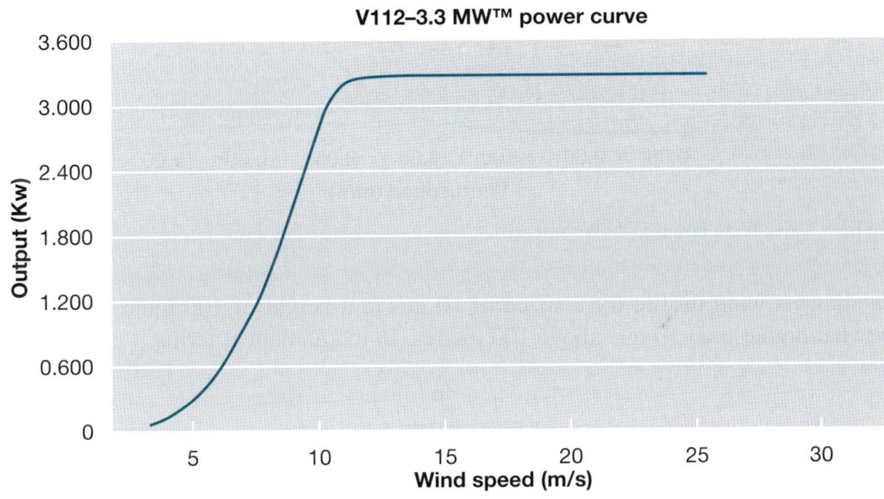

Source: Vestas Wind Systems A/S (2013).

So, the wind turbine is running at about 74 per cent maximum efficiency at $v = 11$ m/s (k = 74 per cent). Looking again at the power curve in figure 11.4, what is the efficiency at around 10 m/s? For $v = 10$ m/s, $P_{max} = 3.49$ MW, the power output is about 3.0 MW and, therefore, the efficiency is about 85 per cent. At a velocity of 7.5 m/s, the efficiency is about 87 per cent. The efficiency factor k varies with wind speed. Wind turbines are deliberately designed to operate at high efficiency at low wind speeds, to extract as much energy as possible. At higher speeds, it is more desirable to stop the turbines from spinning too fast, so the turbine deliberately extracts proportionally less energy from the wind.

Maximising energy production

It is essential to consider how to get the most energy out of the wind farm project. For each wind turbine, you might think choosing the height of the tower would be a simple decision. Yet, this is also a trade-off between cost and energy production. Each extra 10 metres of height costs roughly US$15 000. The benefit is higher wind speed and more electricity production, as will be seen next. The aesthetics of the situation are also important to consider. Towers about the same height as the rotor diameter seem to be aesthetically pleasing and give a useful compromise between extra cost and extra production.

Also consider the design of the towers — most towers are now tapering, tubular columns of steel. They taper because the bending resistance needs to increase linearly from top to bottom. A tapering cross-section reduces the amount of material used — a benefit for the world's limited resources. Alternative designs could include lattice (three-dimensional truss) towers, guyed towers or a hybrid of tubular segments and lattice construction. Clearly, there is scope for optimisation in design and construction that balances material used, cost of assembly, cost of construction (erection) and aesthetics.

Consider the wind profile at any location. The wind swirls around the world and, at 1 kilometre above the surface, it is largely unaffected by obstructions on the surface (assuming reasonably flat terrain). The logarithmic relationship of a wind profile is represented in figure 11.5.

FIGURE 11.5 Logarithmic relationship of a wind profile

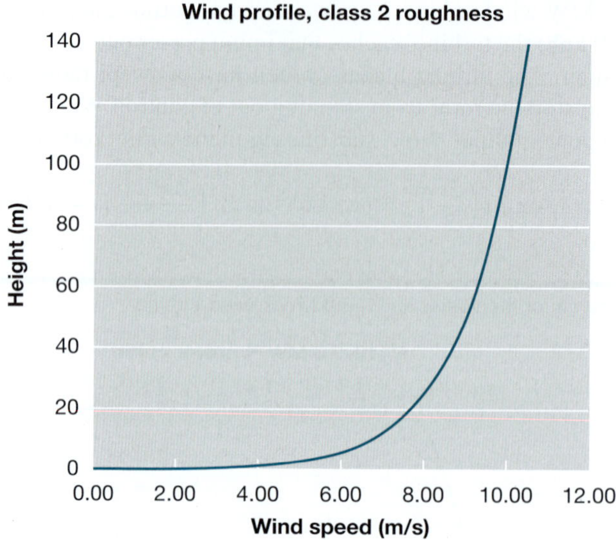

Source: Adapted from Swiss Wind Power Data Website (n.d.).

This figure shows a wind profile for a speed of 10 m/s at a height of 100 metres. Note that the wind speed is somewhat higher than 10 m/s above 100 metres. In mathematical terms:

$$v = v_{\text{ref}} \frac{\ln\left(\frac{z}{z_0}\right)}{\ln\left(\frac{z_{\text{ref}}}{z}\right)}$$

For the example above, the reference height, $z_{\text{ref}} = 100$ metres and the reference velocity, $v_{\text{ref}} = 10\,\text{m/s}$. If the roughness height is $z_0 = 0.1$ metres, then velocities at other heights can be calculated as follows in table 11.8.

TABLE 11.8 **Wind speed, *v* (m/s), versus elevation, *z* (m)**

z (m)	*v* (m/s)
0.1	0.00
1	3.33
5	5.66
10	6.67
20	7.67
40	8.67
60	9.26
80	9.68
100	10.00
120	10.26
140	10.49

There are several conditions that can influence the ideal vertical wind speed profile. One is the effect of Earth's terrain or surface roughness. This variation in wind speed has critical influences on the assessment of wind energy resources and therefore the design of the overall system.

The roughness height is the height above the ground where the wind speed is theoretically zero. This parameter is dependent on the actual terrain of the land in the vicinity. It is an empirical parameter that characterises the influence of surface irregularities on the vertical wind speed profile; the rougher the terrain, the thicker will be the affected layer.

Consider the effect of the surface roughness, z_0, mentioned previously. A roughness of 0.1 metres represents 'Agricultural land with some houses and 8-metre-tall sheltering hedgerows with a distance of approx. 500 metres'. For other terrains, the Swiss Wind Power Data Website (n.d.) or Spera (2009) provide guidance on roughness heights.

Safety

So, what can go wrong with a wind turbine? A risk assessment (as discussed in the chapter on the engineering method) would produce a list detailing the potential concerns as:

- overspeed damage
- blade failure
- vibration failure.

As the wind velocity increases, the rotors are designed to adapt so that they quickly reach a maximum speed. This is done in two common ways: pitch-controlled or stall-controlled. In the former, the blades have a tilting mechanism that reduces their angle with respect to the wind. This reduces the power coming from each blade and stabilises the speed with which the rotor turns. In stall-controlled rotors, the blades are fixed. However, they are designed so that as the wind velocity increases, turbulence is created, which reduces the blades aerodynamic efficiency. This stops the rotor from spinning too quickly.

From a risk point of view, the stall-controlled blade would seem to be a safer design. It does not depend on an active mechanism that might go wrong. The pitch-controlled blades are dependent on such a mechanism, which could fail for a variety of reasons. Either way, it is mandatory to have an overspeed control mechanism, which effectively acts as a braking mechanism.

A blade failure could be potentially catastrophic because it would result in significant out-of-balance forces on the rotor. There would need to be some kind of detection device, which would, in turn, activate the braking mechanism already discussed.

Vibration could be caused by blade failure, as discussed. However, it might also be caused by other factors such as a bird strike, or turbulent wind conditions, or a range of other factors. Each wind turbine is equipped with a detection device that activates the braking mechanism under these conditions.

Further information about the design and safety of wind turbines can be found at Windpower Engineering & Development (2015).

Checking

It is essential that models are checked to ensure calculations are correct. For example, when translating equations into a spreadsheet, hand calculations or a textbook could be used to check the accuracy of the formulae. This is model **verification** — checking that the model is right. These verification tests should be carefully documented in a logbook and made available to other users of the spreadsheet.

It is also important to check that you have the right model. This is model **validation** — checking the model output against real data. If the model does not provide predictive capability, a better model needs to be developed. If you have a number of data sets, you should use some data sets to build the model and other data sets to validate the model.

In the wind turbine example, if power output data did not match our theoretical model, then either the data is unreliable, or the model is not predicting the data very well. When translating the power formula into a spreadsheet or computer program, it would be important to do some hand calculations on a couple of cases to make sure that it is coded correctly.

Hierarchy of models

It is also important to build models of appropriate sophistication. Very simple models (e.g. simple averages or a line of best fit) may be appropriate for some problems or designs. For problems or designs that involve complex systems and/or require very high degrees of accuracy to be obtained, it will be necessary to build models of much greater complexity — which will of course require more time and money.

Sometimes it is necessary to give a quick answer to a problem. For example, which wind turbine should be chosen for this location? A rule of thumb might be used based on average wind speed, or what has been previously chosen for a similar location. However, that recommendation should be verified by using detailed power calculations. If you worked for the manufacturer designing the rotor blades, you would use computational fluid dynamic models of great complexity.

Likewise, if the model is used for making decisions of high value (millions or billions of dollars), then you are more likely to have the time and resources to build models of greater complexity and, hopefully, accuracy. This is the idea of a hierarchy of models. You operate at the level of sophistication that makes sense in the situation. Think about the design of the rotor blades. Large wind turbine companies may sell thousands of each model. Improving its performance by just 1 per cent can make a huge impact to the payback period for the turbine, so it is worth investing large amounts of time to constantly improve the design through sophisticated modelling.

Simpler models can also be used to check the output of more sophisticated models, where it may be impossible to solve the equations by hand, or where there are no simple textbook solutions. Similarly, it is often possible to substitute output from the model into the governing equations to make sure that the values make sense. For example, basic force and moment equilibrium should hold for a structure. This means that the vector sum of all forces on a body is zero and so is the vector sum of all resultant moments. A mass balance should exist across a hydrological, chemical, or electrical system.

SPOTLIGHT

Software checking

Jasmine, a junior engineer, worked in an office where finite element modelling was being used to calculate stresses in a radial gate for a major water reservoir. Radial gates sit on top of a dam spillway and allow a large amount of water to be released from a reservoir in times of major flood. Radial gates are a complex solution to the need for greater flood flow capacity. The simpler solution is a static spillway, but a much wider one. The gates also give operators much greater control over how a flood peak can be reduced as it passes through the reservoir. Attenuation means reducing the maximum flood flow to minimise flood damage to towns and cities downstream.

Jasmine performed basic checks on the computer program's output — checking that the sums of the forces in the X, Y and Z directions were zero (force equilibrium). However, they were not zero. Something was being miscalculated inside the program. After much discussion around the office, it was suggested that the problem might be 'round off' errors, which accumulate when millions of calculations are performed inside a computer with limited precision (they were using a computer with 32-bit 'words', which provide about 8–9 decimal digits of accuracy).

The program was converted to run in 'double precision' (64 bits), which gave about 16 decimal digit accuracy. This solved the problem. From this experience, Jasmine learned about the importance of checking the output from computer programs. With incorrect calculation of stresses, the structure could have been incorrectly designed, possibly leading to collapse and loss of life.

CRITICAL THINKING

When you perform complex calculations in other subjects, how do you check that they are correct? Can you find a simpler method that would confirm a more complex calculation? Find one example of a simple check procedure for a more complex calculation.

SUMMARY

This chapter has explored the evaluation process of engineering problem solving in detail. The overall process was introduced in the chapter on the engineering method. It has concentrated on economic modelling to check the economic feasibility of a project, using a wind farm as an example. Mathematical modelling has been discussed as a process for checking technical feasibility. Other chapters have discussed issues of risk, safety and quality in the design process. We will now briefly revisit each of the learning objectives.

11.1 Evaluate solutions in terms of basic economic principles and consider non-economic criteria in the selection of project options.
The complex part of engineering is usually the evaluation phase of the design process. This is where your technical engineering skills are required to compute performance measures of the various solutions you have identified, for example, power output, battery life, antenna range, braking capacity or stress. When these technical issues have been estimated, the economic assessment of the product can be completed, using techniques such as NPV, IRR, ROI, cost–benefit and payback period.

During the evaluation of the various options for an engineering project, non-economic criteria such as environmental, social, institutional and system reliability need to be considered.

11.2 Consider the technical feasibility of engineering projects.
Most engineering projects require complex calculations that go beyond the scope of this text to discuss. This chapter has indicated how basic fluid mechanics and engineering mechanics are used in the design of wind farms and jet boats.

11.3 Apply principles of mathematical modelling to conceptual and detailed design.
In the process of evaluating the solutions, you will need to use your mathematical modelling skills. These apply as much to the economic analysis as they do to the physical and biological modelling of your proposed solutions. You will draw upon data collected and various existing mathematical models to predict the performance of each of your solutions. In some cases, you will need to develop your own models from first principles. Your various engineering subjects will provide you with the skills for mathematical modelling of engineering systems.

KEY TERMS

cost–benefit (CB) A systematic approach for comparing project options with fluctuating cash flows calculated as the present value stream of net revenues divided by the present value stream of investments.

internal rate of return (IRR) The interest rate at which the NPV is zero, and an estimate of the upper limit on a viable interest rate.

net present value (NPV) The current value of future costs and revenues, required to calculate the capital investment required for a project.

payback period (PBP) The period of time required to recoup the funds expended in an investment/project, or to reach the break-even point, usually expressed in years.

return on investment (ROI) A performance measure used by businesses to identify the efficiency of an investment by calculating the ratio of net revenue over investment.

sensitivity analysis The examination of the outcomes from a model by systematically varying the input parameters, usually by checking reasonable upper and lower bounds on each parameter.

validation Checking the model output against real data. Is this the right model?

verification Checking that the model has been correctly implemented. Is the model right?

working drawings Detailed drawings used for construction or manufacturing.

EXERCISES

1 Briefly explain both the NPV and IRR methods of financially evaluating solutions to a lay person (e.g. a family member).

2 Consider the following financial calculations.

(a) Determine the present value of a single payment of \$1000 in five years' time at a discount rate of 10%.

(b) Which has greater present value, A: \$200 at the end of this year or B: \$500 at the end of ten years? Assume a discount rate of 7%. At what discount rate are they equal?

(c) If a project will return a net profit of \$1 million at the end of each year for ten years, then what is its NPV for a range of discount rates from 5% to 10%? (Use a spreadsheet.)

(d) You are advising your family on the sale of an engineering asset. The asset currently generates annual income of \$100 000. What is a reasonable sale price? Estimate and justify your choice of a discount rate based on current market rates.

3 Rework the wind farm case where the wind farm is to be sold after ten years. What should the sale price be to maintain the NPV as determined above?

4 You have been asked to recommend a choice of wind turbines for a new wind farm. Obtain some wind data from a nearby location and build a computer model that calculates power for the range of wind velocities. *Hint:* You can do this in a couple of ways. One is the brute force solution of simply computing power for every hour of data that you have, which may be many. Another is to sieve the data into categories and compute the frequency of each category; then just compute the average power output for the category and multiply by its frequency.

5 Predict your city's water usage over the next 20 years. Take a hierarchical approach. Can you make this estimate in less than one hour? A good internet search or a call to your local authority might be enough. To verify this result, collect some basic data and fit a line of best fit. How confident do you feel about this approach? Finally, build a more sophisticated model, including estimates of births and deaths and changes in water consumption per head of population over this period.

6 Imagine that a company has the following average daily costs and revenues associated with operating a tunnel.

- Average daily fixed costs of \$80 000.
- Average daily variable costs of five cents per vehicle using the tunnel (for costs such as administration and cleaning).
- Average toll price per vehicle of \$2 (for the sake of simplicity, assume all vehicles regardless of size are charged the same toll).

Based on this information:

(a) calculate the daily number of vehicles the tunnel would need to attract to break even (i.e. cover costs) each day

(b) if the average daily costs remained the same, but the toll price for using the tunnel was \$2.95, what is the new break-even number of vehicles per day?

PROJECT ACTIVITY

Basic human needs include food and water, shelter, energy, communication and transport. To these, health services, community services, mining, manufacturing and entertainment can be added. You may be able to think of others.

Choose an engineering opportunity in one of these areas, which fits your discipline. Some opportunities for the future include better means of growing food, new water sources, renewable energy, new communication devices or services, electric cars, public transport improvements, a bionic eye, online communities, new mining technologies, environmentally sensitive manufacturing and new entertainment media and services.

Your team's challenge is to put together a business proposal for your chosen project. The proposal should demonstrate your capability across the range of techniques discussed in this chapter.

You approach this activity by evaluating the economic feasibility of one or more potential solutions. Try to obtain as much of a realistic cost estimate as possible. What are the development costs? What is the expected income each year? How many devices will you sell?

REFERENCES

Acciona n.d., 'Wind power evolved: AW1500', http://meridian.allenpress.com/jfwm/article-supplement/138729/pdf/10_3996_032012-jfwm-024_s9

Australian Energy Regulator 2023, 'Retail data shows continued support needed for energy consumers', *News release*, 30 November, https://www.aer.gov.au/news/articles/news-releases/retail-data-shows-continued-support-needed energy-consumers

Betz, A 1926, *Wind-Energie und ihreAusnutzungdurchWindmühlen (Wind energy and its use by windmills)*, Vandenhoek & Ruprecht, Göttingen, reprinted in 1982, ÖKO-Buchverlag, Kassel.

Blewett, D 2023, 'Wind turbine cost: How much? Are they worth it in 2023?' *Weather Guard*, 20 March, https://weatherguardwind.com/how-much-does-wind-turbine-cost-worth-it

Clean Energy Regulator 2023, 'Small-scale technology certificates', https://cer.gov.au/schemes/renewable-energy-target/small-scale-renewable-energy-scheme/small-scale-technology-1

——— 2024, 'Renewable Energy Target', https://cer.gov.au/schemes/renewable-energy-target

HamiltonJet 2024, https://www.hamiltonjet.com

Hayes, S & Hargreaves, D 2021, *FEIAP Guidebook on Infrastructure Sustainability*, FEIAP, The Hague, https://www.ies.org.sg

Koen, B 2003, *Discussion of the method*, Oxford University Press, New York.

Martland, CD 2006, 'Project evaluation', lecture notes, MIT OpenCourseWare, Massachusetts, Institute of Technology, Cambridge, https://ocw.mit.edu/courses/civil-and-environmental-engineering/1-201j-introduction-to-transportation-systems-fall-2006/lecture-notes/lect19.pdf

NSW Government 2010, 'The wind energy fact sheet', Department of Environment, Climate Change and Water, https://www.environment.nsw.gov.au/resources/communities/100923-wind-facts.pdf

Spera, DA 2009, *Wind turbine technology*, 2nd edn, ASME Press, New York.

Swiss Wind Power Data Website n.d., 'Wind profile calculator', https://wind-data.ch/tools/profile.php?lng=en

Taiga Motors 2024, 'Orca', https://www.taigamotors.com/en/products/orca

Thamhain, HJ 2013, 'Contemporary methods for evaluating complex project proposals', *Journal of Industrial Engineering International*, vol. 9, https://doi.org/10.1186/2251-712X-9-34

Vestas Wind Systems A/S 2013, '3MW platform', https://www.vestas.com

——— 2024, 'Wind turbine product portfolio', https://www.vestas.com/en/products

Windpower Engineering & Development 2015, 'Designing wind turbines for safety and compliance in a global marketplace', 17 November, https://www.windpowerengineering.com/designing-wind-turbines-for-safety-and-compliance-in-a-global-marketplace

ACKNOWLEDGEMENTS

Photo: © Fokke Baarssen / Adobe Stock Photo

Photo: © MNBB Studio / Shutterstock

Photo: © ChameleonsEye / Shutterstock

Photo: © Liubomir / Adobe Stock Photo

Figure 11.1: © Martland, CD 2006, 'Project evaluation', lecture notes, MIT OpenCourseWare, Massachusetts Institute of Technology, https://ocw.mit.edu/courses/civil-and-environmental-engineering/1-201j-introduction-to-transportation-systems-fall-2006/lecture-notes/lect19.pdf

Figure 11.2: © Acciona, 'Wind power evolved: AW1500', http://meridian.allenpress.com/jfwm/article-supplement/138729/pdf/10_3996_032012-jfwm-024_s9

Figure 11.4: © Vestas Wind Systems A/S 2013, '3MW platform', https://www.vestas.com

Text: © Andrew Schroder, Lead Process Engineer, Bechtel Mining & Metals

CHAPTER 12

Engineering decision making

'When you have to make a choice and don't make it, that is in itself a choice.'

William James

LEARNING OBJECTIVES

After studying this chapter, you should be able to:

12.1 discuss the elements of a decision-making process

12.2 articulate the complexity of a decision-making process

12.3 use the capabilities of a team to improve the decision-making process

12.4 assess the ethics, safety and quality of a decision

12.5 discuss characteristics of decision support tools, techniques and systems.

Introduction

Sometimes big decisions are made with a fanfare of headlines, and sometimes they are made quietly with no more drama than a puff of air. At a dinner party in Tokyo in the summer of 1950, 21 of Japan's most influential corporate leaders, who accounted for some 80 per cent of the country's industrial capacity, listened to W Edwards Deming, an obscure American statistician who had been to Japan only once before. He was certain that he knew how to solve postwar Japan's economic problems. The pursuit of quality, Deming said, was the key to higher productivity, bigger profits, more jobs and, therefore, a richer society. But quality had to be pursued along every link of the supply chain, with the active cooperation of everyone from suppliers to the humblest worker on the factory floor. If Japanese companies followed his 14 points, Deming promised his dinner companions, their goods would be world-class in five years. The notion seemed ridiculous.

At the time, the term 'Made in Japan' was such a joke that some factory owners set up operations in the village of USA so that they could mark their products 'Made in USA'. Japan's industrialists didn't have any other ideas, so they decided to take up Deming's challenge. Deming was a classic prophet without honour in his own country. It wasn't until the 1980s that US manufacturers adopted his principles, largely to meet the competition from Japan. The formalisation of these principles became known as total quality management (TQM). TQM is now, and has been for quite some time, a very important management approach used by many organisations around the world.

Engineers make decisions every day of their professional life; some are 'small', while others are 'large' and can be very complex. This chapter introduces the processes and structure related to decision making, including the elements involved in decision making and the decision-making pyramid. Decision making can relate to issues that are static (little change with time) or dynamic (significant change with time). These challenges can be classified as simple, complicated, complex or chaotic/wicked. As the complexity increases, it is crucial that effective teams are used to determine the final decision. Understanding team dynamics is, therefore, vital for the engineer in these circumstances. There is a range of decision support systems available to assist the team (or individual) to make effective decisions.

12.1 Engineering decision making

LEARNING OBJECTIVE 12.1 Discuss the elements of a decision-making process.

As outlined in the chapter on the engineering method, a combination of constraints and weighted criteria can be used as an aid in decision making by eliminating unreasonable solutions and highlighting the most preferred options. In the chapter on evaluating options, we considered the design of a wind farm. Criteria to be weighted in this example might include issues such as environmental impact; construction time; short-term and long-term costs, such as capital and maintenance; responsiveness to certain wind speeds; and capacity (power output). These criteria fall within the basic categories of technical, social, economic and environmental issues.

Even after detailed analysis, the decision about which solution to choose from a range of feasible alternatives may not be straightforward. For example, while a range of design solutions may be technically feasible, there may still be technical differences to consider. One proposed solution might use technology unfamiliar to local contractors. Another might have a significantly longer design life. Environmental and social effects that may influence which solution is chosen may not be easy to build into an economic model.

For example, some people consider wind turbines to be noisy, and there have been problems with birds being killed by the rotating blades. Such issues could change the solution that is chosen, and this is all part of the decision-making process. At the detailed design stage, issues such as safety, quality and risk must all be considered. We will now discuss these in more detail.

KEY POINT

Risk, safety and quality issues associated with potentially feasible problem solutions must be evaluated as part of the engineering decision-making process.

Engineering decision support

Ullman (2001) sets out key criteria for effective engineering decision making as a key step in problem solving: 'Problem solving is generating and refining information, punctuated by decision making'. Ullman (2001) points out that most of us have never learned effective decision-making skills, particularly when operating as teams or groups. Many people spend their time pushing their own agendas at meetings, trying to have their views adopted. As Ullman (2001) says, a lack of useful decision-making processes means that the skills and knowledge of the team participants are poorly used, and this may explain why many people hate meetings so much.

One key issue of decision making is to recognise that all decisions imply a commitment to spend resources. In an engineering project, this will mean resources in design, construction, manufacturing, maintenance, operations, and so on. This underpins the importance of a *life cycle approach* to engineering decision making. Design decisions have implications throughout the life of a product, which might be 100 years or more. The product cost should be minimised over its lifetime. Life cycle assessment (LCA) was discussed in some detail in the chapter on sustainable engineering, which might be useful to revisit.

Likewise, issues of design for safety and for quality and risk must also be essential components of design thinking. These are considered later in the chapter.

The decision-making pyramid

In the chapter on understanding the problem, the link between data, information, knowledge and wisdom was discussed. Putting these into the context of decision making, we get the **decision-making pyramid** (Ullman 2001). In figure 12.1, wisdom is replaced by judgement and the decision, and information is replaced by models.

FIGURE 12.1 The decision-making pyramid

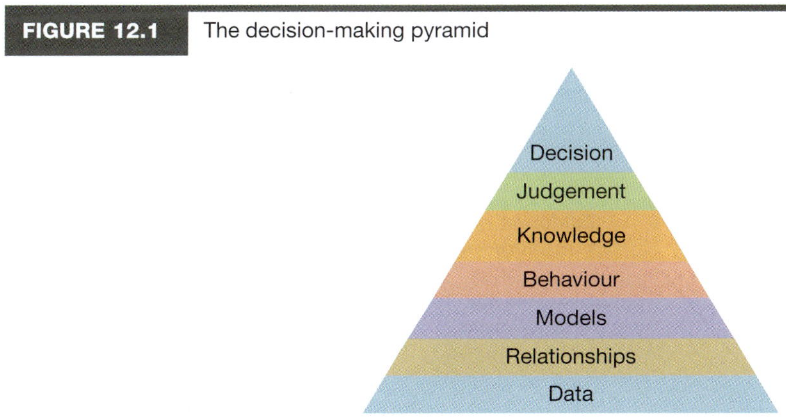

Source: Ullman (2001).

In the twenty-first century, we are all overwhelmed by data, and with the data we need to make some decisions. To do that, we need to make sense of it, and a first step is recognising relationships in the data. This leads to models, as discussed in the chapter on understanding the problem. For example, we might have data about internet traffic. We notice that it varies predictably through the day and across the week, with less traffic when people are asleep or getting ready for work and peaks during the day and again at night. Once the raw data is validated and enriched with metadata it can be interpreted to provide information that can be added to other data/information, modelled and analysed, thereby building a more comprehensive knowledge of the system. We could then develop statistical models to predict traffic by hour throughout the week. As we explore the behaviour of the models, we begin to extract *knowledge* of the system. We begin to understand the users of the system, when they are online, and when they are not.

Of course, there are also times of unpredictable behaviour, for example, a major event such as the release of a new version of an operating system when there might be millions of users wanting to upgrade. Our knowledge of the system, and our wisdom and judgement, could help us *decide* to schedule the release of the operating system upgrade at a time of otherwise low internet usage, for example, over a weekend when business traffic is low. This would avoid clogging major internet links.

Elements of decisions

Decision making in engineering requires those decisions to be systematic and well documented. For this to happen, there needs to be some clear protocols for tracking the decisions made, such as controversial situations where public consultation with a range of stakeholders and a range of conflicting views are involved.

Rittel and Webber (1973) outlined a basic methodology called Issue Based Information System, or IBIS, that they believed would help people address the 'wicked problems' they discussed. They recognised that every decision springs from a set of *issues* that need to be resolved. On each issue, there can be a range of *positions,* and, for each position, there can be a range of *arguments*. Consequently, it should be possible to systematically document the issues, positions and arguments that would lead to an effective decision.

Consider how this might be done in the case of climate change. What are the issues? Pick one, such as rising sea levels. What are the positions? The sceptics might say there will be minimal sea level rise. Others might say 0.5 metres over the next 50 years. Others might say between 2 and 3 metres by the end of the century. What are the arguments that are presented for each of these positions? You might like to investigate some of the claims and counterclaims for each of these positions.

Ullman (2001) builds on the work of Rittel and Webber, starting with an issues-based approach and proposing a shift to a *decision*-oriented view. The *issue* is the commencing point for the designer. It might be the design of a new product (or a component of a new product) or the resolution of a problem in an existing system. Each issue has an associated set of design criteria (specifications, requirements and goals) and design alternatives (options, proposals, ideas), as discussed in the chapter on engineering design and shown schematically in figure 12.2.

FIGURE 12.2 Key elements of decision making

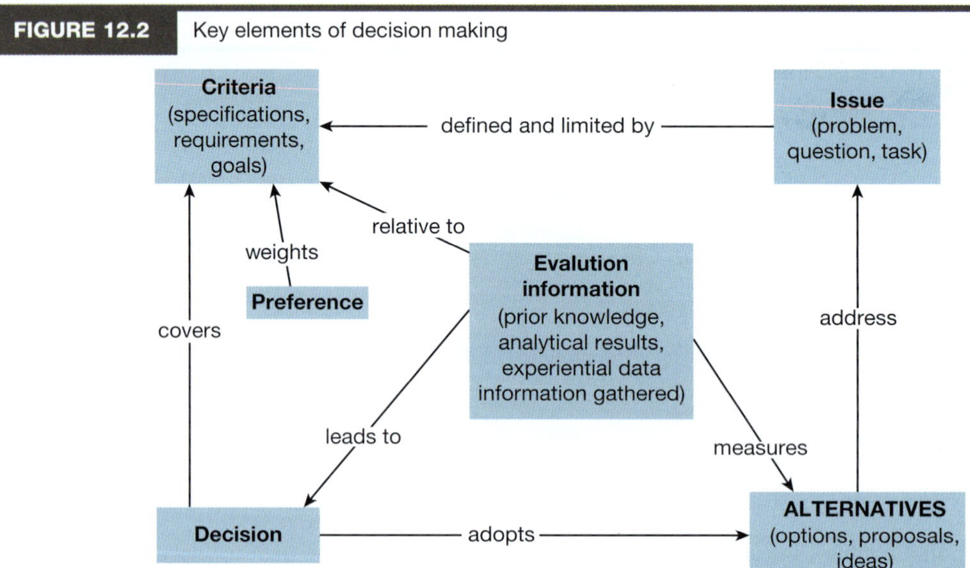

Source: Ullman (2001).

Each criterion is an issue, requiring a process of resolution by carefully documenting the propositions and arguments. For example, design of a new public transport component should include disability access. What is the nature of this access? How can it be codified for the designers? What standards exist? Who needs to be consulted?

Likewise, each alternative has positives and negatives, as discussed in the chapter on the engineering method. Some alternatives can be quickly eliminated because they violate design criteria, for example, size, weight, cost, efficiency, environmental impact, noise and safety. After some research, a design team will have developed a shortlist of *feasible* alternatives. The task then is to *decide* which of them should be implemented.

The weighted rating system described in the chapter on the engineering method is one way to weigh up the shortlist by rating each alternative against the design criteria. This can involve a range of data gathering, including modelling system performance as described in the chapter on evaluating options — both technical performance and economic performance. Measures of social and environmental performance are also necessary. These are often grouped together within an environmental impact assessment or EIA (discussed in more detail in the chapter on sustainable engineering).

This all looks easy enough. So, why is decision making so difficult? Well, some of the confounding *issues* are:

- reaching consensus in a team (e.g. when team members rate and weight factors differently)
- making ethical decisions
- making environmentally responsible decisions
- acknowledging dominance and power relationships in the team
- coping with complicated, complex and wicked situations and systems
- working with intercultural differences (and language differences).

The following sections explore these issues in more detail, providing a range of examples of effective engineering decision making.

12.2 Complexity

LEARNING OBJECTIVE 12.2 Articulate the complexity of a decision-making process.

Payne et al. (1993) describe the difficulties of making complex decisions. They highlight three important aspects: the problem, the person and the social context. Decision making is an adaptive approach to the complexities of the interactions between these three aspects. Decisions are not just about the problem; they must be acceptable to the group, and they must be cognitively accessible to all those involved. This likely means that perfect information is not available, either because it is too difficult or too expensive to obtain. Therefore, the decision must be made with incomplete and uncertain information. There is also a tendency to take shortcuts, to hasten a decision where the cost of additional information exceeds the group's willingness to commit to obtaining that information.

Static and dynamic problems

Problems or issues can be categorised as *static* or *dynamic*. **Static problems** are those where the alternatives and criteria are reasonably known and unchanging. These constitute many engineering problems, and much effort in the last 50 years has been devoted to understanding and optimising the choices. Examples include design of a wheel assembly for a car or an antenna for a mobile phone.

Dynamic problems, on the other hand, are those where the alternatives and criteria are changing with time in a way that needs to be considered. For example, the design of cars has been reasonably static over the last 50 years at least. Although the designs have changed, the key parameters have not changed substantially. All cars have been based on the internal combustion engine, four wheels, braking system, luggage space and so on. At regular intervals, innovations have occurred, such as those stemming from the desire for greater safety. Airbags have become standard components, in even the smallest cars.

Compare that to the current situation, where there is a surge in the use of hybrids, plug-in hybrids and fully electric cars, which are coming to the market from most major suppliers (Winton 2022). There is also a lot of research and development of driverless cars and cars fuelled by hydrogen (Winton 2022). If you were working for one of the major car manufacturers, what would you be designing for the future?

At the extreme end of dynamic problems are 'wicked problems', discussed in the chapter on sustainable engineering. Rittel and Webber (1973) were the first to formally name such problems. Climate change is a wicked problem, as is global poverty, providing clean water and food to all humanity, world peace, and so on. At a city scale, wicked problems can include public transport, water and wastewater, poverty, global pandemics and public health more broadly. The provision of affordable and reliable energy that is generated primarily from renewable resources across a major city is also a wicked problem. Another example might be the provision of effective transport systems in a major city where the vehicles are a mixture of cars as we know them today, driverless cars, trucks and buses.

Ritchey (2007) defined ten characteristics of wicked problems.

1. There is no definitive formulation of a wicked problem (defining wicked problems is itself a wicked problem).
2. Wicked problems have no stopping rule. That is, they are problems that are ongoing. They need to be managed, not solved.
3. Solutions to wicked problems are not true or false, right or wrong, but better or worse.
4. There is no immediate and no ultimate test of a solution to a wicked problem.
5. Every solution to a wicked problem is a 'one-shot operation' because there is no opportunity to learn by trial and error; every attempt counts significantly.

6. Wicked problems do not have an enumerable (or an exhaustively describable) set of potential solutions, nor is there a well-described set of permissible operations that may be incorporated into the plan.
7. Every wicked problem is essentially unique.
8. Every wicked problem is a symptom of another problem.
9. The existence of a discrepancy representing a wicked problem can be explained in numerous ways. The choice of explanation determines the nature of the problem's resolution. That is, the problem looks different to every stakeholder.
10. The planner has no right to be wrong (planners are liable for the consequences of their actions).

Examples for managing wicked problems are given later in the chapter.

Strategies for static problems

There are two classes of strategies for dealing with static problems: *compensatory* and *non-compensatory* strategies. A decision matrix is one example of a compensatory method. In these methods, stronger performance on some measures compensates for weaker performance on others. For example, in choosing a car, good technical performance (speed, economy, etc.) might outweigh poorer performance in the sound system. Someone else might have exactly the opposite set of priorities. This is the role that weighting factors play, allowing the decision maker to adjust the importance of different criteria. The decision matrix approach was originally proposed by Pugh (1990). A more detailed treatment of decision making for multiple objectives is provided by Keeney and Raiffa (1993).

Non-compensatory methods use an elimination approach, and this is often used in the early stages of gathering a list of suitable alternatives. If an alternative fails against a criterion, then it is eliminated from consideration. This might occur for various reasons, such as 'that'll never work!' Whether this is a robust analysis of the alternative is questionable. Reasons to eliminate alternatives might include a lack of familiarity with the technology, cost or environmental impact.

Non-compensatory methods are often used to simplify the choice, reducing the number of feasible alternatives to, in turn, reduce the cognitive load (Ullman 2001). In *The paradox of choice*, Schwartz (2004) points out that most of us do poorly when confronted with too many choices. Consequently, eliminating obviously poor choices can be a useful strategy. The danger is that it is easy to quickly eliminate an option when a more careful analysis would show that it is the best option available. This has been confirmed by research reported by Ullman (2001).

Generally, most people are not good decision makers. Keeney (2004) makes the point that there is a need to equip everyone with better decision-making skills, and that tertiary education is not yet doing that. (We will discuss this issue in more detail later in the chapter.) A further complication is that time pressure usually means that the engineer is still learning about the problem, as well as the range of alternative solutions available, when the decision needs to be made and the project moved to the next stage. People trade accuracy for time saved, preferring less accuracy in shorter time (Ullman 2001), and this works against methodical decision making. An industrial experiment showed that a team of managers made poor intuitive choices but much better decisions when using a more time-intensive, structured decision support process (Naude et al. 1997).

What is needed is the *accuracy* of a compensatory method (e.g. the decision matrix) with the *effort* of a non-compensatory method (such as focusing on key attributes).

Strategies for dynamic problems

Dynamic problems (and wicked problems) are problems in motion. They change with time, fast enough to influence the solutions available. In these cases, the winning strategy seems to be spending most time on exploring the design space (Stauffer & Ullman 1991). This includes understanding the problem, clarifying the design criteria and exploring alternative solutions. This process usually continues until a decision must be made. At that point, the compensatory methods for static problems should be applied unless there is clearly one stand-out solution.

12.3 Team-based decision making

LEARNING OBJECTIVE 12.3 Use the capabilities of a team to improve the decision-making process.

Olson (1996) studied software teams to investigate how they reached their design decisions. After studying ten different teams and ignoring the time spent on project management, there was a surprisingly similar pattern to the design decisions. He found that almost 90 per cent of the discussion related to issues, criteria and alternatives, with the latter two consuming most of the time. For novice teams such as student designers, then, concentrating on these three things clearly makes sense.

Nevertheless, at the pointy end of completing a student project, decisions between alternatives (hopefully thoroughly explored, as discussed earlier) must be made. The Foundation Coalition (2001) outlines seven methods for decision making in teams, adapted from Johnson and Johnson (2000) as follows.

1. Decision made by authority without group discussion — good for quick decisions but poor for using the intellectual resources of the group.
2. Decision by expert — useful when one or two people in the group are the acknowledged experts at this particular issue (e.g. what software should be used for a particular task).
3. Decision by averaging individuals' opinions — useful for ensuring some kind of consensus, but best left to low-value decisions made in a hurry.
4. Decision made by authority after group discussion — gets the decision made, but the group may not be committed to implementing the decision.
5. Decision by minority — often used in executive roles where a subset of the group makes the decision (with similar difficulties as 4).
6. Decision by majority vote — useful for including everyone but can lead to sub-optimal decisions or compromises to ensure the majority vote.
7. Decision by consensus — ensures everyone participates in the process, and all issues, criteria and alternatives should be fully explored. This option is preferable to methods 1–6 but requires more time.

Even with decision by consensus, it is quite likely that decisions between alternative solutions to design problems could be made in simplistic ways (the non-compensatory methods described above). In **team-based decision making**, there needs to be enough time to use a method like the decision matrix to compare the alternatives. This is easily done on a whiteboard, with a row for each alternative and a column for each criterion. One more row is required for the weights assigned to the criteria.

It will quickly become clear in this process that every cell in this table is a decision. The group is required to assign a rating to each criterion for each alternative, plus a set of weights to each criterion. There will almost certainly be differences of opinion on these matters. However, this might be a good time to use method 3 from the list. First, ask each person to fill in their own table, assigning ratings and weights individually. Then, a scribe at the whiteboard can ask for these to be called out and then the values in each cell can be averaged. Alternatively, this could all be done in a spreadsheet.

This process is completed quickly, and everyone has an equal say. There should then be some robust discussion to decide whether the group agrees with the results. Some adjustment of ratings or weights is common as the group converges on consensus.

It is also advisable to do some *sensitivity analysis* to test whether small changes to any of the ratings or weights alter the final decision. Perhaps there are two alternatives quite close together. Can they be effectively separated? Do the weights need more careful thought? Have the most important criteria received sufficient weight? If the decision was based on just the two or three most important criteria, would you make a different decision? Are the option(s) provided by the discipline expert(s) still in the mix of alternatives or have they been discarded due to the team's weighting system?

Effective team environment

Of course, effective team decision making assumes an effective team, one where there is trust and respect for each other, as described in the chapter on working with people. Kline (1999) describes a range of circumstances that should be present if a team is to be effective in decision making: 'everything we do depends on the thinking we do first'. The following ten components are required to create the best thinking environment for a team.

1. *Place.* Choose a physical environment that is welcoming and creative (but not too noisy).
2. *Ease.* Ease creates; urgency destroys — leave time to do good work rather than rushing at the end.
3. *Diversity.* Honour the differences within your team and take advantage of them.
4. *Attention.* Listen with respect, interest and fascination to your team members.
5. *Information.* Share your information and knowledge with the group in a timely fashion.
6. *Incisive questions.* Challenge each other's assumptions that limit ideas and solutions.
7. *Appreciation.* Practise a 5:1 ratio of praise to criticism of your teammates and their ideas.
8. *Encouragement.* Move beyond competition to collaboration with your teammates.
9. *Feelings.* Allow sufficient emotional release to restore thinking; honour other members' feelings.
10. *Equality.* You will have your turn later; listen attentively now.

If you practise these ten steps, then you can move your team into the performing range where decisions become easy. That doesn't mean to say that there will not be disagreements or differences of opinion. It just means that when these disagreements occur, you will listen to each other with an open mind, with the intent of understanding the difference of opinion and with the purpose of finding a resolution. If all else fails, you might need to agree to disagree on the point and let the group majority decide.

Dominance and power

One challenge for student teams is the dominant individual (or individuals) who think it's 'my way or the highway'. This can cause huge problems in consensus decision making. It is all too easy to allow a dominant individual to take charge of a group early. However, this can create a situation where that person believes they will make all the key decisions and badgers others into going along with their views. If the project lasts for many weeks, the situation can become very strained if no-one is willing to call the behaviour and ask for the group to move to collective decision making.

An effective leader becomes a true group facilitator. They act as project manager, making sure that tasks are completed on time. They can chair meetings, organising the agenda beforehand. When it comes to decision making, however, they become consensus builders, drawing out those who have not spoken and making sure that the wisdom of the group becomes available to all members. A good motto is 'No-one knows as much as all of us'.

Another important role for such a group facilitator is to make sure that everyone is learning from the project and developing their skills. One way to ensure this is to identify those with special skills, such as drawing skills or computer skills, and then have another group member who needs to develop those skills become their assistant and understudy. Each person also needs to take responsibility for their own development to assist in this process. For example, you might say 'I really want to improve my Python skills, so can I assist with the modelling?' Develop a learning portfolio and keep track of skills as you master them. Collect evidence to take to a job interview to show your capabilities.

Diversity in teams

There are many other differences in teams, as the discussion of Belbin's team roles in the chapter on working with people indicated. Diversity includes age, gender, nationality, religion, educational background, personality, sexual preference and so on. The challenge, and opportunity, in all teams is to use this diversity to produce better outcomes through better decision making and more cooperative work. So, what are some of the differences that you'll notice? Table 12.1 indicates some pairs of attitudes that you may observe in your student teams (Guyon 2005; McDermott 2011).

TABLE 12.1 Diversity axes in teams

I push my own opinion.	I help the team collectively make the decision.
I focus on the longer term (big picture).	I focus on the immediate issue.
I am influenced by emotion, narrative, relationship.	I am influenced by objective facts and figures.
I communicate verbally.	I communicate pictorially.

Think about each row of the table. Where do you sit on each scale? Where do you think other group members sit on each scale? How will these differences make it easier or harder to reach a decision? Perhaps a group discussion about these differences would be a first step to better understand what influences your decision making.

One obvious difference in engineering teams is gender. For a start, there is rarely a gender balance in engineering teams, since the percentage of women studying and working in engineering is low. There are often articles produced in popular magazines and newspapers that assert a range of differences between male and female decision making that reflect the individualist (male) versus collectivist (female) points of view. However, these stereotypical assertions are not necessarily supported by research. What research in science and engineering institutions has shown is that all-women and mixed teams outperform all-men teams (Rosser 2004).

Differences in decision-making processes may sometimes be influenced by gender, but they will also be influenced by factors such as past experiences, family backgrounds, and cultural and religious backgrounds. The value of diversity, including gender diversity, is that the team has access to all these modes of decision making, which means that a balanced decision is more likely. So, value the contributions of all your team members in the decision-making process.

Cultural differences

Engineering is increasingly a global profession. Around half the engineers in Australia were trained overseas. Much has been written about the differences between cultures (for example, Hofstede 2000). Of all the cultural differences identified, the one that has received the greatest attention is collectivist versus individualist cultures (Güss 2004). Western countries tend to be individualist, where people are inclined to make decisions independently of others; while Asian countries tend to be collectivist, where people make decisions for the common good.

You might imagine, then, that collectivists are better at stakeholder engagement, because they have a genuine interest in arriving at a decision that suits the majority, and that individualists are better at driving through a decision that needs to be made. Of course, each point of view has strengths and weaknesses. The collectivists might get stuck if there is no solution acceptable to all, and the individualists can easily opt for a decision that is ultimately very unpopular. Is it possible to get the best (or worst) of both worlds?

Güss (2004) describes a series of research studies on collectivist versus individualist decision making and comes to the following conclusions. Individualist cultures tend to prefer exploration and information gathering, are achievement-oriented and more willing to take risks. They prefer active and assertive (confrontational) strategies to resolve conflict and they have more confidence in their personal decisions. On the other hand, collectivist cultures tend to pay more attention to the social aspects of the problem. Because they are more sensitive to the social consequences of their actions, they follow a more defensive and incremental strategy, avoiding risk.

Of course, these are generalisations, and, in any culture, both ends of the spectrum exist. So, in any team, expect that you might have the full range of collectivist to individualist points of view. Consider that each person's point of view will enrich the group's decision-making processes. While some members will be brash and assertive, others will be considering the decision from the whole-of-society point of view. If your group is to make good decisions, all these points of view must be integrated.

The current reality in Australia is that you will most likely be working with colleagues whose cultural background will be significantly different from yours. Without knowledge of cultural differences and skills to deal with them you might find this a challenge. Each person may try to do what they intuitively feel is appropriate, but because they might be drawing on different cultural assumptions and expectations about how to behave, they may end up feeling at cross-purposes.

These diverging interpretive frames can produce frustration, resentment and negative evaluations on all sides. The problem is that management techniques developed for the relatively homogeneous workforce of the past generally fail to deliver positive results today.

You might be wondering why all this matters — beyond the obvious values of equity and fairness. It matters because there is overwhelming evidence that diverse teams are more innovative than *either* talented individuals working alone *or* teams where everyone is pretty much the same. Diversity outperforms sameness (Rock & Grant 2016).

Making better decision makers

According to Keeney (2004), most people are not good decision makers, despite some basic training in decision making. Keeney summarises the key steps in decision making (based on Hammond et al. 1999) as follows.

1. *Problem.* Define your problem.
2. *Objectives.* Specify what you are really trying to achieve with your decision.
3. *Alternatives.* Create better alternatives to choose from.
4. *Consequences.* Describe how well each alternative meets your objectives.
5. *Trade-offs.* Balance pros and cons of different alternatives for meeting your objectives.
6. *Uncertainty.* Identify and quantify the major uncertainties affecting your decision.
7. *Risk tolerance.* Account for your willingness to accept risks.
8. *Linked decisions.* Plan.

Of these, steps 1–3 are the ones that need most work, as has been discussed earlier. Solving the right problem is the point of the exercise. Often what presents is a symptom rather than the problem. Investigate and discuss with stakeholders until this is clear. Often some criteria are more important than others and a weighting system can be used effectively to ensure that the critical criteria remain critical, but care should be exercised to ensure that the weighting system does not dilute the importance of the critical criteria.

Objectives are the intended outcomes of the decision. These are usually defined by the client and stakeholders, and some of these may be strict legal conditions. They set the constraints (hard and soft) around the decision. They are *not* under the control of the decision maker in most cases (except in personal decision making).

Alternatives are choices that *are* under the control of the decision maker. They usually represent ways in which the problem can be solved or managed by meeting the required objectives.

Consequences represent the match between the alternatives and the objectives. We have used ratings from 1 to 5 or 1 to 10 to describe such behaviour in the 'Step 3. Evaluating alternative solutions' section in the chapter on the engineering method (in relation to buying a car). This allows for each possible solution to be evaluated against the performance criteria. These could include performance measures, such as fuel economy, weight or environmental footprint.

Trade-offs are usually necessary in multi-criteria decision making. Pugh's decision matrix, discussed earlier, is one way of handling these trade-offs. We discussed earlier compensatory and non-compensatory methods for such decisions. Research shows that, faced with such complex decisions, humans will opt for simple and fast methods rather than accurate ones.

What makes such complex decision even more difficult is that the consequences are often uncertain. In some cases — for example, the design of structures — we have reasonable data to quantify the uncertainty (such as wind speed versus likelihood). In many cases, it is more likely that all we can do is give rough estimates of uncertainty, as we did with the risk assessment in the chapter on the engineering method.

Estimating uncertainty is one issue, and deciding how much risk to accept is another. For example, the design wind speed for a normal building might be a 1-in-50-year storm. For a hospital, the design speed might be 1 in 200 years, because we want the hospital to endure storms that might destroy minor buildings, on the assumption that the hospital will be needed to treat those injured in other storm events. The community accepts that this will cost extra, but it is worth this extra cost. So, risk, which equals probability times consequence, matters in our decision making.

Keeney (2004) outlines several issues in decision making that will help make you a better decision maker. In defining the problem, *find out what matters to people*. Different stakeholders will clearly have different priorities. Spend enough time to find out what these perspectives are. *Understand your own stake in the decision*. Does your future job security depend on this decision? Does that make you risk averse? Will that matter? *Clearly differentiate objectives and alternatives*. Objectives are defined by stakeholders; they act as constraints on the solution space. Alternatives are possible solutions that are available to the decision maker. *Watch for psychological traps*. A common trap is called anchoring. When presented with one possible solution, then other solutions are measured against it. It can be difficult to break out of this mindset to consider radically different solutions. This is a potential pitfall in brainstorming. Finally, *carefully weigh up trade-offs, uncertainties and risks* in the decision.

Beyond rationality

An early model of decision making was *utility theory*, which assumes that people make decisions to maximise their utility (happiness, satisfaction, wellbeing). Utilitarianism became an ethical framework for the organisation of society, with the intention to achieve the greatest wellbeing for the greatest number (or the greatest overall wellbeing). Utilitarian philosophers include David Hume, Jeremy Bentham, John Stuart Mill and Peter Singer (Stafforini 2011). (See the chapter on professional responsibility and ethics for further discussion.) As Bertrand Russell (1998) stated:

> It appeared to me obvious that the happiness of mankind should be the aim of all action, and I discovered to my surprise that there were those who thought otherwise. Belief in happiness, I found, was called Utilitarianism, and was merely one among a number of ethical theories. I adhered to it after this discovery.

Subsequent research has shown that this simple model of utility is too simplistic, and that there are interesting non-linearities in how we actually make decisions. Tversky and Kahneman (1974) and Kahneman and Tversky (1979) found that in decision making 'losses loom larger than gains', and that people focus more on *changes* in their utility states than they focus on the absolute utilities. For example, a \$100 loss has about twice the effect of a potential \$100 gain. They proposed **prospect theory** to recognise the asymmetrical nature of decision making. In this theory, utility is adjusted by multiplying by a value function (see figure 12.3), representing the asymmetrical view of gains and losses.

FIGURE 12.3 Prospect theory value function

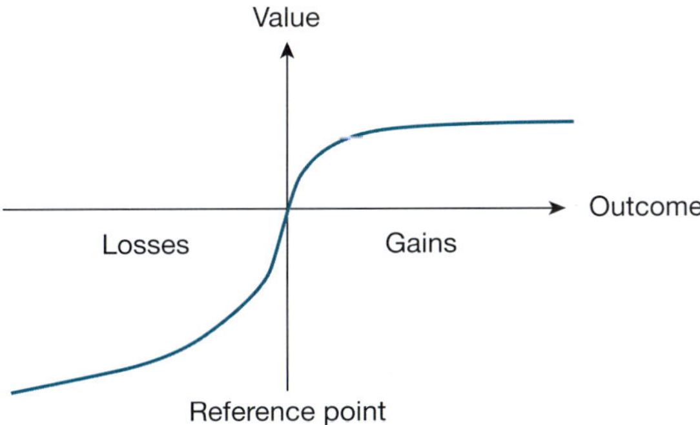

Source: Marc Oliver Rieger (2006), Creative Commons.

One clear illustration of this value function is the relationship between happiness and income. In the United States, studies have shown that above an income of $60 000, happiness does not change significantly (Kahneman 2011). However, below $60 000, it decreases with income, matching the value function shown in figure 12.3.

How does this affect our decision making? For many, it makes us risk averse, avoiding losses rather than seeking gains. This makes change in organisations difficult, for instance, because people fear loss of the current state (the status quo), rather than opting for an uncertain but more appealing future, as neatly summarised by Niccolo Machiavelli (1532):

> . . . there is nothing more difficult to take in hand, more perilous to conduct, or more uncertain in its success, than to take the lead in the introduction of a new order of things, because the innovator has for enemies all those who have done well under the old conditions, and lukewarm defenders in those who may do well under the new. This coolness arises partly from fear of the opponents, who have the laws on their side, and partly from the incredulity of men, who do not readily believe in new things until they have had a long experience of them.

SPOTLIGHT

Understanding complex systems

Design in engineering is a complex business. It helps to understand the type of systems we are trying to influence so that we can better understand the ways in which we might grapple with them. After all, some problems look routine and are quickly solved with standard solutions; other problems are more complex. Is there a way of understanding the difference?

Cynefin is one such methodology, proposed by Snowden (2000). Cynefin is a Welsh word, roughly translated as *place* or *habitat*, but more fully translated as the combination of place, culture, geography and so on — all our past experiences that influence our decision making. Snowden was interested in how people respond to complex systems and posed four main categories.

- *Simple.* This is routine design and decision making. The system is one we have seen before, and standard solutions apply. Best practice can be identified. An example might be design of a mobile phone tower: the technology is straightforward; we've built 500 of these and know just what needs to be done.
- *Complicated.* This is adaptive design. The system has many interacting parts, but it is quite well understood. Nevertheless, there are many available good solutions, so good practice can be identified, but it's difficult to point to a single best practice solution. Designing a mobile phone handset fits this category: there are many competing good designs.
- *Complex.* The system is made up of many interacting components, and the behaviour of the total system cannot be easily understood or predicted. A mobile phone system for a city is an example. It will

have emergent behaviour that cannot be predicted from the behaviour of the individual components. For example, the popularity of text messages (SMS) was emergent behaviour in the early days of mobile phones.

- *Chaotic.* These systems have no apparent relationship between cause and effect at the system level. Social systems sometimes fall into this category. Predictions of weather can be considered a chaotic problem (very small changes in the starting position of a chaotic system can make a big difference as time progresses. This is why the prediction of weather can be for only a few days in advance).

Understanding the kind of problem we are solving, and the kind of decisions that need to be made, is of great assistance. Simple problems will have routine decisions, and these are likely to be available in textbooks, manuals and other documentation. Talking with others in the office can point you towards best practice. Solutions are achieved by *sense, categorise* and *respond* — sense the problem, find what category of standard problem it is, and respond according to best practice guidelines.

Complicated systems are often also well understood in a specialised organisation (e.g. mobile phone companies have a great deal of in-house expertise for designing new mobile phone handsets and their user interfaces). Specialised conferences and journals are also places to access this kind of expertise. This knowledge is often changing rapidly, so there are many good solutions, which quickly become mediocre solutions as technology and user preferences change. Decision making is aided by access to this cutting-edge information: *sense* the problem, *analyse* it and *respond* accordingly.

There are many *complex* problems in engineering (e.g. the design of urban infrastructure). The system (the city) is constantly changing. The best that engineers can manage is to *act*, observe the changes (*sense*) and *respond* further.

Chaotic problems are very difficult to manage because it is not clear to us how we should respond. There may be aspects of these systems that we understand, but most of the system behaviour is not understood due to its complexity. For example, student performance at university is a combination of attitude, aptitude, interest, intention, learning strategies, work ethic, family background, financial means and many other factors. Predicting student success at university is not at all simple. Improving educational systems (and other seemingly chaotic systems) requires us to *act, sense* the system behaviour and *respond* further.

CRITICAL THINKING

Can you think of another example in each category from your discipline of engineering? Can you identify examples of best practice, good practice and emergent behaviour?

12.4 Reviewing key decision-making criteria

LEARNING OBJECTIVE 12.4 Assess the ethics, safety and quality of a decision.

When your team has reached a decision and is ready to communicate and discuss it with the client and other key stakeholders, it is worth reviewing the decision against the key criteria, which are discussed briefly as follows.

Uncertainty

Has your decision taken account of the uncertainty in your data? There is a range of probabilistic methods to aid decision making, such as utility theory, expected value theory and prospect theory. Have you used these tools?

Environment

Typically, we now expect our objective function to include some variant of the triple bottom line. Measures should include economic benefit and impact, environmental effects and social effects, which also includes political challenges. Engineers, of course, have, as a primary concern, the technical adequacy of the solution. For example, a mobile phone must be useful for making phone calls no matter how useful it is for playing MP3 files and browsing the internet.

Ethics

Ethics are a way of formalising social impact and benefits. What ethical frameworks have you used for your decision making? How does this affect your measure of the value of your decision? Ethical frameworks are discussed in the chapter on professional responsibility and ethics, and are neatly summarised by the Markkula Center for Applied Ethics (2010), which suggests that options be evaluated by asking the following questions.

- Which option will produce the most good and do the least harm? (Utilitarian approach)
- Which option best respects the rights of all who have a stake? (Rights approach)
- Which option treats people equally or proportionately? (Justice approach)
- Which option best serves the whole community, not just some members? (Common good approach)
- Which option leads me to act as the sort of person I want to be? (Virtue approach)

These questions enable a robust discussion of the alternatives available in any decision-making process.

Safety

As outlined in the chapter on professional responsibility and ethics, safety is another aspect of social impact, and it is of vital importance in all engineering disciplines. Safety needs to be considered in design, construction and manufacture of artefacts, and in operational contexts.

What are the safety risks during construction? Typically, construction companies are aware of the need to provide a safe workplace. Safe Work Australia provides advice on occupational workplace health and safety. Their website (https://www.safeworkaustralia.gov.au) provides codes of practice (as well as other important information) for a wide range of work activities for each state in Australia.

More complex issues relate to safety during operation. A specialist wind turbine manufacturer can be expected to design equipment based on much experience and safety analysis. Problems experienced in the past include towers toppling and rotor blades snapping off.

In chemical engineering, and in other engineering disciplines where human life is potentially at risk, there are well-developed approaches that can be used to assess hazards in processes and procedures, such as HAZOP — HAZard and OPerability study (CSIRO Minerals 2004). HAZOP considers safety at various stages of a process, such as at the design stage, construction stage and operation stage, as well as by examining equipment, procedures and instrumentation. Consequently, safety should be considered at the conceptual design stage, during detailed design, construction, commissioning and testing, and during operation — when safety will likely be most critical.

In the wind turbine example in the chapter on evaluating options, consider the long-term use of the facility. At regular intervals, there will need to be access to the tower and to the generator and other equipment at the top (90 metres above the ground). What are the regulations for ladders, climbing cages, platforms, tethers and so on? What sort of training will be required? Will this be part of normal induction for any new staff? Will staff need to be licensed to work 90 metres above the ground?

At the design stage of a project, it is crucial for engineers in Australia and New Zealand to comply with any relevant standard from Standards Australia (https://www.standards.org.au) or Standards New Zealand (https://www.standards.govt.nz). These standards are agreed specifications for products, processes, services or performance. They play a vital role in ensuring public safety by preventing accidents and injuries, and minimising the impact of disasters such as earthquakes, fires or electrical hazards. Standards Australia and Standards New Zealand create many standards in consultation with each other, and these standards are denoted by 'AS/NZS' in their title (e.g. AS/NZS 4240.1:2009, which is the Australian and New Zealand Standard for remote controls for mining equipment). Both Standards Australia and Standards New Zealand are members of the International Organization for Standardization (ISO).

Review and improve — quality assurance

In the chapter on evaluating options, the need for checking mathematical models was highlighted. This checking process needs to occur at every level of a project. In professional practice, engineers should check in regularly with their clients. Appropriate times for client engagement include:

- the initial stage (defining the problem and the project — what is required?)
- the research stage (determining whether the problem needs to be redefined considering new information)
- the alternative solutions stage (getting the client's view on the range of options)
- the decision stage (determining whether the recommendation suits the client's needs)
- the final report delivery
- in a follow-up meeting on quality assurance (Was it done right? How could the service have been improved?).

In engineering practice, this final stage is about making sure the client intends to maintain an on-going business relationship. This encourages continuous quality improvement for the organisation. Organisations generally want to know, 'How can we deliver our services more efficiently and profitably, while ensuring satisfied customers?'

This same reflective process is required at every level in an organisation and can be applied to the projects you complete as part of your engineering studies. Three questions that may be asked are as follows.

1. What is going well?
2. What is not going well?
3. What can be done about it?

Learning organisations, or organisations that learn from their past performance, use reflective questions to improve their future performance. They are 'where people continually expand their capacity to create the results they truly desire, where new and expansive patterns of thinking are nurtured, where collective aspiration is set free, and where people are continually learning to see the whole together' (Senge 1990). This consideration of risk and quality is well illustrated by the following spotlight.

SPOTLIGHT

Dubai's Burj Khalifa

Dubai's Burj Khalifa is the tallest free-standing structure in the world. It is 828 metres high, with more than 160 stories. What design criteria had to be considered as part of the overall design? The foundations obviously had to be sufficient not just to support the massive dead weight of the building, but also to resist the very significant wind loads. What other construction methods had to be carefully planned? How was the concrete used for the building superstructure lifted in place, especially when the building neared its completed height? How did the crane drivers get up to their operating cabins when they began their workday? Other considerations were related to the many building services, such as lifts, water supply to and sewage removal from the upper levels. For example, how should the building be evacuated in the case of any emergency? There were other risk and safety issues that needed to be considered in the design of the tower; fire is a problem in any tall building with many people in the building at any time of the day.

The foundations of Burj Khalifa consist of 192 piles, each 50 metres deep, and the total volume of concrete used was 45 000 m^3. For the whole construction, 330 000 m^3 of concrete and 39 000 tonnes of steel reinforcing bar were used. There are 57 elevators and eight escalators in the building. About 946 000 litres of water are supplied to the building every day. Dubai's hot and humid climate means that cooling the building is important. The peak electrical load is 36 megawatts (similar to the power requirements for a town of 20 000 people). It would have been important for the designers to have been confident that enough electricity will always be available from the electrical grid. A look at the Burj Khalifa website (www.burjkhalifa.ae/en) will provide many more technical details of the building and its construction.

The design and construction of this tower was clearly very complex. There were about 30 on-site contracting companies from nations all around the world working on it. Collaboration between all of these companies was crucial for the overall success in completing the building. Successful communication, in all its forms, would have been absolutely vital.

CRITICAL THINKING

You have been asked to review the Burj Khalifa's safety. What are the hazards and vulnerabilities in the tower design? Would you recommend any changes? Make a list of the key design criteria that designers would have needed to consider in designing this tower.

12.5 Decision support systems, tools and techniques

LEARNING OBJECTIVE 12.5 Discuss characteristics of decision support tools, techniques and systems.

A theme repeated through much of this text is that engineers work in the real world, and the real world is complex. Real-world decisions are rarely black and white; they call on an engineer to apply their specialist knowledge and may need to be made quickly. It is also likely that the engineer will be juggling many variables in any sizeable decision situation. In these complex, high stakes, professional settings, decision making is enabled by decision support tools and techniques. The most sophisticated of these are known as decision support systems (or DSSs). Throughout this section, you will come to understand some of the human, data, logic and modelling decision support tools and techniques, and how they are incorporated together in a DSS.

Decision support tools, techniques and systems came about because the human brain has limited processing capacity. When we make decisions, we rely on our 'working memory' to temporarily store and manipulate the relevant information. The problem is that working memory is finite — psychological research suggests working memory can hold 7±2 items, although this may be an over-estimation (Farrington 2011). In complex decision making, we often need to store and process more information than the working memory can hold. We've already seen that most of us respond by simply ignoring many options; sometimes the best options are the ones that are discarded.

Complex concept maps (such as those available online) illustrate how the number of factors and their interrelationship would quickly overwhelm a limit of 7±2. So, how does an engineer mitigate against the sad limitations of human processing capacity? One way is by using decision support tools. Such tools also increase the quality and consistency of decisions by supporting the decision maker with the best information and structure to make that decision.

A quick Google search will reveal a wide range of decision support tools, techniques and systems, from very basic diagrams or flowcharts to the most sophisticated web-linked systems incorporating artificial intelligence and taking account of a mind-boggling array of data and assumptions. Engineers in a particular discipline or a particular company may use specific decision support tools or proprietary DSSs. The pace of DSS development means that keeping up to date with the options and possibilities in your field may prove difficult. In this section, we provide an overview of some decision support tools, and you will garner insights into how engineers use these and how they contribute to the development of more sophisticated DSSs.

'Pen and paper' decision support tools

This class of decision support tools are reasonably unsophisticated and rarely rely on software packages. The amount of data considered in these scenarios may be modest enough that the human brain can keep track of most of it or may rely on a series of 'yes' or 'no' responses from the decision maker. One example of a simple decision support tool is a decision tree, such as Rashidi and Lemass's (2011) on bridge remediation, shown in figure 12.4. **Decision trees** can assist with problems ranging from the simple to the complex — such as how to best travel to university, or how to tackle a difficult engineering project. Such models can include numeric data, economic considerations, context and conditions.

While the kinds of decisions made using 'pen and paper' type decision support tools may be relatively simple, the use of basic tools can lead to more desirable, cheaper and reliable outcomes. Agricultural engineers are often employed to design or review irrigation infrastructure on farms. A New Zealand firm, Irrigation New Zealand Incorporated, offers farmers a 'pen and paper' style decision support tool that allows the farmer to clearly communicate the nature of their farm and operations (e.g. crops, livestock, soil, climate, acreage), their irrigation needs (e.g. flow rates, water quality, water storage, existing infrastructure), the design outputs like system parameters (e.g. net pump suction head required, outlet pressure minimum) and the economic variables (e.g. maximum capital cost of the new irrigation system, maximum annual operating and maintenance costs). The Irrigation New Zealand Incorporated decision support template gathers all of the information needed for an agricultural engineer or irrigation specialist to specify a system that will most likely meet the farmer's needs. However, things can go very wrong for farmers where inadequate information is supplied to the irrigation designer, and where a contract is negotiated that lacks important elements, such as training and post-installation evaluation.

FIGURE 12.4 Decision tree for possible bridge remediation courses of action

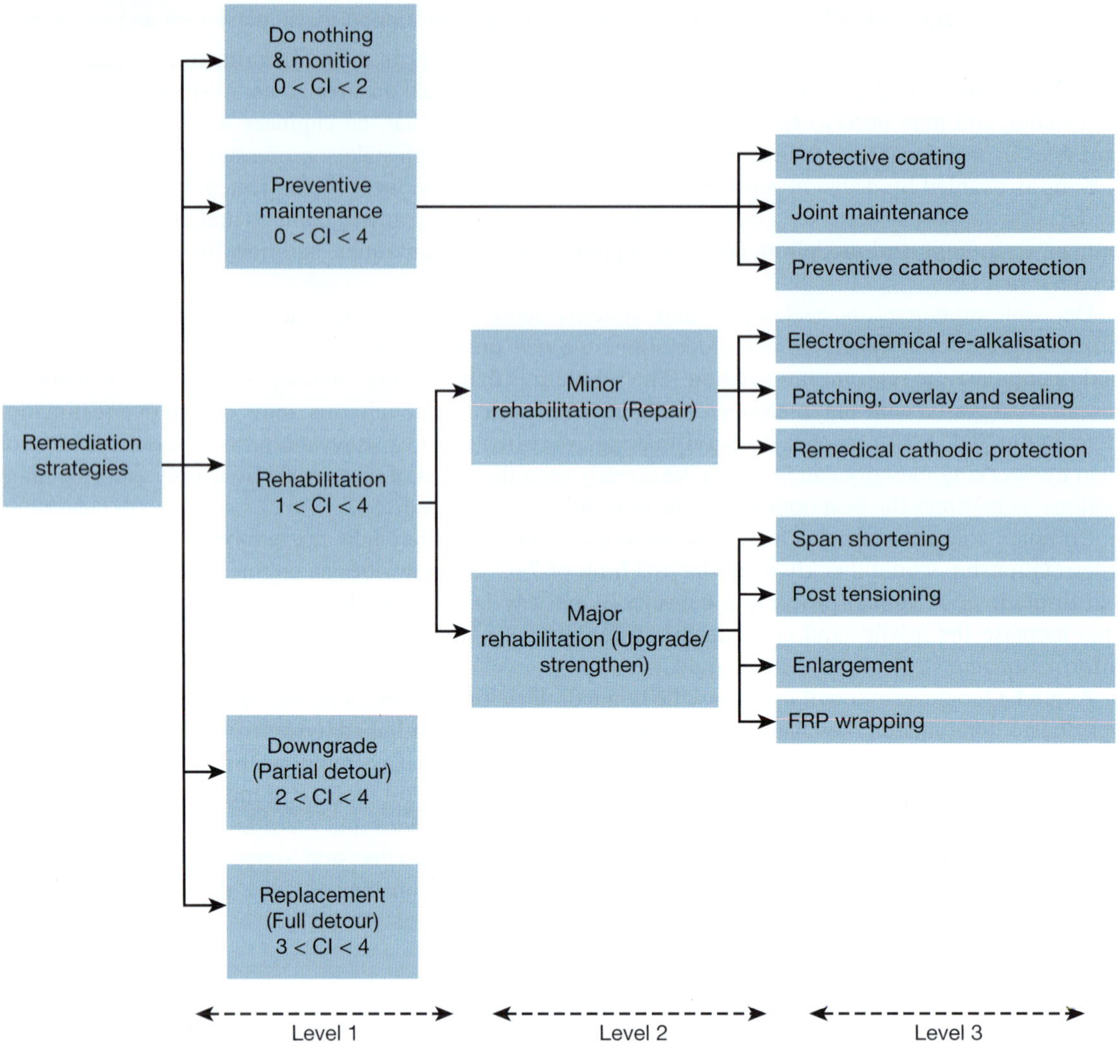

- Remediation strategies
  - Do nothing & monitior 0 < CI < 2
  - Preventive maintenance 0 < CI < 4
    - Protective coating
    - Joint maintenance
    - Preventive cathodic protection
  - Rehabilitation 1 < CI < 4
    - Minor rehabilitation (Repair)
      - Electrochemical re-alkalisation
      - Patching, overlay and sealing
      - Remedical cathodic protection
    - Major rehabilitation (Upgrade/strengthen)
      - Span shortening
      - Post tensioning
      - Enlargement
      - FRP wrapping
  - Downgrade (Partial detour) 2 < CI < 4
  - Replacement (Full detour) 3 < CI < 4

Level 1 | Level 2 | Level 3

Note: CI = Condition Index.
Source: Rashidi and Lemass (2011, p. 6).

SPOTLIGHT

Designing a natural air-conditioning system

In the central business district of cities located in tropical climates, where there may be several skyscrapers, it is very often uncomfortably hot and sticky for people as they go about their normal business. This can be the result of what is known as the heat island effect. These tall buildings effectively trap the heat and then radiate it to all the spaces between them, thereby making it uncomfortable for people walking in those spaces. The sunlight is reflected off the surface of the streets and is absorbed into each building's outside structure. These structures retain that heat and then release it back into the spaces surrounding the building. If you touch the outside of such a building at night you will notice that it remains quite warm and certainly warmer than the surrounding air.

Those designing the 25-storey NBF (formerly Sony) Osaki Building in Tokyo, Japan in 2011 recognised this and decided to install a lattice of pipes on the outside of the building and run rainwater collected from the roof through them. The system is called BioSkin and effectively cancels out the heat island effect. The rainwater is collected, filtered and sterilised before being pumped through the network of pipes. The heat of the building evaporates the water in the pipes (and so uses up heat energy). The temperature of the surface of the building can be reduced by up to 12 °C, and the surrounding air by 2 °C. If all buildings in the vicinity of this building adopted the same technology, then the effect at street level could be significant and would make life more bearable for those walking in this area.

This ability to affect the microclimate was the reason the Council on Tall Buildings and Urban Habitat (CTBUH) recognised this creative approach by awarding the designers its 2014 Innovation Award. 'The potential implications of this are substantial,' the CTBUH says. 'If a large number of buildings in a city used such a system, ambient air temperature could be reduced to the point that cooling loads for many buildings, even those without the system installed, could be reduced.'

CRITICAL THINKING

What sort of decision support tool do you think would have been used during the conceptual design of this building?

Computer-based DSS

Decision support tools and techniques like a decision tree or the very thorough data and needs specification are relatively straightforward and serve as an introduction to more complex **decision support systems (DSSs)**.

In broad terms, a DSS helps professionals to make complex decisions. Rashidi and Lemass (2011, p. 2) define a DSS as 'an interactive computer-based system that uses a model to identify relevant data in order to make decisions'. These authors point out that the use of the term 'system' in DSS implies that a DSS is a set of inter-related components.

DSSs have their origins in the 1960s in the emergence of enabling technology (digital computing), developments in theoretical research (artificial intelligence) and growing industry need (more complex manufacturing and business environments). The earliest DSSs were so-called 'model-driven'. Following these, successive innovation and emerging needs drove the development of 'data-driven', 'communication-driven', 'document-driven' and 'knowledge-driven' systems (figure 12.5) (Power 2007).

FIGURE 12.5 Different types of decision support systems

- **Model-driven DSSs** allow decision makers to access and manipulate financial, optimisation and simulation models. They work with limited data and parameters that are provided by the decision maker to analyse a situation or predict the outcome from a decision. The evolution and accessibility of easy-to-use spreadsheet programs has enabled development and widespread uptake of model-driven DSSs.
- **Data-driven DSSs** were initially focused on access to and manipulation of time-series, real-time or historical data sets. The purpose was to make data more useful by generating summaries and displaying them graphically, hence rendering vast data sets interpretable to decision makers. Recent developments include data mining and machine learning. AI techniques are primarily data-driven (Poniecki-Klotz 2023).
- **Communications-driven DSSs** were developed to aid group decision processes. They harness network and communications technologies to facilitate collaboration and communication for groups that are tasked with agreeing on a course of action. Tools used in these systems include groupware, video conferencing and computer-based bulletin boards (Power 2002).
- **Document-driven DSSs** provide document retrieval and analysis. Web-based search engines are the breakthrough technology that enabled document-driven DSS to emerge as a powerful means of locating or collating key information from document databases, including, for example, scanned or hypertext documents or media files.
- **Knowledge-driven DSSs** are the next frontier in decision support systems. Knowledge-driven DSSs are person-computer systems with specialised problem-solving expertise that can suggest or recommend actions to the decision maker. This class of DSSs is also known as 'intelligent DSS' or IDSS.

Source: Adapted from Power (2007).

We saw earlier in figure 12.1 that there are four key elements that make up a decision: data, models (information), knowledge, decision. As is apparent from figure 12.1, a DSS needs access to *data*, and a capacity to acquire and manipulate that data and transform it via *models* or algorithms into useable *knowledge*.

The DSS also needs a means to *communicate* the knowledge it has produced to a human user in a way that is easily comprehensible (figure 12.6). In any DSS, a range of data from many sources must be integrated in a single location, as will be seen in the following examples. The data is analysed through modelling, data mining, statistical methods and so on. The results are then reported through a graphical user interface so that decision makers are appropriately informed. Some decisions can be recommended by the system, for example, 'the data indicates that this bridge is in a critical state, requiring urgent repair work'. In other systems, the user is presented with the results of complex simulations (e.g. the flood warning system for the Gold Coast, discussed later in the chapter). The user may need to draw their own conclusions; for example, 'we need to evacuate this group of streets or that aged care facility'.

FIGURE 12.6 From data to decision making: a methodology for a decision support system

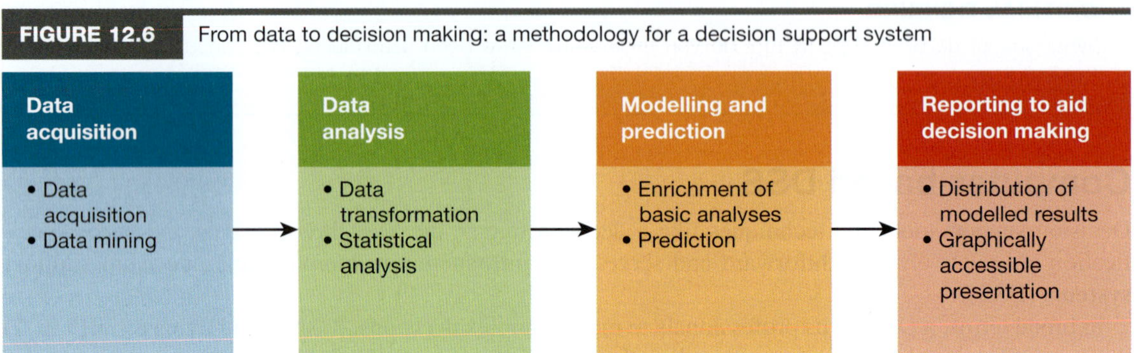

As you will learn in the next section, more sophisticated DSSs may go beyond this model and incorporate additional functions. For example, some IDSSs incorporate the capacity to interact with and learn from human experts and to generate expert interpretation of modelling results.

Underpinning the general schema of functions shown in figure 12.6 are a range of capacities that are essential in the design, development and implementation of a useful, functional DSS. Ullman (2001) listed eleven capacities he viewed as essential to an ideal DSS; these are described in the following subsections.

Support inconsistent, incomplete, uncertain and evolving decision-making information

A DSS must have the ability to process the type of information that is typically available and interpreted during an engineering decision process. Such information and its interpretation may be as follows.

- *Inconsistent.* Information may be interpreted from different life experience or analytical viewpoints. A mix of quantitative (numbers-based) and qualitative (subjective) information can lead to inconsistent interpretation amongst a decision-making group.
- *Incomplete.* Options and the criteria for evaluating them tend to evolve during a decision process, so not all information on all options will be to hand from the outset of the decision process. The team may lack the expertise or inclination to fully evaluate all options against all criteria.
- *Uncertain.* 'Even in the most refined engineering models there are inaccuracies, variations and noise that cannot be ignored' (Ullman 1997).
- *Evolving.* 'Decisions are not made after gathering all the facts, but rather constructed through an incremental process of planning by successive refinement' (Kuipers 1988, cited in Ullman 2001).

Support the building of a shared vision

The quality of a team decision is largely contingent on the amount of time and commitment that a team has for sharing information, and building a shared and agreed sense of the information that is available and its pertinence to the problem. An ideal decision support tool fosters the team to build that shared understanding and vision of the information.

Calculate alternative ranking, rating and risk

An ideal DSS should help users rank a list of alternative decision options, and ideally assist the decision makers to rate those alternatives against a level of satisfaction, thereby separating options from each other in a logical manner. Further benefit is gained where a DSS can provide a measure or indication of the level and nature of risk associated with the various options under consideration.

Suggest what to do next and additional work

There is usually not enough time or other resources to gather the complete, consistent and fully developed information needed to make decisions. One challenge faced by decision makers is to decide which alternatives to eliminate from consideration and which attribute(s) of the remaining alternative(s) to refine and evaluate more thoroughly. Within constraints of time, current knowledge and resources to develop increased knowledge, research should be undertaken to develop the information needed to make a more informed decision. So, an ideal decision support tool will offer sensitivity analysis that permits exploration of the impact of changes to alternative options and the criteria used to assess them. It will show changes that may affect the ranking, rating and risk for each alternative.

Require low cognitive load

This simply means that the decision support tool should make the decision process simpler and easier for the team, rather than more mentally taxing and complicated.

Support rational strategies

It almost goes without saying that a DSS must encapsulate rational strategies. Systems are usually built on carefully assembled rules of practice (heuristics) that have been proven to work. It is also important that the systems *appear* rational. That is, they should recommend decisions that look reasonable to the user, and to the expert user.

Leave a traceable logic trail and information record for justification and reuse

Aside from leaving a history of the decision-making process for later use, there is often the need to show the customer, management and outside reviewers a record of what was done and what was not done to reach a decision. A recorded logical methodology for reaching a decision is important in these cases.

Support a distributed team

An ideal system will support a team of people, complete with their inconsistent, incomplete, uncertain, evolving input as they build a shared vision. Additionally, decision making is becoming increasingly distributed, thus, it is mandatory that a system be capable of supporting people separated by time and distance.

The next sections discuss some examples of systems that embody the use of data, models, communications (networked), documents and knowledge capture.

Networked DSS relying on communications technology

In the section on 'pen and paper' decision support tools we discussed how decisions about farm irrigation can be complex, requiring integration and interpretation of a wide range of data types including: physical (soil, climate, plant or animal physiology), economic (costs of water and equipment, income from produce), environmental (erosion, salinity) and regulatory (environmental flows, water buy-back) parameters. Due to the tensions and complexities associated with irrigation in Australia and New Zealand, the field has been fertile ground for the development of decision support tools and systems. These range from reasonably simple tools to more complex decision support systems developed by software engineers and used by agricultural engineers. Some of these more sophisticated irrigation DSSs have begun to tap into the data opportunities offered by the internet.

DSS can be classified or described on the basis of their **network paradigm** (Car et al. 2007), which means the degree to which they are connected to local, proprietary or wider networks. Car and colleagues (2007) describe five network paradigm categories for irrigation-related DSSs (figure 12.7) and the configuration of each is depicted in figure 12.9. While the examples presented here are focused on irrigation, a 'network paradigm' could be used to classify any number of other currently used DSSs, such as those supporting medication and medical treatment decisions, natural resource management and development planning.

FIGURE 12.7 Five network paradigm categories

- **Category 1** DSSs have no networking abilities and typically consist of desktop applications on a home computer. An example of a category 1 DSS is the crop management tool APSIM 5.3 by APSRU, which is an agricultural production system simulator that models different factors affecting crop performance using data entered into and stored on the DSS user's personal computer.
- **Category 2** DSSs have direct network links to local equipment, such as soil moisture sensors in a paddock or crop of beans. An example of category 2 is IrriMAX™ by Sentek (figure 12.8), which collects soil moisture information from field sensors and creates a user-friendly graphic overview that the farmer can view and interact with on their home PC.
- **Category 3** DSSs use local area networks to access data from such sources as databases and networked sensors. An example of a category 3 DSS is Probe for Windows by Research Services New England, which presents soil moisture data from multiple, network-connected probes.
- **Category 4** DSSs utilise large, proprietary and purpose-built networks, like supervisory control and data acquisition (SCADA) networks, to collect data, as well as using resources available to category 3 DSSs (e.g. LAN-connected probes). An example is ET Drive, which was developed by Micromet and uses local evapotranspiration values, connected via a radio link, in conjunction with database information to control urban irrigation systems.
- **Category 5** DSSs rely on the internet to widely access the kinds of resources available to and used by category 4 DSSs. An example is WaterSense, a sugarcane irrigation scheduling tool developed by the CSIRO that uses the internet to present its user interface to cane farmers, and to download remotely calculated evapotranspiration data.

Source: Adapted from Car et al. (2007).

FIGURE 12.8 Sentek's IrriMAX™ graphic options overview

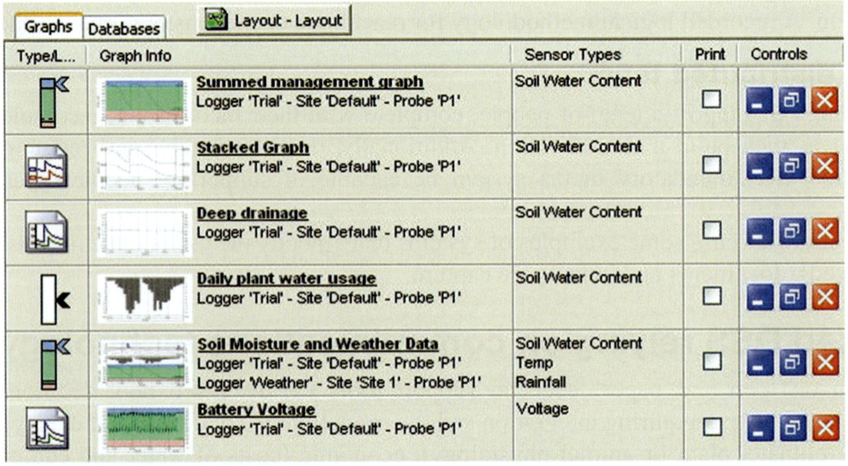

Graphs | Databases | Layout - Layout

Type/L...	Graph Info	Sensor Types	Print	Controls
	Summed management graph Logger 'Trial' - Site 'Default' - Probe 'P1'	Soil Water Content		
	Stacked Graph Logger 'Trial' - Site 'Default' - Probe 'P1'	Soil Water Content		
	Deep drainage Logger 'Trial' - Site 'Default' - Probe 'P1'	Soil Water Content		
	Daily plant water usage Logger 'Trial' - Site 'Default' - Probe 'P1'	Soil Water Content		
	Soil Moisture and Weather Data Logger 'Trial' - Site 'Default' - Probe 'P1' Logger 'Weather' - Site 'Site 1' - Probe 'P1'	Soil Water Content Temp Rainfall		
	Battery Voltage Logger 'Trial' - Site 'Default' - Probe 'P1'	Voltage		

Source: Sentek Technologies.

The evolution of irrigation DSSs to category 5 has been enabled by so-called 'Web 2.0 technologies', which allow users with limited technical knowledge to use and interact with remote computing resources such as servers, databases and other networked information (Car et al. 2007) (see figure 12.9). Advances in connectivity that will be delivered by telecommunications engineers (e.g. the Australian National Broadband Network) and further development of supporting computing platforms may enable a category 6 irrigation DSS (Car et al. 2007). This futuristic category 6 could use machine-to-machine communications over the internet for (artificially) intelligent and potentially 'farmer-free' irrigation management.

The addition of communications technology is one mechanism to enable a DSS, as shown in the irrigation example. The following spotlight looks at a system to schedule bridge maintenance. This system captures an expert's thinking processes for others to use — a simple knowledge capture system. We'll then look at intelligent decision support systems in more detail.

FIGURE 12.9 Categories of networked DSSs

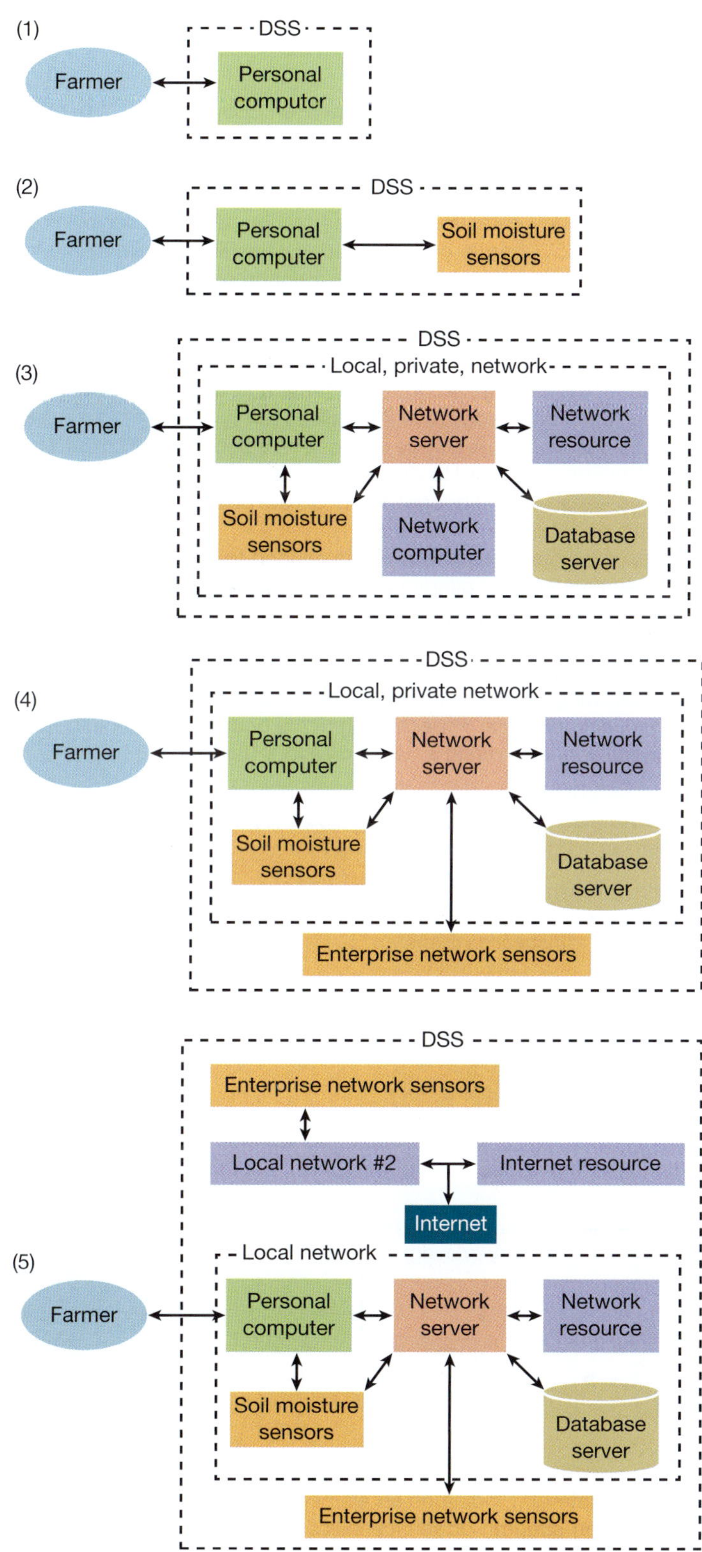

Source: Car et al. (2007).

SPOTLIGHT

SMART decisions for bridge maintenance

Australia is a vast and sunburnt country, criss-crossed with road and rail links. Our extensive transport infrastructure traverses some wild terrain and includes an estimated 33 000 bridges. Not all these bridges are of the scale and grandeur of the Sydney Harbour Bridge; in fact, many are downright basic. Unfortunately, many are also in poor repair. Analysis undertaken in the early 2000s suggested 40 per cent of our bridges are in 'fair' condition, and 15 per cent are in 'poor' condition.

While Australia has avoided any catastrophic bridge failures due to poor maintenance to date, the graphic 2007 collapse of the I35W bridge over the Mississippi River in Minnesota in the United States claimed over a dozen lives and sharpened worldwide attention on the danger of poor upkeep of these important public assets. In August 2018, the viaduct in Genoa, Italy, collapsed killing 43 people and leaving 600 homeless (Mattioli 2019). It was built in the 1960s and Italy's Autostrade indicated they were constantly monitoring and supervising this structure. Indeed, Autostrade were maintaining (strengthening the road foundations of the bridge) when it collapsed. It seems clear that maintenance of the bridge, particularly the steel cables, likely led to the bridge collapse (Mattioli 2019).

The Board of Inquiry into the Minnesota Bridge collapse revealed that the failure was a design fault and not a maintenance fault. Nevertheless, it led to a review of bridges across the United States and elsewhere. It is now estimated that more than 18 000 bridges in the US are operating below their design capacity due to poor maintenance (Transportation for America 2011).

Determining what level of maintenance and repair is needed to achieve the most economical lifespan from a bridge has long been a source of dilemma for asset managers and owners. A recently developed DSS, however, allows remediation strategies to be logically and transparently prioritised. Using the Simple Multi-Attribute Rating Technique (SMART), engineers were able to demonstrate a DSS-supported remediation strategy for a 39-year-old concrete bridge situated on the coast 10 kilometres south of Wollongong, NSW. Following the general bridge remediation flowchart shown in figure 12.10, an inspection database (1) triggers physical inspection (2) of the bridge to gather information and generate a condition rating (3). A bridge condition rating is based on a four-point scale of 'condition rates' for concrete bridge elements.

FIGURE 12.10 General bridge remediation flowchart

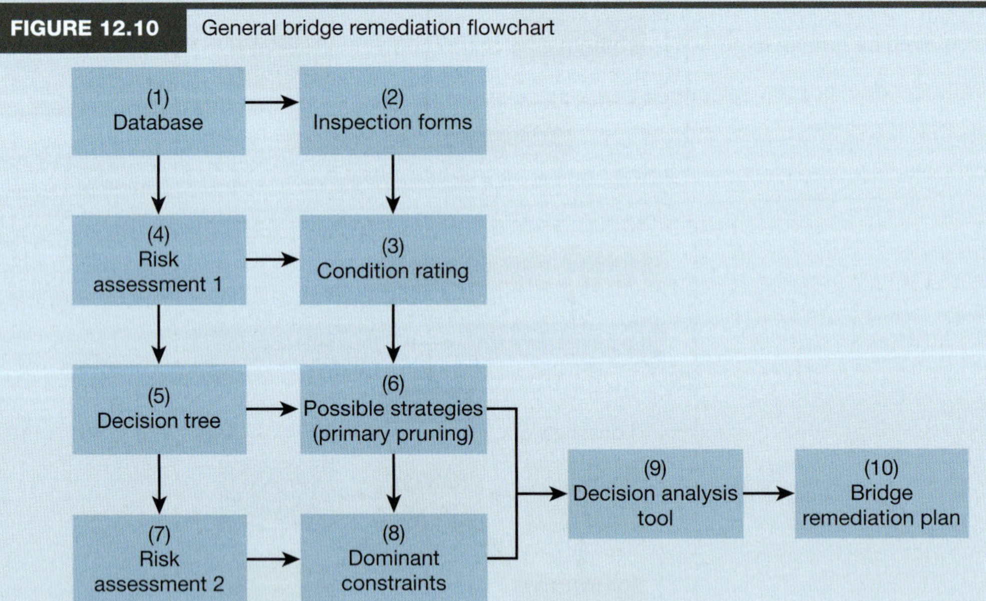

Source: Rashidi and Lemass (2011).

For example, a condition rate of 1 indicates an element might show 'discolouration ... or superficial cracking but without effect on strength or serviceability', whereas condition rate 4 may show 'delaminations, spalls and corrosion of non-prestressed reinforcements' or 'exposure and deterioration of the prestress system' (Rashidi & Lemass 2011, p. 3). With the Wollongong bridge, 102 m^2 out of the total concrete pile area of 695 m^2 were judged to be at condition rate 4. This condition rating was partly explained by the bridge piers being founded in saline water and suffering very high chloride contamination and advanced corrosion. Given the results of risk assessment 1 (4) undertaken based on the bridge's poor condition and analysis of options using a bridge remediation decision tree (5), three possible strategies (6) emerged for repairing the bridge piers: recasting with concrete, surface coating and cathodic protection.

The engineers then considered the wider risks in risk assessment 2 (7) such as cost and public safety, disruption to road users, environmental damage and vulnerability of the remediation project to political pressure. This risk analysis revealed the dominant constraints (8) to the three possible strategies. It is only after all this preliminary data gathering and analysis that the decision analysis tool is employed (9).

The SMART tool uses a ranking method based on Multi-Attribute Utility Theory (MAUT). MAUT allows decision makers to identify the overall goal of their ranking exercise (e.g. remediate the Wollongong bridge), the objectives underpinning that goal (e.g. minimise safety risks, maximise bridge functionality) and the constraints for assessing those objectives (e.g. traffic disruption, expenditure, political pressure). Decision makers then assign each objective and constraint a value or importance ranking, and SMART applies an algorithm to calculate an overall ranking value for each alternative remediation option (e.g. recasting with concrete, surface coating, cathodic protection). In the example of the Wollongong bridge, SMART showed cathodic protection to be the highest ranked option, and this concurred with the expert decision that had been reached by the Roads and Traffic Authority for the bridge remediation plan (10).

CRITICAL THINKING

What is the advantage of incorporating asset managers' or asset owners' values and perception of constraints into a decision process?

Intelligent DSS (IDSS)

As stated earlier, knowledge-driven DSS are also known as **intelligent DSS** (IDSS), fuelled by the artificial intelligence revolution. Power (2007, 2002) describes this class of DSS as having 'expertise' in the form of knowledge about a particular domain, understanding of problems within that domain, and skill at solving some of those problems. Early development of IDSS rested on computer programs to support rule-based reasoning.

Advances in artificial intelligence (AI) have taken IDSS far beyond simple rule-based reasoning to the point where IDSSs are currently developed that can 'learn' by observing the logic and process of decision making employed by human experts in each field. The main immediate goal of IDSS research is to develop DSSs that can make fast judgments (in very complex settings and with large datasets) that look like the decisions that would be made by 'the very best human technicians' (Guerlain et al. 2000, p. 1935).

As Rashidi and Lemass (2011) observed, the partial cloning of human knowledge and support of this with deep algorithmic knowledge in IDSS could lead to improved user understanding, reduced user uncertainty and anxiety, and improvement to the productivity of engineers via highly efficient decision processes. Given that IDSSs improve the efficiency of engineering work, they could 'preserve the valuable knowledge of experts in short supply' (Rashidi & Lemass 2011, p. 2).

Figure 12.11 demonstrates why few IDSSs are currently operational in engineering workplaces in Australia and New Zealand. Designing an IDSS to meet the criteria laid out in figure 12.11 is highly challenging for software engineers and instrumentation specialists. Further, IDSSs need to be customised to the specific context of their use and require extensive co-learning with experts and with the users who will interact with the IDSS on an ongoing basis in a real-life setting. Three packages developed and used in the United States offer actual examples of IDSS in use and provide insight into some of the algorithms and capabilities that are embedded in this technology (Guerlain et al. 2000, p. 1934).

- *Transfusion medicine tutor (TMT).* A DSS that uses an expert knowledge base to critique blood bank workers as they perform medical laboratory analysis tasks. This system acts as an 'intelligent tutor' for medical technologists.
- *Regional crime analysis program (RECAP).* A DSS that mines large crime databases and uses intelligent clustering and other statistical techniques, combined with geographic information system (GIS) information, to help crime analysts discover patterns. The RECAP system was developed to inform decision making in policing and crime prevention.
- *River flooding forecasting system.* A DSS that uses probabilistic (Bayesian) reasoning to predict the likelihood of river flooding based on weather forecasts. This system was used by the US National Weather Service.

FIGURE 12.11 Characteristics of an effective IDSS

(A) Interactivity
- Works well with other databases and with human users who work with it
- Allows users to explore the 'space of possibilities' within constraints, instead of just providing one 'optimal' solution

(B) Information out of data
- Uses intelligent algorithmic techniques to massage data and generate information
- Extracting information (from large amounts of data, and tools for handling small data sets, outliers, and other sources of error and confusion)

(C) Error detection and recovery
- Checks for typical reasoning errors made by people
- The system has some knowledge of its own limitations and checks its own capability for dealing with the given situation

(D) Representation aiding
- Shows and communicates information outputs in a compelling, informative and human-centered way

(E) Event and change detection
- Recognises and effectively communicates important changes and events

(F) Predictive capabilities
- Can predict the effect of actions on future performance (this means predicting both future environmental state and the change in states caused by different decisions)

Source: Adapted from Guerlain et al. (2000).

Kaklauskas (2015) presents several IDSSs, such as text analytics and mining-based DSSs; ambient intelligence and the internet of things (IoT) DSSs; biometrics-based DSSs; recommender, advisory and expert systems; data mining, data analytics, neural networks, remote sensing and their integration with decision support systems and other IDSSs. These other IDSSs include GA-based DSS; fuzzy sets DSS; rough sets-based DSS; intelligent agent-assisted DSS; process mining integration to decision support, adaptive DSS; computer vision–based DSS; sensory DSS and robotic DSS.

Access to geographical and topographical data is one capability that we now take for granted because of Google Maps. There is an increasingly sophisticated set of decision support systems being built on top of this universal access to geographic information systems.

GIS-based DSS

DSSs have been extended to include spatial information resulting in spatial decision support systems (SDSS). The chapter on understanding the problem introduced geographic information systems (GIS), which allow spatial features to be documented and presented on digital maps. GIS are important as standalone tools in engineering decisions, but, in combination with DSS, they represent powerful decision support mechanisms that can condense and present vast quantities of data and modelled outcomes in ways that are uniquely comprehensible. Multi-criteria GIS models can bring earth observation together with predictive modelling and decision theory to enable rich visualisation of current and future spatial development.

An interesting example of GIS combined with predictive modelling is presented in the chapter on your engineering future in the context of climate change adaptation. When reading that chapter you may like to think through how GIS-linked inundation mapping, from modelled rises in sea-level, supports local government development planning.

The following spotlight describes the system used to predict flooding on the Gold Coast to enable emergency services personnel to issue evacuation warnings in plenty of time to move residents to safer locations.

SPOTLIGHT

Flood emergency DSS for the Gold Coast

The Gold Coast City Council (GCCC) employed top-quality decision support in preparing to manage any future catastrophic flooding event. During the late 2000s, the GCCC cooperated with the firm Patterson, Britton and Partners to develop a flood emergency DSS (see figure 12.12) that provides real-time 'flood intelligence' derived from four key datasets (McConnell et al. 2011):

- a GIS database
- a pre-calculated library of flood surfaces
- the capacity for real time flood modelling (based on real-time flood hydrographs from the Bureau of Meteorology)
- a land use planning module.

FIGURE 12.12 Gold Coast City Council flood emergency DSS

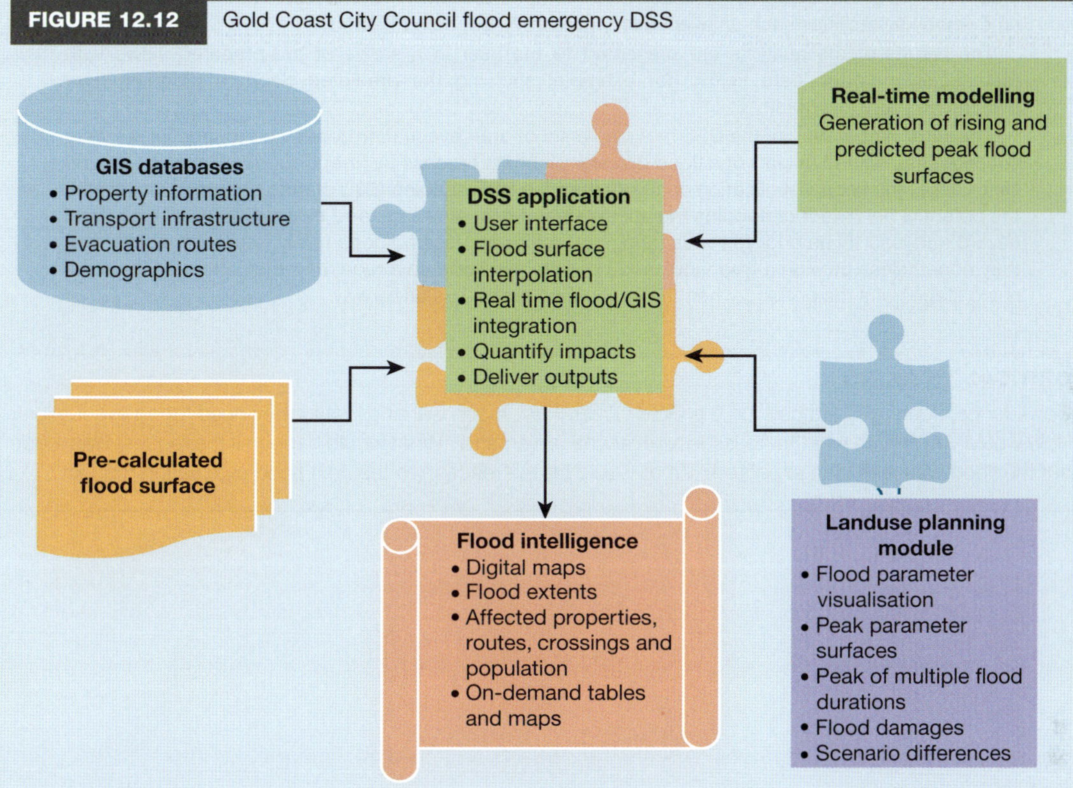

Source: McConnell et al. (2011).

McConnell et al. (2011) outline that:

> [i]n essence, the DSS is a Windows application that connects the property, infrastructure and community GIS databases to the predicted flood surface and updates the various flood-affected fields within those databases. This information can then be presented to support emergency response or planning decisions as thematic maps and tabular summaries. A key element in the DSS is the simplicity of use for emergency response personnel, while maintaining flexibility in the variety of views and integration with any number of GIS layers.

The DSS makes use of the following datasets.

- Baseline topographic data in the form of a digital terrain model plus aerial photos.
- Baseline GIS data made up of the underlying map data (the cadastre), gauging stations, waterways, evacuation centres and evacuation zones.
- Flood data as a library of design flood surfaces, an interpolated flood surface calculated by the DSS, plus a time series of flood surfaces calculated by the flood model for the real-time hydrograph or for design hydrographs.
- Integrated GIS data — property floor levels, bridges, evacuation routes, aged care, childcare, schools.

McConnell et al. (2011) further outline that:

> [t]he DSS integrates all this information, including the automatic execution of flood models based on hydrographs predicted by the Bureau of Meteorology, to provide the flood intelligence needed to respond to a flood emergency. The static pre-calculated library of flood surfaces is built in parallel to the real-time system so that it can be used in case the real-time system malfunctions.
>
> ... In its real-time mode, the DSS reads the predicted rainfall runoff hydrograph provided electronically by [the Bureau of Meteorology] and converts it to a suitable format for the MIKE-11 flood model. The flood model is then automatically run by the DSS and when finished, the DSS then converts the results onto the flood surface spatial framework. ... [a]ll time steps at a pre-selected interval are converted to generate a time series of the evolving flood surface.
>
> Once the peak flood surface has been generated, either from the library or through real-time modelling, the fields in the various integrated GIS layers can then be updated. This analysis typically includes the transfer of flood levels; the determination of flood depths relative to say a floor level or a road level, the determination of air space under bridges, and with the time series data, the anticipated time when key facilities such as evacuation routes, will become affected.
>
> The results of the analysis are presented to the user in a series of pre-prepared views with thematically mapped data fields, [for example] showing the predicted depth of flooding over evacuation route low points...
>
> The initial preparation of the DSS included a set of draft evacuation zones and routes. These were subsequently refined as part of a flood disaster planning process by [the] Council's Counter Disaster Unit [CDU]. The flood visualisation component of the DSS enabled CDU staff to become familiar with the potential behaviour of flooding including rates of rise, evolving flood extents, areas of high flood hazard and lead times prior to roads being cut. Although design floods rarely represent real events, they nonetheless provide a good indication of potential flood coverage, and they often replicate the rising stages of a major flood when the evacuation process is underway.

CRITICAL THINKING

Consider providers of aged care on the Gold Coast. How much warning would they need for an evacuation? How would you decide whether to issue an evacuation demand? What would happen if the request was made but the flood was less than expected? What would be the reaction to the next flood evacuation request?

SUMMARY

This chapter has explored the process of engineering decision making in some detail. The overall process of engineering problem solving was introduced in the chapter on the engineering method and discussed in the chapters on engineering design and evaluating options. This chapter has discussed decision-making processes and systems, with links to issues of ethics, the environment, safety and quality. Some decision support systems have also been discussed to show ways in which expertise is captured to enable others to make more effective decisions. We will now briefly revisit each of the learning objectives.

12.1 Discuss the elements of a decision-making process.

Decisions need to be made about issues (or problems). Issues have a range of criteria, which can be goals and constraints on future performance. Decisions rely on data, information (via models) and knowledge about the issue. Much of the problem-solving process concerns collecting more data, information and knowledge to inform the decision-making process.

12.2 Articulate the complexity of a decision-making process.

Static problems have stable solutions that can be explored carefully and reasonably thoroughly. Complex problems have complex solutions that can often only be explored through implementation, although simulation models can often provide some indications of system behaviour. Design-based research methods can be helpful to evaluate the implementation as it proceeds to help inform future planning choices.

12.3 Use the capabilities of a team to improve the decision-making process.

Teams are a wonderful source of diversity that can improve the quality of the decision-making process. However, this same diversity can get in the way of decision making. The human tendency is to take shortcuts, eliminating as many options as possible until the decision-making process becomes simpler. Research shows that this can eliminate very effective options. Use a systematic voting process using weighted ratings to arrive at the best solution.

12.4 Assess the ethics, safety and quality of a decision.

After the solutions have been evaluated against the system goals (which include technical, economic, environmental and social criteria), one or more recommended solutions need to be identified. This decision-making process must also include an assessment of ethics, safety and quality. Are there any ethical issues in the solution? How will safety be designed into the solution, including its implementation and operation phases? How will quality be managed throughout the project?

12.5 Discuss characteristics of decision support tools, techniques and systems.

There are many problems that are solved repeatedly, but in different circumstances. We have looked at irrigation management, bridge maintenance and flood evacuation. In each case, we want to provide the decision maker with the best data, information and knowledge available in real-time. These systems rely on GIS data, computational models of system behaviour, computer networking and knowledge capture to ensure that the best decision can be made, sometimes in short timeframes (as in disaster management — a good example of a complex decision-making environment).

KEY TERMS

decision support system (DSS) An interactive computer-based system that uses a model to identify relevant data to make decisions.

decision tree A tree-shaped flow chart showing multiple, successive decision points and their likely consequences.

decision-making pyramid Data, models and knowledge inform decision making.

dynamic problems Problems that change quickly with time.

intelligent DSS DSSs with ‘expertise’ in the form of knowledge about a domain, understanding of problems in that domain and skill at solving those problems.

network paradigm The degree to which a DSS is connected to local, proprietary or wider networks.

prospect theory A theory that acknowledges the non-linear way in which we value losses and gains.

static problems Problems that change slowly with time.

team-based decision making Diversity works when used positively. Develop collaborative approaches. Include diversity, such as culture and gender.

EXERCISES

1 Choose one of the wicked problems described earlier and create a mind map of the issues involved. You might find this easier as a group exercise. Next, create a mind map of the range of 'solutions' available. For at least one of these solutions, consider its advantages, disadvantages and side effects. How would you go about introducing such a solution? How would you determine whether it has been successful?
2 Outline the complexity of engineering decision making in relation to assessing safety, quality and risk issues in design in your favourite discipline.
3 Research a decision support system in your discipline. There are many available in biomedical engineering and in computer engineering, for instance.

PROJECT ACTIVITY

You are a member of a planning team for a new mining project in a remote area. One key decision required is how to develop a new township to house the workers for the project (about 500 at any one time). The mine operates 24 hours per day, seven days per week. At one extreme is a construction camp style of accommodation, where workers fly in and fly out; and at the other end is a fully functioning township with a range of facilities.

Your group's challenge is to put together a business proposal for your chosen alternative. The proposal should demonstrate your capability across the range of techniques discussed in this chapter. Here are some reminders to help you approach this activity.

1 Clearly define the problem. What *issue* are you trying to solve?
2 What *criteria* matter in this case?
3 What *alternatives* are available to you?
4 Make an assessment of the *triple bottom line* — social, environmental and economic issues of your solutions at various stages of implementation — such as design, construction/manufacturing and operation.
5 Assess the *potential safety, risk* and *quality* issues of your solutions at various stages of implementation — design, construction/manufacturing and operation.
6 Choose a *preferred solution* using a justifiable decision-making process. Prepare your presentation to demonstrate your recommendation to the evaluation committee.

REFERENCES

Car, NJ, Christen, EW, Hornbuckle, JW & Moore, G 2007, *Towards a new generation of irrigation decision support systems — irrigation informatics?*, International Congress on Modelling and Simulation, Christchurch, New Zealand.

CSIRO Minerals 2004, 'HAZOP (hazard and operability study)', https://www.csiro.au/en/about/people/business-units/mineral-resources

Farrington, J 2011, 'Seven plus or minus two', *Performance Improvement Quarterly*, vol. 23, no. 4, pp. 113–116, https://doi.org/10.1002/piq.20099

Guerlain, S, Brown, D & Mastrangelo, C 2000, 'Intelligent decision support systems', *Proceedings of the IEEE Conference on Systems, Man, and Cybernetics*, Nashville, TN, pp. 1934–1938.

Güss, CD 2004, 'Decision making in individualistic and collectivistic cultures', *Online Readings in Psychology and Culture*, vol. 4, no. 1, https://doi.org/10.9707/2307-0919.1032

Guyon, J 2005, 'The art of the decision', *Fortune*, vol. 144, 14 November.

Hammond, JS, Keeney, RL & Raiffa, H 1999, *Smart choices: A practical guide to making better decisions, volume 226*, Harvard Business School Press, Boston, MA.

Hofstede, G 2000, *Culture's consequences: Comparing values, behaviors, institutions, and organizations across nations*, Sage, Thousand Oaks, CA.

Johnson, DW & Johnson, FP 2000, *Joining together: Group theory and group skills*, Allyn and Bacon, Boston.

Kahneman, D 2011, *Thinking, fast and slow*, Penguin, London.

Kahneman, D & Tversky, A 1979, 'Prospect theory: An analysis of decision under risk', *Econometrica*, vol. 47, no. 2, pp. 263–291, https://doi.org/10.2307/1914185

Kaklauskas, A 2015, *Biometric and intelligent decision making support*, Springer International Publishing, Switzerland.

Keeney, RL 2004, 'Making better decision makers', *Decision Analysis*, vol. 1, no. 4, pp. 193–204, https://psycnet.apa.org/doi/10.1287/deca.1040.0009
Keeney, RL & Raiffa, H 1993, *Decisions with multiple objectives: Preferences and value tradeoffs*, Cambridge University Press, Cambridge.
Kline, N 1999, *Time to think: Listening to ignite the human mind*, Ward Lock Wellington House, London.
Machiavelli, N 1532, *The prince*, https://constitution.org/2-Authors/mac/prince.pdf
Markkula Center for Applied Ethics 2010, 'A framework for thinking ethically', https://www.scu.edu/ethics
Mattioli, G 2019, 'What caused the Genoa bridge collapse — and the end of an Italian national myth?' *The Guardian*, 27 February, https://www.theguardian.com/cities/2019/feb/26/what-caused-the-genoa-morandi-bridge-collapse-and-the-end-of-an-italian-national-myth
McConnell, DD, Druery, BM & Rahman, K 2011, *Flood emergency decision support system for Gold Coast City Council*, https://rous.nsw.gov.au
McDermott, D 2011, 'Gender roles in decision making', *Decision Making Confidence*, https://www.decision-making-confidence.com
Naude, P, Lockett, G & Holmes, K 1997, 'A case study of strategic engineering decision making using judgemental modeling and psychological profiling', *Transactions on Engineering Management*, vol. 44, no. 3, pp. 237–247, https://doi.org/10.1109/17.618075
Olson, GM 1996, 'The structure of activity during design meetings', in TP Moran & JM Carroll (eds), *Design rationale: Concepts, techniques and use*, Lawrence Erlbaum Associates, Mahwah, NJ, pp. 217–244.
Payne, J, Bettman, J & Johnson, E 1993, *The adaptive decision maker*, Cambridge University Press, Cambridge.
Poniecki-Klotz, B 2023, 'Unleashing the power of AI in decision support systems', *Medium*, 1 February, https://medium.com/ubuntu-ai/unleashing-the-power-of-ai-in-decision-support-systems-cb95c7594177
Power, DJ 2002, *Decision support systems: Concepts and resources for managers*, Greenwood/Quorum, Westport, CT.
——— 2007, 'A brief history of decision support systems', *DSS Resources*, 10 March, https://dssresources.com
Pugh, S 1990, *Total design*, Addison-Wesley, Wokingham, UK.
Rashidi, M & Lemass, B 2011, 'A decision support methodology for remediation planning of concrete bridges', *Journal of Construction Engineering and Project Management*, vol. 1, no. 2, pp. 1–10, https://doi.org/10.6106/JCEPM.2011.1.2.001
Ritchey, T 2007, 'Wicked problems: Structuring social messes with morphological analysis', *Swedish Morphological Society*, https://www.swemorph.com/wp.html
Rittel, HW & Webber, MM 1973, 'Dilemmas in a general theory of planning', *Policy Sciences*, vol. 4, pp. 155–169, https://doi.org/10.1007/BF01405730
Rock, D & Grant, H 2016, 'Why diverse teams are smarter', *Forbes*, 4 November, https://hbr.org/2016/11/why-diverse-teams-are-smarter
Rosser, S 2004, 'Gender issues in teaching science', in S Rose & B Brown (eds), *Report on the 2003 Workshop on Gender Issues in the Sciences*, Colby College, Waterville, Maine.
Russell, B 1998, *Autobiography*, Routledge, London.
Schwartz, B 2004, *The paradox of choice*, Harper Perennial, New York.
Senge, P 1990, *The fifth discipline: The art & practice of the learning organization*, Doubleday, New York.
Snowden, D 2000, 'Cynefin: A sense of time and space, the social ecology of knowledge management', in C Despres & D Chauvel, *Knowledge horizons: The present and the promise of knowledge management*, Butterworth Heinemann, Oxford.
Stafforini, P 2011, *Utilitarian philosophers*, https://utilitarianism.net
Stauffer, LA & Ullman, DG 1991, 'Fundamental processes of engineering designers based on empirical data', *Journal of Mechanical Design*, vol. 2, no. 2, pp. 113–125, https://doi.org/10.1080/09544829108901675
The Foundation Coalition 2001, 'Methods for decision making', https://www.coursesidekick.com/management/2296067
Transportation for America 2011, 'New report ranks deficient bridges by metro areas', 19 October, https://t4america.org
Tversky, A & Kahneman, D 1974, 'Judgment under uncertainty: Heuristics and biases', *Science*, vol. 185, no. 4157, pp. 1124–1131, https://doi.org/10.1126/science.185.4157.1124
Ullman, DG 1997, *The mechanical design process*, 2nd edn, McGraw Hill, New York.
——— 2001, 'The ideal engineering decision support system', https://www.semanticscholar.org/paper/The-Ideal-Engineering-Decision-Support-System-Ullman/230749eda976dff8850618f12e2e1129aa1a7e94
Winton, N 2022, 'If the electric car revolution sputters, hydrogen could still succeed', *Forbes*, 23 November, https://www.forbes.com/sites/neilwinton/2022/11/23/if-the-electric-car-revolution-sputters-hydrogen-could-still-succeed

ACKNOWLEDGEMENTS

Photo: © Jacob Lund / Adobe Stock Photo
Photo: © Tomasz Czajkowski / Shutterstock
Photo: © Robert Gilhooly / Alamy Stock Photo
Figures 12.1 and 12.2: © Ullman, DG 2001, 'The ideal engineering decision support system', https://www.semanticscholar.org/paper/The-Ideal-Engineering-Decision-Support-System-Ullman/230749eda976dff8850618f12e2e1129aa1a7e94
Figure 12.3: © Marc Oliver Rieger (2006), Creative Commons.
Figures 12.4, 12.10 and text: © Rashidi, M & Lemass, B 2011, 'A decision support methodology for remediation planning of concrete bridges', *Journal of Construction Engineering and Project Management*, vol. 1, no. 2, pp. 1–10, https://doi.org/10.6106/JCEPM.2011.1.2.001

Figure 12.8: © Sentek Technologies

Figure 12.9: © Car, NJ, Christen, EW, Hornbuckle, JW & Moore, G 2007, *Towards a new generation of irrigation decision support systems — irrigation informatics?*, International Congress on Modelling and Simulation, Christchurch, New Zealand.

Figure 12.12 and text: © McConnell, DD, Druery, BM & Rahman, K 2011, *Flood emergency decision support system for Gold Coast City Council*, https://rous.nsw.gov.au

CHAPTER 13

Managing engineering projects

'A goal without a plan is just a wish.'

Antoine de Saint-Exupéry

LEARNING OBJECTIVES

After studying this chapter, you should be able to:

13.1 provide an overview of project management

13.2 plan the stages of an engineering project

13.3 develop a risk-management plan for a project

13.4 develop a knowledge management plan for a project

13.5 develop a quality plan for a project.

Introduction

In the chapter on the engineering method, project management was introduced in very simple terms, with a discussion on how it relates to the engineering method of problem solving. In this chapter, your understanding of project management will be expanded to cover key skills in scoping projects, being clear and explicit about project goals and the accountability of team members to deliver project elements to time and within budget. You'll also explore important project planning skills like risk analysis and management, information and communication planning, and quality management.

Contemporary engineering teams use proprietary software to plan and run projects, and some of these will be introduced in the coming chapter, as will the process that engineers use to recognise, analyse and rectify projects that are failing. To keep it practical, we'll use a paper-based tool called the One Page Project Manager to show the most fundamental elements of project management by working through an example design and construction of a 'green-star' building.

Complex engineering projects can cost billions and run over several decades — for example, the International Space Station, which has cost around US$150 billion for an ongoing build that started in 1998 (Mosher 2017) and will end its life in 2031 (McKie 2023). Understanding key principles of project management will prepare you for the more complex projects that you'll tackle through the remaining years of your engineering course and as a professional engineer. Engineers complete much of their work through projects, usually with diverse teams of people contributing. Projects can be small in scope and employ just one engineer (e.g. designing and prototyping an app) or large and involved (e.g. 'mega projects' such as Australia's National Broadband Network, the NBN, which commenced in 2009).

Consider the building of the Gold Coast University Hospital (GCUH), delivered in 2013. This $1.8 billion project took seven years to complete (including planning and design) and involved the collaboration of architects with engineers from a range of fields: civil and traffic, structural, mechanical, electrical, hydraulic, fire, ICT and security, in addition to specialists in instrumentation, medical gas, specialised services and ecologically sustainable development (ESD). The GCUH is the largest hospital development to be completed in Australia to date. It consists of six separate buildings giving a total floor space of 170 000 m^2 and provides 750 overnight beds, 20 operating theatres, tertiary-level hospital facilities, pathology and education, and mental health facilities. The hospital continues to expand, with three new wards due in 2024 (Gold Coast Health 2024).

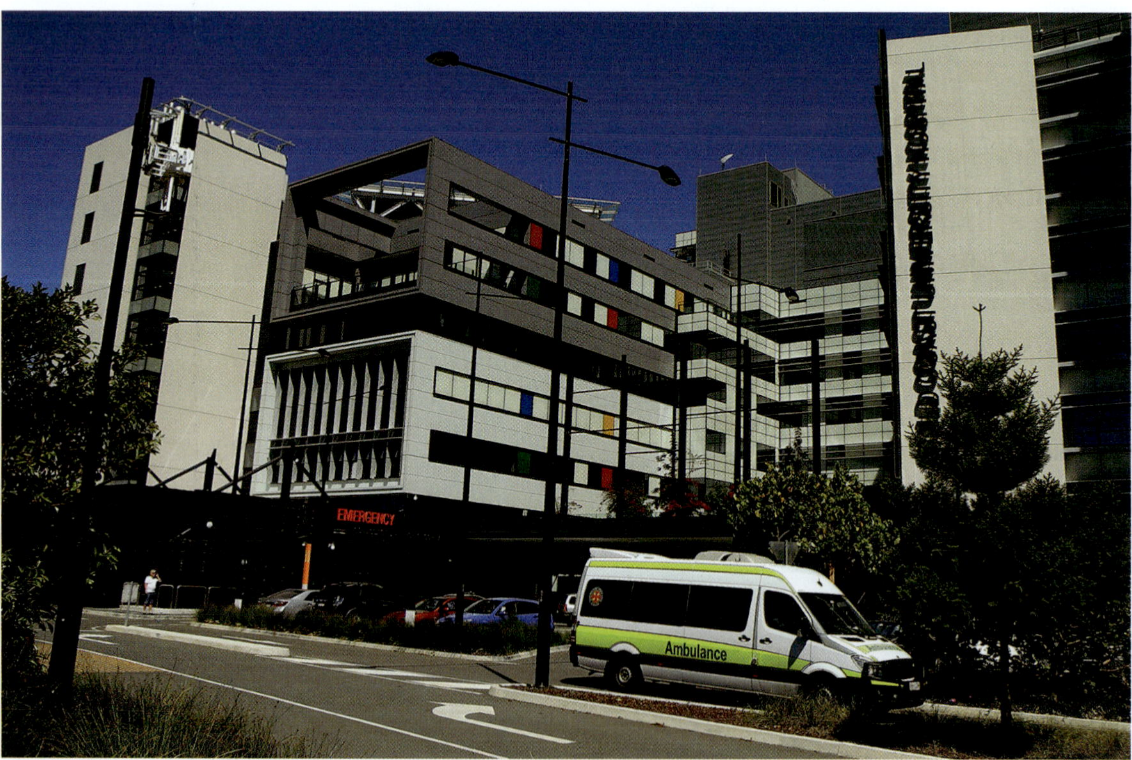

Gold Coast University Hospital, Southport, Queensland, Australia

Mega projects profiled in this chapter show how project management is crucial to the successful delivery of very big and complex projects.

The GCUH project illustrates the subtle elements of project management that can be the key to a project's success. The GCUH project team held weekly multidiscipline coordination workshops with ground rules for 'open and constructive' communication. These meetings were held throughout the design and construction phases of the project and expanded into more comprehensive peer review workshops at each design milestone — for example, at project definition stage, schematic design stage and detailed design stage.

Design for sustainable operation was a key objective throughout the whole GCUH project. The sustainability focus was on building temperature control using computational façade modelling for optimal energy efficiency, to improve façade thermal performance, and engineering highly efficient air-conditioning systems using a cascade of heat transfer systems. The cascade included heat exchange between incoming and exhaust air to gather any remaining 'cool' from the air being expelled, high efficiency chillers and smart variable control for the speed of fans.

Smart buildings, like the GCUH, increasingly rely on sensors and feedback loops to adjust fan speed in real-time response to building temperature changes, thereby eliminating heat accumulation and decreasing unintended temperature fluctuation. Computational modelling of air flows within the building complex was important for effective air-conditioning but also to manage the risk of a fire in the building. If a fire were to break out, sensors can be programmed to control air flow so as not to accelerate the spread of fire. Computational modelling was also used to optimise daylight utilisation and control glare inside the building. The use of volatile organic compounds and formaldehyde in fixatives, paints and other construction materials was minimised. Water sustainability was important given the GCUH's location in Queensland, so the design team included rainwater harvesting and a tank system to capture and store up to one million litres of water, which is used for garden irrigation.

The success of this planned, designed, constructed and commissioned project was dependent on highly effective project management.

This chapter will describe how to break a project into tasks, assign them to people and monitor progress. Projects pass through a series of stages from conception, through detailed design, construction or manufacturing, to operations and maintenance. Each of these stages requires risk management, knowledge management and quality management. These important topics will all be covered in this chapter, in terms of how they relate to project management.

13.1 Understanding project management

LEARNING OBJECTIVE 13.1 Provide an overview of project management.

Most organisations conduct much of their work in **projects** — any work that has a clear purpose that will have agreed beginning and end dates. Change processes are an example. Launching a satellite is an aerospace project as well as a construction project. Rolling out a new customer relationship management (CRM) package is both a software and staff training project. How does project work differ from other work? According to the Project Management Institute (PMI), a project is a 'temporary endeavour undertaken to create a unique product or service' (Klastorin 2004).

A project has a set of definable **tasks** that must be fulfilled to complete the project. These tasks have identifiable dependencies; in other words, some tasks need to be completed before others. Within a project, most tasks can be completed independently, subject to these dependencies. For big projects, this allows a divide-and-conquer strategy to produce a set of parallel teams who work on these various tasks simultaneously.

Project management is the collection of processes used to manage a project to completion. Projects typically have an objective, specifications, start and end dates, funding limits and allocated human and non-human resources (Kerzner 2003). Consider designing and implementing a new light rail system for a city centre. Before implementation can commence, rail corridors need to be identified and approved, appropriate technology designed, costed and manufactured or procured, and disruption to city traffic and other services needs to be scheduled and managed during the construction phase. Each of those steps will require a specific defined set of tasks to be completed, prior to the next step commencing.

The subdivision of a project into tasks allows individuals to comprehend all the parts of very complex projects, and focus on simpler tasks that will eventually unite to make a complex artefact. Consider, for example, assembling a large passenger jet from millions of parts. This is managed by dividing the aircraft into some major components (e.g. wings, engines, tail assembly, fuselage and cockpit) that can be assembled somewhat independently before they are brought together for final assembly.

Project management applies to design processes, changes to operations or staff behaviour, as well as to construction and assembly processes, as we will see in this chapter.

KEY POINT

Engineers complete much of their work through projects, usually with teams of people contributing.

Key factors in project management

Unlike many functions within an organisation, projects are finite in terms of time. This is an important distinction to make as it affects how they need to be managed. A basic idea in project management is the constant tension between three key factors: (1) the *scope* of the project, (2) the *time* available and (3) the *budget*. In the middle of this 'tug-of-war' is the *quality* of the work and the *risks* associated with the project. This relationship is shown in figure 13.1. It is crucial for project managers to consider the interaction of each of these factors.

FIGURE 13.1 The quality and risk of a project depends on its scope, time and budget. If a project's time or budget shrinks, or its scope expands, its quality may reduce and/or its potential risk may increase.

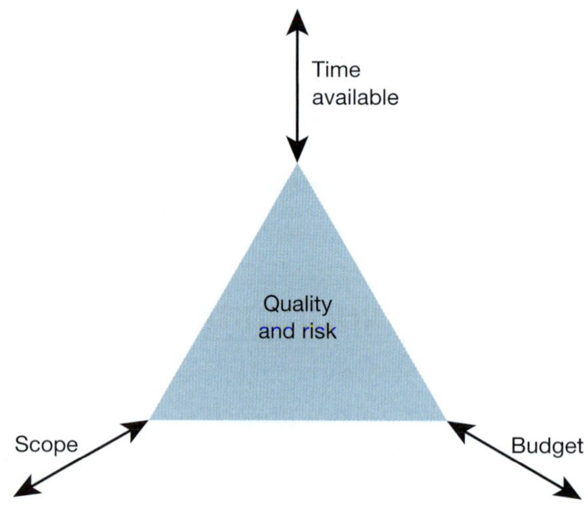

Project management tools help manage the tension between these three factors — time, scope, budget — and ensure that risks are minimised, and quality is maximised.

Project management is itself a cost to a project and, consequently, should not be a burden on the project. There should be 'just enough' project management to ensure that the project is successfully completed, with all objectives satisfied. Projects should be neither over-planned (which is wasteful of resources) nor under-planned (which leads to poor control of the project).

The Project Management Body of Knowledge (PMBOK) and project management standards

The Project Management Institute (PMI) was established in 1969 in the United States in response to the emergence of professional project manager roles in many industries. The PMI created the Project Management Body of Knowledge (PMBOK) and certifies project managers against this standard (PMI 2024). PMI's publication *A Guide to the Project Management Body of Knowledge (PMBOK Guide)* has been widely adopted by many bodies as their project management standard.

The European equivalent to the PMI is the International Project Management Association (IPMA). IPMA was founded in 1967, with a similar competency approach to PMI's PMBOK. The European framework is called the IPMA Competence Baseline (ICB).

The two organisations supported the development of an international standard in project management, which was launched in 2012 by the International Organization for Standardization (ISO). That standard is ISO 21500:2022 Project, Programme and Portfolio Management — Guidance on Programme Management, and it clearly describes processes and concepts that are agreed to constitute 'good practice' in project management (Standards Australia 2022).

Because ISO 21500 covers many processes, it is customary to cluster those processes within several project management concepts. For clarity, you can imagine those concepts as sitting inside certain organisational boundaries. For example, the project you work on as a graduate engineer will sit within a *project* boundary like 'Construction of the Southern Bypass'. That project boundary will contain the *project management, product* and *support processes* to complete the Southern Bypass Project. The concepts that will have preceded your Southern Bypass Project gaining approval will include the *opportunity* to undertake the project, with an assessment by the organisation as to whether the Southern Bypass Project was aligned with the *organisational strategy* and whether the *business case* stacked up. Your project team's objective in completing the Southern Bypass Project will be to produce the project *deliverables* on time and within budget, and those feed into the organisation's demonstrated *operations*. The final key concept that ties together the concepts already mentioned is *benefits*; benefits that flow from completion of the Southern Bypass Project feed through into *organisational strategy*, thereby completing the looped interrelationship of ISO 21500 concepts. You may consider drawing a flow diagram to represent the way these project management concepts link together. Employers value engineering graduates who can understand and articulate how a specific project they are working to complete is integral to the organisational strategy driving the success of the business.

There are 39 project management processes described in ISO 21500, presented under the following subheadings that layout the ten project management subject groups.

1. Integration
2. Stakeholders
3. Scope
4. Resource
5. Time
6. Cost
7. Risk
8. Quality
9. Procurement
10. Communication

For students who learn with a global view in mind, the detailed input/output model in figure 13.2 represents all the complexity of ISO 21500 in an interconnected graphic. You might like to identify and connect the components of some of your previous work experiences or group assessment tasks to this model. Mapping your previous experiences allows you to see which aspects of project management you are already familiar with.

The Australian Institute of Project Management (AIPM) and the New Zealand Project Management Institute are the professional bodies responsible for project management professional development in Australasia. Membership of both bodies is available to both individuals and corporations. Australasian Standards in project management are available from SAI Global (online) and align with the ISO standards.

You will encounter many **standards** in your engineering study and future work. Standards define accepted good practice in a domain of work, such as project management, risk management or knowledge management. Standards are not legal requirements; they are advisory good practice. Other methods and guidelines can be used if they are considered more appropriate. In a legal conflict, however, not following the required standard will raise questions of professional competence, so there must be very good reasons to ignore a standard's recommendations. A **code**, on the other hand, *is* a legal document. It defines *required* practice. An example is the Australian Uniform Building Code, which defines several performance measures that must be followed.

Project management is an essential skill for all knowledge workers in the twenty-first century. You will almost certainly undertake an advanced course in project management within your university degree. This chapter covers basic tools and ideas you can use within your first- and second-year projects.

FIGURE 13.2 Input/output model of ISO 21500

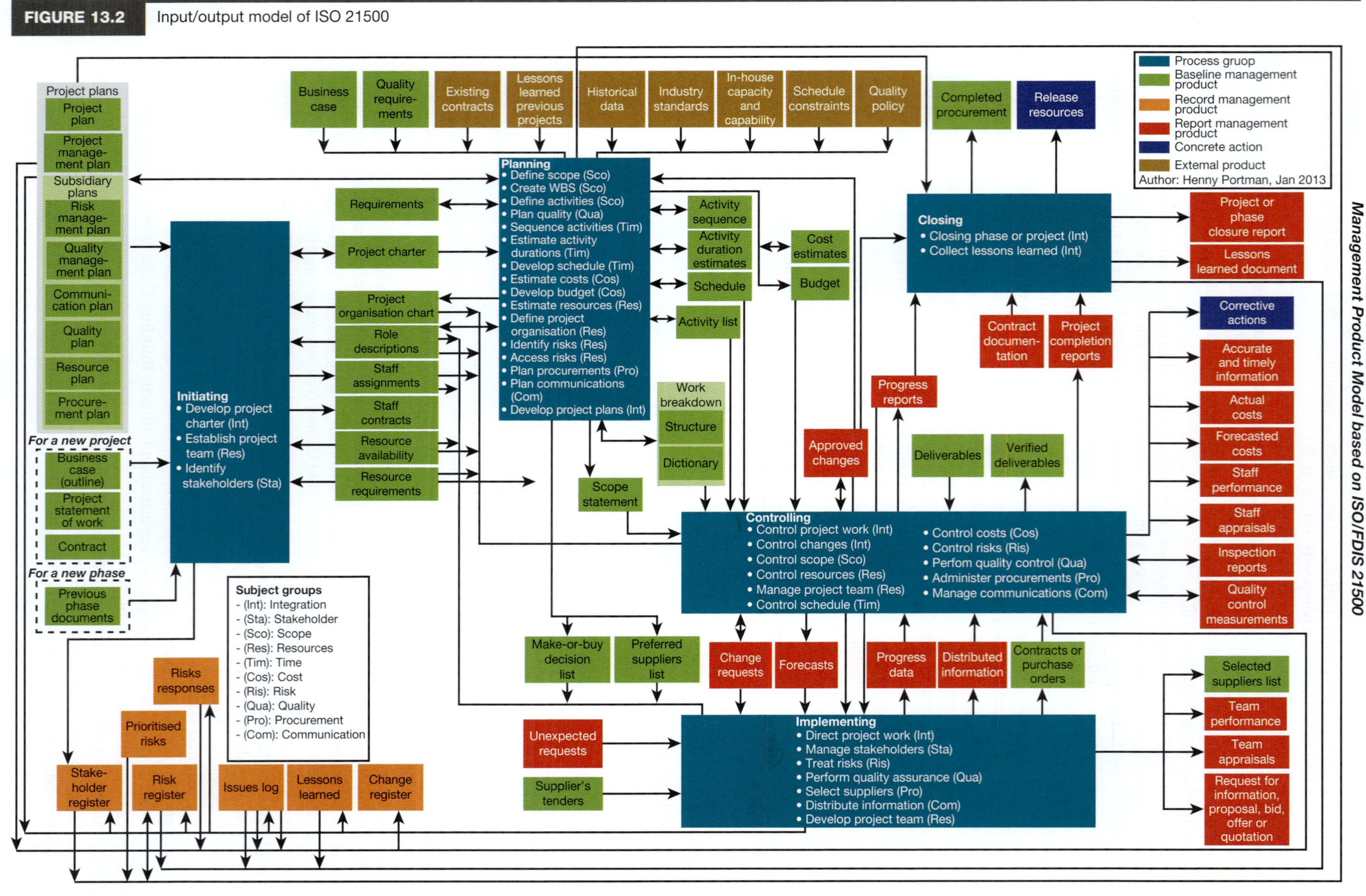

Source: Portman (2013).

SPOTLIGHT

Winning project management in engineering

Each year, the AIPM recognises best practice through a prestigious competitive awards process. Here are three examples of 2018 engineering projects recognised for excellence in the challenge of delivering innovative engineering projects safely on time, within the scope and within budget.

PMAA national winner 2018
Category: Construction\Engineering
Project: Lithium Carbonate Demo Plant
Winners: Rincon Ltd and Turnkey Innovative Engineering Pty Ltd (NSW)

> The Lithium Carbonate Demo project was an innovative, novel and complex undertaking to engineer, construct and commission a demonstration plant within one-year that effectively modelled a new sustainable end-to-end lithium-extraction process that produces high-grade lithium carbonate with low-waste from unconcentrated brine from the Salta-Province in Argentina. The project sought to develop local capability to operate the plant, optimise the new lithium-extraction process and demonstrate commercial viability for investors. This multinational project was highly complex, delivered at a remote site at altitude. This innovative mining capability could revolutionise the industry, boost the region and advance the global transition to renewable energy (Australian Institute of Project Management 2018).

PMAA national winner 2018
Category: Defence\Aerospace
Project: Qantas 787-9 Program
Winners: Qantas Airways Ltd (NSW)

> The Qantas 787-9 Program has successfully introduced a new aircraft type, the Boeing 787-9, to Qantas's international network, an aircraft that has made aviation history by enabling the only non-stop passenger service between Australia and the UK.
>
> . . .
>
> The 787-9 has allowed Qantas to adjust capacity from over-served markets to the strong markets in the US and further enables entirely new non-stop flights to the US.
>
> . . .
>
> This program has introduced innovative customer products suitable for 17-hour flights, trained all affected staff, set-up required infrastructure and managed the operational readiness and introduction of the aircraft into the network (Australian Institute of Project Management 2018).

PMAA national winner 2018
Category: National Project of the Year
Project: Hornsby Track Slab Reconstruction, Sydney Trains
Winners: Major Works Division (NSW)

> The Hornsby Track Slab Reconstruction Project was a high-profile construction project with political sensitivity as it was an enabling project for Transport for NSW's 'more trains, more services' 2018 Timetable. The location of the work was shed 16 of Hornsby Maintenance Centre, which is used for maintaining, cleaning and stabling trains. The project scope was the complete renewal of a 340 m × 20 m concrete track slab with four roads of embedded rail as well as the trade waste drainage system beneath the track slab, required for train cleaning and graffiti removal operations. Planning and delivering a project of this scale would ordinarily take 2 years. This project however was completed in less than 12 months, for 23% less than budget, with no injuries or incidents. In this period, around 25 000 tonnes of spoil was removed and replaced with engineered backfill, nearly 600 m of trade waste drainage was constructed, over 6000 tonnes of concrete was laid and 1.2 km of embedded rail was installed. This was an impressive achievement considering the challenges — the scale of the works, the compressed planning timeframe, working from incomplete

designs, the severely constrained site access for work trains and construction plant, the resource constraints and the requirement to plan all works around the continuing operations of the maintenance centre (Australian Institute of Project Management 2018).

CRITICAL THINKING

Each of the project descriptions in this spotlight mentions the wider context that made each project challenging beyond the technical engineering design and execution. For example, 'political sensitivity' of the Hornsby Track work would have meant the project manager needed to pay particular attention to communicating progress and delays to state government stakeholders and needed to be ready to deal carefully with media enquiries. Identify one or two 'wider context' factors for each project and consider how the project manager would have needed to incorporate them into managing the project.

Project management tools

Common features

One of the key skills for engineers as they move into managerial roles in the years after graduation is to be able to maintain an overview of the work their teams are undertaking. The manager not only needs to know how to check and archive the fine detail, they also need to keep a global view of the complex projects they are running. The global view informs decision making and is important to present interim progress reports to time-poor general managers, CEOs or company boards.

In order to move from a fine-scale to a global view and keep track of projects in real time, most graduate engineers begin to use proprietary software packages for project management soon after graduation. The package your employer uses may be unfamiliar to you because there are dozens to choose from. For example, the following are the project management tools favoured by some of the world's largest and most complex businesses (Robinson 2021).

1. Figma uses Asana
2. CNet uses Planio
3. Dribble uses Flow
4. Dropbox uses Paper
5. Kickstarter uses Trello
6. Amazon and Ghostery use Roadmunk
7. WeWork uses Jira
8. Groupon uses Basecamp
9. Netflix uses Confluence
10. Backpack Health uses Clubhouse
11. Airbnb uses Wrike

We'll investigate the project management tool features these teams require and that are included in almost every premium project management package currently available.

1. Scalable task management

In project management it is important to be able to know and see what everyone is working on so you can fit the pieces together as your project progresses. To achieve that global view, a tool with task management that is organised, highly configurable and scalable as your projects expand is required. Figure 13.3 shows the task management display from the project management package Trello.

Some desirable task management features are:

- flexibility in how you configure and assign tasks including setting deadlines, priority and tagging teammates
- control over who can work on what tasks through role-based permissions
- clear workflows defined by roles and projects to make sure the right people know what to do so that they can do their work.

2. Flexible planning for structuring project tasks

How projects are managed — whether taking a non-linear approach (known as 'agile') or scheduling tasks in a sequential way using a traditional Gantt chart — shouldn't be limited by the project management package. The software you use for project management shouldn't dictate or limit the structure of tasks.

Many engineering project managers prefer a tool that provides simple ways to design and track 'sprints' (subsets of tasks that are completed in sequence). Engineers also tend to like visual methods to track the progress of projects, like the Kanban-style board shown in figure 13.4. The Kanban board shows work as either pending (to do), in progress (working and waiting) or completed (done). It also indicates the urgency or priority of each task (expedite is for urgent tasks, whereas default tasks are not time critical).

FIGURE 13.3 Task management display in project management package Trello

Trello
Quit the tab tango
Add new Trello cards to boards directly from Slack without needing to hop through the app-switching hoop.
beth
chris
matt
trello
+ Invite people
Join
Subscribe
brian
invite @amy @beth @chris @trello
amy
brian
Question: How do we perfect a taco delivery service?
beth
It's all about speed to make sure they're fresh!
chris
But traffic = soggy tortillas
amy
What about a taco drone delivery service?
chris
Love it! Taco drone deliveries or ?

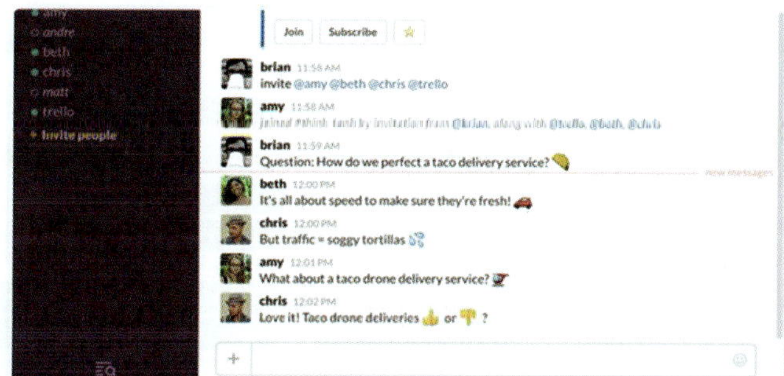

Source: Trello (2024).

FIGURE 13.4 A Kanban board giving a quick visual view of workflow and priority of tasks

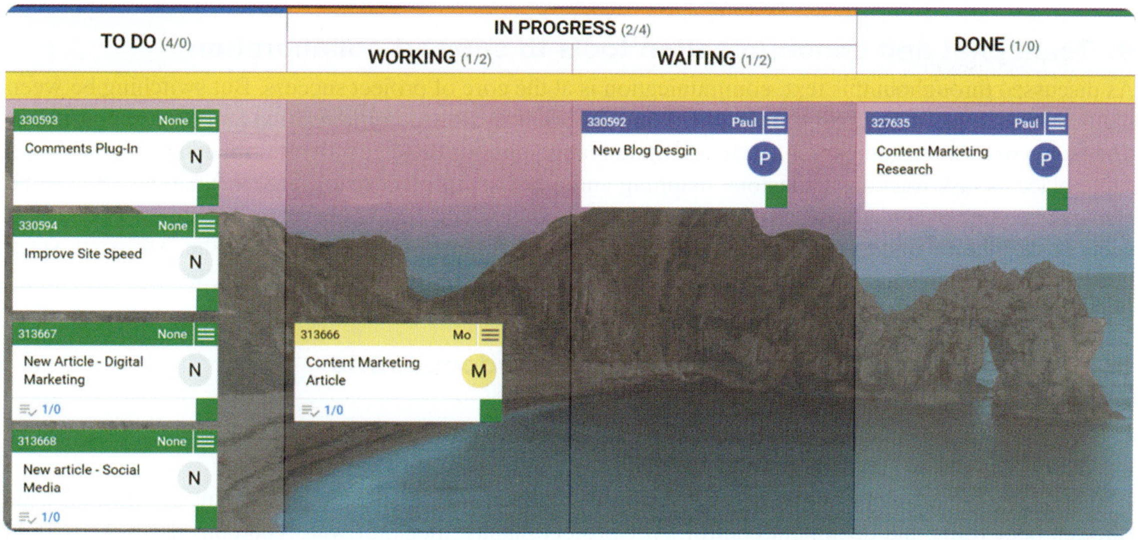

TO DO (4/0)
IN PROGRESS (2/4)
WORKING (1/2)
WAITING (1/2)
DONE (1/0)
330593 None
Comments Plug-In
N
330594 None
Improve Site Speed
N
313667 None
New Article - Digital Marketing
N
1/0
313668 None
New article - Social Media
N
1/0
313666 Mo
Content Marketing Article
M
1/0
330592 Paul
New Blog Desgin
P
327635 Paul
Content Marketing Research
P

Source: Kanbanize (2019).

3. Clear milestones to track progress

Complex engineering projects require a range of tasks to be finished on time to meet a final deadline. Therefore, complex projects are often 'chopped up' into discrete blocks of work. For each block of work, the contract for a project will often set out **milestones**, which are clear, measurable outputs that prove a block of work has been successfully completed. A project contract will often attach a progressive schedule of project payments to the acceptance by a client of the milestone report, or a set of milestone measures. The other benefit of milestones is that a missed milestone gives early indication that a project may be going off-track. Project management software should offer a clear visual 'roadmap' of key project milestones and a way to see real-time progress toward meeting milestones. Figure 13.5 shows software package Roadmunk's timeline page with milestones.

FIGURE 13.5 Software package Roadmunk's milestone indicators provide clarity on project goals when they are added to a roadmap timeline.

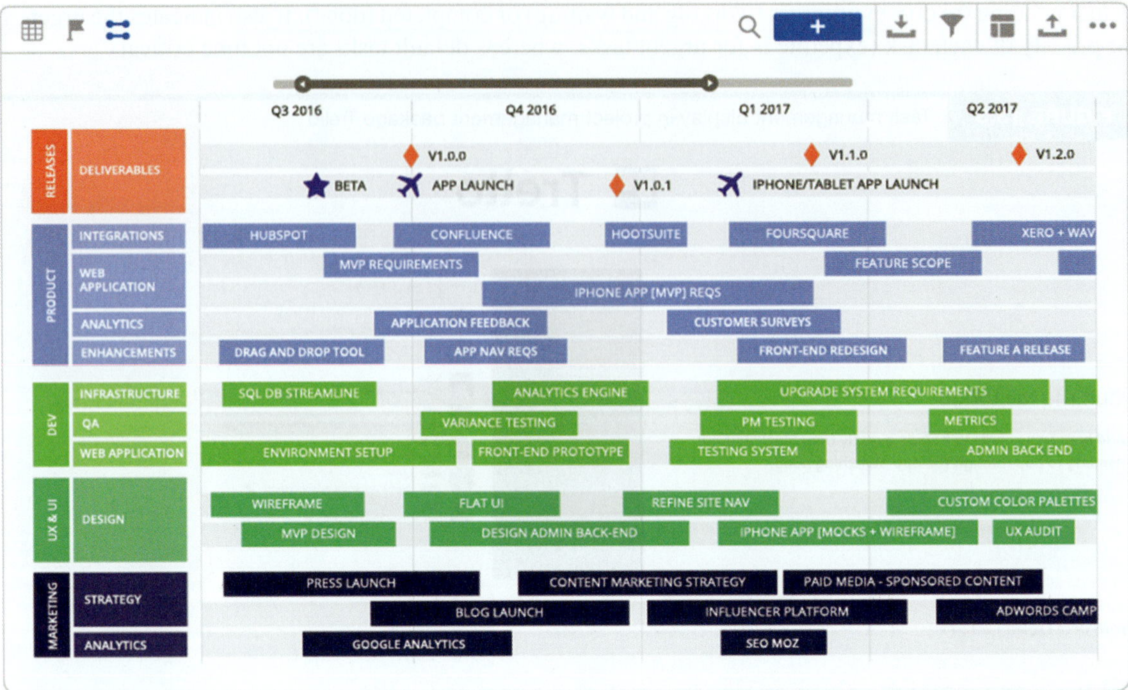

Source: Roadmunk (2019).

4. Team chat and communication tools to support collaboration

As discussed throughout this text, communication is at the core of project success. But switching between platforms, such as from email to task tracking software, can slow communication and lead to confusion. The best project management tools offer communication options that keep all relevant conversations in the same place as task tracking, milestone mapping and other key project management functions. Real-time chat can become overwhelming, so it is useful to set up knowledge channels with clear subscriber listings such as the one in figure 13.6, which features a channel titled 'help-bancly-it' on the Atlassian Slack integration. Clear subscriber listings allow members who need to, remain up to date, without disrupting the workflow and concentration of other sub-teams.

5. Well-organised files and easy access to documents

- 'Where is the project brief again?'
- 'Does anyone know where the photo of our second prototype is stored?'
- 'Who is working on the latest version of the report, and when can I work on it?'

Work teams and organisations, as well as project teams, lose a substantial amount of working time through poor filing and document control. Any project management software you consider using should have the capacity to support teams in setting up and using simple clear filing systems for all key files. It is also advisable to consider seamless archiving for files because engineering projects need to have the capacity to be revived and understood long after the project is completed. This is in case there is a failure associated with the project and to create the possibility for others to learn from any successful approaches to problem solving.

6. Developer-friendly integrations

As you become familiar with different platforms, you will notice some are more geared towards technical work, while others are better for design, content or product work. To maintain good communication and coordination across all sub-teams, it is best to choose a tool that can support all these types of work. The most broadly useable platforms will offer good integration, hosted Git and Subversion repositories and role permissions to limit access to team members who know what they're working with.

FIGURE 13.6 Team chat in Slack

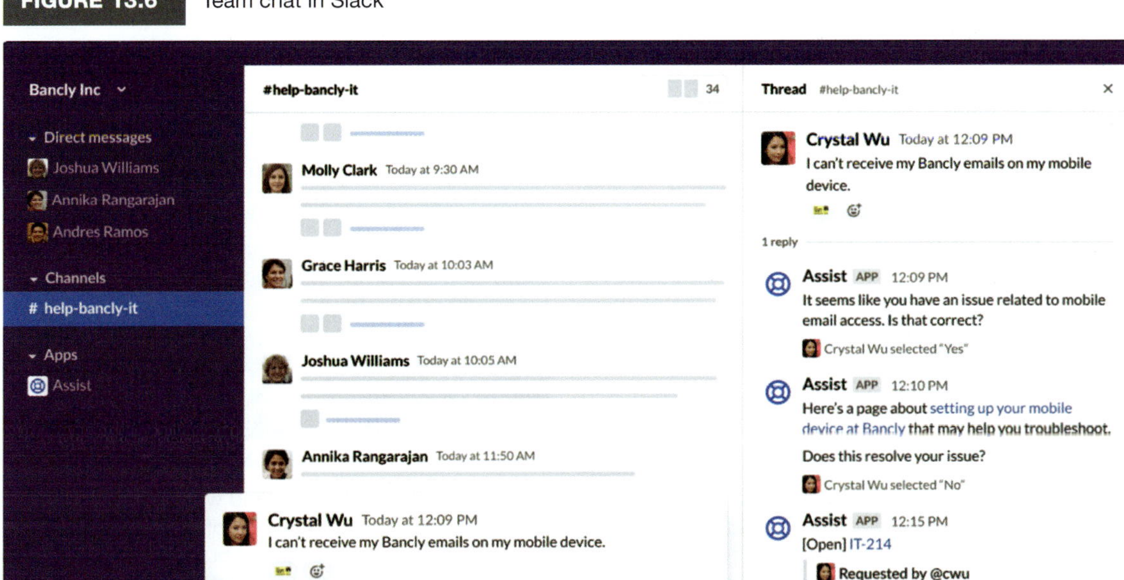

Source: Atlassian (2024).

7. Time tracking for progress feedback and forecasting

Accurate time tracking is essential for projects where a client is being billed by the hour, and the best packages will offer a clear graphical interface so you can communicate openly with a client about where and how each of the hours they paid for was used.

For teams and businesses that tender for project work, accurate and detailed time tracking allows the business to grow a clear sense of how long things take. This may seem strange or obvious but bidding for a project requires an estimate of how long each task or subproject will take and costing that estimate. An incorrect estimate of the time each stage will take can result in a contract that ties an engineering business to delivering a project at a loss. Alternatively, if the tender 'overcooks' the budget with grossly large time estimates, the business is likely to lose the tender to a competitor with a cheaper bid.

13.2 Planning the stages of an engineering project

LEARNING OBJECTIVE 13.2 Plan the stages of an engineering project.

Project management is a key skill for engineers and other professionals, whether managing the completion of work within their own team or managing the delivery of complex projects for external clients. Planning is a crucial aspect of project management. 'Under-planning' will likely lead to poor job control and outcomes; on the other hand, 'over-planning' can be a costly waste of resources.

The aim of project management is to define, plan and manage a series of tasks to achieve an over-all outcome. Projects invariably differ in their scope, complexity, and resourcing and therefore in the amount of planning required. In some cases, there is a legal contract that dictates how a project needs to be managed.

This section presents the design and construction of an environmentally sustainable building as an example of how a complex project can be broken down into many tasks for diverse teams to work on.

KEY POINT

Complex engineering projects can be decomposed into sub-systems to enable completion of the work by specialised teams.

CASE STUDY

Designing and constructing a 'green-star' building

This case study runs through the next several sections and involves the design and construction of a 'green-star' building. Such buildings use natural ventilation and a range of technologies to reduce both water and energy consumption. One example is the PIXEL Building in Melbourne, which has a six-star energy rating and claims to operate with no net release of carbon dioxide into the atmosphere (i.e. is 'carbon neutral') (Introba n.d.).

The PIXEL Building in Carlton, Melbourne

Green buildings are rated by the Green Building Council of Australia and the New Zealand Green Building Council. Typical systems included in these buildings are rainwater capture, grey-water recycling, sewer mining, use of natural light and ventilation, night-time purging of heat from the building, solar panels for water heating and wind turbines. Not all green buildings use all these features. The combination of features leads to ratings of 4, 5 and 6 stars.

The skills of many engineers are required in the design and construction of these buildings, as green buildings contain many active systems — water recycling, convection-based ventilation systems and automatically controlled systems for shading and heating the building. These active systems make the building into a machine rather than a static structure and they must be controlled by electrical, mechanical and software systems. The engineers who work on these buildings work with other professionals, such as architects, in the completion of these complex projects. In such a process, it is common to adopt knowledge and skills from other disciplines. Being widely knowledgeable makes an engineer more valuable in the workplace. This is why many people suggest that the future of engineering is multidisciplinary and multicultural.

Consider an engineering company that is bidding to design such a building. An alliance has been formed with an architectural consultant as well as an established construction company. The work might be distributed as follows.

- Architects will do overall design and layout of the structure, as well as primary liaison with the client and definition of the scope of the work.
- Environmental engineers will consider the overall design of the environmental systems, including maintaining the living systems in the building.
- Structural engineers will handle the building frame and floor slab modelling.
- Geotechnical engineers will deal with the foundations, piling, retaining walls below the ground and the underpinning of adjacent structures and dewatering.
- Geomatic engineers (surveyors) will deal with overall data gathering, monitoring of building movements (e.g. of surrounding structures) and the location and verticality of the construction.
- Electrical engineers will deal with power and lighting, including a new substation in the building, communications (phone, data, wireless) and controllers for the building's many systems (such as heat, lighting, security and water).
- Chemical engineers will design the wastewater treatment plant to recycle water for flushing toilets and irrigation of the green roof (this may include sewer mining).
- Mechanical engineers will deal with pumps for the wastewater treatment plant, grey-water treatment and use on each site, water delivery to all floors and air-conditioning (heat and cooling).
- Software engineers will deal with coordination, control and monitoring of all services from a central control station and smart building control.

At first sight, the scope of the project is bewildering. Where should the engineers start? What crucial decisions need to be made first? What dependencies exist between the work of the various disciplines?

The art of complex project management is to reduce complex problems into simple problems that have a recognisable solution. Thus, complex projects are divided into simpler sub-projects that can be easily completed with known methods, resources and time frames.

Using a tool to plan the project stages

In this section, the One-Page Project Manager (OPPM) will be introduced as a tool for managing this green-star building project (Campbell 2007). While most contemporary engineering consultancies use project management software to plan, pitch and manage complex projects, the OPPM provides a simple view of the skeleton or fundamental parts of managing projects. We use it over the next few sections to illustrate the foundational principles of project planning.

The OPPM focuses on the following six key elements of project management.

1. *Objectives.* Why are you doing this project and what needs to be done?
2. *Major tasks.* What do you need to do to achieve the objectives?
3. *Owners.* Who is responsible for each task?
4. *Target dates.* What are the beginning and end dates for each task?
5. *Costs.* What are the major costs?
6. *Summary and forecast.* What is the current state of each task and what are the costs?

A basic template for this style of project management is shown in figure 13.7 and the six elements will now be explained in the context of the green-star building case study project.

Objectives

A project needs a clear **objective**. An objective is the intended outcome of planned activities, within specific time and other constraints. For the case study, the objective could be to design, within three months, a five-star green energy building with 10 000 square metres (m^2) of floor space. This design project could be followed by a construction project, with the objective to construct the green-star building within two years.

The primary project objective listed in the header of the template in figure 13.7 is broken into component objectives, which are entered in the bottom left corner of the template. These could be the stages of the project, including project researched, documented and agreed upon; alternative solutions identified; alternatives analysed; and final recommendations made.

Other key data recorded in the header are the project name, the project leader completion date and the current date.

Major tasks

Once the objectives are clear, the major tasks can be defined. The one-page planner template in figure 13.7 has space for twenty tasks. If a task is complex, another one-page planner could be used to break it down into subtasks. According to Campbell (2007), tasks need to:

- be a manageable size
- be *distinct*, so they are clearly separable from other tasks, even if one task depends on another
- be *measurable* in terms of progress (e.g. references collected, pages written, number of drawings completed)
- have *commitment* from the owners, although each task could be owned by two people for backup and knowledge sharing.

The progress of each task in the template would be updated every month. If it is complete, it would be shown as a solid dot.

Subjective tasks

The panel of figure 13.7 with rows labelled A–E is a space for listing *subjective tasks* — tasks that it is harder to verify have been completed. Campbell (2007) lists software performance in this category; speed may be desirable but not always attainable at the level required. Weight and speed might be desirable criteria for a new car project. These tasks and/or criteria link to the objectives and to the owners, so that it is clear how they fit into the project and who is responsible for which parts of the project. This space may also be used to summarise progress on each of the sub-objectives in the left-hand column. Colours could be used to show performance in the subjective tasks; for example, green = adequate performance, yellow = worrying performance and red = major concern over performance (which may affect the project budget and/or timeline).

Align tasks and objectives

The next step is to indicate how the tasks link to delivering the objectives. There may be a very simple alignment; for example, the first five tasks may deliver the first objective. Alternatively, a research task might serve three objectives; for example, in the design of a mobile phone, research would be a task that belongs to three different objectives — user interface design, antenna design and camera capability.

The panel on the left in figure 13.7 allows connection between the objectives and the major tasks. A dot in this panel indicates connection between what needs to be done (task) with why it needs to be done (objective).

FIGURE 13.7 One-Page Project Manager (OPPM) template

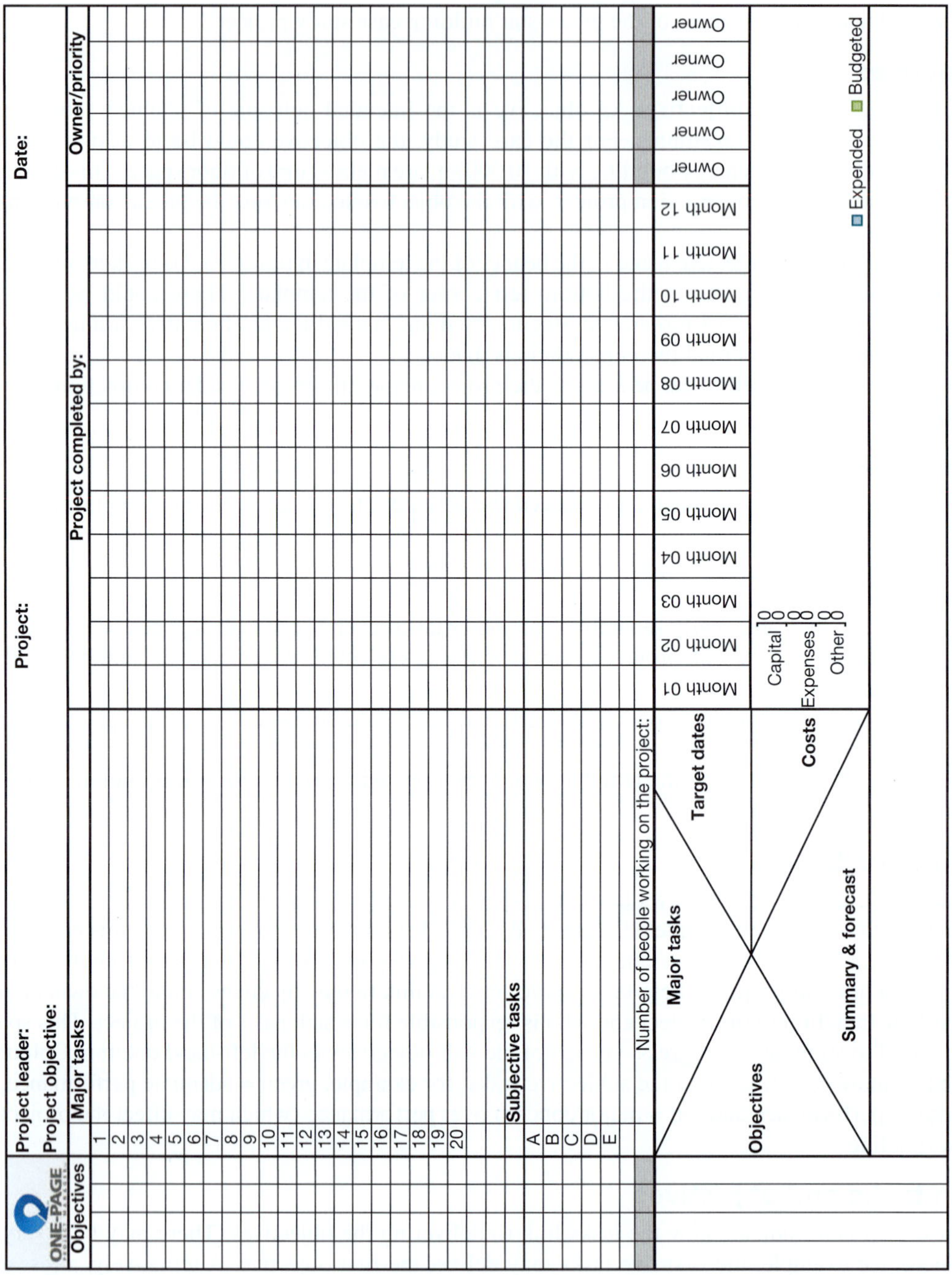

ONE-PAGE PROJECT MANAGER

Project leader:	Project:	Date:
Project objective:		

Objectives		Major tasks	Project completed by:	Owner/priority
	1			
	2			
	3			
	4			
	5			
	6			
	7			
	8			
	9			
	10			
	11			
	12			
	13			
	14			
	15			
	16			
	17			
	18			
	19			
	20			
		Subjective tasks		
	A			
	B			
	C			
	D			
	E			
		Number of people working on the project:		

Major tasks / Target dates / Objectives / Costs / Summary & forecast	Month 01	Month 02	Month 03	Month 04	Month 05	Month 06	Month 07	Month 08	Month 09	Month 10	Month 11	Month 12	Owner	Owner	Owner	Owner	Owner

Costs: Capital 0 0; Expenses 0 0; Other 0 0

■ Expended ■ Budgeted

Source: Campbell (2007).

Target dates

The major constraints on any project are time, resources and scope. The scope is defined by the objectives and the major tasks. The resources include the owners of those tasks. The target dates then define the time available. With an agreed start and end date for the entire project, each task must be fit within these time constraints in a way that respects the dependencies between the tasks. Each task could then be drawn onto the OPPM template shown in figure 13.7. An open circle could represent each time period that would be occupied by each task. The circle could then be filled in (made into a solid dot) when that task is complete. This would enable the project's progress to be tracked and would also provide a simple communication tool between team members. See also figure 13.8 for an example.

Owners

In teams, owners are team members who will do the work or ensure that the work is done. Their names go in the panel at the bottom right of figure 13.7. The owners may be team leaders of their own sub-projects, for example, being responsible for designing or implementing the electrical or mechanical subsystems of the green-star building. Owners are held accountable to ensure tasks are completed. The priorities for tasks are listed in the columns above the owners' names, on the right side of figure 13.7.

It is important to get to know team members so that you can make best use of their skills. In student teams at university, some questions you could ask your fellow group members to open this discussion are as follows.

- What are your strengths, skills and knowledge?
- What skills and knowledge would you like to develop through the current project?
- What roles do you like to play in teams?

Align tasks to owners

Assigning tasks to owners is an important aspect of managing any team project, making sure the distribution of work is reasonable across all team members. Often, there will be one owner per task. Sometimes, there may be two. Using the template from figure 13.7 as a guide, an 'A' could be placed in the cell for the principal owner of a task and a 'B' in the cell for the secondary owner, or backup person. There should only be one principal owner; this person then has clear responsibility for the task. See figure 13.8 as an example.

Costs

A range of costs need to be estimated for components of a project. For example, these might include building costs, software costs and training costs. These numbers can be entered in the figure 13.7 template and represented as a simple bar chart. At the beginning of a project, the bar chart will show estimated costs and actual costs should be shown for reporting purposes.

In student projects, your main resources are the hours that each of you can spend on a project. It is important to allocate realistic hours of effort to each task. The more realistic you make the estimates, the more efficiently work can be allocated around the team. Keeping track of how long it takes you to do project tasks as a student will help you improve your abilities to estimate the time required to complete tasks as a practicing engineer. As an engineer, you will likely have to quote on jobs and will need to have a sense of how to effectively estimate your time.

Summary and forecast

At regular intervals, the state of a project needs to be summarised. Critical issues that should be summarised include the schedule (Are there particular delays?), the resources (Are they adequate?), the people (Are there poorly performing team members/owners of tasks?) and the costs (Is the project running within budget?).

Planning the stages of the green-star building project

The OPPM tool will now be used to demonstrate planning the various stages of the green-star building project, such as:

- bidding for the project
- planning the project
- planning the design of the project
- planning the workload among the team
- developing the conceptual design.

Similar processes would also be used for the detailed design phase and for construction. A plan such as OPPM would be developed for each of these stages.

Bidding for the project

Developing the bid documents will cost more than $20 000, mostly in staff time. A decision must be made on whether this is a good use of company resources.

This is an example of a **feasibility investigation**. Does the company have the required expertise, either in-house or by subcontracting work to other more specialised companies?

The first stage is to scope the bid. This has been done using the OPPM in figure 13.8, where the project has been given the name 'Greenways Building'.

In the absence of better information, two 'working weeks', each of five days, have been allocated to completing tasks. Five team members will contribute to the development of the bid documents. Two extra days have been allocated as slack to account for tasks that take longer than expected. The project has been roughly costed at $20 000 ($2000 salary × five staff × two weeks). This is likely an overestimate of staff time. Another $1000 has been allocated for locating information (e.g. search fees) and copying and printing costs.

To bid or not to bid?

The next stage is to take this proposal to senior management to decide whether to bid for the particular job. The task needs to fit into other work within the office. Some questions to consider include the following.

- Can the required people be spared to work on this job for two weeks?
- Could the work be done more quickly than two weeks?
- Does the company want the job?
- What amount of work will be required to deliver the actual project?
- What will it cost to deliver the project?
- What will the income, or profit, be from the contract?

An overall project plan is needed to address many of these questions. The next stage is to plan the delivery of the project.

Planning the project

Assume senior management gave the green light for the bid to go ahead. Now it becomes necessary to plan the overall project. Project management occurs on several levels, as described earlier. The planning performed at one level is helpful in the next, more detailed, plan. The various subsystems influence the planning process at every level and it is necessary to consider how these subsystems are linked through the stages of construction. For example, what building services can be installed as the building frame goes up?

An overall project plan is shown in figure 13.9. Note that the project is expected to take nine quarters (two and one-quarter years) to implement. There will be four distinct phases (the objectives), being: (1) a design phase, (2) a site and building frame phase, (3) a fit-out phase and (4) a commissioning phase. The subsystem designers must agree about what they will need to do in the different stages of the project (e.g. confirming with the chemical engineers the wastewater treatment plant will be built in the third phrase and commissioned in the fourth phase).

Critical path method (CPM) view

In the middle blocks of figure 13.9, tasks are listed against a timeline in a format known as a Gantt Chart. The critical path method (CPM) takes the Gantt Chart and adds dependencies between the tasks. The key question for preparing a CPM view is: what tasks need to be completed before another can start? CPM capability is a standard feature of project management software packages (Lewis et al. 2019). The method highlights the **critical path**, which is the shortest time in which a project can be completed. Any delays in the activities listed on the critical path will delay subsequent task completion and represent a threat to timely completion of the project. There is no slack in that path through the network of activities. **Slack** is the amount of time a task can be delayed without its completion time affecting other tasks on the critical path.

FIGURE 13.8 Scoping the bid process for the green-star building project

ONE-PAGE PROJECT MANAGER

Project leader: Bob Smith **Project: Greenways Building — The bid** **Date: 27/07/2025**

Project objective: Prepare a bid for the design and supervision work for the Greenways Building

Objectives					**Major tasks**	**Project completed by: 31/08/2025**												**Owner/priority**				
Community and environmental issues	Architectural and structural	Building services				Day 01	Day 02	Day 03	Day 04	Day 05	Day 06	Day 07	Day 08	Day 09	Day 10	Day 11	Day 12	Bob	Mary	Tran	Sue	Arun
O				1	Community assessment	O	O	O	O	O								A	B			
O				2	Environmental assessment	O	O	O	O	O									A	B		
O	O			3	Planning and layout		O	O	O	O	O								B	A		
	O			4	Structural frame			O	O	O	O	O								A	B	
		O		5	Water supply			O	O											B	A	
O		O		6	Water reuse				O	O	O								B		A	
O		O		7	Electrical supply and lighting			O	O	O	O	O							B		A	
O		O		8	Heating and cooling					O	O	O	O	O				B				A
O		O		9	Communications					O	O	O	O	O				B				A
O		O		10	Building control			O	O	O	O	O							B			A
O	O	O		11	Prepare documents	O	O	O	O	O	O	O	O	O	O			A	B	C	D	E
				12	Allowance for the unexpected											O	O	A				
					# People working on the project:																	

Major tasks / **Target dates** / **Objectives** / **Costs** / **Summary & forecast**

Costs	Expended	Budgeted
Capital	0	0
Expenses	0	1000
Salaries	0	20 000

□ Expended □ Budgeted

The plan is to make sure that each subsystem has been scoped for the bid. Some innovative thinking will be required to win this bid. Hence, some investigation and preliminary thinking around the alternatives will be required.

Source: Adapted from Campbell (2007).

FIGURE 13.9 Overall project plan for the green-star building project

ONE-PAGE PROJECT MANAGER

Project leader: Bob Smith **Project: Greenways Building — Overall project plan** **Date: 27/07/2025**

Project objective: Design and supervision work for the Greenways Building

Design phase	Site works and building frame	Building services	Commissioning		Major tasks	Quarter 01	Quarter 02	Quarter 03	Quarter 04	Quarter 05	Quarter 06	Quarter 07	Quarter 08	Quarter 09	Quarter 10	Quarter 11	Quarter 12	Owner/priority
O				1	Community engagement	O												
O				2	Architectural and planning design	O												
O				3	Environmental assessment	O												
O				4	Call and let tenders	O												
O	O			5	Excavation and stabilisation	O	O	O										
O	O			6	Building frame (shell)	O		O	O	O								
O		O	O	7	Water supply	O			O	O	O							
O		O	O	8	Recycled water supply	O			O	O	O							
O		O	O	9	Stormwater collection system	O					O	O						
O		O	O	10	Wastewater treatment plant	O						O	O	O				
O		O	O	11	Electrical supply and lighting	O			O	O	O			O				
O		O	O	12	Communications	O				O	O	O		O				
O		O	O	13	Heating and cooling	O			O	O	O			O				
O		O	O	14	Smart building control	O			O			O	O	O				
O			O	15	Commissioning and hand-over	O								O				
					# People working on the project:													

Major tasks / **Target dates** / **Objectives** / **Costs** / **Summary & forecast**

Costs:

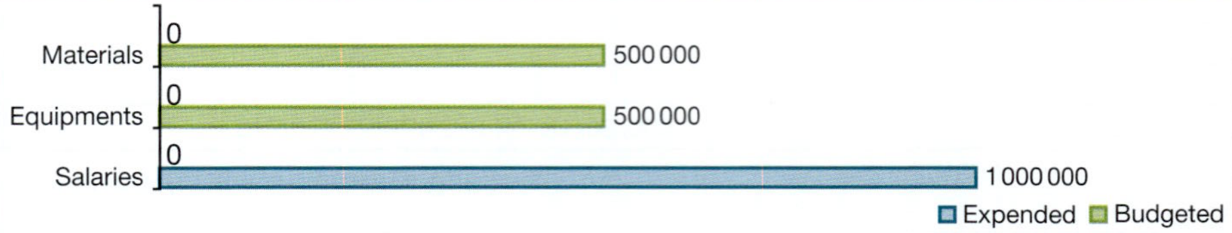

	Start	Budgeted
Materials	0	500 000
Equipments	0	500 000
Salaries	0	1 000 000

Expended / Budgeted

The plan is to implement the project over 9 quarters, with the first quarter being the design stage. Construction and fit-out will take two years. Note that the building frame is complete within one year and final fit-out takes another year.

Source: Adapted from Campbell (2007).

Figure 13.9 shows the overall project plan for the green-star building project completion. What does this look like as a CPM chart? Consider the main activities in the CPM in figure 13.10.

- *Design* occupies the first quarter and must be complete before other activities can begin. This step also includes calling tenders to begin construction.
- The *excavation* and *stabilisation* of the site requires two quarters and must precede construction.
- The *building frame* follows excavation and site works.
- Other activities can begin after the frame has commenced (e.g. *water, electrical, communications*).
- *Commissioning* the building for its occupation is the final activity.

Figure 13.10 shows a CPM plan with three stages (blue bars) and tasks of each stage (green bars). It demonstrates the following different forms of *dependency*.

- Finish to start (e.g. environmental assessment to tender to contract) — activity B starts when activity A is complete.
- Start to start (e.g. frame to water supply) — activity B can start as soon as activity A has started, with the option of a delay (e.g. activity B starts ten days after activity A starts).
- Finish to finish — activity B cannot finish until X days after activity A has finished (e.g. commissioning cannot be completed until ten days after the installation of services).

What is the critical path for this project? This graphical view makes it easy to spot. As we read left to right, then it is clear that the critical path comprises: *design and tendering + site excavation and stabilisation + building structure + services + commissioning and handover*.

Planning the design

Now that the overall green building project has been planned — including design and construction — consider what the design process will resemble. Earlier in figure 13.9, the design is shown to be completed in the first quarter. Is this realistic? Figure 13.11 shows a more detailed project plan for the design process. Take note of the assumed task durations and the general sequence of activities (from general conceptual design through to detailed systems such as the smart building control). Most of the technical subtasks are expected to take six weeks to complete, and these will be staggered.

Planning the workload

Once a plan has been drafted, it may need to be modified based on the availability of resources. Figure 13.11 shows the plan for the design. The main resources for the detailed design are the people who are required to complete the design. The five main contributors are shown as owners in the project plan (Bob, Mary, Tran, Sue and Arun). We will now check to see how their work is distributed across the 12 weeks that are specified for the design.

Figure 13.12 provides some rough workload estimates. Note that some assumptions have been made. If a team member has an 'A' against a task it counts as 1, while a 'B' counts as 0.5 of a task. In week 1, Bob is involved in task 1 only at priority level A, which means a score of 1 for that week. In weeks 6 to 9, he has 6 'B' tasks, for a score of $6 \times 0.5 = 3$. Likewise, for Sue in week 5, she is involved in activities 5, 6 and 7, all of which she leads (priority A tasks). That makes a score of 3 in weeks 5, 6 and 7. The average workload over the 12-week period is 1.9 tasks per person, per week. Since many team members seem to have significant peaks during the period, is it possible to level off the demands on their time to reduce their workload stress?

A quick solution is to look at the overall distribution of tasks and to allocate tasks to reduce the peak workloads and spread the tasks over more weeks, starting many of them sooner. This is shown in figure 13.13, and the corresponding new work schedule is shown in figure 13.14. Whether or not this is a better resource allocation is a matter for debate.

Another solution is to break up some tasks over a longer period. Most project management programs provide resource levelling, which is the automatic distribution of resources across tasks. These diagrams provide the opportunity to anticipate difficulties before they arise and to plan around them. Tasks can be redistributed across the project team or redesigned to complete over a longer time. This is the essence of project management — making sure a complex problem is delivered with the fewest number of surprises, by planning ahead and identifying tasks and the resources required to complete them.

Developing the conceptual design

The pitch has been made to the management team and a decision has been made to go ahead with the bid for the green-star building. There are two weeks to scope out each major element and to make suitable costings. This is the conceptual design stage. The team will come together to work intensively, using quick methods to estimate sizes. From this, the team will estimate design and construction costs.

FIGURE 13.10 Critical path method for the green-star building project

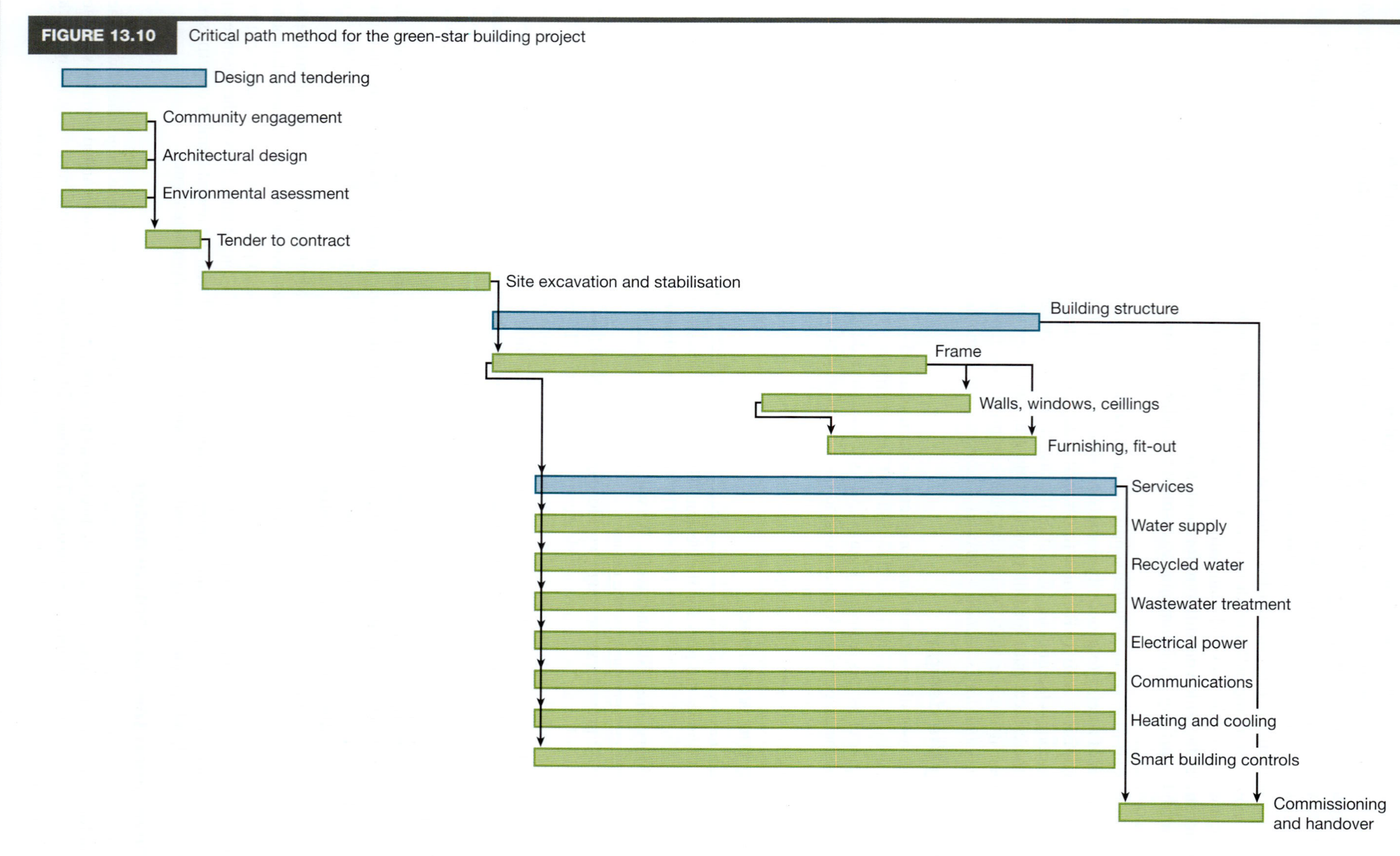

FIGURE 13.11 Detailed design phase of the green-star building project (the first three months)

ONE-PAGE PROJECT MANAGER

Project leader: Bob Smith | **Project: Greenways Building — Design plan** | **Date: 27/07/2025**

Project objective: Design work for Greenways Building

Objectives					Major tasks	Project completed by: 31/08/2025												Owner/priority				
Functionality specification	Site works and building frame	Building services	Commissioning			Week 01	Week 02	Week 03	Week 04	Week 05	Week 06	Week 07	Week 08	Week 09	Week 10	Week 11	Week 12	Bob	Mary	Tran	Sue	Arun
O				1	Community engagement	O	O	O										A	B			
O				2	Architectural and planning design	O	O	O	O	O	O								A	B		
O				3	Environmental assessment	O	O	O	O	O	O								B	A		
O	O			4	Call and let tenders											O	O	A	B			
O	O			5	Geotechnical design			O	O	O	O									B	A	
O	O			6	Building frame (shell)			O	O	O	O	O	O						B		A	
O		O		7	Water supply				O	O	O	O	O	O					B		A	
O		O		8	Recycled water supply				O	O	O	O	O	O				B		A		
O		O		9	Stormwater collection system						O	O	O	O				B		A		
O		O		10	Wastewater treatment plant				O	O	O	O	O	O					B	A		
O		O		11	Electrical supply and lighting				O	O	O	O	O	O				B				A
O		O		12	Communications				O	O	O	O	O	O				B				A
O		O		13	Heating and cooling				O	O	O	O	O	O				B				A
O		O		14	Smart building control					O	O	O	O	O				B				A
O			O	15	Commissioning and hand-over plan								O	O	C			A	B			
					# People working on the project:								O	O	C							

Major tasks / Target dates / Objectives / Costs / Summary & forecast

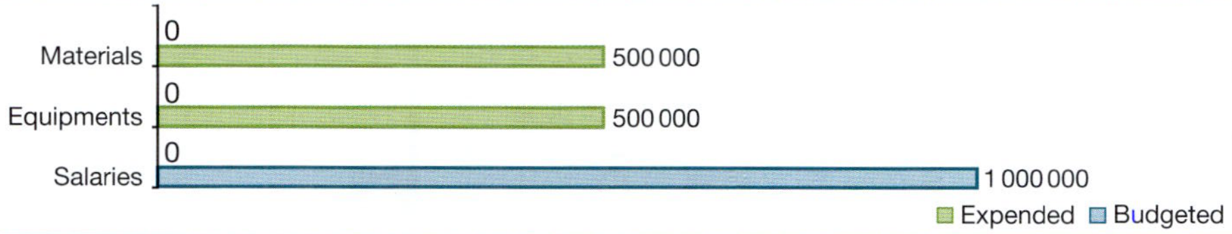

Costs

	Expended	Budgeted
Materials	0	500 000
Equipments	0	500 000
Salaries	0	1 000 000

■ Expended ■ Budgeted

The design will take 10 weeks with tendering occupying the last two weeks.

Source: Adapted from Campbell (2007).

FIGURE 13.12 Initial workload estimates for the green-star building project

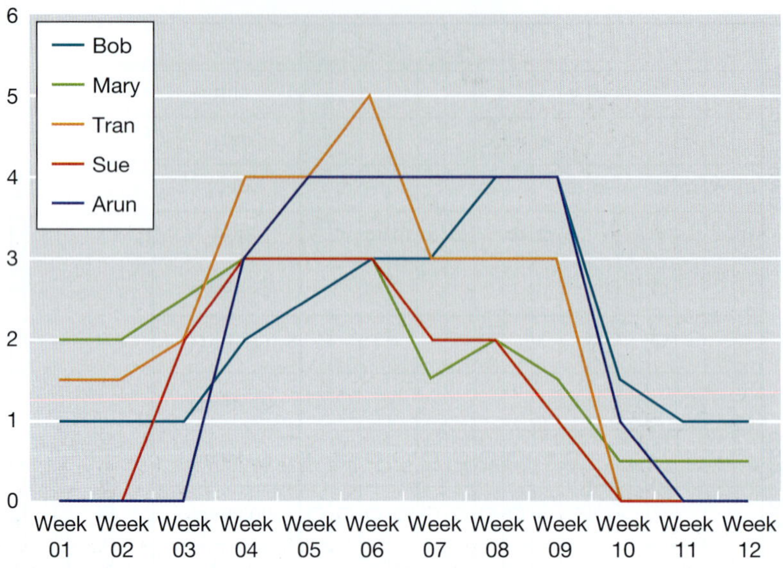

FIGURE 13.13 Potential reallocation of tasks to spread individual team member workload more evenly across the green-star building project

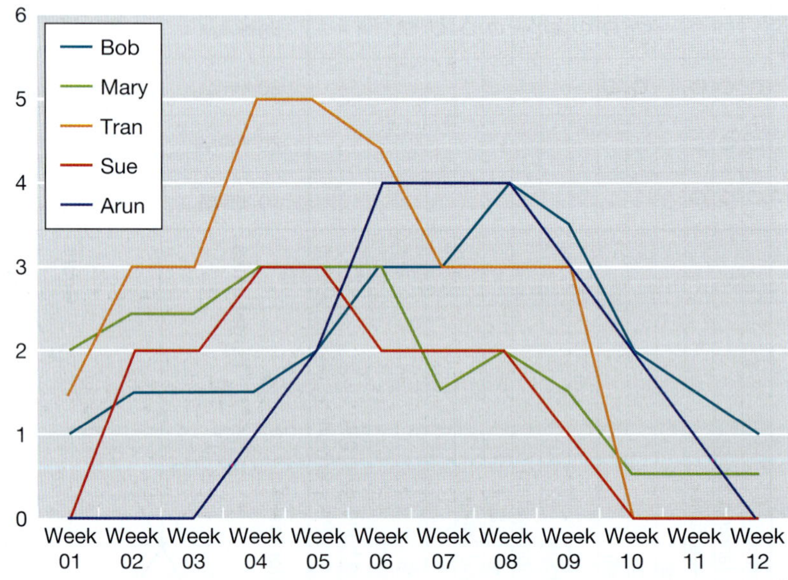

At this stage, average statistics will suffice. Some questions that the team needs to consider are as follows.

- How many people will be in the building?
- How many square metres of floor space are required?
- Roughly how large will the building frame be?
- How many kilometres of items such as electrical and communication cables, heating and cooling ducts, water supply and wastewater collection will be needed?

The client will be interested to see approximate costs for all these factors.

Each of the major design tasks shown in figure 13.11 represents a mini-project. These may be distributed across several people, possibly in different offices. The design process was discussed in the chapters on the engineering method and engineering design. It is as follows.

1. Start with a need with articulated goals and criteria.
2. Conduct research to properly understand and identify the problem.
3. Articulate a range of possible solutions.

4. Analyse the solutions against technical, economic, social and environmental criteria.
5. Check that the recommendations satisfy all the original goals and criteria.
6. Recommend one or more suitable solutions to discuss with the client.

The design of a major building is influenced by many factors, including its purpose, the type of business it will house, the shape of the land and variations in elevation, the climate, preferred orientation, neighbouring developments including farmland, zoning restrictions and height limitations. With such a project, these factors will shape the range of options available to the planner or designer.

The overall layout will define the location of building services such as electricity and communication cables, heating and cooling, water supply, wastewater sewers and stormwater drainage. The developer or owner will want to maximise their profit, so the architects and engineers will be pressured to create as much useable space as possible within the building. This will be traded-off with volumetric space, such as an atrium to provide social space as well as passive cooling.

Stormwater sewers operate under gravity (as do wastewater sewers) and must be oriented downhill to a collection point. Collection and storage must be considered for feeding waste into the wastewater treatment plant. Pumps will be required to lift the treated water throughout the building. All these complications must be reflected in the project plan, including the dependencies.

Figure 13.11 can now be adapted to take account of the dependencies between activities, with the modified plan shown as figure 13.14. This plan recognises that the conceptual design will require full team involvement in the first week to ensure the interdependencies between the subsystems are incorporated in the overall layout. After these key design decisions are made, the subsystems can be developed more independently; however, regular project meetings will still be needed to make sure that changes are known by all team members.

There are natural interdependencies between building design and service design; for example, the layout of the building will determine where services can be located and, of course, how much of each service is required on each floor and how it will be distributed, usually from a central services duct or ducts within the central lift core.

Consider one of the subsystems that matches your own discipline interests. Can the key design decisions be made in isolation from the other subsystems? Questions that can be asked from different engineering discipline perspectives include the following.

- *Electrical.* How is the electricity supply system dependent on other subsystems? How might you estimate its capacity? Is this likely to change over the next 25 years? Why? What telecommunications and data cabling is required? What will be needed in 25 years? How will you provide for capacity upgrade over this time? What will be required to run new data cables?
- *Civil.* Wastewater sewers are often blocked by careless users and need regular maintenance. How will your design make this easier or less frequent, producing lower long-term costs?
- *Mechanical.* There is often poorly distributed heating and cooling within buildings, with some rooms being too hot and others too cool. How will you design a system to provide more even temperature across each floor and between floors?

What has been demonstrated in this section is how project management principles can be applied at many levels of a project. At the overall project level, the plan shows how the project will be delivered. This is the construction or implementation view. Preceding this is the design stage. Some key questions at this stage are as follows.

- What size team is needed to get the design drawings to a stage where they are ready to go out to tender?
- How many people will be required?
- Before the design stage is the bid stage. What effort is required to bid for the job in the first place? How many people are required?
- What needs to be done to prepare the bid?

In the design stage, each task will be further divided into smaller subtasks. These can be allocated around the office or even outside the organisation.

So, project planning and management is a hierarchical process that uses the 'divide and conquer' approach to complex work. The examples of the OPPM show how complex tasks can be divided into simpler ones and how the progress of each task can be reviewed quickly and regularly.

Human resources

People are the key ingredient in successful project completion. In the design of new artefacts, people are also the key resources, together with computer software and hardware, which enable the design process. Clearly, then, building effective project teams is a vital step in project management.

FIGURE 13.14 Version 2 of the design plan for the green-star energy building project

ONE-PAGE PROJECT MANAGER

Project leader: Bob Smith **Project: Greenways Building — Design plan** **Date: 27/07/2025**

Project objective: Design work for Greenways Building

Functionality specification	Site works and building frame	Building services	Commissioning	#	Major tasks	Week 01	Week 02	Week 03	Week 04	Week 05	Week 06	Week 07	Week 08	Week 09	Week 10	Week 11	Week 12	Bob	Mary	Tran	Sue	Arun
Objectives					**Major tasks**	**Project completed by: 31/08/2025**												**Owner/priority**				
O				1	Community engagement	O	O	O										A	B			
O				2	Architectural and planning design	O	O	O	O	O	O								A	B		
O				3	Environmental assessment	O	O	O	O	O	O								B	A		
O	O			4	Call and let tenders											O	O	A	B			
O	O			5	Geotechnical design	O		O	O	O	O									B	A	
O	O			6	Building frame (shell)	O		O	O	O	O	O	O						B		A	
O		O		7	Water supply	O			O	O	O	O	O	O					B		A	
O		O		8	Recycled water supply	O			O	O	O	O	O	O				B		A		
O		O		9	Stormwater collection system	O					O	O	O	O				B		A		
O		O		10	Wastewater treatment plant	O			O	O	O	O	O	O					B	A		
O		O		11	Electrical supply and lighting	O			O	O	O	O	O	O				B				A
O		O		12	Communications	O			O	O	O	O	O	O				B				A
O		O		13	Heating and cooling	O			O	O	O	O	O	O				B				A
O		O		14	Smart building control	O				O	O	O	O	O	O			B				A
O			O	15	Commissioning and hand-over plan								O	O	O			A	B			
					# People working on the project:																	

Major tasks / **Target dates** / **Objectives** / **Costs** / **Summary & forecast**

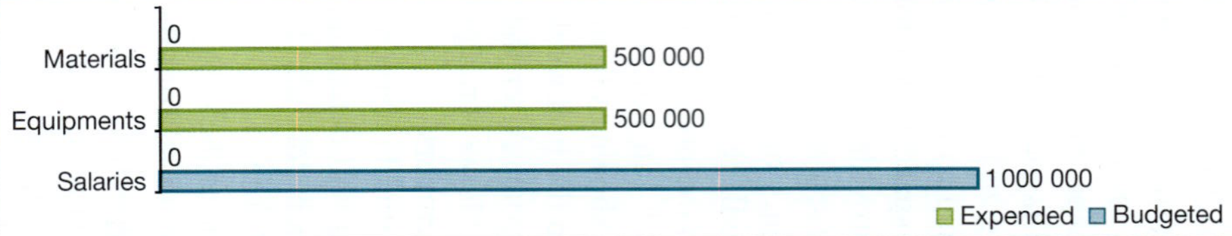

Costs:

	Expended	Budgeted
Materials	0	500 000
Equipments	0	500 000
Salaries	0	1 000 000

■ Expended ■ Budgeted

The design will take 10 weeks with tendering occupying the last two weeks.

Source: Adapted from Campbell (2007).

The chapter on communication skills has already discussed how to build effective teams. Effective communication underpins any successful team. This is particularly true in a project management context, where there might be hundreds of people involved from many different companies. How do they each find the information that they need? Meetings can be effective or a big waste of time in terms of information delivery. Effective meetings are described in the chapters on communication, emphasising such issues as the need for a clear purpose, appropriate timing, and documentation through agendas and minutes. Teams are most effective when they are made up of a diverse collection of talents, as recommended by the Belbin model. Project teams need effective leadership, ideas, delegation and attention to detail, as well as caring roles to maintain team function.

As you might imagine, negotiation is a critical element of project success. Where there are many companies involved, there are likely to be disputes between parties.

Financial resources

Many project disputes centre on the financial aspects of the project. The chapter on evaluating options considered the financial modelling of projects. That modelling concerns the long-term success of a facility or a product. Will it pay for itself and provide a suitable profit over its lifetime?

In terms of project management, a key role for the project management team is to make sure that the project is delivered on budget. The bid for a project must include realistic costing of all the major activities. Where parts of the project will be delivered by external contractors, those estimates need to reflect current market conditions. A shortage of projects in a particular sector might mean that certain services can be bought quite cheaply. Six months later, as new projects come online, the same contractors may be busier and will have increased their prices. So, the estimating and tendering for projects is a complex activity and specialist estimators are usually involved. Quantity surveying is one profession that deals with project cost estimating.

Once the project begins, then the project accountants are required to track actual expenditure against the estimates. Weekly or monthly accounting is required to make sure that the project is running to budget.

SPOTLIGHT

Lean management and Fast Jet

Aircraft modification and maintenance requires significant technical aerospace engineering knowledge and involves considerable expense. The Australian Defence Force (ADF) launched a process in 2009 to change their approach to modification and maintenance on the Hawk Lead-In Fighter Program and the F/A-18 Hornet Program (BAE Systems 2019). The Hawk Lead-In Fighter is a training aircraft that teaches pilots to operate the F/A-18 Hornet. The ADF introduced an approach to management that focused on eliminating all forms of waste from maintenance activity processes while improving quality. They used a process of lean project management, which focuses on identifying the following seven forms of waste in any process (Smartsheet 2019).

Hawk Lead-in Fighter

1. Overproduction — excess production
2. Waiting — time lost to backlogs
3. Transportation — inefficiencies caused by excess moving
4. Overprocessing — unnecessary work
5. Inventory — surplus material
6. Motion — too many steps
7. Defects — quality issues (Smartsheet 2019).

The problem of aircraft maintenance can be understood by mapping the value stream for maintenance activity. Value stream is the sequence of material flows and information flows over the time taken from commencing a maintenance task, until it is successfully completed. In aircraft servicing, aircraft are subjected to line maintenance (pre-flight, between flight, and post-flight system checking and servicing) and base maintenance (full service and checking of aircraft components based on flight hours or calendar life) (Kolanjiappan & Maran 2011). Through value stream mapping that engaged all Hawk Lead-In Fighters and F/A-18 Hornets maintenance staff, ADF were able to achieve a:

- 10 per cent cost reduction on Hawk Lead-In Fighter Program
- 7 per cent cost reduction on Hornet deeper maintenance (BAE Systems 2019).

These cost savings are important given the annual sustainment budget for ADF's Tactical Fighter Systems Program Office is over $200 million and the budget for enhancement of the ADF's fleet of fighter planes usually exceeds $1 billion per annum (Engineers Australia 2014).

Lean strategies continue to allow the ADF to hone and refine modification projects and maintenance activities. Since the 2009 introduction of lean processes, they have introduced phase-based servicing, on-going workforce consultation and engagement, and the use of integrated value stream monitoring, review and evaluation (BAE Systems 2019).

CRITICAL THINKING

Preparing for your exams at the end of a university semester could be considered one of the more important 'projects' in your life at this time. How would lean thinking change the way you approach the project of exam preparation? According to lean principles, identify four examples of waste in your approach to exam preparation and explain what you could do to eliminate them.

13.3 Creating a risk-management plan

LEARNING OBJECTIVE 13.3 Develop a risk-management plan for a project.

Expect the best, plan for the worst, and prepare to be surprised.

Denis Waitley, US motivational speaker and author

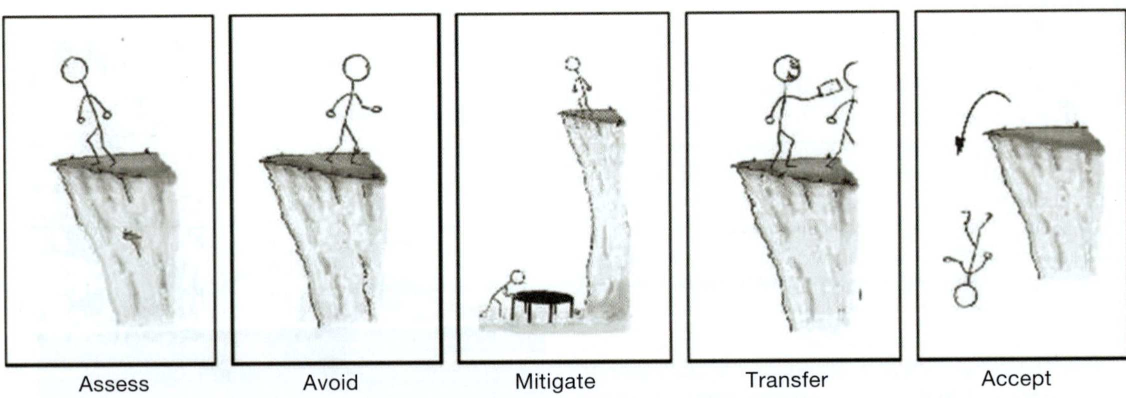

Source: Barron and Barron (2009).

While analysing and managing risk can be challenging and complex, fortunately there is an ISO Standard for risk management, ISO 31000:2018. ISO 31000 is recognised in Australia and New Zealand, and engineering project managers use it to guide their planning and activities on projects. Like the ISO project management standard, the ISO 31000 has a series of interrelated concepts and processes. Figure 13.15 shows the relationships between different topics covered by the standard. During the 'risk identification' stage of developing a risk management plan, an engineer will consider different types of risk. This section will discuss these risks, dependency, design and construction.

KEY POINT

Complementary with project management is the careful risk assessment of all the tasks within the project.

Dependency risks

Project management is about getting the sequence of sub-tasks clear so that each sub-task can be completed at the right time, without interference and without wasting people's time or using equipment unnecessarily. This expertise is developed by companies over many years. Nevertheless, there are still conflicts on construction sites and in manufacturing processes when dependent processes have not produced the components necessary for other processes to be completed. In civil engineering, this can happen because of things such as weather delays, equipment failure, strikes and company bankruptcy. In mechanical engineering and electrical engineering, a company's manufacturing capacity may be compromised by another company's inability to supply essential components for a product.

FIGURE 13.15 Inter-relationship between topics covered by ISO 31000 standard for risk management

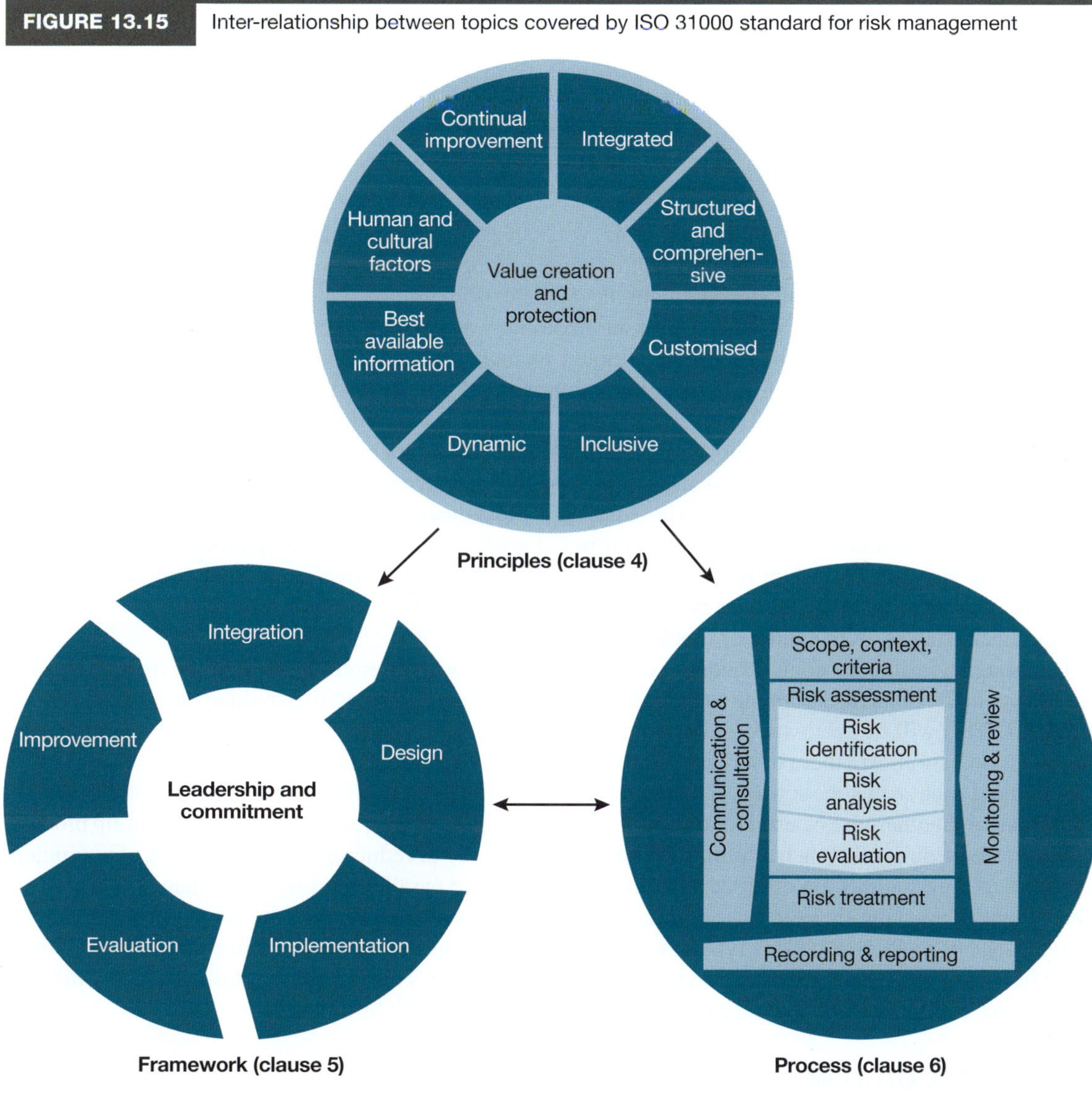

Source: ISO (2018).

Consider the risks associated with contractors running behind schedule and delaying a project. How can project managers protect themselves against these? For example, it is important to complete critical tasks during the dry season — if there is a dry season. Some tasks could be moved off-site to reduce weather disruptions. There is a trend in the construction industry to manufacture wall and floor panels in factories and deliver them to the site for assembly, rather than rely heavily on concrete pours on-site, which are easily disrupted by wet weather.

Project managers can protect themselves against incompetent contractors. The awarding of a contract is not necessarily based on the cheapest price. It is standard practice for companies to deal with contractors they have done business with before, who have a good track record of delivering work to a satisfactory standard and a set deadline.

Developing relationships with other companies who may become reliable contractors is an important part of professional networking. Many aspects of design are often subcontracted to companies with specialised expertise. For example, a project manager might decide the design of a wetland and an associated stormwater drainage system should be outsourced to a specialised consultant. Knowing reliable consultants is extremely valuable and building up those connections is an important role of professional organisations, such as Engineers Australia or Engineering New Zealand. These dependency risks can be managed via methods such as CPM, described earlier in this chapter.

Design risks

Consider the green-star building project. What risks need to be considered in its design? What about during construction? What about during the lifetime of the building? Could the design decisions leave the company vulnerable to litigation? First, consider design risks such as:

- extreme wind events on the building
- loss of energy supply
- incompatible communications providers
- lack of water supply availability
- internal flooding from a burst water main.

Some of these can be quantified. For example, you can estimate the likelihood of an electrical, water or gas supply failure based on past statistics from a supply authority. Likewise, wind velocities are specified in AS/NZS 1170.2:2021 and all structural engineers are expected to be familiar with it. As an engineer, it might be your decision to specify the size of an incoming water main, electricity substation, gas pipeline or optical fibre connection. You could be tempted to make these as small as possible to reduce cost. What would happen if the building population increased and the demand for energy or communications bandwidth surpassed the original capacity? Is this foreseeable? Should you adopt short-term thinking (i.e. aim to save money now)? Who will meet the cost of the expansion of incoming capacity? There can be similar problems with outgoing capacity of stormwater sewers and wastewater sewers that need to be connected to existing infrastructure. These are all critical design decisions.

Future water problems have been anticipated in the green-star building example because the team has opted for on-site reuse and water recycling to flush toilets. The on-site treatment plant may be a risk. If it is overloaded, a spill may occur with possible major consequences to public health. Spills can be caused by equipment failure, overloading or blockage.

The designers will have control over the design of the wastewater system, including the treatment plant. For the treatment plant, it is possible to make sure that there is excess capacity to reduce the risk of overload problems. This will be easy when the building is new because there will be fewer people than the final building population. Are there other ways to manage this risk? It might be possible to include some sort of containment zone so that the sewage can later be recovered and treated. However, this will require basement space that might otherwise be used for different purposes. It will also require an expensive clean-up process.

Duplicating equipment, such as pumps, is quite a normal way of reducing some of this risk of equipment failure. Pumping stations are often designed with two or more parallel pumps, rather than one big pump. This allows for varying flows to be accommodated and allows single pumps to be taken out of service for maintenance or replacement. This type of replication comes at a cost. However, if one pump breaks down, a parallel pump will be automatically switched in to take its place. This is a *cost versus reliability trade-off*, where extra cost is added to increase reliability and to lower the risk of failure.

Construction risks and safety

Workplace safety is a huge issue in all industries. In New Zealand, WorkSafe NZ inspects workplaces to check on safety and health arrangements, investigate accidents at work, and make sure employers and employees comply with health and safety legislation. In Australia, the federal government's Safe Work Australia provides advice on work health and safety and workers compensation issues in terms of national regulatory frameworks. In addition, each state of Australia has its own work health and safety authority that oversees the safety of workers in relation to state issues. These national and state bodies are listed in table 13.1.

For construction projects, compliance with relevant codes and standards is mandatory and non-discretionary. Any risk assessment must be thorough and methodical, and based on a recognised standard. Failure to do this could lead to criminal proceedings. In terms of environmental damage, the Environmental Protection Agency (EPA) regulates soil wash-off from building sites. For example, temporary barriers such as straw bales are required to be used across flow paths to catch sediment washed from sites. This limits sediment from ending up in adjacent streams or in stormwater pipes.

TABLE 13.1 National and state work health and safety bodies — Australia and New Zealand

Body	Country/state
WorkSafe New Zealand	New Zealand
Safe Work Australia	Australia
SafeWork NSW	New South Wales
WorkSafe Victoria	Victoria
Workplace Health and Safety Queensland	Queensland
SafeWork SA	South Australia
WorkSafe WA	Western Australia
WorkSafe Tasmania	Tasmania
WorkSafe ACT	Australian Capital Territory
NT WorkSafe	Northern Territory

For the green-star building case study, construction would introduce a series of new risks. These would include workplace accidents (e.g. electrocution, trench collapses, falling building materials and overturning equipment), environmental damage (e.g. pollution of stormwater systems and dust blowing off the site), productivity failures (e.g. subcontractors failing to complete tasks on time) and resource shortages (such as the unavailability of equipment or materials). How could the engineering company, the project and the client be protected from such risks?

We will analyse work health and safety risks using the guidelines of WorkSafe Victoria as an example. Similar guidelines and legislation exist throughout Australia and in New Zealand, for example, *How to Manage Work Health and Safety Risks Code of Practice* (Safe Work Australia 2018). Construction regulations in Victoria (WorkSafe Victoria 2008) state that construction project responsibilities must be clearly assigned, beginning with the principal contractor whose four responsibilities are to:

1. display a sign with contact details
2. prepare a health and safety coordination plan and keep it up to date — this health and safety plan must contain:
   (a) the names, positions and responsibilities of all people who have specific responsibilities for health and safety
   (b) the arrangements for coordinating the health and safety of everyone who is engaged to do construction work
   (c) the arrangements for managing OHS incidents (now referred to as WHS incidents)
   (d) any site safety rules, with the arrangements for ensuring that everyone at the workplace is informed about the rules
3. make the coordination plan available for inspection
4. make sure that new starters are aware of the coordination plan.

Some amendments to this regulation came into effect on 18 July 2017 (WorkSafe Victoria 2017). While the principles and specifications for a construction project have not been amended, the definition of a construction project has been modified so that a project is defined as having a value greater than $250 000.

Many risks are the responsibility of subcontractors. If you work as a designer or a project manager, you will need to make sure your responsibilities are clearly outlined in your contract. This applies for workplace safety. While contractors are responsible for their own workforce, engineers may be responsible for providing an overall safe site. This may include safe traffic flows of construction vehicles and safe electricity supply. With a construction site, part of the area may need to be stabilised before other contractors are allowed to operate. It is reasonable to require engineers to ensure the work of one contractor does not create safety hazards for others.

Individual contractors or employers on a site have specific responsibilities. For example, in Victoria, employers must work through the following list to control any risks.

1. Wherever it is reasonably practicable, the employer must eliminate any risks to health and safety arising from construction work.
2. If it is not reasonably practicable to eliminate a risk, the risk must be reduced so far as is reasonably practicable by using one of the following control measures (or two or three of them in combination).
   (a) Substituting the hazard with a safer activity, procedure, plant, process or substance.
   (b) Using engineering controls, such as mechanical or electrical devices (e.g. lifting devices).
   (c) Isolating the hazard from people, such as barricading, fencing or guard railing.
3. If there is still any risk, administrative controls must be used to reduce the risk so far as is reasonably practicable (e.g. through effective training).
4. If there is still any risk, suitable personal protective equipment (PPE), such as safety helmets, protective clothing and sunscreen must be used to control the risk. The use of protective equipment is the last line of defence — not the first.

In summary, there are four steps in the process: (1) eliminate, (2) reduce, (3) control, (4) protect.

WorkSafe Victoria (2008) provides the following questions to decide whether it is reasonably practical to eliminate, reduce or control a risk.

- How likely is it that there will be some harm?
- How serious could the consequences be?
- What do you, or should you know about the hazard or risk and ways of eliminating or reducing it (taking into account what is generally known in the industry and information such as material safety data sheets and manufacturers' instructions)?
- Are suitable ways to eliminate or reduce the hazard or risk available?
- How much will it cost to eliminate or reduce the hazard or risk?

In high-risk construction in Victoria, a Safe Work Method Statement (SWMS) must be prepared for all tasks for which someone will be at risk. Similar documentation is required in other Australian states and in New Zealand. In Victoria, the work must be supervised to make sure that it is done according to the SWMS and a copy of the SWMS must be kept as part of the construction documentation. A SWMS:

- lists the types of high-risk construction work being done
- states the health and safety hazards and risks arising from that work
- describes how the risks will be controlled
- describes how the risk control measures will be put in place.

As well as the responsibilities of contractors (employers), individuals are also responsible to work safely so as not to put the safety of others on-site at risk. These include self-employed tradespersons.

Induction training must be provided by a Registered Training Organisation (RTO) and a Construction Induction Card (previously a Red Card) is used to provide proof of training (WorkSafe Victoria 2008). Other states and New Zealand have other forms of proof of training. Induction training is essential for everyone on site so that they understand:

- their rights and responsibilities under WHS law
- common hazards and risks in the construction industry
- basic risk-management principles
- the standard of behaviour expected of workers on construction sites.

This section has considered risks involved in construction, using the green-star building and Victorian construction industry guidelines as an example. For any construction projects, relevant state or national legislation and guidelines need to be carefully consulted.

These controls affect all disciplines in construction, including electrical and mechanical services, communication services and more obvious forms of construction such as earthworks, pouring concrete and assembling steel building frames. Similar controls must be put in place in manufacturing sectors, such as automotive assembly and computer or mobile phone assembly. An advantage of a manufacturing environment is that it changes less frequently than a construction site, which is always changing. Thus, manufacturing risks can be more easily controlled.

Internal project risks

So far, the risks in the project have been looked at from design and construction perspectives. What about internal risks? From an engineer's perspective, what can go wrong within an engineering company to jeopardise the project? The internal risks for the project include:

- loss of key staff through retirement, resignation, death or injury
- lack of expertise in some areas

- not enough people to do the work
- financial failure of the company or contractor (due to other projects) that stops the work
- a major fire in the company offices that destroys project records
- a computer hacker deleting company files on the server.

How can the company minimise these risks? A risk assessment is shown in table 13.2. It highlights critical issues, including maintaining key staff who will do the project and maintaining data; that is, protecting against catastrophic loss of information (e.g. through fire or theft).

TABLE 13.2 **Sample assessment of internal risks — green-star building project**

Cause	Likelihood	Consequence	Risk level	Mitigation
Loss of key staff	Possible	Major	Extreme	Knowledge of project is spread across many staff and stored within the knowledge management system (KMS), which stores all project documentation.
Lack of expertise	Possible	Moderate	High	Company has strong relationships with other consulting companies to provide additional expertise as required.
Not enough staff	Likely	Major	Extreme	Company has a well-developed recruitment program. Outsourcing some work to other companies is possible, though may run into the same problem. Chief executive officer is exploring the possibility of exporting work to an engineering company in China.
Financial failure	Unlikely	Major	High	Company is currently in good condition and financial failure is not expected due to excellent accounting practices and rigorous oversight by both the general manager and the board of management.
Loss of records	Possible	Major	Extreme	Data backups occur every night to an off-site data storage site. Wind back procedures are in place and have been tested. Recovery within 24 hours is assured, subject to acquisition of new hardware, if required.

Long-term risks for the green-star building

Consider the longer term risks involved in the case study. Even after the construction is complete, the building is sold and people have settled into their new environment, potential risks may be realised. These are known as long-term risks and should be anticipated in the design process as much as possible. Examples of long-term risks for the development might include:

- child safety around water bodies, such as a water wall
- health and safety issues around water features on the green roof, through accumulation of various toxins and encouragement of insects, such as mosquitoes
- outbreaks of Legionnaires disease associated with air-conditioning equipment
- safety of stairways, walkways and handrails
- safety of installed equipment, such as signs, lighting and rubbish bins
- electrical safety throughout the building.

How do you think the project team can manage these risks during the design process?

13.4 Developing a knowledge management plan

LEARNING OBJECTIVE 13.4 Develop a knowledge management plan for a project.

In risk analysis, one significant risk is the loss of information and knowledge when key staff leave a project. A knowledge management plan provides for the development of a knowledge management system (KMS) as a strategy to control this risk. We will now look at this important topic in more detail.

We live in a knowledge society, with much wealth now created through innovation. Engineers are primary drivers in this process, creating new electronic consumer products; transport devices such as planes, trains and automobiles; manufacturing and chemical processing facilities that produce mobile phones, computers, fuels and paints; and infrastructure, such as buildings and bridges. Knowledge is critical to an engineering organisation, and it is easily disrupted through loss of staff, inability to get new staff with the right skills (a skills shortage) and the loss of information storage and computer networks.

Most engineering organisations now rely on computer networks for the storage of vital documentation about existing projects. This can include simple documentation, such as emails about the project (often the most difficult documents to track and archive), and complex documentation, such as advanced engineering analysis of electrical components or mechanical components within a proposed system. These documents include information from past projects (at least project information that has been transferred to an electronic format). The documentation may include multiple gigabytes or even terabytes of information.

An Australian/New Zealand Standard exists for knowledge management as a guide to good practice: AS 5037:2005.

KEY POINT

Knowledge management systems help to control risk by making projects less vulnerable to the potential loss of key staff or information.

Document storage, archiving and data mining

Systems have evolved to capture and 'make searchable' these large collections of computer files. Search engines, such as Google, provide the capability to index huge document collections and retrieve information using keywords. Contextual searching is vital if you do not want to sift through thousands of hits when searching for a particular document. In response to this demand, specialised companies have created tools that can precisely locate documents. AI tools, such as ChatGPT, are being used to search large document bases for key insights.

Such tools are only useful if there are strict protocols for the storage of documents within a central repository. If team members store documents on an individual hard drive (which may not automatically back up documents) there is more of a risk of losing the information. These documents cannot be searched centrally and can be lost through a hard disk crash or the theft of a laptop from a work environment. Therefore, it is important that information is stored centrally, with an off-site backup copy and with links to data searching tools. Careful cataloguing of files according to the project will also localise the search process and increase the likelihood a particular document will be located more quickly.

There are also legal requirements for data and documents associated with a project. Data and documents need to be kept for a minimum time in case legal liability issues arise after the project has been completed. Once these measures are in place, the document collection and archive can simultaneously be used for data mining operations, such as looking for trends in company performance, costs and other key performance indicators.

Maintaining the integrity and security of company files has grown in importance and difficulty since the advent of the internet of things and the age of Industry 4.0. Currently, sensing, data sharing and device interconnection mean that company files and data transfers are vulnerable. Figure 13.16 describes some risk mitigation steps engineers can use to maintain data security and integrity in this hyperconnected age.

Sharing knowledge

There has been a long tradition of engineering organisations encouraging knowledge sharing within their organisation through activities such as staff seminars and workshops, and through procedure manuals and other forms of documentation. Word processing, computer networks and the internet have simplified the process of creating a new procedure and having it reviewed and approved. Procedures can also be posted on company servers. This process is now so easy that many engineers are overwhelmed by information, being asked to constantly read more and more pages of new procedures. Nevertheless, access to good information-sharing tools has meant that the cost of sharing information (at least, of making it available) is now quite low.

The challenge is how to encourage people to participate in sharing their knowledge with other people in a company. For many, knowledge is power. Staff may believe they are paid for their knowledge. On this basis, they may believe that the company will employ someone else to do their job for less money if they share their knowledge. Ways of encouraging knowledge-sharing include building trust through face-to-face meetings and providing performance incentives for knowledge-sharing and collaboration (Davenport & Prusak 1998).

FIGURE 13.16 Assessing and maintaining data security and integrity

Physical security
A critical part of cyber defence is to evaluate who has access to your site, what they have access to, and how to manage and control that access to prevent unauthorised physical access to data, workstations and networks.

Information security
Compile a picture of the organisation's risk profile and current resilience. Evaluate and quantify risks to data, software and intellectual property, by assessing inherent risk (number of sites, site location, industry), current security controls and cyber resilience (how prepared the organisation is to respond to and recover from a cyber incident).

System assessment and actions
Create a cyber risk action list through an intensive analysis that prioritises recommendations for improvement across four areas: governance, IT security, insider threat management, and response and recovery.

Communities of practice

Most engineering jobs are rarely routine, which is one of the reasons engineering is such an interesting profession. Engineers build on existing skills and knowledge and stretch their knowledge and skills base with each new problem or project on which they work. Traditionally, engineers learned by asking people in the office around them — usually people who had a bit more experience. By sharing these experiences, a rich collection of knowledge was shared across organisations. This addressed the risk identified earlier of the loss of key staff.

Altshuller (cited in Mazur 1995) suggests 32 per cent of solutions come from personal experience and another 45 per cent of solutions come from members of the same company. According to Altshuller's research, engineers in similar companies could solve another 18 per cent of problems. This research supports the presence of professional bodies such as Engineers Australia and Engineering New Zealand as they provide broad networking opportunities.

Tom Davenport is considered by some to be one of the gurus of knowledge management and he wrote what has become an often-quoted piece more than two decades ago about Hewlett Packard, then the largest technology company in the world — '*If only HP knew what HP knows*' (Davenport 1996). Davenport recognised the challenges of sharing knowledge within large, complex, decentralised, and international organisations and has continued to write about it ever since.

How do organisations know what they know? One way is to get their people to write it down, as has been discussed. The internet has also provided collaborative tools, such as email and email lists, instant messaging, video and teleconferencing, chat rooms, discussion forums and social networking sites. All these tools can be used in various, complementary ways to build a community that then becomes a learning community. Discussion forums are useful community tools that are effective for large numbers of participants. A discussion forum limits participants from getting dozens or even hundreds of personal emails in one day, but a downside is that participants must check the discussion forum or get a daily or weekly digest for information updates. The widespread adoption of social networking sites such as Facebook and X (formerly Twitter) demonstrates that opportunities to network extend beyond workplaces. The internet involves millions of networks of people who share knowledge about different human activities. YouTube and TikTok are two services through which people can share their expertise on a myriad of topics.

Consider the following example: A biomechanics specialist within an international automotive company is investigating and modelling trauma events on human physiology caused by vehicle collisions. Within the company, there are several biomechanics specialists, located at many different offices (in Australia, the United States, the United Kingdom, Sweden, Germany and Japan). These specialists meet regularly, either face to face or via videoconference, and frequently email each other. The group uses an email list to share queries and expertise and a discussion forum in their company's knowledge management system. The forum allows new members to simply subscribe to various lists and older members to unsubscribe when their interests change. The biomechanics specialists use wikis to store recently published reports, to highlight new books in the field and to provide other useful web links for the members in their group.

In Australia, medical and biological engineers network through groups such as the Society for Medical and Biological Engineers Australia (SMBE). Each state group maintains its own website, with meeting and conference announcements, as well as reviews of recent publications and current controversies and notices for workshops and job advertisements. The Australian group is affiliated with the International Federation for Medical and Biological Engineering (IFMBE), which is primarily a federation of national bodies such as the SMBE. In this way, individual engineers are able to keep up to date about the most recent developments in their field, no matter where it happens across the world.

Student knowledge management

So, what does all this mean for you and your next student project? It is possible to build a basic knowledge management system with free public tools offered by companies such as Google and Dropbox. Use of file storage in the 'cloud' is becoming increasingly common. It is also likely that your university is using a common learning management system, such as Canvas or Moodle. These systems offer all the tools you need for knowledge management. For example, your lecturers will have already posted many useful documents for you. They may also have provided a wiki or another space you can use to create documents to share with your group members. There may be a discussion forum capability that you and your classmates can use to share questions and answers with each other. There may also be a chatroom facility for discussion of general topics related to each subject.

Another valuable feature in these systems is the ability for each group in a class to have its own file storage, discussion forum and email list. This allows each group to do all the basic knowledge management activities and it protects the group against the risks identified earlier, such as losing files, hardware failing or a team member being sick on the day of an important presentation. Important files should be safely stored on a central server that is backed up. These can then be easily downloaded when needed. Ask your lecturer for access to this facility. Alternatively, you may like to set up your own site.

13.5 Quality management and its relationship to project management

LEARNING OBJECTIVE 13.5 Develop a quality plan for a project.

At the beginning of this chapter, project management was described as a constant tension between three basic factors: scope (what is to be done), budget (how to pay for it) and time (required to complete the project). These factors lead to risks and impacts on the quality of the final product. Earlier sections have considered the management of resources, time and scope, and the risks that emerge in the process. We will now consider how quality is managed within a project.

Quality management is a *continuous* process within an organisation. Project management consists of achieving tasks within a *finite* time frame. Quality management tools, and the policies and practices of an organisation, are normally used within a project to guarantee its outcome. So, what is quality? Quality can mean *excellence* — very high quality and, perhaps, very high price — such as a Mercedes Benz car. It can also mean reliable, well made and aimed at a different price point, such as a Toyota Corolla. That is, an artefact should be manufactured with exacting tolerances, to function as designed. Thus, *fitness for purpose* is a key element of quality. In fact, engineering is more often focused on fitness for purpose than on extravagance.

It was the Japanese who led the quality movement in the post–World War II era. Before this time, Japanese goods were widely considered unreliable. Using the guidance of quality experts such as Edwards Deming (Deming 1986), Japanese manufacturers transformed the quality of their goods, such as cars, to the point that Toyota became the top-selling car brand in Australia (Blackburn 2006).

Japanese car manufacturers focused on delivering what the customer wanted or needed and did this by transforming their work practices to engage all their people in improving the quality of the systems that produce their goods. Quality circles were established to improve quality within a work group. Rewards encouraged everyone to suggest opportunities to improve quality. Thus, quality shifted from *quality control* (rejecting defective products at the end of the production line, which leads to considerable waste) to *quality assurance*, which is the process of guaranteeing quality products at the end of the production line, with the aim of zero rejection or rework (fixing defective products). It is quality assurance that is applied to engineering projects.

KEY POINT

The degree to which a project meets the client's requirements can be considered a measure of its quality.

Key quality management principles

The ISO Standard 9000:2015, which also serves as the Australian and New Zealand standards for quality management (Standards Australia 2015), provides eight quality management principles, which are:

1. customer focus
2. leadership
3. involvement of people
4. process approach to activities and resources
5. systems approach to management
6. continual improvement of overall performance
7. factual approach to decision making (using data and information)
8. mutually beneficial supplier relationships.

Many of these principles have been described already within a project management framework. Project management must have clear objectives delivered by a well-organised team of people led by group leaders (principles 1, 2 and 3). Projects are ultimately about delivering customer satisfaction. For principle 4, project management techniques are used to organise activities and resources. Systems thinking is used to break down complex problems into simpler ones (principle 5). At each phase of a project, it is necessary to consider how to improve team performance (principle 6). Knowledge management underpins the availability of data and information (principle 7). Finally, mutually beneficial supplier relationships help manage the risks involved in delivering projects (principle 8). These principles also apply to other company activities, not just to project management.

The last principle was considered in some detail in the context of risk management. The need for specialist input from other companies was identified as a key risk-management strategy. These relationships need to be fostered and nurtured, because the company may be dependent on other companies at short notice, just as these companies might be dependent on this company at other times. Likewise, employing contractors with whom the company has worked before is another risk-management strategy. It is all about forming and enhancing relationships that are mutually beneficial.

This is quite different from a commonly held belief that a company or individual should always maximise their own return in any transaction. These ideas were discussed in more detail in the chapter on engineering design. Self-interest may work for one-off transactions, but in an industry where a company will deal many times with the same companies and the same people, it is mutual benefit that works better. In a future project, it might be necessary to work for one of those other companies or an employee may move to this company.

These principles become clearer when viewed from a project perspective. The focus is on building a product with the intent of satisfying the client's need. For that to happen in a reliable way, it is also necessary to measure the performance of a product against the client's original need. There must also be clear management responsibility for resource allocation to ensure quality and continuous quality improvement. Any company (or student group) should be constantly looking to improve its processes, *including* its quality management processes. This means that there are two cycles in operation at the same time. One cycle is producing a quality product and making sure that it meets the client's need. The second is making sure that the project team is continually improving its own internal processes so that each new project is performed more efficiently and cost effectively.

Engineering quality management

Consider a typical engineering project. It begins with a client need, which might be a re-design to upgrade an existing product such as a mobile phone, or it may be the creation of a new artefact such as a bionic eye. The client should have defined clear goals and criteria for the performance of the device. For example, an artificial heart may be required to have a design life of 20 years with no external power supply.

The engineering team will use a design process to produce a set of design drawings and specifications. This may include very complex 3D models, as well as simpler 2D drawings. The engineers will, hopefully, be constantly checking with the client that they have delivered a design that meets their criteria.

Normally, the engineers will not be the manufacturers of the product they designed. This usually happens in other manufacturing or construction companies. In software companies, the line between design and construction can be blurred: systems analysts normally do the design and pass the code-writing phase to teams of programmers. Increasingly, these teams of programmers might be in another part of the world (Friedman 2006), so design and construction are clearly separated — as in other engineering domains.

So, there are three basic organisations: (1) clients, (2) engineers and (3) constructors. What are the quality processes that need to be considered? Consider firstly the relationship between the client and the engineer as shown in figure 13.17.

FIGURE 13.17 Client–engineer quality assurance

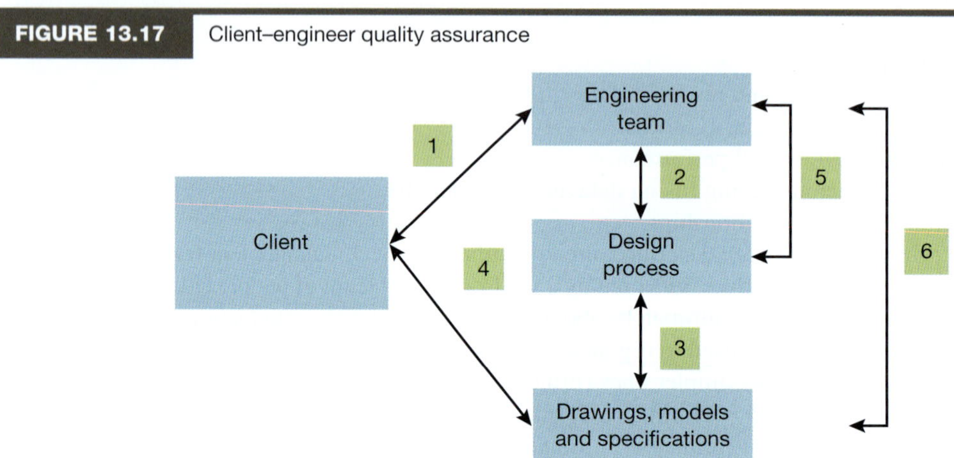

Figure 13.17 shows six quality relationships.

1. Is the client need clearly articulated? Is the problem defined? Is the scope agreed?
2. Will the intended design process deliver the required outcomes?
3. Have all required aspects of the design been clearly articulated in the design products (drawings, models, specifications)?
4. Does the design product satisfy the original client need?
5. Has the design process been as effective as it could have been? Could it be done better next time?
6. What are the learning opportunities for client and engineer?

There is a similar quality relationship between the engineers and the constructors shown in figure 13.18.

These issues are very similar to those in figure 13.17, but an additional issue has been added to explicitly represent the need to revisit the design during the build process. Issues will arise that have not been anticipated in the design. This will then require all the previous steps in figure 13.17, as changes to the design will require checks back against the original client need.

1. Are the drawings, models and specifications adequate for the purposes of construction or manufacture?
2. What project management processes are in place to ensure that the build process will deliver the required outcomes?
3. Is the build process producing the right product? (These are the site supervision aspects.)
4. Does the product match the original drawings and specifications?
5. Are the build processes adequate? Can they be improved for the next time?
6. What are the learning opportunities for all teams?
7. Do the drawings (possibly modified during the build process) properly represent the intended design outcomes?

FIGURE 13.18 Engineer–constructor quality assurance

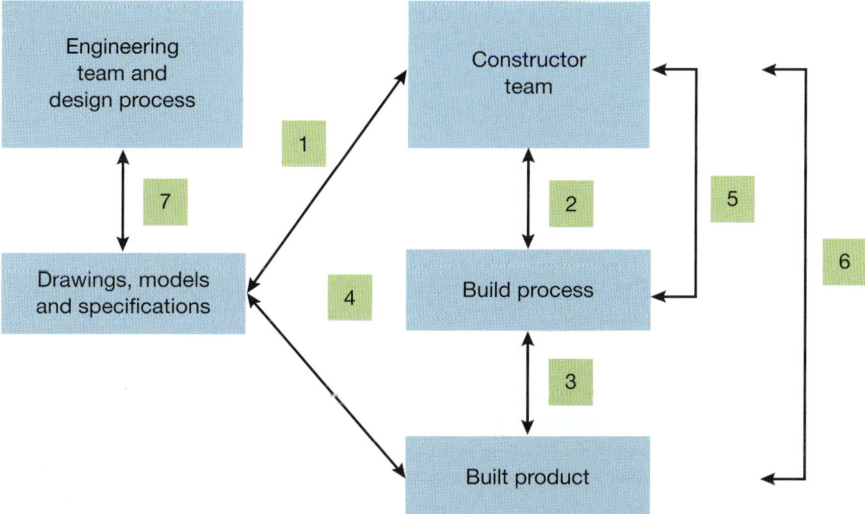

Quality plans and engineering

It is clear from these diagrams that there is a range of quality management processes that are needed in the context of engineering work. Project management processes are important in the delivery of engineering projects and products. There are also other relationships that must be managed, such as:

- getting the scope right (at both design and construction stages). This requires careful negotiation between parties to ensure that the final contract defines what is required to be achieved. Arguments over scope can become very costly, as can variations to the contract
- project management processes
- process quality assurance
- product quality assurance
- project, review, evaluation and learning.

Quality management plans have a number of key components and these can be adapted to the level of detail required for the task. The ISO Standards provide the full picture of rigorous quality management. The components are the six Ps.

1. *Purpose.* Clearly articulate why this is the goal. For example, the purpose may be to ensure that the product is properly defined and both the client and contractor are satisfied. This is important for the contract to be completed with the minimum of fuss and bother to both parties and for each to feel satisfied in the end. One will get a product fit for purpose and the other will get appropriate compensation.
2. *Process.* What needs to be done to ensure that this occurs? Agreement needs to be reached on the scope of the product.
3. *Procedures.* How does this occur? Meetings may be necessary to reach agreement.
4. *Performance indicators.* How will it be known that the goal has been achieved? Certainly, a signed contract should indicate agreement. Minutes of meetings could also document individual points of agreement ready for the final contract.
5. *Improvement Plans.* How can the process be improved? These plans are at two levels. At the immediate level, the requirement is to keep negotiating until agreement is reached. On a meta-level, it may be necessary to rethink how meetings are being conducted. For example, have they become too large? Are there communication difficulties? These questions should help to improve the process for the current project and for future ones.
6. *People.* At the most fundamental level, all the previous five steps will be executed by people: people representing the engineering firm and those representing the client. Are the people involved equipped with the skills for the task or is training required? Is everyone communicating effectively? Is the quality management plan providing effective guidance for quality outcomes? A significant feature of almost any engineering project is the diversity of (human) resources involved. From concept through to installation, commissioning and operation of a design, a mix of skill sets and relationships will come to the fore.

A crucial aspect of this is managing the information flow; that is, the information and protocols necessary for engineers to design in detail based on architectural specifications and drawings, and the information necessary for tradesmen and technicians to construct or install an engineering design.

SPOTLIGHT

Project management oversights and the Australian eCensus fail

The 2016 Australian eCensus was a highly anticipated experiment in online data collection. It turned out to be a very valuable learning experience for all concerned and generated a lively series of memes on social media (try an online image search for 'Census 2016 memes' to see some examples).

The greatest controversy of the national online survey came after a series of service denial attacks caused the website to be taken down on the night of the study and become unavailable to the public for about 40 hours (Australian Broadcasting Corporation 2016).

During a subsequent inquiry into the eCensus failure, much of the blame was placed on lead technology supplier IBM, which was said to have failed to properly plan a response and recovery strategy that could have correctly responded to about a series of denial of service attacks from hackers (Dudley-Nicholson & Bickers 2016). The inquiry found the ABS was right to take the website down due to security concerns, but that IBM had failed to properly plan for how to bring systems back up again (Senate Economics References Committee 2016). This included a failure to undertake an Information Security Registered Assessors Program (IRAP) assessment, which would assess the implementation, appropriateness and effectiveness of an information security system's security controls.

IBM failed to test a router restart or have a backup synchronised in place, both of which were seen to be significant contributing factors to the failure of the eCensus, according to a Senate Committee hearing (Senate Economics References Committee 2016).

> It is impossible to say with certainty and hindsight what would have occurred had the ABS made different decisions, [but] allowing IBM to undertake their own testing and the failure to complete an IRAP assessment appear to be significant oversights in project management (Senate Economics References Committee 2016).

In his report, Allistair MacGibbon (2016), special adviser to the Prime Minister on cybersecurity, found also that the long-term, almost exclusive, relationship between the ABS and IBM had contributed to the problems as any external verification or oversight was lacking.

MacGibbon (2016) said procurement practices had fallen short and echoed a Senate Committee recommendation that the ABS take a more proactive role in validating the resilience of the eCensus application for the 2021 Census.

CRITICAL THINKING

There is an old saying that 'if you fail to plan, you plan to fail'. If you were employed as a project manager for the 2026 Census, what are the main quality priorities you would plan for a 'new and improved' eCensus roll out?

Quality and the engineering student

What does all this mean for you as an engineering student? In the future, you will work with clients to agree upon the scope of your work. In your university projects, you are in a similar relationship with your lecturers; they specify what they want done and together you need to reach an agreement and understanding of what the scope of that work is. Sometimes this is well defined and sometimes there are deliberate options where you can take a project or assignment in different directions. Sometimes there are ambiguities that arise as you get into a project. These are all part of the scope of the work and need to be agreed between you, the consultant and your client (your lecturer). The following questions provide a useful starting point.

- Why do you need to clarify scope with your lecturer? It will save you time by making sure that you are doing what the client (your lecturer) wants. You will not spend precious time doing things that are not required. Don't assume! Clarify your understanding with your lecturer.
- What do you need to do to clarify the lecturer's objectives? You would normally have some discussions, which might be in a class, or via email or in a discussion forum online.
- How will you confirm your understanding and indicate agreement with the scope of the work? A confirming email or discussion forum entry should be adequate. Verbal agreements should be confirmed

immediately in writing. You also need to make sure that these agreements are carefully filed in your design file in a section for project scope. With the multitude of emails that many people now receive, failing to do so may mean you are not able to find the email later if there is a query about what was agreed. Each point of agreement can be tracked via one or more performance indicators.

The following guidelines will also assist you to achieve quality work in your undergraduate projects.

- *Keep the purpose clearly in mind.* Ask yourself what you are trying to achieve. One example might be to learn fluid mechanics well enough to pass the exam.
- *Determine the process.* To learn fluid mechanics, you need to understand basic principles of mass and energy conservation, boundary layer theory and pipe flow.
- *Follow procedures.* Going to lectures, reading the notes and textbook, and completing all the required homework problems is what you need to do.
- *Determine performance indicators.* These could be the number of problems you have been able to complete. Another measure is how quickly you can complete each new problem. For example, an exam is a timed test, so you may need to be able to complete each problem in less than 30 minutes.
- *Identify an improvement plan.* If your time to complete each problem is an hour and you need to be able to complete them in 30 minutes in an exam, you need to do more practice. An appropriate plan would be to set aside another two hours each week to get more practice.

Quality and student team projects

Often at university, you will be required to work in teams on projects. This provides you with an opportunity to implement quality principles to manage the project performance of the team, which you will likely be required to do often in a future engineering career. A quality management plan for a student or other project needs to address the same six issues as previously identified.

1. *Purpose.* Why have you come together as a team? For example, it may be to complete a project to a required standard. What standard is required? Is the team all agreed on this standard? Is the team aiming for a pass or a higher grade?
2. *Process.* What will the team do to achieve the intended outcomes? What are the project and quality management processes that will be implemented?
3. *Procedures.* What procedures will the team follow (such as meeting as a team each week to ensure tasks are complete and updating the project plan)?
4. *Performance indicators.* What measures will the team use to ensure it is on target to achieve its desired outcomes (such as measuring completion of tasks, number of pages written or drawings completed)?
5. *Improvement plans.* What will the team do if it is not on track? How will the potential issues of non-performing or dominating team members be addressed?
6. *People.* What new learning will the team need to complete the task?

This section has provided an overview of quality management and its relationship to project management, both in an engineering student and a professional engineering environment. Key quality principles have been articulated, with a focus on customer satisfaction delivered by empowered people and backed up by processes, procedures, performance indicators and improvement plans.

SUMMARY

This chapter has introduced project management for first-and second-year undergraduate engineering students. A straightforward tool (OPPM) has been outlined that connects the basic elements of scope (the project objectives), the resources (which are mostly people in student projects) and time (within which a project must be completed). The chapter has also described the important topics of risk, quality and knowledge management as they relate to project management. We will now briefly revisit each of the chapter's learning objectives.

13.1 Provide an overview of project management.

A project is any work that has a clear set of tasks and objectives and is to be completed between an agreed beginning and end date. Project management has a long history in the construction of society's infrastructure. In the last century, project management became more formalised and project management is used in many disciplines other than in engineering. Gantt charts and CPM are tools that have been developed to assist with project management.

13.2 Plan the stages of an engineering project.

Planning a project requires systems thinking and a 'divide and conquer' approach. The project manager needs to be able to divide a complex project into simpler tasks or activities that can be assigned to independent teams. The interactions between these activities need to be considered in the allocation to teams and also in the project schedule. Resource restrictions can also have an impact on the scheduling of tasks. Key resources include people as well as items of equipment in limited numbers. Projects pass through several stages. The first is usually feasibility, where the requirement is to assess whether the project should proceed. The next stage is usually conceptual design, in which the basic technology for the project is decided. Detailed design follows, in which the broad technology decisions from the conceptual stage are translated into detailed specifications, drawings and prototypes. The implementation stage follows design. This requires construction or manufacture. Each of these stages requires a new project plan, which means that project management occurs in a hierarchical way at these various levels of detail.

13.3 Develop a risk-management plan for a project.

Project management is fundamentally a risk-management strategy. By carefully planning what will be done within the project, costs can be controlled, resources allocated and delivered on time, and the necessary teams of people and equipment can be assembled at the appropriate moment. Complementary with project management is the careful risk assessment of all the tasks within the project. Any risk must be assessed in terms of its likelihood and consequence, as well as the control measures put in place. Risks can include the interdependency of the project activities, design risks, construction risks and environmental risks. Work health and safety is a crucial issue in the workplace, and hence in project management. Legislation and guidelines are provided by state and national bodies in Australia and New Zealand, and these must be carefully assessed when planning any project.

13.4 Develop a knowledge management plan for a project.

One kind of risk is the knowledge implicit in any engineering project. This knowledge can be lost through key staff leaving or the failure of critical computer systems. Knowledge management systems control this risk by encouraging sharing of information, making the organisation less vulnerable to key individuals, and ensuring any electronic documents are carefully backed up to avoid data loss.

13.5 Develop a quality plan for a project.

The intent of any project is to meet customer requirements. The degree to which this is achieved can be considered a measure of the project's quality. Quality assurance is a systematic process that constantly tracks how these requirements are being met. Quality assurance also encourages a continuous improvement approach so that the project team becomes a learning team, constantly seeking to improve its performance.

KEY TERMS

code A legal document, legislated to define required practice.

critical path The sequence of activities that deliver shortest completion time for a project.

feasibility investigation An investigation that determines whether a project is worth further consideration.

milestone A clear, measurable output that proves a block of work has been successfully completed.
objective The end result of planned activities, within specific time and other constraints.
project A temporary endeavour designed to produce a specific product or service. It has a clear set of tasks and objectives with agreed beginning and end dates.
project management A suite of processes used to manage a project to completion.
slack The amount of time a task can be delayed without its completion time affecting the overall project completion.
standard A document of defined good practice, developed by an expert group within the profession. It contains recommended practice, not legal requirements.
task A component activity within a project.

EXERCISES

1. Explain the key factors in project management, and briefly describe some of the tools that engineers can use to assist with the process.
2. Think of a project you want to work closely on that is linked to the engineering discipline you are studying. It might be an electric vehicle, an unmanned aircraft or a smartphone. What functionality is required? What subsystems are required? Plan a bid to design one of these products. What design steps are required?
3. Assembling a simple electrical device requires the following steps with the duration and precedence shown.

Activity number	Activity	Duration (mins)	Follows activity	Critical resources
1	Organise work bench	1	–	Person
2	Drill part 1	1	1	Drill
3	Drill part 2	1	2	Drill
4	Solder bracket to part 1	1	–	Soldering iron
5	Solder bracket to part 2	1	–	Soldering iron
6	Clean soldering iron	0.5	5	Soldering iron
7	Arrange case	0.5	–	Person
8	Trim part 3	1	–	Trimmer
9	Trim part 4	1	8	Trimmer
10	Attach part 3 to part 1	1	8	Screwdriver
11	Attach part 4 to part 2	1	9	Screwdriver
12	Assemble components into the case	1	–	Person
13	Tidy workbench	1	12	Person

   (a) What activities lie on the critical path (the activities that, if delayed, will delay the whole assembly)?
   (b) What is the minimum time required to assemble the device?
   (c) Is the process faster if two people are assigned to the job and, if so, how would you divide the tasks to achieve the minimum time?
   (d) Does this reduce the cost of assembly?
   (e) How else could you speed up this assembly process?
4. Consider any product of interest to you. It could be something simple, such as a clock, or something complex, such as a car. How many companies contribute components to this product (e.g. look under the bonnet of a car)? List ways these components could fail to arrive at the manufacturing facility. As a product manager, what would you do to protect yourself against these risks? Can you remember an example of product failure that has been covered by the media during the last year?

5 What construction or manufacturing risks can you think of related to your favourite discipline? Consider a mobile phone, an artificial kidney, a GPS device or any other device of interest to you.
6 Locate the website for your local work health and safety authority. You may find Safe Work Australia (https://www.safeworkaustralia.gov.au) or WorkSafe New Zealand (https://www.worksafe.govt.nz) good places to start. Find regulations that apply to your university environment and your workplace, if you are currently employed (e.g. this may be a retail, hospitality or service location). Which of these were you unaware of? What changes to your workplace induction would you recommend?
7 What are (or will be) your data backup and recovery processes for your individual and/or team university assignments? What would you do if you lost your laptop, your backup drive or your memory stick? How would you complete your current assignments? Where would this leave the group members who depend on you? Is this something you need to systematically address in your group?
8 Think of something you are quite good at, for example, a subject at university, a sport or a musical instrument. How would you explain it to someone else? List all of the subtleties of the skills that are difficult to put in writing. What other media might you use to convey your expertise? How would you use a knowledge management system to pass your expertise to others who are interested in the same topic?
9 Outline how quality management principles can be used in your next student team project.

PROJECT ACTIVITY

Choose a project for this chapter that matches your disciplinary interests. Your tutor or lecturer may help in this process, or they may have already proposed a project to which you can apply your project management skills. Map out how you will apply project management principles to various stages of the project. This chapter describes how this occurs within a green-star building design project, which includes civil, environmental, electrical, mechanical, chemical and communication engineering aspects. What process is required to convince a supervisor that there is a need for this new product?

Electronic engineers may like to think about bringing a consumer item to the market, for example, a phone, a GoPro-style camera or a drone. For a civil engineer, the project could be the design of a four-storey building, including structural framing, floor slabs and foundations (including underground parking) and an economic evaluation of the alternatives. For a mechanical engineer, the project could be an aid for the elderly (e.g. a stair climber), a new bicycle, a ski lift or a wind turbine.

Alternatively, the project may be career related. How would you use project management techniques to plan your career? What are the activities involved? If you choose this option, you will need to map out various career directions, locate job advertisements that inspire you, consider the directions of the local and global economies, and think about what skills you will need in the future. You will need to investigate potential employers, reviewing the content on employer websites and networking with staff (if this is viable). Keep at least three career options available and map out the skills you will need in more detail. How do these align with your current program? What studying will you need to do? How will this contribute to a portfolio of work that you could take to your first job interview?

REFERENCES

Australian Broadcasting Corporation 2016, 'Census 2016: ABS website crashes in #censusfail', *ABC News*, 10 August, https://www.abc.net.au/news/2016-08-09/abs-website-inaccessible-on-census-night/7711652

Australian Institute of Project Management 2018, 'Current PMAA winners', https://aipm.com.au

BAE Systems 2019, 'Lean delivers fast jet savings', https://www.baesystems.com

Barron, M & Barron, AR 2009, *Project management for scientists and engineers*, Connexions, Rice University, Houston, Texas.

Blackburn, R 2006, 'Myth busted: Australia's favourite car', https://www.drive.com.au

Campbell, CA 2007, *The one-page project manager*, John Wiley & Sons, Hoboken.

Davenport, TH 1996, *If only HP knew what HP knows*, https://www.academia.edu/33151614/If_Only_HP_Knew_What_HP_Knows

Davenport, TH & Prusak, L 1998, *Working knowledge — how organizations manage what they know*, Harvard Business School Press, Boston.

Deming, WE 1986, *Out of the crisis*, MIT Center for Advanced Engineering Study, Cambridge.

Dudley-Nicholson, J & Bickers, C 2016, 'Australia's 2016 Census had "significant and obvious oversights," report finds', *News.com.au*, 25 November, https://www.news.com.au/technology/online/australias-2016-census-had-significant-and-obvious-oversights-report-finds/news-story/6edcf8f897b2361965bd72683ee6edbe

Engineers Australia 2014, 'RAAF FA-18 Super Hornet tactical fighter: Australian input to upkeep and modifications', https://web.aeromech.usyd.edu.au/yabbfiles/Attachments/asdeseminar_4sept2014_0.pdf
Friedman, TL 2006, *The world is flat*, Farrar, Straus and Giroux, New York.
Gold Coast Health 2024, 'Gold Coast University Hospital sub-acute expansion', https://www.goldcoast.health.qld.gov.au/about-us/initiatives/building-future/accelerated-infrastructure-delivery-program-aidp/gold-coast-university-hospital-sub-acute-expansion
International Organization for Standardization (ISO) 2018, *ISO 31000:2018(en)*, https://www.iso.org/obp/ui#iso:std:iso:31000:ed-2:v1:en
Introba n.d., 'Pixel Building', https://www.introba.com/work/projects/pixel-building
Kerzner, H 2003, *Project management — a systems approach to planning, scheduling and controlling*, John Wiley & Sons, Hoboken.
Klastorin, T 2004, *Project management — tools and trade-offs*, John Wiley & Sons, Hoboken.
Kolanjiappan, S & Maran, K 2011, 'Lean philosophy in aircraft maintenance', *Journal of Management Research and Development*, vol. 1, no. 1, January–December, pp. 27–41.
Lewis, CM, Chatfield, C & Johnson, T 2019, *Microsoft Office Project 2019 step by step*, Microsoft Press, Seattle.
MacGibbon, A 2016, *Review of the events surrounding the 2016 eCensus*, Department of the Prime Minister and Cabinet, Canberra.
Mazur, G 1995, 'TRIZ — Glen Mazur lecture series', https://www.qfdi.org/intro-to-triz-glenn-mazur-lecture-series
McKie, R 2023, '"It only makes the news when the toilets stop working": Has the 25-year-old International Space Station been a waste of space?', *The Guardian*, 30 October, https://www.theguardian.com/science/2023/oct/29/international-space-station-25-years
Mosher, D 2017, 'Congress and trump are running out of time to fix a $100-billion investment in the sky, NASA auditor says', *Business Insider*, 3 November, https://www.businessinsider.com
Portman, H 2013, 'ISO 21500, a quick reference card', https://hennyportman.wordpress.com
PMI 2024, 'Project Management Institute', https://www.pmi.org
Robinson, R 2021, 'The 11 best project management tools used by top technical teams from Netflix to Airbnb (and how to choose the right one for you)', *Planio*, 10 February, https://plan.io/blog/best-project-management-tools-used-by-top-technical-teams
Safe Work Australia 2018, *How to manage work health and safety risks code of practice*, Safe Work Australia, https://www.safeworkaustralia.gov.au/doc/model-code-practice-how-manage-work-health-and-safety-risks
Senate Economics References Committee 2016, *2016 census: Issues of trust*, Commonwealth of Australia, Canberra.
Smartsheet 2019, 'The definitive guide to lean project management', https://www.smartsheet.com
Standards Australia 2015, *AS/NZS ISO 9000:2015 — Quality Management Systems — Fundamentals and Vocabulary*, Standards Australia, Sydney.
——— 2022, *AS ISO 21500:2022 — Project, Programme and Portfolio Management — Guidance on Programme Management*, Standards Australia, Sydney.
WorkSafe Victoria 2017, 'Occupational Health and Safety Regulations 2017: Summary of changes', https://www.worksafe.vic.gov.au
——— 2008, *A handbook for the construction regulations: Working safely in the general construction industry*, https://www.worksafe.vic.gov.au

ACKNOWLEDGEMENTS

Photo: © Chris Hyde / Getty Images
Photos: © Australian Institute of Project Management
Photo: © Grocon Group
Photo: © Ryan Fletcher / Shutterstock
Photo: © Barron, M & Barron, AR 2009, *Project management for scientists and engineers*, Connexions, Rice University, Houston, Texas.
Figure 13.2: © Portman, H 2013, 'ISO 21500, a quick reference card', https://hennyportman.wordpress.com
Figures 13.3 and 13.6: © Atlassian
Figure 13.4: © Kanbanize
Figure 13.5: © Roadmunk
Figure 13.7: © Campbell, CA 2007, *The one-page project manager*, John Wiley & Sons, Hoboken.
Figure 13.15: © Reproduced by John Wiley & Sons Australia Ltd with the permission of Standards Australia Limited under licence CL0624WIL. Copyright in ISO 31000:2018 vests in Standards Australia and ISO/IEC. Users must not copy or reuse this work without the permission of Standards Australia or the copyright owner.
Text: © Australian Institute of Project Management 2018, 'Current PMAA winners', https://aipm.com.au

CHAPTER 14

Communicating information

'The newest computer can merely compound, at speed, the oldest problem in the relations between human beings, and in the end the communicator will be confronted with the old problem, of what to say and how to say it.'

Edward R Murrow

LEARNING OBJECTIVES

After studying this chapter, you should be able to:

14.1 discuss the contexts in which engineering communication occurs

14.2 use the PCR model to develop a communication plan for a project

14.3 plan and prepare written communications in an engineering context

14.4 prepare and give a technical presentation about an engineering project

14.5 discuss the use of visual methods to communicate engineering information

14.6 discuss current approaches that may be used to effectively and ethically implement AI into engineering communication practice.

Introduction

The importance of communication skills for engineers was emphasised in earlier chapters. The chapter on understanding communication described a model that can be used to analyse a communication or to plan a proposed communication. Some important communication skills were discussed in the communication skills chapter. In this chapter, the focus is on the use of the Purpose — Communication — Response (PCR) model to plan communications, and the verbal and visual ways engineers communicate information. A small communication may use only one method of communication, while a larger communication may incorporate a mix of verbal, non-verbal and visual methods.

The following two examples illustrate that during the planning process it is important to get the mix right so that consumers get the messages and the creator receives the desired responses.

The Water Corporation in Western Australia has integrated virtual reality into its Safety in Design Operability Workshops, which are an important part of the engineering design process. These workshops allow plant operators to provide input about a proposed plant design by leveraging their experience to identify potential hazards in the design; explore opportunities for improvements in plant design and efficiency; and provide a user-friendly work environment for plant operators. However, the feedback from operators was that they have difficulty visualising the proposed asset from normal 2D drawings and therefore hazard-causing errors are often missed, meaning the errors are propagated through to the built asset.

While screen-based 3D models have been used in the past, embedding the operator into a virtual representation of the proposed plant took the interaction between operators and design engineers to a whole new level. The operators are able to virtually *walk around* the proposed asset as if it were a built asset. This increases the likelihood of finding design errors or omissions that would have resulted in safety hazards. Operators then provide feedback about the appropriateness of designs (e.g. the location of instrumentation) so that any errors may be corrected, and recommended changes made during the design stages rather than having any defects propagated into the built plant, where it would be significantly more expensive to make corrections.

The virtual reality system was developed in-house by the Water Corporation and set up in a dedicated virtual reality room — called the Holodeck (figure 14.1). The system consists of a virtual reality headset that is coupled with a high-end personal computer loaded with appropriate drafting, design and virtual reality software. Once the system was established, design drafting staff received training on how to export existing 3D models from CAD software into the virtual reality environment. Then the concept was launched when project design managers started using it for real operability studies. The system is able to import 3D models that were already created as part of design work, so there is little extra work required to develop the virtual reality walk through.

To date, the system has been used with the design models for: the Gnarabup (Margaret River) and Bullsbrook waste water treatment plants; the upgrades for the Kellerberrin and Wyalkatchem Effluent Reuse Plants; the Kununurra Diversion Dam Crane; and the Advanced Water Recycling Plant in Craigie, Western Australia. These were cases where existing 3D models were imported into the virtual reality system for operability studies and/or design reviews of these proposed plants to verify the designs before construction.

Stephen Beckwith, Manager, Treatment Operations Engineering at Water Corporation, notes that the benefits to the business of using this system have been:

- improvements in design error identification by both the designer and operator;
- dramatic improvements in the safety in design process through better identification and mitigation of operator hazards or operability concerns through the proactive elimination, substitution, or isolation of the hazard, or through the development of engineering solutions before the infrastructure was built (Beckwith 2024, pers. comm.).

This example highlights the importance of using appropriate communication methods to receive honest and accurate design feedback from future users. Similar virtual reality systems are being used by some other Australian engineering organisations, signalling a change in technology from last century where physical models were used for the same purpose. For example, large physical models were used in the Mt Isa, Queensland mine to view different sections of the mine. These models included ore bodies, existing and proposed shafts, tunnels, train lines, crushers and other plant.

FIGURE 14.1 Using the virtual reality 'Holodeck' for Safety in Design Operability Workshops

(a) Walk through 3D image

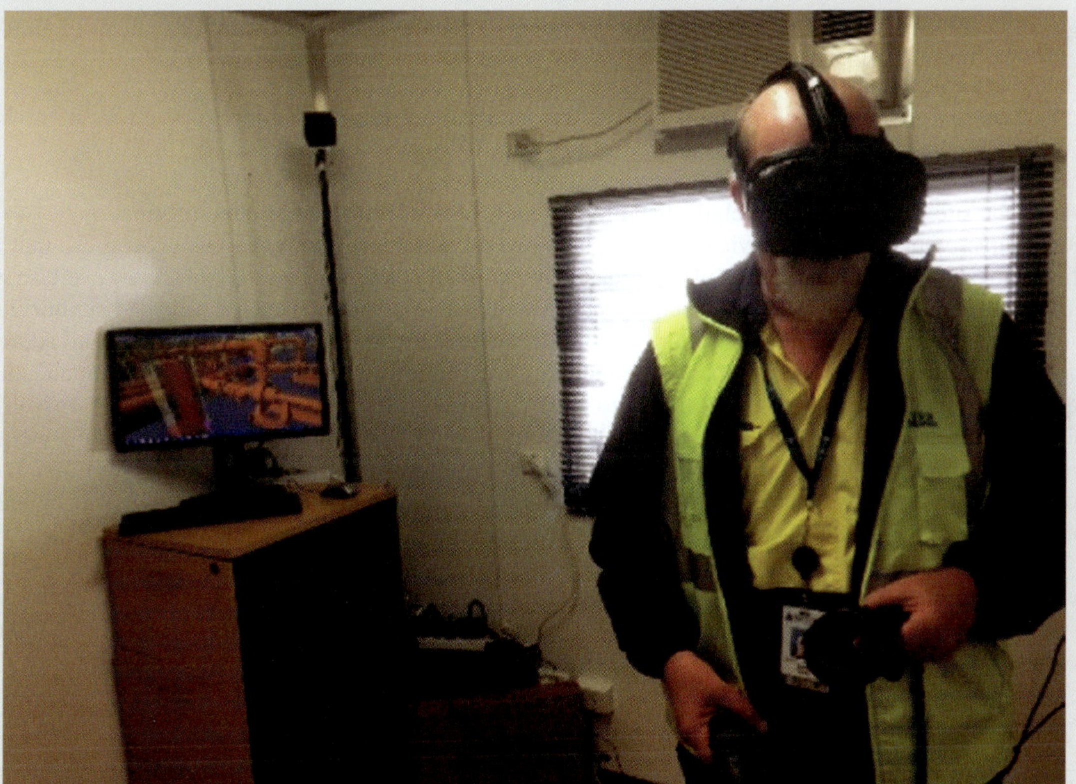

(b) Plant operator using virtual reality headset to explore proposed plant

Source: Stephen Beckwith.

The second example is the Engineers Australia website, which is large and complex as it uses a variety of communication methods to convey specific information to a wide range of users. See https://www.engineersaustralia.org.au.

The home page provides visitors with up-to-date news items and notices about future events, a pop-up menu with a search window, and a menu at the bottom of the page.

The main menu includes the following options, each of which leads to a page with a number of further options in that category: *Membership*, *Credentials*, *For migrants*, *For employers*, *About engineering*, *About us*, *Resources*, *Policy and advocacy*, *News and media*, *Learning and events*, *Engineering communities* and *Contact us*.

The types of information on the site include policy documents, publications, discussion papers, magazine articles, membership criteria and processes, application forms, university program accreditation information and processes, and videos. While much of the information on the site is available to all consumers, a lot of it is password protected for the members of Engineers Australia.

The complexity of the site means that new visitors often need to use the search engine to navigate the site until they become familiar with the layout.

This example shows how a well-planned site may be used to provide the information and services sought by the different types of consumers the organisation hopes will visit the site.

It is common for users to experience one or more of the following problems when visiting large websites:

- difficulty navigating the site
- lack of an efficient and accurate search engine
- out-of-date information
- limited ability to provide feedback on the information they find on the site, its navigability and their overall site experience
- inefficient management of queries and other responses.

A useful exercise would be to carefully explore and review the design of the Engineers Australia website to answer the following questions.

- Is the site easy to navigate?
- Is it easy to find specific information, such as the details about student membership?
- Is it up to date?
- Is the search tool easy to use?
- Does the site provide contact details that allow users to easily request additional information?

Of course, you should also review the design of one or two poorly designed websites, those that are difficult to navigate or that lack key information (or intentionally hide it!). This simple review exercise will help you determine the key design principles and structures that could be used for websites in your field. Some companies do not include any address or contact details, except perhaps for a single email address. You will find that many large companies deliberately use this strategy to minimise the communications they receive from customers. Their aim is to save the time and money they would spend answering questions. Instead they direct people to the frequently asked questions (FAQ) section of their website or other information-rich pages, hoping people will find their answers there. This strategy may save them money but it often leads to a lot of frustration for both existing and potential customers, resulting in the potential loss of some customers. It also wastes customer time instead of company time.

The two examples above demonstrate how well-developed communication plans can be used to deliver messages, and the desired responses, when using a variety of communication methods. An understanding of these issues, and the communication methods discussed in this chapter, will help you to select and use the most appropriate method for a communication and learn to be a more effective communicator.

14.1 Three communication contexts

LEARNING OBJECTIVE 14.1 Discuss the contexts in which engineering communication occurs.

The PCR model highlighted the fact that communication is influenced by the contexts and environments in which it occurs, and the importance of considering these when planning a communication.

There are many contexts in which engineers communicate, but the three most important ones are the business context, the discipline context and the public context. Engineers communicate with members of their engineering team, with other engineers, with other professionals, with business people, with researchers and with members of the public. Because engineering is conducted in the business world, engineers generally communicate using methods that are widely used in business, particularly when communicating with people in other disciplines.

KEY POINT

Engineers should consider the context when planning a communication.

The business context

People who work in business learn to communicate using methods, styles and structures that have developed over time, particularly those that are commonly used in their business sector. Some documents, such as the business letter and company annual report, have become informal standards, while others, such as some specific types of contracts, have evolved into standard templates (e.g. the Standards Australia (2024) series of Australian Standard® Contracts).

When common forms of communication develop and are consistently used by a group, they are known as genres. A **genre** is a form of communication such as a business letter or a novel. More recently the term has also been used to describe the medium of communication such as email or SMS. Thus, a communication would be described as belonging to a particular genre if it generally conforms to the characteristics of the communications in that genre, such as content, language, purpose and style. For example, when a company prepares its annual report in a standard format then it could be said to be representative of the annual report genre.

Of course, this does not mean that all of the communications in a genre are the same. Rather, it means that there will be enough commonality to enable each communication to be effective as the people who regularly work with the genre will be able to access and interpret the information in the communication.

The increasing globalisation of business has meant that business communication methods have evolved and been adapted so that they can be used by business people around the world. While business people may use global communication methods, these will be used in a local context. The term *glocal* is sometimes used to describe this approach (i.e. when a person or organisation is able to 'think global and act local').

The discipline context

The phrase *discourse community* is used to describe the idea of a discipline and its communication culture (Sales 2006). This recognises that the members of the community communicate with each other using verbal, non-verbal and visual genres that are often unique to the community. Swales (1990, pp. 24–27) suggests that there are six defining characteristics that identify a group of individuals as a discourse community. A discourse community:

1. shares a set of common public goals
2. has formal and informal communication mechanisms connecting all of its members
3. requires members to actively exchange information with other members
4. establishes and uses formal and informal conventions to govern participation
5. uses jargon to facilitate communication
6. requires a membership with sufficient expertise to sustain and expand the discourse.

Engineers Australia is a good example of a discourse community. It was established to further the profession of engineering in Australia, build the engineering body of knowledge and facilitate communication between engineers. This is achieved through meetings, conferences, seminars, workshops and publications such as magazines, journals and reports. Some of these activities involve engineers from all branches of engineering, others are discipline specific, and some are for special interest groups that involve engineers from one or more disciplines, for example, the Railway Technical Society of Australasia.

Other professionals, such as accountants, architects, industrial chemists, geologists, information technologists and materials scientists, also have one or more discourse communities within their profession.

When dealing with technical matters, a communication with a professional from another discipline may be considered a *cross-cultural communication*. This is because it involves communicating across a cultural boundary; that is, from one discourse community to another. Obviously, with experience, a person from one discipline can learn to communicate with a person from another discipline about technical matters in that discipline. However, care is required until both people are satisfied that they have a shared understanding about the topic of their communication. For example, a civil engineer working with electrical engineers in the power industry will, over time, acquire a working knowledge of electrical engineering acronyms, terminology and technologies, and will therefore be able to communicate effectively with members of the electrical engineering discipline.

People entering a profession usually receive training in the discourse community's genres so they can communicate effectively with their colleagues (Miller 1994). For this reason, engineering programs are designed to train students in the genres of the engineering discourse community so that when they graduate they will be able to communicate effectively with other engineers in their field. For example, mechanical engineering students will learn about the content, structure and layout of a typical plan that would be used to convey information about the design and manufacture of a mechanical component.

The public context

Increasingly, engineers are having to engage with other professionals, politicians, public servants and members of the public to gain approval for their proposed engineering projects, often during formal public consultations. This means the proposal must be presented in a manner that enables lay people to firstly understand the details and then identify the main issues to be addressed. Only when the details of the proposal are intelligible will people be able to ask questions, challenge concepts and plans, and provide alternative ideas and valuable feedback. This process can be challenging for engineers, particularly when the proposal is based on complex technical engineering concepts and practices.

The following spotlight on the planning discipline in Queensland, provides a good example of how information needs to be intelligible to multiple audiences.

SPOTLIGHT

When public information becomes unintelligible to the public

WITH DR MARITA BASSON

Currently, in Queensland, there is a legal requirement for public consultation associated with the following two major activities undertaken by planners.

- Local government authorities are normally required to display for three months any proposed town planning schemes, strategic plans and regional planning schemes to seek formal input from the public on those plans. Town plans are generally revised every five years and the formal public consultation process is required before the plans can be legally adopted by the relevant state government and implemented by the local authority. Thus, they are 'public domain' documents, and are generally able to be viewed on the internet.
- Town planners prepare development and change-of-land-use applications on behalf of land owners and developers and then lodge them with the relevant local government authority for approval. As part of the approval process, the local authority arranges for notices to be placed on the street boundaries of the relevant parcel of land and in the local paper, informing the public about the application and to invite written comments, support for the proposal or objections to the proposal. Generally, the public have at least one month to respond to the proposal and all of the responses must be formally considered by the local authority before it approves, approves with conditions or rejects the application.

A planning scheme is a legal document that sets the strategic direction for a local government area (LGA) and prescribes the land uses that can be developed on each lot. Land owners normally review the planning scheme before they purchase or develop any land, a task generally undertaken by their legal representative. Apart from planners and the public, planning schemes are used by civil and environmental engineers, builders, surveyors and developers. An understanding of the planning scheme is essential to ensure comprehension of allowable land uses and impacts of future land uses, such as transport corridors.

During the 1960s and early 1970s, most urban and regional planning schemes were prepared by members of what could be described as the technical professions, such as architects, surveyors or civil engineers, rather than by members of the relatively new planning profession. While the public had little input during the plan-making process, there was always a public consultation phase at the end of the process before town planning schemes were approved and implemented.

When a town plan was being prepared by a local authority, it aimed to regulate the existing land uses in the municipality and any expected future land uses, and to identify the sites and corridors for future infrastructure such as highways, schools and shopping centres. The plans prepared by these technical planners were based on the concept of land use zoning, where each allotment, or section, of land was placed in one of the 10–15 land-use zones adopted in the scheme, for example, Residential Zone, Commercial Zone, Park and Recreation Zone and Industrial Zone. The zone allocated to each allotment could be identified from the scheme maps, which showed each allotment in the scheme area and where colours were used to show the various zones. The town planning scheme document was then used to identify the land uses and types of development that were 'allowed', 'allowed with the consent of council' or 'prohibited' in each zone.

Under the then-Local Government Act, the town planning schemes of all of the local government authorities in Queensland — except for the City of Brisbane, which had its own Act — had the same structure and were based on a standard set of land-use definitions, land-use zones, and zoning map colours and legends. When developing a planning scheme, the planners selected the relevant land-use definitions and town planning zones from the standard set defined by the Local Government Department. This meant that, although the town plan for a mainly rural LGA was different to, much smaller and less complex than the town plan for a mainly urban LGA, people were able to find and interpret the required information because the structure, definitions and zones were the same. Importantly, this meant that the planning schemes in Queensland were intelligible to other professionals and members of the public, and they would normally be able to find and interpret the information they required, although some people may have needed assistance from staff at the local authority. Generally, members of the public were also able to prepare and submit the standard forms and documents for planning permits, particularly for routine developments.

The growth to the planning profession in the 1980s and early 1990s resulted in better planning processes, better and more flexible plans and more citizen input into the decision-making processes. But many planners found the planning schemes of this era to be too prescriptive and argued for a change to performance-based planning schemes, where proposed changes in land use could be argued and approved on their merit.

This change occurred in 1997 when the standardised format was discarded when the *Integrated Planning Act 1997* was enacted. Local governments could then prepare their planning scheme in a format they preferred. Planning schemes became less prescriptive and more outcomes-based, resulting in descriptive rather than spatially-based documents. Hence, planning schemes became unintelligible to the public, as well as to many non-planning professionals. Professionals working across two or more local authority areas had to learn how to use multiple planning schemes, each unique in its structure, concepts and maps. The complexity of these planning schemes meant non-planning professionals and members of the public had to employ planning consultants to collect and interpret information and prepare development applications. Thus, the costs of development rose considerably and this affected the public through increased rates and higher prices for land, houses, commercial buildings and units.

The question of the intelligibility of planning schemes was regularly debated during this era. Some people argued that a planning scheme should be intelligible to the public, while others argued that the planning scheme is a legal and highly technical document that should be interpreted by a professional person such as a planner.

Twelve years later, the *Sustainable Planning Act 2009* re-introduced standardised planning schemes and all new planning schemes were based on the state template when they reached the review stage of the five-year cycle.

Further changes occurred with the introduction of the current planning legislation, the *Planning Act 2016*. This Act prescribes some standardised content and format for planning schemes but gives flexibility to local government authorities to cater for the uniqueness of their local government area.

It is clear that, in addition to the challenges resulting from the lack of clarity of the different planning scheme formats, the introduction of performance-based schemes created additional issues. Generally, these schemes are voluminous as they aim to provide the flexibility that will stimulate innovation as well as best practice. These planning schemes encouraged land owners and developers to propose alternative development solutions — a far cry from the prescriptive requirement laid down in earlier schemes.

Through all these changes, the public consultation phase continued to be an important part of the review and development of planning schemes. However, research on the intelligibility of planning schemes suggests that a post-graduate qualification is needed to understand a planning scheme!

CRITICAL THINKING

Review a planning scheme for a house lot you are familiar with by locating the relevant section(s) of the planning scheme on your local authority's website. Are there instructions to guide you on how to use the planning scheme on the council website? Are the written sections of the planning scheme document and the planning scheme maps easy to navigate and read? Could you find the key town planning requirements for your lot? Was the plan intelligible? How big was the planning scheme?

This spotlight highlights the issues surrounding a discipline communicating information to lay people and how it can be hard to find a balance between, firstly, the need for democratic planning processes and easily accessible and comprehensible information and, secondly, the planning discipline objectives such as performance-based planning and modern planning legislation.

This is a case where the outcome of a consultation process would have been improved if the basic principles of the PCR model had been applied. That is, the message needs to be carefully assembled and effectively communicated to the consumers in a package that is easily interpreted by those consumers. It is for this reason that in the engineering disciplines, public consultations often involve a range of media, such as presentations, posters, models, pamphlets, videos, social media and technical reports.

14.2 Planning a communication

LEARNING OBJECTIVE 14.2 Use the PCR model to develop a communication plan for a project.

Using the PCR model to create effective communication

The PCR communication model developed in the chapter on understanding communication can be applied to plan and develop effective communication packages, including business letters, project reports and presentations. When used as a planning tool, the PCR model is applied using the following ten-step process.

1. Define the purpose of the communication, including the desired response from the consumer.
2. Identify the characteristics of the consumer(s).
3. Define the timing and other business requirements for the communication.
4. Consider the contexts in which the communication will occur.
5. Identify and write drafts of the elements of the messages that are to be communicated.
6. Select the most appropriate communication method, channel and media for each element.
7. Assemble the communication package.
8. Forward the package to the gatekeeper for checking and approval.
9. Deliver the package to the consumer.
10. Provide an opportunity for the consumer to respond.

This ten-step checklist may be used for all types of communication; however, the size and type of the communication will dictate the complexity of each step. For example, while the process may be quite straightforward for a business letter, it is likely to be complex for a tender proposal for a multimillion-dollar project.

KEY POINT

Engineers should carefully plan a communication.

SPOTLIGHT

Developing a communication plan for a student project

A team of four first-year engineering students has been assembled to complete a semester-long project. The project specifications stipulated that each team must submit a project report and do a 20-minute presentation at the end of the semester. During their first four meetings the BZ team identified the problems to be solved and then developed strategies to undertake the project and complete the required technical work. At their fourth meeting they decided to develop a communication plan for their final report so that they could identify the information they should gather while undertaking the project. They did this because they thought:

- it would be more efficient to write a draft of each section of work as soon as it was completed
- they should identify opportunities to take photos, videos and other graphics to illustrate the report; this would ensure they captured those images during the relevant stages of the project
- the plan would also help them to identify any additional data they may need for the report.

The team developed the following communication plan.

Step 1. Purpose

(a) The purpose of the project report is to convince the assessors that the team has worked effectively and followed the engineering method to produce an innovative and realistic solution, based on experimental data and supported by relevant literature.

(b) The purpose of the presentation is to outline the key points of the method and outcomes, to convince staff and students that the team worked well together, and that each member of the team was engaged in, and can answer questions about, all aspects of the project.

Step 2. Consumers

The project report will be assessed by the subject lecturer and a tutor; the presentation will be assessed by (a) the lecturer and tutors (60 per cent), and (b) the other student teams (40 per cent).

Step 3. Requirements

(a) The report is to be submitted one week before the end of the semester. It is to be no more than 3000 words in length, including appendices.

(b) All of the members of the team are to participate in the 20-minute presentation, which will be timetabled in the last week of the semester. The team must advise their tutor of any multimedia requirements two weeks before the presentation.

Step 4. Contexts

Each of the team members identified their strengths and preferred roles, as well as their other commitments over the last four weeks of the semester and how these may affect the preparation of the report and the presentation. They also visited and checked the equipment in the multimedia enhanced lecture theatre where they would give their presentation.

Step 5. Draft messages

(a) The requirements for the report are set out in the subject guide. This includes format, content headings and structure. The team planned to write a draft of the relevant section as each project stage is completed.

(b) The team identified seven key messages that they thought they would present.

Step 6. Communication methods

(a) The project report will be in text and include data, graphs and pictures.

(b) The team members will wear BZ T-shirts, jeans and black shoes for the presentation. One member will act as master of ceremonies (MC) to link the other members' presentations. The media will include an A0-sized poster, a slide presentation and a 30-second video of the team in action.

Step 7. Assembling the package

All members were to work on the draft report and identify the key messages for the presentation. To ensure this happened, they agreed it would be an agenda item at every team meeting. The team leader and another member would then complete the report while the other two members scripted and prepared the media for the presentation.

Step 8. Gatekeeper

The team agreed to ask their fourth-year student mentors if they would be the audience for their rehearsal of the presentation. They also asked one of their parents to proofread the final report.

Step 9. Delivery of the communication package

The team members nominated a member to submit a hard copy, or a digital copy of the report to the school office on the due date and another member who would lodge their media requirements with the tutor by the due date.

Step 10. Response

The team members discussed ways in which they could engage the audience and encourage them to respond enthusiastically during their presentation, and give the team a high mark!

The team members used the plan to guide their work while they undertook the project. They regularly reviewed the plan and updated it as their work progressed. This included discussions about the feedback they received from their teachers, and other people, during the life of the project.

CRITICAL THINKING

Use the ten-step PCR process to develop a communication plan for your next project.

Developing a communication plan for an engineering project

All engineering organisations have policies and protocols relating to everyday forms of communication, such as letters, emails, reports, hardcopy and digital copies of plans, and digital models. Larger organisations would also have guidelines and templates for the many documents required during the life cycle of a project, from commencement to completion. The guidelines would normally include document approval processes and quality assurance requirements. The following discussion covers some of the broad principles that should be considered when developing a communication plan at the beginning of a project.

The expected outcomes for a project are normally defined in a document, such as in a client brief, a contract or a set of specifications. Part of the project planning process normally includes the development of a plan setting out the methods that will be used to communicate the project outcomes to the various stakeholders. For example, a project planning team may consider the following questions.

- How will the outcomes and the supporting information be reported to the client?
- What outcomes will be reported to the chief executive officer (CEO) or the company board?
- Are any user or maintenance manuals required?
- What, if any, information must be reported to the government at a local, state or federal level?
- Will any media releases be required and, if so, when will they be required?
- Can the project be used for any excellence awards?
- Will the project incorporate any innovative solutions? If so, should this be communicated to the profession at a conference or in a refereed journal article?
- Will there be any intellectual property (IP) that should be retained or protected by a patent?

A communication plan will usually include regular project reports to some or all of the stakeholders. On smaller projects, this information may be reported at weekly or monthly team meetings. One of the benefits of developing a communication plan early in the life of a project is that the required information can be collected and stored along the way so it is ready when required. This will save a lot of time and effort when the final report has to be prepared, which often occurs at the busiest phase of a project.

Using the model

The ten-step process used for the student project can be used to develop plans for many types of communication, although the prompt questions may change to suit the type of documents or presentations being prepared. The length of time spent on the planning process will also vary, as it will depend on the type, size and complexity of the communication. It may take only a few moments, or many weeks.

The model may also be used to analyse a previous communication, to understand why it was successful or why it failed to achieve the desired response. For example, you could analyse some final-year student reports on topics that interest you to identify any common communication strategies or approaches used by those teams.

The PCR process can be used to identify the information that should be communicated, the contexts and environments in which the communication will occur, the methods of communication, and the formats, styles and media that will be used to communicate the information. The communication approach that will be used is also decided during this planning phase.

Approach

The approach that is adopted during the planning stage may be influenced by the type of information being communicated, particularly when the information relates to a decision, an outcome or a recommendation based on an opinion. The approach will also be influenced by the purpose of a communication. For example, a persuasive approach may be adopted if the purpose of a communication is to persuade a prospective client to accept a proposed design.

KEY POINT

Engineers should carefully consider the approaches and methods that will be used for a communication.

A factual communication

Although it is difficult to identify a single engineering discourse, there are many common features in the genres used by engineers. You have already learned about many of these features, and in table 14.1 they have been summarised into seven principles that underpin the majority of communications. To help you remember these key principles, they have been arranged to form the acronym FACTUAL.

TABLE 14.1 Key principles for developing an engineering communication

Focused	A communication should be focused on its purpose. Only the required information should be conveyed so that it is not lost amongst irrelevant information. The key criteria are effectiveness and efficiency.
Accurate	A communication should be accurate and should contain any relevant evidence associated with measurements, facts and other information. Information included in the communication should be verifiable and precise. Opinions should only be included when they are required and they should be supported by evidence and sound arguments.
Complete	A communication should contain all of the information the target consumer will need to be able to understand and use the information correctly.
Targeted	A communication should be tailored to meet the needs of the target consumer(s). Their characteristics and their business relationship should help to determine the content and the channels used to convey the information.
Unambiguous	A communication should be clear and unambiguous. Clarity is normally enhanced when simple English and sentence structures are used to convey information.
Accessible	A communication should use appropriate language, style and format to ensure that consumers are able to access and understand the information that is conveyed. For example, a report written for an engineering audience that contains technical jargon and formulas would not be suitable for distribution at a public consultation meeting about a project.
Logical	A communication should be presented in a logical and intuitive format that the target consumer will be able to easily follow and understand.

A professional opinion

Engineers may be asked to provide an *opinion* if there is insufficient evidence to state a fact or to draw a firm conclusion. In this case they would use their professional judgement to develop an opinion. This would be based on their knowledge, skills and experience. A professional opinion is developed after conducting a careful review of the available evidence, identifying the possible conclusions that can be drawn from that evidence, and then deciding the most likely conclusion. Young engineers would not normally be expected to give a professional opinion as considerable experience, expertise and judgement is required to formulate such an opinion.

When providing an opinion, engineers would state the evidence, the possible conclusions and the reasons the adopted conclusion was selected. Opinions are often embedded in qualifying statements, such as 'On the best available evidence . . .' or 'After a careful review of the available evidence, it is my opinion that . . .' Any such qualifications would be carefully worded and expressed so that the opinion is embedded in the contractual, engineering, project and other relevant contexts. The opinion may also be qualified by statements about the expertise and experience of the engineer who is providing the opinion.

Engineers need to be mindful when giving an opinion that they also consider any potential liability or negligence. If perceived to be an expert, an engineer may have a higher duty of care to those relying on that information, or be required to ensure they provide a higher standard of expertise or judgement.

Finally, it should be noted that in a dispute, equally qualified engineers may provide different opinions about a particular matter. While these opinions may be based on the same evidence, each engineer would interpret the evidence based on their knowledge, skills and experience.

A persuasive communication

Sales (2006) reports that engineers are reluctant to use persuasive text in documents such as tender proposals as this form of writing is not generally accepted as part of the engineering discourse. This may be because engineers are uncomfortable with this style of writing, which they may see as a 'sales pitch'.

However, for those engineers who are involved in tendering for projects or building a client base, this approach will be an important part of their communication strategy. Engineers may also use the persuasive approach to sell their ideas, designs or products to their supervisor or their organisation.

Engineers often use a restrained approach when persuading a consumer, by appealing to their intellect and relying on the logic of their written arguments, or the supporting figures, to demonstrate their case (Sales 2006). This could be described as the 'facts speak for themselves' approach.

There are other approaches; for example, an appropriate analogy may be used to clarify or reinforce the case being presented. In this case, the analogy may be more persuasive or direct than the communication itself. Another approach is to provide simple statements in a document that summarise a key point or highlight an important piece of information. These do not have to be written in persuasive language to have the desired impact on the consumer.

Persuasion is one of the tools of leadership. Leaders use the power of persuasion to argue a case, to convince team members of the path to follow or the design approach to be used on a project. They also use persuasive communication techniques to cast a vision for their organisation.

One of the keys to effective communication is to decide which approach should be used. Generally, one approach would be adopted for a small communication, while two or more approaches may be used in larger documents, in different sections of a document or where the communication plan identified that a range of approaches and media should be used.

Communication methods, styles, formats and media

There are many communication methods, styles, formats and media that can be used to communicate a message. In addition, contexts influence the communication methods engineers use, particularly information formats, styles and media. Getting the mix right is an important part of the planning process.

While you are at university your lecturers and tutors will provide you with the policies, guidelines and templates that you will be required to use when preparing an assignment or a project report, or when giving a presentation. Similarly, engineering organisations have different policies and guidelines; therefore, the structure and type of information that is communicated in each channel, rather than the style or the format, is the focus of this section.

The following spotlight was based on ChatGPT responses to three similar prompts, with only the word count changing: *Write XXX words on the pros and cons of using AI to prepare engineering project reports.* The first prompt requested a 300-word response, the second a 500-word response, and, finally, a 1000-word response, although the response provided was just over 800 words. The results from the three prompts were then reviewed, and merged with minimal editing to produce the spotlight, and finally copyedited for consistency with the surrounding text.

SPOTLIGHT

What does ChatGPT say?

WITH CHATGPT

ChatGPT: Title: Exploring the Pros and Cons of Leveraging AI for Engineering Project Reports

In the contemporary landscape of engineering project management, the integration of artificial intelligence (AI) has become increasingly prevalent, revolutionising the way reports are prepared and analysed. With its ability to streamline processes, enhance accuracy, and offer valuable insights, AI presents a plethora of advantages. However, it also brings along its own set of challenges and limitations. In this discourse, we delve into the nuanced pros and cons of employing AI in the preparation of engineering project reports.

Pros

- *Efficiency*. One of the most significant advantages of utilising AI in engineering project reports is its ability to enhance efficiency. AI-powered tools can automate various tasks involved in report preparation, such as data collection, analysis, identifying key insights, summarisation and formatting. This ability to automate repetitive tasks streamlines the writing process and generates reports in a fraction of the time it would take a human. This efficiency allows engineers and project managers to focus on more critical aspects of the project such as design and problem-solving.
- *Accuracy*. AI algorithms are designed to process vast amounts of data with precision and accuracy, minimising the risk of human error inherent in manual report writing. By leveraging machine learning

models, AI systems can analyse complex datasets and generate accurate insights, leading to more reliable project reports. This accuracy is particularly important in engineering, where precise and reliable information is essential for making informed decisions. It ensures that decision-making processes are based on solid, data-driven information, thereby reducing the risk of costly mistakes.

- *Data visualisation*. AI technologies enable the creation of dynamic data visualisation tools that transform raw data into insightful graphs, charts and interactive dashboards. These visual representations help stakeholders better understand project metrics, trends and patterns, facilitating more informed decision making. Additionally, compelling visualisations enhance the clarity and effectiveness of project reports, making them more engaging and accessible to a wider audience.
- *Predictive analytics*. AI-driven predictive analytics empower engineers and project managers to anticipate potential challenges and opportunities throughout the project lifecycle. By analysing historical data and identifying patterns, AI algorithms can forecast project outcomes, resource requirements and potential risks. This proactive approach enables stakeholders to devise strategies to mitigate risks and optimise project performance, ultimately leading to better outcomes.
- *Customisation*. AI tools can be tailored to suit the specific needs and preferences of engineering projects. Advanced AI systems can be tailored to specific engineering domains, allowing for the generation of reports that align closely with the specific needs and preferences of the engineering team, and with industry standards and best practices. Whether it's generating custom reports, analysing specialised datasets, or implementing unique algorithms, AI solutions offer a high degree of customisation. This flexibility allows organisations to adapt AI technologies to their workflows seamlessly, maximising their utility and effectiveness in generating project reports tailored to specific requirements.
- *Consistency*. AI ensures consistency in the format, tone, style and language of project reports across different stages and team members. This consistency helps to improve readability and comprehension facilitating effective communication within project teams and for stakeholders making it easier for them to understand the information presented in the report. Additionally, consistent formatting can help to maintain the professional appearance of the report, which is important in industries where attention to detail is critical.
- *Accessibility*. AI-powered systems can generate reports in multiple languages and formats, making them accessible to a wider audience. This is particularly important in global projects where stakeholders may speak different languages or have different cultural backgrounds. By providing reports in multiple languages, AI can help to promote collaboration and communication among team members from diverse backgrounds
- *Scalability*. AI-powered report writing scales effortlessly to accommodate projects of varying sizes and complexities. Whether it's a small-scale initiative or a large-scale engineering endeavour, AI can adapt to the scope and requirements, delivering consistent results without compromising quality.

Cons

- *Dependency on data quality.* The effectiveness of AI-powered project reporting heavily relies on the quality and integrity of the underlying data. If the data fed into the AI system is incomplete, inaccurate or biased, it can compromise the integrity of the generated reports, potentially leading to erroneous conclusions or recommendations, undermining the reliability of project reports. Ensuring data quality and consistency remains a significant challenge, requiring meticulous data governance practices and continuous monitoring.
- *Lack of contextual understanding*. One of the main drawbacks of using AI for writing engineering project reports is its lack of contextual understanding. While AI can process large amounts of data and identify patterns, it may struggle to understand the nuances of a particular project or industry. This can lead to reports that are inaccurate or irrelevant. Engineering projects are often situated in specialised fields where complex multidimensional factors, including technical specifications, regulatory requirements, and stakeholder preferences, may not always be adequately captured by AI algorithms. These limitations can result in oversimplified analyses, inaccuracies, misinterpretations that lead to recommendations that overlook critical contextual nuances, particularly in situations where contextual understanding is crucial for accurate reporting.
- *Ethical and bias concerns*. The use of AI in writing engineering project reports raises a number of ethical concerns, including issues of accountability, transparency and bias. For example, if an AI system produces a report that contains inaccuracies or biases, who is responsible for correcting them? If the data fed into the AI system is incomplete, inaccurate or biased, it can compromise the integrity of the generated reports, potentially leading to erroneous conclusions or recommendations. This is because AI algorithms are susceptible to biases inherent in the data they are trained on, leading to skewed outcomes and discriminatory practices. In the context of engineering project reports, biased AI models may inadvertently perpetuate inequalities or reinforce existing disparities, particularly concerning resource allocation, environmental impact assessment and community engagement. Addressing ethical concerns and mitigating biases in AI systems requires transparent data sources, diverse training datasets, and robust algorithmic fairness measures.

- *Limited creativity and innovation*. While AI excels at processing and analysing existing data, it inherently lacks the creativity and intuition of human writers and decision makers. As a result, AI-generated reports may sometimes lack nuanced insights or innovative solutions that human authors can provide. Engineering projects often require innovative problem solving and out-of-the-box thinking, which may not be fully replicable by AI systems. Relying solely on AI-generated reports may stifle creativity and innovation within project teams, hindering the exploration of unconventional solutions and novel approaches.
- *Security and privacy risks*. The integration of AI in engineering project reporting introduces potential security and privacy risks associated with data storage, processing and sharing. AI algorithms require access to sensitive project data, raising concerns about unauthorised access, data breaches and privacy violations. Safeguarding confidential project information against cyber threats and ensuring compliance with data protection regulations are paramount considerations when deploying AI-powered reporting systems.
- *Maintenance and updates*. AI systems require regular maintenance and updates to remain effective and relevant. This includes updating algorithms, fine-tuning parameters, and incorporating new data sources or features. Failure to maintain AI systems adequately can result in outdated or suboptimal performance, undermining the value of AI-generated reports.

In conclusion, the utilisation of AI in engineering project reporting offers a multitude of benefits, including enhanced efficiency, accuracy, data visualisation, predictive analytics and customisation. However, it also presents challenges such as dependency on data quality, lack of contextual understanding, ethical and bias concerns, limited creativity and security risks. To harness the full potential of AI while mitigating its drawbacks, organisations must implement robust data governance practices, address ethical considerations, foster interdisciplinary collaboration and prioritise cybersecurity measures. By striking a balance between the strengths and limitations of AI, engineering project teams can leverage its transformative capabilities to drive innovation, improve decision making and achieve successful project outcomes.

CRITICAL THINKING

Carefully review the 'pros' and 'cons' in the spotlight and consider what changes you should make to your current AI methodology.

This spotlight highlighted the advantages and disadvantages of using AI and the need for engineers to always carefully review and verify AI responses before editing and situating them into the project report and, where necessary, populating them with personal and project-specific information. The use of AI in engineering communications will be discussed in more detail in section 4.6 of this chapter.

It is likely you will have to use most of the following communication methods during your professional career: writing, speaking, listening, charting, graphing, drawing, modelling, demonstrating and photography. This list may be daunting but, as you will see in the following sections, there are many tools and techniques you can use to help you develop these skills. Some of these were discussed in the chapter on communication skills, and the remaining topics will be discussed in this chapter. We will also see how these communication skills and techniques are applied in typical engineering and business contexts. Once again, they are grouped into three main categories: verbal communication, visual communication and non-verbal communication.

14.3 Writing in the engineering workplace

LEARNING OBJECTIVE 14.3 Plan and prepare written communications in an engineering context.

While engineering organisations generally have communication policies, guidelines, conventions and templates for many of the written documents they prepare, they are unlikely to have guidelines that cover grammar and style because they assume that their staff have previously acquired the relevant knowledge and skills. Therefore, it is important that you acquire these skills at university. There are many books on these topics in university and public libraries, and most universities have online resources and tutorials available for students, for example, the Australian Government Style Manual: https://www.stylemanual.gov.au.

Susan Conrad and Timothy Pfeiffer researched a continuing problem in engineering education: the mismatch between the writing skills of engineering graduates and the demands of writing in the workplace. Susan is a professor of applied linguistics at Portland State University in Oregon and Timothy is a senior engineer and manager in Portland. The aim of their project was to develop instructional materials that will provide students with the skills needed to write as a practitioner. They hope this will limit the amount of 'red ink' supervisors write on the first reports that graduates submit when they commence practice.

Rather than assuming what the problem was, they began with an empirical analysis of the language features in a large collection of texts. This involved the analysis of 661 documents (382 from students and 279 from practitioners) such as reports, technical memos, proposals, project emails, lab reports and site visit reports. In all, 100 practitioners and 335 students contributed to the writing of the documents. Some of their findings are reported in the following spotlight.

KEY POINT

Engineers should use an appropriate format, structure and style for a written communication.

SPOTLIGHT

The differences between academic and practitioner writing

WITH SUSAN CONRAD, PORTLAND STATE UNIVERSITY
AND TIMOTHY PFEIFFER, PE, FOUNDATION ENGINEERING INC.

One of our initial observations was that the practitioner texts were easier to understand than the student texts even though they contained more complicated technical information. While the information in practitioner texts is dense, *each sentence normally expresses only one idea*. Our grammatical analysis also found that practitioners tend to *use simpler sentence structure*, with their sentences often containing just one clause, without subordinate or embedded clauses.

Let's look at an example of practitioner writing:

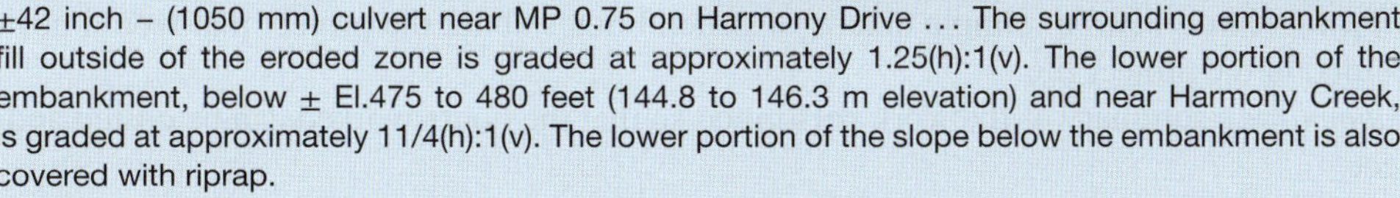

> A section of the road embankment has eroded directly beneath and around the ±42 inch – (1050 mm) culvert near MP 0.75 on Harmony Drive ... The surrounding embankment fill outside of the eroded zone is graded at approximately 1.25(h):1(v). The lower portion of the embankment, below ± El.475 to 480 feet (144.8 to 146.3 m elevation) and near Harmony Creek, is graded at approximately 11/4(h):1(v). The lower portion of the slope below the embankment is also covered with riprap.

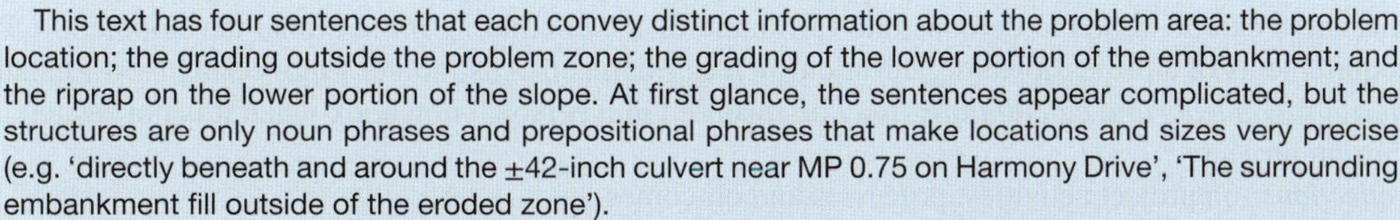

This text has four sentences that each convey distinct information about the problem area: the problem location; the grading outside the problem zone; the grading of the lower portion of the embankment; and the riprap on the lower portion of the slope. At first glance, the sentences appear complicated, but the structures are only noun phrases and prepositional phrases that make locations and sizes very precise (e.g. 'directly beneath and around the ±42-inch culvert near MP 0.75 on Harmony Drive', 'The surrounding embankment fill outside of the eroded zone').

On the other hand, students tend to write more complex sentences, with subordinate and embedded clauses that cover multiple ideas in a single sentence. The information is less dense overall, and the students' noun phrases and prepositional phrases are often ambiguous.

Let's look at an example of a student's writing:

> The width of the existing landslide is approximately 70 to 100 feet (20 to 30 metres) and at the top of the landslide is the existing road with exposed foundation to the scarp. The highest elevation part of the scarp has the inclination of approximately 80 degrees while the rest of the scarp inclination varies between 40 and 60 degrees.

The student text covers multiple ideas in each sentence. For example, the first sentence is a compound sentence that describes the width of the landslide and then attempts to express an idea about the location of the road and scarp. The idea is unclear due to a vague prepositional phrase ('with exposed foundation to the scarp'). The student's literal description of the slide width is vague; they probably mean that 'The landslide varies from approximately 70 to 100 feet wide'. The result of the grammatical choices ▶

is that the student writing is less precise. In fact, in sentences such as these, the students' meaning often becomes inaccurate.

We also found that when practitioners write for clients, their grammar choices exhibit some striking differences from students' choices. Practitioners establish their firm's responsibility for certain work or decisions by using active voice (e.g. 'we drilled', 'we recommend'). Once responsibility is established, they may also use passive voice ('a hole was drilled'). Students, in contrast, tend to use passive voice indiscriminately.

So, what is the take-home message from this aspect of our work? Well, when practitioners discuss their writing, they make repeated reference to two concerns: the first is *the need for precise information* and the second is *the ease of reading for their clients*. They strive for sentences that clients can read quickly with no questions about the meaning. Relatively simple sentence structures help to achieve this. The precise, accurate information and easy-to-understand sentences fulfil a need to communicate with as little ambiguity as possible. Not only are the clients happy, but practitioners noted that this limits a firm's liability.

For students, the key message is that accurate, precise and unambiguous writing is as much engineering as design and calculation are; in other words, *writing is engineering*.

CRITICAL THINKING

Review a recent lab report you wrote and analyse the complexity of the sentence structures you used, the voice you used, and the accuracy and precision of your writing. Then, turn it into a technical memo for a client and rewrite the report using the practitioner's principles outlined in the spotlight.

Practice notes

The ability to write accurate and efficient notes is a skill engineers use in many spheres of their work — for example, to record the outcomes of meetings with clients, government organisations, contractors or other members of their team. Depending on the work they are undertaking, they may record brief notes about their activities in either a personal notebook or a project notebook. Some examples of the types of personal information they record are:

- *site visits:* notes about conversations, site conditions, problems encountered, solutions and decisions
- *formal office meetings:* notes about relevant issues, decisions or actions discussed at the meeting
- *meetings or phone conversations:* notes about the topics discussed, key pieces of information and decisions
- *general project information:* simple details or general notes about the project that would not normally be recorded in the project file
- *work summary:* a summary of the work performed so that the hours worked can be costed and billed to a project
- *professional development:* a summary of any education, training or other activities undertaken so that these can be tallied to demonstrate fulfilment of the continuing professional development (CPD) requirements of a professional organisation such as Engineers Australia or Engineering New Zealand
- *designs, concepts and calculations:* may be recorded in a personal notebook when it is not related to a current project
- *reflections:* on project activities, processes and outcomes.

When they are working on some types of projects or sites, engineers may be required to record additional information about project or site activities, often on a daily basis. For example, on a construction site they may have to record the weather, the personnel on site, the activities undertaken by those people, and any problems or issues, decisions, actions, design variations and deliveries. Engineers use a range of media to record these notes. For example:

- a standard appointments diary. A vertical line can be drawn down the middle of the page with the left-hand side used to record 'planned' activities and the right-hand side to record the actual activities undertaken.
- a spiral-bound or hard-cover A4 notebook. A hard-cover notebook is obviously more durable than a spiral-bound format and, from an evidence perspective, pages cannot be removed without leaving evidence of their removal.

Some organisations produce their own versions of site diaries, notebooks or logbooks to facilitate the recording of the required information in a consistent manner. Notes of meetings and conversations are then placed in the relevant project file. In other cases the notes of a meeting or conversation are typed and then circulated so that all parties can record their agreement by signing and dating the notes. This is done to ensure that all parties agree that the notes are a true and correct record of a meeting or conversation. The fully signed copy would then be filed.

While the notes engineers write are, from a quality assurance perspective, an important record of their activities or conversations, they may also prove to be invaluable when a dispute arises. They may even be used as evidence when an engineer is required to testify in a court case, or is called to testify as an expert witness. For example, the meeting notes prepared by Underground Mine Manager, Mr Pat Ball, became an important piece of evidence during the Coroners inquiry into the 2006 Beaconsfield mine collapse, in which one miner was killed and two other miners were trapped nearly a kilometre below the surface for two weeks before being miraculously rescued (Denholm 2008). The subsequent coronial inquiry into the tragedy highlighted the need for engineers and other professionals to write and archive accurate notes about their work activities, and to have a highly efficient system that facilitates the retrieval of those notes.

Some of the common types of documents used by engineers to formally communicate information are discussed in the following sections.

Business correspondence

Most day-to-day communication in the business world is conducted face to face, by telephone or through *correspondence* (communication through one of the forms of written communication). Often the outcomes of a meeting or telephone conversation lead to an exchange of letters or emails. These forms of communication should be written using formal language, as they are generally taken to be a record of the information that was communicated from one party to another.

The standard components that are normally found in business correspondence are as follows.

- *The company letterhead.* The name and contact details of the organisation are included and generally the name and contact details of the person sending the letter. The contact details section may include information such as street address, postal address, telephone number, mobile phone number, email address and Australian Business Number (ABN) or, in New Zealand, the GST number. In cases where there is more than one office, the contact details for all of the offices may be included. In this case, the office where the letter originated will also be identified.
- *The date of the communication.* The day, month and year, and, in some cases, the time.
- *The recipient's name and address.* The postal address would normally be included here, unless the recipient has indicated that the street address is the preferred address or that the letter is to be conveyed by a courier.
- *Salutation.* The greeting should be formal, for example: 'Dear Sir/Madam', 'Dear Dr Loader' or 'Dear Ms Taylor'. The first name of a person should only be used for friends or business colleagues.
- *Subject line.* The topic or main purpose of the letter is identified here. The subject line is an optional component. When the letter is being written in response to an earlier letter, the same subject line would be used along with any project or file reference information.
- *Body.* The information to be communicated, which may vary from a single paragraph to multiple pages.
- *Close.* The phrase would be selected from common phrases such as 'Yours faithfully', 'Yours sincerely' or 'Kind regards'.
- *Signature block.* The signature and printed name of the writer. This section may also include the writer's position or role in the organisation.

Some optional components normally printed after the signature block are as follows.

- *Document/project reference.* The number or code of the document so that the letter can be filed and, if necessary, located at a later date.
- *Enclosures.* The documents that are to be attached to the letter are listed here. This helps to ensure that they are attached before the letter is sent. This also advises the reader of the presence of attachments if they become separated from the letter, or if a copy is made of the letter, but not the attachments.
- *Copy details.* The names of the recipients are listed if a copy of the letter is sent to other people in the organisation, or other organisations.

Great care should be exercised when writing and selecting the content of a written communication, particularly when using email and other forms of electronic communication. This is because emails can easily be circulated widely by a single keystroke, both within an organisation and beyond. This can be intentional or unintentional (perhaps through the selection of an incorrect address). The consequences of the circulation of commercially sensitive or offensive material can be devastating for the writer and those that are the subject of the email (or other form of communication), as well as those who may have 'forwarded' the email. Personal lives, business relationships and careers can be affected, and there have been cases where the writers of offensive emails have ended up in court.

Mobile phone text messages should only be used to communicate information in a business context when no other suitable method is available. This is because the language used is highly abbreviated and subject to misinterpretation. In addition, it is difficult to obtain a suitable record of a text message for filing in a project file. It should be noted that police and other law enforcement agencies are able to seize phones, tablets and computers to access emails, images, messages and other information as evidence for their investigations.

Some of the business documents regularly used by engineers are discussed in the following sections.

Business emails

An email quickly conveys information to people within an organisation, external to an organisation and in other locations. The convenience of email technology often means that an informal style is used and the wording is not checked as thoroughly as it would be if it was in a formal letter. This is fine when emailing friends or co-workers; however, it may not be appropriate when writing to the CEO, clients, colleagues or other organisations. Business emails should use formal language and be prepared in the same manner — and with as much care — as a letter or a memo. Generally, an email should be formal if it is likely to be filed by the creator or the consumers.

Many people (Krotz n.d.; Martin 2000; Shea 1994; Stahl 2004) have written about email etiquette and the following 'do' and 'don't' list is representative of good email etiquette.

- Never forget the individual reading your email is a person. Don't be offensive.
- Never email anything you wouldn't say to the recipient's face. Don't write an email when you are angry.
- Avoid using block capitals in an email as this is generally considered to be 'shouting'.
- Proofread your emails. Check grammar, spelling, names and information.
- Be professional. Try to respond to an email within one working day and be ethical in your use of emails.
- Know when email is not the best form of communication. Email may not be appropriate for sensitive information, and complex information may be better communicated by telephone or at a face-to-face meeting. Talking directly enables each party to clarify details and ask follow-up questions.
- Manage the size of your emails. Don't ramble, and be careful about the size of any attachments.
- Email is not chain mail. Don't circulate an email you have received without the writer's permission. Remember, once you have sent your email you have lost control of it.
- Count to ten before pressing SEND. Check the recipient details and the contents of the email, and verify any attachments by opening them within the email. Then decide whether you really do want to send the email.
- Manage your email use as emails can disrupt your work and become unmanageable. One strategy is to look at or respond to emails only two or three times each day.
- Back up your emails. Business emails should be electronically filed in a project file like any other form of business communication.

One advantage of using emails is that it is easy to acknowledge receipt of a message by answering it directly or replying with a simple message, such as 'Thanks for the email. I expect to be able to respond to your questions on Friday.' A message like this is good practice as it informs the sender you have received their email and shows them you have prioritised their request.

Business memos

A memorandum (memo) is often used instead of a letter for inter-office communication within an organisation. A memo can also be written to a job file to record the key facts from a telephone call or other communication. A formal memo contains many of the components of a business letter. In recent years, formal memos have been replaced by emails in many organisations, although the same language and style is normally used.

A memorandum should begin with a short sentence or two that clearly state the purpose of the memorandum. This should help the reader to quickly identify the importance of the communication and any actions that they may need to take. The next section of the memorandum should provide the reader with the background to the document. This is followed by the main section of the memorandum, which provides information, discusses issues and outlines any decisions that have been made. The final section summarises the information in the document and lists the key outcomes and any requested actions.

A sample memorandum is presented in figure 14.2. It highlights the standard and optional components of the memorandum.

FIGURE 14.2 Sample memorandum (optional components are identified by dotted lines)

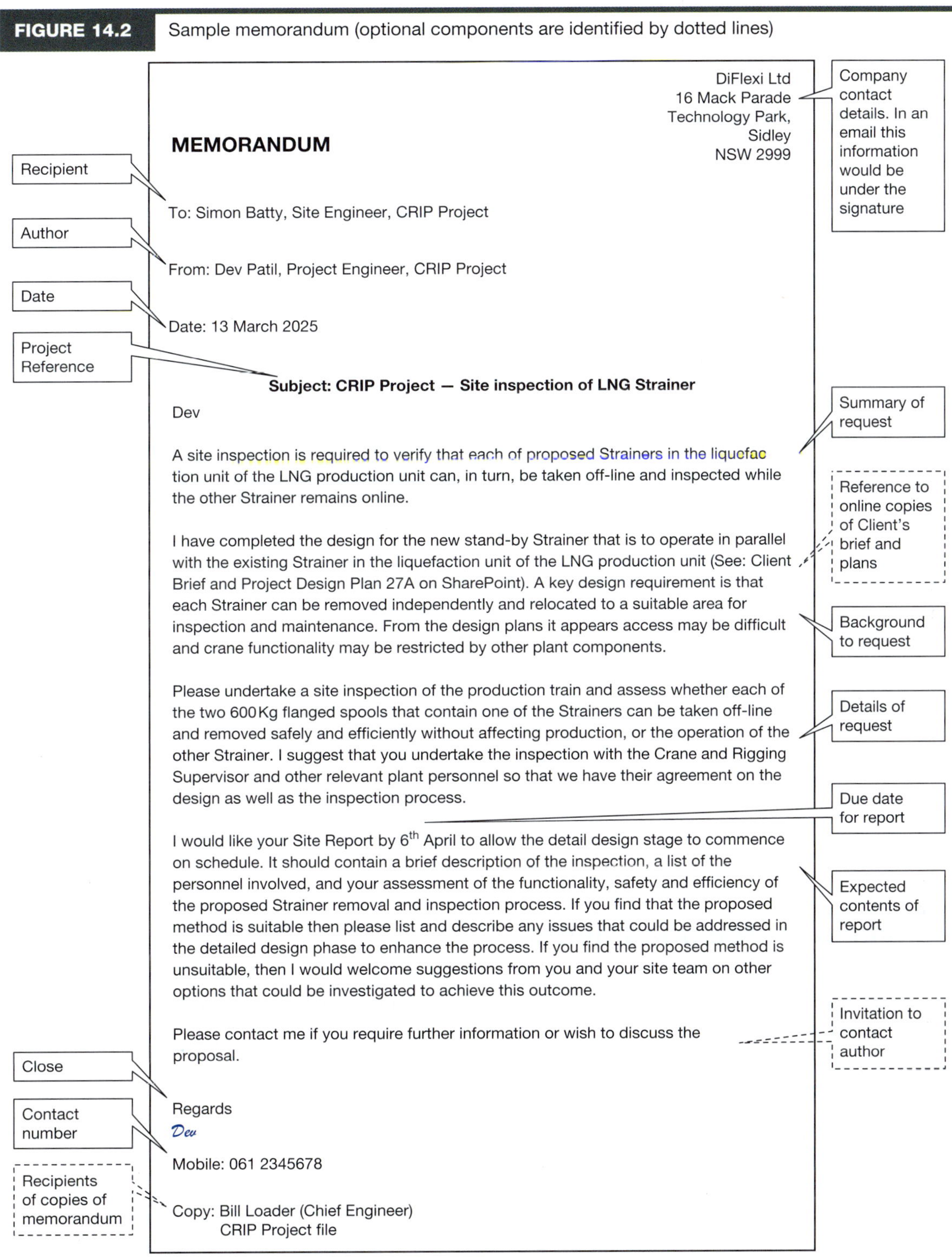

DiFlexi Ltd
16 Mack Parade
Technology Park,
Sidley
NSW 2999

MEMORANDUM

To: Simon Batty, Site Engineer, CRIP Project

From: Dev Patil, Project Engineer, CRIP Project

Date: 13 March 2025

Subject: CRIP Project — Site inspection of LNG Strainer

Dev

A site inspection is required to verify that each of proposed Strainers in the liquefaction unit of the LNG production unit can, in turn, be taken off-line and inspected while the other Strainer remains online.

I have completed the design for the new stand-by Strainer that is to operate in parallel with the existing Strainer in the liquefaction unit of the LNG production unit (See: Client Brief and Project Design Plan 27A on SharePoint). A key design requirement is that each Strainer can be removed independently and relocated to a suitable area for inspection and maintenance. From the design plans it appears access may be difficult and crane functionality may be restricted by other plant components.

Please undertake a site inspection of the production train and assess whether each of the two 600Kg flanged spools that contain one of the Strainers can be taken off-line and removed safely and efficiently without affecting production, or the operation of the other Strainer. I suggest that you undertake the inspection with the Crane and Rigging Supervisor and other relevant plant personnel so that we have their agreement on the design as well as the inspection process.

I would like your Site Report by 6th April to allow the detail design stage to commence on schedule. It should contain a brief description of the inspection, a list of the personnel involved, and your assessment of the functionality, safety and efficiency of the proposed Strainer removal and inspection process. If you find that the proposed method is suitable then please list and describe any issues that could be addressed in the detailed design phase to enhance the process. If you find the proposed method is unsuitable, then I would welcome suggestions from you and your site team on other options that could be investigated to achieve this outcome.

Please contact me if you require further information or wish to discuss the proposal.

Regards

Dev

Mobile: 061 2345678

Copy: Bill Loader (Chief Engineer)
CRIP Project file

Business letters

A letter is the traditional means of formal business communication; however, it is a comparatively slow means of communication, particularly when sent by post. Business letters are normally written in a formal style, with a standard format, and are used to convey information to a person or another organisation.

A sample business letter showing the standard and optional components is presented in figure 14.3.

FIGURE 14.3 Sample business letter (optional components are identified by dotted lines)

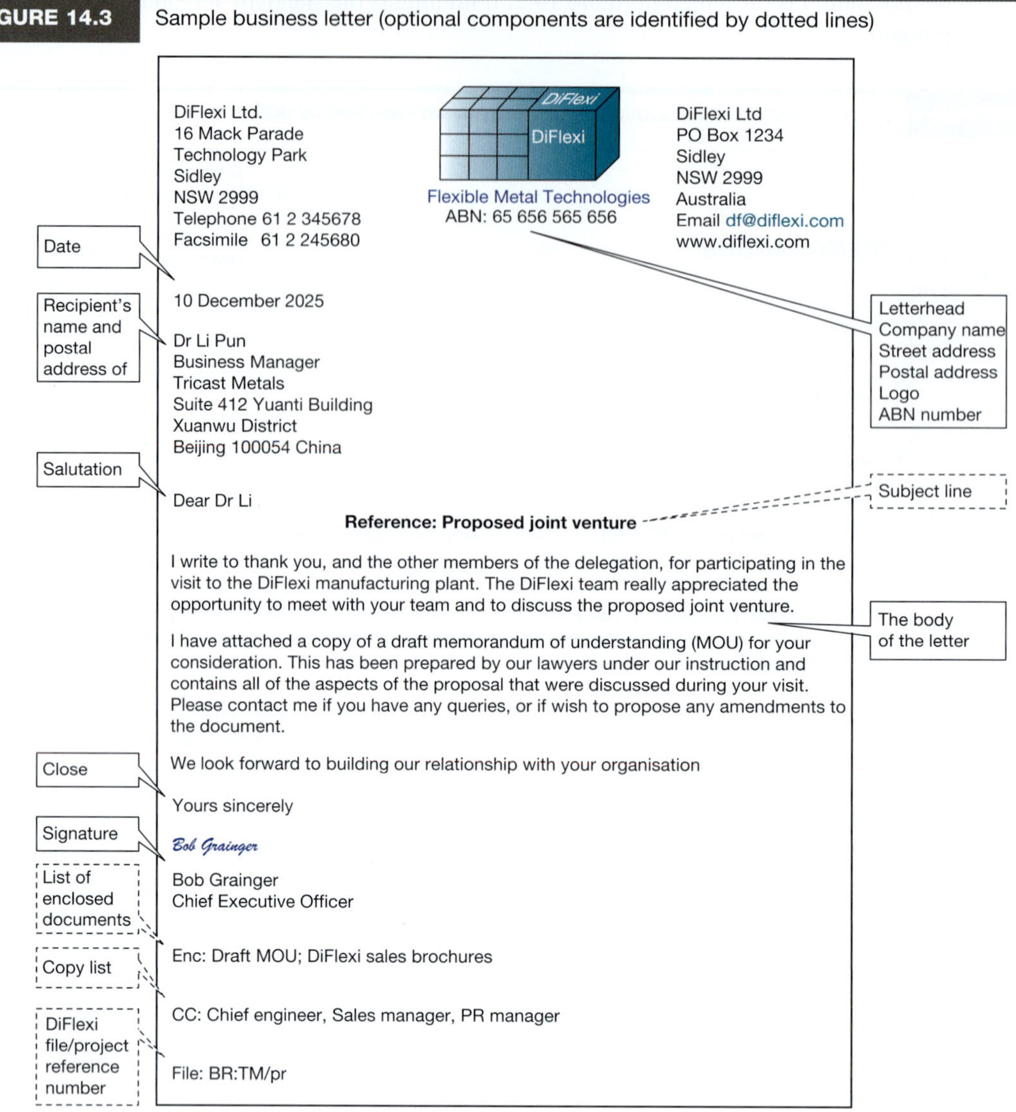

DiFlexi Ltd.
16 Mack Parade
Technology Park
Sidley
NSW 2999
Telephone 61 2 345678
Facsimile 61 2 245680

DiFlexi
Flexible Metal Technologies
ABN: 65 656 565 656

DiFlexi Ltd
PO Box 1234
Sidley
NSW 2999
Australia
Email df@diflexi.com
www.diflexi.com

10 December 2025

Dr Li Pun
Business Manager
Tricast Metals
Suite 412 Yuanti Building
Xuanwu District
Beijing 100054 China

Dear Dr Li

Reference: Proposed joint venture

I write to thank you, and the other members of the delegation, for participating in the visit to the DiFlexi manufacturing plant. The DiFlexi team really appreciated the opportunity to meet with your team and to discuss the proposed joint venture.

I have attached a copy of a draft memorandum of understanding (MOU) for your consideration. This has been prepared by our lawyers under our instruction and contains all of the aspects of the proposal that were discussed during your visit. Please contact me if you have any queries, or if wish to propose any amendments to the document.

We look forward to building our relationship with your organisation

Yours sincerely

Bob Graínger

Bob Grainger
Chief Executive Officer

Enc: Draft MOU; DiFlexi sales brochures

CC: Chief engineer, Sales manager, PR manager

File: BR:TM/pr

Human resource documents

There are many documents that support the management of people (human resources, or HR) in an organisation. Larger organisations normally develop templates for the documents that support HR activities, such as:

- position descriptions
- job advertisements
- job applications
- letters of appointment
- performance reviews
- terminations.

While some sections of the templates may be modified to suit specific cases, they retain core elements, formats and styles. Generally, these are factual documents that have been carefully crafted to ensure they accurately reflect the intended purpose. They will often be approved by a company's legal team prior to deployment, as they will be a key part of the evidence used to resolve any disputes. Two human resource documents are described in the following sections to illustrate the types of information that are normally included in these documents.

Position descriptions

A job or position description is normally developed for every job in an organisation. It provides detailed information about key aspects of the job.

- *Position identification information.* This can include position title, classification under the relevant industrial award, salary scales, job status (full time, part time or casual) and the department or section of the organisation.
- *General information for a position of this type and at this level.* This includes information on matters such as scope, authority, responsibility, accountability, staff development and workplace health and safety.
- *Specific duties and responsibilities.* This section lists the key duties and responsibilities for the specific position being described, such as job objectives, scope, specific duties, responsibilities, financial and other accountabilities, positions that report to this position and the position(s) that this position reports to. An 'Other duties as required ...' clause is often included in the list to provide managers with some flexibility.
- *Knowledge and skill requirements.* This section can include qualifications, licences, capabilities and experience. These will be used to assess applications for the position.

The latest version of the Australian and New Zealand Standard Classification of Occupations (ANZSCO) (Australian Bureau of Statistics 2022), is a good starting point for information when developing a new position description. It classifies and provides general information for all occupations and jobs in Australia and New Zealand — information that is used for statistical and other types of analysis. A skill and attributes-based hierarchical system is used to classify the occupations, beginning with broad groups and cascading down to individual occupations. The entry for electronics engineers is reproduced in figure 14.4.

FIGURE 14.4 ANZSCO listing for the position of Electronics Engineer

2334 Electronics Engineers

ELECTRONICS ENGINEERS design, develop, adapt, install, test and maintain electronic components, circuits and systems used for computer systems, communication systems, entertainment, transport and other industrial applications.

Indicative Skill Level:

In Australia and New Zealand:

- Most occupations in this unit group have a level of skill commensurate with a bachelor degree or higher qualification. In some instances relevant experience and/or on-the-job training may be required in addition to the formal qualification (ANZSCO Skill Level 1).
- Registration or licensing may be required.

Tasks Include:

- designing electronic components, circuits and systems used for computer, communication and control systems, command and warfare systems, and other industrial applications
- designing software, especially embedded software, to be used within such systems
- developing apparatus and procedures to test electronic components, circuits and systems
- supervising installation and commissioning of computer, communication and control systems, and ensuring proper control and protection methods
- establishing and monitoring performance and safety standards and procedures for operation, modification, maintenance and repair of such systems
- designing communications bearers based on wired, optical fibre and wireless communication media
- analysing communications traffic and level of service, and determining the type of installation, location, layout and transmission medium for communication systems
- designing and developing signal processing algorithms and implementing these through appropriate choice of hardware and software
- designing the architecture, modelling and integration of communication and control systems, command and warfare systems, and other industrial applications.

Occupation:

233411 Electronics Engineer

233411 Electronics Engineer

Designs, develops, adapts, installs, tests and maintains electronic components, circuits and systems used for computer, command and warfare systems, communication systems, entertainment, transport and other industrial applications. Registration or licensing may be required.

Skill Level: 1

Specialisation:

- Combat System Engineer

Source: Australian Bureau of Statistics (2022).

In the figure, the classification number (233411) defines the place in the hierarchy for the occupation Electronics Engineer. This occupation is in:

- Major Group: 2 Professionals
- Sub-Major Group: 23 Design, Engineering, Science and Transport Professionals
- Minor Group: 233: Engineering Professionals
- Unit Group: 2334: Electronics Engineers
- Occupation: 233411: Electronics Engineer; Specialisation: Combat System Engineer.

The other Engineering Professionals in the Minor Group 233 are: 2331 (Chemical and Materials Engineers); 2332 (Civil Engineering Professionals); 2333 (Electrical Engineers); 2335 (Industrial, Mechanical and Production Engineers); 2336 (Mining Engineers); and 2339 (Other Engineering Professionals). The 'Other Engineering Professionals' category is used for the engineering disciplines that have not been allocated a code — usually because they are relatively new disciplines, for example, Environmental Engineering, Biomedical Engineering and Mechatronic Engineering.

The other members of the engineering team are classified under Major Group 3 — Technicians and Trade Workers.

- The occupations of Electronic Engineering Draftperson and Electronic Engineering Technician are described in the position description for Unit Group 3124 — Electronic Engineering Draftpersons and Technicians.
- The occupations of Business Machine Mechanic, Communications Operator, Electronic Equipment Trades Worker, Electronic Instrument Trades Worker (General) and Electronic Instrument Trades Worker (Special Class) are described in the position description for Unit Group 3423 — Electronic Engineering Draftpersons and Technicians.

These documents can be adapted to list the specific duties and other details for a proposed electronics engineering, technician or trade position.

Letter of appointment

A letter of appointment would normally include:

- a greeting and congratulatory statement
- position identification and location
- commencement date
- hours of work
- commencing salary
- position description
- information about other entitlements, such as leave (annual, long-service, maternity, personal and sick), superannuation, salary sacrificing and personal use of a company car
- termination notification requirements for both parties.

Letters of appointment should be filed for safe-keeping, as they represent the contract between the employee and the employer, and would be important evidence if a dispute arises.

Financial documents

The business sector in most countries is highly regulated to protect the interests of the various stakeholders, and to build investor confidence in the marketplace. The sources of regulation in Australia include the Corporations Act, the Australian Competition and Consumer Commission (ACCC) and the Australian Taxation Office (ATO). Accounting standards are defined by the Australian Accounting Standards Board, in consultation with the relevant Australian professional bodies and similar agencies in other countries. They are implemented by accountants who, in turn, are regulated by the code of conduct of the professional organisation they belong to, such as CPA Australia.

Many engineers deal with financial documents on a daily basis, particularly those in management, project management or procurement roles. For example, they may prepare, receive or review:

- balance sheets
- bank statements
- budgets
- income statements
- monthly activity statements
- receipts
- tax invoices

- tender documents
- quotations.

The content of these documents is normally stipulated by regulation, although the format and style may vary. While there are many standard templates available on the internet or through office supply companies, it is advisable to seek the advice of the company accountant to ensure that all of the required information is included in a document. The tax invoice in figure 14.5, is used to illustrate this point, as it is one of the most used financial documents, both in business and in everyday life.

FIGURE 14.5 Sample tax invoice for services supplied

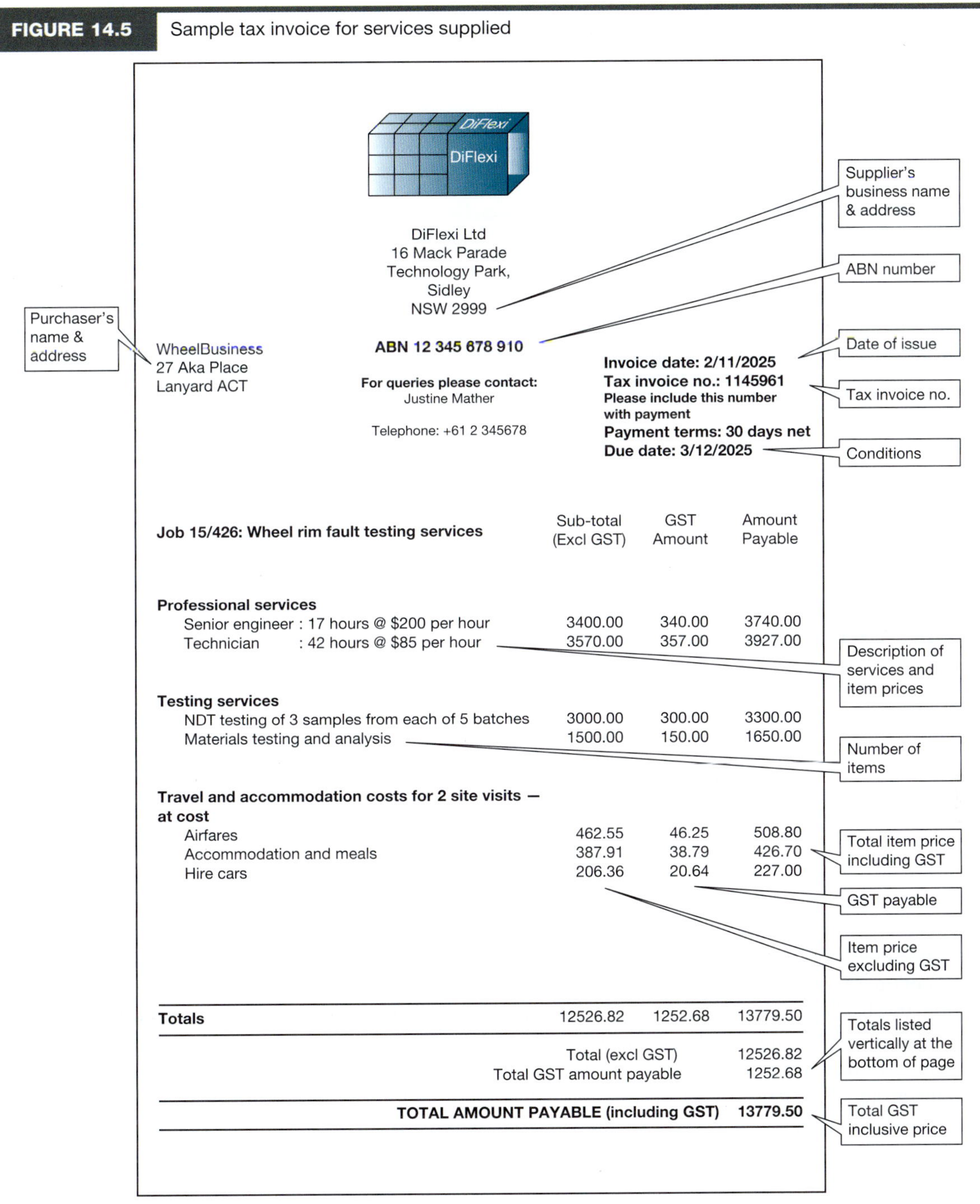

DiFlexi

DiFlexi Ltd
16 Mack Parade
Technology Park,
Sidley
NSW 2999

ABN 12 345 678 910

WheelBusiness
27 Aka Place
Lanyard ACT

For queries please contact:
Justine Mather

Telephone: +61 2 345678

Invoice date: 2/11/2025
Tax invoice no.: 1145961
Please include this number with payment
Payment terms: 30 days net
Due date: 3/12/2025

Job 15/426: Wheel rim fault testing services	Sub-total (Excl GST)	GST Amount	Amount Payable
Professional services			
Senior engineer : 17 hours @ $200 per hour	3400.00	340.00	3740.00
Technician : 42 hours @ $85 per hour	3570.00	357.00	3927.00
Testing services			
NDT testing of 3 samples from each of 5 batches	3000.00	300.00	3300.00
Materials testing and analysis	1500.00	150.00	1650.00
Travel and accommodation costs for 2 site visits — at cost			
Airfares	462.55	46.25	508.80
Accommodation and meals	387.91	38.79	426.70
Hire cars	206.36	20.64	227.00
Totals	12526.82	1252.68	13779.50

Total (excl GST)	12526.82
Total GST amount payable	1252.68
TOTAL AMOUNT PAYABLE (including GST)	**13779.50**

Both the ATO and New Zealand Inland Revenue have published guidelines on their websites that define the required content and layout of tax invoices and invoices. In New Zealand, the information that must be shown on a tax invoice depends on the value of the goods or services supplied. A tax invoice is used when the price of some or all of the goods or services include GST. An invoice is used when the goods or services do not include GST. Table 14.2 shows the information requirements for Australian and New Zealand tax invoices for an amount of more than $1000.

TABLE 14.2 **Information required on Australian and New Zealand tax invoices**

Information	Australia	New Zealand
Currency	Australian dollars	New Zealand dollars
Supplier's business name	The seller's identity	Name or trade name
Supplier's business number	The seller's Australian Business Number (ABN)	GST Number
The words 'Tax invoice'	Stated prominently with invoice number	In a prominent place
Date	The date the tax invoice was issued	The date of issue
Purchaser's business name	Buyer's name or ABN	Name and address of recipient
Purchaser's business number	Buyer's name or ABN	
Goods or services	A unit description	A description
Quantity	Quantities	Quantity or volume
Number of items	Optional	
Item price	The price for each item	Price excluding tax
GST	GST for each item, if any	The GST payable
Total item price	GST inclusive price for each item	
Total amount payable	The total GST exclusive price, total GST and total GST inclusive price vertically in the bottom right hand corner	Total amount payable including GST

Source: Adapted from Australian Taxation Office (2023); New Zealand Inland Revenue (2024).

Project initiation documents

There are many ways in which engineers are engaged to undertake a design, project or other task, such as the following examples.

- A prospective client walks into a consultant's office and asks to see an engineer about a specific problem, such as a request for a foundation design for a house. Once the engineering details and fee details have been agreed the client may instruct the engineer to proceed with the work.
- The manager of the sales section of a large manufacturing company requests the information technology (IT) section of the company to develop new sales processing software. In this case the sales manager is the client.
- An existing client engages a consultant to undertake a new project. Consultants would normally have regular clients, with both parties benefiting from their business relationship.
- A large engineering company wins a tender to manage a multimillion-dollar project and the signing of the contract formalises their engagement.

From this list it can be seen that an engineering project normally begins with an engineer (or engineering organisation) responding to a client request (such as a brief or tender). In some fields, the project initiation process is well defined; for example, in the construction and procurement industries. In 2010, a joint working group of the Australian Construction Industry Forum Ltd (ACIF) and the Australian Procurement and Construction Council Inc. (APCC) prepared and published *A Guide to Project Initiation for Project Sponsors, Clients and Owners*. The guide describes the four-step project initiation process as:

1. project ideas sources
2. concept development
3. evaluation
4. definition.

The guide suggests that 'the use of rigorous information analysis at the beginning of a project has the greatest potential to significantly improve the whole project' (ACIF & APCC 2010, p. 1).

Once the details of the project have been defined, a contract is developed and signed by all the relevant parties. Once the project has begun it is likely that other parties will be engaged to undertake defined parts of the project. In this case the engineer (or engineering organisation) may become the client, and prepare the brief, instruction or tender on behalf of the principal client.

To be able to undertake these roles, engineers need to have a good understanding of the business, financial and legal aspects of the various documents used to initiate a project.

A client brief

When a client briefs an engineer about work that is to be undertaken, it is important that the details are carefully reviewed and refined so the final written brief contains all of the information required to define the project, and its outcomes, clearly and accurately. If the brief results from a face-to-face meeting with the client, the brief development process will involve reading, questioning, listening, negotiation and note-taking skills. Many organisations use a checklist to guide this process to ensure that nothing is overlooked. The checklist contains all of the items that are normally set out in a brief. Once the brief is drawn up and finalised, it is checked and signed by all parties.

Depending on the size and type of project, the brief may be integrated into a contract that would be signed by all parties before the commencement of the project.

A design brief

This may be a subset of a client brief. A design brief normally contains much more technical information than a client brief. It contains all of the technical specifications and requirements for a design task. Once again, it should be carefully reviewed to ensure that it includes all of the relevant information that the designer(s) will require to complete the task and meet the required specifications and intended use.

A tender

A tender process is a competitive process that begins when an organisation (Organisation A) invites individuals or organisations to make a formal offer (a tender) to supply goods or undertake works in line with stated terms and conditions. The invitation normally includes a closing date for the submission of tenders. The individuals or organisations that submitted a tender are known as *tenderers* as they *tendered* for the work. After the closing date, Organisation A reviews the submitted tenders and selects one, often the lowest price, and advises the tenderers of the outcome. The winning organisation then supplies the goods or completes the works for the agreed price and under the agreed terms and conditions.

The tender process begins with the preparation of the tender documents, which normally specify the requirements in relation to the:

- terms and conditions to ensure tenderers adhere to the tendering procedures and processes
- contract
- technical documents (detailing the project requirements, specifications and scope of the work that is to be undertaken).

Depending on the size and scope of the project, the tender is then advertised in the local, national, or international media. The advertisement is a *Request to/for Tender* and it invites interested organisations to obtain a copy of the tender documents and to submit a response. The tender process may be managed by the client, or by a third party — known as the principal — who acts on behalf of the client. In many cases prospective tenderers must attend a mandatory briefing and site inspection if their proposal or bid is to be considered in the tender process.

In some cases, the tender process may have two stages, with the first stage being an invitation for interested parties to submit an expression of interest (EOI), or a registration of interest (ROI). The following example illustrates a typical ROI and tender processes.

In an advertisement in The Weekend Australian (8 September 2018), the WA Water Corporation invited an ROI from 'organisations with suitable experience, capability and capacity ... for the delivery of a side-stream demmonification facility at the Beenyup Wastewater Treatment Plant ... to improve water quality for the Advanced Water Recycling Plant'.

The relevant ROI documents were made available from 10 September 2018, via the Suppliers Portal, to companies registered as Water Corporation suppliers and bidders. Organisations seeking registration as a supplier or bidder could apply online, allowing up to five working days to process the application.

A mandatory requirement was that all applicants be granted Water Corporation Health Safety and Environment (HSE) Prequalification, at the level nominated by the ROI, prior to the contract being awarded. This application could be lodged via the Water Corporation website.

ROIs were be submitted in writing by no later than Tuesday 9 October 2018.

Many government and other agencies now use online tendering portals to advertise and accept tenders. For example, the eTendering site of Transport for NSW (https://www.tenders.nsw.gov.au) lists contracts that are proposed, current and closed.

A communication plan is essential when developing large or complex tender documentation. For example, table 14.3 illustrates the size and complexity of the documentation prepared for a tender proposal for a large engineering project.

TABLE 14.3 The tender proposal

Document	Word count	Page count
Executive summary	1 661 words	8 pages
Volume 1: Commercial proposal	5 496 words	57 pages
Volume 2: Project management proposal	59 761 words	248 pages
Volume 3: Technical proposal	58 615 words	305 pages
Volume 4: Tender support technical data	31 153 words	164 pages
Totals	156 686 words	782 pages

Source: Sales (2006, pp. 154–155).

People from a number of disciplines were involved in preparing the documents listed in the table, with the engineers mainly concerned with *Volume 3: Technical proposal*. In tender proposals like this, the documents may include drawings, specifications, photographs, statistics, graphs and charts, as well as other media. The purpose of the visual components would be to clarify and support the text and help persuade the consumer of the technical capability and commercial viability of the proposal.

Often people from many organisations are involved in large proposals like this. Different people may write and prepare parts of the document, and some may never see the full proposal once it is assembled. For large documents, specialist authors (possibly with an engineering background) may design the structure of the documents and produce and publish the final package.

Once the overall communication plan has been defined, a communication strategy for each component can be planned in detail. The strategy should be designed so that the required information is communicated in the most timely and effective manner. The following spotlight reports on a hypothetical explanation of how an engineer might do this for a wind power project.

SPOTLIGHT

Planning a proposal

Cyclone Consulting is preparing a bid for a tender to design and construct a small wind farm to produce power for the Northern Regional Council (NRC). The chief engineer (CE) made the following notes while planning the bid document.

Step 1. Defining the purpose

To persuade the Northern Regional Council that Cyclone Consulting has the technical knowledge, experience and capability to design and construct the wind farm and associated infrastructure in accordance with the specifications and budget.

Step 2. Client

Northern Regional Council owns and operates an old coal-fired power station and a solar farm. This will be their first wind farm. They worked with another company, Energy Options, to prepare the specifications for the project. Cyclone Consulting has built a good relationship with NRC while working on projects over the last ten years.

Step 3. Requirements

The requirements for the proposal documentation are set out in detail in the tender specifications. The CE decided the core pieces of information to be communicated were:

- Cyclone's technical expertise, including the CVs of key staff
- staff recruitment and training strategies

- Cyclone's project management expertise and experience
- a realistic construction plan including workplace health and safety
- management expertise
- a detailed budget
- risk-management strategies.

Step 4. Contexts

The CE noted that they had worked in similar contexts when they developed a solar farm for NRC at a nearby location. However, they will need to enhance their expertise in wind power generation, which will require some research and staff training.

Step 5. Draft messages

The CE decided to use the documents prepared for a previously successful tender as a template.

Step 6. Communication methods

Written documents and a five-minute video about Cyclone Consulting, using an earlier successful project to highlight the company's capabilities.

Step 7. Assembly

The size of the proposal documentation will require the workload to be allocated to a number of staff members and also to some consultants. The CE prepared a schedule showing the authors, timelines and other details for each part of the document. This included a note that the technical team will need outside assistance from Wind Power Pty Ltd and that Imagineering Media will prepare the video and publish the package.

There are eight weeks to complete the documents, leaving two weeks to proof, check, publish and submit. The CEO reviewed the schedule and developed strategies to ensure that each deadline was met.

Part	Title	Author	Approach	Channel	Due
1	Executive summary	Chief engineer	Persuasive	Text and graphics	21 May
2	Technical proposal	Design team and Wind Power	Factual	Text and graphics	21 Apr
3	Project management	Project management team	Persuasive	Text and charts	28 Apr
4	Commercial	Accountant and lawyer	Factual	Text and charts	28 Apr
5	Company profile	Imagineering Media	Persuasive	Video/DVD	21 Apr

Step 8. Gatekeeper

The CE decided who would do the final check of each part and the final package.

Part	Gatekeeper	Completed
1	Lawyer	15 May
2	Chief engineer, lawyer	15 May
3	Lawyer	12 May
4	Company accountant	15 May
5	Chief engineer, lawyer, design team	28 April
Package	Chief engineer, lawyer	17 May

Step 9. Delivery

The CE made the following notes.

- The publisher will take three days to print and collate the documents. The final proofs to be delivered to Imagineering Media on 18 May. Six copies of the documentation are required.
- The tender is due at the head office of Northern Regional Council before 5 pm on 26 May.

Step 10. Response

The CE made the following notes.

- Ensure a receipt is obtained after lodgement of the documents.
- Phone a contact in Council one week after lodgement to check everything is in order.
- Find out when the successful tender will be announced.

CRITICAL THINKING

Use the ten-step PCR process to plan an assignment report you will be preparing this semester.

The design of large documents, or sets of documents, can be a major task because it is important that each component, or section, is designed so that when they are combined, they communicate a consistent message and achieve the purpose of the communication. A good understanding of the communication process helps writers to develop documents that are focused on achieving the purpose of the communication, that is, to communicate information and elicit the desired response from the consumer.

Engineers may participate in both sides of this process but not, of course, in the same tender. They could act as the principal and be responsible for preparing the tender documentation and managing the tender process on behalf of the client, or they could be responsible for preparing a proposal in response to an invitation to tender advertised by another party.

A contract

A **contract** represents a binding agreement, between two or more parties, requiring the parties to perform certain obligations. Generally, a contract will set out the requirements and obligations to be complied with or undertaken by the respective parties and reflects the risk allocation between the parties. Often contracts are the result of a tender process and/or negotiations and may include the tender documents, or part of such documentation. A contract is enforceable and recognised by law.

Typically a contract sets out all of the details and requirements relating to a project. This may include:

- specifications and drawings
- standards and quality requirements
- outcomes or performance criteria
- general terms and conditions.

Preparing a contract requires considerable skill and expertise. When an engineer is responsible for preparing a contract, they may use a standard form contract, for example, a standard form contract produced by Standards Australia (2024) or the International Federation of Consulting Engineers (FIDIC) (2024), which are internationally recognised.

Engineers may engage the services of specialist contract, commercial or project lawyers to prepare or review the contract. Alternatively, a client may have engaged lawyers to provide legal services, and require the engineer (as the client's representative) to confer with the lawyers regarding the contract. Larger organisations may have an in-house counsel or a legal support group to undertake this role. Obtaining legal advice before a contract is signed ensures the parties fully understand the risks and obligations under the contract. Once the contract is signed, the parties will be bound by the terms and conditions. If a party is unable to fulfil or undertake its obligations, that party may face serious commercial and legal ramifications.

SPOTLIGHT

When things go wrong: a legal perspective

Ilsa Kuiper is a qualified lawyer and civil engineer. Before embarking on a career in construction law she practised as a civil engineer in the water resources, environmental and mining engineering industries. Here she gives a legal perspective on contracts.

ILSA KUIPER, LAWYER AND CIVIL ENGINEER

In its simplest form, the contract reflects the intention of the parties to undertake certain tasks or services in consideration of their relative value (generally for the payment of money). If a dispute arises during the course of an engineering project, and cannot be resolved by the parties, it is the contract and the law that can provide the process for the resolution of the dispute.

Disputes can arise out of a wide range of issues in connection with a project. The following scenarios are some simple examples of how disputes can arise, particularly in relation to time, cost and quality.

Example A. Time

Under a contract, Party A and Party B agree for the project to be completed by a certain date. As a result of changes to the scope, Party B claims for an extension of time in accordance with the requirements of the contract.

Party A rejects the claim on the basis that Party B has not demonstrated that the changes to the scope impacted the critical path program and hence is not entitled to claim an extension of time.

If the completion date is not extended, Party B may be subject to paying *liquidated damages* to Party A for failing to complete the works by the completion date. In this case, liquidated damages are payments for the losses Party A would incur as a result of the completion date not being achieved by Party B.

Example B. Cost

Under a contract, Party A and Party B agree for the project to be undertaken for an amount: $X. During the course of the project, Party B claims for amount $Y (in addition to $X), due to variations based upon changes to the scope directed by Party A and the subsequent delays.

In response, Party A rejects the claim on the basis that the work was within the scope under the contract and that Party B caused the delays. Party A's position is that Party B is therefore not entitled to the additional amount of $Y.

Example C. Quality

Under a contract, Party A and Party B agree to construct an element of the project in compliance with a certain standard. During the defects liability period, Party A discovers a defect in the element and notifies Party B of the defect, and requests rectification of the defect in accordance with the contract provisions.

Party B responds by rejecting Party A's assertion that the element contains a defect because Party B believes the defect was caused by Party A's substandard maintenance program, which was not in accordance with the contract requirements.

If parties are unable to resolve their differences, then legal advice should be sought. Although well drafted contracts will generally include clauses about how the parties are to approach disputes, legal guidance should be sought early in the process. This is to ensure a party's rights are maintained (and not waived) and any potential exposure to legal and commercial risks are reduced or avoided.

CRITICAL THINKING

Download a short-form contract from the Engineering New Zealand website (https://www.engineeringnz.org/engineer-tools/engineering-documents/contracts) or an AUS-SPEC Contract from the NATSPEC website (Students and Lecturers, https://natspec.com.au/training-support/students). Review the document and then list the key pieces of information required.

It is clear from the examples listed in the spotlight that although a contract should clearly set out the obligations of the parties involved, disputes may arise that can result in legal action. To minimise risk, many companies use legal advisors to review all of their important documents before they are sent to the relevant parties. In this sense, the legal advisers are acting as gatekeepers.

Consider the case of the Wembley Stadium. Among the many firms that were involved in this massive project were two Australian organisations — Connell Wagner (now called Aurecon) who received an Australian Engineering Excellence Award from Engineers Australia for the structural design of the arch and the roof of the stadium, and construction firm Multiplex. As outlined in the following spotlight, Multiplex subsequently took legal action against the UK engineering consultants involved in the project, with both parties claiming the other was legally responsible for the problems that occurred during the construction phase of the project.

SPOTLIGHT

Legal games at Wembley Stadium

Imagine beginning the construction of a complex £326.5 million project (Quinn 2008). That was the exciting but formidable prospect facing the Australian company Multiplex in 2000 when it tendered for, and won, the contract to build the new Wembley Stadium in the UK (Quinn 2008). The original Wembley Stadium on the same site had closed in that same year. It had been host to the 1948 Olympics, the 1985 Live Aid Concert and hundreds of FA Cup football matches.

The new venue was designed to host high-profile soccer matches, pop concerts and other events. The plan included:

- a 133-metre high, 315-metre wide roof structure, now known as the Wembley Arch
- undercover seating for 90 000 people
- 8 restaurants and 34 bars
- the capacity to serve 40 000 glasses of beer and even more soft drinks during half-time
- 2600 toilets (Mott MacDonald Group Limited 2008).

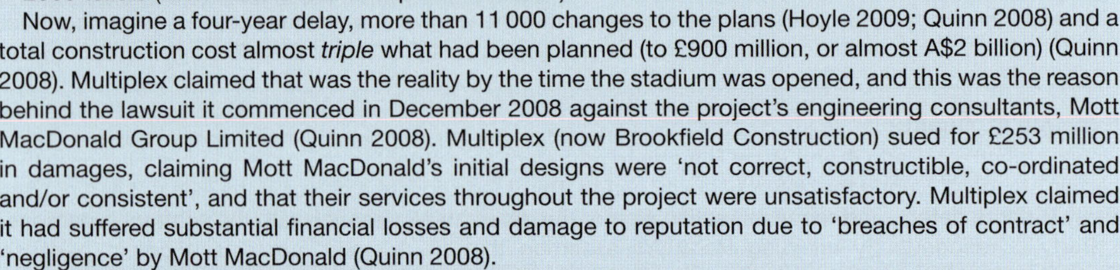

Now, imagine a four-year delay, more than 11 000 changes to the plans (Hoyle 2009; Quinn 2008) and a total construction cost almost *triple* what had been planned (to £900 million, or almost A$2 billion) (Quinn 2008). Multiplex claimed that was the reality by the time the stadium was opened, and this was the reason behind the lawsuit it commenced in December 2008 against the project's engineering consultants, Mott MacDonald Group Limited (Quinn 2008). Multiplex (now Brookfield Construction) sued for £253 million in damages, claiming Mott MacDonald's initial designs were 'not correct, constructible, co-ordinated and/or consistent', and that their services throughout the project were unsatisfactory. Multiplex claimed it had suffered substantial financial losses and damage to reputation due to 'breaches of contract' and 'negligence' by Mott MacDonald (Quinn 2008).

In early July 2009, Mott MacDonald filed its defence papers in preparation for the hearing in the High Court later that month. In April 2010, Justice Coulson advised the parties to seek mediation or face legal costs of more than £74 million (Lynch 2010). Then in June 2010, Lynch (2010) and Donati (2010) reported that the two firms had settled the case and issued a joint statement that said:

> We can confirm that Brookfield Construction and Mott MacDonald as leader of the Mott Stadium Consortium consisting of themselves, Sinclair Knight Merz (Europe) and Aurecon Australia Pty have settled the legal proceedings between Brookfield and Mott over the design and construction of Wembley Stadium.

CRITICAL THINKING

What impacts do you think this legal battle had on (a) the companies involved in the case and (b) the engineers working for those companies?

Risk management documents

A risk management process can be used to identify, assess and manage risk in the workplace. While often associated with work health and safety (WHS) practices — previously called occupational health and safety (OHS) — the risk management process may also be used to assess other aspects of a project such as financial risk and environmental risk. A key component of any risk management process is record keeping, as the relevant documents can be used to help demonstrate compliance. Examples of these documents are discussed in this section.

Under an Intergovernmental Agreement for Regulatory and Operational Reform in Occupational Health and Safety, signed on 3 July 2008 by the Council of Australian Governments (COAG), all states and territories were to adopt the same WHS legislation on 1 January 2012. However, at that time only the Australian Capital Territory, New South Wales, Queensland, the Northern Territory and the Commonwealth had harmonised their WHS laws. This means that businesses, organisations and workers, including volunteers, in these jurisdictions are protected by the same WHS laws. Other states were expected to harmonise their laws and regulations in the following years.

The *model WHS Act* was updated on 24 November 2023 and now includes all amendments made since 2011. For the *model WHS Act* to have effect in a jurisdiction it must be implemented in that jurisdiction. Amendments to the *model WHS Act* do not automatically apply in a jurisdiction.

By March 2024, the model WHS laws had been implemented in all jurisdictions except Victoria. Some jurisdictions have made variations in their respective WHS laws compared with the model WHS laws (Safe Work Australia 2024).

Safe Work Australia develops model Codes of Practice on behalf of the governments that are part of the agreement. Each code is a practical guide on how to achieve the standards defined in the laws and the relevant regulations.

A project manager is normally responsible for implementing the WHS risk management system for a project. As the project management role is normally undertaken by an engineer, it is likely that an engineer will personally undertake this work, or delegate it to another engineer working on the project. On larger projects this may be delegated to a WHS specialist.

When WHS standards and codes are implemented the risk management process is applied to each task being undertaken on a project and the outcomes are documented and archived. For example, there are four steps in Safe Work Australia's risk management process (Safe Work Australia 2018, p. 7).

1. *Identify hazards.* The workplace should be carefully reviewed to identify any hazards that may cause harm in the physical environment, in the tasks being undertaken and the equipment and materials being used.
2. *Assess risks.* The hazards should be carefully assessed to answer the following questions: How may harm occur? What is the nature of the harm that may be caused? How serious could the harm be? What is the likelihood of the harm occurring?
3. *Control risks.* To determine how to control the risks, the following questions need to be asked: What controls can be put in place to mitigate or eliminate the risk? Can the hazard be substituted with something safer? Can people be isolated from the hazard? What would be the most effective control measures that are reasonably practicable in the circumstances? How can these be implemented? What personal protection equipment (PPE) should be worn? What training and supervision are required?
4. *Review hazards and control measures.* Some of the questions to be asked are: Have the hazards changed? Are the control measures working as planned? Are they effective? Can they be improved?

While there are a number of approaches that can be used to assess risk, all involve consultation with the relevant workers. There are also a number of tools that have been developed to help people categorise the risk in a systematic manner. Many of these tools are based on the use of a 2D matrix like that shown in table 14.4.

TABLE 14.4 Using a matrix to assess risk categories

	Likelihood of occurrence			
Severity of harm	**Highly likely**	**Likely**	**Unlikely**	**Highly unlikely**
Life-threatening	High	High	High	Medium–high
Permanent injury	High	High	Medium–high	Medium
Temporary injury	Medium–high	Medium–high	Medium	Low
Discomfort	Medium–high	Medium	Low	Low
Nuisance only	Medium	Medium	Low	Low

Source: CCH (2007, p. 125).

A number of questions can be used to assess the likelihood of harm occurring.

- How often is the task undertaken?
- How many people may be affected?
- How often and for how long might people be exposed to the hazard?
- Has this hazard caused harm before in this or another workplace?
- How effective are the existing controls at reducing the risk?
- Could the way people behave in the workplace, or the differences between people in the workplace, alter the risk?

By assessing the likelihood of an incident occurring, the number of people that may be harmed and the severity of the harm that may result from such an incident, an assessor is able to assess the category of risk associated with a hazard. In table 14.4, four categories were used: high, medium–high, medium and low. Some systems suggest five categories should be used (certain to occur, very likely, possible, unlikely and rare), while others use a numerical scale to rate risk from low to high; for example, risk is rated on a 5-point scale, with 1 for low risk and 5 for high risk.

The outcomes for each risk assessment are documented and issued to the people undertaking the specified tasks. Where appropriate these people also receive training to ensure that they operate in accordance with the defined control measures. A risk assessment document will normally have a tabular format like that shown in table 14.5.

TABLE 14.5 **An example of a WHS risk assessment document**

DiFlexi Metal Technologies WHS risk management			
Project: Installation of new equipment		**Assessors:** JKD	**Date:** 27/11/2025
Activity	**Hazards**	**Risk category**	**Risk control measures**
Unload equipment	Crush injuries from crane dropping load	Low, as crane and lifting equipment are new	Staff to wear hardhats and steel capped shoes Staff to stand clear during unloading operation
Installation on hold-down bolts on pad	1 Crush injuries 2 Back and muscle injuries from nudging load onto hold-down bolts	1 Low, as crane and lifting equipment are new 2 Medium, due to weight of equipment and confined operating space	1 Staff to wear hardhats and steel capped shoes 2 Staff to be trained to use proper lifting and pushing procedures

Once the risk has been assessed, the appropriate level of control measures must be decided by working through a hierarchy of control measures, such as that shown in shown in figure 14.6.

FIGURE 14.6 The hierarchy of control measures

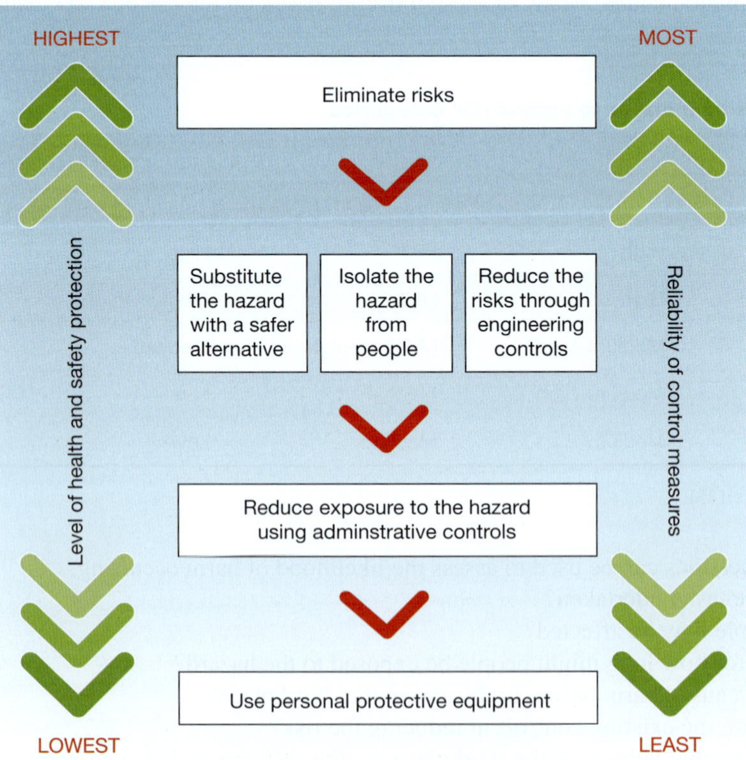

Source: Safe Work Australia (2018, p. 19).

The most effective control is to eliminate the risk; this can only be achieved by eliminating the hazard, which should be the ultimate aim. However, if this is not reasonably practicable, then the risk must be minimised by working through the other alternatives in the hierarchy. The lower levels in the hierarchy are less effective because they can only minimise the risk.

Specific control measures are then developed and implemented and the relevant documentation is filed. The outcome of this process may be included in a risk register, as shown in table 14.6.

TABLE 14.6 **An example of a risk register**

Location: Installation of new lathe onto hold down bolts in machine shop						**Date: 27/11/2025**		
						How will the controls be implemented?		
Hazard	**What is the harm that the hazard could cause?**	**What is the likelihood that the harm will occur?**	**What is the level of risk?**	**How effective are the current controls?**	**What further controls are required?**	**Action by**	**Due date**	**Date completed**
Lowering lathe onto someone's hand or foot	Crush injuries	Unlikely	Low as crane and lifting equipment are new and operators are skilled	—	Staff must wear hardhats and steel-capped shoes	FB	29/11/25	29/11/25
Nudging lathe onto hold-down bolts	Back and muscle injuries	Possible	Medium, due to weight of equipment and confined operating space	—	Staff to be trained to use proper lifting and pushing procedures	FB	29/11/25	29/11/25

An instruction

During the life of a project an engineer may issue written instructions to construction staff, contractors or other project staff. An engineer's *instruction* is a direction for the specified work to proceed. The work specified in an instruction may be part of an existing contract, or a variation of the contract, (i.e. a new task that was not identified when the contract was prepared). An instruction will normally include details of the engineering organisation that issued the instruction, the project name and reference, the date, the contractor's name and contact details, the task specifications, and commencement and completion information. Copies of all of the instructions issued on a project are normally filed in the project records.

The following spotlight highlights the critical importance of carefully documenting all of the agreed changes that occur during the life of a project, clearly communicating those changes to all of the relevant stakeholders (including through written instructions) and ensuring that all of the stakeholders have received, understood and, where necessary, acted on those instructions.

SPOTLIGHT

Death by lack of communication and lack of design

TIM McCARTHY, UNIVERSITY OF WOLLONGONG

In 1995 I met two men with albatrosses* hanging from their necks, two men whose mission was to tell the engineering world what they had been through. Jack Gillum was the older of the two. He had been the principal and owner of a structural engineering consultancy, Jack D. Gillum and Associates in St Louis, Missouri. In 1976 Gillum had hired the younger man, Greg Luth, fresh out of university as one of his design engineers. The pair looked like they might have been father and son and their joint story is a timely reminder to all engineers, young and old.

Gillum told the audience of university students in that 1995 seminar how it felt 15 years earlier when he returned home on a Friday evening to a ringing telephone. The caller told him that there had been a ▶

collapse at the Hyatt Regency Hotel in Kansas City, a construction project that Gillum had completed less than a year before. Gillum told the 1995 gathering that he had felt 'shattered to the core'. There were fatalities and many injuries.

Greg Luth had been the junior engineer on the structural design. The particular component that failed was one that he had designed — or hadn't, as the ultimate investigation determined. There had been a breakdown in communication and misunderstanding about who was doing what in the fast-tracked project. Most of the steel connections had been designed by the steel contractor and signed off by the structural engineer. However, for the walkway, there were a couple of late design changes that were to prove fatal. The walkways were slender structures suspended on hanger rods from the roof of the ceiling of the atrium. There were two walkways hanging from one set of rods. One connected the second floors (first floor in Australian terminology) and one the fourth. Parallel to these walkways was one that spanned the atrium at third-floor level. The architect wanted these to look extra slender and thin and requested that the rod diameters be reduced by using stronger steel. The second change was requested by the steel contractor. The original design used a single 45 mm diameter rod from ceiling to the base of the lower walkway as shown in figure 14.7. This proved too difficult to build as it would entail threading a long rod through the hole in the fourth-floor crossbeam. The solution was to cut the rod at the fourth floor and introduce an extra bolt. This change doubled the force on the bottom bolt at the fourth-floor level. Reducing the diameter from 45 mm to 31 mm and doubling the forces could have easily been accommodated but two crucial mistakes were made (Brady 2018; Gillum 2000; Luth 2000; Montgomery 2001; ThinkReliability 2019).

FIGURE 14.7 Walkway connection detail

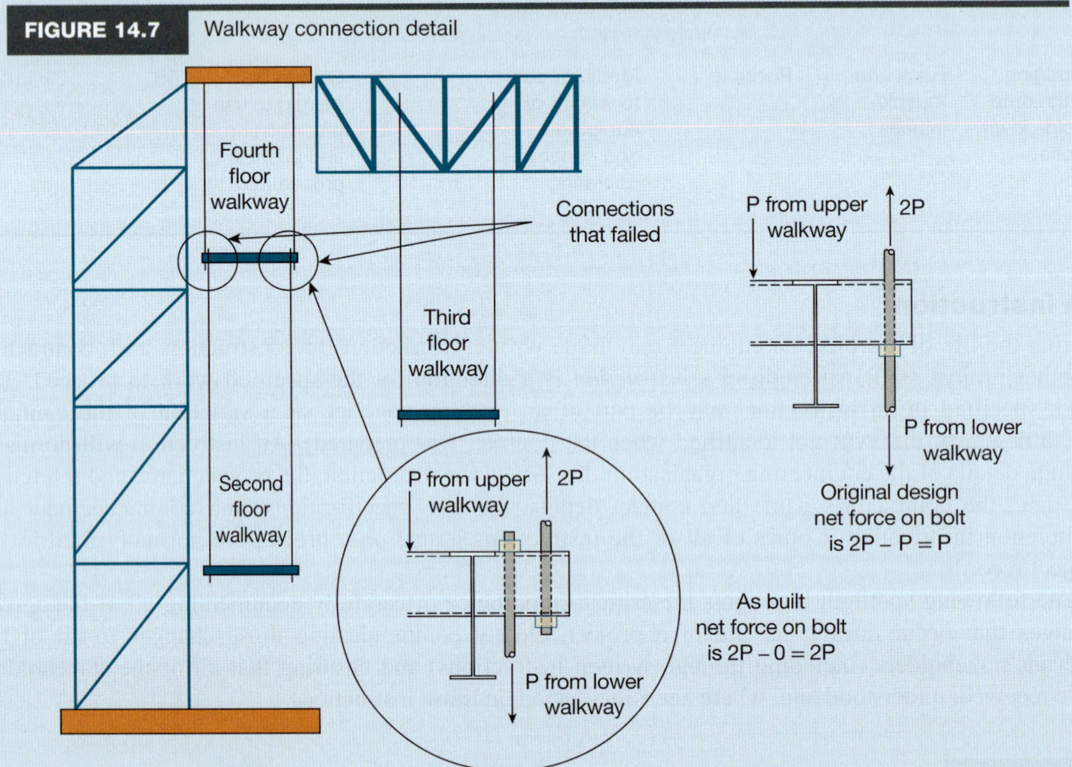

The first mistake was that the specification for the stronger steel appeared on one drawing but not on the critical one sent to the steel contractor, who used the original lower strength rods. The diameter was reduced but the strength was not increased. The second mistake occurred because the structural designer thought the steel contractor had designed the new connection. The steel contractor thought the structural engineer had designed it. In fact nobody designed it and nobody realised that the force had doubled. The hanger connection that was built was guaranteed to fail (Brady 2018; Gillum 2000; Luth 2000; Montgomery 2001).

On 17 July 1981, at approximately 7.05 pm, there was a loud thud as one of the fourth-floor bolts pulled through its hole (Brady 2018). The top walkway dropped an initial six inches, the bolt quickly tore through the top of the crossbeam and each connection failed progressively and rapidly. The two walkways were not crowded and were not overloaded. However, the atrium below was crowded with hundreds of couples attending a regular tea dance. A total of 114 people were killed and 216 were injured (Brady 2018;

Luth 2000; ThinkReliability 2019). It was the worst structural engineering disaster in US history. Forty-five minutes later, Jack Gillum was receiving the news (Gillum 2000).

A major enquiry was held to examine the causes and to determine culpability. This enquiry concluded in 1984 that Jack Gillum, as the Engineer of Record, and Daniel Duncan, as the Project Engineer, were guilty of negligence (Robbins 1985). They lost their licences to practise in the state of Missouri (Montgomery 2001). GCE International (Jack Gillum and Associates changed its name after the collapse) lost its licence to operate as a consultancy (ThinkReliability 2019). The steel contractor was cleared of responsibility even though that contractor had designed the bulk of the connections and had requested this change and had omitted the higher strength specification (Gillum 2000; Luth 2000).

This case clarified that the Engineer of Record holds ultimate responsibility even if someone else does the work (Gillum 2000). Greg Luth, as a junior engineer, was not held responsible, even though he was the one tasked with doing the checks. Luth went back to university and undertook a PhD on structural analysis. Both Gillum and Luth, of their own volition, toured the country giving seminars and presentations about the tragedy and their mistakes, in an effort to educate other engineers. (Gillum 2000; Luth 2000). Luth subsequently established a successful structural engineering and building practice in California (Gregory P Luth & Associates Inc. 2019). Jack Gillum went on to practise outside the state of Missouri and dedicated his work to the victims of the Hyatt Regency collapse. In his own words 'There is hardly a day that goes by that I don't think about the Hyatt collapse, the lives that were lost or marred forever, the relatives that lost their loved ones and the effect it has had on Kansas City, the construction industry and everyone connected with the project' (Gillum 2000). Gillum died in 2012, aged 83 (Horan & McConaty 2019).

Although completely exonerated by the enquiry, the Hyatt Regency Hotel group paid out over $140 million in compensation to the victims of the disaster (ThinkReliability 2019). This sum vastly exceeded the insurance liability policy that Jack Gillum and Associates had in place.

The third-floor walkway remains while the second and fourth floors just have gaping holes. The hangers from the fourth floor to the ceiling dangle.

Source: Dr Lee Lowery Jnr. Creative Commons.

It is clear that the hanger and bolt have pulled through the crossbeam causing the walkway to plummet.

Source: Dr Lee Lowery Jnr. Creative Commons.

**See* The Rime of the Ancient Mariner *by Samuel Taylor Coleridge; the mariner kills an albatross and suffers great remorse seeking to tell his story to each person he meets.*

CRITICAL THINKING

Greg Luth was a young engineer in his twenties, caught up in a major engineering disaster that shaped his life thereafter. You will soon be embarking on your career in engineering. Consider what procedures and processes are required so that your design team will detect errors and omissions before the item is made. As a rookie engineer, how would you challenge norms that you deem incorrect or risky?

Read Luth's paper on the chronology of the collapse. At which point do you think the engineers should have spotted the problem? Apart from the miscommunication what other factors were significant contributors to the failure at the Hyatt Regency Hotel in 1981?

A technical manual

An important deliverable from many projects is a technical manual, which is prepared to guide clients, or product purchasers, about the installation, safe use, maintenance, repair and disposal of a product. It is important that manuals are written using language that is suitable for intended purchasers and users. One way to check this is to ask one or more of the intended users to evaluate the content, structure and style of a manual prior to its publication and distribution.

14.4 Technical presentations

LEARNING OBJECTIVE 14.4 Prepare and give a technical presentation about an engineering project.

The chapter on communication skills introduced some presentation skills. This section builds on those skills and outlines some techniques that can be used to prepare technical presentations, a regular activity for many engineers.

How can poor presentation be avoided?

Many presenters use PowerPoint, or similar software, to present technical information to a range of audiences. A common reason that such presentations fail to communicate the intended message is that each slide is crammed with technical information and presented in a similar hierarchical structure, regardless of the nature of the message on each slide. This approach was common in the early days of PowerPoint use, when people were still learning how to use this media effectively.

Since then, there has been a quiet revolution in the way many astute presenters are using PowerPoint, and other presentation software, to communicate technical information. Their approach is based on three underpinning principles.

1. The presentation must succeed in accurately communicating the desired messages both during the presentation and in the future as an archived record of the presentation.
2. The use of slides affects both the way the speaker prepares a talk and how the talk will be presented.
3. An assertion–evidence approach (Alley & Neeley 2005) is used to design the slides rather than the traditional heading and bullet point format. Using this approach, an assertion is made at the top of a slide and then the speaker provides the evidence that supports the assertion.

An Institute of Electrical and Electronics Engineers book titled *Slide rules: Design, build, and archive presentations in the engineering and technical fields* (Nathans-Kelly & Nicometo 2014) addresses these issues from an engineering perspective and provides hundreds of examples of dos and don'ts to help engineers prepare and present technical presentations. Some of these are outlined in the following spotlight.

SPOTLIGHT

Slide rules

TRACI NATHANS-KELLY, CORNELL UNIVERSITY

AND CHRISTINE GROHOWSKI NICOMETO, UNIVERSITY OF WISCONSIN-MADISON

The practising engineers we have worked with spend between 20 per cent and 80 per cent of their time preparing and delivering communications to colleagues, managers and other stakeholders within their organisations.

We believe that engineering and technical communications need to be as clear, elegant, concise and accurate as the work itself. To help engineers hone their presentation skills, we developed some techniques that fundamentally re-craft the way engineers use slides in their technical work. In our book, we provide guidance for engineers and other technical people who desire to move towards best practices for their complex presentations. Four important principles are discussed in great detail in the book, and they are outlined as follows.

1. *Revisit presentation assumptions*. Cognitive science tells us that when audiences must simultaneously process information presented orally (listening to a speaker) while decoding text (e.g. reading bullet points), the capacity to process the information in each cognitive channel diminishes greatly. Your audiences will retain more information when there is less to process at a given moment. This doesn't mean that you 'dumb it down'; rather, your job as a technical presenter is to control the message. We also know that well-crafted and targeted visuals help us to remember information better than the same information presented as text. Storytelling techniques also support technical talks.
2. *Write sentence headers*. The header on each slide should be used to frame the content, but current practice is to deploy fragmented headers. We often see 'Results' or 'Analysis', for example. However, fragmented headers demand that audiences guess your point. Instead of playing guessing games with your audience, use concise sentences that summarise the content of the slide. Using a full-sentence statement as the header focuses audiences on the main point unambiguously — this is the assertion. The storyline becomes clearer in a presentation when the main points are provided.
3. *Use targeted visuals*. Instead of bullets in the main acreage of your slides, use targeted visuals. Humans are visual beings, and they will respond to well-structured and well-presented graphics, such as pictures, graphs, drawings and cutaways, that build the information as you move through the presentation. This is particularly important when presenting complex technical information as it allows audiences at all competency levels to follow the story and receive your key messages. Remember, the graphics are there to support the header statement and the spoken words.
4. *Archive details for future use*. To ensure that the archived presentation houses a complete set of technical details, you should include additional talking points and supporting information (e.g. data, results, formulae, references, your talk) in the Notes section of each slide. When used well, the Notes provide the full, rich technical account for future users. This includes the evidence that supports the assertion. Copies can also be circulated electronically as a PDF — but remember to save 'Notes Pages' when you create the archival PDF of your slides. These notes provide continuity and documentation for you and your organisation beyond the moments of the live presentation.

Figure 14.8 is an example of 'old' practices.

FIGURE 14.8 A typical bullet-point slide

Analysis: SCADA

- Load profile
  - Summated value
  - Double redundancy system
  - July 2014
- Mission support is key
- Balanced across system, but:
  - Rack at 259kW
  - Mode at 118.5kW
  - Trapped: 1405.kW

Source: Barclay (2014).

Figure 14.9 has the same information but demonstrates how the four principles can transform slides, allowing for a more dynamic and authentic technical presentation. This slide also incorporates all of the technical information in the Notes section for audience information, and for archival purposes.

FIGURE 14.9 An example of a slide that uses the four principles

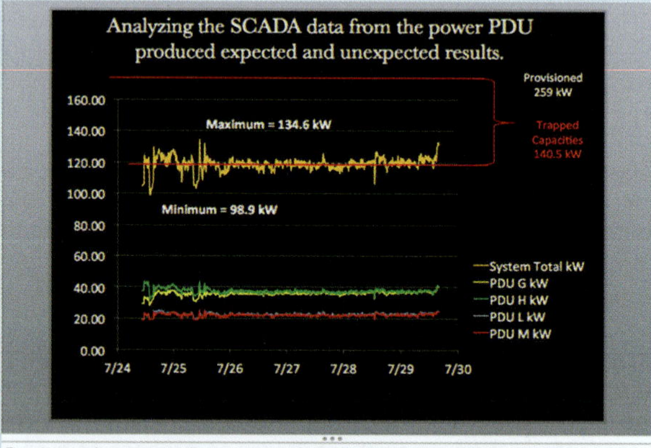

Source: Barclay (2014).

We encourage you to use these principles for your future presentations.

CRITICAL THINKING

Review a presentation that you or your student team prepared this semester. What changes would you need to make to incorporate the principles outlined in the spotlight? Redesign the slide set to incorporate those changes.

14.5 Visual communication

LEARNING OBJECTIVE 14.5 Discuss the use of visual methods to communicate engineering information.

Drawing, modelling, photography and sketching have always been part of the engineer's communication tools and, with the introduction of computer technologies, they have become even more powerful. The advances in computer technologies over the last thirty years have enhanced these tools and led to the introduction of new tools such as animation, videos, 3D printing, and walkthrough and flyover technologies. These tools can give the designer greater understanding of the intricacies of a design, and the ability to create a range of products — such as plans and models — that communicate the design to the people who will implement it. These products provide the client, the manufacturer or constructor with an understanding of the size, shape and location of each component, and the spatial relationships between the components.

Many of the innovations in the visual communications field that have been introduced over the last few decades were originally developed for computer games or movie animations and simulations. These technologies have flowed through to other industries, such as engineering, where they have been adapted to produce a range of new and innovative products. These products may be used by engineers to develop designs or to communicate design and construction information to clients, members of a project team, construction or manufacturing staff or members of the public.

Some of these new technologies became known as 'disruptive technologies' as they radically changed the way we do things and their impact may be greater and more far reaching than the developer first anticipated. 3D printing is an example of a disruptive technology.

Drawings, plans and sketches

While engineers need to be skilled in the preparation of plans, models and sketches in their field, they also need to be proficient at reading and using the information on plans that are prepared by engineers in other fields of engineering and, in some cases, plans prepared by professionals in allied disciplines, such as architects, town planners and surveyors. This is because they need to access and interpret the information on those types of plans, each of which has its own set of conventions, symbols, styles and formats. For example, figure 14.10 is an example of an electronic circuit diagram using the standard set of symbols for that discipline.

FIGURE 14.10 A circuit diagram for a 100 W power amplifier

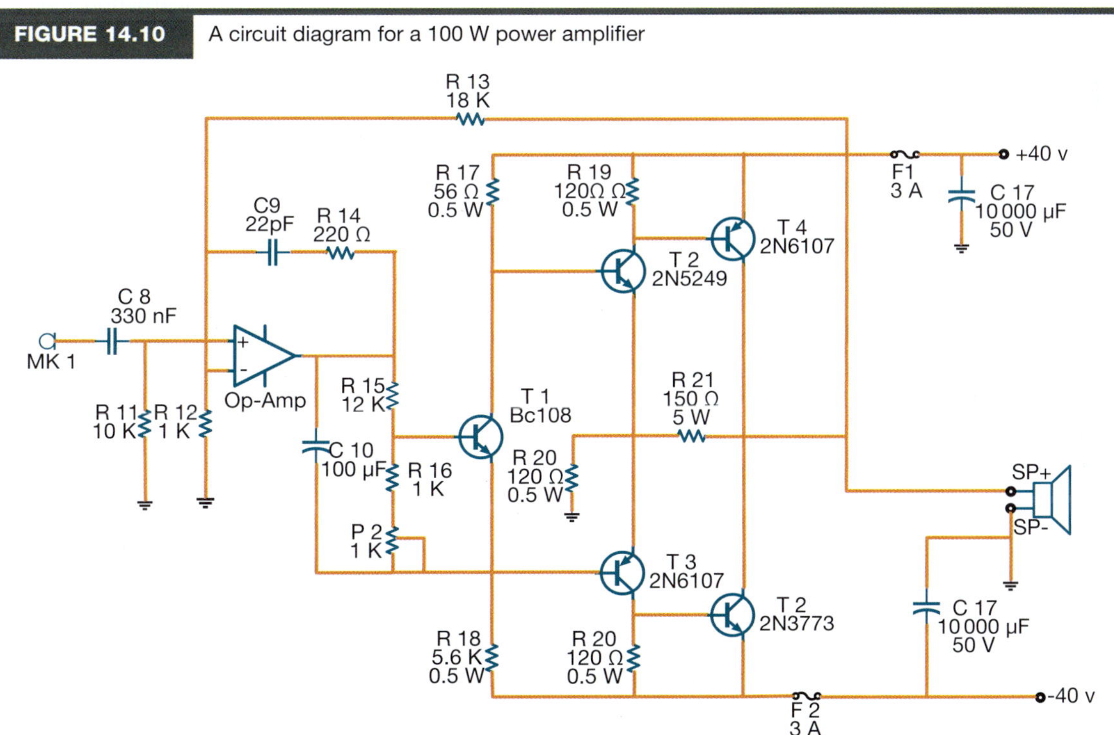

Source: Ali Kharrazi, University of Western Australia (2018).

Before using any information from a plan, you should check that it is accurate enough for your purposes and verify that it is the latest version of the plan. You can check the accuracy of a plan by comparing some of the plan measurements of existing features with your actual measurements of the features. You can also check the internal accuracy of the plan measurements. For example, the part lengths of a component should add up to the total length shown for the same component.

Sketches

Engineers use sketches for a variety of reasons, such as explaining construction details to on-site workers, or sketching preliminary concepts or designs. They are used to aid the thinking process, to test ideas and to guide discussion. While project sketches should be drawn in a project journal so the idea is captured, engineers often get an idea in non-work environments and have to use whatever paper is available (such as the back of an envelope, or a bar coaster) unless, of course, they have their mobile phone or an electronic tablet with them. These sketches should be retained and copied into a journal at the first opportunity.

Some of the latest 3D modelling software packages enable designers to begin designing with a freehand sketch, before transforming this into a scaled model. These sketches may also be used to communicate design concepts to other design staff, or members of a project team.

Two-dimensional plans

Two-dimensional (2D) plans are the mainstay of engineering. They are the media used to capture the concepts and details of designs, as well as detailed construction information and, once completed, the 'as-constructed' information. Generations of engineers and engineering designers have developed highly specialised plan formats that are used to communicate information in such a way that it is easily understood by the people who will use the plans. They use plan views, elevations, cross-sections, longitudinal sections, and isometric and axonometric perspectives to communicate this information. Figure 14.11 is an example of a 2D construction plan for one floor of a multi-storey building.

Importantly, these plans always incorporate relevant metadata. This includes information such as company details; project information; client information; the drawing title; staff information, including who designed, drafted, checked, verified and approved the plan; relevant dates; and key identification and location data, such as the plan number, the version number, the location, the north point, the scales, the sheet size and the legends. Before using a plan, the most critical information to check is, firstly, that the plan has been released for construction or manufacture and, secondly, what the plan version number is. This should be checked with the relevant organisation to ensure that the copy you have is a copy of the latest version. This information is contained on the right-hand side of the sample construction plan in figure 14.11a. That section of the plan is enlarged and annotated in figure 14.11b.

Photography

Photographs have been used by engineers for many years to record events, evidence and progress. The advent of digital photography has meant that:

- images are immediately available for use
- more photographs and videos can be taken at no extra cost
- images can be easily transmitted to other people
- the current status of an artifact can be documented.

The digital camera (including those in tablets and mobile phones) is an important communication tool and you should learn to use it properly so you can take photographs in different conditions and at a range of distances. For example, you should be able to take photographs in different environmental and lighting conditions.

Video footage is also an important medium for recording engineering activities. It is particularly useful when the key information to be captured involves movement. For example, a video of an experiment can be replayed at normal speed, or at slower speeds, to check for any occurrences that may have been missed at the time of the experiment. Like photographs, videos should be archived so they can be used to provide evidence if required. For example, it can be used to show that a specified activity took place, or to show the operational behaviour of a piece of machinery.

Before you take photographs or videos, you should consider what information you are trying to capture and communicate. This will help you to frame the pictures so they show the key pieces of information. In some situations, you may not know what you are looking for, such as when a piece of machinery breaks down but the cause is not immediately apparent. In this case, you should take as many photographs as possible from a variety of angles and distances. This strategy should help to ensure you have captured the crucial pieces of information, even if you do not know it at the time.

The increasing quality of the cameras integrated in mobile phones makes them ideal for capturing images or video and then transmitting these to colleagues in the office or other locations for information or advice.

Engineering models

Engineers have always used models to communicate information about a design. Traditionally they have used physical (i.e. constructed) models of a completed design to communicate information about the design to clients, members of the project team and the general public. These days, they are more likely to use computer-generated models to communicate design information.

FIGURE 14.11a A 2D construction plan

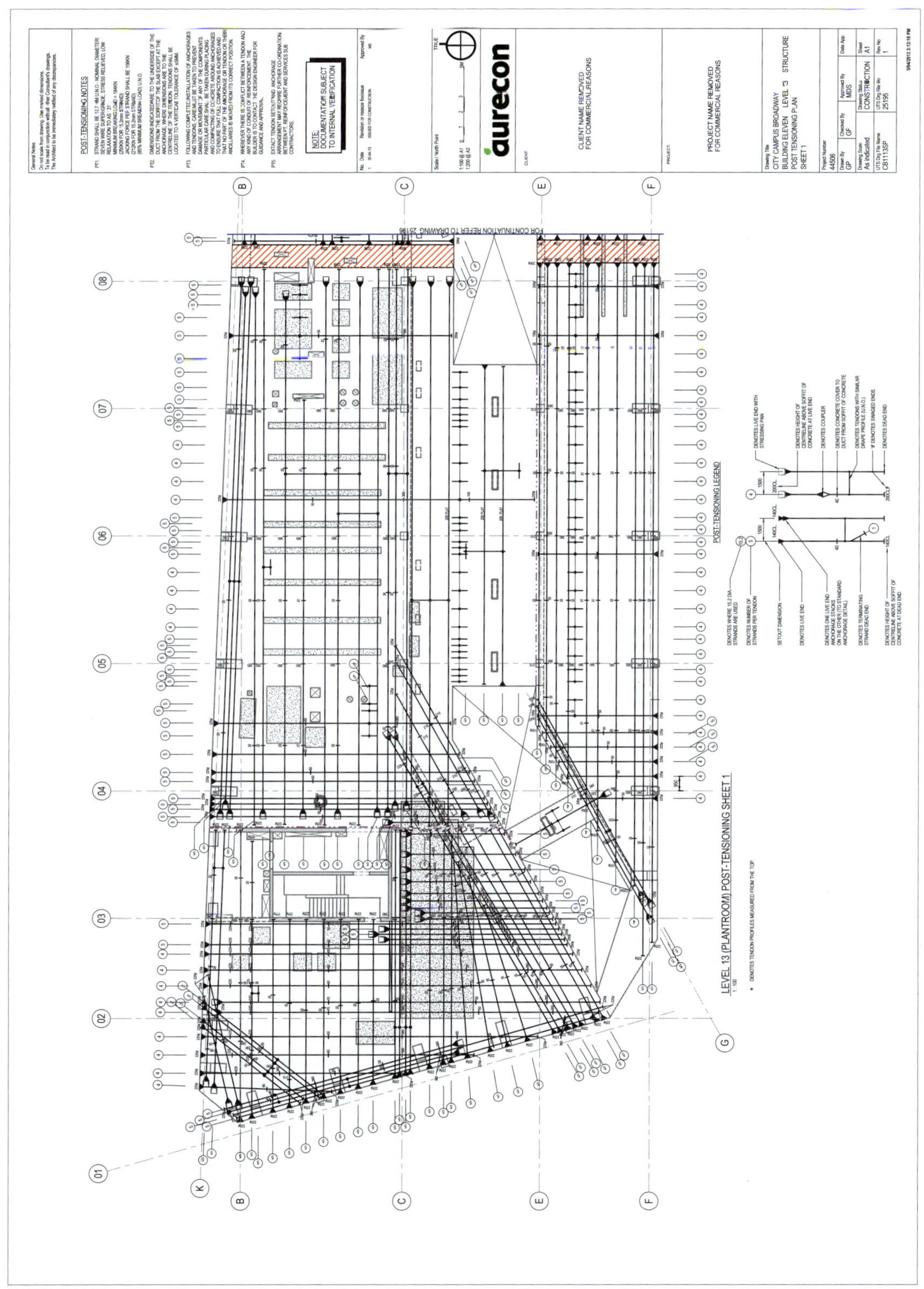

Source: Aurecon (2015).

The drawing–modelling process has also changed. Until the 1980s, engineers used 2D drawings to develop, document and communicate design information and, if necessary, 3D models were then constructed from those drawings. Today, the process is likely to be reversed, with the initial 3D design sketches being developed directly into 3D computer models. Once the 3D design is complete, the computer software enables the designer to generate 2D plans for certification and construction or manufacturing purposes.

FIGURE 14.11b An annotated title block from the 2D construction plan presented in figure 14.11a

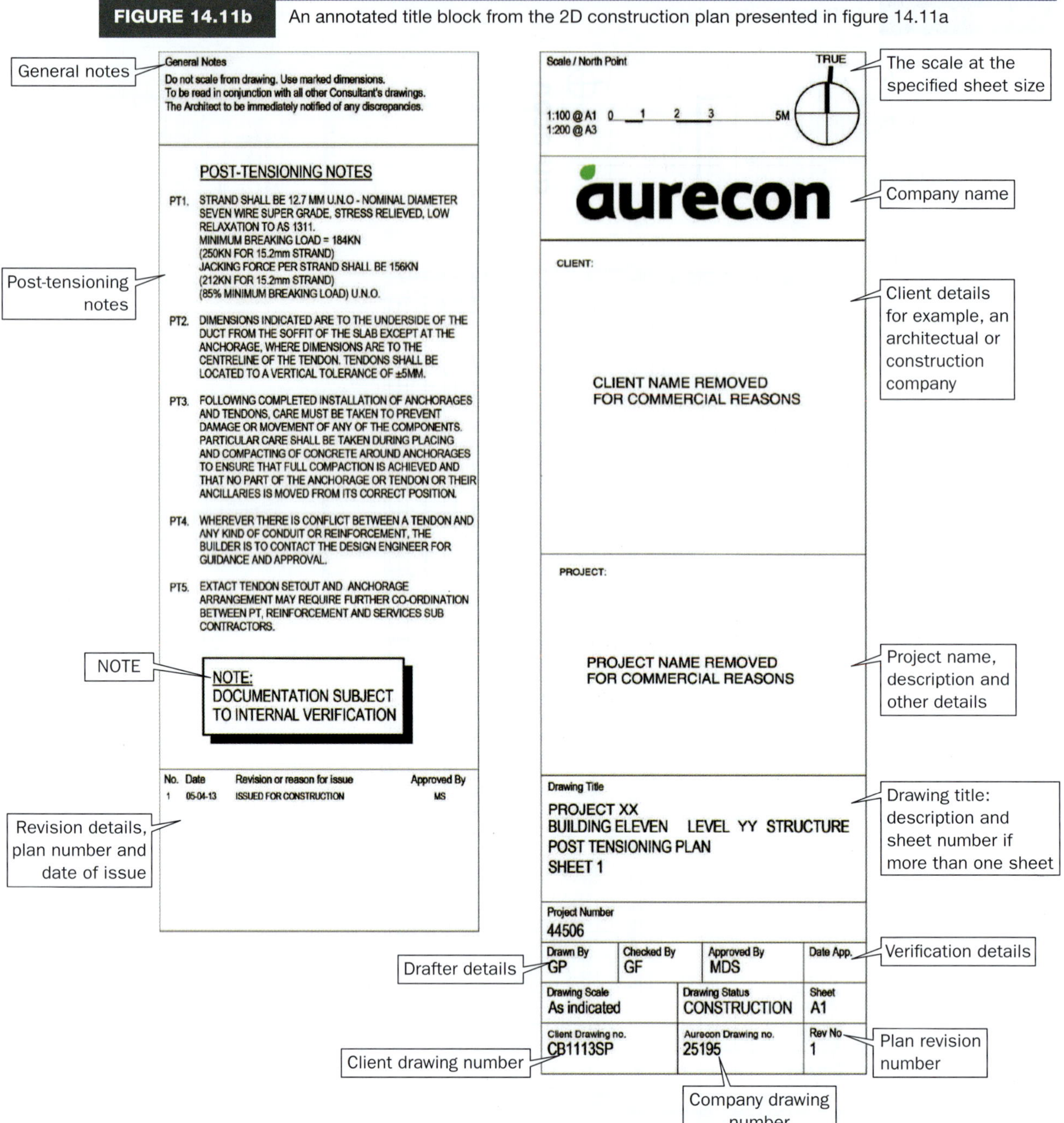

Source: Aurecon (2015).

Physical models

Physical models are often used to give designers and constructors a better understanding of the spatial relationships between components in a complex construction environment (e.g. the construction of a crusher plant in a chamber in an underground mine). Other constructed models are used as training aids and can normally be dismantled and reassembled. For example, a model could be used by a construction team to develop a sequence plan for the construction of the component or infrastructure. Part of this process may involve the creation of *cut-throughs* — 3D sections that enable the engineering team to see inside the components and explore the spatial relationships between them. This enables them to develop an assembly plan for each component, and to identify the spatial requirements of the path that will enable each component to be moved from the assembly area to its final position. It is interesting to note that surgeons use models in a similar way to plan complex surgery. The ability to physically handle and manipulate 3D physical models is one of the reasons they will continue to have a place in the engineering industry.

Computer-generated 3D models

Contemporary 3D design software packages provide the designer with enormous power and flexibility. The model can be rotated in real time, which enables the designer to view and work on the model from the most appropriate viewpoint. The model may also be designed to replicate the planned movements of the actual machine. These capabilities may also be used to communicate information about the design to the client or other members of the project team.

Three-dimensional building design software packages normally include options that enable the designer to walk through rooms and spaces and view them from all angles. Designers can also check the presence of sunlight and shade at various times of the day, or during different times of the year. They are able to locate the 'sun' in real-time by nominating the location of the building, its orientation, and the required time and date.

Figure 14.12 is an example of a 3D-generated perspective view. It is one of many images that were used by the designers to evaluate the design of the Kurilpa Bridge and communicate key aspects of the design to the clients and regulatory bodies. The design elements and shadows can be seen in this image.

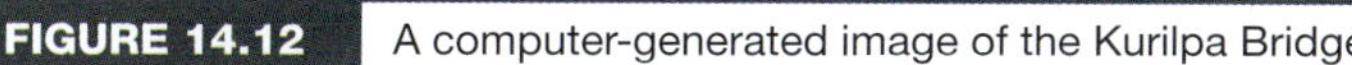

FIGURE 14.12 A computer-generated image of the Kurilpa Bridge

Source: Arup (2011).

The designers of the bridge, which spans the Brisbane River near the Gallery of Modern Art at South Brisbane's Kurilpa Point, used a variety of 3D software packages to undertake the architectural, structural and civil design components for the project. Constructed in 2009, the bridge design won a number of engineering and construction awards, including the World Transport Building of the Year at the World Architecture Festival in November 2011. The architectural design was completed by Cox Rayner Architects, and the structural design by Arup. The structural design is based on the principle of 'tensegrity', a term coined by Richard Buckminster Fuller to describe a system of balanced compressive and tensile forces.

Structural design software packages generally use colour to indicate important design information to the designer, such as deflection, with the colour red normally indicating design 'hotspots' or important features (see figure 14.13).

Fly-over views can also be created for larger construction projects, so that the project team and clients can visually experience the completed project. In a similar way, road design software can be used to create drive-over views of a proposed highway. This enables the designers to 'drive' along the route to check for blind spots and to identify any other design problems. These features can also be used by the design team to create visual communication packages (such as videos) that can be used to gain approval for the design from the client and development assessment agencies. Visual communication packages may also be created to market a project.

FIGURE 14.13 The deformed shape of the structural elements of Kurilpa Bridge. The actual deformation has been magnified 125 times.

Source: Arup (2011).

KEY POINT

Engineers use a range of visual methods to communicate information.

3D printing

For many people, particularly clients and designers, the ability to hold, touch, see, rotate or pull apart a physical model of a product can be an important part of the approval to manufacture decision-making process. Late last century rapid prototyping machines were used to produce 3D models for this purpose. These machines were expensive and were mainly used for prototyping products before manufacture.

Over the last decade, changes in technology have reduced the cost of these machines to the point where they are in common use across a range of industries. Along the way, the name for this type of equipment changed to '3D printer'. There are even desktop models and printers are produced for home use so that artists and even children can produce plastic models. Generally, 3D printers build up the model by 'printing' a series of thin layers of material, one on top of another until the model is complete. Each layer is based on a thin section, or slice, from the 3D computer model. More sophisticated machines enable changes of colour or material from layer to layer.

From a communication perspective, the ability to cheaply produce multicoloured 3D models of complex objects enables engineers and designers to use them as part of their communication strategy for reporting project outcomes to clients, users or the general public. They can also be used to produce promotional products. For example, imagine using a 3D printer to produce high-quality chocolates incorporating words or symbols that promote your company or a project (University of Exeter 2011). That is a communication strategy that may engage all of the senses of the target consumers — and even overload them.

Computer-generated 3D$^+$ models

Many 3D modelling packages include options that enable the user to add one or more dimensions to a 3D model, for example, time or motion. Thus, the resulting models, referred to as 3D$^+$ models in this text, could be described as 4D or even 5D models. These additional dimensions can be used to communicate important information about the movement of a component or a fluid in a process. Seeing a simulated process enables designers, clients and users to evaluate a design prior to manufacture or construction — a cost-effective alternative to prototyping (see figure 14.14).

The options available in each software package vary and, while some packages are used in a number of engineering disciplines, others are discipline specific.

We are very familiar with 3D gaming as a means to experience a fantasy world where there is little risk of real injury or death — apart from the avatars we choose to be or those we encounter! The use and deployment of 3D graphically generated models of spatial domains and engineered systems has grown enormously in the last twenty years, along with the capabilities of the software systems that drive them. The ability to develop and use 3D$^+$ models introduces a temporal dimension to construction and enables the optimisation of, for example, construction scheduling. They are powerful environments for both communication and decision-making purposes.

FIGURE 14.14 A locomotive boiler model developed using Autodesk's Maya software (which has motion capabilities)

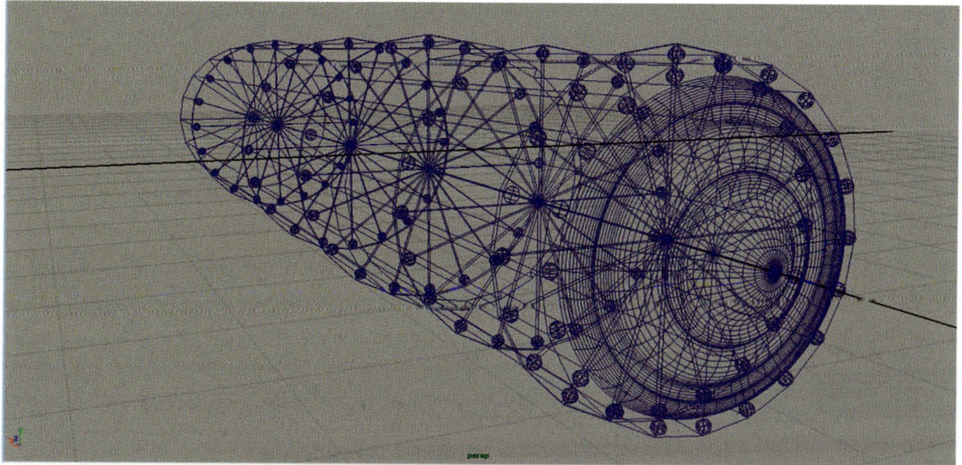

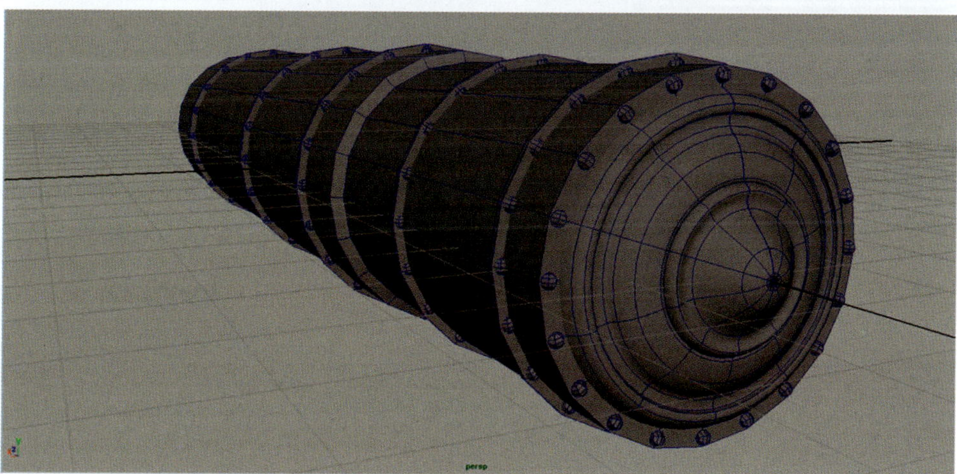

The following spotlights highlight the way engineers in different disciplines are using 3D+ environments to communicate information for education, technical and training purposes. Often these environments communicate the information using many of our senses, for example, sight, sound, touch and emotions.

SPOTLIGHT

Using 3D+ models to communicate design and operational information

WITH PROFESSOR IAN CAMERON, UNIVERSITY OF QUEENSLAND

Beyond the computer graphic worlds is the real 3D or 4D world, where all the fidelity of engineering and the interaction of humans with those systems can be displayed. Using the power of high-resolution 3D photography, real-world situations can be captured and used for a wide range of applications. These applications can provide powerful and rich training environments where people can:

- spatially familiarise themselves with equipment
- gain an appreciation of the scale of equipment and operations
- interact with the technology and conduct operational activities
- build a deeper understanding of the relationships between the engineered system and its many engineering representations, such as design, control and operational drawings.

In these environments, the 'distance' between the in-silico environment and the real world is minimised, through increased 'presence' or 'immersion'. These 'real' 3D and 4D environments are being developed for a wide range of engineering applications, particularly in the process industries, where complex multi-unit plants can be presented to the user in a manner that enhances the in-depth understanding of design and operational principles as well as issues of complexity, the response to system failures and the importance of enhanced diagnosis and decision making. They have applicability well beyond engineering.

These environments allow users to move around the plant through various means, such as isometric mini-maps or selectable hotspots, in the 3D virtual reality (VR) and interactive plot plans. They can also be used to embed a wide range of engineering resources for learning purposes. These include process flow diagrams, piping and instrumentation diagrams, vessel design details, animations of process flows or equipment internal dynamic flows, guided commentaries and much more. Even real plant data are available for analysis and design activities. All these features can be used to situate learning activities in real-world environments and can be used to communicate design intentions, and how they are realised. They can also be used by training staff and operations staff and for work health and safety courses. This makes them powerful education and training tools for both universities and companies.

Figure 14.15 is a screen shot of part of a major oil refinery that is used for separating crude oil into its constituent components. The user can access the detailed engineering drawings of components, as well as an animation of how one of the major separation column internals might behave.

FIGURE 14.15 3D+ modelling for an oil refinery

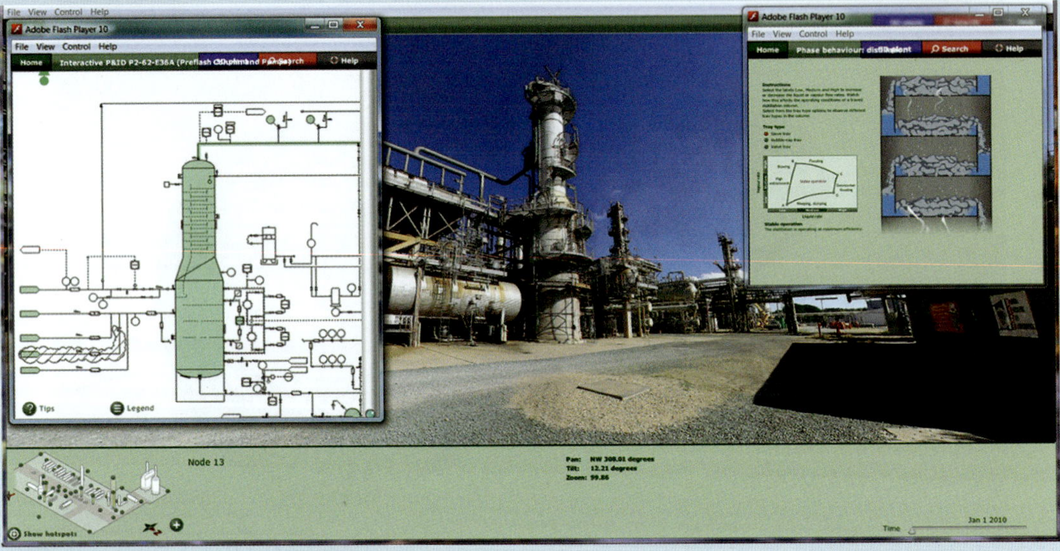

Source: Ian Cameron (2014, pers. comm.).

Other disciplines are seeing the benefits of these environments, for example, veterinary science, nursing and pharmaceuticals as they provide remote-site familiarisation. This means there are plenty of opportunities for real-world 3D+ environments to be developed and deployed for communicating design intentions and operations.

SPOTLIGHT

Using virtual mines to communicate the complexity of mining operations to students

WITH PROFESSOR PETER KNIGHTS, UNIVERSITY OF QUEENSLAND

The University of Queensland jumped into the future in 2014 with the opening of its Immersive Learning Facility. The facility uses three high-powered projectors — each one powered by a separate computer — to display virtual mines on a semi-circular wall with a diameter of eight metres. The room comfortably fits up to 30 students at a time, and software is used to synchronise the images projected by the three computers. Users have the option of viewing the simulation in two dimensions or in three by wearing polarised glasses. These facilities can be used to simulate the real-world environments of other engineering disciplines and are being used in some museums to simulate natural environments and systems (see figure 14.16).

The facility enables mining engineering students to experience the complexity and magnitude of both surface and underground mining operations and link their theoretical knowledge and understanding to real-world mining engineering. They are able to see how all of the mining structures, excavations, machinery and services fit together to form a mining system. In the virtual mine, the operator can drive through the mine and fly over, under and around operating machinery. At the press of a button, the user can jump out of the mine and view it in plan view, which enables them to put everything into context. Pressing another button takes them back to the desired point in the 3D virtual mine where they can continue experiencing the ongoing mining operations.

FIGURE 14.16 Engineers undertake simulation training in an immersion theatre similar to UQ's Immersive Learning Facility.

How real is the experience? Real enough that some people have reported they suffered motion sickness as they travelled over the open pit mine or through the maze of drives and cross-cut tunnels in an underground mine. The virtual mine is proving to be a powerful tool for communicating the reality of mining environments, the complexity of mining operations and the operational characteristics of mining equipment. Importantly, students are likely to have a better understanding of mining operations than they would receive from a site visit. This is because mines, particularly underground mines, can be confusing places and it is difficult to build a comprehensive understanding from just one visit. More importantly, simulations allow students to view, and study, mining operations from perspectives that would not be possible in a real mine because of safety issues or technical limitations.

Importantly, students can experience these environments without having to go through induction processes or be subjected to the physical risks of going underground. Another advantage is that students can make multiple visits into a virtual mine without leaving the campus. These sessions can be timetabled like any other classes — a major consideration for universities when compared to the complexity of arranging visits to mine sites, in particular underground mines, where group sizes, travel and safety issues have to be overcome.

These virtual reality tools are not just used to support mine design and student learning, as the underground coal mine model is also used by the Queensland Government's Safety in Mines Testing and Research Station to train mine deputies for emergency situations and responses. They use a series of hazards that are built into the simulation. For example, a trainer can trigger a fire in a conveyor motor or a roof fall in a drift to test whether the deputies understand and can apply the Trigger Action Response Plans formulated to assist safe operation and emergency responses.

Virtual mines can also be used as 'collaborative design hubs' to design mines by bringing geologists, and mining, geomechanical and metallurgical engineers together into the one design space to visualise, discuss and analyse a mine plan. This process can short-circuit hours of independent efforts and ensure that the mining team have a shared understanding of mining designs and operations.

So, come in, pull up a chair and make yourself comfortable: a tour of a virtual mine or industrial plant is starting at a university near you!

CRITICAL THINKING

3D⁺ environments are becoming ubiquitous in the modern world. Consider a 3D or 3D⁺ learning or simulated environment that you have experienced, and briefly list the advantages and disadvantages of learning in that environment rather than a real-world environment. Compare the pros and cons of the design group using individual virtual headsets with the theatre experience described in the spotlight.

The spotlights in this section emphasise the changing nature of communication strategies and methods. While many of the changes can be attributed to changing technologies, many of the new methods and strategies result from the actions of people who used a systematic planning process to develop their communication strategies. In doing so, they initiated new or improved ways of communicating their messages so that they achieved their purposes. We can expect similar changes, and perhaps disruptions to current best practices, resulting from future developments in AI, drones, 3D printing, VR headsets and associated technologies.

14.6 Integrating AI into engineering communications

LEARNING OBJECTIVE 14.6 Discuss current approaches that may be used to effectively and ethically implement AI into engineering communication practice.

Engineers Australia's submission in response to the Australian Government's *Safe and responsible AI in Australia* discussion paper (Australian Government 2023) included the following statement.

> An AI system is an engineered system. It is defined with inputs and outputs and operations through a programmed, albeit complex, calculation which uses the inputs to generate an output (Engineers Australia 2023, p. 3).

Thus, at its core, any intent an AI system displays has been programmed into the system, and, as we saw in the earlier spotlight, its behaviour is non-deterministic; that is, you will get a different response each time you use the same prompt.

As we shall see throughout this section, the dilemma faced by those who are integrating a large language model (LLM) AI system into their professional practice is how to harness the benefits of AI while minimising the risk of negative outcomes from using the technology.

AI is a rapidly changing field, and what is good practice today may be discarded in two or three years. Thus, the lag time involved in the review process may mean that refereed articles in research and professional journals may be dated by the time they are published. Due to this lag time, some publishers (e.g. SSRN) have begun to publish working drafts of papers online for comment and discussion while the review process is undertaken. These authors may also be given permission to write short articles for professional magazines, highlighting key aspects of their research. These articles are usually written from first-hand research experiences and practitioner perspectives, and the AI experiences they report on are still relevant and relatively up to date. For this reason, some of the articles and papers referred to in this section are drawn from professional magazines.

AI is a disruptive technology, and even those at the forefront of its adoption find it difficult to predict the future impacts of AI on their field. For example, McAfee et al. (2023, p. 47) commented that '[g]enerative AI is still a nascent technology, and we can't predict exactly how it will be put to work in the years ahead'.

While some researchers have made positive comments about their experiences with AI, others have commented on the negative outcomes of AI projects, for example:

> Sadly, beneath the aspirational headlines and the tantalising potential lies a sobering reality: Most AI projects fail. Some estimates place the failure rate as high as 80% — almost double the rate of corporate IT project failures a decade ago (Bojinov 2023, p. 54).

Interestingly, the articles reviewed for this section, which were written by researchers working in the AI field, listed both positive and negative experiences in their published articles. For example, Bojinov (2023, p. 54) commented that 'AI is great at identifying patterns and providing predictions for well-formulated problems, but it fails to practice emotional intelligence and exercise moral and ethical judgement'. Eapen et al. (2023, p. 58) found that:

> There is a tremendous apprehension about the potential of generative AI — technologies that can create new content such as audio, text, images and video — to replace people in many jobs. But one of the biggest opportunities generative AI offers to businesses and governments is to augment human creativity and overcome the challenges of democratising innovation.

McAfee et al. (2023, p. 43) stated that:

> AI has already displayed a breathtaking ability to create new content such as music, speech, text, images, and video and is currently used, for instance, to create software, to transcribe physicians' interactions with their patients, and to allow people to converse with a customer-relationship management system. On the other hand, it is far from perfect: It sometimes produces distorted or entirely fabricated output and can be oblivious to privacy and copyright concerns.

They also noted that while '[t]here are ways to both reap the benefits of generative AI and contain its risks . . . users should take an LLM's output with a grain of salt' (McAfee et al. 2023, p. 47). They provided an example of a task that is too risky for generative AI to be involved in at all, writing that a client:

> was thrilled to learn that ChatGPT could take a list of books, articles and websites and generate a site of properly formatted references for them. Upon checking its output, he was dismayed to find that some of the references were wrong. He found it quicker to create all the references by hand than to check every aspect of the ones generated by the LLM (McAfee et al. 2023, p. 47).

AI and engineering communications

So, what does all this mean for engineers and their communication skills? Eapen et al. (2023) noted that AI may remove communication obstacles facing those who lack the skills to communicate their new ideas in written or visual form. However, to harness the communication power of AI, engineers will need a range of new skills.

It is time to revisit the earlier spotlight, where ChatGPT listed the pros and cons of using AI to write project reports. You will remember that the eight pros were efficiency, accuracy, data visualisation, predictive analytics, customisation, consistency, accessibility and scalability, and that the six cons were dependency on data quality, lack of contextual understanding, ethical and bias concerns, limited creativity and innovation, security and privacy risks, and maintenance and updates. These pros and cons suggest the issues that engineers will need to address to use AI to prepare written messages that enable them to communicate effectively, efficiently and ethically.

Up to this point the discussion in this section has mainly been about AI and written communications. It should be noted, however, that the capabilities of GPT models are rapidly expanding in other modalities such as image and video, and these capabilities are starting to be harnessed in the architecture and conceptual design fields.

Mollick (2023) described some issues that will need to be addressed in addition to those listed above.

- The greatest problems with using AI are that there are no instruction manuals and that no one really knows the full range of capabilities of the most advanced LLMs, the best ways to use them or the conditions under which they fail.
- AI is immensely powerful for some tasks but fails completely or subtly on others. Only through experience will engineers learn where it works and where it doesn't.
- AI is a skills leveller. While AI will boost the performance of all who use it, the increase in performance for skilled employees is likely to be less than that achieved by employees with lower skills.
- People can 'fall asleep at the wheel' and go onto autopilot when using AI. This may mean they often fail to identify mistakes when using AI-generated output (see also Dell'Acqua 2022). This also means individuals and their organisation fail to learn from their experiences with AI.

Budding AI users should take every opportunity to develop their knowledge and skills in using the adopted AI system.

- Undertake every AI training opportunity offered (Leonardi 2023, p. 51).
- Learn as much as possible about how the AI system works, its key uses and its limitations.
- Clearly understand the privacy policies of the AI system to ensure that confidential information is not added to, or downloaded from, the system's knowledge base (McAfee et al. 2023, p. 47).

- Develop sound 'prompt engineering' skills (Leonardi 2023, p. 51), that is, how to create effective prompts when working with AI. Prompt engineering is a new field for all of us, and is still more an art than a science (McAfee et al. 2023, p. 48). Many prompt engineering training opportunities and resources are emerging on the internet.
- Learn how to evaluate and identify mistakes in AI output (Leonardi 2023, p. 51).
- Learn how to train the AI system to contextualise and format output to meet the organisation's requirements (Leonardi 2023, p. 51).

KEY POINT

Engineers should train in prompt engineering.

Some principles for integrating AI into engineering communications

It is important to note that the steps discussed in this section can be applied at the macro level (e.g. the integration of AI into a section of an organisation) or at the micro level (e.g. writing individual reports). However, some may be more appropriate at one level than the other.

As a first step, Leonardi (2023, p. 50) recommends that each user should review the tasks they undertake in their organisation and allocate them into three categories:

1. tasks that AI can't or should not do
2. tasks where AI can be used to augment the user's work
3. tasks that can be automated by AI.

This is not a simple task, as Mollick (2023) suggests that only with experience will users be able to identify tasks that may be undertaken by AI, and those that should not. This review process should be undertaken regularly when new technologies or changes in existing technologies are introduced and as the employee becomes more experienced. Then, McAfee et al. (2023) suggest that the best way for a user to gain experience with an AI system is on a low-risk project where it can be trialled on discreet tasks.

Once a user gains experience with an AI system, they can try other approaches. For example, Mollick (2023) found that a lot of consultants gained the benefits of AI without the disadvantages if they applied one of the following two approaches: become a Centaur or become a Cyborg.

- When the Centaur approach is used there is a clear line between the tasks the user undertakes and the tasks undertaken by the AI system. The user 'switches between AI and human tasks, allocating responsibilities based on the strengths and capabilities of each entity' (Mollick 2023).
- When the Cyborg approach is used the work of the user and the AI system is blended as individual tasks are not delegated to the user or the AI system. Instead, the user works in tandem with the AI system (Mollick 2023).

McAfee et al. (2003, p. 48) recommend that an iterative approach is ideal when using generative AI systems.

> You must figure out how to phrase your prompts to get the most useful responses. You also frequently have to tell the system to try again and give it notes on how to do better. Directing it to assume a persona, or change its tone or style is often effective.

McAfee et al. (2023, p. 45) also suggest that the best way to learn and gain the benefits of AI is through rapid iteration, where the user moves through repeated cycles of Boyd's (1995) OODA loop:

- Observing the situation: Carefully review and reflect on AI output.
- Orienting for action: Consider shifting the focus or direction.
- Deciding what to do: Plan how to achieve that.
- Acting: Implementing the plan (e.g. revising the prompt)

A simple example of the OODA process was used to develop the earlier spotlight on using AI for engineering project reports, as follows.

- Step 1 — The first ChatGPT prompt was: *Write 300 words on the pros and cons of using AI to prepare engineering project reports.* The response included four 'pros' and four 'cons'.
- Step 2 — The second ChatGPT prompt was: *Write 500 words on the pros and cons of using AI to prepare engineering project reports.* The response included five 'pros' and five 'cons'.
- Step 3 — The third ChatGPT prompt was: *Write 1000 words on the pros and cons of using AI to prepare engineering project reports.* The response included five 'pros' and five 'cons'.

- Step 4 — The three responses were carefully reviewed. While the three lists of 'pros' and 'cons' were similar, they were worded differently. Importantly, it was noted that not all of the 'pros' and 'cons' in the 300-word response were included in the 500-word response and, likewise, not all of the 'pros' and 'cons' in the 500-word response were included in the list for the 1000-word response. Finally, it is interesting to note that the 300-word response used simple English and was easier to read and understand. As expected, the 500- and 1000-word responses provided more detail, but the responses were not as readable.
- Step 5 — The key points from each set of 'pros' and 'cons' were combined. While the spotlight is based on the 1000-word response, the 'pros' and 'cons' from each response were carefully reviewed and consolidated to provide clarity and additional detail. Thus, all of the 'pros' and 'cons' were included in the spotlight, resulting in eight 'pros' and six 'cons'. It should be noted, however, that the wording from the original content was retained where possible, and the introduction and conclusion were not rewritten.

This stepped approach can be used by engineers integrating AI into their practice.

The risks of using external AI systems

Many authors have identified the risks of using external AI systems, as shown in the following examples.

- If you use a confidential report to help train a generative AI system, bits of the report's content might later show up in the response to a prompt from someone who shouldn't have access to that information. Consequently, it's important to be clear on the privacy policies of any generative AI you're using (McAfee et al. 2023, p. 47).
- It is crucial to identify the ethical implications of a project and to formulate a plan for addressing them during the development phase. If that doesn't happen, the costs can be high — in terms of reputational harm, government fines, and the engineering required to retrofit the system to deal with them (Bojinov 2023, p. 55).
- Privacy requires that AI models safeguard personal data and provide guarantees that it won't be leaked (Bojinov 2023, p. 55).
- ChatGPT has been trained on enormous amounts of text, some of which is still covered by copyright or other IP rights (McAfee et al., 2023, p. 48).
- 'Garbage in, garbage out' is one of the oldest sayings in the computer era, and its true now more than ever. If a machine learning system is trained on biased data, the results it generates will reflect that bias (McAfee et al. 2023, p. 48).
- Increasing transparency in when and how AI is used will help to build confidence and increase user trust. Transparency is critical at the point of data collection, including what data is being collected, how and where it will be stored, who will have access and how it will be used. It should also be clear as to if the data will be used to train future AI systems. Greater transparency should be required when the systems are used to assist in making decisions related to individual circumstances (Engineers Australia 2023, p. 8).
- For transparency, users need to understand how an AI model works, evaluate its functionality, and know its strengths and limitations (Bojinov 2023, p. 55).

Many of these issues are likely to be addressed at the organisational level. One approach is the adoption of an enterprise system. For example, on 16 August 2023 McKinsey & Company announced that they had launched 'Lilli', their own *internal* generative AI solution, stating that 'it is a platform that provides a streamlined, impartial search and synthesis of the firm's vast stores of knowledge to bring our best insights, quickly and efficiently, to clients' (McKinsey & Company 2023).

McKinsey & Company is a leading global consulting firm that works with clients across many industries and sectors to address complex challenges. It has offices in 70 countries and Lilli draws on the information in more than forty curated knowledge sources, containing a total of more than 100 000 documents and interview transcripts. Erik Roth, who is the senior partner with McKinsey and leads the development of Lilli, stated: 'Lilli aggregates our knowledge and capabilities in one place for the first time and will allow us to spend more time with clients activating those insights and recommendations and maximising the value we can create' (McKinsey & Company 2023).

Wark (2024, pers. comm.) expects most engineering organisations will be using enterprise-based AI systems after developing either their own foundational AI model or a pre-trained model combined with techniques like retrieval augmented generation (RAG) to combine external data with their own enterprise data. AECOM is an American-based global design, engineering, construction and management corporation with offices around the world, including in Australia. It is working on its own enterprise AI system (Wark 2024, pers. comm.).

Using a RAG system to augment an LLM system is more cost effective than developing an enterprise system. It enables an organisation to constrain the LLM to using the organisation's data, while retaining the capacity to use data from its external foundational knowledge base. A RAG system allows users to add in specialised and up-to-date research and data, and to receive accurate source and attribution information from the LLM.

Finally, it is likely that governments around the world will enact legislation to regulate the use of AI. For example, the European Parliament approved the Artificial Intelligence Act on 20 March 2024 — the first AI legislation to be enacted. The new regulations:

- 'ban certain AI applications that threaten citizens' rights, including biometric [data] ... and the untargeted scraping of facial images from the internet'
- list the obligations of the developers and users of high-risk AI systems to protect, for example, the health, education, safety and fundamental rights of its citizens, and the rule of law, infrastructure, essential services and environment of each country in the Union
- list the transparency requirements for general-purpose AI systems
- identify the need for measures to support innovation and small-to-medium enterprises (European Parliament 2024).

In summary, the regulation seeks to mitigate risk while promoting innovation.

It will be two or three years before these regulations are enacted. While the major impact of these regulations will be in Europe, they will also impact developers and users across the globe.

No doubt other governments will use this legislation to inform the development of their own legislation. For example, the Australian Government sought submissions when it published the *Safe and responsible AI in Australia* discussion paper in June 2023 (Australian Government 2023). The following statement was included in Engineers Australia's response to the discussion paper.

> Australia needs to strike a balance in the regulation of AI, to allow the benefits to be realised, while also ensuring the negative aspects of the technology are mitigated. This balance should take a risk-based approach to ensure the safe and responsible use of the technology (Engineers Australia 2023, p. 3).

During the consultation period, the Australian Government published its interim response to the consultation on 17 January 2024 (Australian Government 2024). The interim response covered:

- what it heard from stakeholders
- how the government will ensure AI is designed, developed and deployed safely and responsibly.

The Engineers Australia submission was one of 510 online submissions. The government also held three ministerial round tables with a total of 64 participants, four in-person round tables (59 participants), four virtual round tables (81 participants), and one virtual town hall (345 participants) during the consultation period. The interim response included the following statements on risk.

> A preliminary analysis of submissions found at least 10 legislative frameworks that may require amendments to respond to applications of AI. Many AI risks outlined in submissions were well known before recent advances generative AI. These include:
> - inaccuracies in model inputs and outputs
> - biased or poor-quality model training data
> - model slippage over time
> - discriminatory or biased outputs
> - a lack of transparency about how and when AI systems are being used (Australian Government 2024, p. 4).

The document also highlighted the impact that these real and perceived risks were having.

> Submissions highlighted specific risks associated with newer, very powerful AI models. These 'frontier' models exceed the capacity of previous models and can generate new content quickly and easily. Submissions noted that 'frontier' AI models may require targeted attention ... Responses also indicated public concern that AI systems are not being tested appropriately, and have limited detail on how they function. As a result this is reducing public and business trust and reliability in the outcomes of these systems (Australian Government 2024 , p. 4).

We await the final legislative outcomes of this consultation process.

We are all on an AI learning journey into an AI-driven future that is difficult to predict. It is accepted that generative AI will become ubiquitous and continue to disrupt current communication practices going into the future. The dilemma for us is to ensure we harness the positives and mitigate or minimise the negatives associated with AI. To be able to do this, engineers will need to have the knowledge and skills to ethically evaluate, refine and enhance the quality of AI output while protecting the privacy of data and information. This highlights the importance of both understanding, and having experience with, traditional communication methods, models and practices. Only then will we be able to evaluate new AI-driven communication methods.

Remember, we are professionally responsible for every communication we produce, although our responsibility may be lessened if a communication is signed off by a gatekeeper.

It would be interesting, and perhaps important, for you to start a reflective journal to record your experiences with AI systems. Then, in the future, you will be able to look back on your learning journey and identify the critical shifts in your practice.

This section was designed to provide you with some principles that may be used to integrate AI into your communication practices. However, as the adoption of AI is in its infancy, and because AI technology is rapidly changing, these principles will need to be regularly reviewed and updated as you and others in your field of engineering gain experience in this emerging field. So, watch this space!

SUMMARY

Communicating information is a critical skill for engineers, and this chapter described a wide range of the contexts, approaches, methods, formats and media used by engineers. The importance of careful planning was emphasised and the ten-step PCR planning process was used to illustrate this. The various spotlights and examples in the chapter highlighted different aspects of the communication process and what happens when communication does not occur. Finally, the impacts of AI on communications were discussed, where the emphasis was on harnessing the rewards while mitigating the risks. We will now briefly revisit each of the chapter's learning objectives.

14.1 Discuss the contexts in which engineering communication occurs.
Engineers mainly communicate in the business world where they use the genres, methods, formats and styles that are used by professionals in Australia and around the world. Engineers often use other genres when they communicate with other engineers, particularly those in the same discipline. In this case the communication occurs within an engineering discourse community. For this reason engineers should carefully consider the context when planning a communication.

14.2 Use the PCR model to develop a communication plan for a project.
Engineers (and students) should use their understanding of the communication process to plan a communication. The ten steps in the PCR planning process can be used to develop a communication plan for a range of different types of communication; for example, a student project, a presentation or a tender proposal. The prompt questions may change to suit the type of documents or presentations being prepared. Careful communication planning is vital, although the length of time spent on the planning process will vary depending on the type, size and complexity of the communication. Examples were provided that demonstrate how the PCR model could be used to plan a communication by engineering students and professional engineers. Three communication approaches are factual communication, opinion and persuasive communication. Engineers generally use the factual approach, although the other approaches are often used, particularly by experienced engineers. Most organisations have policies and guidelines relating to the communication methods, styles, formats and media that are commonly used, as well as the approval processes that are to be followed before a communication is delivered. A large communication is likely to have a mix of communication approaches and methods, that is, verbal, visual and non-verbal. The ten-step process from the PCR model can be used to plan the appropriate methods for a communication.

14.3 Plan and prepare written communications in an engineering context.
Engineers can spend a lot of their time at work writing or responding to written documents. This includes correspondence, emails, project initiation documents, financial documents, human resource documents, workplace health and safety documents and project reports. Each of these documents has a purpose, and most have a defined format. Engineers need to consider the legal and regulatory contexts of all the documents they prepare, not just contracts. Engineers should always use an appropriate format, structure and style for a written communication so that the information is accessible by the consumers and achieves its purpose.

14.4 Prepare and give a technical presentation about an engineering project.
Adequately preparing for verbal communication, even informal communication, is important in a professional environment. A variety of tips and techniques can be used to enhance formal presentations, particularly those highlighted in the Slide rules spotlight. Preparing and delivering formal presentations is a routine activity for many professional engineers, and these presentations should be carefully planned so that they achieve their purpose.

14.5 Discuss the use of visual methods to communicate engineering information.
Visual communication methods have always been an important part of engineering. They are used in the design process, and to communicate design information to clients, users and the other members of the engineering team. Engineers use a range of visual communication methods, including sketches, plans, models and photography. The power of contemporary design software packages and 3D printing technology enables sophisticated communication products to be created from 3D designs. These products are then used to communicate design information to clients, project teams, construction teams and members of the public. The rapid development of $3D^{+}$ software has opened up a whole new world of possibilities that engineers can use to communicate their messages.

14.6 Discuss current approaches that may be used to effectively and ethically implement AI into engineering communication practice.

Artificial intelligence (AI) systems are being integrated into the engineering communications practices of many engineering organisations in Australia. AI is a disruptive technology that comes with both risks and rewards. Users must develop sound practices that enable them to harness the advantages provided by AI systems while mitigating or eliminating the risks. To do this they must be well trained and ever vigilant and adopt ethical practices.

KEY TERMS

contract A legally binding agreement between two or more parties that requires the parties to perform certain obligations.

genre A standard form of communication used by a group.

EXERCISES

1. Review a practice-based article about a topic you are studying. Are there any aspects you found difficult to understand? Did the author assume readers would have specific background knowledge? Were there any acronyms or abbreviations you did not understand?
2. Think of a presentation, lecture, speech or interview that you have seen lately that held your interest, such as a speech by the best man at a friend's wedding; a riveting podcast of an interview with your favourite singer, politician or athlete; or a recent engineering lecture. What was it that made it such a good presentation? Did it hold your interest because the topic was interesting or was there more to it than that? Write notes on the elements of the presentation that made it appealing for you.
3. Review the use of visual communication methods in your lectures this semester. Which was the stand-out method? Why?
4. Review the sections on the integration of AI into engineering communications and then develop a list of principles that you will adopt to guide your future use of AI.

PROJECT ACTIVITY

Develop a communication plan that you will use to communicate the outcomes of a university assignment. Your report on this project should include:

- a one-page report on the way you developed your communication plan
- a copy of your communication strategy, including communication methods and channels
- a reflection on the problems you had while preparing the plan, and any assumptions you had to make.

If you have already used the strategy to develop the documents and/or presentations for a university assignment, you should also reflect on the success of the communication.

REFERENCES

ACIF & APCC 2010, *A guide to project initiation for project sponsors, clients and owners*, Canberra.

Alley, M & Neeley, KA 2005, 'Rethinking the design of presentation slides: A case for sentence headlines and visual evidence', *Technical Communication*, vol. 52, no. 4, pp. 417–426.

Australian Bureau of Statistics 2022, *ANZSCO — Australian and New Zealand Standard Classification of Occupations*, 22 November, https://www.abs.gov.au/statistics/classifications/anzsco-australian-and-new-zealand-standard-classification-occupations/latest-release

Australian Government 2023, *Safe and responsible AI in Australia*, Department of Industry, Science and Resources, June, https://storage.googleapis.com/converlens-au-industry/industry/p/prj2452c8e24d7a400c72429/public_assets/Safe-and-responsible-AI-in-Australia-discussion-paper.pdf

—— 2024, *Safe and responsible AI in Australia consultation: Australian Government's interim response,* Department of Industry, Science and Resources, 17 January, https://storage.googleapis.com/converlens-au-industry/industry/p/prj2452c8e24d7a400c72429/public_assets/safe-and-responsible-ai-in-australia-governments-interim-response.pdf

Australian Taxation Office 2023, 'Tax invoices', https://www.ato.gov.au/businesses-and-organisations/gst-excise-and-indirect-taxes/gst/tax-invoices

Barclay, C 2014, 'Quantifying trapped capacities in a cloud environment', unpublished presentation.

Bojinov, I 2023, 'Keep your AI projects on track: Most go off course', *Harvard Business Review*, November–December, Boston, MA.

Boyd, JR 1995, 'The essence of winning and losing', 28 June, https://www.coljohnboyd.com/static/documents/1995-06-28__Boyd_John_R__The_Essence_of_Winning_and_Losing__PPT-PDF.pdf

Brady, S 2018, 'Hyatt Regency: The human price of failure', *Engineers Journal*, 11 December, https://www.engineersireland.ie/Engineers-Journal

CCH 2007, *Australian master OHS & environment guide*, 2nd edn, CCH Australia, Illinois.

Dell'Acqua, F 2022, *Falling asleep at the wheel: Human/AI collaboration in a field experiment on HR recruiters,* working paper, 23 January, https://static1.squarespace.com/static/604b23e38c22a96e9c78879e/t/62d5d9448d061f7327e8a7e7/1658181956291/Falling+Asleep+at+the+Wheel+-+Fabrizio+DellAcqua.pdf

Denholm, P 2008, 'Poisoned politics', *The Australian*, 27 May, https://www.theaustralian.com.au

Donati, M 2010, 'Multiplex and Mott MacDonald settle Wembley legal battle', *Construction News*, 25 June, https://www.constructionnews.co.uk

Eapen, T, Finkenstadt, D, Folk, J & Venkataswamy, L 2023, 'How generative AI can augment human creativity: Use it to promote divergent thinking', *Harvard Business Review*, July–August, Boston, MA.

Engineers Australia 2023, *Responsible AI in Australia: An engineering perspective*, Engineers Australia's submission to the Safe and Responsible AI in Australia discussion paper, July, https://www.engineersaustralia.org.au/publications/responsible-ai-australia-engineering-perspective

European Parliament 2024, *Artificial Intelligence Act: MEPs adopt landmark law*, Press release, 13 March, https://www.europarl.europa.eu/news/en/press-room/20240308IPR19015/artificial-intelligence-act-meps-adopt-landmark-law

Gillum, JD 2000, 'The engineer of record and design responsibilities', *Journal of Performance of Constructed Facilities*, vol. 14, no. 2, pp. 67–70, https://doi.org/10.1061/(ASCE)0887-3828(2000)14:2(67)

Gregory P Luth & Associates Inc. 2019, 'Designing and building', https://www.gplainc.com

Horan & McConaty 2019, *Jack D. Gillum*, obit, https://www.horancares.com

Hoyle, R 2009, 'Mott MacDonald steels itself for high court battle over Wembley Stadium', *The Guardian*, 9 July, https://www.theguardian.com/uk

International Federation of Consulting Engineers 2024, *Publications*, https://www.fidic.org/bookshop

Krotz, JL n.d., 'Got e-mail manners? See these dos and don'ts', https://www.microsoft.com/en-au

Leonardi, P 2023, 'Helping employees succeed with generative AI: How to manage performance when new technology brings constant and unpredictable change', *Harvard Business Review*, November–December, Boston, MA.

Luth, GP 2000, 'Chronology and context of the Hyatt Regency collapse', *Journal of Performance of Constructed Facilities*, vol. 14, no. 2, pp. 51–61, https://doi.org/10.1061/(ASCE)0887-3828(2000)14:2(51)

Lynch, D 2010, 'Multiplex and Mott MacDonald settle Wembley case', *The New Civil Engineer*, 25 June, https://www.newcivilengineer.com

Martin, JA 2000, 'Mind your e-mail manners: Don't contribute to rudeness on the web; learn these ten rules and use them forever', https://www.pcworld.com

McAfee, A, Rock, D & Brynjolfsson, E 2023, 'How to capitalize on generative AI: A guide to realizing benefits while limiting its risks', *Harvard Business Review*, November–December, Boston, MA.

McKinsey & Company 2023, 'Meet Lilli, our generative AI tool that's a researcher, a time saver, and an inspiration', *McKinsey Blog*, 16 August, https://www.mckinsey.com/about-us/new-at-mckinsey-blog/meet-lilli-our-generative-ai-tool

Miller, C 1994, 'Genre in social action', in A Freedman & P Medway (eds), *Genre and the new rhetoric*, Taylor & Francis, London.

Mollick, E 2023, 'Centaurs and cyborgs on the jagged frontier', 16 September, https://www.oneusefulthing.org/p/centaurs-and-cyborgs-on-the-jagged

Montgomery, R 2001, '20 years later: Many are continuing to learn from skywalk collapse', *The Kansas City Star*, 3 October, https://www.kansascity.com

Mott MacDonald Group Limited 2008, 'Wembley Stadium, UK', https://www.mottmac.com

Nathans-Kelly, T & Nicometo, C 2014, *Slide rules: Design, build and archive presentations in the engineering and technical fields*, Institute of Electrical and Electronic Engineers Inc., John Wiley & Sons, Hoboken.

New Zealand Inland Revenue 2024, 'Chapter 2 — Tax invoice requirements', https://www.taxpolicy.ird.govt.nz/publications/2020/2020-ip-gst-issues/chapter-2

Quinn, B 2008, '£253m legal battle over Wembley delays: Under-fire building firm accuses stadium's consultants of negligence', *The Guardian*, 16 March, https://www.theguardian.com/uk

Robbins, W 1985, 'Engineers are held at fault in "81 hotel disaster"', *The New York Times*, 16 November, https://www.nytimes.com/1985/11/16/us/engineers-are-held-at-fault-in-81-hotel-disaster.html

Safe Work Australia 2018, *How to manage work health and safety risks*, https://www.safeworkaustralia.gov.au/system/files/documents/1901/code_of_practice_-_how_to_manage_work_health_and_safety_risks_1.pdf

——— 2024, 'Model WHS laws', https://www.safeworkaustralia.gov.au/law-and-regulation/model-whs-laws

Sales, H 2006, *Professional communication in engineering*, Palgrave and Macmillan, Basingstoke, UK.

Shea, V 1994, *Net etiquette*, Albion Books, San Francisco.

Stahl, A 2004, 'Stand up straight and mind your e-mail manners', https://www.microsoft.com/en-au

Standards Australia 2024, https://www.standards.org.au

Swales, J 1990, *Genre analysis — English in academic and research settings*, Cambridge University Press, Cambridge.

The Weekend Australian 2018, 8–9 September, p. 44.

ThinkReliability 2019, *Root cause analysis of the Hyatt Regency disaster — cautionary tale about assumptions,* https://dev.thinkreliability.com/case_studies/root-cause-analysis-of-the-hyatt-regency-disaster-cautionary-tale-about-assumptions

University of Exeter 2011, 'Chocolate research shapes the future of gift shopping', *Research News*, https://news.exeter.ac.uk

ACKNOWLEDGEMENTS

Photo: © Nestor Costa / Adobe Stock Photo

Photo: © NDABCREATIVITY / Adobe Stock Photo

Photo: © kerkezz / Adobe Stock Photo

Photo: © photo.ua / Shutterstock

Photos: © Dr Lee Lowery Jnr. Creative Commons

Figure 14.1 and text: © Stephen Beckwith

Figure 14.4: © Australian Bureau of Statistics 2022, *ANZSCO — Australian and New Zealand Standard Classification of Occupations*, 22 November, https://www.abs.gov.au/statistics/classifications/anzsco-australian-and-new-zealand-standard-classification-occupations/latest-release

Figure 14.6: © Safe Work Australia 2018, *How to manage work health and safety risks*, https://www.safeworkaustralia.gov.au/system/files/documents/1901/code_of_practice_-_how_to_manage_work_health_and_safety_risks_1.pdf

Figures 14.8 and 14.9: © Barclay, C 2014, 'Quantifying trapped capacities in a cloud environment', unpublished presentation.

Figure 14.10: © Ali Kharrazi, University of Western Australia (2018).

Figures 14.11a and 14.11b: © Aurecon (2015).

Figures 14.12 and 14.13: © Arup (2011).

Figure 14.14: © John Wiley & Sons, Inc.

Figure 14.15: © Ian Cameron

Figure 14.16: © Bloomberg / Getty Images

Text: © McAfee, A, Rock, D, and & Brynjolfsson, E 2023, 'How to capitalize on Generative AI: A guide to realizing benefits while limiting its risks', *Harvard Business Review*, November–December, Boston, MA.

Text: © Bojinov, I 2023, 'Keep your AI projects on track: Most go off course', *Harvard Business Review*, November–December, Boston, MA.

Text: © Eapen, T, Finkenstadt, D, Folk, J & Venkataswamy, L 2023, 'How generative AI can augment human creativity: Use it to promote divergent thinking', *Harvard Business Review*, July–August, Boston, MA.

Text: © Engineers Australia 2023, *Responsible AI in Australia: An engineering perspective*, Engineers Australia's submission to the Safe and Responsible AI in Australia discussion paper, July, https://www.engineersaustralia.org.au/publications/responsible-ai-australia-engineering-perspective

Text: © McKinsey & Company, 2023, 'Meet Lilli, our generative AI tool that's a researcher, a time saver, and an inspiration', *McKinsey Blog*, 16 August, https://www.mckinsey.com/about-us/new-at-mckinsey-blog/meet-lilli-our-generative-ai-tool

Text: © Australian Government 2024, *Safe and responsible AI in Australia consultation: Australian Government's interim response,* Department of Industry, Science and Resources, 17 January, https://storage.googleapis.com/converlens-au-industry/industry/p/prj2452c8e24d7a400c72429/public_assets/safe-and-responsible-ai-in-australia-governments-interim-response.pdf

ACKNOWLEDGEMENTS

Photo: © Nestor Rizhniak / [illegible]
Photo: © [illegible] / Adobe Stock Photo
Photo: © [illegible] / Adobe Stock Photo
Photo: © [illegible] / Shutterstock
[illegible] © Lee Bowers / Creative Commons
[illegible]
Figure 23.1: Australian Bureau of Statistics 2021, ANZSCO – *Australian and New Zealand Standard Classification of Occupations*, 22 November, https://www.abs.gov.au/statistics/classifications/anzsco-australian-and-new-zealand-standard-classification-occupations/latest-release
Figure 23.2: Safe Work Australia 2018, [illegible]
[illegible]
[illegible]
[illegible]
[illegible]
[illegible]
[illegible]
[illegible]

PART 6

PLANNING YOUR CAREER

CHAPTER 15

Your engineering future

'Be the change you want to see in the world.'

Mahatma Gandhi — Indian pacifist and independence leader (1869–1948)

LEARNING OBJECTIVES

After studying this chapter, you should be able to:

15.1 discuss the engineer's role in addressing global challenges

15.2 describe how globalisation may affect you in your future professional life

15.3 identify and discuss some emerging fields in engineering

15.4 manage yourself in your workplace, and develop a plan to meet your career goals.

Introduction

As a first-year student, you might not be focused on your future career. You are probably more concerned about getting through the next few months with decent marks. That is important; however, so is the process of planning out what your future options might be and how you can use the next few months and years to set yourself up for the career/future that you want. What you learn in the coming few years, about engineering and about yourself, can set you on the path to a satisfying, meaningful future. In this chapter, we attempt to see into the crystal ball and think through the main global and social challenges that engineers will work on in coming decades, challenges that the field of engineering faces, some of the remarkable innovation at the leading edge of engineering disciplines, and how you might plan to contribute to society's challenges and succeed in a job that fits whatever engineering passion or purpose you hold highest.

Consider the following scenario a young engineer faced.

CASE STUDY

Jono

Jono pulled another tissue from the box and blew his nose. The flu he had contracted on his overnight flight from Hong Kong had kicked in. He'd done a COVID test and it was looking like 'just' the flu. The drag coefficients for his company's latest catamaran hull were proving really hard to sort out. His head was thumping and the computer screen was a blur. A couple of days off would probably have been wise but there was a deadline to meet and he was the only on-site engineer with the necessary level of skill in fluid dynamics to handle the drag calculations. Jono decided a break from staring at the equations would help. He opened his email and clicked on a message from his manager.

> **From:** Kay_Gruen@acme.com.nz
> **To:** Jono_Westwood@acme.com.nz
> **Date:** Wed, 22 Jan 2024 10:33:49
> **Subject:** RE: Annual Performance Appraisal mtg
>
> Jono,
>
> A reminder — your annual performance appraisal is this Friday (10 am).
>
> I've got your paperwork from last year but have not rec'd your Personal Performance Appraisal for 2023. Will also need your Career Development Plan for 2024 for the meeting. If you get paperwork to me today, I'll look over it and draft Manager's Appraisal for you to review during our meeting. All the forms you need are on the server in HR folder.
>
> Thanks, Kay

Jono suppressed a curse — he had forgotten about his performance appraisal. The paperwork would take half a day and he really needed a clear head to sort out what to say. There were many things he had done 'ok' in the last 12 months, but there were one or two things he had not handled particularly well; like the time he had been part of a shouting match with a technician, and the time he missed a deadline for calculations and held the production team up for three days.

Also, he had been thinking about applying for a job with a different division but there were some questions he needed to clear up about whether this would be a good career move. Maybe he could talk it over with Kay? He wondered if Kay might lose confidence in his commitment to the hull design team though.

A tough week just started looking a bit tougher. Jono was already having a hard time dealing with the flu and a work commitment he felt obliged to meet. Now he faced the task of describing and providing evidence of his performance over the past 12 months, as well as detailing what his career plans were for the next 12 months and explaining how he planned to meet them.

Have you ever been in a similar position to Jono? You may remember times in your personal or work life when you have worked under pressure, with competing deadlines. At times, your physical or mental state may have made it hard for you to perform at your best. Part of planning your future as an engineer involves developing strategies to manage yourself, your work and your workplace. Undergraduate study is an ideal time to develop career skills, like those needed by Jono to deal with the competing priorities and demands of his health, design work and career planning.

In this chapter, we look at the exciting and important work coming out of new and emerging fields of engineering such as biotechnology, animatronics, smart materials and engineering responses to climate change. We will also consider engineering work and careers from the perspective of career planning and workplace management. The chapter will provide insights into how to explain your work performance, handle yourself in the workplace and plan short- and long-term strategies to meet your career goals. Additionally, the chapter reviews some of the forces that will have an impact on your employment and employability in the future, with topics such as the globalisation of engineering work.

Engineering is a profession that has much to offer graduates who are well prepared and confident in explaining who they are, what they can do and where they want to go in their professional life.

15.1 Engineering to meet future global challenges

LEARNING OBJECTIVE 15.1 Discuss the engineer's role in addressing global challenges.

You and your classmates are the future of engineering. Look around the lecture hall, the lab or in the mirror and you will see people who will work on some of the most important, intractable and interesting challenges our society will face in coming years. The work of all engineers is important to society, and the coming decades offer some profound challenges that will be addressed by engineers working in multidisciplinary teams with other specialists (such as policy-makers, economists, social geographers, biologists, communications specialists and architects) and with stakeholders whose lives, environment and livelihood rest on the solutions crafted by such teams.

The potential roles of engineers in meeting two global challenges are reviewed here: accessible health care and climate change adaptation. After reading this section, you may like to think through what engineers in your discipline could contribute to address other current and future global challenges such as information overload, global viral pandemics, social media and privacy, fossil fuel availability, food security, overpopulation, peace-making and national security, and water availability.

Antibiotic resistance

Micro-organisms — such as bacteria, fungi and yeasts — are single-celled entities typically measured in microns (millionths of a metre). Microbes are our constant companions; they are all around us — in the air, on our skin, in our food and nestled in every nook and cranny of our reproductive, respiratory and digestive systems. This might sound alarming, but it is mostly a good thing as micro-organisms in our digestive system keep us healthy by lining our stomachs, effectively guarding against invasion by disease causing bacteria like *Salmonella typhi* (which causes typhoid fever) and *Helicobacter pylori* (which causes stomach ulcers).

Without the micro-organism yeast, beer would be a bittersweet experience — sugary, flat, astringent and entirely lacking in alcohol. Biofilms are made up of millions upon millions of bacteria living, eating and reproducing in their slimy conjoined colonies. They look unpleasant but are responsible for secondary treatment of our sewage, stripping excess nutrients from our wastewater and reducing the incidence of toxic algal blooms in receiving waters.

In short, many micro-organisms are our friends and without them we'd be in trouble. But (and there is always a 'but'), some micro-organisms cause no end of problems, and a particular class of bacteria offer one of the most serious challenges facing humankind. Engineering has a role to play in meeting this challenge and we will review that role after a bit of background on antibiotics, antibiotic resistance and the threat resistance poses to humanity.

Antibiotics, as their name suggests, are chemical agents that act against living cells. The earliest antibiotics used in medicine were chemical compounds that various micro-organisms (e.g. fungi, bacteria) made to combat or kill other micro-organisms — chemical warfare on a tiny scale. Penicillin is a compound produced by fungi of the genus *Penicillium*, and was the first antibiotic to be mass produced and made available for medical use. One of the greatest milestones in medical history, it took over 15 years of exhaustive research by some of the greatest microbiologists, chemists, medical doctors and mycologists of the day to reach the point, in 1943, where penicillin could be mass manufactured and used as a routine medical therapy. Penicillin was a wonder drug that made dramatic improvements to infection control and post-operative survival rates, and drastically reduced mortality rates from previously life-threatening diseases, such as meningitis, gangrene, syphilis and pneumonia. However, as early as 1947, microbes began to appear that could resist penicillin. So began the race by medical researchers and pharmaceuticals manufacturers to discover new antibiotics that would keep us one step ahead of antibiotic resistance, and avoid a return to the days when a septic cut or sore throat could quickly develop into a life-threatening systemic bacterial disease.

Antibiotic resistance is when disease-causing micro-organisms become immune to the effects of an antibiotic that was previously used against them. Resistance to antibiotics can develop through the natural genetic variation that results over cycles of microbial reproduction. However, human activities that expose microbes to low or sub-lethal doses of antibiotic are currently speeding up the development of resistance. Low exposure can leave those microbes that are slightly more resistant to the antibiotic, while killing off microbes that are less resistant to the drug. This means that the 'tougher' microbes are left behind to breed together and genetic variability in their offspring inevitably leads to a proportion of 'extra-tough' microbes in the next generations. In the world of antibiotic resistance, these offspring may include the health professional's next nightmare: a new 'superbug'. Superbug is the name given by the World Health Organization (WHO) and others to describe micro-organisms that are concurrently resistant to many antibiotics. Multi-resistant *Staphylococcus aureus* (MRSA) and vancomycin-resistant *enterococci* (VRE) are of main concern for Australian health authorities, and a quick Google search will bring up reports and newspaper articles on the threat these two superbugs pose for Australia and its health services. The WHO has described antibiotic resistance as one of the world's most critical health challenges, stating that it is 'rising to dangerously high levels in all parts of the world . . . threatening our ability to treat common infectious diseases' (World Health Organization 2019).

The COVID-19 pandemic has added to concerns about antibiotic resistance; an American research team identified secondary bacterial pneumonia as the cause of death for nearly half of the patients they reviewed who had died with COVID-19 (Paul 2023). During the scramble to respond, it became common for medical doctors to prescribe antibiotics on the basis of limited information about the microbes driving secondary infection. Early in the COVID-19 pandemic, prestigious medical journal *Lancet Global Health* published that:

> When faced with patients who are critically ill and hospitalised, and the diagnosis of a potential bacterial superinfection is uncertain, healthcare providers will and should err on the side of treating with broad-spectrum antibiotics.
>
> . . .
>
> Adherence to established antibiotic stewardship programs has declined as providers struggle to save the lives of patients with COVID-19 (Ginsburg & Klugman 2020).

The Lancet published sobering examples of the impact of antibiotic resistance (Murray et al. 2022), outlining that:

- in 2019, there were an estimated 4.95 million deaths globally associated with antimicrobial resistance (AMR)
- six pathogens with resistance were implicated in 73 per cent of global AMR-associated deaths, with resistant *Escherichia coli* and *Staphylococcus aureus* the leading killers.

The WHO identified methicillin-resistant *Staphylococcus aureus* (MRSA) as particularly lethal; people with MRSA 'are estimated to be 64 per cent more likely to die than people with a non-resistant form of the infection' (World Health Organization 2014).

But what does all this mean to an engineer? Apart from the generic concern that engineers might have about falling foul of an antibiotic resistant microbe themselves, what role might engineers play in the fight against these devastating microbes, and in current efforts to safeguard the effectiveness of our antibiotic frontline? As at 2024, 137 countries were enrolled to participate in the WHO-led Global Antimicrobial Resistance and Use Surveillance System (GLASS) (World Health Organization 2024). The use of antibiotics in ways that reduce development of AMR have been enshrined in the Sustainable Development Goals, specifically SDG3, which makes the issue of AMR very relevant for engineers working in agriculture, water, wastewater, sustainable development and population health, for example (World Health Organization 2021).

Distinguished Professor Amy Pruden, an environmental engineering researcher at Virginia Tech in the United States, has been investigating levels of antibiotic resistance in receiving waters downstream from municipal sewage treatment plants and feedlot operations. Feedlots are high-density animal production facilities where pigs, poultry and cattle are raised for meat. One of the reasons for the development of resistance among microbes is the inappropriate or incorrect use of antibiotics, and WHO cites antibiotic use in some feedlots as an example. Sub-therapeutic doses of antibiotics are often included in feed to promote growth or prevent disease among animals that are being raised in close contact. Animals raised in feedlots need antibiotics to combat routine health problems, and health effects related to crowding; transportation stress; inappropriate handling; procedures with possible pain impact, such as branding, dehorning, debeaking or castration; restriction of innate behaviours; or adjustments to diet (Mench et al. 2008). A concern with the routine sub-therapeutic use of antibiotics in animal feed is that oral antibiotics can pass through the digestive system of humans and animals and be excreted in active form. Conventional sewage treatment (activated sludge) has been shown to reduce the concentration of antibiotics in wastewater by only 92 per cent (Watkinson et al. 2007), leaving 8 per cent in treated effluent.

Dr Pruden and her colleagues tested receiving waters, and found antibiotic resistance in 100 per cent of receiving waters downstream of sewage treatment plants and feedlots, compared with just one incidence of antibiotic resistance in pristine stretches of river (upstream of receiving waters) (Storteboom et al. 2010). Regular, sub-therapeutic use of antibiotics in feedlots, excretion of active antibiotics in faecal waste and inadequate wastewater treatment meant that receiving waters had become a breeding ground for antibiotic resistant micro-organisms.

In Australia, the cattle feedlot industry has Antimicrobial Stewardship Guidelines (Meat and Livestock Australia 2018) which support feedlot operators towards best-practice management of antimicrobials. Research supported by Meat and Livestock Australia (MLA) showed an increase from 1.5 per cent ampicillin resistant faecal *E. coli* at entry into the feedlot, to 9.6 per cent ampicillin resistance for faecal *E. coli* collected after cattle had been slaughtered. At slaughter, faecal samples yielded 20.7 per cent tetracycline resistance (Messele et al. 2023). Both of these antibiotics are in current use in human health care for combating bacterial infection.

Pruden has explained that reducing the spread of antibiotic resistance is a critical measure needed to prolong the effectiveness of currently available antibiotics. What a challenge for the engineer!

Amy Pruden, Virginia Tech Associate Professor of Civil and Environmental Engineering

To address antibiotic residue in wastewaters, chemical and civil engineers might need to design wastewater treatment facilities to remove antibiotics. Reverse osmosis, nanofiltration, constructed wetlands and ozonation have all been evaluated as effective in antibiotic removal from wastewater, although the technology is novel and may be viewed by water authorities and feedlot owners as unacceptably expensive. If antibiotic pollution is a regular component of wastewater, what then are the implications for engineers tasked to design treatment of wastewater for potable reuse? Should environmental engineers include antibiotic monitoring in regular water quality assessments? How might agricultural engineers redesign intensive meat production systems to alleviate animal stress, reduce health problems and decrease the need for widespread routine use of antibiotics in animal production? These are just a few of the challenges that engineers might tackle in the fight against antibiotic resistance. Can you think of any other roles that engineers, such as those in your discipline, might take in addressing this complex problem?

In addition to antibiotic pollutants in wastewater causing a buildup of microbial resistance in the environment, antibiotic resistance can also be caused by incorrect use of antibiotic medicines. Many engineering disciplines — chemical, pharmaceutical, material, mechatronic, biomedical and bioprocess — have knowledge and skills to apply to the problem. These engineers could assist in the design of more effective antibiotic 'delivery systems' (i.e. how medicines are administered). More effective delivery systems might meet a specification that no antibiotic emerge from the human and animal digestive system in active condition. Alternatively, a slow-release coating on antibiotic tablets could reduce the number of tablets patients need to take; or implantable or remotely activated delivery devices could be developed to reduce patients' responsibility for administering their own medication.

The following spotlight shows how engineering research is working towards better health outcomes in another area where the delivery and control of medication can be challenging for health professionals.

SPOTLIGHT

Digital diagnosis: how your smartphone or wearable device could forecast illness

CALEB FERGUSON, WESTERN SYDNEY UNIVERSITY AND SALLY INGLIS, UNIVERSITY OF TECHNOLOGY SYDNEY

What if you could forecast sickness, before you even had any symptoms? Your smartphone and digital data might be able to help.

If you carry your smartphone with you everywhere, then it's probably already tracking a lot of data about you and your behaviour. If you have a wearable device, you'll be generating a multitude of fitness data too. Put it all together and you get your 'digital phenotype' — a comprehensive picture of your health and well-being, kind of like a digital jigsaw puzzle.

When collated with data obtained from clinical assessment, such as blood tests and diagnostic imaging, it could generate valuable insights for health professionals to inform care planning. And there is increasing evidence that web data can be useful to digitally detect disease, and predict suicide, influenza outbreaks or new cases of HIV.

What do your digital devices know about you?

As you get out of bed, your smartphone captures your usual time of waking, and possibly the amount and quality of sleep you had that night. If you commute to work, it captures your journey — as well the amount of time spent sitting or standing during that travel period — via GPS and geographical data.

Over the course of the day, it tracks your physical activity — distance walked or climbed, standing time, and an estimate of the exercise undertaken and energy expended — via accelerometers.

You might have logged your medical history, diet and weight data through an app or a smart scale. Your heart rate and rhythm could be recorded via your smart watch. Your device usage, including screen time and internet browsing history, is captured by your phone itself, or via various apps you may have downloaded.

How could this information help?

Take the example of chronic heart failure. An estimated 480 000 Australians are living with this common and burdensome cardiovascular condition, which is increasingly prevalent among older adults. It's also common for these individuals to live with other health conditions.

While heart failure is a progressive and terminal condition, there are often signs and symptoms preceding periods of deterioration. These include ankle or leg swelling (oedema), shortness of breath (dyspnoea), fatigue and reduced capacity for exercise.

If someone living with heart failure was being monitored via a smartphone, the data could highlight increasingly disrupted sleep patterns, reduced physical activity, changes in weight and irregularity in vital signs, such as heart rate. These changes in a person's daily functioning would clearly indicate a deteriorating condition that should, ideally, trigger a visit to their nurse, GP or emergency department.

Taking it a step further, the changes could trigger an alert to a health professional, allowing timely clinical assessment and intervention. This would not only save significant health costs, but improve symptom management and unwanted deterioration.

Are health professionals using this information?

Globally, the population is living longer and with multiple chronic health conditions. But few health care professionals are currently tapping into the digital phenotype to inform clinical decisions.

There isn't a lot of education about how to access these data, or how to routinely apply it within a practice, at the bedside or in the clinic. Protocols, procedures and platforms for collecting, analysing and integrating these data into medical record systems aren't common either.

Some forms of remote health monitoring already exist. For example, telehealth and telemonitoring systems allow clinicians to remotely deliver health care via telephone and web-based platforms. But their uptake in routine clinical care has been relatively slow — perhaps because it requires both patients and clinicians to manually log data. This is in contrast to the kind of passive data collection that occurs with smartphone and wearable devices.

The next frontier of health care

We urgently need innovative models of health care that support patient monitoring in those with chronic conditions, and the early detection of symptoms. That means establishing data feedback loops between patients and clinicians. Passive data collection via smartphones and wearable devices may provide monitoring technology as a low cost, accessible solution that doesn't put a huge burden on the provider and patient for data input.

It is critical to embed these technologies within the existing medical data and electronic medical record systems. But these technologies present complex clinical, legal, ethical and health systems challenges.

Patients must consent to health professionals accessing and using their personal digital data to inform their healthcare. There must be data security all along the chain. And we must not overlook value of a direct and personal relationship between the patient and clinician.

Monitoring data alone cannot improve outcomes, but it may be used to identify deterioration in the patient's condition and thereby enable timely clinical assessment and intervention.

Source: Ferguson and Inglis (2018). Originally published on *The Conversation*.

CRITICAL THINKING

Can you imagine other support that might now be possible with new wearable technologies and wireless communication?

Climate change adaptation

As an engineering undergraduate student, your introductory studies in physics (electromagnetic radiation) and thermodynamics (heat transfer, conservation of energy) will provide good grounding to understand and explain the general principles that govern our planet's climate.

What is climate change?

Key to explaining climate change is the 'greenhouse effect'. Geoscience Australia points out that the 'greenhouse effect' is an important and natural part of the Earth's climate and, without it, our planet would be a far colder place (Geoscience Australia 2011). When sunlight hits the surface of the earth it is absorbed, and the visible light (short-wave radiation) is converted into heat (infra-red or long-wave radiation), which is radiated back into the atmosphere towards space. Figure 15.1 shows the electromagnetic spectrum and highlights the sun's energy spectrum with its peak in the visible light section of the spectrum.

Figure 15.2 shows the greenhouse effect. As Geoscience Australia explains, sunlight is converted to heat at the Earth's surface and radiated back into the atmosphere where so-called 'greenhouse gases' (e.g. molecules of carbon dioxide, methane, water vapour) absorb some of the radiated heat (infra-red or long-wave radiation) (Geoscience Australia 2011). Eventually, these molecules emit heat back into the atmosphere as infra-red radiation. Some of this infra-red radiation is absorbed by other greenhouse gases, some is emitted to space, and some is absorbed at the Earth's surface; the cycles of absorption, conversion and emission are then repeated (figure 15.2).

Essentially, these cycles slow the loss of heat to space, keeping the Earth's surface warmer than it would be without our blanket of greenhouse gases. Without the greenhouse effect, the atmosphere would be about 30 °C cooler, and life on Earth would be a very difficult prospect indeed (Geoscience Australia 2011). However, as Goldilocks found, there is a sweet spot in terms of temperature.

FIGURE 15.1 A schematic of the electromagnetic spectrum, showing the sun's energy output (yellow line) in relation to wavelength

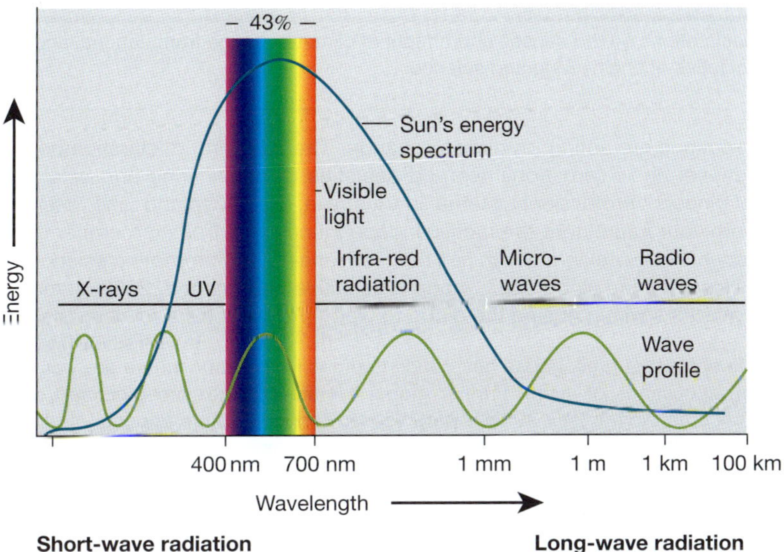

Source: Geoscience Australia (2011).

FIGURE 15.2 An overview of the greenhouse effect

Source: Reproduced from IPCC Working Group Second Assessment Report (1996) by Geoscience Australia (2011).

It is widely agreed that human activity is the cause of an enhanced greenhouse effect through excessive emissions of carbon dioxide and other greenhouse gases (see figure 15.3). As is clear from figure 15.2, an enhanced greenhouse effect increases the retention of heat within the Earth's atmosphere. While the impact of increased heat retention is highly variable across the globe, the net result is global climate change in the form of higher average global temperatures and increased climate variability.

FIGURE 15.3 Declarations on the causes and effects of climate change from national and international learned societies of engineering and science

Mainstream organisations are acting on climate change. The following declarations on climate change come from the national and international learned societies of engineering and science which are the gatekeepers of engineers' professional status in Australia, New Zealand and internationally. These organisations govern your future right to practise independently as a trusted member of society.

The Royal Society of New Zealand, which has been advancing and promoting science, technology and the humanities in New Zealand since 1867, made a definitive statement in 2008, stating that increasing greenhouse gas emissions were causing the globe to warm (Royal Society of New Zealand 2008):

> Measurements show that greenhouse gas concentrations in the atmosphere are well above levels seen for many thousands of years. Further global climate changes are predicted, with impacts expected to become more costly as time progresses. Reducing future impacts of climate change will require substantial reductions of greenhouse gas emissions.

The international Institution of Chemical Engineers (IChemE) reviewed and updated their position on climate change in 2022, stating:

> Climate science is established — we are in the midst of a climate emergency. Human activity is causing the climate to change, with significant adverse consequences. IChemE accepts the veracity of the science and its conclusions published by the Intergovernmental Panel on Climate Change (IPCC). To avoid irreparable social, economic and environmental damage, it is essential that we accelerate our efforts to decarbonise our economic systems (p. 2).

The following photo illustrates how extreme weather events (e.g. storm surges) and sea-level rise are likely to affect coastal infrastructure. Damage to coastal infrastructure is a serious concern in Australia, where around 85 per cent of the population live within 50 kilometres of the coast (Australian Bureau of Statistics 2004).

Some of the impacts of climate change are already being felt including more extreme weather events, ocean acidification and coral bleaching, sea-level rise and changes to the intensity and distribution of rainfall.

In the *State of the Climate 2022* report, the Commonwealth Scientific and Industrial Research Organisation (CSIRO) and Bureau of Meteorology (BOM) reported the following (Bureau of Meteorology & CSIRO 2022).

- Australia's climate has warmed by an average of 1.47 °C since 1910.
- Around Australia, sea levels continue to rise, sea surface temperatures have increased by an average of 1.05 °C since 1900, and seawater is becoming increasingly acidic.
- May to July rainfall has declined by 19 per cent since 1970 in the southwest of Australia.
- Across large parts of Australia, extreme fire weather has increased since the 1950s, with longer fire seasons.
- Globally, concentrations in the atmosphere of long-lived greenhouse gases including carbon dioxide are still increasing, with global annual mean carbon dioxide (CO_2) concentrations reaching 414.4 parts per million (ppm) in 2021.

If you are interested in learning more, you can consult a range of reputable and comprehensive resources via the websites of the Intergovernmental Panel on Climate Change (IPCC), CSIRO, US EPA Climate Change Section, the ARC Centre of Excellence for Climate System Science (a coalition of five Australian universities with research expertise in climate systems) or Geoscience Australia.

Engineers and climate change

Many engineers are already contributing to reductions in greenhouse gas emissions. In New Zealand, engineers from a range of disciplines are responding to the emissions trading scheme that was enacted in 2008 (Leining & Kerr 2018) and assisting industries, schools, councils and communities to reign in their energy use and resulting carbon footprint. In the chapter on sustainable engineering, the Z-Mag spotlight provided insight into the very practical action that mechanical engineers can take to reduce carbon emissions (and, in that case, product defects and energy costs). Policies related to climate change are examples of how engineering work is influenced by politics.

The Engineers Australia Code of Ethics mandates that professional engineers registered with Engineers Australia 'practise within areas of competence' (Engineers Australia 2022). For those registered with Engineering New Zealand, the Code states that engineers shall 'only undertake engineering activities that are within your competence' (Engineering New Zealand 2016). This means that engineers without specialist knowledge of climate science need to be cautious in commenting on climate change. As professional engineers trained and registered to practise in your specialist discipline area, your climate change adaptation role will likely be to react to the recommendations of specialist climate scientists, and the policy-makers and politicians who decide how our society responds to threats and changes of this nature. Coastal engineering is one of the many disciplines that will be at the forefront of the response if current climate models are accurate. The following spotlight shows the kind of technically robust and socially acceptable response that coastal engineers and engineers in the maritime and local government sectors might be asked to plan.

SPOTLIGHT

Climate change adaptation and maritime engineering

In 2011, Geoscience Australia released maps illustrating the potential effect of sea-level rise on key urban regions of the Australian coast. The maps modelled likely impacts for the period around 2100 with low, medium and high sea-level rise assumptions of 50 centimetres, 80 centimetres and 110 centimetres respectively. The maps revealed a 110-centimetre sea-level rise inundation map for Newcastle, New South Wales, where the Newcastle Port Corporation operates one of the major dispatch points for Australia's mineral exports (e.g. coal and iron ore), agricultural products (e.g. wheat, barley and canola), containerised goods and general cargo. Port facilities in the suburb of Carrington are likely to be substantially affected, including several berths, coal and grain operations, and ancillary services such as a helicopter base, NSW Maritime Police and NSW Water Police.

Dr Melissa Nursey-Bray from the University of Adelaide's Department of Geography and Environmental Studies and her colleagues investigated the likely impact of climate change on coastal infrastructure, including ports (Nursey-Bray et al. 2012). Some of the findings from this study are shown in table 15.1 and highlight the challenges facing future coastal, maritime, ocean, civil, mechanical, environmental and supply chain engineers and naval architects who will work to adapt our ports, shipping and coastal infrastructure to withstand the impacts of climate change.

TABLE 15.1 **Observed and projected changes in some climate change observations and their likely impact on the port sector**

Climate change observation	Observed changes (to date)	Projected changes: from present to 2100*	Likely (direct) impact on Australian ports
Sea level	Rise of 1.8 $\pm$ 0.5 mm yr^{-1} since 1960	Rise of 0.18–0.59 m, least increase in the southern oceans and parts of North Atlantic Ocean	Storm surges, inundation and flooding of ports, insurance costs, WHS issues
Ocean acidity	Increase in acidic (surface pH decrease of 0.1 units)	Increase in acidity (surface pH decrease by 0.14–0.35 pH units)	Increased corrosion, biodeterioration resulting in higher maintenance costs
Tropical storms (e.g. cyclones, hurricanes, typhoons)	Increase in intensity	Increase in intensity — but fewer of them — further penetration at higher latitudes	Increased wind speeds = damage to structures, engineering upgrades to structures and cargo handling equipment required
Heavy precipitation events	Increase in number	Increase in heavy daily rainfall events in many regions	Coastal flooding in port cities, increased run-off, disruption of transport (road and rail) to ports; siltation, heavy metals and pollutants entering ports, increased dredging

* Solomon et al. (2007).

Source: Nursey-Bray et al. (2012).

Engineers will need to take local and international action to minimise and respond to the impacts of climate change. Civil engineer, Sue Murphy, led the Western Australian Water Corporation for ten years, as CEO from 2008 (Water Corporation of Western Australia 2018). The Water Corporation is the main supplier of safe and reliable drinking water, waste water treatment and stormwater drainage in WA, which has experienced dramatic rainfall changes since the 1970s. With significant stakeholder consultation, the Water Corporation has been proactive in sustainably managing water, based on climate observations and modelling, by reducing consumption, accessing groundwater, establishing desalination plants and recycling treated wastewater through monitored groundwater replenishment (Water Corporation of Western Australia 2018).

Engineering and ecological restoration

Agricultural engineer, Andrew Guzzomi, works with agricultural scientists and botanists to address problems such as landcare and feeding the world's population. Restoring native vegetation after mining, erosion, clearing, pests or heavy stocking is important. Native vegetation can help protect against erosion and salination, and provide food and habitat for native fauna. Additionally, maintaining biodiversity is important for protecting our resilience against climate change.

Unfortunately, many native Australian seeds are more difficult to handle in bulk using machinery than common commercially used food grains. This is because the native seeds often have husky uneven exteriors. To improve the process, engineering final-year project student Alan Ling and seed scientists Todd Erickson, David Merritt and Kingsley Dixon developed a seed flamer (see figure 15.4). Their patented device repeatedly exposes seeds of wild plant species to a flame, under precise control (see figure 15.5). It systematically and efficiently improves wild seed geometry, transforming the ability to store and handle native seeds that are needed to revegetate the vast degraded global environments also typical of remote Australia such as mine sites and degraded agricultural lands.

FIGURE 15.4 Andrew Guzzomi with a seed flamer

Source: Andrew Guzzomi.

FIGURE 15.5 Flaming seeds

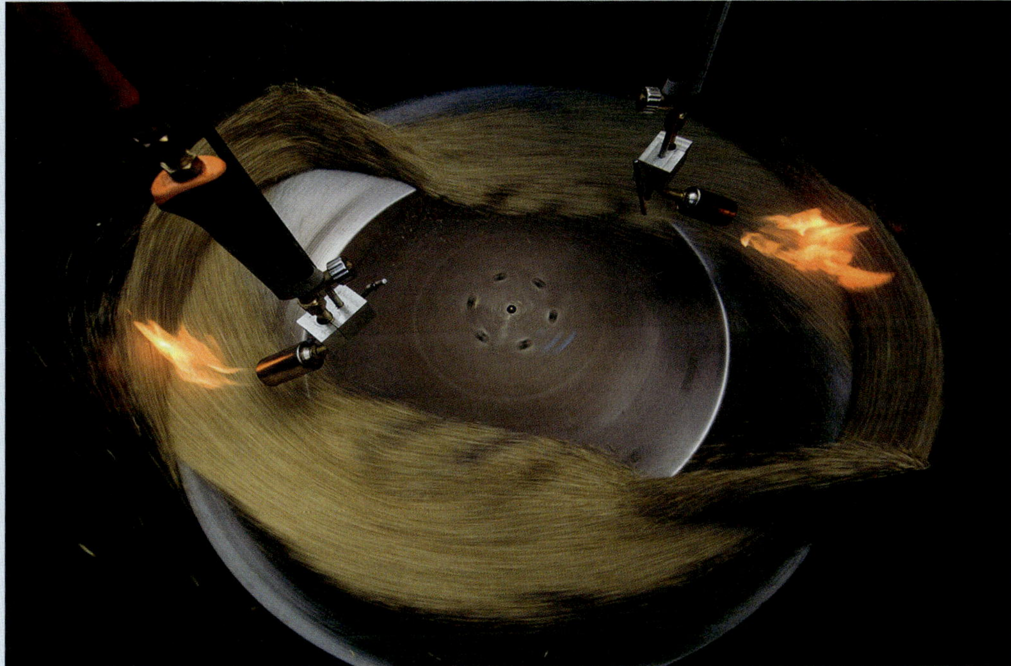

Source: Andrew Guzzomi.

CRITICAL THINKING

Thinking of the likely impacts of climate change, how could your discipline of engineering contribute to addressing those impacts? How would those sorts of impacts affect the place you call home?

15.2 Engineering and globalisation

LEARNING OBJECTIVE 15.2 Describe how globalisation may affect you in your future professional life.

Fast forward five years from now — you will be a qualified engineer working in one of many different types of activities and occupations. You may even use your engineering degree in creative ways that we haven't yet thought of. The skills you develop during this time will be invaluable for a wide variety of career paths. It is also worth noting that the term 'engineer' is used in many different ways in everyday speech. The classic misconception in some countries is that if you are an engineer you are considered to be a train driver, while in countries where the term is more closely connected to the term 'ingenious' engineers are thought to be highly innovative creators. There are several different levels of engineers designated by Engineers Australia (and other international professional bodies), which include professional engineers, engineering technologists and engineering associates. If you are studying for a four-year degree or masters program then you will be aiming for professional engineer status, which means that you are responsible for designing the systems that many others will be using and addressing previously unseen problems. This brings with it much excitement and creativity and a huge degree of responsibility, as discussed in the sections on ethics within this text. In this section we are going to look at some of the most recent trends in globalisation as they relate to engineers and engineering — from a personal level of employment to a broader level of the systems and processes that you as engineers will develop and implement and how these are affected by, and in turn influence, globalisation.

Employment in Australia

Governments, employers and those who pay for engineering services or products need engineers. However, the numbers of engineers, the industries, locations and specific capabilities needed change over time. Graduates and experienced engineers adapt and move in order to work and contribute to society. Qualified engineers have capabilities that are valuable in many fields and they often work in roles other than as professional engineers.

During the mid-2000s, key groups and individuals in Australia and New Zealand realised that each country was heading for serious and ongoing shortfalls in the number of graduate engineers entering the labour market. In 2023, CEO of Engineers Australia Romilly Madew AO launched the most recent *Statistical Overview of the Engineering Profession* report which suggested the engineering skills crisis has deepened. She pointed out that 'as the world continues to embrace technology and systems, becoming more sophisticated and interdependent, our economy and society are more reliant than ever on the engineering profession' (Engineers Australia 2023a). The report laid out the demographics of the Australian engineering workforce. Can you see yourself in the statistics in figure 15.6?

FIGURE 15.6 Australian engineering workforce statistics 2023

Engineering graduates

- Around 25 per cent of Bachelor of Engineering (Honours) students graduate in the minimum time for full-time study (many students study part time).
- Around 75 per cent of engineering students graduate within six years.
- Over 75 per cent of domestic commencements complete a degree in nine years.
- There is a 5 per cent attrition rate in the first year, with a further 20 per cent attrition over later years.
- Low growth in domestic commencements relative to the population will limit the supply of Australian engineers over the short term.

Women in engineering

- Only 17.7 per cent of engineering graduates are women, with 18.9 per cent enrolment.
- 16 per cent of qualified engineers in Australia are women, with 76 per cent of them born overseas. Australian-born women make up just 3.8 per cent of the total engineering qualified population in Australia.
- Female engineers make up 14 per cent of the engineering workforce.
- The share of women in engineering occupations increased in most industries, except for a decline from 7.3 per cent to 6 per cent in agriculture, forestry and fisheries.
- At the current rate (2.4 per cent increase every five years), it will take 70.8 years for female engineers to reach equal representation in the profession.
- Lack of awareness about engineering is a common reason for women not considering it as a career.

Engineers not working in engineering

- Over 20 per cent of Australia's qualified engineers are not in the labour force.
- It's expected that up to 68 133 engineers will retire over the next 15 years, with 25 000 retiring in the next five years.
- There's a retention problem, with approximately 3200 engineers leaving the profession for other sectors annually.

Migrant engineers

- The majority of Australia's engineering workforce is born overseas.
- Overseas-born engineers contributed to 70 per cent of the growth in the engineering labour force from 2016 to 2021.
- Engineers born overseas comprise 62.7 per cent of the qualified engineer population, 62 per cent of the engineer-qualified labour force, and 55.8 per cent of the population in engineering occupations.
- Australia competes with other countries for skilled migrant engineers.

Millennial engineers

- Nearly 50 per cent of Australia's qualified engineers in the labour force are under 40 years of age.
- Millennials are the largest generation in the engineering profession.
- The proportion of qualified engineers undertaking part-time work is increasing faster than those in full-time positions.

First Nations engineers

- While First Nations qualified engineers and those working in engineering occupations increased in all industries, they remain a small proportion of the engineering population at just 0.3 per cent.

Source: Adapted from Engineers Australia (2023b).

For many decades, engineers in Australia and elsewhere have worked in professions outside engineering (Trevelyan & Tilli 2010). As the above statistics show, many engineers continue to find opportunities outside engineering, even early in their careers. In Australia the end of the mining boom was associated with reduced employment of engineers. Using Australian Bureau of Statistics Education and Work Survey data related to all qualified engineers (associate, technologist and professional), Engineers Australia reported a reduction in the demand for engineers to undertake engineering work of 7.3 per cent in 2016, compared with 4.6 per cent drop in the years of the global financial crisis (Kaspura 2017, p. 32). Palmer and Campbell (2018) analysed Australian census data and reported that in 2016 of the 18 580 citizens or permanent residents aged 20 to 24 years with bachelor engineering degrees, 24 per cent were working as professional engineers, 39 per cent were otherwise employed and 36 per cent were not working. Engineering graduates are valuable in engineering roles and their skills are also sought after for in other work. Fluctuations in the economy, beyond the control of the individual, mean you may be forced to seek work outside your immediate areas of specialism or even outside engineering.

KEY POINT

Engineering graduates work in engineering and non-engineering roles and industries.

Globalisation

Globalisation is a complex term used by different people in different ways. It is commonly used to explain what happens when companies or individuals begin to operate internationally due to liberalisation of trade and commercial restrictions, but it means much more than this. The COVID-19 pandemic showed some of the many facets of globalisation. High availability and use of international travel allowed the virus to disseminate rapidly to all corners of the globe in a matter of weeks. The *Harvard Business Review* reported that the peak of the crisis and public health response resulted in:

> the largest and fastest decline in international flows in modern history [including] a 13–32% decline in merchandise trade, a 30–40% reduction in foreign direct investment, and a 44–80% drop in international airline passengers (Altman 2020).

However, the COVID-19 pandemic period saw unprecedented global cooperation on the manufacture and distribution of health products (e.g. rapid development, mass manufacture and distribution of new vaccines).

There are those who see globalisation as positive, a merging of cultures and systems, a progression that is inevitable as we move increasingly into a global way of living. Others are highly critical, and those who support the 'anti-globalisation movement' criticise the rapid introduction of market-driven processes to countries whose economies are not strong enough to withstand globalisation's severe backlash that affects vulnerable populations. They are also concerned that Western culture will gradually wipe out other less dominant ways of living, indigenous ways of being will be eroded globally, we will increasingly see the same shops and brands (often US or European) in villages all over the world, we will have less choice in what we buy and that minority languages — which hold within them certain ways of expressing ideas and thoughts — will disappear.

A major issue is considered to be the decreasing diversity of food sources (we have reduced from several thousand to just a handful of potato varieties in some countries), which will render us defenceless against resistant diseases. The local food movement supports locally grown and distributed food to combat some of these issues.

Companies that operate in a globalised way are sometimes referred to as multinational or transnational companies. Examples of multinational companies include Microsoft, McDonald's and BHP Billiton. Globalisation might be viewed through economic, political, geographical, philosophical, technological or sociological lenses. The way we view something changes what we focus on and consider important. Consider the lens of engineering: how might engineers think about this tricky concept; what has it got it do with us? It turns out that we have much to do with it, both in terms of input and output.

Engineers create the technologies that allow faster or cheaper transportation or communication across national boundaries and are therefore drivers of these global changes. They also take advantage of the potential for outsourcing production of components to countries where labour is cheaper. This can have the positive effect of generating jobs in areas of low employment, but it can also have devastating effects where factories close down and move to chase lower wages. Australia is suffering from a decimation of its manufacturing industry as a result of labour costs, due in part to the high cost of living. Furthermore, outsourcing of labour can introduce ethical issues where worker rights are abused (in sweatshops, for example) under circumstances in which the work is done outside the national boundary, and codes of ethics or conduct cannot be monitored.

KEY POINT

Globalisation offers future engineers great opportunities to work across national borders and within diverse cultures, while including substantial associated challenges such as cross-cultural communication, adherence to diverse ethical and legislative frameworks, and managing work–life balance.

Development and post development

Globalisation is connected to another critical concept, which is crucial to the role engineers play on a local and international level: **international development**. Like globalisation, international development has many mixed definitions and interpretations. For some it is simply inevitable 'progress', while for others it is a system whereby one more dominant nation 'helps' another to reach some global standard or way of living. Critics argue that globalisation today is a contemporary version of the 'development' of colonial times. Mike Davis has studied India under British rule and is highly critical of strategies that were promoted as 'developing' a nation but that he believes had more selfish motives (Davis 2002). He suggests that 'Londoners were eating India's bread' (Davis 2002, p. 26). He notes that between 1875 and 1900 'enough grain for 25 million people was exported from a country suffering the worst famines in history' (Davis 2002, p. 26). He blames engineering in part for this, 'the newly constructed railroads, lauded as institutional safeguards against famine, were used by merchants to ship grain inventories from outlying drought stricken districts to central depots for hoarding' (Davis 2002, p. 26). Historians and theorists such as Davis maintain that the reason nations are 'third world' or 'developing' or 'global south' is largely due to the actions of 'first world' nations in past years. Hence, they are highly critical of engineers in developed countries 'helping' other nations, which they have inadequate understanding of, potentially causing yet more (albeit unintended) problems.

All of this makes for a very complicated set of ideas to help you consider your future. We have tried, in the chapter on ethics, to help you think about *macro ethics* and the role that engineers play in the broader society. As you learn more about social and economic processes, you will discover that engineering is not apolitical; it is highly enmeshed with global and political changes. You are not expected to know about all of these trends in detail but the more you know about what you don't know, the more this will help you to ask the right questions of the right people at the right time.

SPOTLIGHT

On mining, poverty and development

Mining has a rather turbulent history in Peru. One of the key operators is Newmont Mining Corporation, which also operates in Australia. In 2007, Newmont became the first gold company to be part of the Dow Jones Sustainability World Index. Newmont also has a suite of policies and standards related to corporate social responsibility. Their code of conduct (Newmont 2018) includes, among others, tenets on topics such as:

- safety
- sustainability
- inclusion and respect
- fair hiring
- collaboration
- honesty
- established standards of operation and resource development
- relationships based on integrity
- honesty in the marketplace
- speaking up and cooperating candidly in investigations.

Mural in Celendin near the Conga extension site, commemorating local community members who died in the 2012 conflict

In a 2014 study on mining and community engagement in the Newmont Yanacocha and Conga mining areas in Cajamarca, by one of this text's authors, one of the local community members described his view of development as follows.

> I have one vision of development, and I think that the first clash starts between what the community wants and what the company wants. They come here imposing their type of development, that we are not used to here and we believe that it is not the best type of development, right? Because they come and install their supermarkets, they come and impose new forms of living that completely contradict with the one we have ... before we used to live well, and it's like that they say, 'no you are poor'. I have been in different countries, right? I have seen poverty in Europe, I have seen poverty in the US, and those poor, they are truly poor because they do not have anywhere to wash their hands. The

poor that we have in our communities they have land and we can sow and we have something to eat. But in the US, those poor people do not eat anything right? (Baillie et al. 2014)

CRITICAL THINKING

Newmont has a sustainability and community engagement plan, and yet in Peru they have experienced years of ongoing conflict with local community members. In the views of the person interviewed, one of the reasons for this clash is that the company has different views of development than the locals. Is there any way of reconciling this?

15.3 Futuristic engineering: emerging fields

LEARNING OBJECTIVE 15.3 Identify and discuss some emerging fields in engineering.

In the introduction to engineering chapter, we speculated on the future of engineering. We discussed how new social demands are driving engineers to deploy their technical skills and knowledge in new ways. It was suggested future engineers may fall into one of five categories: (1) technical specialist, (2) integrator, (3) change agent, (4) project manager and (5) asset maintainer (Henley Management College 2006; King Report 2008). Future engineers will also need a range of sociotechnical skills to operate effectively as professionals who serve society. The focus of the discussion in the introduction to engineering chapter was on how engineers will need to practise in the future. In this section, what engineers will do in the future is discussed. We will look at some emerging fields of engineering in which engineering science is being applied in new ways to address problems and create new technological opportunities.

Engineering is a field in which innovations in technology and the shifting demands of society mean constant change. The fundamental concepts of engineering tend to be relatively stable; for example, the conservation of energy, corrosion chemistry and fluid mechanics. However, technological innovation means such fundamentals are applied in new and creative ways in order to generate new technologies and applications (e.g. solar photovoltaic cells with nanotechnology components, new alloys with different corrosion characteristics, machine learning with new computing technology, autonomous vehicles with improved sensors and control systems, smart devices with new information and communication technology, new design techniques with improved visualisation technology such as virtual and augmented reality, electric vehicles and increased use of renewable energy with improved energy storage technology). The need for constant innovation means engineering fields are constantly expanding and diversifying. This phenomenon is described as the emergence of new fields in engineering. While it is difficult to accurately predict what the future holds, 'hot spots' for engineering innovation can develop in established engineering disciplines, in which future engineers may find satisfying careers. The next sections provide information about some different emerging engineering fields.

KEY POINT

New fields of engineering are continuously emerging in response to technological and scientific breakthroughs and new expectations and needs of society.

Industrial biotechnology

Industrial biotechnology is the development and use of technology for commercial or industrial-scale production of biological products. Biological processes such as cell reproduction, photosynthesis, enzyme generation and antibiotic synthesis are increased ('upscaled') to industrial proportions, for mass production of living tissue, photo-catalysed energy, enzymes and therapeutic drugs.

Some engineers who specialise in industrial biotechnology improve outcomes in the textiles industry by developing technology for mass production of bio-products that result in more efficient and environmentally sound processes and products. For example, catalase enzymes are mass produced in systems designed and run by biotechnology engineers.

Catalase enzymes are used in the textiles industry to degrade residual hydrogen peroxide, which is used to bleach raw cotton. Enzyme breakdown of hydrogen peroxide reduces water use in preparing bleached cotton to receive dye, and improves subsequent dye uptake, thereby improving the efficiency of dye usage. A second textiles example is the design by biotechnology engineers of systems to mass produce and apply cellulase enzymes to remove surface fuzz from finished textiles ('bio-polishing'), or to give finished textiles a stonewashed appearance ('bio-stoning'). Bio-polishing and bio-stoning cause less physical damage to textile fibres than the traditionally used abrasive polishing and stoning tools, and so result in more durable and robust textiles (Ikbal et al. 2024).

Materials science

Materials science investigates how the chemical and physical structure of materials influences how they function. Materials science is mostly focused on materials that have engineering application, such as steel, wood, copper, cotton, fibreglass and carbon. Materials scientists also develop and investigate new composite materials, such as polymers and alloys, and are at the forefront of development of **nanotechnology** (construction of particles in the nanometre size range that are designed to react or respond to cues and perform tasks in the sub-microscopic realm). Materials science is an exciting field where innovation is impacting in diverse areas; these include development of smart fabrics for clothing and defence technologies research. Smart fabrics are found in apparel such as jackets with micro-electronic components to regulate the jacket's warmth according to the wearer's body temperature; fabric that responds to increased ultraviolet (UV) radiation by decreasing UV permeability and super skins for use in the pool that reputedly decrease drag on swimmers. Miniaturised optics have numerous emerging applications in fields such as medicine, agriculture and bio-security.

SPOTLIGHT

Microscope in a needle

Breast cancer surgery is traumatic and expensive. It can be difficult for the surgeon to identify the edge of the tumour during an operation, in order to know exactly how much tissue to remove. If surgeons could precisely identify the edge of a tumour, then additional traumatic and costly surgery could be avoided, and risks reduced. To this end, a large interdisciplinary team of engineers, scientists and medical researchers, led by David Sampson, has developed a microscope in a needle (see figure 15.7) (Optical+Biomedical Engineering Laboratory 2019).

David and his team (Sampson 2019, pers. comm.) recognised the need for a hand-held device that surgeons could use to sense and visualise the edge of cancer tumours during surgery. They then set about determining the feasibility and conceiving a plan for developing the microscope in a needle.

In 2007, an engineering team was formed to build a 3D optical coherence tomography (OCT) in a needle. OCT senses different reflections of near infra-red light from different tissues (Zysk et al. 2007). The team developed a system with the capacity to miniaturise the optics and scan tissues that could fit into hypodermic needles with an outer diameter of 310 micrometres (Zysk et al. 2007). To ensure the scanning for the images occurs at the correct site in the patient's body, the team developed tracking and guidance technologies (Lau et al. 2010).

FIGURE 15.7 Microscope in a needle

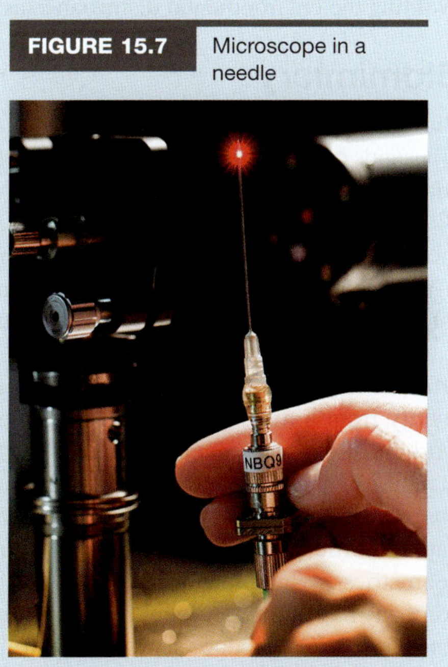

Source: University of Western Australia.

Building on this existing technology, Sampson and his team first tested their technology with lung tissue and lymph nodes, before testing with breast tumours. The microscope in a needle allows surgeons to visualise the margins of tumours and other disease to perform safer surgery (Allen et al. 2016).

CRITICAL THINKING

Many new materials are emerging. List some applications for which they could be used.

Phytomining

Consider the process of mining gold and other hard rock ore resources. During the 1800s, Australia and New Zealand reaped bountiful riches through gold mining in Northern Tasmania, Ballarat in Victoria and Central Otago. In those days, hard rock gold mining involved a pick, a wheelbarrow and the occasional stick of dynamite (for those who were wealthy, brave or foolhardy enough to use it). A lot of hard work was involved and the conditions were dangerous as well as physically exhausting (Blainey 2003). Today, hard rock gold mining utilises sophisticated drilling plant and blasting techniques. However, it is still not easy.

Engineering researchers have been investigating a form of ore extraction called **phytomining**. Phytomining involves using metal-accumulating plants to extract ore from ore bodies deemed 'sub-economic' (i.e. the deposit is too small or diffuse to offer a good return on investing in extracting it).

Phytomining researchers select and grow plants known as 'metal hyper-accumulating' species in soils that are rich in the target metal. These plants, as the name suggests, preferentially take up and sequester the target metal into their biomass (i.e. their leaves and stem). Once the crop is fully grown, phytomining researchers harvest the plant and burn it to produce a 'bio-ore' that can be further refined to produce useable ore products.

Early research in phytomining by the US Bureau of Mines showed the nickel hyper-accumulator milkwort jewelflower (*Streptanthus polygaloides*) could yield 100 kilograms per hectare of sulfur-free nickel. Milkwort jewelflower has been surpassed in the nickel phytomining stakes by more effective hyper-accumulators in the weed species *Alyssum bertolonii* and *Berkheya coddii*. On some soils, *Berkheya coddii* was shown to accumulate more than 20 tonnes per hectare (Institute of Terrestrial Ecosystems 2006).

While this approach to mining is in its research phase, the application of phytomining offers the possibility of exploiting ores that are not financially viable to mine by conventional methods; bioremediating soils and sludges where mixed toxins or the concentration of metal pollutants makes conventional, physical treatment impractical; conserving substantial energy via low temperature burning for ore recovery (as opposed to smelting); recovering ores from dumpsites as opposed to virgin mine activity; and, potentially, avoiding the environmental disturbance associated with open-cut mining of new ore bodies.

Biomimicry

Biomimicry is observing the forms and functioning of biological systems (which have evolved through thousands of years of natural selection) as the basis for designing engineered solutions. In biomimicry, the anatomical or physiological function of living things is imitated, adapted and replicated in technologies. Biomimicry in engineering ranges from the simple — Velcro™, which was inspired by the sticking mechanism of hooks used by plant burrs and prickle seeds to stick to animal fur for the purpose of dispersal — to the complex — Nissan Motors' Advanced Technology Center researching bumblebee flight as a basis for new generation vehicular crash-avoidance systems (Nissan 2008).

An Australian naturalist and entrepreneur, Jay Harman, has used biomimicry to engineer remarkably efficient impellers, pumps and fans. An impeller is a rotating device inside a closed tube that transfers energy from a motor to increase the speed of fluid flow through the tube. Jay Harman spent 12 years working in the Australian outback for the Department of Fisheries and Wildlife and observed a repetition in nature of spirals — in stream eddies, smoke rising from his campfire, flowers and in shells. Based on his observation of nature, he developed the PAX Streamlining Principle, which is a guideline for translating nature's efficiencies into industrial applications (Pax Scientific 2004). The result was 'streamlining geometries', which underpinned the design of energy-efficient and quiet fluid technologies. Harman's company has produced impellers, pumps and fans inspired by natural forms that require up to30 per cent less energy, and produce less noise and heat than parts made by giants such as General Electric, using traditional engineering techniques (Watters 2007). Biomimicry is discussed in further detail in the chapter on engineering design.

An impeller designed by PAX Scientific that draws inspiration for its form from the calla lily flower

SPOTLIGHT

Mimicking the retina

Electronic engineer Jason Eshraghian is mimicking retinal cells and brain functions for vision to develop improved computer vision and computer processing. He merged aspects from computational neuroscience, integrated circuit design and machine learning to develop a programmable retinal simulator and silicon retina image chip. The research could advance the design of smart image sensors that operate at high resolutions with ultra-low power, much like the retina.

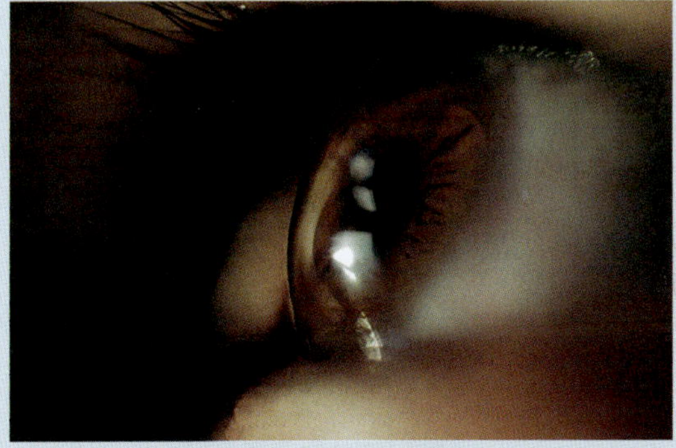

The human eye, used in biomimicry

The approach to computer vision uses machine learning, a numerical process involving reaching a decision by comparing many examples in large datasets. The team used datasets collated from a range of current- and voltage-clamp experiments performed on various retinal cells of animals by other research groups over the past seven decades (e.g. tiger salamander, giant squid, goldfish, mice, cats). However, machine learning involves a very large number of computations and requires parallelisation of these numerous calculations in order to run efficiently.

Rather than relying on continuous analogue signals or clock-based digital approaches, commonly used in computing, Jason and collaborators developed a neuromorphic vision chip, inspired by retinal cells, which gives a spike-based response to a change in scene. The chip functions similarly to the retina in that it only consumes power when there are changes in a scene, otherwise known as 'event-driven' sensing. The chip generates a signal when there is a change of light input and intensity is encoded in the frequency of spike firing.

With further work on interfacing, the vision chip could be used in therapeutic health care by passing the sequence of spike signals through the visual cortex and enabling the brain to construct meaningful visual information, which could potentially assist the visually impaired.

Jason and collaborators also developed a chip that encodes information using spikes instead of binary signals and have shown that it can be used in other demanding processing such as machine learning applications, significantly reducing the amount of information required to be processed.

CRITICAL THINKING

This spotlight includes some examples of biomimicry. Other studies are applied with less innocent aims, for example, studying the movement of dolphins through water in order to facilitate the movement of torpedoes through the sea to locate enemy vessels. What are your thoughts about studying animal behaviour in order to create weapons?

Animatronics

Animatronics is the design and creation of robotic or mechanised puppets used mostly in the entertainment industry. The process involves the creation of an accurate scale model of an animatronic creature and scanning by a 3D digitiser to guide milling of full-sized blocks of rigid polyurethane foam. The full-sized foam model serves as the base for moulding the creature's skin, so it is ready for application of the rigid members (e.g. structural elements such as bones), mechanical devices (e.g. gears and hydraulics) and electronic components (e.g. circuit boards, actuators and telemetry devices). The sophistication of the electronic components and associated control mechanisms (e.g. software or remote control) affect how natural and believable creature movements are to an intended audience.

SPOTLIGHT

Weta Workshop and Weta Digital

There continue to be improvements in the technology that is used in animatronic and CGI work but who can forget the scene in *The Lord of the Rings (LOTR)* trilogy, which was directed by New Zealand's Peter Jackson, in which the viewer meets the gruesome CGI (computer-generated imagery) Gollum — a mutated, human-like creature with bulbous eyes, a pale complexion and an unpleasant appetite for raw meat?

A Weta animatronic yellow-eyed penguin created for a television program. Photo: Steve Unwin.

The team responsible for the special effects, prosthetics, costumes and make up in the *LOTR* trilogy are from sister companies Weta Workshop and Weta Digital in Wellington, New Zealand. The multiple Academy Award–winning companies have developed a range of sophisticated engineered solutions, creating frighteningly life-like computer graphics and animatronic creatures that have terrified and entranced moviegoers in films including the *LOTR* trilogy, *King Kong* and *Black Sheep*. The crew at Weta Workshop also developed killer animatronic sheep for the 2006 film *Black Sheep*, and some of the guns for the 2009 film *Avatar*. Following the success of the *LOTR* trilogy, Peter Jackson was made director and executive producer for three films based on JRR Tolkien's book *The Hobbit*. Discussing the content of the films in an interview with TheOneRing.net (2008), Hobbit-franchise screenwriter Guillermo del Toro commented that he would use different technology for *The Hobbit* than that used in the *LOTR* trilogy, with making more use of:

> ... full animatronics and animatronic creatures enhanced with CGI, as opposed to CGI creatures themselves. We really want to take the state-of-the-art animatronics and take a leap ten years into the future with the technology we will develop for the creatures in the movie.

Although animatronics is mostly used to create great movies to entertain the masses, there are other, more serious applications for the technology. The Weta Workshop has assisted research into the mating habits of the endangered New Zealand tuatara. Tuatara are a unique lizard-like creature that used to be found throughout New Zealand. Their numbers have been reduced so much on mainland New Zealand that they are now restricted in range to a few tiny off-shore islands and have a precarious conservation status. Because of this, scientists in New Zealand are keen to establish new breeding colonies. Unfortunately, the tuatara only breeds once every four years and research suggests that only one in four males gets to breed at all. To ensure the success of breeding in a new colony, it is important that male tuataras who are motivated and successful breeders are selected. This is where animatronics from the Weta Workshop came into play. The workshop created an animatronic tuatara named 'Robo-Ollie' to help researchers investigate the breeding behaviour of male tuataras (Catalyst 2008).

CRITICAL THINKING

In the introduction to engineering chapter we claimed that engineering work is ubiquitous. This spotlight demonstrates you can't even escape engineering on a trip to The Shire. Is there any activity you undertake that does not involve engineering? Think of some and challenge your classmates to prove you wrong.

In this section, we have explored some of the novel and emerging engineering fields that will contribute to human development, environmental protection and quality of life in the coming decades. As a first-year student, your undergraduate years will be a time to decide which field will provide the stimulation and rewards to make working life meaningful for you. But choosing an ideal field or job is only half of the equation. The other half is in developing the skills, 'smarts' and attitude to survive and thrive in the workplace. The following section focuses on career development for engineers, and the skills undergraduate engineers need to develop to thrive in their early career, particularly the challenging professional task of managing their own managers.

15.4 An engineering career

LEARNING OBJECTIVE 15.4 Manage yourself in your workplace, and develop a plan to meet your career goals.

Consider the case study scenario raised at the outset of this chapter. Jono was battling the flu, struggling with some complicated drag coefficients for a ship's hull and attempting to think through some important possible career changes. The reality of working life for many engineers is one of multiple pressures and competing deadlines. With looming skills shortages and the challenges of practising engineering in a globalised workplace, career management for engineers is likely to remain complex for the foreseeable future. Because of this, you need to cultivate career management as a conscious practice and a habit from your undergraduate years.

In order to thrive as a professional engineer and to attain a desired job or career path, it is important to plan for success in the recruitment process (curriculum vitae, selection criteria, interview). Part of this is learning how to present your skills and experience effectively in a *curriculum vitae*, e-portfolio or via professional networking sites like LinkedIn. The spotlight about communicating our brand (in the chapter on understanding communication) highlights how important it is to take control of your personal and professional brand, particularly in the online environment. Once you have been offered and have accepted a job, it is important to appraise your work environment from time to time. This means taking the time to observe the work culture — such as taking notice of who appears to be successful in this environment and how they work; what kind of hours are worked; and how people behave towards one another, including support staff and managers. This is not to say that it is essential to conform to the workplace culture; rather, it is important to work out if you can (and want to) succeed within it — and if so, how?

KEY POINT

Being aware of your interactions with others at work will allow you to make the most of your job and your career prospects.

Work to rule — or be inspired to work towards a fulfilling career

What does it mean to *work to rule*? This approach to work involves being conscientious; that is, doing what is asked of you, but nothing more, and achieving a high degree of accuracy with your work. As a student, this is the equivalent of aiming for 50 per cent and living by the 'Ps (passes) mean degrees' maxim. This approach will probably get you a degree at the end of your time at university, but it is questionable how well equipped you will be to excel in your chosen career.

For professional engineers, work to rule is the minimum acceptable approach. If, when you graduate, you are happy in the role you are offered and not particularly ambitious, this may be a satisfactory strategy for you. However, some employers hope for more from their employees. Being motivated to *be yourself* and find a career that suits you, and your values and needs is the only way to realise this — and thereby be more effective in your work. However, it is important to be realistic about your abilities and knowledge. One characteristic that engineering employers value in an employee is the ability to know when to ask for help or seek more information. While it is good to be self-motivated and independent at work, the potential cost of a new engineer not seeking help can be high in terms of financial loss or workplace health and safety risk. It is important, particularly for young or inexperienced engineers, to be willing and able to get on with work when they can, but also to seek advice when they are unsure.

Management approaches

Consider the following scenario. Miranda, a graduate electrical engineer skilled in design, divided her working week between two small engineering consultancies. At one of the consultancies, the manager would brief Miranda on the project and leave her to get on with the job. The manager considered Miranda to be a 'star performer' and was delighted with the accuracy and innovation of her design work. At the other consultancy, the manager leaned over Miranda's shoulder, checking her preliminary sketches, calculations and circuits as she worked. A working day at the second consultancy started with a detailed review of the progress on Miranda's project. In a short time, Miranda's work at the second consultancy became slow, stilted and substandard, and her performance suffered because her second manager was practising micro-management.

Micro-management is an approach in which the manager retains the right to make most work-related decisions. This is often at the expense of staff autonomy. People who work for a micro-manager become starved of the pleasure and self-confidence that comes from being trusted to make decisions about work tasks. If you are micro-managed, you are unlikely to learn to take responsibility for your work. Micro-managers also suffer from their own behaviour. There are not enough hours in a day for a person to handle the myriad of decision-making opportunities involved in managing and organising an engineering team. Micro-management is one of a range of different approaches to management that can make a work environment challenging.

Another difficult approach is reactive management. **Reactive management** is when a manager tends to focus more time and attention on immediate tasks, and fails to tackle longer term tasks. Some longer term tasks, such as establishing time management systems, and staff and business development, might not be urgent. However, if unresolved, such tasks can threaten the long-term viability of a group or company. Figure 15.8 shows the Time Management Matrix that has been developed to explain reactive management (Covey 2004). Reactive managers tend to expend most of their energy in dealing with tasks listed in the urgent column (important and less important) of the matrix; whereas a proactive manager tends to focus their attention on tasks in the important row (urgent and not urgent) of the matrix.

To thrive in the workplace, it is useful to objectively appraise your manager's approach to management. The employee who can diagnose their manager's approach will be in a good position to assess how it fits with their own preferred style of working and being managed. If there is a serious mismatch between an employee's and a manager's management styles, communication, negotiation and adjustment may be required.

Different types of engineering organisations

There are many different structures of organisations that undertake engineering. Some of the more typical (sometimes expressed differently in different countries) are:

- charity or not-for-profit organisations
- cooperatives — owned and managed by workers who share profit between themselves
- micro entities — less than 10 members
- SMEs — small to medium sized enterprises with 50 to 250 members
- corporations — usually large companies with public shareholders
- government utilities — overseeing public infrastructure and operations such as main roads.

It doesn't have to be the case that on graduation you go to work for a large corporation. You could think about establishing your own business or a cooperative with some friends and colleagues. Is there something you are passionate about?

FIGURE 15.8 The Time Management Matrix

	Urgent	Not urgent
Important	I – Crisis – Pressing issues – Deadlines – Meetings	II – Preparation – Planning – Prevention – Relationship building – Personal development
Not important	III – Interruptions – Some mail and email – Many popular activities	IV – Trivial emails – Some phone calls – Time wasters

Source: Covey (2004).

Efficiency and respect in the workplace

As an engineering graduate, you will likely move into a management role reasonably early in your career. This may happen sooner than it is comfortable, and possibly before you have developed a strong, personal understanding of what it means to be a good manager. Undoubtedly, some experiences of participating in and leading team projects during your undergraduate career will give you first-hand experience of the joys and trials of trying to organise the work and personalities of a group of strong-minded individuals! While some organisations offer explicit support for the development of management skills, many will expect you to learn 'on the job', which may involve a degree of trial and error. One ethic to consider in moving into a management role is that of respect. Individuals in any workplace are likely to have skills, knowledge and experiences that are worthy of your respect, regardless of their place in the workplace hierarchy. A smart manager conveys respect partly because the sense of being respected can engender a better work ethic in some workers; and also because tapping effectively into the skills, knowledge and experience of the full workforce can lead to profound improvements in productivity, safety and workplace cohesion.

Lean thinking or 'Lean' takes a mindset of respect and applies it to the optimisation of business operations. Lean is a set of improvement philosophies and tools that have been developed by Toyota since the 1940s. The increasing application of Lean in the analysis and optimisation of production and manufacturing mean that it is an increasingly important analytical tool for graduate engineers.

Lean is based on:

- understanding how the customer defines 'value' — a process that will physically transform the raw material into the finished product the customer wants
- creating flow — streamlining the movement of people, materials and information so as to reduce production lead-time (the time taken to transform raw materials into consumer product) and reduce costs to the organisation
- removing the non-valued added (the waste component of all functions).

According to the Lean way of analysing production systems, there are seven wastes that each add to a product's lead-time:

1. transport
2. inventory
3. motion
4. waiting
5. overprocessing
6. overproduction
7. defects.

Generally, the personnel who add value (i.e. transform raw materials to product) in a factory are the operators who run machinery. Lean thinking focuses organisations and engineers on more actively supporting operators in their frontline work. Lean is a structured process that allows organisations to support operators through continuous improvement and in implementing a rapid response to the elimination of waste in all its forms. Waste elimination via Lean requires analysis and cooperative action by *all* personnel in the factory (e.g. manager, engineers, operators).

Continuing professional development (CPD)

As a professional engineer, you will have a formal responsibility to continue to learn and update your skills. This process is called **continuing professional development (CPD)**.

Continuing professional development is part of the minimum requirement for registration as a professional engineer with Engineers Australia's voluntary National Engineering Register (NER) (Engineers Australia 2023c). Although conditions and requirements for professional practice vary between Australian states, becoming chartered and registered is now even more important as legislation requiring registration to practise is becoming widespread in Australia. Engineering New Zealand (2024) recommend CPD as a learning mix for success and recommend:

> . . . gaining the right breadth and depth of continuing professional development (CPD). Completing a mix of learning activities each year that cover both business and technical skills will ensure you stay relevant and up to date. As part of your Engineering New Zealand membership, you need to complete 40 hours CPD/year.

The aim of CPD is to maintain or develop your capacity to perform all the diverse tasks that are a part of your work as an engineer. Specifically, Engineers Australia (2024) states that CPD should:

- maintain your technical competence
- retain and enhance your workplace effectiveness
- learn good leadership skills and how to mentor others
- successfully deal with career challenges
- better serve the community.

According to Engineers Australia's CPD policy, engineers listed on the National Professional Engineers Register and engineers applying for or maintaining Chartered status with Engineers Australia must participate in and gather evidence of a minimum of 150 hours of structured CPD over each three-year period. The types of CPD described and recognised by Engineers Australia (2024) include:

- any tertiary course taken either as an individual course or for a formal post-graduate award
- short courses, workshops, seminars and discussion groups, conferences, technical inspections and technical meetings
- workplace learning activities that extend your area of practice or area of engineering
- private study that extends your knowledge and skill
- service to the engineering profession
- preparation and presentation of material for courses and conferences
- tertiary teaching or academic research
- structured activities that meet the objectives of the [Engineers Australia's] CPD policy.

When you enrolled in an undergraduate engineering program, did you realise your university studies would only be the beginning of a lifetime commitment to learning about engineering? The Engineers Australia list discusses a range of different learning experiences, opportunities and activities you may choose to do in the future. You will need to choose the CPD activities and approaches that best suit your needs as a life-long learner.

Career planning

While it is possible to pursue and document CPD in order to satisfy the requirements of being a professional engineer, an engineer gains a great deal more from CPD if they approach this requirement as a strategic opportunity. Career planning is the process of thinking through and mapping out a path from a current working situation to a preferred working situation. There are four main considerations in a career development plan:

1. identification of goals
2. auditing skills and competencies
3. planning and maximising development activities
4. setting a realistic timeframe.

Questions you might ask yourself at each stage of a career development planning process are outlined in figure 15.9.

FIGURE 15.9 The four stages of a career development plan

Goals. What are your goals over the next five years? Your answer to this question should mainly focus on your work-related goals, but should also encompass or take account of personal goals or constraints that you want to fit with your work life (e.g. desires to take time out for travel, family commitments and so on). Importantly, goals should be optimistic and realistic.

Skills and competencies. Given your goals, what skills and competencies do you already have? What skills and competencies do you need to develop? In earlier chapters, you spent time considering your existing competencies and areas in which you were not so strong. This is an important, ongoing activity for engineers who commit to career development planning.

Development activities. What are the main or priority skills and competencies you need in order to reach your career goals? What development activities, experiences or support can you obtain to develop the necessary skills? How can you target your CPD accordingly? Become strategic about your CPD as an engineer — there is little point in participating in CPD that merely reinforces skills or competencies you already have. As a student, you may plan your periods of work experience so they maximise career development.

Time frame. What are your highest priorities for development? How can development activities be accommodated within your normal workload? It is important to set up a time frame for achieving goals. This is particularly the case for a new engineer, as the immediate demands of work sometimes make it much easier to continue to delay career development planning and being strategic about CPD.

This general schema for career development planning should look fairly familiar to you if you have read the chapter on managing engineering projects. A career development plan can be viewed as a project to be managed, with the project being your career and the main manager being you.

Career development planning may seem a long way off with years of study between you and your first professional job. However, decisions you make in the next year or two may influence your future career prospects. During undergraduate study, it is important to begin building a network of contacts in engineering, partly to get a feel for how real engineers do their jobs, and partly because employment opportunities and career progression can sometimes be influenced by who you know (in addition to what you know). Nurture your career prospects by being well prepared, interacting with companies you contact for summer work or practicum, discussing your plans and aspirations with engineers you meet on a social basis, and planning questions for those engineers who provide guest lecturers in your course. Being strategic in building your network of contacts and working hard over the coming years of your undergraduate degree should ensure that your engineering future offers the deep satisfaction of a career that is just right for you.

SUMMARY

In this chapter consideration has been given to your future as an engineer in a changing world. In the coming years of undergraduate study, you will have the opportunity to develop skills in managing yourself in the workplace, strategies for identifying and moving towards your career goals, and a clear sense of which of the many and varied roles and fields you may wish to fill as a graduate engineer. We will now briefly revisit each of the chapter's learning objectives.

15.1 Discuss the engineer's role in addressing global challenges.
There are a range of challenges currently facing humanity. These include global climate change, depletion of antibiotic defences, food security and water supply for a growing world population, and managing information and safeguarding privacy. Because engineering is ubiquitous in modern life, engineers have an important role to play in addressing these challenges and possibly in pre-empting unforeseen global challenges.

15.2 Describe how globalisation may affect you in your future professional life.
Globalisation is the complex concept describing the trade, economies, lifestyles, cultures and traditions that are increasingly passing across international boundaries. Most large corporations already work across borders and this brings with it the excitement of working in different countries and the challenge of doing so with respect for the traditions and cultures of the local communities in offshore locations.

15.3 Identify and discuss some emerging fields in engineering.
Emerging and innovative fields in engineering include industrial biotechnology, animatronics, materials science and phytomining, just to name a few. Several of these fields blend existing fields of engineering with other disciplines (e.g. plant science + chemical engineering + mining engineering) or are fields in which innovation is particularly rich (such as materials science and mechatronics).

15.4 Manage yourself in your workplace, and develop a plan to meet your career goals.
Techniques and strategies you might use in the workplace to advance in your career as an engineer include managing yourself in the workplace and understanding your manager's approach to management. You will have an obligation when you are a registered professional engineer to engage in CPD. All of these factors should assist you in meeting your career goals.

KEY TERMS

animatronics The design and creation of robotic or mechanised puppets used mostly in the entertainment industry.

antibiotic resistance When disease-causing micro-organisms become immune to the effects of an antibiotic that was previously used against them.

biomimicry Drawing inspiration for engineered solutions from the forms and functioning of biological systems.

continuing professional development (CPD) A formal responsibility of professional engineers to continue to learn and update their knowledge and skills.

globalisation When companies or individuals begin to operate across national boundaries.

industrial biotechnology The development and use of technology for commercial or industrial-scale production of biological products.

international development Development of countries on an international scale, usually according to Western standards of progress.

lean thinking A structured thinking process for continuous improvement and the elimination of waste in all its forms.

materials science The investigation of how the chemical and physical structure of materials influences the way in which they function.

micro-management When the manager retains the right to make most work-related decisions.

nanotechnology Construction of particles in the nanometre size range that are designed to react or respond to cues and perform tasks in the sub-microscopic realm.

phytomining The use of metal hyper-accumulating plants to extract metals from soils.

reactive management When a manager focuses mostly on immediate tasks, and fails to tackle longer term tasks.

EXERCISES

1. Search the internet for globalised companies that operate in the field of engineering you want to work in. Try to find out if the company does different work in different countries (e.g. if they are involved in raw materials extraction in one country and downstream processing in another). In which city and country is the main research and development office for the company? Where are the consumers of the majority of the goods or services of the company based? What are the social considerations that these companies have had to consider when working in these locations?
2. Look through the 'Engineering spotlights: at a glance' grid in the front matter of the text to refresh your memory about the spotlight sections covered throughout all of the chapters in this text. What are the jobs and technologies you can discern for your favourite engineering discipline? Which ones look the most interesting to you? Why?
3. Consider a recent technical challenge you encountered. What was the main problem or function required? See if you can find out through biomimetics research how nature has solved a problem similar to this.
4. It is probable that whatever fusion of engineering and other disciplines you can think of; there will be someone — somewhere in the world — doing research and development in this area. Is there a fusion field you are interested in researching or perhaps even working in? Think of activities you enjoy doing outside of engineering, for example, bird watching, ballet, pyrotechnics or martial arts. Do an online keywords search on your hobby and your enrolled field of engineering. What did you come up with? Is there a fusion field that is researching or developing engineering innovation that links your enrolled field of engineering with your hobby? Try searching some odd combinations; for example, mechanical engineering and hip hop, sheep and electronics, civil engineering and asthma, agricultural engineering and trousers, robotics and fish, or daisies and mining. What might be the practical application of such research? How may developments in highly unusual fusion fields lead into new and innovative solutions to societal problems?
5. Can you imagine yourself as the director of a start-up company or cooperative? What would your business be? Where would you start up? Whom with? Create a business plan for your start-up. Research what legal steps you need to take to incorporate your business.
6. Form teams in your classroom and draw a table using a whiteboard or butcher's paper. The row headings should list four different scale lengths: (1) nano/molecular/chemical, (2) individual/human, (3) plant/factory/local and (4) national/international/global. Three of the columns should focus on different engineering disciplines. Set a timer to five minutes and see how many different processes, products or impacts your team can brainstorm for each of the different scales. Then, consider your team's results. Are there any boxes in the matrix that are blank? Why is this? Might the field concerned have little or no relevance to the corresponding scale length, or does your group need to learn more about a particular field or scale length? Has your team profiled the field you intend to work in? Have you considered the range of scale lengths your intended field impacts on before? What does this mean for your short- and long-term learning needs?
7. Have you ever worked for or with a micro-manager or a reactive manager (or even a lecturer, tutor or teacher)? How did you perform under either of these management approaches? You might be able to think of times when the behaviour of a manager, coach, mentor or teacher has substantially influenced your performance — for better or for worse. Either on your own or with one or two classmates, identify particular managers or leaders whose management style has affected your work. Use Blake and Mouton's (1985) Managerial Grid III, shown in figure 15.10, to diagnose the management style of the manager. Based on this diagnosis, what would you say is the management style you respond best to? Which style makes it hard for you to perform? How would you explain the best management style for you to a prospective team leader or manager?
8. Go to Engineers Australia's or Engineering New Zealand's webpage and find the section devoted to CPD. If you are currently in the workforce, investigate the approach your organisation takes to helping you document and present CPD. The Engineers Australia website has an Excel spreadsheet you can download and use to record CPD. An online option is also available. Using the spreadsheet, document the learning activities or situations you can list as CPD, based on your life experience. You may like to do this in pairs. If you have no activities or situations to list, research and write a list of CPD opportunities that may help you to achieve your career goals.

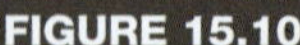

FIGURE 15.10 Managerial grid

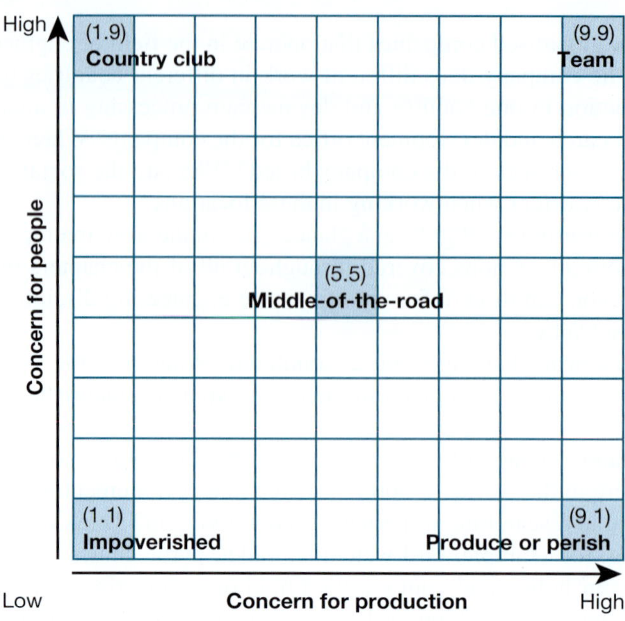

Source: Blake and Mouton (1985).

PROJECT ACTIVITY

Your task is to develop paperwork to take to an annual performance appraisal. This will involve a performance appraisal (PA) of your work as a learner over the past semester, and a career development plan (CDP) that explains your short-term (over the next 12 months) and longer term (over the next three years) development goals.

To complete the PA, you will need to gather examples of your learning over the last semester to support your PA claims. These may include assignment reports, marks you have gained in this subject or other relevant subjects, and personal testimonials from tutors and classmates. You might like to include some forms of evidence from alternative sources, such as written references from employers, records showing participation in social or sporting clubs, certificates of completion in night classes and conformation of volunteer work you have done.

Think widely and be creative. Almost anything you do in your life might count as evidence of knowledge, a skill or an attitude that is unique or valuable. The challenge for you will be in presenting the evidence and explaining it in a compelling way.

REFERENCES

Allen, WM, Chin, L, Wijesinghe, P, Kirk, RW, Latham, B, Sampson, DD, Saunders, CM & Kennedy, BF 2016, 'Wide-field optical coherence micro-elastography for intraoperative assessment of human breast cancer margins', *Biomedical Optics Express*, vol. 7, no. 10, pp. 4139–4153, https://doi.org/10.1364%2FBOE.7.004139

Altman, S 2020, 'Will COVID-19 have a lasting impact on globalisation?', *Harvard Business Review*, 20 May, https://hbr.org/2020/05/will-covid-19-have-a-lasting-impact-on-globalization

Baillie, C, Armstrong, R, Rondon, G & Fourie, A 2014, *Community engagement experiences: Communities telling their stories to inform future practice*, IM4DC final report, University Western Australia, Crawley.

Blainey, G 2003, *The rush that never ended: A history of Australian mining*, 5th edn, Melbourne University Press, Victoria.

Blake, RR & Mouton, JS 1985, *The managerial grid III: The key to leadership excellence*, Gulf Publishing Co., Houston.

Bureau of Meteorology & CSIRO 2022, *State of the climate 2022*, Commonwealth of Australia, https://www.csiro.au/en/research/environmental-impacts/climate-change/state-of-the-climate

Catalyst 2008, 'Robot tuatara', https://www.abc.net.au

Covey, SR 2004, *The 7 habits of highly effective people*, Free Press, New York.

Davis, M 2002, *Late Victorian holocausts: El Nino famines and the making of the third world*, Verso Press, Brooklyn.

Engineering New Zealand 2016, *Code of Ethical Conduct*, https://www.engineeringnz.org/engineer-tools/ethics-rules-standards/code-ethical-conduct

——— 2024, 'Professional development', https://www.engineeringnz.org/cpd

Engineers Australia 2022, *Code of Ethics and Guidelines on Professional Conduct*, https://www.engineersaustralia.org.au/publications/code-ethics

——— 2023a, 'New report reveals deepening engineering skills crisis', Media release, 16 November, https://www.engineersaustralia.org.au/news-and-media/2023/12/new-report-reveals-deepening-engineering-skills-crisis

——— 2023b, *The engineering profession: A statistical overview*, 15th edn, https://www.engineersaustralia.org.au/publications/engineering-profession-statistical-overview-15th-edition

——— 2023c, 'Minimum requirements for registration for independent practice as a professional engineer in Australia', https://www.engineersaustralia.org.au/sites/default/files/2023-02/GUIDE-minimum-registration-requirements.pdf

——— 2024, 'CPD requirements', https://www.engineersaustralia.org.au/membership/cpd-requirements

Ferguson, C & Inglis, S 2018, 'Digital diagnosis: How your smartphone or wearable device could forecast illness', *The Conversation*, 23 October, https://theconversation.com/digital-diagnosis-how-your-smartphone-or-wearable-device-could-forecast-illness-102385

Geoscience Australia 2011, 'The enhanced greenhouse effect (global warming)', *OzCoasts: Coastal indicators*, https://ozcoasts.org.au

Ginsburg, AS & Klugman, KP 2020, 'COVID-19 pneumonia and the appropriate use of antibiotics', *Lancet Global Health*, vol. 8, no. 12, pp. E1453–E1454, https://doi.org/10.1016/S2214-109X(20)30444-7

Henley Management College 2006, *Educating engineers for the 21st century: The industry view*, report for the Royal Academy of Engineering by Nigel Spinks, Nick Silburn and David Birchall, 8 March, United Kingdom.

Ikbal, MS, Tisha, FA, Asheque, A, Hasnat, E, Uddin, MA 2024, 'Eco-friendly biopolishing of cotton fabric through wasted sugarcane bagasse-derived enzymes', *Heliyon,* vol. 10, no. 4, https://doi.org/10.1016/j.heliyon.2024.e26346

Institute of Terrestrial Ecosystems 2006, 'Soil protection group: Phytomining', https://ites.ethz.ch

Institution of Chemical Engineers (IChemE) 2022, 'IChemE position on climate change', https://www.icheme.org/media/14873/icheme-climate-change-statement.pdf

Kaspura, A 2017, *The engineering profession: A statistical overview*, 13th edn, Engineers Australia, Canberra.

King, R 2008, *Addressing the supply and quality of engineering graduates for the new century*, report on Engineering Education in Australia for the Carrick Institute for Teaching and Learning in Higher Education, Sydney.

Lau, B, McLauglin, RA, Curatolo, A, Kirk, RW, Gerstmann, DK & Sampson, DD 2010, 'Imaging true 3D endoscopic anatomy by incorporating magnetic tracking with optical coherence tomography: Proof-of-principle for airways', *Optics Express*, vol. 18, no. 26, pp. 27173–27180, https://doi.org/10.1364/OE.18.027173

Leining, C & Kerr, S 2018, *A guide to the New Zealand Emissions Trading Scheme*, report prepared for the Ministry for the Environment, Motu Economic and Public Policy Research, https://environment.govt.nz

Mench, JA, James, H, Pajor, EA & Thompson, PB 2008, 'The welfare of animals in concentrated animal feeding operations', in *Report to the Pew Commission on Industrial Farm Animal Production*, Pew Commission on Industrial Farm Animal Production, Washington, DC.

Meat and Livestock Australia 2018, *Antimicrobial stewardship guidelines for the Australian cattle feedlot industry*, March, https://www.mla.com.au/globalassets/mla-corporate/research-and-development/program-areas/animal-health-welfare-and-biosecurity/mla_antimicrobial-stewardship-guidelines.pdf

Messele, YE, Khallawi, MA, Veltman, T, Trott, D, Kidd, SP, Low, WY & Petrovski, KR 2023, *Final report: Longitudinal analysis of antimicrobial resistance of E.coli, Salmonella, and Enterococcus species during pre-feedlot, feedlot and slaughter periods*, Meat and Livestock Australia, https://www.mla.com.au/contentassets/a8d5c9d64cc641668c97aa3459524d2c/b.flt.3003-final-report.pdf

Murray, C 2022, 'Global burden of bacterial antimicrobial resistance in 2019: A systematic analysis', *The Lancet*, vol. 399, no. 10325, pp. 629–655, https://doi.org/10.1016/S0140-6736(21)02724-0

Newmont 2018, *Our code of conduct*, Newmont, Colorado, https://www.newmont.com

Nissan 2008, 'Crash avoidance robotic car inspired by flight of the bumblebee', https://www.nissan-global.com/EN

Nursey-Bray, M, Blackwell, B, Brooks, B, Campbell, ML, Goldsworthy, L, Haugstetter, H, Rodrigues, I, Roome, M, Wright, JT, Francis, J & Hewitt, CL 2012, 'Vulnerabilities and adaptation of ports to climate change', *Journal of Environmental Planning and Management*, vol. 56, no. 7, pp. 1021–1045, https://doi.org/10.1080/09640568.2012.716363

Optical+Biomedical Engineering Laboratory 2019, 'Microscope-in-a-needle', https://obel.ee.uwa.edu.au

Palmer, S & Campbell, M 2018, 'Using census data to better understand engineering occupational outcomes', *Proceedings of the Australasian Association for Engineering Education Conference*, 9–12 December, Hamilton, New Zealand.

Paul, M 2023, 'Secondary bacterial pneumonia drove many COVID-19 deaths', *News Center*, 5 May, Northwestern University Feinberg School of Medicine, https://news.feinberg.northwestern.edu/2023/05/05/secondary-bacterial-pneumonia-drove-many-covid-19-deaths

PAX Scientific 2004, 'Jay Harman', https://paxscientific.com

Royal Society of New Zealand 2008, 'Climate change statement from the Royal Society of New Zealand', 10 July, https://www.royalsociety.org.nz

Solomon, S, Qin, D, Manning, M, Chen, Z, Marquis, M, Averyt, KB, Tignor, M & Miller, HL (eds) 2007, *Climate change 2007: The physical science basis*, Cambridge University Press, Cambridge and New York.

Storteboom, H, Arabi, M, Davis, JG, Crimi, B & Pruden, A 2010, 'Tracking antibiotic resistance genes in the South Platte River basin using molecular signatures of urban, agricultural, and pristine sources', *Environmental Science & Technology*, vol. 44, no. 19, p. 7397, https://doi.org/10.1021/es101657s

TheOneRing.net 2008, 'Guillermo del Toro chats with TORN about *The Hobbit* films!', 25 April, https://www.theonering.net/torwp/2008/04/25/28747-guillermo-del-toro-chats-with-torn-about-the-hobbit-films

Trevelyan, JP & Tilli, S 2010, 'Labour force outcomes for engineering graduates in Australia', *Australasian Journal of Engineering Education*, vol. 16, no. 2, pp. 101–122, https://doi.org/10.1080/22054952.2010.11464047

Water Corporation of Western Australia 2018, *Annual report 2018*, https://www.watercorporation.com.au

Watkinson, AJ, Murby, EJ & Costanzo, SD 2007, 'Removal of antibiotics in conventional and advanced wastewater treatment: Implications for environmental discharge and wastewater recycling', *Water Research*, vol. 41, no. 18, pp. 4164–4176, https://doi.org/10.1016/j.watres.2007.04.005

Watters, E 2007, 'Product design, nature's way', *Business 2.0 Magazine*, 12 June, https://edition.cnn.com/business

World Health Organization 2014, 'WHO's first global report on antibiotic resistance reveals serious, worldwide threat to public health', 30 April, https://www.who.int/news/item/30-04-2014-who-s-first-global-report-on-antibiotic-resistance-reveals-serious-worldwide-threat-to-public-health

——— 2019, 'World Antibiotic Awareness Week', November, https://www.who.int/campaigns/world-amr-awareness-week/2019

——— 2021, 'Antimicrobial resistance: Fact sheet on Sustainable Development Goals (SDGs): Health targets', *Fact sheet*, 15 April, https://www.who.int/europe/publications/i/item/WHO-EURO-2017-2375-42130-58025

——— 2024, 'AMR and AMC open call: WHO invites Member States to participate in GLASS', https://www.who.int/initiatives/glass/country-participation

Zysk, AM, Adie, SG, Armstrong, JJ, Leigh, MS, Paduch, A, Sampson, DD, Nguyen, FT & Boppart, S 2007, 'Needle-based refractive index measurement using low-coherence interferometry', *Optics Letters*, vol. 32, no. 4, pp. 385–387, https://doi.org/10.1364/OL.32.000385

ACKNOWLEDGEMENTS

Photo: © Amy Pruden

Photo: © Kite_rin / Shutterstock

Photo: © katacarix / Shutterstock

Photo: © Eric Feinblatt

Photo: © PAX Scientific, Inc

Photo: © Jason Eshraghian

Photo: © Weta

Figure 15.1: © Geoscience Australia 2011

Figure 15.2: Reproduced from IPCC Working Group Second Assessment Report (1996) by Geoscience Australia (2011).

Figures 15.4 and 15.5: © Andrew Guzzomi

Figure 15.7: © University of Western Australia

Figure 15.8: © Covey, SR 2004, *The 7 habits of highly effective people*, Free Press, New York.

Figure 15.10: © Blake, RR & Mouton, JS 1985, *The managerial grid III: The key to leadership excellence*, Gulf Publishing Co., Houston.

Table 15.1: © Nursey-Bray, M et al. 2012, 'Vulnerabilities and adaptation of ports to climate change', *Journal of Environmental Planning and Management*, vol. 56, no. 7, pp. 1021–1045, https://doi.org/10.1080/09640568.2012.716363

Text: © Ferguson, C & Inglis, S 2018, 'Digital diagnosis: How your smartphone or wearable device could forecast illness', *The Conversation*, 23 October, https://theconversation.com/digital-diagnosis-how-your-smartphone-or-wearable-device-could-forecast-illness-102385

Text: © Institution of Chemical Engineers 2022, 'IChemE position on Climate Change', https://www.icheme.org/media/14873/icheme-climate-change-statement.pdf

Text: © Baillie, C, Armstrong, R, Rondon, G & Fourie, A 2014, *Community engagement experiences: Communities telling their stories to inform future practice*, IM4DC final report, University Western Australia, Crawley.

Text: © Engineers Australia 2024, 'CPD requirements', https://www.engineersaustralia.org.au/membership/cpd-requirements

INDEX